Life in the Universe

Fourth Edition

Jeffrey Bennett

University of Colorado at Boulder
Big Kid Science

Seth Shostak

SETI Institute

PEARSON

Boston Columbus Indianapolis New York San Francisco Upper Saddle River
Amsterdam Cape Town Dubai London Madrid Milan Munich Paris Montreal Toronto
Delhi Mexico City Sao Paulo Sydney Hong Kong Seoul Singapore Taipei •Tokyo

Editor-in-Chief: Jeanne Zalesky
Executive Editor: Nancy Whilton
Project Managers: Tema Goodwin/Lizette Faraji
Program Manager: Mary Ripley
Director of Product Management Services: Erin Gregg
Program/Project Management Team Lead: Kristen Flathman
Production Management: Lifland et al., Bookmakers
Copyeditor: Quica Ostrander
Compositor: Cenveo Publisher Services
Design and Cover Manager: Mark Ong
Interior and Cover Designer: John Walker
Illustrations: Rolin Graphics Inc.
Rights and Permissions Project Manager: Maya Gomez
Photo Researcher: Amy Dunleavy
Manufacturing Buyer: Maura Zaldivar-Garcia
Director of Marketing: Christy Lesko
Marketing Manager: Elizabeth Ellsworth
Printer/Binder/Cover Printer: LSC Communications
Cover Photo Credits: NGC 281, Martin Pugh; Earth, NASA; Illustration: Graham Johnson of www.fiVth.com

Library of Congress Cataloging–in–Publication Data

Names: Bennett, Jeffrey O. | Shostak, G. Seth.
Title: Life in the universe / Jeffrey Bennett, Seth Shostak.
Description: Fourth edition. | San Francisco : Pearson, 2016. | Includes index.
Identifiers: LCCN 2015039754 | ISBN 9780134089089
Subjects: LCSH: Exobiology. | Life--Origin.
Classification: LCC QH327 .B45 2016 | DDC 576.8/39--dc23
LC record available at http://lccn.loc.gov/2015039754

ISBN-10: **0-13-408908-1**
ISBN-13: **978-0-13-408908-9**

Dedication

The quest to understand life on Earth and the prospects for life elsewhere in the universe touches on the most profound questions of human existence. It sheds light on our origins, teaches us to appreciate how and why our existence on Earth became possible, and inspires us to wonder about the incredible possibilities that may await us in space. We dedicate this book to all who wish to join in this quest, with the sincere hope that knowledge will help our species act wisely and responsibly.

All this world is heavy with the promise of greater things, and a day will come, one day in the unending succession of days, when beings, beings who are now latent in our thoughts and hidden in our loins, shall stand upon this earth as one stands upon a footstool, and shall laugh and reach their hands amidst the stars.

H. G. Wells (1866–1946)

Brief Contents

Detailed Contents

Preface

To the Reader

Few questions have so inspired humans through the ages as the mystery of whether we are alone in the universe. Many ancient Greek philosophers were confident that intelligent beings could be found far beyond Earth. When the first telescopes were trained on the Moon in the seventeenth century, some eminent astronomers interpreted lunar features as proof of an inhabited world. A little over a century ago, belief in a civilization on Mars became so widespread that the term *martian* became synonymous with *alien*. But despite this historical interest in the possibility of extraterrestrial life, until quite recently few scientists devoted much effort to understanding the issues surrounding it, let alone to making a serious search for life.

In the past few decades, however, a remarkable convergence of biology, geology, astronomy, and other sciences has brought the issue of extraterrestrial life to the forefront of research. Advances in our understanding of the origin of life on Earth are helping us predict the conditions under which life might arise in other places. Discoveries of microbes thriving under extreme conditions (at least by human standards) on Earth have raised hopes that life might survive even in some of the harsh environments found elsewhere in our own solar system. Proof that planets exist around other stars—first obtained in the 1990s—has given added impetus to the study of the conditions that might allow for life in other star systems. Technological advances are making it possible for us to engage in unprecedented, large-scale scrutiny of the sky for signals from other civilizations, spurring heightened interest in the search for extraterrestrial intelligence (SETI). Perhaps most important, scientists have found the interdisciplinary study of issues related to the search for life beyond Earth to have intrinsic value, independent of whether the search is ultimately successful.

Given the intense research efforts being undertaken by the scientific community and the long-standing public fascination with the search for life, it should be no surprise that the study of life in the universe—also known as *astrobiology*—has become one of the most publicly visible sciences. Colleges, too, have recognized the growing importance of this discipline, and many have established courses in astrobiology. This book aims to serve such courses by offering a comprehensive introduction to the broad science of life in the universe.

Although this is a textbook, it is designed to be of interest to *anyone* with a desire to learn about the current state of research in astrobiology. No special scientific training or background is assumed, and all necessary scientific concepts are reviewed as they arise. If you have a basic high school education and a willingness to learn, you are capable of understanding every topic covered in this book. We wish you well in your efforts.

Jeffrey Bennett
Seth Shostak

To Current or Prospective Instructors

The rest of this preface is aimed primarily at current or prospective instructors teaching courses on life in the universe. Students and general readers might still find it useful, because it explains some of the motivation behind the pedagogical features and organization of this book and may thereby help you get the most from your reading.

Why Teach a Course on Life in the Universe?

By itself, the rapid rise of research interest in astrobiology might not be enough to justify the creation of new courses for non-science majors. But the subject has at least three crucial features that together make a strong case for adding it to the standard science offerings:

1. For students who take only one or a few required science courses, the interdisciplinary nature of the study of life in the universe offers a broader understanding of a range of scientific research than can a course in any single discipline.
2. Public fascination with UFOs and alien visitation offers a unique opportunity to use life in the universe courses as vehicles for teaching about the nature of science and how to distinguish true science from pseudoscience.
3. The science of life in the universe considers many of the most profound questions we can ask, including What is life? How did life begin on Earth? Are we alone? Could we colonize other planets or other star systems? Students are nearly always interested in these questions, making it easy to motivate even those students who study science only because it is required.

These features probably also explain the growing number of life in the universe courses being offered at colleges around the world (as well as some at the high school level). It's worth noting that, besides being fascinating to students, a course on life in the universe can be a great experience for instructors. The interdisciplinary nature of the subject means that no matter what your specific scientific background, you are sure to learn something new when you teach an astrobiology course at any level.

Course Types and Pacing

This book is designed primarily for use in courses for nonscience majors, such as required core courses in natural science or elective follow-up courses for students who lack the preparation needed for more technical offerings in astrobiology. However, past editions have also been used successfully in higher-level courses, often supplemented with journal articles on original research. This book can also be used at the senior high school level, especially for integrated science courses that seek to break down the traditional boundaries separating individual science disciplines.

Although the chapters are not all of equal length, it should be possible to cover them at an average rate of approximately one chapter per week in a typical 3-hours-per-week college course. The 13 chapters in this book should therefore provide about the right amount of material for a typical one-semester college course. If you are teaching a one-quarter course, you might need to be selective in your coverage, perhaps dropping some topics entirely. If you are teaching a yearlong course, you'll have time to go into greater depth as you spread out the material for an average pace of about one chapter every 2 weeks.

The Topical (Part) Structure of
Life in the Universe

The interdisciplinary nature of astrobiology can make it difficult to decide where emphasis should be placed. In this book, we follow the general consensus revealed in discussions with instructors of astrobiology courses, which suggests a rough balance between the different disciplines that contribute to the study of life in the universe. We've therefore developed this book with a four-part structure, outlined below. (See the table of contents for more detail.)

PART I. INTRODUCING LIFE IN THE UNIVERSE (CHAPTERS 1–3). Chapter 1 offers a brief overview of the topic of life in the universe and why this science has moved to the forefront of research. Chapter 2 discusses the nature of science, based on the assumption that this is many students' first real exposure to how scientific thinking differs from other modes of thinking. Chapter 3 presents fundamental astronomical and physical concepts necessary for understanding the rest of the course material, including the formation of planetary systems.

PART II. LIFE ON EARTH (CHAPTERS 4–6). This is the first of three parts devoted to in-depth study of astrobiology issues.

Here we discuss the current state of knowledge about life on Earth. Chapter 4 discusses the geological conditions that have made Earth habitable. Chapter 5 explores the nature of life on Earth. Chapter 6 discusses current ideas about the origin and subsequent evolution of life on Earth.

PART III. LIFE IN THE SOLAR SYSTEM (CHAPTERS 7–10). We next use what we've learned about life on Earth in Part II to explore the possibilities for life elsewhere in our solar system. Chapter 7 discusses the environmental requirements for life and then offers a brief tour of various worlds in our solar system, exploring their potential habitability. Chapters 8 and 9 focus on the places in our solar system that seem most likely to offer possibilities for extraterrestrial life: Mars (Chapter 8) and the jovian moons Europa, Ganymede, Callisto, Titan, Enceladus, and Triton (Chapter 9). Chapter 10 discusses how habitability evolves over time in the solar system, with emphasis on comparing the past and present habitability of Venus and Earth; this chapter also introduces the concept of a habitable zone around a star, setting the stage for the discussion of life beyond our solar system in Part IV.

PART IV. LIFE AMONG THE STARS (CHAPTERS 11–13, EPILOGUE). This final set of chapters deals with the question of life beyond our solar system. Chapter 11 focuses on our rapidly growing understanding of extrasolar planets, including the types of stars they orbit, how we detect them, their similarities to and differences from the planets of our own solar system, and prospects for habitability among the different types of planets. Chapter 12 covers the search for extraterrestrial intelligence (SETI). Chapter 13 discusses the challenge of and prospects for interstellar travel, and then uses these ideas to investigate the Fermi paradox ("Where is everybody?"), the potential solutions to the paradox, and the implications of the considered solutions. The Epilogue is designed as a short wrap-up of the course, focusing on philosophical issues relating to the search for life beyond Earth.

Pedagogical Features of *Life in the Universe*

Along with the main narrative, *Life in the Universe* includes a number of pedagogical devices designed to enhance student learning:

- **Basic Chapter Structure** Each chapter is carefully structured to ensure that students understand the goals up front, learn the details, and pull together all the ideas at the end. In particular, note the following key structural elements:
  - **Chapter Learning Goals** Each chapter opens with a page offering an enticing image and a brief overview of the chapter, including a list of the section titles and associated learning goals. The learning goals are presented as key questions designed to help students both understand what they will be learning about and stay focused on these key goals as they work through the chapter.

- **Introduction and Epigraph** The main chapter text begins with a two- to three-paragraph introduction to the chapter material and an inspirational quotation relevant to the chapter.

- **Section Structure** Chapters are divided into numbered sections, each addressing one key aspect of the chapter material. Each section begins with a short introduction that leads into a set of learning goals relevant to the section—the same learning goals listed at the beginning of the chapter.

- **The Process of Science in Action** The entire book is built around showing that science is a process, helping students understand how scientific ideas arise and how they gain acceptance through careful studies of evidence. To reinforce these ideas, every chapter ends with a final section designated as "The Process of Science in Action," in which we explore one topic in particular depth to show students various aspects of how science works in practice.

- **The Big Picture** Every chapter narrative ends with this feature, designed to help students put what they've learned in the chapter into the context of the overall goal of gaining a broader perspective on ourselves, our planet, and prospects for life beyond Earth.

- **Chapter Summary** The end-of-chapter summary offers a concise review of the learning goal questions, helping reinforce student understanding of key concepts from the chapter. Thumbnail figures are included to remind students of key illustrations and photos in the chapter.

- **End-of-Chapter Exercises** Each chapter includes an extensive set of exercises that can be used for study, discussion, or assignment. The end-of-chapter exercises are organized into the following subsets:

  - **Review Questions:** Questions that students should be able to answer from the reading alone

  - **Does It Make Sense?** (or similar title): A set of short statements that students are expected to evaluate, determining whether each statement makes sense and explaining why or why not; these exercises are generally easy once students understand a particular concept, but very difficult otherwise, making them an excellent probe of comprehension

  - **Quick Quiz:** A short multiple-choice quiz that allows students to check their progress

  - **Process of Science Questions:** Essay or discussion questions that help students focus on how science progresses over time

  - **Group Work Exercise:** A suggested activity designed for collaborative learning in class

  - **Short-Answer/Essay Questions:** Questions that go beyond the Review Questions in asking for conceptual interpretation

  - **Quantitative Problems:** Problems that require some mathematics, usually based on topics covered in the Cosmic Calculations boxes

  - **Discussion Questions:** Open-ended questions for class discussions

  - **Web Projects:** A few suggestions for additional web-based research

- **Additional Features** You'll find a number of other features designed to increase student understanding, both within individual chapters and at the end of the book, including the following:

  - **Annotated Figures** Key figures in each chapter use the research-proven technique of annotation—the placement on the figure of carefully crafted text (in blue) to guide students through interpreting graphs, following process figures, and translating between different representations.

  - **Think About It** This feature, which appears throughout the book in the form of short questions integrated into the narrative, gives students the opportunity to reflect on important new concepts. It also serves as an excellent starting point for classroom discussions.

  - **Cosmic Calculations Boxes** These boxes contain optional mathematics exercises. Many of the quantitative exercises at the ends of chapters are based on these boxes.

  - **Special Topic Boxes** These boxes contain supplementary discussion topics related to the chapter material but not prerequisite to the continuing discussion.

  - **Movie Madness Boxes** These boxes contain brief discussions of popular movies that deal with various aspects of life in the universe, presented in a way designed to be both humorous and informative.

  - **Cross-References** When a concept is covered in greater detail elsewhere in the book, we include a cross-reference in brackets to the relevant section (e.g., [Section 5.2]).

  - **Glossary** A detailed glossary makes it easy for students to look up important terms.

  - **Appendixes** The appendixes contain a number of useful references and tables including key constants (Appendix A), key formulas (Appendix B), key mathematical skills (Appendix C), the periodic table (Appendix D), and a summary of key solar system facts (Appendix E).

  - **MasteringAstronomy® Resources** New to this edition, *Life in the Universe* now has a dedicated MasteringAstronomy site, with numerous resources including online quizzes, Interactive Figures and Photos, Self-Guided Tutorials, and much more.

New for the Fourth Edition

Astrobiology is a fast-moving field, and there have been many new developments since we wrote the third edition. You will therefore find many sections of the book almost entirely rewritten, though we have retained the basic organization of the text. Here, briefly, is a list of some of the most important changes and updates we have made:

- Every chapter has undergone at least some substantial change, in order to bring the scientific material fully up-to-date with recent research.

- In Chapter 2, we have enhanced the discussion of the nature of science with the new Table 2.1, which summarizes how the same terms often have different meanings in science and in everyday usage. We've also reorganized and rewritten Section 2.4.

- Chapter 3 has been significantly reorganized and rewritten, particularly in Sections 3.3 through 3.5, in order to reflect new understanding of extrasolar planetary systems.

- Section 5.5 on extremophiles has been almost completely rewritten in light of new discoveries.

- Chapter 6 has been heavily rewritten, particularly in Sections 6.1 and 6.2, in light of new research concerning the most ancient fossils of life.

- Chapter 7 has been significantly revised, particularly in Sections 7.3 and 7.4, so that it now covers several new spacecraft missions, including relevant results from *Dawn* at Ceres and *New Horizons* at Pluto.

- Chapter 8 on Mars has been nearly entirely rewritten to reflect the latest results from *Curiosity*, *MAVEN*, and more.

- Chapter 9 has several important scientific updates, particularly with regard to recent results from the *Cassini* mission in its studies of Titan and Enceladus.

- In Chapter 10, we have significantly updated and revised the final section on global warming.

- Chapter 11 has been almost completely reorganized and rewritten in light of the rapidly advancing study of extrasolar planets.

- In every chapter, we have added one or two "process of science" questions and one new group work question to the exercise set.

- We've replaced three Movie Madness boxes entirely and updated numerous others.

Supplements and Resources

In addition to the book itself, a number of supplements are available to help you as an instructor. The following is a brief summary; contact your local Pearson representative for more information.

- **MasteringAstronomy®** (www.masteringastronomy. com). MasteringAstronomy is the most widely used and most advanced astronomy tutorial and assessment system in the world. By capturing the step-by-step work of students nationally, MasteringAstronomy has established an unparalleled database of learning challenges and patterns. Using these student data, every activity and problem has been refined. The result is a library of activities of unique educational effectiveness and assessment accuracy. MasteringAstronomy provides students with two learning systems in one: a dynamic self-study area and the ability to participate in online assignments.

  MasteringAstronomy provides instructors with a fast and effective way to give uncompromising, wide-ranging online homework assignments of just the right difficulty level and duration. Tutorials built around text content are available in MasteringAstronomy. The tutorials coach 85 percent of students to the correct answer with specific wrong-answer feedback. Powerful post-diagnostics enable instructors to assess the progress of their class as a whole and to quickly identify an individual student's areas of difficulty. A media-rich self-study area is included that students can use whether or not the instructor assigns homework.

- *Pearson eText 2.0* (ISBN 978-0-13-408002-4). An interactive Pearson eText will be available for this edition.

  - Now available on smartphones and tablets

  - Seamlessly integrated videos and other rich media

  - Accessible (screen-reader ready)

  - Configurable reading settings, including resizable type and night-reading mode

  - Instructor and student note-taking, highlighting, bookmarking, and search

- *Life in the Universe Activities Manual*, Second Edition, by Ed Prather, Erika Offerdahl, and Tim Slater (ISBN 978-0-80-531712-1). This manual provides creative projects that explore a wide range of concepts in astrobiology. It can be used as a laboratory component for a life in the universe course or as a source for group activities in the classroom.

- **Additional Instructor Resources.** This instructor resource area residing in MasteringAstronomy includes jpegs of all figures from the text, PowerPoint® Lecture Outlines that incorporate figures, photos, and multi-media, and the Test Bank in both Word and Testgen® formats. TestGen is an easy-to-use, fully networkable program for creating tests ranging from short quizzes to long exams. Questions from the Test Bank are supplied, and professors can use the Question Editor to modify existing questions or create new questions.

Acknowledgments

A textbook may carry the names of its authors, but it is the result of the hard work of a long list of committed individuals.

We could not possibly name everyone who has had a part in this book, but we would like to call attention to a few people who have played particularly important roles. First, we thank the friends and family members who put up with us during the long hours that we worked on this book. Without their support, this book would not have been possible.

At Pearson, we offer special thanks to our editors Nancy Whilton, Tema Goodwin, and Lizette Faraji (who put in countless hours to make this book meet its schedule). Many others have also helped make this book happen, including Adam Black, Joan Marsh, Mary Douglas, Margot Otway, Claire Masson, Michael Gillespie, Debbie Hardin, Sally Lifland, Mark Ong, and many more.

We've also been fortunate to be able to draw on the expertise of several other Pearson authors, in some cases drawing ideas and artwork directly from their outstanding texts. For their gracious help, we thank the authors of the Campbell *Biology* textbooks and the authors of *The Cosmic Perspective* astronomy texts. And very special thanks go to Bruce Jakosky, who was our coauthor on the first edition and provided much of the vision around which this book has been built.

Finally, we thank the many people who have carefully reviewed portions of the book in order to help us make it both as scientifically up-to-date and as pedagogically useful as possible:

Wayne Anderson, Sacramento City College
Timothy Barker, Wheaton College
Wendy Hagen Bauer, Wellesley College
Laura Baumgartner, University of Colorado, Boulder
Jim Bell, Cornell University
Raymond Bigliani, Farmingdale State University of New York
Janice Bishop, SETI Institute
Sukanta Bose, Washington State University
Greg Bothun, University of Oregon
Paul Braterman, University of North Texas
Juan Cabanela, Haverford College
Christopher Churchill, New Mexico State University
Leo Connolly, San Bernardino State
Manfred Cuntz, University of Texas at Arlington
Alfonso Davila, SETI Institute
Steven J. Dick, U.S. Naval Observatory
James Dilley, Ohio University
Anthony Dobrovolskis, SETI Institute
Alberto G. Fairén, Cornell University
Jack Farmer, Arizona State University
Steven Federman, University of Toledo
Eric Feigelson, Penn State University
Daniel Frank, University of Colorado School of Medicine
Richard Frankel, California Polytechnic State University

Rica S. French, MiraCosta College
Tracy Furutani, California Polytechnic State University
Bob Garrison, University of Toronto
Harold Geller, George Mason University
Perry A. Gerakines, University of Alabama at Birmingham
Donna H. Gifford, Pima Community College
Kevin Grazier, Santa Monica College
Bob Greeney, Holyoke Community College
Bruce Hapke, University of Pittsburgh
William Hebard, Babson College
Beth Hufnagel, Anne Arundel Community College
James Kasting, Penn State University
Laura Kay, Barnard College
Jim Knapp, Holyoke Community College
David W. Koerner, Northern Arizona University
Karen Kolehmainen, California State University, San Bernardino
Kenneth M. Lanzetta, Stony Brook University
Kristin Larson, Western Washington University
James Lattimer, Stony Brook University
Jack Lissauer, NASA Ames Research Center
Abraham Loeb, Harvard University
Bruce Margon, Space Telescope Science Institute
Lori Marino, Emory University
Christopher Matzner, University of Toronto
Gary Melcher, Pima Community College
Stephen Mojzsis, University of Colorado, Boulder
Michele Montgomery, University of Central Florida
Ken Nealson, University of Southern California
Norm Pace, University of Colorado, Boulder
Stacy Palen, Weber State University
Robert Pappalardo, Jet Propulsion Laboratory, California Institute of Technology
Robert Pennock, Michigan State University
James Pierce, Minnesota State University at Mankato
Eugenie Scott, National Center for Science Education
Beverly J. Smith, East Tennessee State University
Inseok Song, University of Georgia
Charles M. Telesco, University of Florida
David Thomas, Lyon College
Glenn Tiede, Bowling Green State University
Gianfranco Vidali, Syracuse State University
Fred Walter, Stony Brook University
John Wernegreen, Eastern Kentucky University
William Wharton, Wheaton College
Nicolle Zellner, Albion College
Ben Zuckerman, University of California, Los Angeles

About the Authors

Jeffrey Bennett

Jeffrey Bennett, a recipient of the American Institute of Physics Science Communication Award, holds a B.A. in biophysics (UC San Diego) and an M.S. and Ph.D. in astrophysics (University of Colorado). He specializes in science and math education and has taught at every level from preschool through graduate school. Career highlights include serving 2 years as a visiting senior scientist at NASA headquarters, where he developed programs to build stronger links between research and education, and proposing and helping to develop the Voyage scale model solar system on the National Mall (Washington, D.C.). He is the lead author of textbooks in astronomy, astrobiology, mathematics, and statistics, and of critically acclaimed books for the public including *Beyond UFOs* (Princeton University Press, 2008/2011), *Math for Life* (Big Kid Science, 2014), *What Is Relativity?* (Columbia University Press, 2014), and *On Teaching Science* (Big Kid Science, 2014). He is also the author of six science picture books for children, including *Max Goes to the Moon*, *The Wizard Who Saved the World*, and *I, Humanity*; all six have been launched to the International Space Station and read aloud by astronauts for NASA's Story Time from Space program. Dr. Bennett lives in Boulder, Colorado, with his wife, children, and dog. His personal website is www.jeffreybennett.com.

Seth Shostak

Seth Shostak earned his B.A. in physics from Princeton University (1965) and a Ph.D. in astronomy from the California Institute of Technology (1972). He is currently a senior astronomer and Director of the Center for SETI Research at the SETI Institute in Mountain View, California, where he helps press the search for intelligent cosmic company. For much of his career, Seth conducted radio astronomy research on galaxies and investigated the fact that these massive objects contain large amounts of unseen mass. He has worked at the National Radio Astronomy Observatory in Charlottesville, Virginia, as well as at the Kapteyn Astronomical Institute in Groningen, the Netherlands (where he learned to speak bad Dutch). Seth also founded and ran a company that produced computer animation for television. He has written more than four hundred popular articles on various topics in astronomy, technology, film, and television. A frequent fixture on the lecture circuit, Seth gives approximately 70 talks annually at both educational and corporate institutions; he is also a frequent commentator on astronomical matters for radio and television. His book *Confessions of an Alien Hunter: A Scientist's Search for Extraterrestrial Intelligence* (National Geographic, 2009) details the latest ideas, as well as the personal experience of his day job. When he's not trying to track down aliens, Seth can often be found behind the microphone, as host of the SETI Institute's weekly one-hour radio show about science, *Big Picture Science*.

How to Succeed in Your Astrobiology Course

If Your Course Is	Times for Reading the Assigned Text (per week)	Times for Homework Assignments (per week)	Times for Review and Test Preparation (average per week)	Total Study Time (per week)
3 credits	2 to 4 hours	2 to 3 hours	2 hours	6 to 9 hours
4 credits	3 to 5 hours	2 to 4 hours	3 hours	8 to 12 hours
5 credits	3 to 5 hours	3 to 6 hours	4 hours	10 to 15 hours

The Key to Success: Study Time

The single most important key to success in any college course is to spend enough time studying. A general rule of thumb for college classes is that you should expect to study about 2 to 3 hours per week *outside* of class for each unit of credit. For example, based on this rule of thumb, a student taking 15 credit hours should expect to spend 30 to 45 hours each week studying outside of class. Combined with time in class, this works out to a total of 45 to 60 hours spent on academic work—not much more than the time a typical job requires, and you get to choose your own hours. Of course, if you are working while you attend school, you will need to budget your time carefully.

As a rough guideline, your study time might be divided as shown in the table above. If you find that you are spending fewer hours than these guidelines suggest, you can probably improve your grade by studying longer. If you are spending more hours than these guidelines suggest, you may be studying inefficiently; in that case, you should talk to your instructor about how to study more effectively.

Using This Book

Each chapter in this book is designed to make it easy for you to study effectively and efficiently. To get the most out of each chapter, you might wish to use the following study plan.

- A textbook is not a novel, and you'll learn best by reading the elements of this text in the following order:

  1. Start by reading the Learning Goals (in the form of key questions) and the introductory paragraphs at the beginning of the chapter so that you'll know what you are trying to learn.
  2. Get an overview of key concepts by studying the illustrations and their captions and annotations. The illustrations highlight most major concepts, so this "illustrations first" strategy gives you an opportunity to survey the concepts before you read about them in depth. You will find the two-page Cosmic Context figures especially useful.
  3. Read the chapter narrative, trying the Think About It questions as you go along, but save the boxed features (e.g., Cosmic Calculations, Special Topics, Movie Madness) to read later. As you read, make notes on the pages to remind yourself of ideas you'll want to review later. Take notes as you read, but avoid using a highlight pen (or a highlighting tool if you are using an e-book), which makes it too easy to highlight mindlessly.
  4. After reading the chapter once, go back through and read the boxed features.
  5. Review the Chapter Summary, ideally by trying to answer the Learning Goal questions for yourself before reading the given answers.

- After completing the reading as outlined above, test your understanding with the end-of-chapter exercises. A good way to begin is to make sure you can answer all of the Review and Quick Quiz Questions; if you don't know an answer, look back through the chapter until you figure it out.
- Visit the MasteringAstronomy® site and make use of resources that will help you further build your understanding. These resources have been developed specifically to help you learn the most important ideas in your course, and they have been extensively tested to make sure they are effective. They really do work, and the only way you'll gain their benefits is by going to the website and using them.

General Strategies for Studying

- Budget your time effectively. Studying 1 or 2 hours each day is more effective, and far less painful, than studying all night before homework is due or before exams.
- Engage your brain. Learning is an active process, not a passive experience. Whether you are reading, listening to a lecture, or working on assignments, always make sure that your mind is actively engaged. If you find your mind drifting or find yourself falling asleep, make a conscious effort to revive yourself, or take a break if necessary.
- Don't miss class. Listening to lectures and participating in discussions is much more effective than reading someone else's notes. Active participation will help you retain what you are learning. Also, be sure to complete any assigned reading *before* the class in which it will be discussed. This is crucial, since class lectures and discussions are designed to help reinforce key ideas from the reading.
- Take advantage of resources offered by your professor, whether it be email, office hours, review sessions, online

chats, or other opportunities to talk to and get to know your professor. Most professors will go out of their way to help you learn in any way that they can.

- Start your homework early. The more time you allow yourself, the easier it is to get help if you need it. If a concept gives you trouble, do additional reading or studying beyond what has been assigned. And if you still have trouble, ask for help: You surely can find friends, peers, or teachers who will be glad to help you learn.
- Working together with friends can be valuable in helping you understand difficult concepts, but be sure that you learn *with* your friends and do not become dependent on them.
- Don't try to multitask. A large body of research shows that human beings simply are not good at multitasking: When we attempt it, we do more poorly at all of the individual tasks. And in case you think you are an exception, the same research found that those people who believed they were best at multitasking were actually the worst! So when it is time to study, turn off your electronic devices, find a quiet spot, and concentrate on focusing your efforts.

Preparing for Exams

- Study the Review Questions, and rework problems and other assignments; try additional questions to be sure you understand the concepts. Study your performance on assignments, quizzes, or exams from earlier in the term.
- Work through the relevant chapter quizzes and other study resources available at the MasteringAstronomy® site.
- Study your notes from lectures and discussions. Pay attention to what your instructor expects you to know for an exam.
- Reread the relevant sections in the textbook, paying special attention to notes you have made on the pages.
- Study individually *before* joining a study group with friends. Study groups are effective only if every individual comes prepared to contribute.
- Don't stay up too late before an exam. Don't eat a big meal within an hour of the exam (thinking is more difficult when blood is being diverted to the digestive system).
- Try to relax before and during the exam. If you have studied effectively, you are capable of doing well. Staying relaxed will help you think clearly.

Presenting Homework and Writing Assignments

All work that you turn in should be of *collegiate quality:* neat and easy to read, well organized, and demonstrating mastery of the subject matter. Future employers and teachers will expect this quality of work. Moreover, although submitting homework of collegiate quality requires "extra" effort, it serves two important purposes directly related to learning:

1. The effort you expend in clearly explaining your work solidifies your learning. In particular, research has shown that writing and speaking trigger different areas of your brain. Writing something down—even when you think you already understand it—reinforces your learning by involving other areas of your brain.
2. If you make your work clear and self-contained (that is, make it a document that you can read without referring to the questions in the text), you will have a much more useful study guide when you review for a quiz or exam.

The following guidelines will help ensure that your assignments meet the standards of collegiate quality:

- Always use proper grammar, proper sentence and paragraph structure, and proper spelling. Do not use texting shorthand.
- Make all answers and other writing fully self-contained. A good test is to imagine that a friend will be reading your work and to ask yourself whether the friend will understand exactly what you are trying to say. It is also helpful to read your work out loud to yourself, making sure that it sounds clear and coherent.
- In problems that require calculation:

  1. Be sure to *show your work* clearly so that both you and your instructor can follow the process you used to obtain an answer. Also, use standard mathematical symbols, rather than "calculator-ese." For example, show multiplication with the symbol (not with an asterisk), and write 10^5, not 10^5 or 10E5.
  2. *Check that word problems have word answers.* That is, after you have completed any necessary calculations, make sure that any problem stated in words is answered with one or more *complete sentences* that describe the point of the problem and the meaning of your solution.
  3. Express your word answers in a way that would be *meaningful* to most people. For example, most people would find it more meaningful if you expressed a result of 720 hours as 1 month. Similarly, if a precise calculation yields an answer of 9,745,600 years, it may be more meaningfully expressed in words as "nearly 10 million years."

- Include illustrations whenever they help explain your answer, and make sure your illustrations are neat and clear. For example, if you graph by hand, use a ruler to make straight lines. If you use software to make illustrations, be careful not to make them overly cluttered with unnecessary features.
- If you study with friends, be sure that you turn in your own work stated in your own words—you should avoid anything that might give even the *appearance* of possible academic dishonesty.

Credits

1 A Universe of Life?

▲ **About the photo:** Earth is home to an abundance of life, making us wonder if other worlds might also be home to life.

The night sky glitters with stars, each a sun, much like our own Sun. Many stars have planets, some of which may be much like Earth and other planets of our own solar system. Among these countless worlds, it may seem hard to imagine that our world could be the only home for life. But while the possibility of life beyond Earth might seem quite reasonable, we do not yet know whether it actually exists.

Learning whether the universe is full of life holds great significance for the way we view ourselves and our planet. If life is rare or nonexistent elsewhere, we will view our planet with added wonder. If life is common, we'll know that Earth is not quite as special as it may seem. If civilizations are common, we'll be forced to accept that humanity is just one of many intelligent species inhabiting the universe. The profound implications of finding—or not finding—extraterrestrial life make the question of life beyond Earth an exciting topic of study.

The primary purpose of this book is to give you the background needed to understand new and exciting developments in the human quest to find life beyond Earth. We'll begin in this chapter with a brief introduction to the subject and to why it has become such a hot topic of scientific research.

1.1 The Possibility of Life Beyond Earth

Aliens are everywhere, at least if you follow the popular media (Figure 1.1). Starships on television, such as the *Enterprise* or *Voyager,* are on constant prowl throughout the galaxy, seeking out new life and hoping it speaks English (or something close enough to English for the "universal translator"). In *Star Wars,* aliens from many planets gather at bars to share drinks and stories, and presumably to marvel at the fact that they have greater similarity in their level of technology than do different nations on Earth. Closer to home, movies like *Independence Day, Men in Black,* and *War of the Worlds* feature brave Earthlings battling evil aliens—or, as in the case of *Avatar,* brave aliens battling evil humans— while numerous websites carry headlines about the latest alien landings. Even serious newspapers and magazines run occasional articles about UFO sightings or about claims that the U.S. government is hiding alien corpses at "Area 51."

Scientists are interested in aliens too, although most scientists remain deeply skeptical about reports of aliens on Earth (for reasons we'll discuss later in the book). Scientists are therefore searching for life elsewhere, looking for evidence of life on other worlds in our solar system, trying to learn whether we should expect to find life on planets orbiting other stars, and scanning for signals broadcast by other civilizations. Indeed, the study of life in the universe is one of the most exciting fields of active scientific research, largely because of its clear significance: The discovery of life of any kind beyond Earth would forever change our perspective on how we fit into the universe as a whole, and would undoubtedly teach us much more about life here on Earth as well.

What are we searching for?

When we say we are searching for *life* in the universe, just what is it that we are looking for? Is it the kind of intelligent life we see portrayed in science fiction TV shows and films? Is it something more akin to the plants and animals we see in parks and zoos? Is it tiny, bacteria-like microbes? Or could it be something else entirely?

The simple answer is "all of the above." When we search for **extraterrestrial life,** or life beyond Earth, we are looking for any sign of life, be it simple, complex, or intelligent. We don't care if it looks exactly like life we are familiar with on Earth or if it is dramatically different. However, we can't really answer the question of what we are looking for unless we know what life *is*.

Unfortunately, defining life is no simple matter, not even here on Earth where we have bountiful examples of it. Ask yourself: What common attributes make us think that a bacterium, a beetle, a mushroom, a tumbleweed, a maple tree, and a human are all alive, while we think that a crystal, a cloud, an ocean, or a fire are not? If you spend just a little time considering this question, you'll begin to appreciate its difficulty. For example, you might say that life can move, but the same is true of clouds and oceans. You might say that life can grow, but so can crystals. Or you might say that life can reproduce and spread, but so can fire. We will explore in Chapter 5 how scientists try to answer this question and come up with a general definition of life, but for now it should be clear that this is a complicated question that affects how we search for life in the universe.

Because of this definitional difficulty, the scientific search for extraterrestrial life in the universe generally presumes a search for life that is at least somewhat Earth-like and that we could therefore recognize based on what we know from studying life on Earth. Science fiction fans will object that this search is far too limited, and they may be right—but we have to start somewhere, so we begin with what we understand.

Think About It Name a few recent television shows and movies that involve aliens of some sort. Do you think any of these shows portray aliens in a scientifically realistic fashion? Explain.

FIGURE 1.1
Aliens have become a part of modern culture, as illustrated in this movie poster.

Is it reasonable to imagine life beyond Earth?

The scientific search for life in the universe is a relatively recent development in human history, but the idea of extraterrestrial life is not. Many ancient cultures told stories about beings living among the stars and, as we'll discuss in Chapter 2, the ancient Greeks engaged in serious philosophical debate about the possibility of life beyond Earth.

Until quite recently, however, all these ideas remained purely speculative, because there was no way to study the question of extraterrestrial life scientifically. It was always possible to *imagine* extraterrestrial life, but there was no scientific reason to think that it could really exist. Indeed, the relatively small amounts of data that might have shed some light on the question of life beyond Earth were often misinterpreted. Prior to the twentieth century, for example, some scientists guessed that Venus might harbor a tropical paradise—a guess that was based on little more than the fact that Venus is covered by clouds and closer than Earth to the Sun. Mars was the subject of even more intense debate, largely because a handful of scientists thought they saw long, straight canals on the surface [Section 8.1]. The canals, which don't really exist, were cited as evidence of a martian civilization.

Today, we have enough telescopic and spacecraft photos of the planets and large moons in our solar system to be quite confident that no civilization has ever existed on any of them. The prospect of large animals or plants seems almost equally improbable. Nevertheless, scientific interest in life beyond Earth has exploded in the past few decades. Why?

We'll spend most of the rest of the book answering this question, but we can summarize the key points briefly. First, although large, multicellular life in our solar system seems unlikely anywhere but on Earth, new discoveries in both planetary science and biology make it seem plausible that simpler life—perhaps tiny microbes—might yet exist on other planets or moons of the solar system. Second, while we've long known that the universe is full of *stars*, we've only recently gained concrete evidence telling us that it is also full of *planets*, which means there are far more places where we could potentially search for life. Third, advances in both scientific understanding and technology now make it possible to study the question of life in the universe through established techniques of science, something that was not possible just a few decades ago. For example, we now understand enough about biology to explore the conditions that might make it possible for life to exist on other worlds, and we know enough about planets, and many of their moons, to consider which ones might be capable of harboring life. We are also rapidly developing the spacecraft technology needed to search for microbes on other worlds of our solar system and the telescope technology needed to look for signs of life among the stars.

The bottom line is that while it remains possible that life exists only on Earth, we now have plenty of scientific reasons to think that life might be widespread and that we might detect it if it is.

1.2 The Scientific Context of the Search

Almost every field of scientific research has at least some bearing on the search for life in the universe. Even seemingly unrelated fields such as mathematics and computer science play important roles. For example, we use mathematics to do the many computations that help us understand all other areas of science, and we use computers to simulate everything from the formation of stars and planets to the way in which the molecules of life interact with one another. However, three disciplines play an especially important role in framing the context of the scientific search for life: astronomy, planetary science (which includes geology and atmospheric science), and biology.

How does astronomy help us understand the possibilities for extraterrestrial life?

For most of human history, our conception of the cosmos was quite different from what it is today. Earth was widely assumed to be the center of the universe. Planets were lights in the sky, named for ancient gods, and no one had reason to think they could be *worlds* on which we might search for life. Stars were simply other lights in the sky, distinguished from the planets only by the fact that they remained fixed in the patterns of the constellations, and few people even considered the possibility that our Sun could be one of the stars. Moreover, with the Sun and planets presumed to be orbiting around Earth, there was no reason to think that stars could have planets of their own, let alone planets on which there might be life.

When you consider the dominance that this Earth-centered, or **geocentric,** view of the universe held for thousands of years, it becomes obvious that astronomy plays a key role in framing the context of the modern search for life. We will discuss in Chapter 2 how and why the human view of the cosmos changed dramatically about 400 years ago, and we'll consider the modern astronomical context in some detail in Chapter 3. But the point should already be clear: We now know that Earth is but one tiny world orbiting one rather ordinary star in a vast cosmos, and this fact opens up countless possibilities for life on other worlds.

Astronomy provides context to the search for life in many other ways as well, but one more is important enough to mention right now: By studying distant objects, we have learned that the physical laws that operate in the rest of the universe are the same as those that operate right here on Earth. This tells us that if something happened here, it is possible that the same thing could have happened somewhere else, at least in principle. We are not the center of the universe in location, and we have no reason to think we are "central" in any other way, either. To summarize, the astronomical context makes it clear that the universe holds an enormous number of stars that could potentially be orbited by planets with life (Figure 1.2).

How does planetary science help us understand the possibilities for extraterrestrial life?

Planetary science is the name we give to the study of almost everything having to do with planets. It includes the study of planets themselves, as well as the study of moons orbiting planets, the study of how planets form, and the study of other objects that may form in association with planets (such as asteroids and comets). Planetary science helps set the context for the search for life in the universe in several different ways, but two are especially important.

First, by learning how planets form, we develop an understanding of how common we might expect planets to be. Until just about the middle of the last century, we really had no basis for assuming that many other stars would have their own planets. Some scientists thought this likely, while others did not, and we lacked the data needed to distinguish between the two possibilities [Section 3.5]. But during the latter half of the twentieth century, a growing understanding of the processes by which our own solar system formed—much of it based on evidence obtained through human visits to the Moon and spacecraft visits to other planets—gradually made it seem more likely that other stars might similarly be born with planetary systems.

Still, as recently as 1995, no one was sure whether planets encircled other stars like the Sun. That was the year in which the first strong evidence was obtained for the existence of **extrasolar planets,** or planets orbiting stars other than our Sun. Since that time, additional discoveries of extrasolar planets have poured in at an astonishing rate, so that the number of known extrasolar planets now far exceeds the number of planets of our solar system (Figure 1.3). Based on the statistics of these discoveries, it now seems likely that most stars have planets and, as we'll discuss in Chapter 11, it seems reasonable to imagine that life—and

FIGURE 1.2
The astronomical context tells us that our Sun is an ordinary star in a vast universe, implying that there could be an enormous number of stars with planets that might potentially host life. This Hubble Space Telescope photo shows a cluster of young, massive stars (NGC 3603) surrounded by a gas cloud in which Sun-like stars may still be forming. Careful study of distant stars and gas clouds shows that they are made of the same basic chemical elements and obey the same physical laws that we are familiar with on Earth.

FIGURE 1.3
Artist's conception of the planet "Kepler 11f," which is about twice the mass of Earth and orbits its star with at least five other planets. Kepler 11f is one of thousands of planets discovered by the *Kepler* spacecraft.

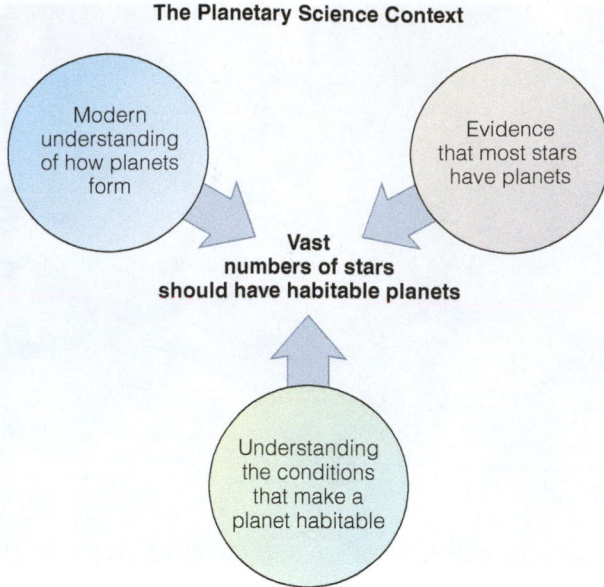

The Planetary Science Context

Modern understanding of how planets form

Evidence that most stars have planets

Vast numbers of stars should have habitable planets

Understanding the conditions that make a planet habitable

FIGURE 1.4
The astronomical context showed us that vast numbers of stars could be hosts to planets, and planetary science suggests that these stars are indeed orbited by planets, many of which should be habitable.

possibly even civilizations—could exist on at least some of these planets or their moons.

A second way in which planetary science shapes the context for the search for life is by helping us understand why planets differ. For example, by studying planets and comparing them to one another, we have learned why some planets are rocky like Earth while others, like Jupiter, contain vast amounts of hydrogen and helium gas. We've also learned why Venus is *so* much hotter than Earth despite the fact that, in the scheme of our solar system, it is only slightly closer to the Sun. Similarly, we can now explain why the Moon is desolate and barren even though it orbits the Sun at essentially the same distance as Earth, and we have a fairly good idea of why Mars is cold and dry today, when evidence shows that it was warmer and wetter in the distant past.

This understanding of how planets work gives us deeper insight into the nature of planetary systems in general. More important to our purposes, it also helps us understand what to look for as we search for **habitable worlds**—worlds that have the ingredients and conditions necessary for life. After all, given that there are far more worlds in the universe than we can ever hope to study in detail, we can improve our odds of success in finding life by constraining the search to those worlds that are the most promising. Be sure to note that when we ask whether a world is habitable, we are asking whether it offers environmental conditions under which life *could* arise or survive, not whether it actually harbors life.

Also keep in mind that when we say a world is habitable, we do not necessarily mean that familiar plants, animals, or people could survive there. For much of Earth's history, nearly all life was microscopic, and even today, the total mass of microbes on Earth is greater than that of all plants and animals combined. The search for habitable worlds is primarily a search for places where microbes of some kind might survive, though we might find larger organisms as well. To summarize, the planetary science context suggests that most of the stars in the universe should indeed have planets and that we should expect many of them to be at least potential homes for life (Figure 1.4).

How does biology help us understand the possibilities for extraterrestrial life?

Astronomy, planetary science, and other science disciplines play important roles in shaping the context for the search for life in the universe, but since we are searching for *life*, the context of biology is especially important. Just as you wouldn't look for a house to buy without knowing something about real estate, it would make no sense to search for life if we didn't know something about how life functions. The key question about the biological context of the search revolves around whether we should expect biology to be rare or common in the cosmos.

Wherever we have looked in the universe, we have found clear evidence that the same laws of nature are operating. We see galaxies sprinkled throughout space, and we see that the same stellar processes that occur in one place also occur in others. In situations in which we can observe orbital motions, we find that they agree with what we expect from the law of gravity. These and other measurements make us confident that the basic laws of physics that we've discovered here on Earth also hold throughout the universe.

We can be similarly confident that the laws of chemistry are universal. Observations of distant stars show that they are made of the same chemical elements that we find here in our own solar system, and that interstellar gas clouds contain many of the same molecules we find on Earth. This provides conclusive evidence that atoms come in the same types and combine in the same ways throughout the universe.

Could biology also be universal? That is, could the biological processes we find on Earth be common throughout the cosmos? If the answer is yes, then the search for life elsewhere should be exciting and fruitful. If the answer is no, then life may be a rarity.

Because we haven't yet observed biology anywhere beyond Earth, we can't yet know whether biology is universal. However, evidence from our own planet gives us at least some reason to think that it might be. Laboratory experiments suggest that chemical constituents found on the early Earth would have combined readily into complex organic (carbon-based) molecules, including many of the building blocks of life [Section 6.2]. Indeed, scientists have found organic molecules in meteorites (chunks of rock that fall to Earth from space) and, through spectroscopy [Section 3.3], in clouds of gas between the stars. The fact that such molecules form even under the extreme conditions of space suggests that they form quite readily and may be common on many worlds.

Of course, the mere presence of organic molecules does not necessarily mean that life will arise, but the history of life on Earth gives us some reason to think that the step from chemistry to biology is not especially difficult. As we'll discuss in Chapter 6, geological evidence tells us that life on Earth arose quite early in Earth's history, at least on a geological time scale. If the transition from chemistry to biology were exceedingly improbable, we might expect that it would have required much more time. The early origin of life on Earth therefore suggests—but certainly does not prove—that life might also emerge quickly on other worlds with similar conditions.

Think About It Microbial life on Earth predates intelligent life like us by at least 3 to 4 billion years. Do you think this fact tells us anything about the likelihood of finding *intelligent* life, as opposed to finding any life, on extrasolar planets? Explain.

If life really can be expected to emerge under the right conditions, the only remaining question is the prevalence of those "right" conditions. Here, too, recent discoveries give us reason to think that biology could be common. In particular, biologists have found that microscopic life can survive and prosper under a much wider range of circumstances than was believed only a few decades ago [Section 5.5]. For example, we now know that life exists in extremely hot water near deep-sea volcanic vents, in the frigid conditions of Antarctica, and inside rocks buried a kilometer or more beneath the Earth's surface. Indeed, if we were to export these strange organisms from Earth to other worlds in our solar system—perhaps to Mars or Europa—it seems possible that at least some of them would survive. This suggests that the range of "right" conditions for life may be quite broad, in which case it might be possible to find life even on planets that are significantly different in character from Earth.

In summary, we have no reason to think that life ought to be rare and several reasons to expect that it may be quite common (Figure 1.5). If life is indeed common, studying it will give us new insights into life on Earth, even if we don't find other intelligent civilizations. These enticing

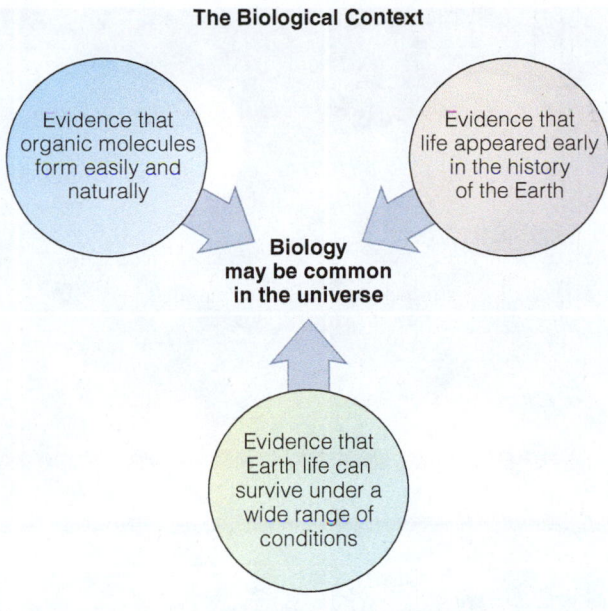

The Biological Context

FIGURE 1.5
Evidence from the study of life on Earth gives us at least some reason to think that biology may be common among the many potentially habitable worlds in the universe.

FIGURE 1.6
A "family portrait" of the eight official planets that orbit our Sun, along with the dwarf planets Ceres (located in the asteroid belt between Mars and Jupiter) and Pluto (located in the Kuiper belt beyond Neptune). Going across the top row and then the bottom row, the planets and dwarf planets are shown in order of increasing average distance from the Sun; the photos are *not* shown to scale.

prospects have captured the interest of scientists from many disciplines and from around the world, giving birth to a new science devoted to the study of, and search for, life in the universe.

1.3 Places to Search

The study of life in the universe involves fundamental research in all the scientific areas we have already mentioned, and others as well. Indeed, as you'll see throughout this book, the study of extraterrestrial life goes far beyond simply searching for living organisms. Still, all of this study is driven by the possibility that life exists elsewhere, so before we dive into any details, it's worth a quick overview of the places and methods we use in the search.

Where should we search for life in the universe?

The search for life in the universe takes place on several different levels. First, and foremost in many ways, it is a study of life right here on Earth. As we discussed earlier, we are still learning about the places and conditions under which terrestrial life exists, and many scientists are busy searching for new species of life on our own world. After all, the more we know about life here, the better we'll be able to search for it elsewhere.

SEARCHING OUR OWN SOLAR SYSTEM Turning our attention to places besides Earth, the first place to search for life is on other worlds in our own solar system. Our solar system has a lot of worlds: It has the planets and dwarf planets (including Ceres, Pluto, and Eris) orbiting the Sun, moons orbiting planets, and huge numbers of smaller objects such as asteroids and comets.

Figure 1.6 shows some of our best current views of the planets (and two of the five currently identified dwarf planets) in our solar system. Note that it is *not* to scale, since its purpose is to show each world as we know it today from spacecraft or through telescopes; you can turn to Figure 3.3 to see the sizes correctly scaled.

The photos alone make clear how different Earth is from every other planet in our solar system. Ours is the only planet with oceans of liquid water on its surface, a fact that provides an instant clue about why Earth is home to so much life: Water is crucial to all terrestrial life. Indeed, as we'll discuss in Chapters 5 and 7, we have good reason to think that liquid water is always a requirement for life, though it's possible that a few other liquids might work in place of water.

Given that we are primarily looking for life that is at least somewhat Earth-like, the need for water or some other liquid places constraints on where we might find life. Among the planets, Mars is the most promising candidate. As we'll discuss in detail in Chapter 8, strong evidence tells us that the now-barren surface of Mars (Figure 1.7) once had flowing water, making it seem reasonable to imagine life having arisen on Mars at that time. Mars still has significant amounts of water ice, so it is even possible that life exists on Mars today, perhaps hidden away in places where volcanic heat keeps underground water liquid. Past or present life seems much less likely on any of the other planets, though we can't rule it out completely; we'll discuss these dim prospects for planetary life in Chapter 7.

Aside from the planets, the most promising abodes for life in the solar system are a few of the large moons. At least five moons are potential candidates for life, including Jupiter's moon Europa (Figure 1.8). Current evidence strongly suggests that Europa hides a deep ocean of liquid water under its icy crust. Indeed, if we are interpreting the evidence correctly, the Europan ocean may have considerably more water than all of Earth's oceans combined [Section 9.2]. Because we suspect that life on Earth got started in the deep oceans [Section 6.1], Europa may well have all the conditions needed both for life to have arisen and for its ongoing survival. Two other moons of Jupiter—Ganymede and Callisto—also show some evidence for subsurface oceans, though the evidence is less strong and other considerations (primarily availability of energy) make them poorer prospects for harboring life. Other candidates for life include Saturn's moons Titan, which has a thick atmosphere and lakes of liquid methane, and Enceladus, which appears to have a subsurface ocean from which we observe fountains of ice spraying out into space [Section 9.3].

SEARCHING AMONG THE STARS In terms of numbers, there are many more places to look for life on planets and moons around other stars than in our own solar system. However, the incredible distances to the stars [Section 3.2] make searches of these worlds much more difficult. All stars are so far away that we will need great leaps in technology to have any hope at all of sending spacecraft to study their planets up close; for example, with current spacecraft, the journey to even the nearest stars would take close to 100,000 years.

With visits out of reach, telescopic searches represent our only hope of finding life on extrasolar worlds. Current telescope technology is able to detect extrasolar planets only under certain conditions. But the technology is advancing rapidly, and within a couple of decades we may have telescopes that are able to obtain moderate-resolution pictures and spectra of planets and moons around other stars. As a result, one important area of research is trying to figure out the photographic or spectral "signatures" that would tell us we are looking at a world with life.

FIGURE 1.7
The surface of Mars, photographed by NASA's *Curiosity* rover. The martian surface is dry and barren today, but strong evidence points to liquid water on its surface in the distant past.

FIGURE 1.8
This photograph shows Jupiter and two of its moons: Io is the moon in front of Jupiter's Great Red Spot, and Europa is to the right. Scientists suspect that Europa has a deep ocean beneath its surface of ice, making it a prime target in the search for life in our solar system.

Could aliens be searching for us?

So far we have talked about searching for life that is not searching for us—that is, life that we could identify only by seeing it with our spacecraft or telescopes. But if life really is common in the universe, there could be other places like Earth where life has evolved to become intelligent enough to be interested in searching for life beyond its home world. In that case, it is possible that other civilizations might actually be broadcasting signals that we could detect. The **search for extraterrestrial intelligence,** or **SETI,** which we'll discuss in Chapter 12, focuses on the search for such signals from alien civilizations (Figure 1.9). Although we don't know whether the search will meet with success, we can be sure that the unambiguous receipt of an alien message would be one of the most significant discoveries in human history—not to mention the fact that it would also probably answer many of our other questions about life in the universe.

1.4 The New Science of Astrobiology

We have seen that the study of life in the universe is a multidisciplinary field of scientific research, involving scientists with training in many different specialties. Nevertheless, because it has become a prominent and important area of study, it would be good to give the science of life in the universe its own name. A number of different names are in use, including "exobiology" and "bioastronomy," but in this book we follow the lead of NASA and call it **astrobiology.** This term is meant to invoke the combination of astronomy (the study of the universe) and biology (the study of life), so *astrobiology* literally means "the study of life in the universe."

How do we study the possibility of life beyond Earth?

Because astrobiology is a young science, scientists are still working to decide where to focus their research efforts. One major player in this effort is the NASA Astrobiology Institute, a collaboration involving scientists from NASA

FIGURE 1.9

This 140-foot radio telescope in West Virginia was used in 1996 to search for signals from extraterrestrial civilizations.

Movie Madness CINEMA ALIENS

Aliens should probably join the Screen Actors Guild. Every year, Hollywood reliably cranks out a handful of films in which visitors from distant star systems mess with our minds, our bodies, or our entire planet.

Cinema aliens are typecast, usually available in only two flavors: good and bad. A few, like loveable, wrinkly-faced little E.T., are willing to make a field trip of a few million light-years simply to pick some plants and hang with the kids. But most of these uninvited guests are cranky: They spend their time either dithering with our personal lives or blowing up famous landmarks just because they can.

Extraterrestrials didn't snag many movie roles until after the Second World War, when the rapid development of rocketry seemed to suggest that we'd soon be taking rides to the Moon, to Mars, and beyond. For the popcorn-eating public, it seemed inevitable that our descendants would visit other worlds as casually as you might head for the mall. And if we could do this, then it seemed only reasonable that advanced aliens were already roaming space, like motorcycle gangs on a Sunday afternoon.

The movie moguls studiously ignored the fact (which you'll encounter later in this book) that traveling between the stars is enormously more difficult than checking out the planets of your own solar system. The aliens won't do it just to share play time with the neighborhood children, or abduct you for unauthorized breeding experiments.

But the really big problem with Hollywood aliens, other than the fact that they seldom wear clothes, is that these frequently nasty visitors are inevitably portrayed as being close to our own level of technical development. We can engage the bad ones in aerial dogfights, or challenge them to a manly light-saber duel. But the reality is somewhat different. As we'll discuss in Chapter 13, if we ever make contact with actual aliens, their culture will almost certainly be thousands, millions, or billions of years beyond ours.

Of course, an invasion by hostile aliens with a million-year head start on *Homo sapiens* wouldn't make for an interesting movie. It would be Godzilla versus the chipmunks. But you don't mistake the movies for reality, do you?

and more than a dozen other research institutions across the United States. Similar efforts are under way in other countries, including the United Kingdom, Sweden, France, Spain, Russia, and Australia. These collaborations are among the most interdisciplinary in any area of science, bringing together astronomers, biologists, geologists, chemists, and many others seeking to understand the prospects of finding life beyond Earth.

Although different groups concentrate on different problems, most astrobiology research is concentrated in the following three areas:

1. Studying the conditions conducive to the origin and ongoing existence of life
2. Looking for such conditions on other planets in our solar system and around other stars
3. Looking for the actual occurrence of life elsewhere

Astrobiology therefore includes much more than simply searching for extraterrestrial life or civilizations. At a fundamental level, astrobiology research seeks to reveal the connections between living organisms and the places where they reside. In this sense, finding no life (on Mars, for example) is just as significant a result as finding life, because either way we learn about the conditions that can lead to the presence of life, about how life evolves in conjunction with planets, and about whether life is likely to be rare or common throughout the universe.

In the rest of this book, we will focus on the three areas listed above. After discussing the scientific context of the search in greater detail in Chapters 2 and 3, we'll turn our attention in Chapters 4 through 6 to the nature, origin, and evolution of life on Earth. This study of the history of life on our planet will help us understand the conditions under which we might expect to find life elsewhere. We'll then discuss prospects for life elsewhere in our solar system in Chapters 7 through 10, and the prospects for finding life—including intelligent life—beyond our solar system in Chapters 11 through 13. Along the way, we'll also learn what science can currently say about the future of life on Earth, we'll consider possible futures for our own species, and we'll discuss the philosophical implications of the search for—and potential discovery of—life beyond Earth.

The Big Picture

PUTTING CHAPTER 1 IN PERSPECTIVE

This chapter has offered a brief overview of the ideas we will cover in more depth in the rest of the book, primarily so that you will have a sense of what to expect in the rest of your study of life in the universe. As we will do in every chapter, we conclude with a brief "big picture" recap of how these ideas fit into the overall goals of the scientific study of life in the universe:

- Despite the abundance of aliens in popular media, we don't yet have any convincing evidence for life beyond Earth. Nevertheless, current understanding of astronomy, planetary science, and biology gives us good reason to think that it is at least reasonable to imagine that life may be widespread, and the discovery of extraterrestrial life of any kind would have profound significance to our understanding of life in the universe.

- It's conceivable that life may exist on any of several worlds in our own solar system, but it's extremely unlikely that any of this life is intelligent. However, we find many more possibilities when we consider life on planets or moons around other stars. And, through the search for extraterrestrial intelligence (SETI), it is even possible that we could receive a signal from an advanced civilization.

- The prospect that life may be common in the universe has given rise to the new science of *astrobiology*, an exciting and interdisciplinary topic of research that focuses both on understanding the possibility of finding life elsewhere and on the actual search for life beyond Earth.

Summary of Key Concepts

1.1 The Possibility of Life Beyond Earth

What are we searching for?
The search for **extraterrestrial life** is in principle a search for *any* kind of life. However, the difficulty of clearly defining life means that it's easier to focus the search on life that is at least somewhat similar to life here on Earth. This still opens a wide range of possibilities, from bacteria-like microbes to complex plants and animals.

Is it reasonable to imagine life beyond Earth?

People have long considered the possibility of life beyond Earth, but only recently have we been able to examine this possibility through the lens of science. While we have no evidence at this time of actual life beyond Earth, our scientific understanding of the possibilities makes it reasonable to think that life could exist elsewhere.

1.2 The Scientific Context of the Search

How does astronomy help us understand the possibilities for extraterrestrial life?
Astronomy tells us that we live on just a tiny planet orbiting one rather ordinary star in a vast cosmos, and that the same physical laws that operate here also operate throughout the universe. Together these ideas suggest that there could be many other worlds with life.

How does planetary science help us understand the possibilities for extraterrestrial life?

Based on current understanding of how planets form, we expect planets to be common around other stars—an idea that has been confirmed by discoveries of **extrasolar planets.** By learning how planets work, we learn the conditions that might make a **habitable world,** meaning a world that has the basic necessities for life, even if it does not actually have life.

How does biology help us understand the possibilities for extraterrestrial life?
Modern biology provides three lines of evidence suggesting that life *might* be common on other habitable worlds: (1) The fact that life arose quickly on Earth suggests that it might occur on any world that has the "right" conditions. (2) We know from observations of meteorites and interstellar clouds that organic molecules are common throughout the galaxy, suggesting that we'll find them on many other worlds. (3) The fact that some life on Earth survives even under seemingly extreme conditions suggests that life is hardy enough to survive in many other places as well.

1.3 Places to Search

Where should we search for life in the universe?

The search begins right here on Earth, as we seek to learn more about the life on our own planet. Elsewhere in our solar system we can search many planets and moons, but current understanding suggests that the most promising candidates for life are Mars and a few moons, including Jupiter's moon Europa or Saturn's moon Enceladus. In the future, we should be able to conduct telescopic searches for life around other stars.

Could aliens be searching for us?

If life is common in the universe, civilizations might also be common, in which case other civilizations might be conducting their own searches and broadcasting signals of their existence. We look for such signals from alien civilizations through the **search for extraterrestrial intelligence,** or **SETI.**

1.4 The New Science of Astrobiology

How do we study the possibility of life beyond Earth?
The science of life in the universe, or astrobiology, focuses on three major areas: (1) studying the conditions conducive to the origin and ongoing existence of life; (2) looking for such conditions on other planets in our solar system and around other stars; and (3) looking for the actual occurrence of life elsewhere. Together, these studies should help us understand the connections between living organisms and the places where they reside.

Exercises and Problems

MasteringAstronomy® For instructor-assigned homework and other learning materials, go to MasteringAstronomy®.

REVIEW QUESTIONS

Short-Answer Questions Based on the Reading

1. Why are scientists interested in the possibility of life beyond Earth?

2. People have long been interested in life beyond Earth. What is different today that makes this possibility seem scientifically reasonable?

3. What do we mean by a *geocentric universe?* In general terms, contrast a geocentric view of the universe with our modern view of the universe.

4. What are *extrasolar planets?* In what way does their discovery make it seem more reasonable to imagine finding life elsewhere?

5. What do we mean by a *habitable* world? Does a habitable world necessarily have life?

6. What do we mean by the "universality" of physics and chemistry? Although we don't know yet whether biology is similarly universal, what evidence makes it seem that it might be?

7. Besides Earth, what worlds in our solar system seem most likely to have life? Why?

8. Could we actually detect life on an extrasolar planet and moon with current technology? Explain.

9. What is the *search for extraterrestrial intelligence (SETI)?*

10. What do we mean by *astrobiology?* What are the major areas of research in astrobiology?

TEST YOUR UNDERSTANDING

Quick Quiz

Choose the best answer to each of the following. Explain your reasoning with one or more complete sentences.

11. An *extrasolar planet* is (a) a planet that is larger than our Sun; (b) a planet that orbits a star other than our Sun; (c) a planet located in another galaxy.

12. A *habitable planet* is (a) a planet that has oceans like Earth; (b) a planet that has life of some kind; (c) a planet that may or may not have life, but that has environmental conditions under which it seems that life could arise or survive.

13. By a *geocentric* view of the universe, we mean (a) the ancient idea that Earth resided at the center of the universe; (b) the idea that Earth is the only planet with life in the universe; (c) a view of the universe shaped by current understanding of geological science.

14. According to current scientific understanding, life on Earth (a) was exceedingly improbable; (b) arose quite soon after conditions allowed it; (c) may have been inevitable, but took billions of years to develop.

15. The correct order for the eight official planets in our solar system, from closest to farthest from the Sun, is (a) Mercury, Venus, Earth, Mars, Saturn, Jupiter, Neptune, Uranus; (b) Mercury, Venus, Earth, Mars, Jupiter, Uranus, Neptune, Saturn; (c) Mercury, Venus, Earth, Mars, Jupiter, Saturn, Uranus, Neptune.

16. Today, the research known as SETI is conducted primarily by (a) scanning the skies for signals from alien civilizations; (b) sending spacecraft to the planets; (c) using telescopes to observe extrasolar planets.

17. If we sent one of our current spacecraft to a nearby star (besides the Sun), the trip would take about (a) a decade; (b) a century; (c) 100,000 years.

18. Scientists today are interested in searching for life on Mars because (a) we see clear evidence of a past civilization on Mars; (b) Mars contains frozen water ice at its polar caps; (c) evidence suggests that Mars had liquid water on its surface in the distant past.

19. Based on current evidence, the object in our solar system most likely to have a deep, subsurface ocean of liquid water is (a) Mars; (b) Europa; (c) Neptune.

20. Based on the way scientists view the study of astrobiology, failure to find life on any other world would mean (a) the whole subject has been a waste of time; (b) we must have done something wrong, since life has to exist beyond Earth; (c) we have learned important lessons about the conditions that made life on Earth possible.

PROCESS OF SCIENCE

21. *Universal Laws.* Briefly discuss how the idea that the laws of nature are universal is important to the study of astrobiology. Based on what you know about the universality of the laws of physics and chemistry, do you think it is likely that there are also universal laws of biology? Defend your opinion.

22. *The Science of Astrobiology.* The study of astrobiology is sometimes criticized as being the study of something for which we have no data, since we do not yet have evidence of life beyond Earth. Is astrobiology a science or speculation? Defend your opinion.

GROUP WORK EXERCISE

23. *Aliens in the Movies.* **Roles:** *Scribe* (takes notes on the group's activities), *Proposer* (proposes suggestions/explanations to the group), *Skeptic* (points out weaknesses in proposed suggestions/explanations), and *Moderator* (leads group discussion and makes sure everyone contributes). **Activity:** Come up with a list of as many movies as possible that involve aliens. Then come to a group consensus in ranking the top five such movies of all time, giving a brief reason for why you like each movie in your top five. Compare your top five list with the lists made by other groups.

Chapter 1   A Universe of Life?   **13**

INVESTIGATE FURTHER

In-Depth Questions to Increase Your Understanding

Short-Answer/Essay Questions

24. *Aliens Among Us.* Conduct an informal poll of your friends or classmates. How many believe we have already been visited by aliens? On what do they base their beliefs? How strong are their convictions on this issue? Summarize your survey results, and based on these results, write a paragraph or two discussing whether public interest in aliens visiting Earth has any bearing on the scientific study of astrobiology.

25. *Conducting the Search.* Given the large number of possible places to look for life, how would you prioritize the search? In other words, where would you look first for life on other worlds in our own solar system, and how would you come up with a search strategy for other star systems? Make a list of priorities and write a few sentences to explain your search strategy.

26. *Funding for Astrobiology.* Imagine that you are a member of Congress, so it is your job to decide how much government funding goes to research in astrobiology. What factors would influence your decision? Make a brief list of at least five important factors, then write a paragraph summarizing whether you would increase or decrease such funding from the current level and why.

WEB PROJECTS

27. *Astrobiology News.* Go to NASA's Astrobiology home page and read some of the recent news from astrobiology research. Choose one recent news article, and write a one- to two-page summary of the research and how it fits into astrobiology research in general.

28. *The NASA Astrobiology Institute.* Visit the site for the NASA Astrobiology Institute (NAI) and learn how it is organized and the type of research it supports. Also learn whether your school or any nearby institutions participate in the NAI. In one page or less, describe the NAI and its work and discuss the particular contributions of any institutions located near you.

29. *International Astrobiology.* Search the Web for information on astrobiology efforts outside the United States. Learn about the effort in one particular country or group of countries. What areas of research are emphasized? How do the researchers involved in the effort collaborate with other international astrobiology efforts? Write a one- to two-page report on your findings.

30. *The Search for Extraterrestrial Intelligence.* Go to the home page for the SETI Institute. Learn more about how SETI is funded and how the institute does its work. Summarize your findings in about one page.

2

The Science of Life in the Universe

▲ **About the photo:** Perspective view of an ancient riverbed on Mars (named Reull Vallis), created using data from the *Mars Express* orbiter. Evidence like this lies at the heart of the modern scientific search for life beyond Earth.

Extraterrestrial life may sound like a modern idea, but stories of life beyond Earth reach far back into ancient times. Many of these stories concerned mythical or supernatural beings living among the constellations, but some were not so different from the ideas we consider today. Nevertheless, the present-day search for life in the universe differs from ancient speculations in an important way: While ancient people could do little more than guess about the possibility of finding life elsewhere, we can now study this possibility with the powerful methods of modern science.

Given that we don't yet know of any life beyond Earth, you might wonder how we can make a science of life in the universe. The answer is that we use science to help us understand the conditions under which we might expect to find life, the likely characteristics of life elsewhere, and the methods we can use to search for it. Because the methods of science are so integral to the search for life beyond Earth, we devote this chapter to understanding those methods and how they developed.

2.1 The Ancient Debate About Life Beyond Earth

More than 2300 years ago, scholars of ancient Greece were already engaged in a lively debate about the possibility of life beyond Earth. Some scholars argued that there *must* be life elsewhere, while others argued the opposite. This impassioned debate may in some ways seem a historical curiosity, but the mere fact that it occurred tells us that a major change in human thinking was already underway.

Deeper in the past, our ancestors looked at the sky and attributed what they saw to the arbitrary actions of mythological beings, an idea still reflected in the fact that the planets carry the names of mythological gods. In contrast, the Greek scholars sought rational explanations for what they could observe in the universe around them. As far as we know, these Greek efforts marked the first attempts to understand the universe through methods closely resembling the ones we use in science today. Therefore, if we want to understand how modern science works—and how we can use it to study the possibility of life beyond Earth—we must begin by peering into the past, to see how observations of the sky started humanity on the road to modern science and kindled interest in the question of whether the universe is ours alone.

How did attempts to understand the sky start us on the road to science?

Imagine living in ancient times, looking up at the sky without the benefit of our modern knowledge. What would you see?

Every day, the Sun rises in the east and sets in the west, its precise path varying with the seasons. At night, the stars circle the sky (Figure 2.1), with different constellations prominent at different times of year. The Moon goes through monthly phases, from new to full and back again, while the planets gradually meander among the stars in seemingly mysterious ways. All the while, the ground beneath you feels steady and solid. It would be quite natural to assume—as did people of many early cultures—that Earth is a flat, motionless surface under a domelike sky across which the heavenly bodies move.

The story of how we progressed from this simple, intuitive view of Earth and the heavens to our modern understanding of Earth as a tiny planet in a vast cosmos is in many ways the story of the development of science itself. Our ancestors were curious about many aspects of the world around them, but astronomy held special interest. The Sun clearly plays a central role in our lives, governing daylight and darkness and marking the progression of the seasons. The Moon's connection to the tides would have been obvious to people living near the sea. The evident power of these celestial bodies probably explains why they attained prominent roles in many early religions and may be one reason why it seemed so important to know the sky. Careful observations of the sky also served practical needs by enabling ancient peoples to keep track of the time and the seasons—crucial requirements for agricultural societies.

As civilizations rose, astronomical observations became more careful and elaborate. In some cases, the results were recorded in writing. The ancient Chinese kept detailed records of astronomical observations beginning some 5000 years ago. By about 2500 years ago, written records allowed the Babylonians (in the region of modern-day Iraq) to predict eclipses with great success. Halfway around the world (and a few centuries later), the Mayans of Central America independently developed the same ability.

These ancient, recorded observations of astronomy represent databases of facts—the raw material of science. But in most cases for which we have historical records, it appears that these facts were never used for much beyond meeting immediate religious and practical needs. An exception was ancient Greece, where scholars attempted to use them to understand the architecture of the cosmos.

EARLY GREEK SCIENCE Greece gradually rose as a power in the Middle East beginning around 800 B.C. and was well established by about 500 B.C. Its geographical location placed it at a crossroads for travelers, merchants, and armies of northern Africa, Asia, and Europe. Building on the diverse ideas brought forth by the meeting of these many cultures, ancient Greek philosophers began to move human understanding of nature from the mythological to the rational.

We generally trace the origin of Greek science to the philosopher Thales (c. 624–546 B.C.; pronounced "THAY-lees"). Among his many accomplishments, Thales was the first person known to have addressed the question "What is the universe made of?" without resorting to supernatural explanations. His own guess—that the universe fundamentally consisted of water and that Earth was a flat disk on an infinite ocean—was not widely accepted even in his own time, but his mere asking of the question helped set the stage for all later science. For the first time, someone had suggested that the world was inherently understandable and not just the result of arbitrary or incomprehensible events.

The scholarly tradition begun by Thales was carried on by others, perhaps most famously by Plato (428–348 B.C.) and his student Aristotle (384–322 B.C.). Each Greek philosopher introduced new ideas, sometimes in contradiction to the ideas of others. None of these ideas rose quite to the level of modern science, primarily because the Greeks tended to rely more on pure thought and intuition than on observations or experimental tests. Nevertheless, with hindsight we can see at least three major innovations in Greek thought that helped pave the way for modern science.

FIGURE 2.1

This photograph, taken at Arches National Park with a 6-hour exposure, shows daily paths of stars in the sky. Notice that stars near the North Star (Polaris) make complete daily circles, while those farther from the North Star rise in the east and set in the west. Ancient people were quite familiar with patterns of motion like these.

First, the Greek philosophers developed a tradition of trying to understand nature without resorting to supernatural explanations. For example, although earlier Greeks might simply have accepted that the Sun moves across the sky because it is pulled by the god Apollo in his chariot—an idea whose roots were already lost in antiquity—the philosophers sought a natural explanation that caused them to speculate anew about the construction of the heavens. They were free to think creatively because they were not simply trying to prove preconceived ideas, and they recognized that new ideas should be open to challenge. As a result, they often worked communally, debating and testing each other's proposals. This tradition of challenging virtually every new idea remains one of the distinguishing features of scientific work today.

Second, the Greeks developed mathematics in the form of geometry. They valued this discipline for its own sake, and they understood its power, using geometry to solve both engineering and scientific problems. Without their mathematical sophistication, they would not have gone far in their attempts to make sense of the cosmos. Like the Greek tradition of challenging ideas, the use of mathematics to help explore the implications of new ideas remains an important part of modern science.

Third, while much of their philosophical activity consisted of subtle debates with little connection to observations or experiments, the Greeks also understood that an explanation about the world could not be right if it disagreed with observed facts. This willingness to discard explanations that simply don't work is also a crucial part of modern science.

THE GEOCENTRIC MODEL Perhaps the greatest Greek contribution to science came from the way they synthesized all three innovations into the idea of creating **models** of nature, an idea that is still central to modern science. Scientific models differ somewhat from the models you may be familiar with in everyday life. In our daily lives, we tend to think of models as miniature physical representations, such as model cars or airplanes. In contrast, a scientific model is a conceptual representation whose purpose is to explain and predict observed phenomena. For example, a model of Earth's climate uses logic, mathematics, and known physical laws in an attempt to represent the way in which the climate works. Its purpose is to explain and predict climate changes, such as the changes that may occur with global warming. Just as a model airplane does not faithfully represent every aspect of a real airplane, a scientific model may not fully explain all our observations of nature. Nevertheless, even the failings of a scientific model can be useful, because they often point the way toward building a better model.

Think About It Conceptual models aren't just important in science; they often affect day-to-day policy decisions. For example, economists use models to predict how new policies will affect the federal budget. Describe at least two other cases in which models affect our daily lives.

In astronomy, the Greeks constructed conceptual models of the universe in an attempt to explain what they observed in the sky, an effort that quickly led them past simplistic ideas of a flat Earth under a dome-shaped sky to a far more sophisticated view of the cosmos. One of the first crucial steps was taken by a student of Thales, Anaximander (c. 610–547 B.C.). In an attempt to explain the way the northern sky appears to turn around the North Star each day (see Figure 2.1), Anaximander suggested that the heavens must form a complete sphere—the **celestial sphere**—around Earth (Figure 2.2). Moreover, based on how the sky varies with latitude, he realized that Earth's surface must be curved, though he incorrectly guessed Earth to be a cylinder rather than a sphere.

The idea of a round Earth probably followed soon, and by about 500 B.C. it was part of the teachings of Pythagoras (c. 560–480 B.C.). He and his followers most likely adopted a spherical Earth for philosophical reasons: The Pythagoreans had a mystical interest in mathematical perfection, and they considered a sphere to be geometrically perfect. More than a century later, Aristotle cited observations of Earth's curved shadow on the Moon during lunar eclipses as evidence for a spherical Earth. Greek philosophers adopted a **geocentric** (Earth-centered) **model** of the universe, with a spherical Earth at the center of a great celestial sphere.

Incidentally, this shows the error of the widespread myth that Columbus proved Earth to be round when he sailed to America in 1492. Not only were scholars of the time well aware of Earth's round shape; they even knew Earth's approximate size: Earth's circumference was first measured (fairly accurately) in about 240 B.C. by the Greek scientist Eratosthenes. In fact, a likely reason why Columbus had so much difficulty finding a sponsor for his voyages was that he tried to argue a point on which he was dead wrong: He claimed the distance by sea from western Europe to eastern Asia to be much less than many scholars had estimated it to be. His erroneous belief would almost certainly have led his voyage to disaster if the Americas hadn't stood in his way.

THE MYSTERY OF PLANETARY MOTION If you watch the sky closely, you'll notice that while the patterns of the constellations seem not to change, the Sun, the Moon, and the five planets visible to the naked eye (Mercury, Venus, Mars, Jupiter, and Saturn) gradually move among the constellations from one day to the next. Indeed, the word *planet* comes from the Greek for "wanderer," and it originally referred to the Sun and Moon as well as to the five visible planets. Our seven-day week is directly traceable to the fact that seven "planets" are visible in the heavens (Table 2.1).

The wanderings of these objects convinced the Greek philosophers that there had to be more to the heavens than just a single sphere surrounding Earth. The Sun and Moon each move steadily through the constellations, with the Sun completing a circuit around the celestial sphere each year and the Moon completing each circuit in about a month (think

FIGURE 2.2
The early Greek geocentric model consisted of a central Earth surrounded by the celestial sphere, which is shown here marked with modern constellation borders and a few reference points and circles. We still use the idea of the celestial sphere when making astronomical observations, but we no longer imagine that it reflects reality.

TABLE 2.1 The Seven Days of the Week and the Astronomical Objects They Honor

The seven days were originally based on the seven visible "wanderers" of the sky. In English, the correspondence is still obvious for Sunday, "Moonday," and "Saturnday" (other days take names from Germanic gods); other connections are clearer in languages such as French and Spanish.

Object	English	French	Spanish
Sun	Sunday	dimanche	domingo
Moon	Monday	lundi	lunes
Mars	Tuesday	mardi	martes
Mercury	Wednesday	mercredi	miércoles
Jupiter	Thursday	jeudi	jueves
Venus	Friday	vendredi	viernes
Saturn	Saturday	samedi	sábado

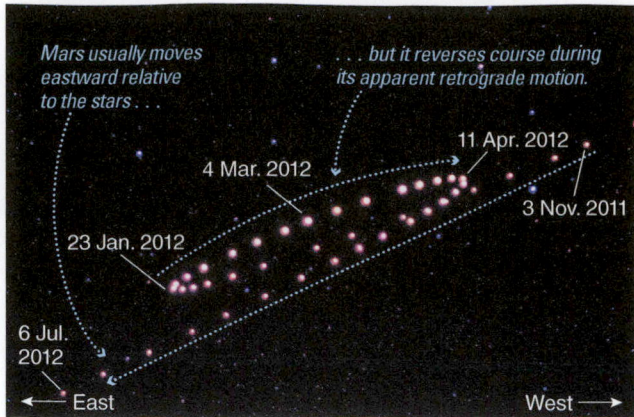

Mars usually moves eastward relative to the stars . . .

. . . but it reverses course during its apparent retrograde motion.

11 Apr. 2012

4 Mar. 2012

3 Nov. 2011

23 Jan. 2012

6 Jul. 2012

← East

West →

FIGURE 2.3

This composite of individual photos (taken at 5- to 7-day intervals in 2011 and 2012) shows a retrograde loop of Mars. Note that Mars is biggest and brightest in the middle of the retrograde loop, because that is where it is closest to Earth in its orbit.

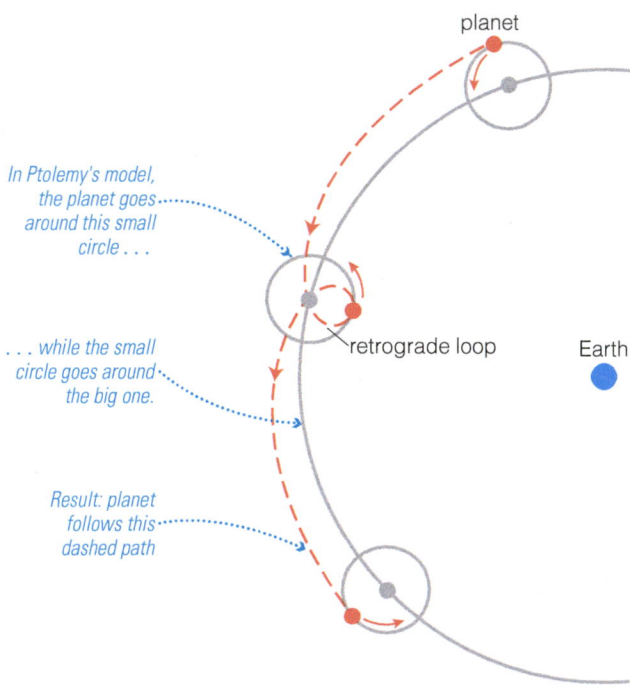

planet

In Ptolemy's model, the planet goes around this small circle . . .

. . . while the small circle goes around the big one.

retrograde loop

Earth

Result: planet follows this dashed path

FIGURE 2.4 INTERACTIVE FIGURE

This diagram shows how the Ptolemaic model accounted for apparent retrograde motion. Each planet is assumed to move around a small circle that turns on a larger circle. The resulting path (dashed) includes a loop in which the planet goes backward as seen from Earth.

"moonth"). The Greeks could account for this motion by adding separate spheres for the Sun and Moon, each nested within the sphere of the stars, and allowing these spheres to turn at different rates from the sphere of the stars. But the five visible planets posed a much greater mystery.

If you observe the position of a planet (such as Mars or Jupiter) relative to the stars over a period of many months, you'll find not only that its speed and brightness vary considerably but also that its direction of motion sometimes changes. While the planets usually move eastward relative to the constellations, sometimes they reverse course and go backward (Figure 2.3). These periods of **apparent retrograde motion** (*retrograde* means "backward") last from a few weeks to a few months, depending on the planet.

This seemingly erratic planetary motion was not so easy to explain with rotating spheres, especially because the Greeks generally accepted a notion of "heavenly perfection," enunciated most clearly by Plato, which demanded that all heavenly objects move in perfect circles. How could a planet sometimes go backward when moving in a perfect circle? The Greeks came up with a number of ingenious ideas that preserved Earth's central position, culminating with a complex model of planetary motion described by the astronomer Ptolemy (c. A.D. 100–170; pronounced "TOL-e-mee"); we refer to Ptolemy's model as the **Ptolemaic model** to distinguish it from earlier geocentric models. This model reproduced retrograde motion by having planets move around Earth on small circles that turned around larger circles. A planet following this circle-on-circle motion traces a loop as seen from Earth, with the backward portion of the loop mimicking apparent retrograde motion (Figure 2.4).

The circle-on-circle motion may itself seem somewhat complex, but Ptolemy found that he also had to use many other mathematical tricks, including putting some of the circles off-center, to get his model to agree with observations. Despite all this complexity, he achieved remarkable success: His model could correctly forecast future planetary positions to within a few degrees of arc—roughly equivalent to holding your hand at arm's length against the sky. Indeed, the Ptolemaic model generally worked so well that it remained in use for the next 1500 years. When Arabic scholars translated Ptolemy's book describing the model in around A.D. 800, they gave it the title *Almagest,* derived from words meaning "the greatest work."

AN ALTERNATIVE MODEL In about 260 B.C., the Greek scientist Aristarchus (c. 310–230 B.C.) offered a radical departure from the conventional wisdom: He suggested that Earth goes around the Sun, rather than vice versa. Little of Aristarchus's work survives to the present day, so we do not know exactly how he came up with his Sun-centered idea. We do know that he made measurements that convinced him that the Sun is much larger than Earth, so perhaps he simply concluded that it was more natural for the smaller Earth to orbit the larger Sun. In addition, he almost certainly recognized that a Sun-centered system offers a much more natural explanation for apparent retrograde motion.

You can see how the Sun-centered system explains retrograde motion with a simple demonstration (Figure 2.5a). Find an empty area (such as a sports field or a big lawn), and mark a spot in the middle to represent the Sun. You can represent Earth, walking counterclockwise around the Sun, while a friend represents a more distant planet (such as Mars, Jupiter, or Saturn) by walking counterclockwise around the Sun

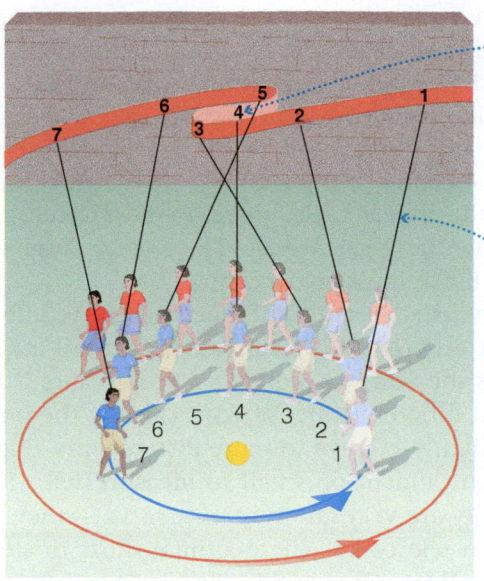

Apparent retrograde motion occurs between positions 3 and 5, as the inner person (planet) passes the outer person (planet).

Follow the lines of sight from inner person (planet) to outer person (planet) to see where the outer one appears against the background.

East

West

Leo

Gemini

Cancer

Earth orbit

Mars orbit

a The retrograde motion demonstration: Watch how your friend (in red) usually appears to move forward against the background of the building in the distance but appears to move backward as you (in blue) catch up to and pass her in your "orbit."

b This diagram shows the same idea applied to a planet. Follow the lines of sight from Earth to Mars in numerical order. Notice that Mars appears to move westward relative to the distant stars (from points 3 to 5) as Earth passes it by in its orbit.

FIGURE 2.5 INTERACTIVE FIGURE
Apparent retrograde motion—the occasional "backward" motion of the planets relative to the stars—has a simple explanation in a Sun-centered solar system.

at a greater distance. Your friend should walk more slowly than you, because more distant planets orbit the Sun more slowly. As you walk, watch how your friend appears to move relative to buildings or trees in the distance. Although both of you always walk in the same direction around the Sun, your friend will appear to move backward against the background during the part of your "orbit" at which you catch up to and pass him or her. To understand the apparent retrograde motions of Mercury and Venus, which are closer to the Sun than is Earth, simply switch places with your friend and repeat the demonstration. The demonstration applies to all the planets. For example, because Mars takes about 2 years to orbit the Sun (actually, 1.88 years), it covers about half its orbit during the 1 year in which Earth makes a complete orbit. If you trace lines of sight from Earth to Mars from different points in their orbits, you will see that the line of sight usually moves eastward relative to the stars but moves westward during the time when Earth is passing Mars in its orbit (Figure 2.5b). Like your friend in the demonstration, Mars never actually changes direction. It only appears to change direction from our perspective on Earth.

Despite the elegance of this Sun-centered model for the universe, Aristarchus had little success in convincing his contemporaries to accept it. Some of the reasons for this rejection were purely philosophical and not based on any hard evidence. However, at least one major objection was firmly rooted in observations: Aristarchus's idea seemed inconsistent with observations of stellar positions in the sky.

To understand the inconsistency, imagine what would happen if you placed the Sun rather than Earth at the center of the celestial sphere, with Earth orbiting the Sun some distance away. In that case, Earth would be closer to different portions of the celestial sphere at different times of year. When we were closer to a particular part of the sphere, the stars on that part of the sphere would appear more widely separated than

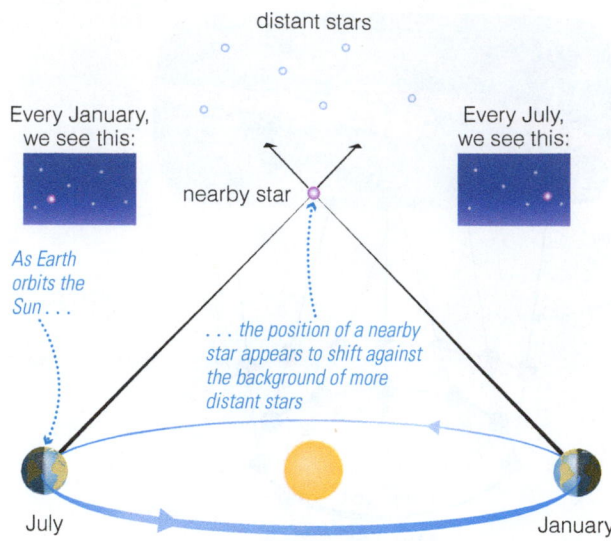

distant stars

Every January, we see this:

nearby star

Every July, we see this:

As Earth orbits the Sun...

...the position of a nearby star appears to shift against the background of more distant stars

July

January

FIGURE 2.6 INTERACTIVE FIGURE

If Earth orbits the Sun, then over the course of each year we should see nearby stars shift slightly back and forth relative to more distant stars (*stellar parallax*). The Greeks could not detect any such shift, and used this fact to argue that Earth must be at the center of the universe. Today, we *can* detect stellar parallax with telescopic observations, proving that Earth does orbit the Sun. (This figure is greatly exaggerated; the actual shift is far too small to detect with the naked eye.)

they would when we were farther from that part of the sphere, just as the spacing between the two headlights on a car looks greater when you are closer to the car. This would create annual shifts in the separations of stars—but the Greeks observed no such shifts. They knew that there were only two possible ways to account for the lack of an observed shift: Either Earth was at the center of the universe or the stars were so far away as to make the shift undetectable by eye. To most Greeks, it seemed unreasonable to imagine that the stars could be *that* far away, which led them to conclude that Earth must hold a central place.

This argument about stellar shifts still holds when we allow for the reality that stars lie at different distances rather than all on the same sphere: As Earth orbits the Sun, we look at particular stars from slightly different positions at different times of year, causing the positions of nearby stars to shift slightly relative to more distant stars (Figure 2.6). Although such shifts are much too small to measure with the naked eye—because stars really are very far away [Section 3.2]—they are easily detectable with modern telescopes. These annual shifts in stellar position, called **stellar parallax,** now provide concrete proof that Earth really does go around the Sun.

THE ROOTS OF MODERN SCIENCE Although the Greeks ultimately rejected the correct idea—that Earth orbits the Sun—we have seen that they did so for reasons that made good sense at the time. Not all of their reasons would pass the test of modern science; for example, their preference for motion in perfect circles came only from their cultural ideas of aesthetics and not from any actual data. But they also went to a lot of effort to ensure that their models were consistent with observations, and in that way they laid the foundation of modern science. And while Aristarchus may not have won the day in his own time, his idea remained alive in books. Some 1800 years after he first proposed it, Aristarchus's Sun-centered model apparently came to the attention of a Polish astronomer named Nicholas Copernicus (1473–1543), who took the idea and ran with it in a way that led directly to the development of modern science. We'll return to this story shortly.

Why did the Greeks argue about the possibility of life beyond Earth?

Almost from the moment that Thales asked his question of what the universe was made of, the Greeks realized that the answer would have bearing on the possibility of life elsewhere. This might seem surprising in light of their geocentric beliefs, because they didn't think of the planets or stars as worlds in the way we think of them today. Instead, the Greeks generally considered the "world" to include both Earth and the heavenly spheres that they imagined to surround it, and they were at least open to the possibility that other such "worlds" might exist.

As we noted earlier, Thales guessed that the world consisted fundamentally of water, with Earth floating on an infinite ocean, but his student Anaximander imagined a more mystical element that he called *apeiron,* meaning "infinite." Anaximander suggested that all material things arose from and returned to the apeiron, which allowed him to imagine that worlds might be born and die repeatedly through eternal time. So even though he made no known claim of life existing elsewhere in the present, Anaximander essentially suggested that other Earths and other beings might exist at other times.

Other Greeks took the debate in a slightly different direction, and eventually a consensus emerged in favor of the world's having been built from four elements: fire, water, earth, and air. However, two distinct schools of thought emerged concerning the nature and extent of these elements:

- The *atomists* held that both Earth and the heavens were made from an infinite number of indivisible atoms of each of the four elements.
- The *Aristotelians* (after Aristotle) held that the four elements—not necessarily made from atoms—were confined to the realm of Earth, while the heavens were made of a fifth element, often called the *aether* (or *ether*) or the *quintessence* (literally, "the fifth essence").

The differences in the two schools of thought led to two fundamentally different conclusions about the possibility of extraterrestrial life.

Think About It Look up the words *ethereal* and *quintessence* in the dictionary. How do their definitions relate to the Aristotelian idea that the heavens were composed of an element distinct from the elements of Earth? Explain.

The atomist doctrine was developed largely by Democritus (c. 470–380 B.C.), and his views show how the idea led almost inevitably to belief in extraterrestrial life. Democritus argued that the world—both Earth and the heavens—had been created by the random motions of infinite atoms. Because this idea held that the number of atoms was infinite, it was natural to assume that the same processes that created our world could also have created others. This philosophy on life beyond Earth is clearly described in the following quotation from a later atomist, Epicurus (341–270 B.C.):

> There are infinite worlds both like and unlike this world of ours ... we must believe that in all worlds there are living creatures and plants and other things we see in this world.*

Aristotle had a different view. He believed that each of the four elements had its own natural motion and place. For example, he believed that the element earth moved naturally toward the center of the universe, an idea that offered an explanation for the Greek assumption that Earth resided in a central place. The element fire, he claimed, naturally rose away from the center, which explained why flames jut upward into the sky. These incorrect ideas about physics, which were not disproved until the time of Galileo and Newton almost 2000 years later, caused Aristotle to reject the atomist idea of many worlds. If there was more than one world, there would be more than one natural place for the elements to go, which would be a logical contradiction. Aristotle concluded:

> The world must be unique.... There cannot be several worlds.

Interestingly, Aristotle's philosophies were not particularly influential until many centuries after his death. His books were preserved and valued—in particular, by Islamic scholars of the late first millennium—but they were unknown in Europe until they were translated into Latin in the twelfth and thirteenth centuries. St. Thomas Aquinas (1225–1274)

---

*From Epicurus's "Letter to Herodotus"; the authors thank David Darling for finding this quotation and the one from Aristotle, both of which appear in Darling's book *The Extraterrestrial Encyclopedia*, Three Rivers Press, 2000.

integrated Aristotle's philosophy into Christian theology. At this point, the contradiction between the Aristotelian notion of a single world and the atomist notion of many worlds became a subject of great concern to Christian theologians. Moreover, because the atomist view held that our world came into existence through random motions of atoms, and hence without the need for any intelligent Creator, atomism became associated with atheism. The debate about extraterrestrial life thereby became intertwined with debates about religion. Even today, the theological issues are not fully settled, and echoes of the ancient Greek debate between the atomists and the Aristotelians still reverberate in our time.

2.2 The Copernican Revolution

Greek ideas gained great influence in the ancient world, in large part because the Greeks proved to be as adept at politics and war as they were at philosophy. In about 330 B.C., Alexander the Great began a series of conquests that expanded the Greek Empire throughout the Middle East. Alexander had a keen interest in science and education, perhaps because he grew up with Aristotle as his personal tutor. Alexander established the city of Alexandria in Egypt, and his successors founded the renowned Library of Alexandria. Though it is sometimes difficult to distinguish fact from legend in stories of this great Library, there is little doubt that it was once the world's preeminent center of research, housing up to a half million books written on papyrus scrolls. While the details of the Library's destruction are hazy and subject to disagreement among historians, the Library was ultimately destroyed, and most of its books were lost forever.

The relatively few books from the Library that survive today were preserved primarily thanks to the rise of a new center of intellectual inquiry in Baghdad (in present-day Iraq). As European civilization fell into the Dark Ages, scholars of the new religion of Islam sought knowledge of mathematics and astronomy in hopes of better understanding the wisdom of Allah. The Islamic scholars translated and thereby saved many of the remaining ancient Greek works. Building on what they learned from the Greek manuscripts, they went on to develop the mathematics of algebra as well as many new instruments and techniques for astronomical observation.

The Islamic world of the Middle Ages was in frequent contact with Hindu scholars from India, who in turn brought ideas and discoveries from China. Hence, the intellectual center in Baghdad achieved a synthesis of the surviving work of the ancient Greeks, the Indians, the Chinese, and the contributions of its own scholars. This accumulated knowledge spread throughout the Byzantine Empire (the eastern part of the former Roman Empire). When the Byzantine capital of Constantinople (modern-day Istanbul) fell in 1453, many Eastern scholars headed west to Europe, carrying with them the knowledge that helped ignite the European Renaissance. The stage was set for a dramatic rethinking of humanity and our place in the universe.

How did the Copernican revolution further the development of science?

In 1543, Nicholas Copernicus published *De Revolutionibus Orbium Coelestium* ("Concerning the Revolutions of the Heavenly Spheres"), launching what we now call the **Copernican revolution.** In his book, Copernicus revived Aristarchus's radical suggestion of a Sun-centered solar system and described the idea with enough mathematical detail to make it a valid competitor to the Earth-centered, Ptolemaic model. Over the next century and a half, philosophers and scientists (who were often one and the same) debated and tested the Copernican idea. Many of the ideas that now form the foundation of modern science first arose as this debate played out. Indeed, the Copernican revolution had such a profound impact on philosophy that we cannot understand modern science without first understanding the key features of this revolution.

COPERNICUS—THE REVOLUTION BEGINS By the time of Copernicus's birth in 1473, tables of planetary motion based on the Ptolemaic model had become noticeably inaccurate. However, few people were willing to undertake the difficult calculations required to revise the tables. Indeed, the best tables available were already two centuries old, having been compiled under the guidance of the Spanish monarch Alphonso X (1221–1284). Commenting on the tedious nature of the work involved, the monarch is said to have complained that "If I had been present at the creation, I would have recommended a simpler design for the universe."

Copernicus began studying astronomy in his late teens. He soon became aware of the inaccuracies of the Ptolemaic predictions and began a quest for a better way to predict planetary positions. He adopted Aristarchus's Sun-centered idea, probably because he was drawn to its simple explanation for the apparent retrograde motion of the planets (see Figure 2.5). As he worked out the mathematical details of his model, Copernicus discovered simple geometric relationships that allowed him to calculate each planet's orbital period around the Sun and its relative distance from the Sun in terms of Earth–Sun distance. The success of his model in providing a geometric layout for the solar system further convinced him that the Sun-centered idea must be correct. Despite his own confidence in the model, Copernicus was hesitant to publish his work, fearing that the idea of a moving Earth would be considered absurd.* However, he discussed his system with other scholars, including high-ranking officials of the Church, who urged him to publish a book. Copernicus saw the first printed copy of his book on the day he died—May 24, 1543.

Publication of the book spread the Sun-centered idea widely, and many scholars were drawn to its aesthetic advantages. However, the Copernican model gained relatively few converts over the next 50 years, for a good reason: It didn't work all that well. The primary problem was that while Copernicus had been willing to overturn Earth's central place in the cosmos, he held fast to the ancient belief that heavenly motion must occur in perfect circles. This incorrect assumption forced him to add numerous complexities to his system (including circles on circles much like those used by Ptolemy) to get it to make decent predictions. In the end, his complete model was no more accurate and no less complex than the Ptolemaic model, and few people were willing to throw out thousands of years of tradition for a new model that worked just as poorly as the old one.

TYCHO—A NEW STANDARD IN OBSERVATIONAL DATA Part of the difficulty faced by astronomers who sought to improve either the Ptolemaic or the

---

*Indeed, in the Preface of *De Revolutionibus*, Copernicus offered a theological defense of the Sun-centered idea: "Behold, in the middle of the universe resides the Sun. For who, in this most beautiful Temple, would set this lamp in another or a better place, whence to illumine all things at once?"

FIGURE 2.7

Tycho Brahe in his naked-eye observatory, which worked much like a giant protractor. He could sit and observe a planet through the rectangular hole in the wall as an assistant used a sliding marker to measure the angle on the protractor.

Copernican model was a lack of quality data. The telescope had not yet been invented, and existing naked-eye observations were not particularly accurate. In the late sixteenth century, Danish nobleman Tycho Brahe (1546–1601), usually known simply as Tycho (commonly pronounced "TIE-koe"), set about correcting this problem.

Tycho was an eccentric genius who, at age 20, lost part of his nose in a sword fight with another student over who was the better mathematician. Taking advantage of his royal connections, he built large naked-eye observatories (Figure 2.7) that worked much like giant protractors, and over a period of three decades he used them to measure planetary positions to within 1 minute of arc ($\frac{1}{60}$ of 1°)—which is less than the thickness of a fingernail held at arm's length.

KEPLER—A SUCCESSFUL MODEL OF PLANETARY MOTION Tycho never came up with a fully satisfactory explanation for his observations (though he made a valiant attempt), but he found someone else who did. In 1600, he hired a young German astronomer named Johannes Kepler (1571–1630). Kepler and Tycho had a strained relationship,* but in 1601, as he lay on his deathbed, Tycho begged Kepler to find a system that would make sense of his observations so "that it may not appear I have lived in vain."

Kepler was deeply religious and believed that understanding the geometry of the heavens would bring him closer to God. Like Copernicus, he believed that planetary orbits should be perfect circles, so he worked diligently to match circular motions to Tycho's data. After years of effort, he found a set of circular orbits that matched most of Tycho's observations quite well. Even in the worst cases, which were for the planet Mars, Kepler's predicted positions differed from Tycho's observations by only about 8 arcminutes.

Kepler surely was tempted to ignore these discrepancies and attribute them to errors by Tycho. After all, 8 arcminutes is barely one-fourth the angular diameter of the full moon. But Kepler trusted Tycho's careful work. The small discrepancies finally led Kepler to abandon the idea of circular orbits—and to find the correct solution to the ancient riddle of planetary motion. About this event, Kepler wrote,

If I had believed that we could ignore these eight minutes [of arc], I would have patched up my hypothesis accordingly. But, since it was not permissible to ignore, those eight minutes pointed the road to a complete reformation in astronomy.

Kepler's decision to trust the data over his preconceived beliefs marked an important transition point in the history of science. Once he abandoned perfect circles, he was free to try other ideas and he soon hit on the correct one: Planetary orbits take the shapes of the special types of ovals known as *ellipses.* He then used his knowledge of mathematics to put his new model of planetary motion on a firm footing, expressing the key features of the model with what we now call **Kepler's laws of planetary motion:**

- **Kepler's first law:** *The orbit of each planet about the Sun is an ellipse with the Sun at one focus* (Figure 2.8). This law tells us that a planet's distance from the Sun varies during its orbit. Its closest point is called **perihelion** (from the Greek for "near the Sun") and its farthest point is called **aphelion** ("away from the Sun"). The *average*

---

*For a particularly moving version of the story of Tycho and Kepler, see *Cosmos*, by Carl Sagan, Episode 3.

of a planet's perihelion and aphelion distances is the length of its **semimajor axis** (which we will refer to simply as the planet's average distance from the Sun).

- **Kepler's second law:** *A planet moves faster in the part of its orbit nearer the Sun and slower when farther from the Sun, sweeping out equal areas in equal times.* As shown in Figure 2.9, the "sweeping" refers to an imaginary line connecting the planet to the Sun, and keeping the areas equal means that the planet moves a greater distance (and hence is moving faster) when it is near perihelion than it does in the same amount of time near aphelion.

- **Kepler's third law:** *More distant planets orbit the Sun at slower average speeds, obeying the precise mathematical relationship $p^2 = a^3$; p is the planet's orbital period in years and a is its average distance (semimajor axis) from the Sun in astronomical units.* [One **astronomical unit (AU)** is defined as Earth's average distance from the Sun, or about 149.6 million kilometers.] The mathematical statement of Kepler's third law allows us to calculate the average orbital speed of each planet (Figure 2.10).

Kepler published his first two laws in 1609 and his third in 1619. Together, they made a model that could predict planetary positions with far greater accuracy than Ptolemy's Earth-centered model. Indeed, Kepler's model has worked so well that we now see it not just as an abstract model, but instead as revealing a deep, underlying truth about planetary motion.

GALILEO—ANSWERING THE REMAINING OBJECTIONS The success of Kepler's laws in matching Tycho's data provided strong evidence in favor of Copernicus's placement of the Sun, rather than Earth, at the center of the solar system. Nevertheless, many scientists still voiced reasonable objections to the Copernican view. There were three basic objections, all rooted in the 2000-year-old beliefs of Aristotle:

- First, Aristotle had held that Earth could not be moving because, if it were, objects such as birds, falling stones, and clouds would be left behind as Earth moved along its way.
- Second, the idea of noncircular orbits contradicted the view that the heavens—the realm of the Sun, Moon, planets, and stars—must be perfect and unchanging.
- Third, no one had detected the stellar parallax that should occur if Earth orbits the Sun.

Galileo Galilei (1564–1642), nearly always known by only his first name, answered all three objections.

Galileo defused the first objection with experiments that almost single-handedly overturned the Aristotelian view of physics. In particular, he used experiments with rolling balls to demonstrate that a moving object remains in motion *unless* a force acts to stop it (an idea now codified in Newton's first law of motion). This insight explained why objects that share Earth's motion through space—such as birds, falling stones, and clouds—should *stay* with Earth rather than falling behind as Aristotle had argued. This same idea explains why passengers stay with a moving airplane even when they leave their seats.

The notion of heavenly perfection was already under challenge by Galileo's time, because Tycho had observed a supernova and proved

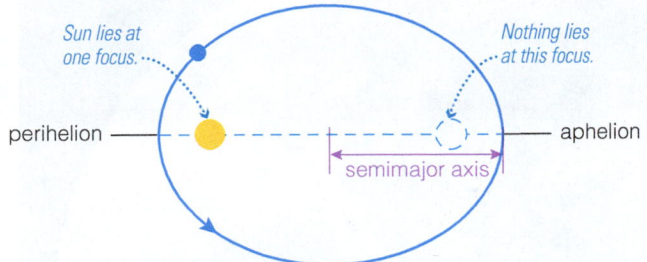

FIGURE 2.8 INTERACTIVE FIGURE
Kepler's first law: The orbit of each planet about the Sun is an ellipse with the Sun at one focus. (The ellipse shown here is more eccentric, or "stretched out," than any of the actual planetary orbits in our solar system.)

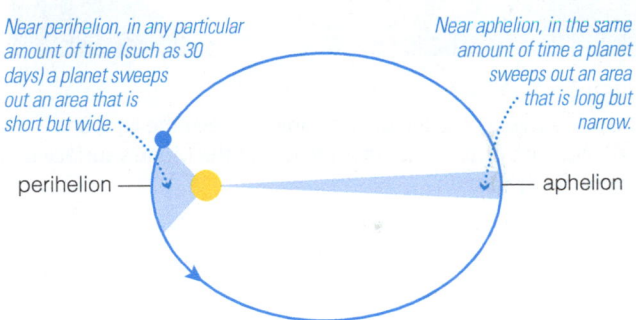

The areas swept out in 30-day periods are all equal.

FIGURE 2.9 INTERACTIVE FIGURE
Kepler's second law: As a planet moves around its orbit, it moves faster when closer to the Sun than when farther away, so that an imaginary line connecting it to the Sun sweeps out equal areas (the shaded regions) in equal times.

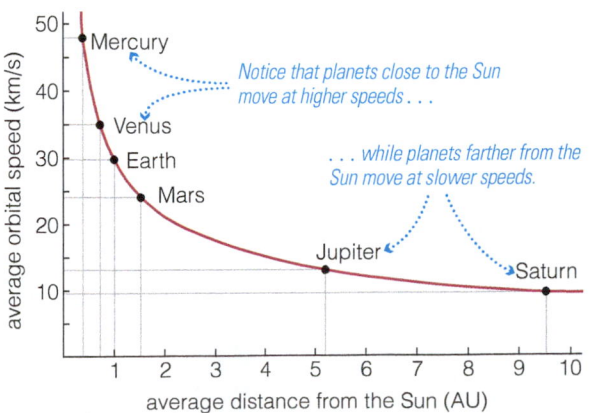

FIGURE 2.10
This graph, based on Kepler's third law ($p^2 = a^3$) and modern values of planetary distances, shows that more distant planets orbit the Sun more slowly.

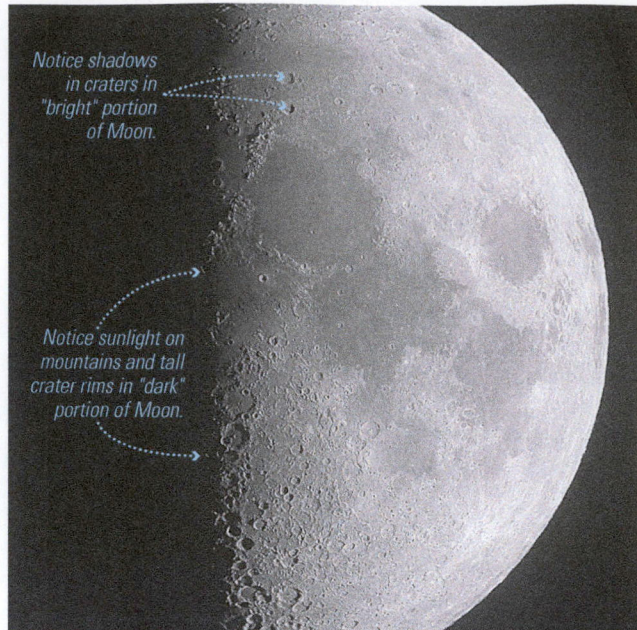

Notice shadows in craters in "bright" portion of Moon.

Notice sunlight on mountains and tall crater rims in "dark" portion of Moon.

FIGURE 2.11
Shadows visible near the dividing line between the light and dark portions of the lunar face prove that the Moon's surface is not perfectly smooth.

FIGURE 2.12 INTERACTIVE FIGURE
Galileo's telescopic observations of Venus proved that it orbits the Sun rather than Earth.

that comets lie beyond the Moon; these observations showed that the heavens *do* sometimes undergo change. But Galileo drove the new idea home after he built a telescope in late 1609. (Galileo did not invent the telescope, but his innovations made it much more powerful.) Through his telescope, Galileo saw sunspots on the Sun, which were considered "imperfections" at the time. He also used his telescope to prove that the Moon has mountains and valleys like the "imperfect" Earth by noticing the shadows cast near the dividing line between the light and dark portions of the lunar face (Figure 2.11). If the heavens were not perfect, then the idea of elliptical orbits (as opposed to "perfect" circles) was not so objectionable.

The third objection—the absence of observable stellar parallax—had been a particular concern of Tycho's. Based on his estimates of the distances of stars, Tycho believed that his naked-eye observations were sufficiently precise to detect stellar parallax if Earth did in fact orbit the Sun. Refuting Tycho's argument required showing that the stars were more distant than Tycho had thought and therefore too distant for him to have observed stellar parallax. Although Galileo didn't actually prove this fact, he provided strong evidence in its favor. For example, he saw with his telescope that the Milky Way resolved into countless individual stars. This discovery helped him argue that the stars were far more numerous and more distant than Tycho had believed.

In hindsight, the final nails in the coffin of the Earth-centered universe came with two of Galileo's earliest discoveries through the telescope. First, he observed four moons clearly orbiting Jupiter, not Earth. Soon thereafter, he observed that Venus goes through phases in a way that proved that it must orbit the Sun and not Earth (Figure 2.12). Together, these observations offered clear proof that Earth is *not* the center of everything.*

---

*While these observations proved that Earth is not the center of everything, they did not by themselves prove that Earth orbits the Sun; direct proof of that fact did not come until later, with measurements of stellar parallax and of an effect known as the *aberration of starlight* that also occurs only because of Earth's motion. Nevertheless, the existence of Jupiter's moons showed that moons can orbit a moving planet like Jupiter, which overcame some critics' complaints that the Moon could not stay with a moving Earth, and the proof that Venus orbits the Sun provided clear validation of Kepler's model of Sun-centered planetary motion.

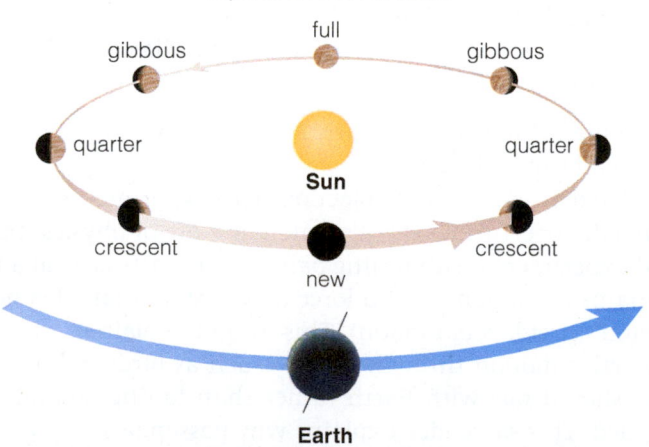

a In the Ptolemaic system, Venus orbits Earth, moving around a small circle on its larger orbital circle; the center of the small circle lies on the Earth-Sun line. Thus, if this view were correct, Venus's phases would range only from new to crescent.

b In reality, Venus orbits the Sun, so from Earth we can see it in many different phases. This is just what Galileo observed, allowing him to prove that Venus really does orbit the Sun.

Although we now recognize that Galileo won the day, the story was more complex in his own time, when Catholic Church doctrine still held Earth to be the center of the universe. On June 22, 1633, Galileo was brought before a Church inquisition in Rome and ordered to recant his claim that Earth orbits the Sun. Nearly 70 years old and likely fearing for his life, Galileo did as ordered. However, legend has it that as he rose from his knees, he whispered under his breath, *Eppur si muove*—Italian for "And yet it moves." (Given the likely consequences if Church officials had heard him say this, most historians doubt the legend.)

The Church did not formally vindicate Galileo until 1992, but the Church had given up the argument long before that. Today, Catholic scientists are at the forefront of much astronomical research, and official Church teachings are compatible not only with Earth's planetary status but also with the theories of the Big Bang and the subsequent evolution of the cosmos and of life.

Think About It Although the Catholic Church today teaches that science and the Bible are compatible, not all religious denominations hold the same belief. Do *you* think that science and the Bible are compatible? Defend your opinion.

NEWTON—THE REVOLUTION CONCLUDES Kepler's model worked so well and Galileo so successfully defused the remaining objections that by about the 1630s, scientists were nearly unanimous in accepting the validity of Kepler's laws of planetary motion. However, no one yet knew *why* the planets should move in elliptical orbits with varying speeds. The question became a topic of great debate, and a few scientists even guessed the correct answer—but they could not prove it, largely because the necessary understanding of physics and mathematics didn't exist yet. This understanding finally came through the remarkable work of Sir Isaac Newton (1642–1727), who invented the mathematics of calculus and used it to explain and discover many fundamental principles of physics.

In 1687, Newton published a famous book usually called *Principia*, short for *Philosophiae Naturalis Principia Mathematica* ("Mathematical Principles of Natural Philosophy"). In it, he laid out precise mathematical descriptions of how motion works in general, ideas that we now describe as **Newton's laws of motion.** For reference, Figure 2.13 illustrates the three laws of motion, although we will not make much use of them in this book. (Be careful not to confuse *Newton's* three laws, which apply to all motion, with *Kepler's* three laws, which describe only the motion of planets moving about the Sun.)

Newton continued on in *Principia* to describe his universal law of gravitation (see Section 2.4), and then used mathematics to prove that Kepler's laws are natural consequences of the laws of motion and gravity. In essence, Newton had created a new model for the inner workings of the universe in which motion is governed by clear laws and the force of gravity. The model explained so much about the nature of motion in the everyday world, as well as about the movements of the planets, that the geocentric idea could no longer be taken seriously.

LOOKING BACK AT REVOLUTIONARY SCIENCE Fewer than 150 years passed between Copernicus's publication of *De Revolutionibus* in 1543 and Newton's publication of *Principia* in 1687, such a short time in the scope of human history that we call it a revolution. A quick look back shows that

Cosmic Calculations 2.1
KEPLER'S THIRD LAW

When Kepler discovered his third law ($p^2 = a^3$), he knew only that it applied to the orbits of planets about the Sun. In fact, it applies to any orbiting object as long as the following two conditions are met:

1. The object orbits the Sun *or* another object of precisely the same mass.
2. We use units of *years* for the orbital period and *AU* for the orbital distance.

(Newton extended the law to *all* orbiting objects; see Cosmic Calculations 7.1.)

Example 1: The largest asteroid, Ceres, orbits the Sun at an average distance (semimajor axis) of 2.77 AU. What is its orbital period?

Solution: Both conditions are met, so we solve Kepler's third law for the orbital period p and substitute the given orbital distance, $a = 2.77$ AU:

$$p^2 = a^3 \Rightarrow p = \sqrt{a^3} = \sqrt{2.77^3} \approx 4.6$$

Ceres has an orbital period of 4.6 years.

Example 2: A planet is discovered orbiting every three months around a star of the same mass as our Sun. What is the planet's average orbital distance?

Solution: The first condition is met, and we can satisfy the second by converting the orbital period from months to years: $p = 3$ months $= 0.25$ year. We now solve Kepler's third law for the average distance a:

$$p^2 = a^3 \Rightarrow a = \sqrt[3]{p^2} = \sqrt[3]{0.25^2} \approx 0.40$$

The planet orbits its star at an average distance of 0.40 AU, which is nearly the same as Mercury's average distance from the Sun.

Newton's first law of motion:
An object moves at constant velocity unless a net force acts to change its speed or direction.

Example: A spaceship needs no fuel to keep moving in space.

Newton's second law of motion:
Force = mass × acceleration

Example: A baseball accelerates as the pitcher applies a force by moving his arm. (Once the ball is released, the force from the pitcher's arm ceases, and the ball's path changes only because of the forces of gravity and air resistance.)

Newton's third law of motion:
For any force, there is always an equal and opposite reaction force.

Example: A rocket is propelled upward by a force equal and opposite to the force with which gas is expelled out its back.

FIGURE 2.13
Newton's three laws of motion.

the revolution not only caused a radical change in human perspective on our place in the universe—shifting Earth from a central role to being just one of many worlds—but also altered our ideas about how knowledge should be acquired. For example, while previous generations had tolerated inaccuracies in the predictions of the Ptolemaic model, Copernicus and his followers felt compelled to find models of nature that could actually reproduce what they observed.

The eventual success of Kepler's model also led to a new emphasis on understanding *why* nature works as it does. Past generations had relied almost solely on their cultural senses of aesthetics in guessing that the world was built with perfect circles and spheres and indivisible atoms, and they seemed content to accept these guesses even without any evidence of their reality. By Newton's time, guessing was no longer good enough: Instead, you had to present hard evidence, backed by rigorous mathematics, to convince your colleagues that you'd hit on something that truly brought us closer to understanding the nature of the universe.

How did the Copernican revolution alter the ancient debate on extraterrestrial life?

The Copernican revolution did not deal directly with the question of life in the universe, but it had a major effect on the way people thought about the issue. You can see why by thinking back to the ancient Greek debate.

Recall that while the atomists believed that there were many worlds, Aristotle held that this world *must* be unique and located in the center of everything, largely because his ideas of physics convinced him that all the "earth" in the universe would have naturally fallen to the center. The Copernican revolution therefore proved that Aristotle was wrong: Earth is not the center of the universe, after all.

Of course, the fact that Aristotle was wrong did *not* mean that the atomists had been right, but many of the Copernican-era scientists assumed that they had been. Galileo suggested that lunar features he saw through his telescope might be land and water much like that on Earth. Kepler agreed and went further, suggesting that the Moon had an atmosphere and was inhabited by intelligent beings. Kepler even wrote a science fiction story, *Somnium* ("The Dream"), in which he imagined a trip to the Moon and described the lunar inhabitants. The Dominican friar and philosopher Giordano Bruno was convinced of the existence of extraterrestrial life, a belief that contributed to battles with authorities that ultimately got him burned at the stake (see Special Topic 2.1).

Later scientists took the atomist belief even further. William Herschel (1738–1822), most famous as co-discoverer (with his sister Caroline) of the planet Uranus, assumed that all the planets were inhabited. In the late nineteenth century, when Percival Lowell (1855–1916) believed he saw canals on Mars (despite the fact that other astronomers wielding even bigger telescopes could not) [Section 8.1], it's quite likely that he was still being influenced by the philosophical ruminations of people who had lived more than 2000 years earlier.

If this debate about extraterrestrial life shows anything, it's probably this: *It's possible to argue almost endlessly, as long as there are no actual facts to get in the way.* With hindsight, it's easy for us to see that everything from the musings of the ancient Greeks to Lowell's martian canals was based more on hopes and beliefs than on any type of real evidence.

Nevertheless, the Copernican revolution really did mark a turning point in the debate about extraterrestrial life. For the first time, it was possible to test one of the ancient ideas—Aristotle's—and its failure caused it to be discarded. And while the Copernican revolution did not tell us whether the atomists had been right about life, it did make clear

Special Topic 2.1 GEOCENTRISM AND THE CHURCH

The case of Galileo is often portrayed as having exposed a deep conflict between science and religion. However, the history of the debate over geocentrism shows that the reality was much more complex, with deep divisions even within the Church hierarchy.

Perhaps the clearest evidence for a more open-minded Church comes from the case of Copernicus, whose revolutionary work was supported by many Church officials. A less well known and even earlier example concerns Nicholas of Cusa (1401–1464), who published a book arguing for a Sun-centered solar system in 1440, more than a century before Copernicus's book. Nicholas even weighed in on the subject of extraterrestrial life, writing

Rather than think that so many stars and parts of the heavens are un-inhabited and that this earth of ours alone is peopled ... we will suppose that in every region there are inhabitants, differing in nature by rank and allowing their origin to God ...

Church officials were apparently so untroubled by these radical ideas that they ordained Nicholas as a priest in the same year his book was published, and he later became a Cardinal. (Copernicus probably was not aware of this earlier work by Nicholas of Cusa.)

Many other scientists received similar support within the Church. Indeed, for most of his life, Galileo counted Cardinals—and even the pope who later excommunicated him—among his friends. Some historians suspect that Galileo got into trouble less for his views than for the way he portrayed them. For example, in 1632—just a year before his famous trial—he published a book in which two fictional characters debated the geocentric and Sun-centered views. He named the character taking the geocentric position Simplicio—essentially "simple-minded"—and someone apparently convinced the pope that the character was meant to be the pope. Moreover, as described by the noted modern author Isaac Asimov,

The book was all the more damaging to those who felt themselves insulted, because it was written in vigorous Italian for the general public (and not merely for the Latin-learned scholars) and was quickly translated into other languages—even Chinese!

If it was personality rather than belief that got Galileo into trouble, he was not the only one. The Italian philosopher Giordano Bruno (1548–1600), who had once been a Dominican monk, became an early and extreme supporter not only of the Copernican system but also of the idea of extraterrestrial life. In his book *On the Infinite Universe and Worlds*, published in 1584, Bruno wrote,

[It] is impossible that a rational being ... can imagine that these innumerable worlds, manifest as like to our own or yet more magnificent, should be destitute of similar or even superior inhabitants.

Note that Bruno was so adamant in his beliefs that he claimed that no "rational being" could disagree with him, so it's unsurprising that he drew the wrath of conservative Church officials (on numerous issues, not just extraterrestrial life). Bruno was branded a heretic and burned at the stake on February 17, 1600.

Perhaps the main lesson to be drawn from these stories is that while science has advanced dramatically in the past several centuries, people remain much the same. The Church was never a monolithic entity, and just as different people today debate the meaning of words in the Bible or other religious texts, Church scholars also held many different opinions at the time of the Copernican revolution. The political pendulum swung back and forth—or perhaps even chaotically—between the geocentric and Copernican views. Even when the evidence became overwhelming, a few diehards never gave in, and only the passing of generations finally ended the antagonism that had accompanied the great debate.

that the Moon and the planets really are other *worlds,* not mere lights in the sky. That fact alone makes it plausible to imagine life elsewhere, even if we still do not have the data necessary to conclude whether such life actually exists.

2.3 The Nature of Modern Science

The story of how our ancestors gradually figured out the basic architecture of the cosmos exhibits many features of what we now consider "good science." For example, we have seen how models were formulated and tested against observations, and then modified or replaced if they failed those tests. The story also illustrates some classic mistakes, such as the apparent failure of anyone before Kepler to question the belief that orbits must be circles. The ultimate success of the Copernican revolution led scientists, philosophers, and theologians to reassess the various modes of thinking that played a role in the 2000-year process of discovering Earth's place in the universe. Now, let's examine how the principles of modern science emerged from the lessons learned in the Copernican revolution.

How can we distinguish science from nonscience?

It's surprisingly difficult to define the term *science* precisely. The word comes from the Latin *scientia,* meaning "knowledge," but not all knowledge is science. For example, you may know what music you like best, but your musical taste is not a result of scientific study.

APPROACHES TO SCIENCE One reason science is difficult to define is that not all science works in the same way. For example, you've probably heard that science is supposed to proceed according to something called the "scientific method." As an idealized illustration of this method, consider what you would do if your flashlight suddenly stopped working. You might *hypothesize* that the flashlight's batteries have died. This type of tentative explanation, or **hypothesis,** is sometimes called an *educated guess*—in this case, it is "educated" because you already know that flashlights need batteries. Your hypothesis allows you to make a simple prediction: If you replace the batteries with new ones, the flashlight should work. You can test this prediction by replacing the batteries. If the flashlight now works, you've confirmed your hypothesis. If it doesn't, you must revise or discard your hypothesis, usually in favor of some other one that you can also test (such as that the bulb is burned out). Figure 2.14 illustrates the basic flow of this process.

The scientific method can be a useful idealization, but real science rarely progresses in such an orderly way. Scientific progress often begins with someone going out and looking at nature in a general way, rather than conducting a careful set of experiments. For example, Galileo wasn't looking for anything in particular when he pointed his telescope at the sky and made his first startling discoveries. We still often approach science in this way today, such as when we build new telescopes or send missions to other worlds. Science often depends on exploration.

We must also use alternative approaches when we attempt to understand past events, such as the history of Earth or the origin and evolution of life on Earth. We cannot repeat or vary the past, so we must instead rely on careful study of evidence left behind by past events. For

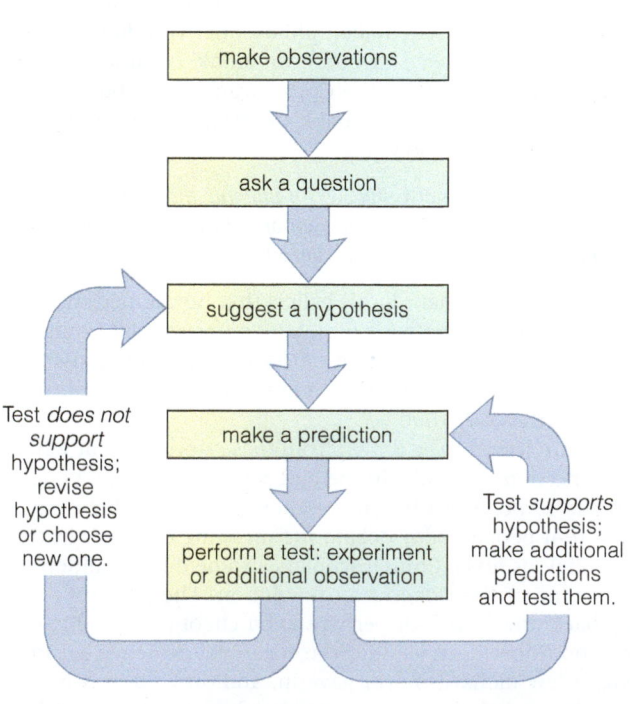

FIGURE 2.14

This diagram illustrates what we often call the *scientific method.*

make observations

ask a question

suggest a hypothesis

Test *does not support* hypothesis; revise hypothesis or choose new one.

make a prediction

perform a test: experiment or additional observation

Test *supports* hypothesis; make additional predictions and test them.

example, we learn about early life on Earth not by observing it directly but by piecing together its story from an examination of fossils and other evidence that we can find today. Nevertheless, we can still apply at least some elements of the scientific method. For example, when scientists first proposed the idea that a massive impact may have been responsible for the death of the dinosaurs [Section 6.4], they were able to predict some of the other types of evidence that should exist if their hypothesis was correct. These predictions allowed other scientists to plan observations that might uncover this evidence, and when they succeeded—such as in discovering an impact crater of the right age—support for the impact hypothesis grew much stronger.

A further complication in describing how science works comes from the fact that scientists are human beings, so their intuitions and personal beliefs inevitably influence their work. Copernicus, for example, adopted the idea that Earth orbits the Sun not because he had carefully tested this idea but because he believed it made more sense than the prevailing view of an Earth-centered universe. While his intuition guided him to the right general idea, he erred in the specifics because he still held Plato's ancient belief that heavenly motion must be in perfect circles.

Given the variety of ways in which it is possible to approach science, how can we identify what is science and what is not? To answer this question, we must look a little deeper at the distinguishing characteristics of scientific thinking.

HALLMARKS OF SCIENCE One way to define scientific thinking is to list the criteria that scientists use when they judge competing models of nature. Historians and philosophers of science have examined (and continue to examine) this issue in great depth, and different experts express somewhat different viewpoints on the details. Nevertheless, everything we now consider to be science shares the following three basic characteristics, which we will refer to as the *hallmarks of science* (Figure 2.15):

- Modern science seeks explanations for observed phenomena that rely solely on natural causes.
- Science progresses through the creation and testing of models of nature that explain the observations as simply as possible.
- A scientific model must make testable predictions about natural phenomena that would force us to revise or abandon the model if the predictions do not agree with observations.

Each of these hallmarks is evident in the story of the Copernican revolution. The first shows up in the way Tycho's careful measurements of planetary motion motivated Kepler to come up with a better explanation for those motions. The second is evident in the way several competing models were compared and tested, most notably those of Ptolemy, Copernicus, and Kepler. We see the third in the fact that each model could make precise predictions about the future motions of the Sun, Moon, planets, and stars in our sky. Kepler's model gained acceptance because it worked, while the competing models lost favor because their predictions failed to match the observations. Figure 2.16 summarizes the Copernican revolution and how it illustrates the hallmarks of science.

OCCAM'S RAZOR The criterion of simplicity in the second hallmark deserves additional explanation. Remember that Copernicus's original model did *not* match the data noticeably better than Ptolemy's model. If

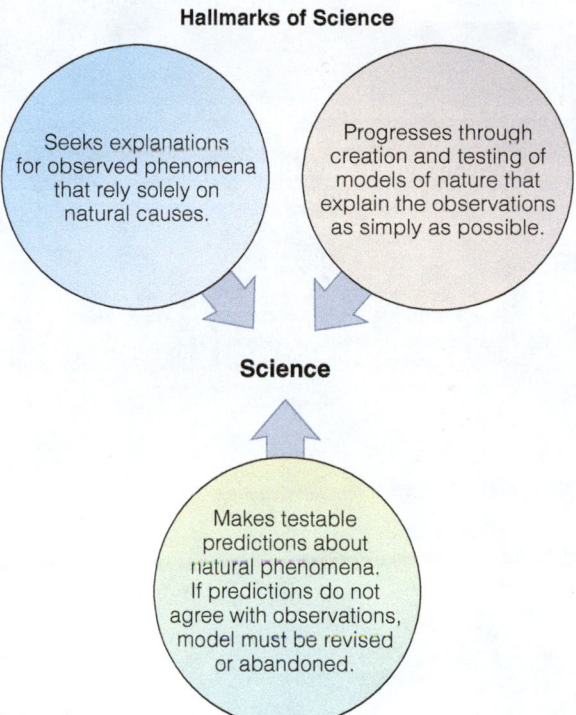

Hallmarks of Science

Seeks explanations for observed phenomena that rely solely on natural causes.

Progresses through creation and testing of models of nature that explain the observations as simply as possible.

Science

Makes testable predictions about natural phenomena. If predictions do not agree with observations, model must be revised or abandoned.

FIGURE 2.15
Hallmarks of science.

Ancient Earth-centered models of the universe easily explained the simple motions of the Sun and Moon through our sky, but had difficulty explaining the more complicated motions of the planets. The quest to understand planetary motions ultimately led to a revolution in our thinking about Earth's place in the universe that illustrates the process of science. This figure summarizes the major steps in that process.

(1) Night by night, planets usually move from west to east relative to the stars. However, during periods of *apparent retrograde motion,* they reverse direction for a few weeks to months [Section 2.1]. The ancient Greeks knew that any credible model of the solar system had to explain these observations.

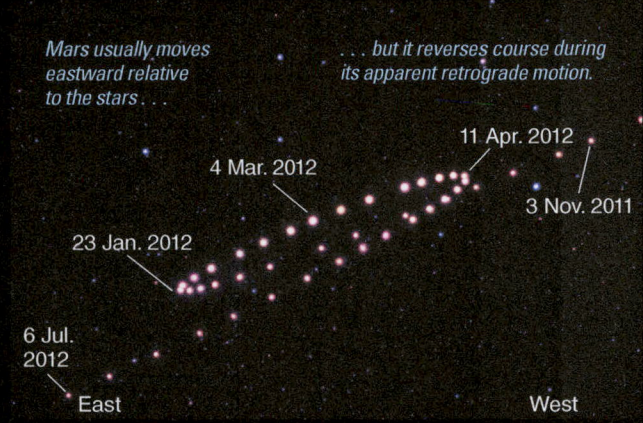

Mars usually moves eastward relative to the stars . . .

. . . but it reverses course during its apparent retrograde motion.

11 Apr. 2012

4 Mar. 2012

3 Nov. 2011

23 Jan. 2012

6 Jul. 2012

East   West

This composite photo shows the apparent retrograde motion of Mars.

(2) Most ancient Greek thinkers assumed that Earth remained fixed at the center of the solar system. To explain retrograde motion, they therefore added a complicated scheme of circles moving upon circles to their Earth-centered model. However, at least some Greeks, such as Aristarchus, preferred a Sun-centered model, which offered a simpler explanation for retrograde motion.

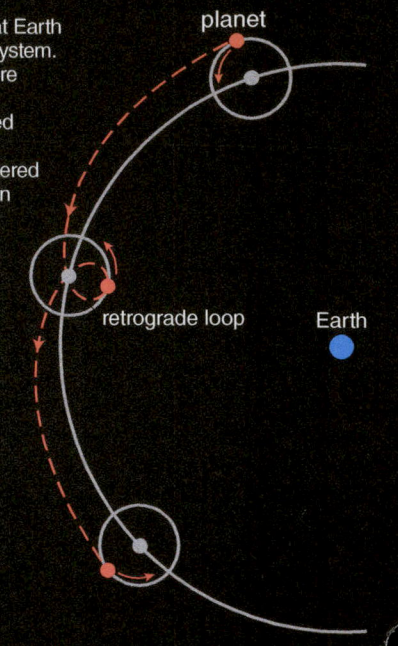

planet

retrograde loop   Earth

The Greek geocentric model explained apparent retrograde motion by having planets move around Earth on small circles that turned on larger circles.

HALLMARK OF SCIENCE   **A scientific model must seek explanations for observed phenomena that rely solely on natural causes.** The ancient Greeks used geometry to explain their observations of planetary motion.

(Left page)
A schematic map of the universe from 1539 with Earth at the center and the Sun (Solis) orbiting it between Venus (Veneris) and Mars (Martis).

(Right page)
A page from Copernicus's De Revolutionibus, *published in 1543, showing the Sun (Sol) at the center and Earth (Terra) orbiting between Venus and Mars.*

LIBRI COSMO.   Fo.V.

Schema huius præmiſſæ diuiſionis

Sphærarum.

net, in q
diximu
deniʠ l
currens

EMPIREVM   HABITACVLVM

Decimum Coelum   Primũ Mobile

Nonũ Coelum   Criſtallinũ

Octauum   Firmamentũ

SATVRNI

COELV   IOVIS

MARTIS

SOLIS

VENERIS

MERCVRII

LVNÆ

COELVM   DEI

③ By the time of Copernicus (1473–1543), predictions based on the Earth-centered model had become noticeably inaccurate. Hoping for improvement, Copernicus revived the Sun-centered idea. He did not succeed in making substantially better predictions because he retained the ancient belief that planets must move in perfect circles, but he inspired a revolution continued over the next century by Tycho, Kepler, and Galileo.

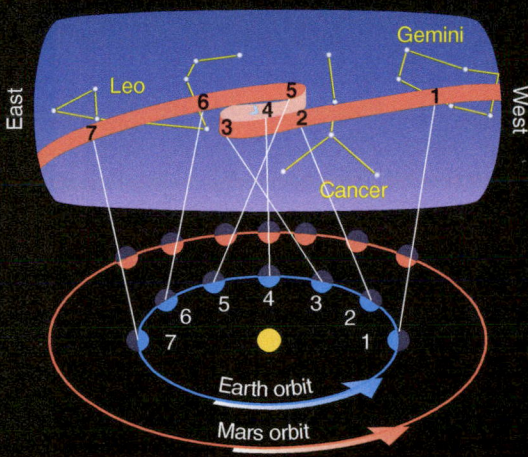

Apparent retrograde motion is simply explained in a Sun-centered system. Notice how Mars appears to change direction as Earth moves past it.

HALLMARK OF SCIENCE Science progresses through creation and testing of models of nature that explain the observations as simply as possible. Copernicus developed a Sun-centered model in hopes of explaining observations better than the more complicated Earth-centered model.

④ Tycho exposed flaws in both the ancient Greek and Copernican models by observing planetary motions with unprecedented accuracy. His observations led to Kepler's breakthrough insight that planetary orbits are elliptical, not circular, and enabled Kepler to develop his three laws of planetary motion.

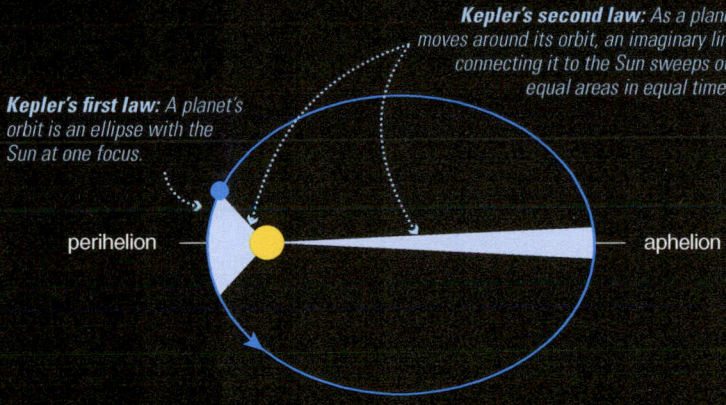

Kepler's first law: A planet's orbit is an ellipse with the Sun at one focus.

Kepler's second law: As a planet moves around its orbit, an imaginary line connecting it to the Sun sweeps out equal areas in equal times.

perihelion

aphelion

Kepler's third law: More distant planets orbit at slower average speeds, obeying $p^2 = a^3$.

HALLMARK OF SCIENCE A scientific model makes testable predictions about natural phenomena. If predictions do not agree with observations, the model must be revised or abandoned. Kepler could not make his model agree with observations until he abandoned the belief that planets move in perfect circles.

⑤ Galileo's experiments and telescopic observations overcame remaining scientific objections to the Sun-centered model. Together, Galileo's discoveries and the success of Kepler's laws in predicting planetary motion overthrew the Earth-centered model once and for all.

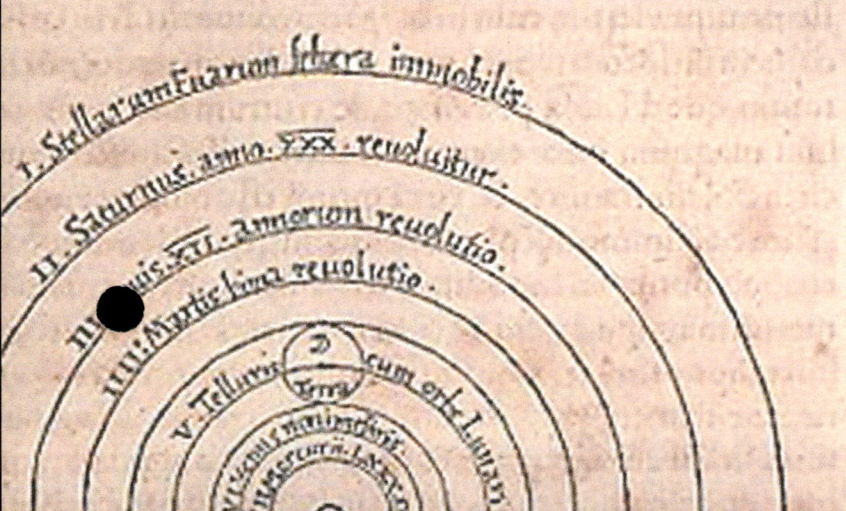

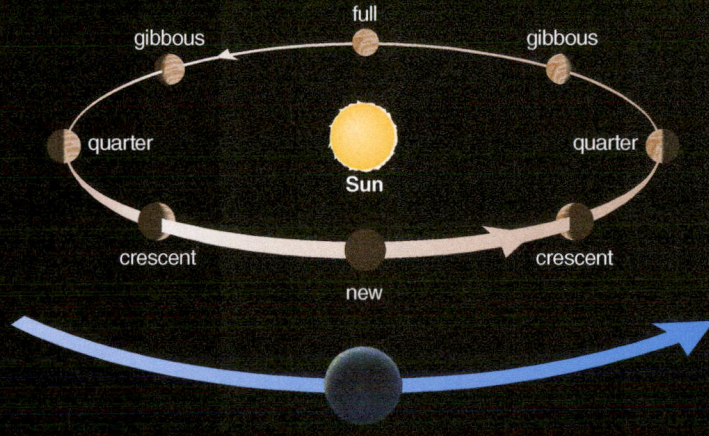

With his telescope, Galileo saw phases of Venus that are consistent only with the idea that Venus orbits the Sun rather than Earth.

scientists had judged this model solely on the accuracy of its predictions, they might have rejected it immediately. However, many scientists found elements of the Copernican model appealing, such as its simple explanation for apparent retrograde motion. They therefore kept the model alive until Kepler found a way to make it work.

If agreement with data were the sole criterion for judgment, we could imagine a modern-day Ptolemy adding millions or billions of additional circles to the geocentric model in an effort to improve its agreement with observations. A sufficiently complex geocentric model could in principle reproduce the observations with almost perfect accuracy—but it still would not convince us that Earth is the center of the universe. We would still choose the Copernican view over the geocentric view because its predictions would be just as accurate but follow a much simpler model of nature. The idea that scientists should prefer the simpler of two models that agree equally well with observations is called *Occam's razor,* after the medieval scholar William of Occam (1285–1349).

VERIFIABLE OBSERVATIONS The third hallmark of science forces us to face the question of what counts as an "observation" against which a prediction can be tested. Consider the claim that aliens are visiting Earth in UFOs. Proponents of this claim say that thousands of eyewitness reports of UFO encounters provide evidence that it is true. But do these personal testimonials count as *scientific* evidence? On the surface, the answer isn't obvious, because all scientific studies involve eyewitness accounts on some level. For example, only a handful of scientists have personally made detailed tests of Einstein's theory of relativity, and it is their personal reports of the results that have convinced other scientists of the theory's validity. However, there's an important difference between personal testimony about a scientific test and a UFO: The first can be verified by anyone, at least in principle, while the second cannot.

Understanding this difference is crucial to understanding what counts as science and what does not. Even though you may never have conducted a test of Einstein's theory of relativity yourself, there's nothing stopping you from doing so. It might require several years of study before you have the necessary background to conduct the test, but you could then confirm the results reported by other scientists. In other words, while you may currently be trusting the eyewitness testimony of scientists, you always have the option of verifying their testimony for yourself.

In contrast, there is no way for you to verify someone's eyewitness account of a UFO. Without hard evidence such as clear photographs or pieces of the UFO, there is nothing that you could evaluate for yourself, even in principle. (And in those cases where "hard evidence" for UFO sightings has been presented, scientific study has never yet found the evidence to be strong enough to support the claim of alien spacecraft [Section 12.4].) Moreover, scientific studies of eyewitness testimony show it to be notoriously unreliable. For example, different eyewitnesses often disagree on what they saw even immediately after an event has occurred. As time passes, memories of the event may change further. In some cases in which memory has been checked against reality, people have reported vivid memories of events that never happened at all. This explains something that virtually all of us have experienced: disagreements with a friend about who did what and when. Since both people cannot be right in such cases, at least one person must have a memory that differs from reality.

The demonstrated unreliability of eyewitness testimony explains why it is generally considered insufficient for a conviction in criminal court; at least some other evidence is required. For the same reason, we cannot accept eyewitness testimony by itself as evidence in science, no matter who reports it or how many people offer similar testimony.

SCIENCE AND PSEUDOSCIENCE It's important to realize that science is not the only valid way of seeking knowledge. For example, suppose you are shopping for a car, learning to play drums, or pondering the meaning of life. In each case, you might make observations, exercise logic, and test hypotheses. Yet these pursuits clearly are not science, because they are not directed at developing testable explanations for observed natural phenomena. As long as nonscientific searches for knowledge make no claims about how the natural world works, they do not conflict with science.

However, you will often hear claims about the natural world that seem to be based on observational evidence but do not treat evidence in a truly scientific way. Such claims are often called **pseudoscience,** which means "false science." To distinguish real science from pseudoscience, a good first step is to check whether a particular claim exhibits all three hallmarks of science. Consider the example of people who claim a psychic ability to "see" the future and use it to make specific, testable predictions. In this sense, "seeing" the future sounds scientific, since we can test it. However, numerous studies have examined the predictions of "seers" and have found that their predictions come true no more often than would be expected by pure chance. If the seers were scientific, they would admit that this evidence undercuts their claim of psychic abilities. Instead, they generally make excuses, such as saying that the predictions didn't come true because of some type of "psychic interference." Making testable claims but then ignoring the results of the tests marks the claimed ability to see the future as pseudoscience.

OBJECTIVITY IN SCIENCE The idea that science is objective, meaning that all people should be able to find the same results, is important to the validity of science as a means of seeking knowledge. However, there is a difference between the overall objectivity of science and the objectivity of individual scientists.

Science is practiced by human beings, and individual scientists may bring their personal biases and beliefs to their scientific work. For example, most scientists choose their research projects based on personal interests rather than on some objective formula. In extreme cases, scientists have even been known to cheat—either deliberately or subconsciously—to obtain a result they desire. For example, in the late nineteenth century, astronomer Percival Lowell claimed to see a network of artificial canals in blurry telescopic images of Mars, leading him to conclude that there was a great Martian civilization. But no such canals actually exist, so Lowell must have allowed his beliefs about extraterrestrial life to influence the way he interpreted what he saw. A more deliberate—and much more damaging—case of cheating occurred in 1998, when British physician Andrew Wakefield published results claiming a link between childhood vaccines and autism, but follow-up research revealed the claim to be fraudulent.

Bias can sometimes show up even in the thinking of the scientific community as a whole. Some valid ideas may not be considered by any scientist because they fall too far outside the general patterns of thought, or **paradigm,** of the time. Einstein's theory of relativity provides an example. Many scientists in the decades before Einstein had gleaned hints of the theory but did not investigate them, at least in part because the ideas seemed too outlandish.

The beauty of science is that it encourages continued testing by many people. Even if personal biases affect some results, tests by others should eventually uncover the mistakes. Similarly, if a new idea is correct but falls outside the accepted paradigm, sufficient testing and verification of the idea should eventually force a paradigm shift. In that sense, *science ultimately provides a means of bringing people to agreement*, at least on topics that can be studied scientifically.

What is a scientific theory?

The most successful scientific models explain a wide variety of observations in terms of just a few general principles. When a powerful yet simple model makes predictions that survive repeated and varied testing, scientists elevate its status and call it a **theory.** Some famous examples are Isaac Newton's theory of gravity, Charles Darwin's theory of evolution, and Albert Einstein's theory of relativity.

THE MEANING OF *THEORY* AND OTHER SCIENTIFIC TERMS The scientific meaning of the word *theory* is quite different from its everyday meaning, in which we equate a theory more closely with speculation or a hypothesis. In everyday life, someone might say, "I have a new theory about why people enjoy the beach." Without the support of a broad range of evidence that others have tested and confirmed, this "theory" is really only a guess. In contrast, Newton's theory of gravity qualifies as a scientific theory because it uses simple physical principles to explain many observations and experiments. "Theory" is just one of many terms that are used with different meaning in science than in everyday life. Table 2.2 summarizes some of these terms, emphasizing those that you'll encounter in this book.

Despite its success in explaining observed phenomena, a scientific theory can never be proved true beyond all doubt, because future observations may disagree with its predictions. However, anything that qualifies as a scientific theory must be supported by a large, compelling body of evidence.

In this sense, a scientific theory is not at all like a hypothesis or any other type of guess. We are free to change a hypothesis at any time, because it has not yet been carefully tested. In contrast, we can discard or replace a scientific theory only if we have a better way of explaining the evidence that supports it.

Again, the theories of Newton and Einstein offer great examples. A vast body of evidence supports Newton's theory of gravity, but by the late nineteenth century scientists had begun to discover cases where its predictions did not perfectly match observations. These discrepancies were explained only when Einstein developed his general theory of relativity, which was able to match the observations. Still, the many successes of Newton's theory could not be ignored, and Einstein's theory would not have gained acceptance if it had not been able to explain these successes equally well. It did, and that is why we now view Einstein's theory as a broader theory of gravity than Newton's theory. As we will discuss in

TABLE 2.2 Scientific Usage Often Differs from Everyday Usage

This table lists some words you will encounter in this book that have a different meaning in science than in everyday life. (Adapted from a table published by Richard Somerville and Susan Joy Hassol in Physics Today, *Oct. 2011.)*

Term	Everyday Meaning	Scientific Meaning	Example
model	something you build, like a model airplane	a representation of nature, sometimes using mathematics or computer simulations, that is intended to explain or predict observed phenomena	A model of planetary motion can be used to calculate exactly where planets should appear in our sky.
hypothesis	a guess or assumption of almost any type	a model that has been proposed to explain some observations, but which has not yet been rigorously confirmed	Scientists hypothesize that the Moon was formed by a giant impact, but there is not enough evidence to be fully confident in this model.
theory	speculation	a particularly powerful model that has been so extensively tested and verified that we have extremely high confidence in its validity	Einstein's theory of relativity successfully explains a broad range of natural phenomena and has passed a great many tests of its validity.
bias	distortion, political motive	tendency toward a particular result	Current techniques for detecting extrasolar planets are biased toward detecting large planets.
critical	really important; involving criticism, often negative	right on the edge	A boiling point is a "critical value" because above that temperature, a liquid will boil away.
deviation	strangeness or unacceptable behavior	change or difference	The recent deviation in global temperatures compared to their long-term average implies that something is heating the planet.
enhance/enrich	improve	increase or add more, but not necessarily making something "better"	"Enhanced color" means colors that have been brightened. "Enriched with iron" means containing more iron.
error	mistake	range of uncertainty	The "margin of error" tells us how closely measured values are likely to reflect true values.
feedback	a response	a self-regulating (negative feedback) or self-reinforcing (positive feedback) cycle	Gravity can provide positive feedback to a forming planet: Adding mass leads to stronger gravity, which attracts more mass, and so on.
state (as a noun)	a place or location	a description of current condition	The Sun is in a state of balance, so it shines steadily.
trick	deception or prank	clever approach	A mathematical trick solved the problem.
uncertainty	ignorance	a range of possible values around some central value	The measured age of our solar system is 4.55 billion years with an uncertainty of 0.02 billion years.
values	ethics, monetary value	numbers or quantities	The speed of light has a measured value of 300,000 km/s.

the next section, some scientists today are seeking a theory of gravity that will go beyond Einstein's. If any new theory ever gains acceptance, it will have to match all the successes of Einstein's theory as well as work in new realms where Einstein's theory does not.

Think About It When people claim that something is "only a theory," what do you think they mean? Does this meaning of *theory* agree with the definition of a theory in science? Do scientists always use the word *theory* in its "scientific" sense? Explain.

THE QUEST FOR A THEORY OF LIFE IN THE UNIVERSE We do not yet have a theory of life in the universe, because we do not yet have the data to distinguish between many different hypotheses, which range from the hypothesis of no life anyplace else to the hypothesis that civilizations are abundant in our own galaxy. But thanks to the historical process that gave us the principles of modern science, we have a good idea of what we need to do if we ever hope to verify one of those hypotheses and turn it into a broad-based theory of life in the universe. That is why we can now make a modern science of astrobiology: not because we actually understand it yet but because we now know how to choose appropriate research projects to help us learn about the possibility of finding life elsewhere and how to go out and search for life that might exist within our solar system or beyond.

✳ THE PROCESS OF SCIENCE IN ACTION

2.4 The Fact and Theory of Gravity

We've completed our overview of the nature of modern science and its historical development. We've discussed the general process by which science advances, a process that is crucial to all sciences but is particularly important in astrobiology where, for example, widespread belief in aliens sometimes makes it difficult to separate fact from fiction. Because of its importance, we will continue to focus on the process of science throughout the book. In addition, in the final numbered section of this and all remaining chapters, we will take one topic and explore it in more depth, using it to illustrate some aspect of the process of science in action.

In this chapter, we focus on gravity. Gravity is obviously important to life in the universe. On a simple level, life would float off its planet without gravity. On a deeper level, stars and planets could never have been born in the first place without gravity, so we presume that life could not start in a universe in which gravity were absent or in which it worked significantly differently than it does in our universe. Gravity also

Movie Madness GRAVITY

Want to go into space? Of course you do, and you can. Simply write a big check to a private rocket company and wait in a long line for a short ride.

Alternatively, you can put on your 3D glasses and watch the movie *Gravity*, which took home seven Oscars.

It's an "incident" story. A small handful of astronauts are on a servicing mission for the Hubble Telescope (apparently they didn't get the memo that repair efforts ended in 2009), and as the film opens they're busy torqueing up bolts on a solar panel while engaging in the witty banter that characterizes all movie astronauts. This could get boring, but fortunately the situation quickly turns uglier than monkfish.

The Russians, for inscrutable reasons, have blown up a satellite somewhere, creating lots of hi-tech shrapnel in space. Now, that's bad manners, and self-destructive too, as there's a chance that the resulting junk will hit one of their own space assets. And this sudden trove of trash is threatening the repair crew with catastrophe 350 miles above the Earth.

Mind you, in reality, there wouldn't be much chance that the debris would actually hit them (or anything else). Space is big—really big—and space stations and orbiting telescopes don't follow one another

around like circus elephants. They're slaves to Kepler's laws, and they careen about the planet at different heights and in different directions. (The exceptions are satellites in geosynchronous orbit, but that's not where the Hubble Telescope, the Space Station, or anything else in this film is located.) The chance that you'd be hit even once by a freshly formed cloud of satellite debris is not much different from the odds of being beaned by a meteor in your backyard.

But in *Gravity*, getting hit once is just the initial round, and the astronauts are compelled to save themselves by repeatedly jet-packing their way to new orbiting oases, a tactic about as plausible as rescuing yourself at sea by backstroking from one island to another. In truth, you'd need real rockets to get to that next astronaut safe house.

Of course, it's easy to nitpick about the goofy orbital mechanics in *Gravity*. But what's really exceptional about this film is that it conveys the sensation of being in space a whole lot better than all those NASA videos you watched as a kid. Shot in large format and 3D, this movie is as stunning as a taser and will boldly take you where you've never been before: into orbit.

And you and your friends or family will be back, safe and sound, by bedtime.

provides a great example of the distinction in science between a "fact" and a "theory," which is the idea we will focus on here.

How does the fact of gravity differ from the theory of gravity?

Gravity is clearly a fact: Things really do fall when you drop them, and planets really do orbit the Sun. However, despite our daily experience with gravity, an adequate *theory* of gravity took a long time to develop. In ancient Greece, Aristotle imagined gravity to be an inherent property of heavy objects and claimed that heavier objects would fall to the ground faster than lighter-weight objects. Galileo put this idea to the test in a series of experiments that supposedly included dropping weights from the Leaning Tower of Pisa. His results showed that all objects fall to the ground at the same rate, as long as air resistance is unimportant. Aristotle was therefore wrong about gravity, but Galileo's ideas about gravity still fell short of being a useful theory.

Think About It Find a piece of paper and a small rock. Hold both at the same height, one in each hand, and let them go at the same instant. The rock, of course, hits the ground first. Next, crumple the paper into a small ball and repeat the experiment. What happens? Explain how this experiment suggests that, without air resistance, gravity causes all falling objects to fall at the same rate.

NEWTON'S THEORY OF GRAVITY The breakthrough in our understanding of gravity came from Isaac Newton. By his own account, he experienced a moment of inspiration in 1666 when he saw an apple fall to the ground. He suddenly realized that the gravity making the apple fall was the same force that held the Moon in orbit around Earth. With this insight, Newton eliminated the long-held distinction between the realm of the heavens and the realm of Earth. For the first time, the two realms were brought together as one *universe* governed by a single set of principles.

Newton worked hard to turn his insight into a theory of gravity, which he published in 1687 (in his book *Principia*). Newton expressed the force of gravity mathematically with his **universal law of gravitation.** Three simple statements summarize this law:

- Every mass attracts every other mass through the force called *gravity*.
- The strength of the gravitational force attracting any two objects is *directly proportional* to the product of their masses. For example, doubling the mass of *one* object doubles the force of gravity between the two objects.
- The strength of gravity between two objects decreases with the *square* of the distance between their centers. That is, the gravitational force follows an **inverse square law** with distance. For example, doubling the distance between two objects weakens the force of gravity by a factor of 2^2, or 4.

These three statements tell us everything we need to know about Newton's universal law of gravitation. Mathematically, all three statements can be combined into a single equation, usually written like this:

$$F_g = G\frac{M_1 M_2}{d^2}$$

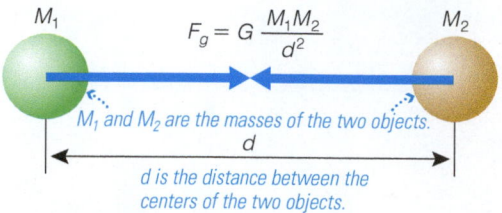

The **universal law of gravitation** tells us the strength of the gravitational attraction between the two objects.

$$F_g = G \frac{M_1 M_2}{d^2}$$

M_1 and M_2 are the masses of the two objects.

d is the distance between the centers of the two objects.

FIGURE 2.17
The universal law of gravitation is an inverse square law, which means the force of gravity declines with the square of the distance d between two objects.

where F_g is the force of gravitational attraction, M_1 and M_2 are the masses of the two objects, and d is the distance between their centers (Figure 2.17). The symbol G is a constant called the **gravitational constant,** and its numerical value has been measured to be $G = 6.67 \times 10^{-11}\,\mathrm{m^3/(kg \times s^2)}$.

Think About It How does the gravitational force between two objects change if the distance between them triples? If the distance between them drops by half?

Newton's theory of gravity gained rapid acceptance because it explained a great many facts that other scientists had already discovered. For example, it explained Galileo's observations about falling objects and Kepler's laws of planetary motion. Even more impressively, it led to new predictive successes. Shortly after Newton published his theory, Sir Edmund Halley used it to calculate the orbit of a comet that had been seen in 1682, from which he predicted the comet's return in 1758. Halley's Comet returned on schedule, which is why it now bears his name. In 1846, after carefully examining the orbit of Uranus, the French astronomer Urbain Leverrier used Newton's theory to predict that Uranus's orbit was being affected by a previously undiscovered eighth planet.* He predicted the location of the planet and sent a letter suggesting a search to Johann Galle of the Berlin Observatory. On the night of September 23, 1846, Galle discovered Neptune within 1° of the position predicted by Leverrier. It was a stunning triumph for Newton's theory.

A PROBLEM APPEARS Today, we can apply Newton's theory of gravity to objects throughout the universe, including the orbits of extrasolar planets around their stars, of stars around the Milky Way Galaxy, and of galaxies in orbit of each other. There seems no reason to doubt the universality of the law. However, we also now know that Newton's law does not tell the entire story of gravity.

The first hint of a problem with Newton's theory arose not long after Leverrier's success in predicting the existence of Neptune. Astronomers discovered a slight discrepancy between the observed characteristics of the orbit of Mercury and the characteristics predicted by Newton's theory. The discrepancy was very small, and Mercury was the only planet that showed any problem, but there seemed no way to make it go away: Unless the data were wrong, which seemed highly unlikely, Newton's theory was giving a slightly incorrect prediction for the orbit of Mercury.

Leverrier set to work on this new problem, suggesting it might be solved if there were yet another unseen planet, this one orbiting the Sun closer than Mercury. He even gave it a name—Vulcan. But searches turned up no sign of this planet, and we now know that it does not exist. So why was there a discrepancy in Mercury's orbit? Albert Einstein (1879–1955) provided the answer when he published his *general theory of relativity* in 1915.

EINSTEIN'S SOLUTION To understand what Einstein did, we need to look a little more deeply at Newton's conception of gravity. According to

---

*The same idea had been put forward a few years earlier in England by a student named John Adams, but he did not succeed soon enough in convincing anyone to search for the planet; Leverrier was apparently unaware of Adams's work.

Newton's theory, every mass exerts a gravitational attraction on every other mass, no matter how far away it is. If you think about it, this idea of "action at a distance" is rather mysterious. For example, how does Earth "feel" the Sun's attraction and know to orbit it? Newton himself was troubled by this idea. A few years after publishing his law of gravity in 1687, Newton wrote:

> That one body may act upon another at a distance through a vacuum, ... and force may be conveyed from one to another, is to me so great an absurdity, that I believe no man, who has ... a competent faculty in thinking, can ever fall into it.*

This type of "absurdity" was troubling to Einstein, whose scientific career can in many ways be viewed as a quest to find simple principles underlying mysterious laws. Although we will not go into the details, Einstein discovered that he could explain the mysterious action at a distance by assuming that all objects reside in something known as four-dimensional *spacetime*. Massive objects curve this spacetime, and other objects simply follow the curvature much like marbles following the contours of a bowl. Figure 2.18 uses a two-dimensional analogy to illustrate the idea, showing how planetary orbits are the straightest paths allowed by the structure of spacetime near the Sun. Einstein removed the mystery of "action at a distance" by telling us that gravity arises from the way in which masses affect the basic structure of the universe; in other words, he told us that gravity *is* "curvature of spacetime."

When Einstein worked out the mathematical details of his theory, he found that it predicted an orbit for Mercury that matched the observations. Not long after, astronomers put Einstein's new theory to the test during a solar eclipse, finding that it successfully predicted the precise positions of stars visible near the blocked disk of the Sun, while Newton's theory gave a prediction that was slightly off. Scientists have continued to test both Newton's and Einstein's theories ever since. In every case in which the two theories give different answers, Einstein's theory has matched the observations while Newton's theory has not. That is why, today, we consider Einstein's general theory of relativity to have supplanted Newton's theory as our "best" theory of gravity.

Does this mean that Newton's theory of gravity was "wrong"? Remember that Newton's theory successfully explains nearly all observations of gravity in the universe, and it works so well that we can use it to plot the courses of spacecraft to the planets. Moreover, in all cases in which Newton's theory works well, Einstein's theory gives essentially the same answers. The differences in the predictions between the two theories are noticeable only with extremely precise measurements or in cases where gravity is unusually strong. We therefore do not say that Newton's theory was wrong, but rather that it was only an *approximation* to a more exact theory of gravity— Einstein's general theory of relativity. Under most circumstances, the approximation is so good that we can barely tell the difference between the two theories of gravity, but in cases of strong gravity, Einstein's theory works and Newton's fails.

While Einstein's theory of gravity has so far passed every test that it has been subjected to, most scientists suspect that we'll eventually find

The mass of the Sun causes spacetime to curve . . .

. . . so freely moving objects (such as planets and comets) follow the straightest possible paths allowed by the curvature of spacetime.

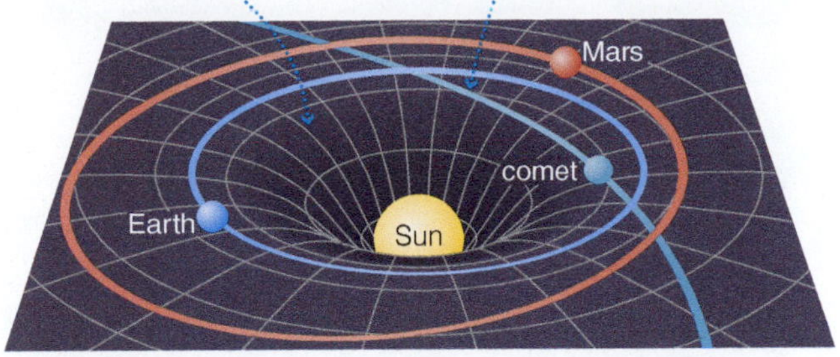

FIGURE 2.18

According to Einstein's general theory of relativity, the Sun curves spacetime much as a heavy weight curves a rubber sheet, and planets simply follow this curvature in their orbits.

---

*Letter from Newton, 1692–1693, as quoted in J. A. Wheeler, *A Journey into Gravity and Spacetime*, Scientific American Library, 1990, p. 2.

an even better theory of gravity. The reason is that for the most extreme possible case of gravity—that which occurs at the infinitely small and high-density center of a *black hole*—Einstein's theory of relativity gives a different answer than the equally well tested theory of the very small (known as the theory of quantum mechanics). Because these two theories contradict each other in this special case, scientists know that one or both will ultimately have to be modified.

THE BOTTOM LINE Gravity is both a fact and a theory. The *fact* of gravity is obvious in the observations we make of falling objects on Earth and orbiting objects in space. The *theory* of gravity is our best explanation of those observations, and we can use it to make precise predictions of how objects will behave due to gravity. In the future, our theory of gravity may be further improved, but gravity remains a fact regardless of how we revise the theory. Note that gravity is not unique in this way: Scientists make the same type of distinction in many other cases, such as when they talk about the fact of atoms being real and the atomic theory used to explain them, and when they talk about fact of evolution revealed in the fossil record and the theory used to explain how evolution occurs.

The Big Picture

PUTTING CHAPTER 2 IN PERSPECTIVE

In this chapter, we've explored the development and nature of science, and how thoughts about life in the universe changed with the development of science. As you continue your studies, keep in mind the following "big picture" ideas:

- The questions that drive research about life in the universe have been debated for thousands of years, but only recently have we begun to acquire data that allow us to address the questions scientifically. In particular, the fundamental change in human perspective that came with the Copernican revolution had a dramatic impact on the question of life in the universe, because it showed that planets really are other *worlds* and not mere lights in the sky.

- The ideas that underlie modern science—what we've called the "hallmarks of science"—developed gradually, and largely as a result of the attempt to understand Earth's place in the universe. Science always begins by assuming that the world is inherently understandable and that we can learn how it works by observing it and by examining the processes that affect it. All of science, therefore, is based on observations of the world around us.

- Science is not the only valid way in which we can seek knowledge, but it has proved enormously useful, having driven the great progress both in our understanding of nature and in the development of technology that has occurred in the past 400 years.

Summary of Key Concepts

2.1 The Ancient Debate About Life Beyond Earth

How did attempts to understand the sky start us on the road to science?

 The development of science began with Greek attempts to create **models** to explain observations of the heavens. Although most Greek philosophers favored a **geocentric model**, which we now know to be incorrect, their reasons for this choice made sense at the time. One of the primary difficulties of the model was that it required a complicated explanation for the **apparent retrograde motion** of the planets, with planets going around small circles on larger circles that went around Earth, rather than the much simpler explanation that we find with a Sun-centered model.

Why did the Greeks argue about the possibility of life beyond Earth?

Some Greek philosophers (the *atomists*) held that our world formed among an infinite number of indivisible atoms, and this infinity implied the existence of other worlds. In contrast, Aristotle and his followers (the *Aristotelians*) argued that all earth must have fallen to the center of the universe, which rationalized the belief in a geocentric universe and the belief that the heavens were fundamentally different from Earth. This implied that Earth must be unique, in which case no other worlds or other life could exist.

2.2 The Copernican Revolution

How did the Copernican revolution further the development of science?

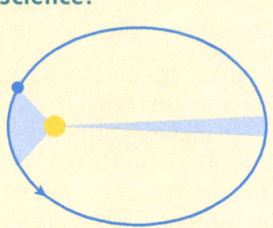 During the **Copernican revolution,** scientists began to place much greater emphasis on making sure that models successfully reproduced observations, and learned to trust data even when it contradicted deeply held beliefs. This willingness to let data drive the development of models led Kepler to develop what we now call **Kepler's laws of planetary motion,** and later led to the deeper understanding that came with **Newton's laws of motion** and the **universal law of gravitation.**

How did the Copernican revolution alter the ancient debate on extraterrestrial life?

The Copernican revolution showed that Aristotle's Earth-centered beliefs had been incorrect, effectively ruling out his argument for Earth's uniqueness. Many scientists of the time therefore assumed that the atomists had been correct, and that other worlds and life are widespread. However, the data didn't really support this view, which is why we still seek to learn whether life exists elsewhere.

2.3 The Nature of Modern Science

How can we distinguish science from nonscience?

Science generally exhibits these three hallmarks: (1) Modern science seeks explanations for observed phenomena that rely solely on natural causes. (2) Science progresses through the creation and testing of models of nature that explain the observations as simply as possible. (3) A scientific model must make testable predictions about natural phenomena that would force us to revise or abandon the model if the predictions do not agree with observations.

What is a scientific theory?

A scientific **theory** is a simple yet powerful model that explains a wide variety of observations in terms of just a few general principles, and has attained the status of a theory by surviving repeated and varied testing.

✳ **THE PROCESS OF SCIENCE IN ACTION**

2.4 The Fact and Theory of Gravity

How does the fact of gravity differ from the theory of gravity?

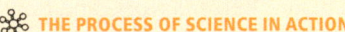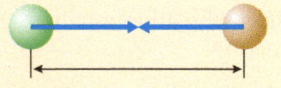 Gravity is a *fact* in that objects really do fall to the ground and planets really do orbit the Sun. The *theory* of gravity is used to explain why gravity acts as it does. While the fact of gravity does not change, the theory can be improved with time, as Einstein's general theory of relativity improved on Newton's theory of gravity.

Exercises and Problems

MasteringAstronomy® *For instructor-assigned homework and other learning materials, go to MasteringAstronomy®.*

REVIEW QUESTIONS

Short-Answer Questions Based on the Reading

1. Describe at least three characteristics of Greek thinking that helped pave the way for the development of modern science.

2. What do we mean by a *model* of nature? Summarize the development of the Greek *geocentric model*, from Thales through Ptolemy.

3. What is *apparent retrograde motion*, and why was it so difficult to explain with the geocentric model? What is its real explanation?

4. Who first proposed the idea that Earth is a planet orbiting the Sun, and when? Why didn't this model gain wide acceptance in ancient Greece?

5. Briefly describe and contrast the different views of the atomists and the Aristotelians on the subject of extraterrestrial life.

6. What was the *Copernican revolution*, and how did it change the human view of the universe? Briefly describe the major players and events in the Copernican revolution.

7. Why didn't Copernicus's model gain immediate acceptance? Why did some scientists favor it, despite this drawback?

8. State and explain each of *Kepler's laws of planetary motion*. Why did they gain acceptance?

9. Briefly describe three reasonable objections to the Sun-centered model that still remained even after Kepler's work, and how Galileo's work overcame each of these objections.

10. How did Newton's discoveries about the laws of motion and the universal law of gravitation put the Sun-centered model on an even stronger footing?

11. How did the Copernican revolution affect scholarly thought regarding the question of life beyond Earth?

12. What is the difference between a *hypothesis* and a *theory* in science?

13. Describe each of the three hallmarks of science and give an example of how we can see each one in the unfolding of the Copernican revolution.

14. What is Occam's razor? Give an example of how it applies.

15. Why doesn't science accept personal testimony as evidence? Explain.

16. In what sense is gravity both a fact and a theory? Explain clearly.

17. What is Newton's *universal law of gravitation*? Write it in equation form, and clearly explain what the equation tells us. What do we mean when we say that the law is an *inverse square law*?

18. How did Einstein's general theory of relativity change our view of gravity?

TEST YOUR UNDERSTANDING

Science or Nonscience?

Each of the following statements makes some type of claim. Decide in each case whether the claim could be evaluated scientifically or whether it falls into the realm of nonscience. Explain clearly; not all of these have definitive answers, so your explanation is more important than your chosen answer.

19. Lionel Messi is the best soccer player of his generation.

20. Several kilometers below its surface, Europa has an ocean of liquid water.

21. My house is haunted by ghosts, who make the creaking noises I hear each night.

22. There are no lakes or seas on Mars today.

23. All life in the universe must use DNA as its genetic material.

24. Children born when Jupiter is in the constellation Taurus are more likely to be musicians than other children.

25. Aliens can manipulate time so that they can abduct and perform experiments on people who never realize they were taken.

26. Newton's law of gravity explains the orbits of planets around other stars just as well as it explains the orbits of planets in our own solar system.

27. God created the laws of motion that were discovered by Newton.

28. A huge fleet of alien spacecraft will land on Earth and introduce an era of peace and prosperity on January 1, 2030.

Quick Quiz

Choose the best answer to each of the following. Explain your reasoning with one or more complete sentences.

29. In Ptolemy's geocentric model, the retrograde motion of a planet occurs when (a) Earth is about to pass the planet in its orbit around the Sun; (b) the planet actually goes backward in its orbit around Earth; (c) the planet is aligned with the Moon in our sky.

30. Which of the following was *not* a major advantage of Copernicus's Sun-centered model over the Ptolemaic model? (a) It made significantly better predictions of planetary positions in our sky. (b) It offered a more natural explanation for the apparent retrograde motion of planets in our sky. (c) It allowed calculation of the orbital periods and distances of the planets.

31. Earth is closer to the Sun in January than in July. Therefore, in accord with Kepler's second law, (a) Earth travels faster in its orbit around the Sun in July than in January; (b) Earth travels faster in its orbit around the Sun in January than in July; (c) Earth has summer in January and winter in July.

32. According to Kepler's *third* law, (a) Mercury travels fastest in the part of its orbit in which it is closest to the Sun; (b) Jupiter orbits the Sun at a faster speed than Saturn; (c) all the planets have nearly circular orbits.

33. Tycho Brahe's contributions to astronomy included (a) inventing the telescope; (b) proving that Earth orbits the Sun; (c) collecting data that enabled Kepler to discover the laws of planetary motion.

34. Galileo's contributions to astronomy included (a) discovering the laws of planetary motion; (b) discovering the law of gravity;

(c) making observations and conducting experiments that dispelled scientific objections to the Sun-centered model.

35. Which of the following is *not* true about scientific progress? (a) Science progresses through the creation and testing of models of nature. (b) Science advances only through strict application of the scientific method. (c) Science avoids explanations that invoke the supernatural.

36. Which of the following is *not* true about a scientific theory? (a) A theory must explain a wide range of observations or experiments. (b) Even the strongest theories can never be proved true beyond all doubt. (c) A theory is essentially an educated guess.

37. How did the Copernican revolution alter perceptions of the ancient Greek debate over extraterrestrial life? (a) It showed that Aristotle's argument for a unique Earth was incorrect. (b) It showed that the atomists were correct in their belief in an infinite cosmos. (c) It proved that extraterrestrial life must really exist.

38. When Einstein's theory of gravity (general relativity) gained acceptance, it demonstrated that Newton's theory had been (a) wrong; (b) incomplete; (c) really only a guess.

PROCESS OF SCIENCE

39. *Greek Models.* As we discussed in this chapter, the Greeks actually considered both Earth-centered and Sun-centered models of the cosmos.
    a. Briefly describe the pros and cons of each model as they were seen in ancient times, and explain why most Greeks preferred the geocentric model.
    b. Suppose you could travel back in time and show the Greeks *one* observation from modern times. If your goal was to convince the Greeks to accept the Sun-centered model, what observation would you choose? Do you think it would convince them? Explain.

40. *What Makes It Science?* Read ahead and choose a single idea in the modern view of the cosmos that is discussed in Chapter 3, such as "The universe is expanding," "The universe began with a Big Bang," "We are made from elements manufactured by stars," or "The Sun orbits the center of the Milky Way Galaxy."
    a. Briefly describe how the idea you have chosen is rooted in each of the three hallmarks of science discussed in this chapter. (That is, explain how it is based on observations, how our understanding of it depends on a model, and how the model is testable.)
    b. No matter how strongly the evidence may support a scientific idea, we can never be certain beyond all doubt that the idea is true. For the idea you have chosen, describe an observation that might cause us to call the idea into question. Then briefly discuss whether you think that, overall, the idea is likely or unlikely to hold up to future observations. Defend your opinion.

GROUP WORK EXERCISE

41. *Testing UFOs.* **Roles:** *Scribe* (takes notes on the group's activities), *Proposer* (proposes suggestions/explanations to the group), *Skeptic* (points out weaknesses in proposed suggestions/explanations), and *Moderator* (leads group discussion and makes sure everyone contributes). **Activity:** As a group, search the Web and identify at least one popular claim of alien visitation (such as a claim about the Roswell crash, about an alien abduction, or about aliens among us). Imagine that you had access to all the relevant material on which the claim is based, and create a plan that would allow you to test the validity of the claim. Speculate on what you would expect your test to show.

INVESTIGATE FURTHER

Short-Answer/Essay Questions

42. *Copernican Players.* Using a bulleted list format, write a one-page summary of the major roles that Copernicus, Tycho, Kepler, Galileo, and Newton played in overturning the ancient belief in an Earth-centered universe, along with a brief description of how each individual's work contributed to the development of modern science.

43. *Atomists and Aristotelians.* The ancient Greek arguments about the possible existence of extraterrestrial life continued for centuries. Write a short summary of the arguments, and then write a one- to two-page essay in which you describe how the Greek debate differs from the current scientific debate about extraterrestrial life.

44. *Science or Nonscience?* Find a recent news report from "mainstream" media (such as a major newspaper or magazine) that makes some type of claim about extraterrestrial life. Analyze the report and decide whether the claim is scientific or nonscientific. Write two or three paragraphs explaining your conclusion.

45. *Influence on History.* Based on what you have learned about the Copernican revolution, write a one- to two-page essay about how you believe it altered the course of human history.

46. *Discovery of Neptune.*
    a. In what sense was Neptune discovered by mathematics, rather than by a telescope? How did this discovery lend further support to Newton's theory of gravity? Explain.
    b. According to the idea known as *astrology,* the positions of the planets among the constellations, as seen from Earth, determine the courses of our lives. Astrologers claim that they must carefully chart the motions of *all* the planets to cast accurate predictions (horoscopes). In that case, say skeptics, astrologers should have been able to predict the existence of Neptune long before it was predicted by astronomers, since they should have noticed inaccuracies in their predictions. But they did not. Do you think this fact tells us anything about the validity of astrology? Defend your opinion in a one- to two-page essay.

47. *Biographical Research: Post-Copernican Viewpoints on Life in the Universe.* Many seventeenth- and eighteenth-century writers expressed interesting opinions on extraterrestrial life. Each individual listed below wrote a book that discussed this topic; book titles (and original publication dates) follow each name. Choose one or more individuals and research their arguments about extraterrestrial life. (You can find many of these books online in their entirety.) Write a one- to two-page summary of the person's arguments, and discuss which (if any) parts of these arguments are still valid in the current debate over life on other worlds.
    Bishop John Wilkins, *Discovery of a World in the Moone* (1638).
    René Descartes, *Philosophical Principles* (1644).

Bernard Le Bovier De Fontenelle, *Conversations on the Plurality of Worlds* (1686).

Richard Bentley, *A Confutation of Atheism from the Origin and Frame of the World* (1693).

Christiaan Huygens, *Cosmotheros, or, Conjectures Concerning the Celestial Earths and Their Adornments* (1698).

William Derham, *Astro-Theology: Or a Demonstration of the Being and Attributes of God from a Survey of the Heavens* (1715).

Thomas Wright, *An Original Theory or New Hypothesis of the Universe* (1750).

Thomas Paine, *The Age of Reason* (1793).

48. *Research: Religion and Life Beyond Earth.* Choose one religion (your own or another) and investigate its beliefs with regard to the possibility of life on other worlds. If scholars of this religion have made any definitive statements about this possibility, what did they conclude? If there are no definitive statements, discuss whether the religious beliefs are in any way incompatible with the idea of extraterrestrial life. Report your findings in a short essay.

Quantitative Problems

Be sure to show all calculations clearly and state your final answers in complete sentences.

49. *Sedna Orbit.* The object Sedna orbits our Sun at an average distance (semimajor axis) of 509 AU. What is its orbital period?

50. *Eris Orbit.* The dwarf planet Eris, which is slightly larger than Pluto, orbits the Sun every 557 years. What is its average distance (semimajor axis) from the Sun? How does its average distance compare to that of Pluto?

51. *New Planet Orbit.* A newly discovered planet orbits a distant star with the same mass as the Sun at an average distance of 112 million kilometers. Find the planet's orbital period.

52. *Halley's Orbit.* Halley's Comet orbits the Sun every 76.0 years. (a) Find its semimajor axis distance. (b) Halley's perihelion distance is about 90 million kilometers from the Sun. What is its aphelion distance? (c) Does Halley's Comet spend most of its time near its perihelion distance, near its aphelion distance, or halfway in between? Explain.

53. *Newton's Universal Law of Gravitation.*
    a. How does quadrupling the distance between two objects affect the gravitational force between them?
    b. Suppose the Sun were somehow replaced by a star with twice as much mass. What would happen to the gravitational force between Earth and the Sun?
    c. Suppose Earth were moved to one-third of its current distance from the Sun. What would happen to the gravitational force between Earth and the Sun?

Discussion Questions

54. *Science and Religion.* Science and religion are often claimed to be in conflict. Do you believe this conflict is real and hence irreconcilable, or is it a result of misunderstanding the differing natures of science and religion? Defend your opinion.

55. *The Impact of Science.* The modern world is filled with ideas, knowledge, and technology that developed through science and application of the scientific method. Discuss some of these things and how they affect our lives. Which of these impacts do you think are positive? Which are negative? Overall, do you think science has benefited the human race? Defend your opinion.

56. *Absolute Truth.* An important issue in the philosophy of science is whether science deals with absolute truth. We can think about this issue by imagining the science of other civilizations. For example, would aliens necessarily discover the same laws of physics that we have discovered, or would the laws they observe depend on the type of culture they have? How does the answer to this question relate to the idea of absolute truth in science? Overall, do you believe that science is concerned with absolute truth? Defend your opinion.

WEB PROJECTS

57. *The Galileo Affair.* In recent years, the Vatican has devoted a lot of resources to learning more about the trial of Galileo and understanding past actions of the Church in the Galileo case. Learn more about such studies, and write a short report about the current Vatican view of the case.

58. *Pseudoscience.* Choose a pseudoscientific claim that has been in the news recently, and learn more about it and how scientists have "debunked" it. Write a short summary of your findings.

59. *UFOlogy.* You can find an amazing amount of material about UFOs on the Web. Search for some such sites. Choose one that looks particularly interesting or entertaining and, in light of what you have learned about science, evaluate the site critically. Write a short review of the site.

60. *Gravitational Lensing.* Go to the Hubble Space Telescope website to find out what astronomers mean by "gravitational lensing," and locate at least two pictures that show examples of this phenomenon. How does the existence of gravitational lensing support Einstein's general theory of relativity, and what does it tell us about the idea that gravity works the same way throughout the universe?

3 The Universal Context of Life

LEARNING GOALS

3.1 THE UNIVERSE AND LIFE
- What major lessons does modern astronomy teach us about our place in the universe?

3.2 THE STRUCTURE, SCALE, AND HISTORY OF THE UNIVERSE
- What does modern science tell us about the structure of the universe?
- What does modern science tell us about the history of the universe?
- How big is the universe?

3.3 A UNIVERSE OF MATTER AND ENERGY
- What are the building blocks of matter?
- What is energy?
- What is light?

3.4 OUR SOLAR SYSTEM
- What are the major features of our solar system?
- How does modern science explain the features of our solar system?

✲ THE PROCESS OF SCIENCE IN ACTION

3.5 ONGOING DEVELOPMENT OF THE NEBULAR THEORY
- How did the nebular model gain acceptance?
- How have other planetary systems altered our understanding of the nebular theory?

▲ **About the photo:** This Hubble Space Telescope image shows a portion of the Carina Nebula, an interstellar cloud of gas and dust in which new stars and planets are being born today.

> Do there exist many worlds, or is there but a single world? This is one of the most noble and exalted questions in the study of Nature.
>
> *Saint Albertus Magnus (c. 1206–1280)*

The study of life in the universe brings together many different fields of research, each contributing to different aspects of our understanding. Biology helps us understand the nature of life, so that we know what we are searching for. Chemistry and biochemistry help us understand how life works, and how it might have arisen in the first place. Planetary science, which includes geology and atmospheric science, helps us understand the conditions that might allow life to arise and survive on other worlds. Physics teaches us about the fundamental laws of nature that both enable and constrain the possibilities for life elsewhere. That is why we will study aspects of all of these sciences as we continue our survey of astrobiology in this book.

The universe is the arena in which all these sciences come together, so to gain a true appreciation of the scientific search for life, we first need to discuss fundamental concepts of the universe itself. In this chapter, we'll explore modern understanding of the overall nature of the universe and of our solar system, along with basic properties of the matter and energy that make up all living things.

3.1 The Universe and Life

In Chapter 2, we saw how and why people long assumed that Earth was at the center of the universe and that the Sun, Moon, planets, and stars belonged to a separate realm known as "the heavens." This Earth-centered (geocentric) view of the universe gave our planet a unique place in the cosmos and implied a clear distinction between Earth and anyplace else. Although the geocentric belief did not prevent people from speculating about inhabitants of the heavens—recall the debate between the atomists and the Aristotelians—it certainly limited the possibilities.

The possibilities changed dramatically with the Copernican revolution, when Earth lost its central place and we learned that the planets are *worlds*, not just wandering lights in the night sky. The realization that the Sun is a star added even more possibilities, because it became reasonable to imagine planets around other stars.

We still do not know whether any other planet has life, either in our own solar system or elsewhere. However, we have learned a great deal about the universe in the past 400 years, including much about its size, content, and history. As we will discuss in this chapter, these new discoveries have given us good reason to think that it's worth some scientific effort to search for life beyond Earth.

What major lessons does modern astronomy teach us about our place in the universe?

The study of astronomy is so old that we cannot even pinpoint when it began, but for most of human history this study focused almost exclusively on observing the motions of visible objects in the sky. These observations were enough to meet immediate practical needs, such as being able to tell the time and the seasons from the Sun's path through the sky or being able to predict the tides from the position and phase of the Moon.

The human realm of astronomy expanded when the Greeks began their attempts to *explain* the observed motions by seeking to learn the architecture of the cosmos, an effort that bore fruit some 2000 years later when Kepler finally succeeded in describing the laws by which the planets orbit the Sun. Even then, however, science had advanced only far enough to say *how* the planets move around the Sun, not *why* they move as they do.

The key change in human understanding that allowed the emphasis to shift from "how" to both "how" and "why" occurred when Newton discovered that the planets are held in their orbits about the Sun by the same force of gravity that makes an apple fall to Earth. With this discovery, Newton delivered the final, shattering blow to the Aristotelian conception that Earth must by necessity be unique. The heavens could no longer be considered a separate realm made from different material (the *ether* or *quintessence*) and operating under different laws from Earth. Newton ended the ancient distinction between Earth and the heavens, merging both as parts of one universe.

The modern science of astronomy begins where Newton left off, as we seek to use his and other discoveries about the laws of nature to understand both the history and the workings of the objects we see in the sky. Today, powerful telescopes enable us to study objects whose existence was not even contemplated in Newton's time, and experimental techniques allow us to probe the inner workings of the cosmos at a level far deeper than Newton probably could have imagined. Almost everything we have learned through modern astronomy has at least some importance to the study of life in the universe. For our purposes in this book, however, three ideas are especially important in framing the universal context for everything else we will study:

- *The universe is vast and old.* Its vastness implies an enormous number of worlds on which life might possibly have arisen, and its old age means there has been plenty of time for life to begin and evolve.
- *The elements of life are widespread.* Observations show that the basic chemical elements that make up Earth and life are present throughout the universe.
- *The same physical laws that operate on Earth operate throughout the universe.* Every experiment and observation made to date has given additional support to Newton's conclusion that the laws of nature are the same everywhere. In that case, it is reasonable to think that the same processes that made life possible on Earth have also made life possible on other worlds.

Together, these ideas reinforce the primary lesson of the Copernican revolution: *We are not the center of the universe.* Our planet may be special to us, but it is just one planet orbiting one rather average star in a universe that is home to many similar stars and, probably, to many similar planets. The apparent ordinariness of our circumstances is a major reason that it seems plausible to imagine a universe full of life.

3.2 The Structure, Scale, and History of the Universe

We will devote the rest of this chapter to understanding the three important ideas just listed. They are all interrelated, so rather than examining them one at a time, we'll focus on key features of the universe as we understand it today.

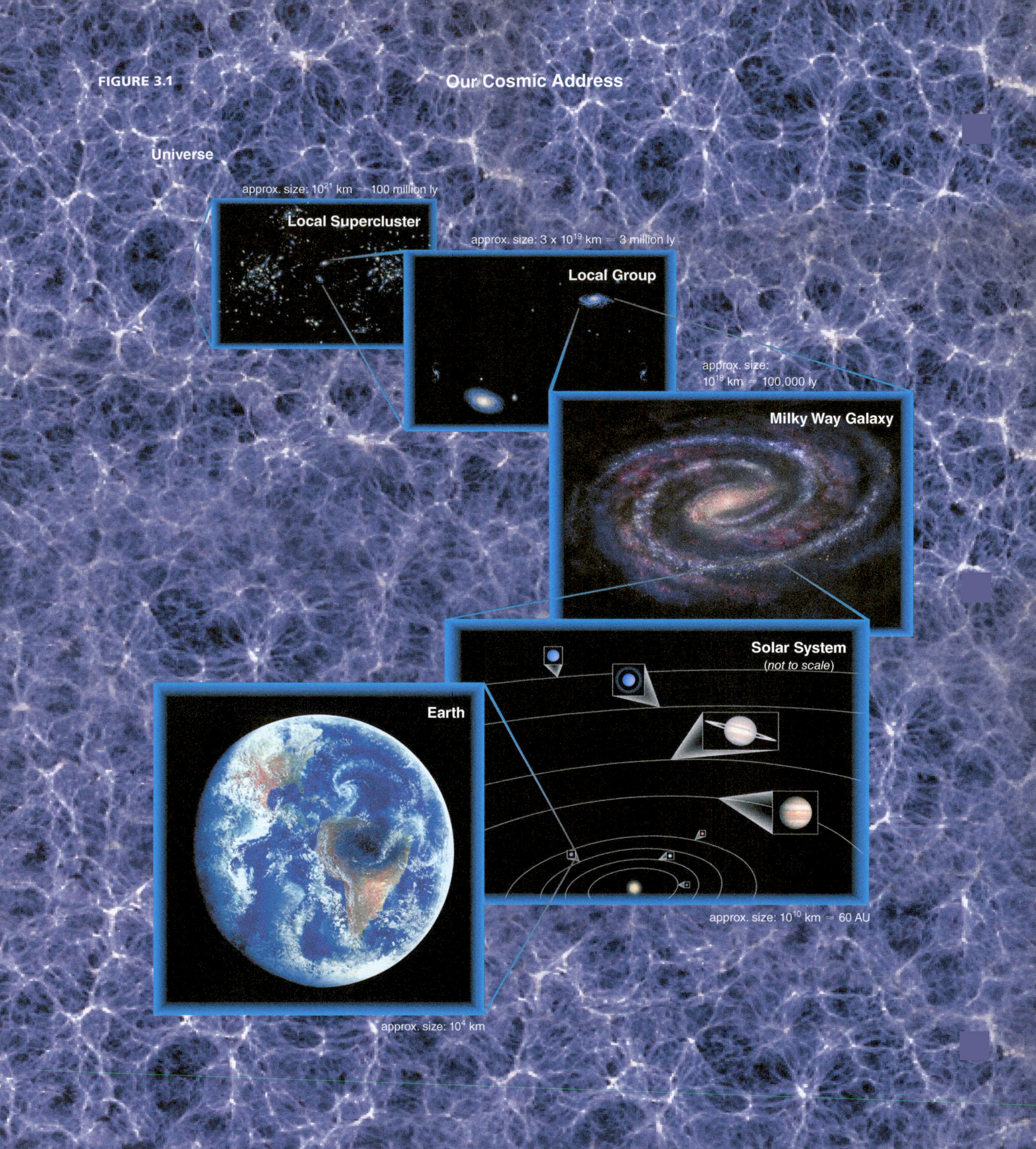

FIGURE 3.1

Our Cosmic Address

Universe

approx. size: 10^{21} km $\approx$ 100 million ly

Local Supercluster

approx. size: 3×10^{19} km $\approx$ 3 million ly

Local Group

approx. size:
10^{18} km $\approx$ 100,000 ly

Milky Way Galaxy

Solar System
(not to scale)

Earth

approx. size: 10^{10} km $\approx$ 60 AU

approx. size: 10^4 km

What does modern science tell us about the structure of the universe?

Let's begin our brief survey of the universe by examining the current state of knowledge about its general makeup.

OUR COSMIC ADDRESS Figure 3.1 illustrates our place in the universe with what we might call our "cosmic address." Earth is a planet in our **solar system,** which consists of the Sun and all the objects that orbit it: the planets and their moons and countless smaller objects including rocky *asteroids* and icy *comets*.

Our Sun is a star, just like the stars we see in our night sky. The Sun and all the stars we can see with the naked eye make up only a small part of a huge, disk-shaped collection of stars called the **Milky Way Galaxy.** A **galaxy** is a great island of stars in space, containing from a few hundred million to a trillion or more stars. The Milky Way Galaxy is relatively large, containing more than 100 billion stars. Our solar system is located about halfway out through the disk of the galaxy, where it orbits the galactic center once every 230 million years.

Billions of other galaxies are scattered through space and are usually found in groups. Our Milky Way, for example, is one of the two largest among more than 70 galaxies in the *Local Group.* Groups of galaxies with more than a few dozen large members are often called *galaxy clusters.*

On a very large scale, observations show galaxies and galaxy clusters to be arranged in giant chains and sheets, with huge voids between them. The regions in which galaxies and galaxy clusters are most tightly packed are called *superclusters,* which are essentially clusters of galaxy clusters. Our Local Group is located in the outskirts of the Local Supercluster (which was recently named *Laniakea,* Hawaiian for "immense heaven").

Together, all these structures make up our **universe.** In other words, the universe is the sum total of all matter and energy, encompassing the superclusters and voids and everything within them.

Think About It Some people think that our tiny physical size in the vast universe makes us insignificant. Others think that our ability to learn about the wonders of the universe gives us significance despite our size. What do *you* think?

THE SCALE OF THE SOLAR SYSTEM It's easy to list the levels of structure shown in Figure 3.1, but it takes additional thought to comprehend the vast scales involved. Let's begin by considering the scale of our own solar system.

Illustrations and photo montages often make our solar system look as if it were crowded with planets and moons, but the reality is far different. One of the best ways to develop perspective on cosmic sizes and distances is to imagine our solar system shrunk down to a scale on which you could walk through it. The Voyage scale model solar system in Washington, D.C., makes such a walk possible (Figure 3.2). The Voyage model shows the Sun and the planets, and the distances between them, at *one ten-billionth* of their actual sizes and distances.

FIGURE 3.2
This photo shows the pedestals housing the Sun (the gold sphere on the nearest pedestal) and the inner planets in the Voyage scale model solar system (Washington, D.C.). The model planets are encased in the sidewalk-facing disks visible at about eye level on the planet pedestals. The National Air and Space Museum is located to the left of the walkway.

a The scaled sizes (but not distances) of the Sun, the planets, and the two largest known dwarf planets.

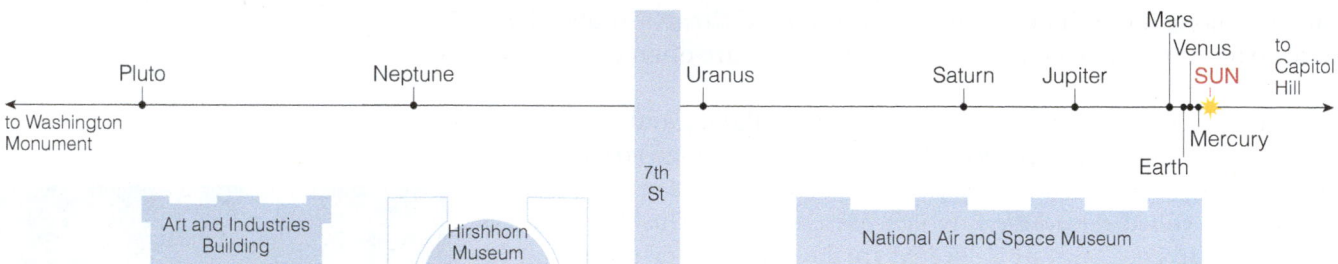

b Locations of the Sun and planets in the Voyage model (Washington, D.C.); the distance from the Sun to Pluto is about 600 meters (1/3 mile). Planets are lined up in the model, but in reality each planet orbits the Sun independently and a perfect alignment never occurs.

FIGURE 3.3 INTERACTIVE FIGURE
The Voyage model represents the solar system at *one ten-billionth* of its actual size. Pluto is included in the Voyage model, which was built before the International Astronomical Union reclassified Pluto as a dwarf planet.

Figure 3.3a shows the Sun and planets at their correct sizes (but not distances) on the Voyage scale: The model Sun is about the size of a large grapefruit, Jupiter is about the size of a marble, and Earth is about the size of the ball point in a pen. You can immediately see some key facts about our solar system. For example, the Sun is far larger than any of the planets, which themselves vary considerably in size: The entire Earth could be swallowed up by the storm on Jupiter known as the Great Red Spot (visible near Jupiter's lower left in the painting).

The scale of the solar system becomes even more remarkable when you combine the sizes shown in Figure 3.3a with the distances illustrated by the map of the Voyage model in Figure 3.3b. For example, the ball point–size Earth is located about 15 meters (16.5 yards) from the grapefruit-size Sun, which means you can picture Earth's orbit by imagining a ball point taking a year to make a circle of radius 15 meters around a grapefruit.

Perhaps the most striking feature of our solar system when we view it to scale is its emptiness. The Voyage model shows the planets along a straight path, so we'd need to draw each planet's orbit around the model Sun to show the full extent of our planetary system. Fitting all these orbits would require an area measuring more than a kilometer on a side—an area equivalent to more than 300 football fields arranged in a grid. Spread over this large area, only the grapefruit-size Sun, the planets and dwarf planets, and a few moons would be big enough to notice with your eyes. The rest of it would look virtually empty (that's why we call it *space!*).

Seeing the solar system to scale can help us understand why the search for life in the solar system is only just beginning. The Moon, the only other world on which humans have ever stepped (Figure 3.4), lies only about 4 centimeters (1½ inches) from Earth in the model. On this scale, the palm of your hand can cover the entire region of the universe in which humans have so far traveled. Our robotic spacecraft have traveled much farther, but these journeys require long travel times. For example, while you can walk from the Sun to Pluto in a few minutes on the Voyage scale, the *New Horizons* spacecraft—which travels nearly 100 times as fast as a commercial jet—took more than 9 years from launch until its 2015 flyby of Pluto.

DISTANCES TO STARS If you visit the Voyage model in Washington, D.C., you can walk the roughly 600-meter distance from the Sun to Pluto in just a few minutes. How much farther would you have to walk to reach the next star on this scale?

Amazingly, you would need to walk to California. If this answer seems hard to believe, you can check it for yourself. We usually measure the distances to stars in units of **light-years;** 1 light-year is the *distance* that light can travel in 1 year, which is about 10 trillion kilometers (see Cosmic Calculations 3.1, page 58). On the 1-to-10-billion Voyage scale, a light-year becomes about 1000 kilometers (because 10 trillion ÷ 10 billion = 1000). The nearest star system to our own, a three-star system called Alpha Centauri, is about 4.4 light-years away. This distance becomes about 4400 kilometers (2700 miles) on the 1-to-10-billion scale, or roughly equivalent to the distance across the United States (Figure 3.5).

The tremendous distances to the stars give us some perspective on the technological challenge of searching for life in other star systems. For example, because the largest star of the Alpha Centauri system is roughly the same size and brightness as our Sun, viewing it in the night sky is somewhat like being in Washington, D.C., and seeing a bright grapefruit in San Francisco (neglecting the problems introduced by the curvature of Earth). It may seem amazing that we can see this star at all, but the blackness of the night sky allows the naked eye to see it as a faint dot of light. Now, consider the difficulty of detecting *planets* orbiting nearby stars, which is equivalent to looking from Washington, D.C., and trying to find ball points or marbles orbiting grapefruits in California or beyond. Remarkably, we now have technology that can detect such planets [Section 11.2], but studying these worlds in detail or searching them for life remains well beyond our current capabilities.

The vast distances to the stars offer a sobering lesson about interstellar travel. Although science fiction shows like *Star Trek* and *Star Wars* make such travel look easy, the reality is far different. Consider the *Voyager 2* spacecraft. Launched in 1977, *Voyager 2* flew by Jupiter in 1979, Saturn in 1981, Uranus in 1986, and Neptune in 1989. It is now bound for the stars at a speed of close to 50,000 kilometers per hour—about 100 times as fast as a speeding bullet. But even at this speed, *Voyager 2* would take about 100,000 years to reach Alpha Centauri if it were headed in that direction (which it's not). Convenient interstellar travel remains well beyond our present technology.

FIGURE 3.4

So far, the Moon is the only other world ever visited by humans. This famous photograph from the first Moon landing (*Apollo 11* in July 1969) shows astronaut Buzz Aldrin, with Neil Armstrong reflected in his visor. Armstrong was the first to step onto the Moon's surface, saying, "That's one small step for a man, one giant leap for mankind."

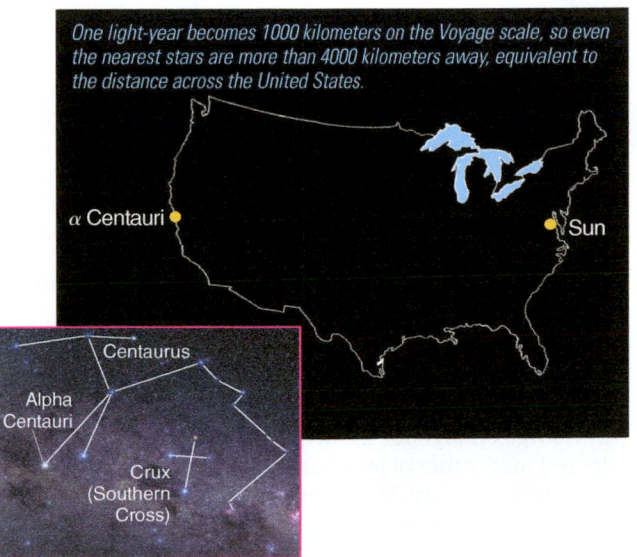

FIGURE 3.5

On the same 1-to-10-billion scale on which you can walk from the Sun to Pluto in just a few minutes, you'd need to cross the United States to reach Alpha Centauri, the nearest other star system. The inset shows the location and appearance of Alpha Centauri in the night sky.

THE SCALE OF THE GALAXY We turn now to the Milky Way Galaxy, which is so vast that only a handful of its more than 100 billion stars could even fit on Earth with the 1-to-10-billion scale. To picture the galaxy, let's reduce our solar system scale by another factor of one billion (making it a scale of 1 to 10^{19}). On this new scale, each light-year becomes 1 millimeter, and the 100,000-light-year diameter of the Milky Way Galaxy becomes 100 meters, or about the length of a football field. Visualize a football field with a scale model of our galaxy centered over midfield. Our entire solar system is a microscopic dot located around the 20-yard line (corresponding to our real distance of about 27,000 light-years from the center of the galaxy). The 4.4-light-year separation between our solar system and Alpha Centauri becomes just 4.4 millimeters on this scale—smaller than the width of your little finger. If you stood at the position of our solar system in this model, millions of star systems would lie within reach of your arms.

Another way to get a handle on the size of the galaxy is to think about light-travel times. Light travels extremely fast by earthly standards. If you could circle Earth at the speed of light of 300,000 kilometers per second, you could complete almost eight circuits in just 1 second. But despite this awesome speed, light requires years to cross the vast chasms between the stars. That is why we measure interstellar distances in light-years. For example, when we say that Alpha Centauri is 4.4 light-years away, we mean that its light takes 4.4 years to reach us. This fact has an astonishing implication: It means that we cannot see what Alpha Centauri looks like today, but can see only what it looked like 4.4 years ago, when the light that is now reaching our eyes and telescopes first left on its journey. It also has an important implication for the possibility of carrying on a conversation with any beings who might happen to live in the Alpha Centauri system. We generally transmit messages over long distances with radio waves, which are a form of light and hence travel at the speed of light (see Section 3.3). If we sent a radio message to Alpha Centauri, the message would take 4.4 years to get there, and any reply would take the same 4.4 years to travel to us. You'd need a lot of patience for a conversation in which it would be almost 9 years from the time you said, "Hello, how are you?" until you heard the reply, "Fine, thanks, and you?"

KEY ASTRONOMICAL DEFINITIONS

star: A large, glowing ball of gas that generates heat and light through nuclear fusion in its core. Our Sun is a star.

planet: A moderately large object that orbits a star and shines primarily by reflecting light from its star. Based on a definition approved in 2006, an object can be considered a planet only if it (1) orbits a star, (2) is large enough for its own gravity to make it round, and (3) has cleared most other objects from its orbital path. An object that meets the first two criteria but has *not* cleared its orbital path, like Pluto, is designated a *dwarf planet*.

extrasolar planet: A planet orbiting a star other than our Sun.

habitable planet (or **habitable world**)**:** A planet (or other type of world) with environmental conditions under which life could potentially arise or survive. (Sometimes called an *exoplanet*.)

moon (or **satellite**)**:** An object that orbits a planet. The term *satellite* can refer to any object orbiting another object.

asteroid: A relatively small and rocky object that orbits a star.

comet: A relatively small and ice-rich object that orbits a star.

solar system: The Sun and all the material that orbits it, including the planets. The term *solar system* technically refers only to our own star system (because *solar* means "of the Sun"), but it is sometimes applied to other star systems.

star system: A star (sometimes more than one star) and any planets and other materials that orbit it.

galaxy: A great island of stars in space, containing from a few million to a trillion or more stars, all held together by gravity.

universe (or **cosmos**)**:** The sum total of all matter and energy—that is, all galaxies and everything within and between them.

The effect becomes more dramatic at greater distances. Consider the Orion Nebula, a giant cloud of gas and dust (meaning tiny solid particles) in which new stars and planets are currently being born (Figure 3.6). The Orion Nebula lies about 1350 light-years away, which means that if we were to receive a radio message from aliens in the Orion Nebula, it would have to have been sent some 1350 years ago. If we sent a message in return, we couldn't expect to hear a reply for at least 2700 years.

The Orion Nebula is still quite near relative to the scale of the galaxy: On our football-field-size scale model, the nebula lies only about 1.5 meters from Earth. It takes light 100,000 years to cross the 100,000-light-year diameter of the Milky Way Galaxy. Given that we are located about 27,000 light-years from the galactic center, a signal now reaching us from the far outer edge of the galaxy would have been sent more than 70,000 years ago. The *search for extraterrestrial intelligence,* or SETI, which listens for signals from alien civilizations, is in essence a search to hear from civilizations that used radio technology some decades, centuries, or millennia in the past.

The number of star systems in the Milky Way Galaxy is no less remarkable than its size. Imagine that you are having difficulty falling asleep at night, perhaps because you are contemplating the vastness of our galaxy. Instead of counting sheep, you decide to count stars. The Milky Way has more than 100 billion stars (perhaps as many as a trillion). If you are able to count about one star each second, on average, how long would it take you to count 100 billion stars? Clearly, the answer is 100 billion (10^{11}) seconds, but how long is that? You can get the answer by dividing 100 billion seconds by 60 seconds per minute, 60 minutes per hour, 24 hours per day, and 365 days per year. If you do this calculation, you'll find that 100 billion seconds is more than 3000 years. In other words, you would need thousands of years just to *count* the stars in the Milky Way Galaxy, let alone to study them or search their planets for signs of life. And this assumes you never take a break—no sleeping, no eating, and absolutely no dying!

THE CONTENT OF THE UNIVERSE The Milky Way Galaxy is just one of billions of galaxies in the universe, and we'll discuss the overall extent of the universe shortly. First, however, it's worth briefly discussing the content of the universe. We've said that the universe is the sum total of all matter and energy, but what exactly is this? Until a few decades ago, astronomers assumed that the matter of the universe was primarily found in stars that were organized into galaxies, while the energy of the universe took the form of light. It now seems that this "visible" matter and energy are just the tip of the iceberg in a universe that remains far more mysterious.

Just as planets orbit the Sun, stars orbit the center of the Milky Way Galaxy. The more massive the galaxy, the stronger its gravity and the faster stars should be orbiting. By carefully studying stellar orbits, astronomers have been able to put together a map of the distribution of matter in the Milky Way. The surprising result is that while most of the matter that we can see consists of stars and gas clouds in the galaxy's relatively flat *disk,* most of the mass lies unseen in a much larger, spherical *halo* that surrounds the disk (Figure 3.7). We don't know the nature of this unseen mass in the halo, so we call it **dark matter** to indicate that we have not detected any light coming from it, even though we have detected its gravitational effects. Studies of other galaxies suggest that they also

FIGURE 3.6 INTERACTIVE PHOTO
The Orion Nebula, located about 1350 light-years away, photographed by the Hubble Space Telescope. The inset shows its location within the constellation Orion.

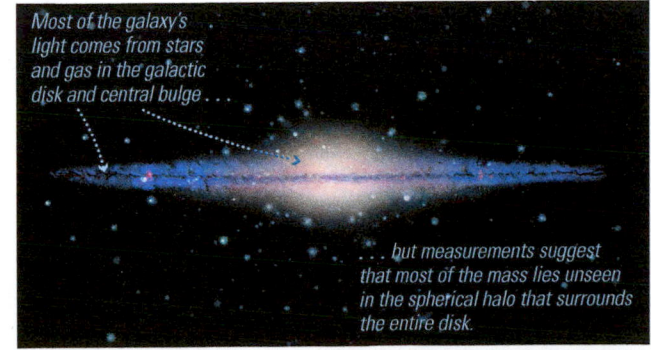

FIGURE 3.7
This painting shows an edge-on view of the Milky Way Galaxy. Study of galactic rotation shows that most of the galaxy's mass lies unseen in the halo that surrounds and encompasses the stars and gas of the disk and bulge. Because this mass emits no light that we have detected, we call it *dark matter.*

Cosmic Calculations 3.1

HOW FAR IS A LIGHT-YEAR?

One light-year (ly) is defined to be the distance that light can travel in 1 year. This distance is fixed because light always travels through space at the *speed of light*, which is 300,000 kilometers per second (186,000 miles per second).

It's easy to calculate the distance represented by a light-year if you recall that

$$\text{distance} = \text{speed} \times \text{time}$$

For example, if you travel at a speed of 50 kilometers per hour for 2 hours, you will travel 100 kilometers. To find the distance represented by 1 light-year, we need to multiply the speed of light by 1 year:

$$1 \text{ light-year} = (\text{speed of light}) \times (1 \text{ yr})$$

Because we are given the speed of light in units of kilometers per second but the time as 1 year, we must carry out the multiplication while converting 1 year into seconds. You can find a review of unit conversions in Appendix C; here, we show the result for this particular case:

$$1 \text{ light-year} = \left(300,000 \, \frac{\text{km}}{\text{s}} \right) \times (1 \text{ yr})$$

$$= \left(300,000 \, \frac{\text{km}}{\text{s}} \right) \times$$

$$\left(1 \text{ yr} \times 365 \, \frac{\text{day}}{\text{yr}} \times 24 \, \frac{\text{hr}}{\text{day}} \times 60 \, \frac{\text{min}}{\text{hr}} \times 60 \, \frac{\text{s}}{\text{min}} \right)$$

$$= 9,460,000,000,000 \text{ km}$$

That is, 1 light-year is equivalent to 9.46 trillion kilometers, or almost 10 trillion kilometers. Be sure to note that a light-year is a unit of *distance*, not time.

---

are made mostly of dark matter. In fact, most of the mass in the universe seems to be made of this mysterious dark matter, which means that its gravity must have played a key role in assembling galaxies.

Evidence of the existence of dark matter has been building for several decades. More recently, scientists have gathered evidence of an even greater mystery: The universe seems to contain a mysterious form of energy—nicknamed **dark energy** by analogy to dark matter—that is pushing galaxies apart even while their gravity tries to draw them together. As is the case with dark matter, scientists have good reason to think that dark energy exists but lack any real understanding of its nature.

In recent years, scientists have been able to conduct a sort of census of the matter and energy in the universe. The results show that dark energy and dark matter are by far the main ingredients of the universe. The ordinary matter—atoms and molecules—that makes up stars and planets and life apparently represents no more than a few percent of all the matter and energy in the universe.

Because they appear to be the dominant constituents of the universe but we don't know much about them, dark matter and dark energy are arguably the biggest mysteries in astronomy today. However, they do not appear to affect the general evolution of stars, planets, or life, so they seem unlikely to affect our study of life in the universe. Still, as we seek to answer questions about life elsewhere, the mysteries of dark matter and dark energy should remind us that nature may still hold surprises that no one has foreseen.

Think About It We generally ignore dark matter and dark energy in discussions of astrobiology, because based on current understanding, they should have no significant effects on chemical or biological processes. Nevertheless, they appear to be the most abundant constituents of our universe, and therefore have had major influence on such things as the birth and evolution of galaxies. Do you think that dark matter and dark energy ought to be discussed further in an astrobiology class? Defend your opinion.

What does modern science tell us about the history of the universe?

Figure 3.8 summarizes the history of the universe according to modern science. Follow the figure as you read through this section. As you will see, this history further reinforces the idea that we live on a world that ought to be similar to many others, giving us additional reason to think that life might exist elsewhere.

THE BIG BANG AND THE EXPANDING UNIVERSE Telescopic observations of distant galaxies show that the entire universe is *expanding*, meaning that average distances between galaxies are increasing with time (see Special Topic 3.1). This fact implies that galaxies must have been closer together in the past, and if we go back far enough, we must reach the point at which the **expansion** began. We call this beginning the **Big Bang,** and scientists use the observed rate of expansion to calculate that it occurred about 14 billion years ago. The three cubes in the upper-left corner of Figure 3.8 represent the expansion of a small piece of the universe over time.

In addition to the observed expansion, two other key lines of evidence support the idea that the universe began in a Big Bang. You can understand them by noting that just as compressing gas inside a car

FIGURE 3.8

Our Cosmic Origins

Birth of the Universe: The expansion of the universe began with the hot and dense Big Bang. The cubes show how one region of the universe has expanded with time. The universe continues to expand, but on smaller scales gravity has pulled matter together to make galaxies.

Galaxies as Cosmic Recycling Plants: The early universe contained only two chemical elements: hydrogen and helium. All other elements were made by stars and recycled from one stellar generation to the next within galaxies like our Milky Way.

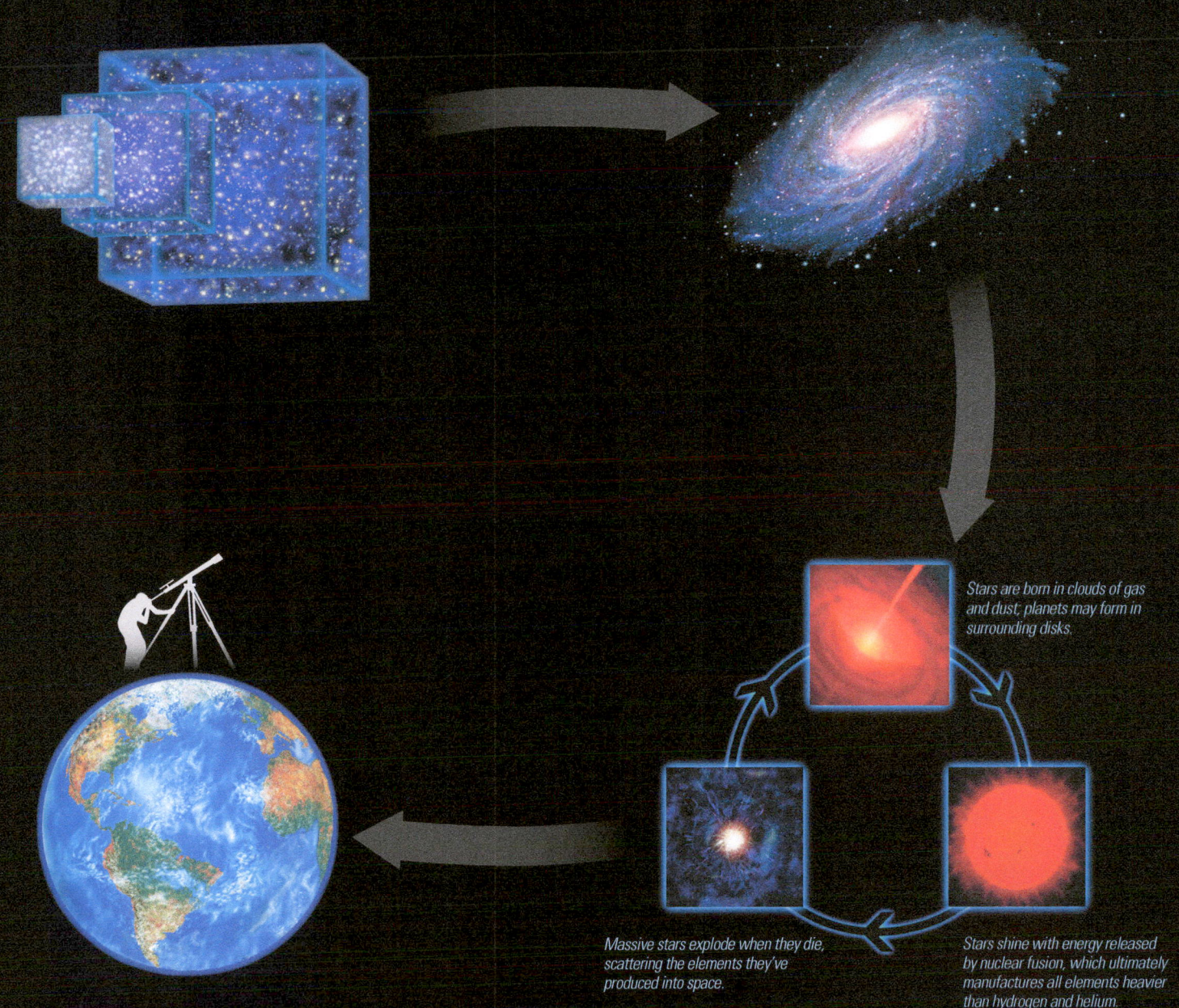

Stars are born in clouds of gas and dust; planets may form in surrounding disks.

Massive stars explode when they die, scattering the elements they've produced into space.

Stars shine with energy released by nuclear fusion, which ultimately manufactures all elements heavier than hydrogen and helium.

Earth and Life: By the time our solar system was born, 4½ billion years ago, about 2% of the original hydrogen and helium had been converted into heavier elements. We are therefore "star stuff," because we and our planet are made from elements manufactured in stars that lived and died long ago.

Life Cycles of Stars: Many generations of stars have lived and died in the Milky Way.

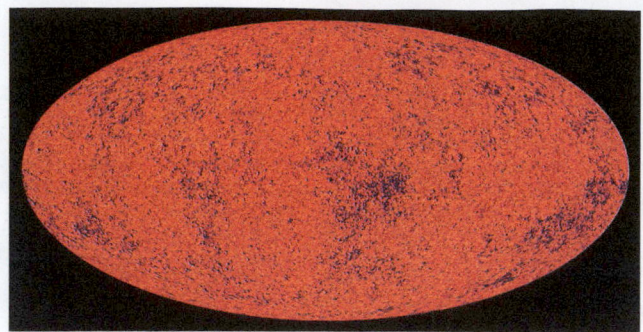

FIGURE 3.9

This all-sky map shows the cosmic microwave background as measured by the *Planck* spacecraft. The spectrum of this radiation shows it to be everywhere representative of a temperature of about 2.73 Kelvin—and this temperature makes sense only if it is leftover heat from a hot Big Bang that occurred about 14 billion years ago. (The color differences correspond to very slight variations in the temperature, which arose from tiny density differences in the early universe that allowed gravity to pull matter together into galaxies.)

engine (the piston compresses gas in a cylinder) makes the gas much hotter and denser, the universe must have been much hotter and denser if it was smaller in the past. At the time of the Big Bang, the universe must have had all its matter compressed to extremely high temperature and density, producing intensely bright radiation (light). Calculations show that as the universe expanded and cooled with time, it should have left behind a faint "glow" of radiation that we could detect with radio telescopes. This radiation, known as the *cosmic microwave background*, has indeed been detected and studied (Figure 3.9), providing strong evidence that the Big Bang really occurred.

The second additional line of evidence comes from the overall chemical composition of the universe. Calculations that run the expansion backward allow scientists to predict exactly when and how the chemical elements should have been born in the early universe. The calculations predict that if the Big Bang occurred, then the chemical composition of the universe should be about three-fourths hydrogen and one-fourth helium (by mass). Observations show that this is indeed a close match to the overall chemical composition of the universe. This excellent agreement between prediction and observation gives additional strong support to the Big Bang theory. Note also that these calculations predict that the universe was born without any elements heavier than hydrogen and helium (except a trace of lithium)—which means the early universe lacked the elements that make life on Earth. As we'll discuss shortly, these elements were made later.

The universe as a whole has continued to expand ever since the Big Bang, but on smaller scales the force of gravity has drawn matter together. Structures such as galaxies and clusters of galaxies occupy regions where gravity has won out against the overall expansion. That is, while the universe as a whole continues to expand, individual galaxies and galaxy clusters do *not* expand. This idea is also illustrated by the three cubes in Figure 3.8. Notice that as the region as a whole grows larger, the matter within it has clumped into galaxies and galaxy clusters. Most galaxies, including our own Milky Way, probably formed within a billion years after the Big Bang.

STELLAR LIVES AND GALACTIC RECYCLING Within galaxies like the Milky Way, gravity drives the collapse of clouds of gas and dust to form stars and planets. Stars are not living organisms, but they nonetheless go through "life cycles," as illustrated in the lower right of Figure 3.8. A star is born when gravity compresses the material in a cloud to the point at which the center becomes dense enough and hot enough to generate energy by **nuclear fusion,** the process in which lightweight atomic nuclei smash together and stick (or fuse) to make heavier nuclei. Planets may be born at the same time. In much the same way that spinning a ball of dough causes it to spread out into a flat pizza, the natural spin of a contracting interstellar cloud keeps some of its gas spread away from its center while shaping it into a flattened disk. (We'll discuss the process in more detail in Section 3.4.) The planets of our own solar system formed in such a disk, which is why they all ended up orbiting the Sun in nearly the same plane.

Once a star is born, it shines with energy released by the nuclear fusion in its core. During most of a star's life, nuclear fusion combines hydrogen nuclei to make helium nuclei (Figure 3.10). It takes four hydrogen nuclei to make one helium nucleus (the process involves several

Hydrogen fusion

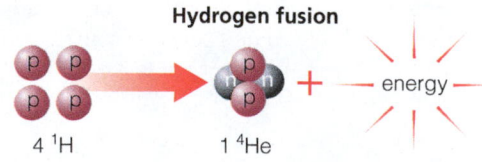

4 ^{1}H   1 ^{4}He

FIGURE 3.10

The hydrogen fusion reaction: Four hydrogen nuclei (protons, in red) fuse to make one helium nucleus (two protons and two neutrons). The helium nucleus has slightly less mass than the four hydrogen nuclei combined (by about 0.7%); this "lost" mass is converted to energy in accord with Einstein's formula $E = mc^2$. This diagram shows the overall fusion reaction; in reality, this reaction proceeds in several steps, with only two nuclei fusing at a time.

At the dawn of the last century, many astronomers assumed that the universe as a whole was permanent and largely unchanging. However, thanks largely to work in the 1920s by Edwin Hubble, we now know that the universe is expanding. That is, average distances between galaxies in the universe are increasing with time, and space itself is growing to account for these larger distances.

Hubble discovered the universal expansion by observing many galaxies and the speeds at which they appear to move relative to Earth. His observations revealed two key facts:

1. Virtually every galaxy in the universe (except those within the Local Group) is moving away from us.
2. The more distant the galaxy, the faster it appears to be racing away.

Figure 1 uses a simple analogy to show how these observations lead to the conclusion that the universe is expanding. Imagine that you make a raisin cake in which the distance between adjacent raisins is 1 centimeter. You place the cake in an oven, where it expands as it bakes. After 1 hour, you remove the cake, which has expanded so that the distance between adjacent raisins has increased to 3 centimeters. From the outside, the expansion of the cake is fairly obvious. But what would you see if you lived in the cake, as we live in the universe?

Pick any raisin (it doesn't matter which one), call it the Local Raisin, and identify it in the pictures of the cake both before and after baking. Figure 1 shows one possible choice for the Local Raisin, with three nearby raisins labeled. The accompanying table summarizes what you would see if you lived within the Local Raisin. Notice, for example, that Raisin 1 starts out at a distance of 1 centimeter from the Local Raisin before baking and ends up at a distance of 3 centimeters after baking, which means it moves a distance of 2 centimeters away from the Local Raisin during the hour of baking. Hence, its speed as seen from the Local Raisin is 2 centimeters per hour. Raisin 2 moves from a distance of 2 centimeters before baking to a distance of 6 centimeters after baking, which means it moves a distance of 4 centimeters away from the Local Raisin during the hour. Hence, its speed is 4 centimeters per hour, or twice as fast as the speed of Raisin 1. Generalizing, the fact that the cake is expanding means that all raisins are moving away from the Local Raisin, with more distant raisins moving away faster. Hubble's discovery that galaxies are moving in much the same way as the raisins in the cake, with most moving away from us and more distant ones moving away faster, implies that the universe in which we live is expanding much like the raisin cake.

Hubble's original measurements of the universal expansion were fairly crude, but they have been greatly improved since then. Today, we know the rate of expansion to within a few percent, and we even have measurements that show roughly how the expansion rate has changed with time. Just as knowing a car's speed and its current distance from home can allow you to determine how long it's been since the car left, knowing the rate of expansion and the current separations of galaxies allows astronomers to determine how long it's been since the expansion began. The answer—about 14 billion years (more precisely, 13.8 billion years)—must represent the age of the universe.

You might imagine that this expansion is similar to an explosion that has, over billions of years, filled a vast "empty room" with stars and galaxies. But that's not correct, as the "empty room"—space itself—didn't exist before the Big Bang. Space is just as much a result of the Big Bang as are matter and energy. It is the expansion of space that causes the distances between galaxies to increase, not that they are racing away in some preexisting space. This also resolves the problem that, if you could only look far enough into the universe, you would find galaxies whose distance from us is increasing faster than the speed of light, apparently violating the theory of relativity. It's true that this theory doesn't permit matter to move through space faster than light speed, but it places no restrictions on the speed at which space itself can expand. Like the raisins in the cake, the galaxies are just being carried along as their "cake"—space—increases in size.

Measurements of the expansion rate are also responsible for one of the biggest mysteries in astronomy: the mystery of *dark energy*, discussed briefly in this chapter. If you throw a ball upward, gravity makes it slow down as it rises. In much the same way, we would expect the mutual gravitational attraction of all the galaxies in the universe to slow the expansion rate with time. However, measurements seem to show just the opposite: The expansion rate has been *increasing* with time, at least for the past few billion years. No one knows what is causing this acceleration of the expansion, but it must be some type of energy that can push galaxies apart. That is where the idea of dark energy comes from, even though we do not yet know what it is.

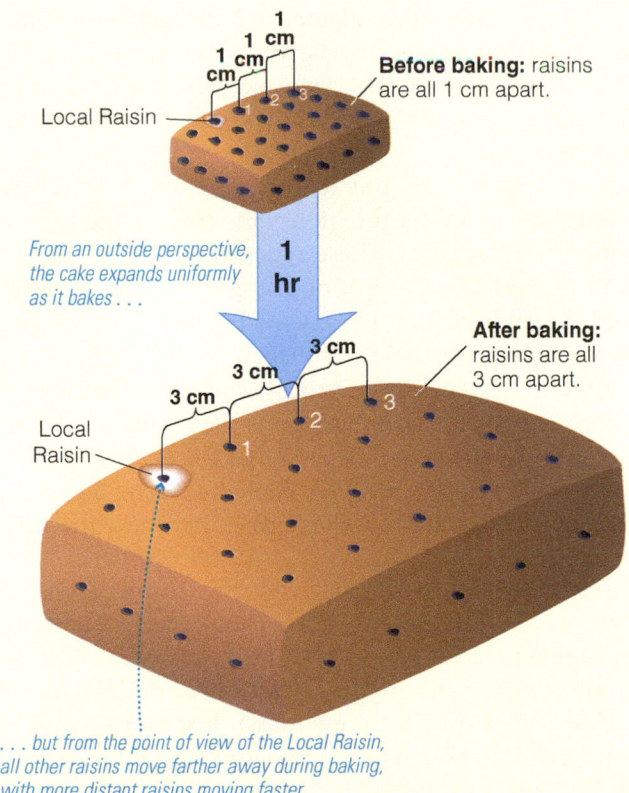

From an outside perspective, the cake expands uniformly as it bakes . . .

1 hr

Before baking: raisins are all 1 cm apart.

After baking: raisins are all 3 cm apart.

. . . but from the point of view of the Local Raisin, all other raisins move farther away during baking, with more distant raisins moving faster.

FIGURE 1 INTERACTIVE FIGURE
An expanding raisin cake offers an analogy to the expanding universe.

Distances and Speeds as Seen from the Local Raisin

Raisin Number	Distance Before Baking	Distance After Baking (1 hour later)	Speed
1	1 cm	3 cm	2 cm/hr
2	2 cm	6 cm	4 cm/hr
3	3 cm	9 cm	6 cm/hr
⋮	⋮	⋮	⋮

FIGURE 3.11 INTERACTIVE PHOTO
The Crab Nebula is the remnant gas from a massive star whose explosion (supernova) was witnessed on Earth in A.D. 1054. The glowing gas is moving outward at high speed from the center, confirming its explosive origin. (The central object, a *pulsar*, offers further confirmation, as pulsars are now known to be remains of stars that have exploded.) In a few tens of thousands of years, the gas will have fully dispersed, mixing the elements forged in the exploded star with other gas in the Milky Way Galaxy.

steps); energy is released because a helium nucleus has slightly less mass than the four hydrogen nuclei. This means that a small amount of the mass of the hydrogen has disappeared and become energy in accord with Einstein's famous formula expressing the equivalence of matter and energy, $E = mc^2$ (where E is the energy, m is the mass, and c is the speed of light). Indeed, that is how our Sun shines today—with energy generated deep in its core by the fusion of hydrogen into helium.

A star lives until it exhausts all its usable fuel for fusion. The rate at which a star burns through its fuel depends on its mass: More massive stars, with much denser and hotter cores, burn through their fuel at far greater rates than less massive stars. In essence, more massive stars have their engines running hotter and therefore faster. This more than makes up for the fact that larger stars have more fuel to burn. The most massive stars live only a few million years, while stars like our Sun live 10 billion years and lower-mass stars can live hundreds of billions of years.

In its final death throes, when its fuel for fusion has been exhausted, a star blows much of its content back out into space. In particular, massive stars die in titanic explosions called *supernovae* (Figure 3.11). The returned matter mixes with other matter floating between the stars in the galaxy, eventually becoming part of new clouds of gas and dust from which new generations of stars can be born. Galaxies therefore function as cosmic recycling plants, reusing material expelled from dying stars to make new generations of stars and planets.

WE ARE STAR STUFF The recycling of stellar material is connected to our existence in an even deeper way. Observations of stars of different ages confirm the Big Bang theory prediction that the universe was born containing only the simplest chemical elements: hydrogen and helium (and a trace of lithium). Living things, and Earth itself, are made primarily of other elements, such as carbon, nitrogen, oxygen, and iron. Where did these elements come from? Evidence shows that these elements were manufactured by stars.

We cannot see inside stars, but we can use the laws of physics to predict what must happen under the high-temperature and high-density conditions found in stellar cores. These types of calculations tell us that stars spend most of their lives generating energy by fusing hydrogen into helium. Toward the ends of their lives, stars like our Sun can fuse helium into carbon: Fusing three helium nuclei together makes one carbon nucleus. The Sun will stop the fusion process there, but the cores of more massive stars can continue on to create many other heavy elements. For example, they can fuse carbon into oxygen and silicon, oxygen into neon and sulfur, and so on up to iron. Still other elements can be produced by nuclear reactions that accompany stellar death. All these manufactured elements then disperse into space after the star dies.

At least three lines of observational evidence confirm this theoretical prediction. First, stars of different ages show the expected pattern in the proportions of elements heavier than helium that they contain. The oldest stars are made of nearly pure hydrogen and helium (heavier elements make up less than about 0.1% of their mass), just as we would expect for objects born before there had been time for stars to make much else. Younger stars, like our Sun, were born with higher proportions (up to about 2%) of their mass in the form of elements heavier than hydrogen and helium, telling us that they were born from gas clouds that contained the elements manufactured and released by earlier generations of stars.

The second line of evidence comes from studies of the overall abundances of chemical elements in the universe today. The theory of nuclear fusion in massive stars makes specific predictions about relative abundances; for example, it predicts that the elements carbon and oxygen should be more abundant than nitrogen and that neon should be more abundant than fluorine. Figure 3.12 shows the observed relative abundances of the elements. Notice, for example, that nitrogen is indeed less abundant than carbon and oxygen. In fact, detailed calculations predict a pattern of abundances that almost perfectly matches these observations, even including all the up and down wiggles that appear on the graph.

The third line of evidence comes from studies of the gas from exploding stars (such as the Crab Nebula shown in Figure 3.11). Models of massive stars and their deaths allow astronomers to calculate the precise makeup expected for these clouds from recently deceased stars, and again, the observations match the predictions quite well.

The importance of this stellar manufacturing should be clear: Without it, our universe would not contain the chemical elements of which we are made. The recycling of matter and the production of heavier elements had already been taking place in the Milky Way Galaxy for billions of years before the Sun and the planets were born. By that time, stars had converted about 2% of the original hydrogen and helium into heavier elements, so the cloud that gave birth to our solar system was made of about 98% hydrogen and helium and 2% of everything else. This 2% may sound small, but it was more than enough to make the small rocky planets of our solar system, including Earth. On Earth, some of these elements became the raw ingredients of life, ultimately blossoming into the great diversity of life we see today.

In summary, most of the material from which we and our planet are made was created inside stars that died before the birth of our Sun. As astronomer Carl Sagan (1934–1996) said, we are "star stuff."

IMPLICATIONS FOR LIFE IN THE UNIVERSE The fact that we are made of "star stuff" has important implications for the possibility of finding life elsewhere in the universe. The processes of stellar and galactic recycling operate throughout the Milky Way Galaxy, as well as in every similar galaxy throughout the universe, so we expect the chemical composition of many other star systems to be quite similar to that of our own. Observations confirm this expectation. While there is some variation in the precise proportions—in particular, stars that were born long ago have much lower proportions of heavy elements—the overall composition of our solar system is typical. We conclude that many and perhaps even most other star systems have the necessary raw materials to build Earth-like planets and life.

Think About It The oldest stars in the galaxy are generally found in the halo, while younger stars are always found in the disk. Identify these regions in Figure 3.7. What does the difference in heavy-element abundance tell you about which region of the galaxy formed first? Do you think the difference affects the likelihood of finding Earth-like planets or life in the halo versus the disk? (*Note:* We'll discuss this topic further in Chapter 11.)

THE SCALE OF TIME When we discussed the structure of the universe, we found that we had to carefully consider scale to understand how greatly one level differs from the next. In much the same way, it's easy to

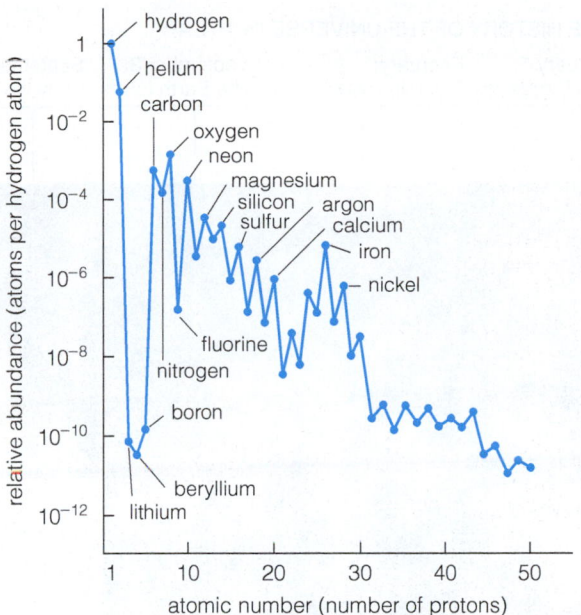

FIGURE 3.12
This graph shows the observed abundances of elements in the Milky Way, relative to the abundance of hydrogen (set to 1 in this comparison). For example, the graph shows a nitrogen abundance of about 10^{-4}, or 1/10,000, which means there are about 10,000 times as many hydrogen atoms as nitrogen atoms in the galaxy. Note that hydrogen and helium are still by far the dominant chemical elements; the overall chemical content of our galaxy is about 98% hydrogen and helium (by mass) and 2% everything else combined.

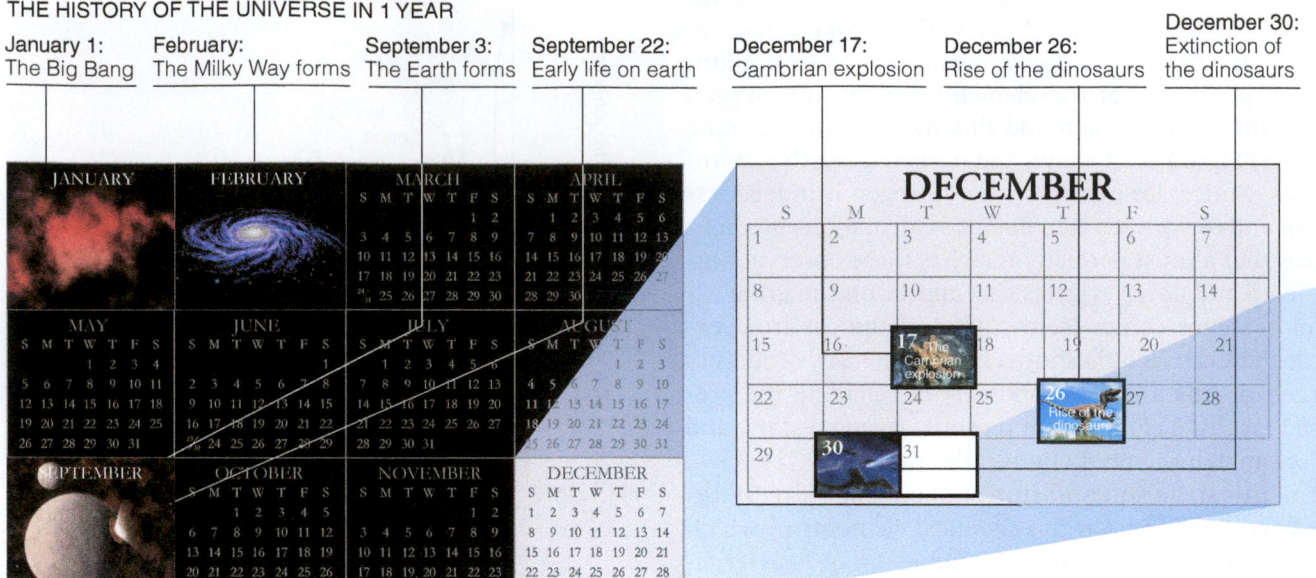

January 1: The Big Bang

February: The Milky Way forms

September 3: The Earth forms

September 22: Early life on earth

December 17: Cambrian explosion

December 26: Rise of the dinosaurs

December 30: Extinction of the dinosaurs

FIGURE 3.13

The cosmic calendar compresses the 14-billion-year history of the universe into 1 year, so that each month represents a little more than 1 billion years (more precisely, 1.17 billion years). This cosmic calendar is adapted from a version created by Carl Sagan.

state that the universe is 14 billion years old, but it requires some deeper thought to begin to grasp the truly astronomical meaning of this age.

You are probably familiar with the use of time lines to represent historical events. We'll use a slight variation on this theme, making a scale for time in which we imagine compressing the entire history of the universe, from the Big Bang to the present, into a single year (Figure 3.13). On this *cosmic calendar*, the Big Bang takes place at the first instant of January 1, and the present is the stroke of midnight on December 31. For a universe that is about 14 billion years old, each month on the cosmic calendar represents a little more than 1 billion years, each day represents about 40 million years, and every second represents more than 400 years.

On this time scale, the Milky Way Galaxy probably formed in February. Many generations of stars lived and died in the subsequent cosmic months, enriching the galaxy with the "star stuff" from which we and our planet are made.

Our solar system and our planet did not form until early September on this scale, or $4\frac{1}{2}$ billion years ago in real time [Section 4.2]. By late September, life on Earth was flourishing. However, for most of Earth's history, living organisms remained microscopic. On the scale of the cosmic calendar, recognizable animals became prominent only in mid-December. Early dinosaurs appeared on the day after Christmas. Then, in a cosmic instant, the dinosaurs disappeared forever—probably because of the impact of an asteroid or a comet [Section 6.4]. In real time, the death of the dinosaurs occurred some 65 million years ago, but on the cosmic calendar it was only yesterday. With the dinosaurs gone, furry mammals inherited Earth. Some 60 million years later, or around 9 P.M. on December 31 of the cosmic calendar, early hominids (human ancestors) began to walk upright.

Perhaps the most astonishing thing about the cosmic calendar is that the entire history of human civilization falls into just the last half-minute. The ancient Egyptians built the pyramids only about 11 seconds ago on this scale. About 1 second ago, Kepler and Galileo proved that Earth

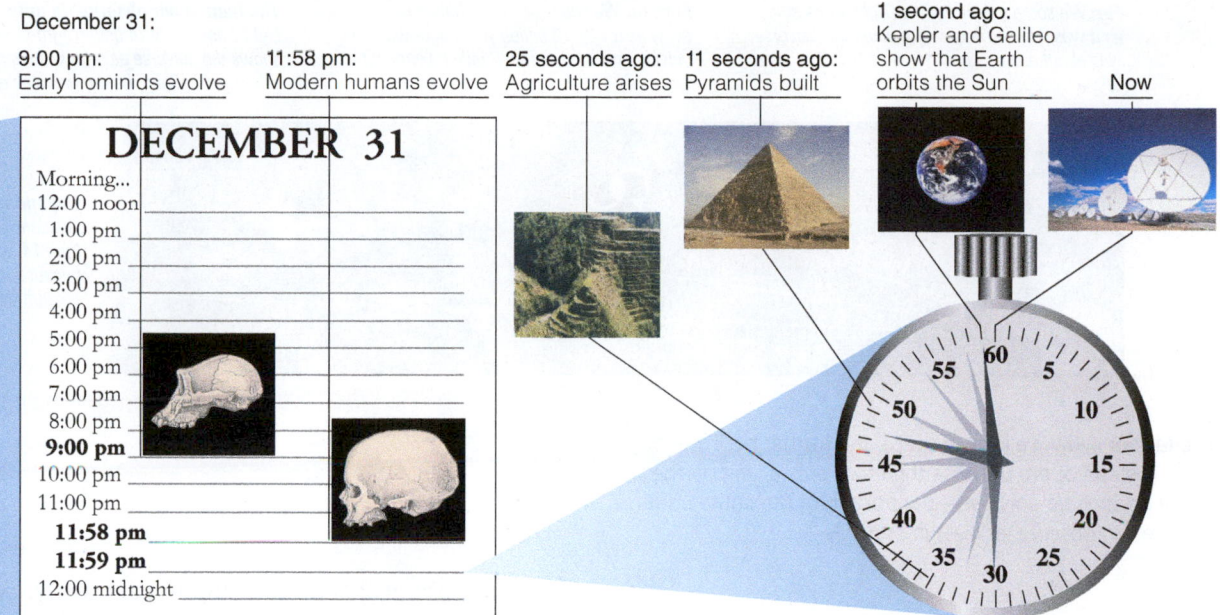

December 31:

9:00 pm:
Early hominids evolve

11:58 pm:
Modern humans evolve

25 seconds ago:
Agriculture arises

11 seconds ago:
Pyramids built

1 second ago:
Kepler and Galileo
show that Earth
orbits the Sun

Now

DECEMBER 31

Morning...
12:00 noon
1:00 pm
2:00 pm
3:00 pm
4:00 pm
5:00 pm
6:00 pm
7:00 pm
8:00 pm
9:00 pm
10:00 pm
11:00 pm
11:58 pm
11:59 pm
12:00 midnight

orbits the Sun rather than vice versa. The average college student was born about 0.05 second ago, around 11:59:59.95 P.M. on the cosmic calendar. On the scale of cosmic time, the human species is the youngest of infants, and a human lifetime is a mere blink of an eye.

Like the scale of space, the fantastic scale of time carries important lessons about extraterrestrial life, if it exists. For example, the fact that the universe is so much older than Earth means that there ought to be many worlds that have had plenty of time for life to arise and evolve. If those worlds have had biological histories similar to Earth's, they might have had civilizations millions or even billions of years ago. We'll explore this idea and its astonishing implications in Chapter 13. The scale of time also holds sobering lessons for our own future. Species have come and gone in the months of the cosmic calendar during which life has flourished on Earth, and there's no special reason to think that our fate should be any different. Unless we learn enough about ourselves and our planet to find ways to survive into the next cosmic year, we will end up as little more than a momentary blip in the long history of the universe.

How big is the universe?

We've stated that there are billions of galaxies in the universe, but can we be any more precise? In fact, when we think of the universe as the sum total of *all* matter and energy, we really don't know how large it is—the universe could well be infinite, in which case it contains an infinite number of galaxies. However, the age of the universe places a fundamental limit on the portion of the universe that we can possibly see, even with the most powerful telescopes imaginable. To understand why, think again about the time it takes light to travel vast distances through the universe.

THE OBSERVABLE UNIVERSE When we look to great distances, we are also looking far back into the past. Figure 3.14 shows the nearest large galaxy to our own—the Great Galaxy in Andromeda, also known as M31. It is

FIGURE 3.14
The Great Galaxy in Andromeda (M31). When we look at this galaxy—which is faintly visible to the naked eye in the location shown in the inset—we see light that has been traveling through space for 2.5 million years.

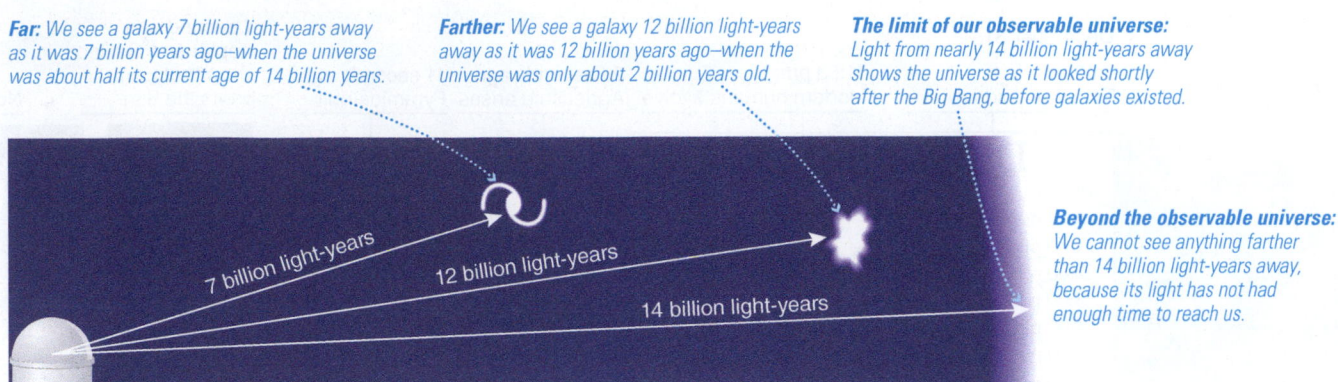

Far: We see a galaxy 7 billion light-years away as it was 7 billion years ago—when the universe was about half its current age of 14 billion years.

Farther: We see a galaxy 12 billion light-years away as it was 12 billion years ago—when the universe was only about 2 billion years old.

The limit of our observable universe: Light from nearly 14 billion light-years away shows the universe as it looked shortly after the Big Bang, before galaxies existed.

Beyond the observable universe: We cannot see anything farther than 14 billion light-years away, because its light has not had enough time to reach us.

7 billion light-years

12 billion light-years

14 billion light-years

FIGURE 3.15

The farther away we look in space, the further back we look in time. The age of the universe therefore puts a limit on the size of the *observable* universe—the portion of the entire universe that we can observe, at least in principle.

located about 2.5 million light-years away, which means the photo in Figure 3.14 shows this galaxy as it was about 2.5 million years ago, long before modern humans even existed. This might seem like a long time in the past, but it's unlikely that the Andromeda galaxy looks significantly different today: The galaxy is so large that it takes some 200 million years just to rotate once, so in 2 million years it barely changes at all. At much greater distances, however, we begin to see back to a time when the entire universe was significantly younger than it is today.

Consider, for example, a galaxy that is 1 billion light-years away. Its light has taken 1 billion years to reach us, which means we are seeing it as it looked 1 billion years ago—when the universe was only 13 billion years old, rather than its current 14 billion years.* Next, consider a galaxy that is 7 billion light-years away. We see this galaxy as it looked 7 billion years ago—which means we see it as it was when the universe was only half its current age. If we look at a galaxy that is 12 billion light-years away, we see it as it was 12 billion years ago, when the universe was only 2 billion years old. And if we tried to look beyond 14 billion light-years, we'd be looking to a time more than 14 billion years ago—which is before the universe existed and therefore means that there is nothing to see. This distance of 14 billion light-years therefore marks the boundary (or *horizon*) of our **observable universe**—the portion of the entire universe that we can potentially observe (Figure 3.15).

The concept of the observable universe has at least two important philosophical implications that are worth keeping in mind. First, the fact that we cannot observe anything more than 14 billion light-years away does *not* mean that nothing exists at such distances. In fact, we have good reason to think that the universe goes on far beyond 14 billion light-years, and some evidence suggests that the universe is infinite in extent. It's just that we have no hope of seeing or studying any objects that lie beyond the bounds of our observable universe. Second, notice that by definition, we are the center of our observable universe, since it

---

*Distances to faraway galaxies become difficult to define in an expanding universe, because the galaxies today are significantly farther away than they were when their light left on its journey to us. In this book, we use distances based on light-travel time (often called the *lookback time*). For example, when we say that a galaxy is 1 billion light-years away, we mean that its light traveled through space for 1 billion years to reach us.

is defined by a light-travel distance in all directions from *us*. However, being in the center of the *observable* universe is very different from being in the center of *the* universe. The latter would imply a special location, and the former does not. Observers on any planet around any star in any galaxy must also be at the center of their own observable universe; for example, people living in a distant galaxy would say the observable universe extends 14 billion light-years in all directions from them rather than from us. (You may realize that this means that they can see at least some galaxies that we cannot, and vice versa, because their observable universe only partially overlaps ours.) While the idea that we lie at the center of our own observable universe may sound like a throwback to pre-Copernican times, it's really not, since it still does not give us any special place in the *whole* universe.

WORLDS BEYOND IMAGINATION Because the observable universe has a finite size, it must contain a finite number of galaxies. We do not know exactly how many, because there are too many to count and because some galaxies are so faint that we cannot see them even with our best telescopes. Nevertheless, we can *estimate* the number of galaxies in the observable universe by counting the number that we can see in pictures made with our most powerful telescopes. Figure 3.16 shows a remarkable photo, taken by the Hubble Space Telescope, that shows a tiny piece of the sky in great detail. By counting the galaxies in Figure 3.16 and multiplying by the number of such photos it would take to make a montage of the entire sky, astronomers estimate that the observable universe contains about 100 billion galaxies. Just as it would take thousands of years to count the more than 100 billion stars in the Milky Way, it would take thousands of years to count all the galaxies in the observable universe.

Now, let's think about the total number of *stars* in all these galaxies. If we assume 100 billion stars per galaxy—similar to the number in the Milky Way—the total number of stars in the observable universe is roughly 100 billion × 100 billion, or 10,000,000,000,000,000,000,000 (10^{22}). How big is this number? Visit a beach. Run your hands through the fine-grained sand. Imagine counting each tiny grain of sand as it slips through your fingers. Then imagine counting every grain of sand on the beach and continuing on to count *every* grain of dry sand on *every* beach on Earth. If you could actually complete this task, you would find that, roughly speaking, the number of grains of sand is comparable to the number of stars in the observable universe (Figure 3.17).

The total number of *worlds*—by which we mean any reasonably large bodies in space, such as planets, moons, or even large asteroids—may be even greater. If planetary systems are as common as recent discoveries suggest, most stars have at least a few planets or moons. Clearly, our universe contains worlds beyond imagination.

Think About It Contemplate the incredible numbers of stars in our galaxy and in the universe, and the fact that each star is a potential sun for a system of planets. How does this perspective affect your thoughts about the possibilities for finding life—or intelligent life—beyond Earth? Explain.

The incredible size of the universe poses a practical challenge to the search for life beyond Earth: We can no more hope to search all the possible places where we might find life than we could hope to study every grain of dry sand on every beach on Earth. We will therefore confine

FIGURE 3.16
This photograph, known as the Hubble Extreme Deep Field, shows thousands of galaxies, some more than 12 billion light-years away. The field of view of this image would fit behind a grain of sand held at arm's length against the sky. Nearly every tiny dot in this photo is an entire galaxy of stars.

FIGURE 3.17
The number of stars in the observable universe is comparable to the number of grains of dry sand on all the beaches on Earth.

our discussions of the search for life to the search within the Milky Way Galaxy, and presume that we'd find similar results if we could study other galaxies.

THE FINE-TUNED UNIVERSE We have briefly surveyed modern understanding of the universe, and have discussed a little bit of the evidence that has given us this picture. But why is the universe like this? This may seem a strange question to ask in a science book, but it is one that many scientists are now asking themselves. The interest in this question has been sparked by the realization that our universe appears to be "fine-tuned" for life.

The logic behind the fine-tuning idea, sometimes known as the *anthropic principle,* goes like this: We are here today, able to study the universe and learn about its basic properties. But if any of those properties were much different, we could never have come to exist in the first place. For example, consider the expansion of the universe. If the universe were expanding much more rapidly, all its matter would have flown apart before gravity could have collected it into galaxies, stars, or planets. And if the universe were expanding much more slowly, gravity would have pulled all the matter back together, causing the universe to collapse before there was time for life to get started. The expansion rate had to be "just right" for galaxies to form, a fact that looks even more remarkable when you take into account the acceleration of the expansion, which also has to be of just the right value so that it would have accelerated neither too much nor too little by now. Similar considerations apply to many other fundamental properties of the universe, including the ratio of the strengths of different forces (such as gravity and the electromag-

---

Movie Madness INTERSTELLAR

It's long been fashionable for Hollywood to depict humanity's future as bleaker than a Siberian winter. In *Interstellar*, the troubles are everywhere, and of our own making: Some unspecified misstep has caused an environmental catastrophe called "the blight." Despite the fact that almost all the people have left their cubicles to take up farming, the world is running out of food.

This is not trivial. Think of all the things humans eat: Just about every variety of plant or animal—from kohlrabi to cockroaches—is on the menu somewhere. But apparently all these comestibles have been wiped out, which makes the blight more deadly than the Permian extinction. There's only one crop that hasn't come a cropper: corn. This is undoubtedly a relief to movie concession stands, which sell the stuff basted in butter at a price that's 30 times more per pound than gasoline (but tastes better).

Someone has to do something, and *Interstellar* follows the exploits of a former NASA pilot who learns of a wormhole in our solar system, apparently put in place by some mysterious beings who are trying to make it easy for humanity to escape our doomed world and seek refuge in another galaxy.

Now that's a plot line you don't see every day. Fortunately, the director has engaged Kip Thorne, a renowned Caltech scientist, to make sure that both the premise and the visuals of the film are semi-plausible. Mind you, in truth wormhole travel happens only on physicists' blackboards, and it's not clear that anyone—blighted or otherwise—is

ever going to drop into one of these Einsteinian spacetime warps and come out at a different cosmic address. But this is sci-fi, with the emphasis on "fi."

Interstellar's plot line is so complicated that many will find it less comprehensible than Latin grammar. To help out, someone has put diagrams on the Internet showing the time sequence of events in this film. (Really, someone has.) Of course, time is a slippery concept when cast members rocket through wormholes or wander around in five-dimensional space. But not to worry—if you don't have a diagram, you can always sit back and groove to the special effects.

Oh, and as for our descendants emigrating to another galaxy, well . . . given that there are about a trillion planets in our own galaxy, this seems like an overreaction (though, in fairness, the plot required a black hole of large enough mass that the tidal forces wouldn't kill you as you crossed the event horizon, and such black holes exist only at the centers of a few other galaxies). It's quite possible that some of us will eventually decamp to other cosmic locales, but those would most plausibly be habitats on the Moon or Mars—or even closer, in giant space colonies orbiting Earth.

Sending humanity to another galaxy is a bit like walking to Timbuktu for lunch, instead of heading for a diner down the street. Doing the latter is easier and cheaper, and offers a better chance that you'll actually make it. But then again, cinema isn't about any of that.

---

netic force), the size scale on which quantum effects become important, and even the fact that we live in three dimensions of space—current theories of physics hold that many more dimensions actually exist, so three was not the only possibility. In every case, a change—perhaps even a very small change—could have prevented us from being here.

There's no doubt that the universe really is fine-tuned in a way that makes our existence possible; the debate is over what this means. Some people argue that this is merely a philosophical question, since we obviously would not be here talking about it if it weren't so, in which case the question falls outside the realm of science. Others argue that the fine-tuning implies some "specialness" to humans in particular, with a few going as far as to claim that we should conclude that we *are* unique and hence the only life in the universe. Many physicists are seeking a natural explanation for the observed fine-tuning. One set of models that physicists have proposed to explain the workings of our universe suggests that in fact there should be a huge number of universes (sometimes called a *multiverse*), each with its own set of parameters such as its number of dimensions, ratio of force strength, and expansion rate. In essence, this viewpoint holds that lots of universes exist, most of which are unsuitable for life, and we live in one of the rare universes suitable for life simply because we can. Other physicists suspect that we'll ultimately find a simpler explanation for why the universe turned out just right, and that once we find this explanation, we'll realize that it had to be this way; as Einstein once put it, perhaps "God had no choice" in setting the parameters of creation. From this viewpoint, our apparent "specialness" may simply be a consequence of our still-incomplete knowledge of physics.

3.3 A Universe of Matter and Energy

Now that we have surveyed modern understanding of the universe on the large scale, our next major goal in this chapter will be to consider the worlds on which we might possibly find life, which means learning about the nature and origin of planetary systems. However, before we can do that, we first need to focus in a little more depth on the basic constituents of the universe— *matter* and *energy*—and on how we learn about distant objects in the universe by studying their light.

Many of the ideas in this section may already be familiar to you (especially if you recall high school chemistry or physics), but don't worry if they are not; this section will provide you with all the basic information needed to be successful as you work through this book. Also note that, as we discussed earlier, we will restrict our focus to "ordinary" matter and energy, ignoring *dark* matter and *dark* energy because they do not appear to be necessary for our discussions of stars, planets, and life.

What are the building blocks of matter?

In Chapter 2, we discussed the ancient Greek idea that matter consists of four elements—fire, water, earth, and air—and the claim by Democritus and his followers (the "atomists") that these elements came in tiny, indivisible pieces that they called *atoms*, a Greek term meaning "indivisible." Today, we have a similar idea, but there are a lot more elements, and fire, water, earth, and air are *not* among them. In addition, we now know that the atoms that make the elements are themselves made from smaller

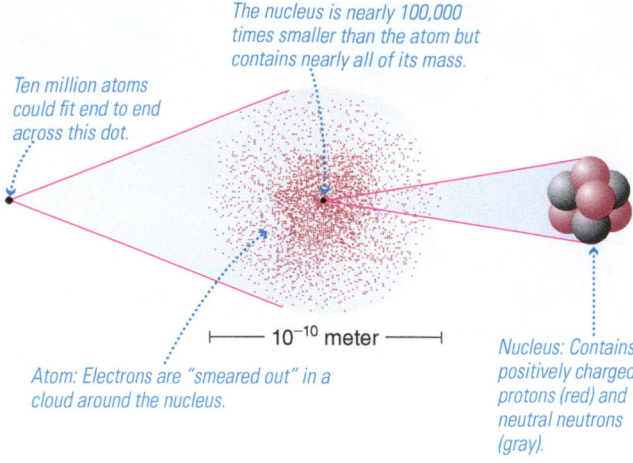

The nucleus is nearly 100,000 times smaller than the atom but contains nearly all of its mass.

Ten million atoms could fit end to end across this dot.

⊢—— 10^{-10} meter ——⊣

Atom: Electrons are "smeared out" in a cloud around the nucleus.

Nucleus: Contains positively charged protons (red) and neutral neutrons (gray).

FIGURE 3.18
The structure of a typical atom. Atoms are extremely tiny: The atom shown in the middle is magnified to about one billion times its actual size, and the nucleus on the right is magnified to about 100 trillion times its actual size.

pieces. Let's take a brief look at our current understanding of atoms and other microscopic forms of matter.

ATOMIC STRUCTURE Atoms come in different types, and each type corresponds to a different chemical **element**. Some of the most familiar chemical elements are hydrogen, helium, carbon, oxygen, silicon, iron, gold, silver, lead, and uranium.

Atoms are made of particles that we call **protons, neutrons,** and **electrons** (Figure 3.18). Protons and neutrons are found in the tiny **nucleus** at the center of the atom. The rest of the atom's volume contains electrons, which surround the nucleus. Although we can think of electrons as tiny particles, they are not quite like tiny grains of sand and they don't orbit the nucleus the way planets orbit the Sun. Instead, the electrons in an atom form a kind of "smeared out" cloud that surrounds the nucleus and gives the atom its apparent size. The electrons aren't really cloudy, but it is impossible to pinpoint their positions in the atom.

Figure 3.18 also shows several other important features of atoms. First, notice that atoms are incredibly small: Millions could fit end to end across the period at the end of this sentence, and the number of atoms in a single drop of water (typically, 10^{22} atoms) may equal the number of stars in the observable universe. At the same time, the electrons give the atom a size far larger than that of its nucleus; if you imagine an atom on a scale that makes its nucleus the size of your fist, its electron cloud would be many kilometers wide. Nevertheless, most of the atom's mass resides in its nucleus, because protons and neutrons are each about 2000 times as massive as an electron.

The properties of an atom depend mainly on the amount of **electrical charge** in its nucleus; an object's electrical charge is a measure of how strongly it will interact with other charged particles. We define the electrical charge of a proton as the basic unit of positive charge, which we write as +1. The electron has an electrical charge that is precisely opposite that of a proton, so we say it has negative charge (−1). Neutrons are electrically neutral, meaning that they have no charge. Oppositely charged particles attract one another, and similarly charged particles repel one another. An atom is held together by the attraction between the positively charged protons in the nucleus and the negatively charged electrons that surround the nucleus.*

Most of the atoms in and around you contain the same number of electrons as protons, making them electrically neutral overall. However, atoms often lose or gain electrons, in which case they obtain a net electrical charge. We call such atoms **ions.** A *positive ion* is an atom that has lost one or more electrons so that it has more positive than negative charge overall; a *negative ion* is an atom that has gained one or more electrons, giving it a net negative charge. The net electrical charge of an atom turns out to be exceedingly important to life: Because the nucleus is buried so deeply inside an atom, interactions between atoms are almost exclusively interactions between their electrons. Indeed, these electrical interactions between atoms essentially make up everything that we think of as *chemistry*—and since chemical reactions are the foundation of all the processes that occur in living organisms, we see that the electrical interactions of atoms underlie everything we know about life.

---

*You may wonder why electrical repulsion doesn't cause the positively charged protons in a nucleus to fly apart from one another. The answer is that an even stronger force, called the strong force, overcomes electrical repulsion and holds the nucleus together.

ATOMIC TERMINOLOGY Figure 3.19 summarizes several pieces of atomic terminology that we will make use of throughout this book. First, each different chemical element contains a different number of protons in its nucleus. This number is its **atomic number.** For example, a hydrogen nucleus contains just one proton, so its atomic number is 1. A helium nucleus contains two protons, so its atomic number is 2. The complete set of the more than 100 known elements is listed in the **periodic table of the elements** (see Appendix D).

The *combined* number of protons and neutrons in an atom is called its **atomic mass number.** The atomic mass number of ordinary hydrogen is 1 because its nucleus is just a single proton. Helium usually has two neutrons in addition to its two protons, giving it an atomic mass number of 4. Carbon usually has six protons and six neutrons, giving it an atomic mass number of 12.

Every atom of a given element contains exactly the same number of protons, but the number of neutrons can vary. For example, all carbon atoms have six protons, but they may have six, seven, or eight neutrons. Versions of an element with different numbers of neutrons are called **isotopes** of the element. Isotopes are named by listing their element name and atomic mass number. For example, the most common isotope of carbon has six protons and six neutrons, giving it atomic mass number $6 + 6 = 12$, so we call it carbon-12. Two other isotopes of carbon are carbon-13 (six protons and seven neutrons give it atomic mass number 13) and carbon-14 (six protons and eight neutrons give it atomic mass number 14). We can also write the atomic mass number of an isotope as a superscript to the left of the element symbol: ^{12}C, ^{13}C, ^{14}C. We read ^{12}C as "carbon-12."

Think About It The symbol ^{16}O represents oxygen with an atomic mass number of 16; it is the most common form of oxygen, containing eight protons and eight neutrons. What does the symbol ^{18}O represent?

MOLECULES The number of different material substances is far greater than the number of chemical elements because atoms can combine to form **molecules.** Some molecules consist of two or more atoms of the same element. For example, we breathe O_2, oxygen molecules made of two oxygen atoms. Other molecules, such as water, are made up of atoms of two or more different elements; these types of molecules are called **compounds.** Water is an example of a compound, because its symbol H_2O tells us that a water molecule contains two hydrogen atoms and one oxygen atom.

The chemical properties of a molecule are different from those of its individual atoms. For example, molecular oxygen (O_2) behaves differently from atomic oxygen (O), and water behaves differently from pure hydrogen or pure oxygen. Life on Earth is based on the complex chemistry of molecules (compounds) containing carbon, which are called **organic molecules** (or **organic compounds**). In diagrams, molecules are often represented with ball-and-stick models that show how their atoms are arranged (Figure 3.20).

PHASES OF MATTER Everyday experience tells us that a substance can behave dramatically differently in different *phases*, even though it is still made of the same atoms or molecules. For example, molecules of H_2O can exist in three familiar phases: as **solid** ice, as **liquid** water, and as the **gas** we call water vapor. How can the same molecules look and act so differently in these different phases?

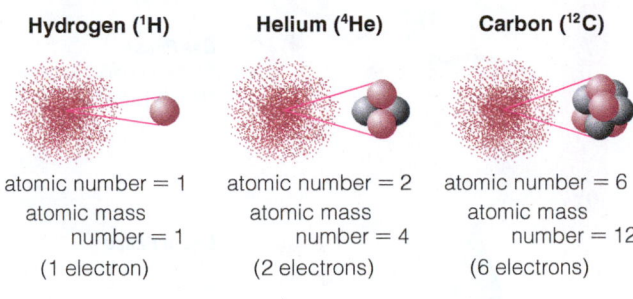

atomic number = number of protons
atomic mass number = number of protons + neutrons
(A neutral atom has the same number of electrons as protons.)

Hydrogen (^{1}H)   **Helium (^{4}He)**   **Carbon (^{12}C)**

atomic number = 1
atomic mass number = 1
(1 electron)

atomic number = 2
atomic mass number = 4
(2 electrons)

atomic number = 6
atomic mass number = 12
(6 electrons)

*Different **isotopes** of a given element contain the same number of protons, but different numbers of neutrons.*

Isotopes of Carbon

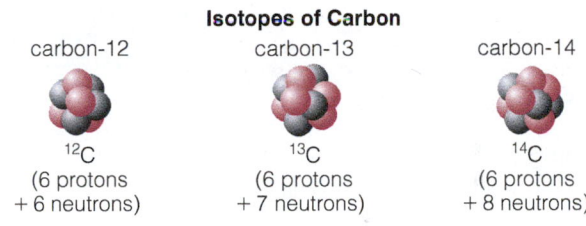

carbon-12   carbon-13   carbon-14

^{12}C
(6 protons
+ 6 neutrons)

^{13}C
(6 protons
+ 7 neutrons)

^{14}C
(6 protons
+ 8 neutrons)

FIGURE 3.19
Terminology of atoms.

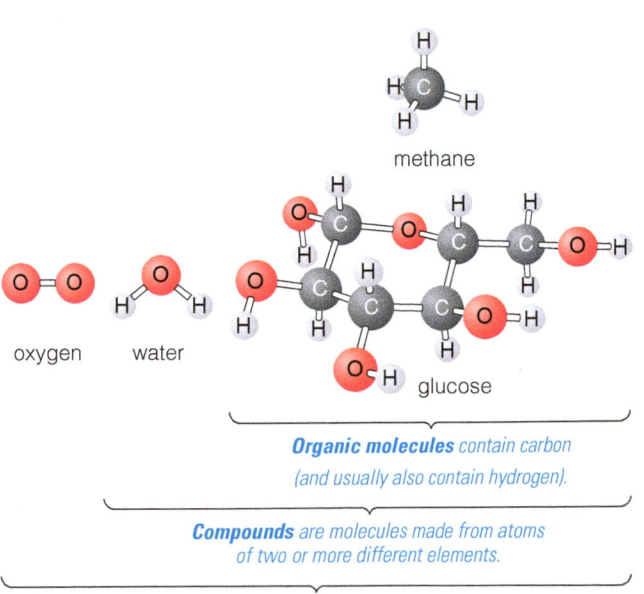

methane

oxygen   water   glucose

Organic molecules contain carbon (and usually also contain hydrogen).

Compounds are molecules made from atoms of two or more different elements.

Molecules consist of two or more atoms.

FIGURE 3.20
Terminology of molecules. (Adapted from Campbell, Reece, Taylor, Simon, *Biology Concepts & Connections*.)

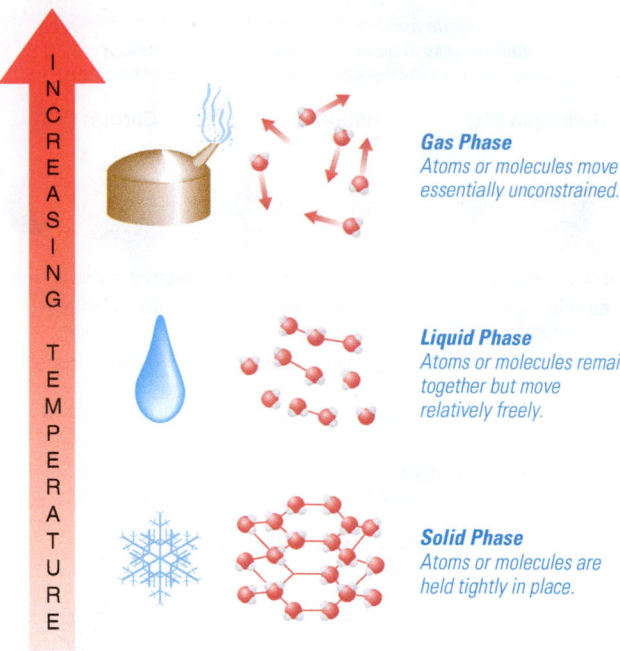

FIGURE 3.21

The basic progression of phase changes in water.

Gas Phase
Atoms or molecules move essentially unconstrained.

Liquid Phase
Atoms or molecules remain together but move relatively freely.

Solid Phase
Atoms or molecules are held tightly in place.

You are probably familiar with the idea of a **chemical bond,** the name we give to the interactions between electrons that hold the atoms in a molecule together. For example, we say that chemical bonds hold the hydrogen and oxygen atoms together in a molecule of H_2O. Similar but much weaker interactions among electrons hold together the many water molecules in a block of ice or a pool of water. We can think of the interactions that keep neighboring atoms or molecules close together as another type of bond.

If we think in terms of bonds, the phases of solid, liquid, and gas differ in the strength of the bonds between neighboring atoms and molecules. Phase changes occur when one type of bond is broken and replaced by another. Changes in either pressure or temperature (or both) can cause phase changes, but it's easier to think about temperature.

Consider water as an example (Figure 3.21). At low temperatures, water molecules are bound tightly to their neighbors, making the *solid* structure of ice. As long as the temperature remains below freezing, the water molecules in ice remain rigidly held together; we often say that they have a *crystal* structure, meaning that the molecules are arranged in a precise geometrical pattern. However, the molecules within this crystal structure are always vibrating, and higher temperature means greater vibrations.

The **melting point** (0°C at sea level on Earth) is the temperature at which the water molecules finally break the solid bonds of ice. The molecules can then move much more freely among one another, allowing the water to flow as a *liquid.* However, the molecules in liquid water are not completely free of one another, as we can tell from the fact that droplets of water can stay intact. Thus, adjacent molecules in liquid water are still held together by a type of bond, though a much looser bond than the one that holds them together in solid ice.

If we continue to heat the water, the average speeds of the water molecules increase, and high enough speeds will ultimately break the bonds between neighboring molecules altogether. The molecules will then be able to move freely, and freely moving particles constitute what we call a *gas.* Above the **boiling point** (100°C at sea level), all the bonds between adjacent molecules are broken so that the water can exist only as a gas.

We see ice melting into liquid water and liquid water boiling into gas so often that it's tempting to think that's the end of the story. However, a little thought should convince you that the reality has to be more complex. For example, you know that Earth's atmosphere contains water vapor that condenses to form clouds and rain. But Earth's temperature is well below the boiling point of water (luckily for us!), so how is it that our atmosphere can contain water in the gas phase?

You'll understand the answer if you remember that temperature is a measure of the *average* motion of the particles in a substance; individual particles may move substantially faster or slower than the average. Even at the low temperatures at which most water molecules are bound together as ice or liquid, a few molecules will always move fast enough to break free of their neighbors and enter the gas phase. In other words, some gas (water vapor) is always present along with solid ice or liquid water. The process by which molecules break free is often called **vaporization,** because the escaped molecules enter the gas (or vapor) phase. More technically, vaporization from a solid is called *sublimation,* while vaporization from a liquid is called *evaporation.* Higher temperatures lead to higher rates of vaporization.

The same basic ideas hold for other substances, but their melting and boiling temperatures differ from those of water. Moreover, although we won't go into detail here, remember that pressure also has an important effect. For example, high pressure can cause a substance to remain in the solid phase even when the temperature is above the low-pressure boiling point.

What is energy?

Energy is what makes matter move. In other words, without energy, nothing would have happened in the universe. In that sense, we can think of everything that ever occurs in the universe, including all the processes that occur in living organisms, as an interplay between matter and energy.

BASIC TYPES OF ENERGY Energy comes in many different types. For example, there is the energy we get from the food we eat, the energy that makes our cars go, and the energy a light bulb emits. Fortunately, we can classify the many types of energy into three major categories (Figure 3.22):

- Energy of motion, or **kinetic energy** (*kinetic* comes from a Greek word meaning "motion"). Falling rocks, orbiting planets, and the molecules moving in the air around us are all examples of objects with kinetic energy.
- Energy carried by light, or **radiative energy** (the word *radiation* is often used as a synonym for *light*). All light carries energy, which is why light can cause changes in matter. For example, light can alter molecules in our eyes—thereby allowing us to see—or warm the surface of a planet.
- Stored energy, or **potential energy,** which might later be converted into kinetic or radiative energy. For example, a rock perched on a ledge has *gravitational* potential energy because it will fall if it slips off the edge, and gasoline contains *chemical* potential energy that can be converted into the kinetic energy of the moving car.

Although all forms of energy fall into one of these three major categories, it's sometimes useful to subdivide these categories further. We've already noted that gravitational potential energy and chemical potential energy are both subcategories of potential energy. For our purposes in this book, it's important to be familiar with two other subcategories of energy: *mass-energy*, which is a type of potential energy, and *thermal energy*, which is a type of kinetic energy.

Mass-energy embodies the idea that mass itself is a form of stored energy. This idea was discovered by Einstein as part of his special theory of relativity and is commonly stated through his famous formula $E = mc^2$, which we discussed earlier in the context of nuclear fusion in stars.

Thermal energy represents the collective kinetic energy of the many individual atoms or molecules moving randomly within a substance like a rock or the air. For a given substance, higher temperature means more rapid motion of its atoms and molecules, and therefore means more thermal energy. This idea explains why many chemical and biochemical reactions proceed more rapidly at higher temperatures: With more thermal energy available, it is possible for this energy to fuel a larger number of individual reactions.

CONSERVATION OF ENERGY It is possible for energy to change from one form to another—indeed, such changes are the primary drivers of life.

Energy can be converted from one form to another.

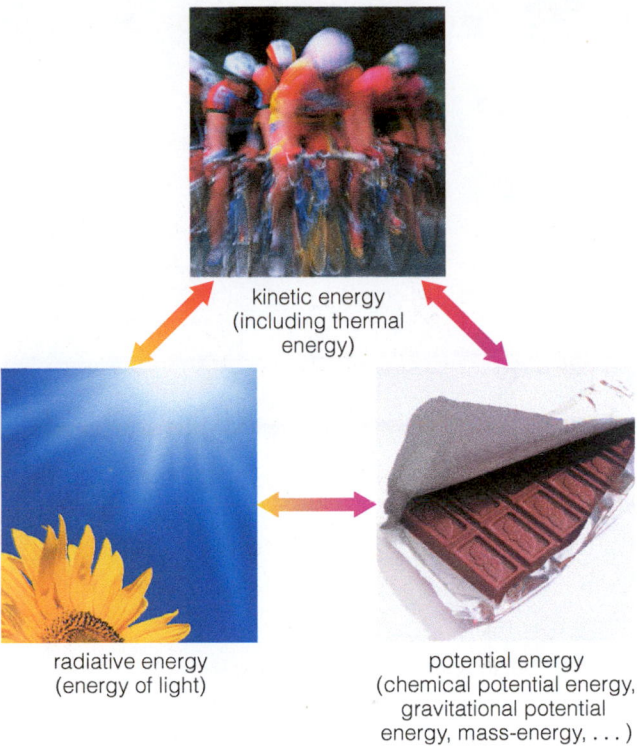

kinetic energy
(including thermal energy)

radiative energy
(energy of light)

potential energy
(chemical potential energy, gravitational potential energy, mass-energy, . . .)

FIGURE 3.22
The three basic categories of energy. Energy can be converted from one form to another, but it can never be created or destroyed, an idea embodied in the law of conservation of energy.

For example, our bodies take the chemical potential energy stored in food and use it to make molecules move in ways that allow our leg muscles to contract for walking, our blood and skin to create scabs over wounds, and neurons in our brains to fire in ways that make thought possible.

However, while energy can *change* from one form to another, it can be neither created nor destroyed. This idea represents what we call the **law of conservation of energy.** As a simple example, imagine that you've just thrown a baseball. The baseball is now moving, so it has kinetic energy. Where did this kinetic energy come from? The baseball got its kinetic energy from the motion of your arm as you threw it. Your arm, in turn, got its kinetic energy from the release of chemical potential energy stored in your muscle tissues. Your muscles got this energy from the chemical potential energy stored in the foods you ate. The energy stored in the foods came from sunlight, which plants convert into chemical potential energy through photosynthesis. The radiative energy of the Sun was generated through the process of nuclear fusion, which releases some of the mass-energy stored in the Sun's supply of hydrogen. The mass-energy stored in the hydrogen came from the birth of the universe in the Big Bang, which in this sense represents the moment in which all matter and energy is thought to have come into existence. The energy transfer will continue after your throw. Air resistance will slow the ball down, and a few bounces will eventually bring it to rest on the ground, which means the ball will have transferred its kinetic energy to molecules in the air and ground. It may be difficult to trace the energy after this point, but it will never disappear.

The law of conservation of energy is critical to understanding life and prospects for life on other worlds. Although some form of stored energy is available almost everywhere, in most cases there is no viable way for life to extract the energy for its own use. Therefore, as we'll discuss in later chapters, the availability of a viable energy source is a crucial factor in determining whether a world is habitable.

What is light?

Nearly all the information we have about distant planets and stars comes from studying their light. Let's briefly examine key properties of light that make it possible to learn so much from it.

BASIC PROPERTIES OF LIGHT As we have already seen, light is a form of energy (radiative energy) that travels through empty space at the high speed of 300,000 kilometers per second. More specifically, light is characterized by rapidly changing electric and magnetic fields, which is why we often call light an **electromagnetic wave** (Figure 3.23). Like other types of waves (such as water waves, sound waves, or waves on a vibrating string), light is characterized by a **wavelength** (the distance between adjacent peaks of the electric or magnetic field) and a **frequency** (the rate at which the electric and magnetic fields change). The standard unit of frequency, *hertz* (Hz), is equivalent to waves (or cycles) per second; for example, 10^3 hertz means that $10^3 = 1000$ wave peaks pass by a point each second.

Unlike most other types of waves, light also exhibits properties that we usually attribute to particles. In particular, light comes in distinct "pieces," called **photons,** that can exert pressure, knock electrons out of atoms, or cause molecules to start rotating and vibrating. In other words, light is both a wave and a particle, so the best way to think of light is as a collection of photons that are each characterized by a wavelength and a frequency.

Wavelength is the distance between adjacent peaks of the electric (and magnetic) field . . .

. . . while frequency is the number of times each second that the electric (and magnetic) field vibrates up and down (or side to side) at any point.

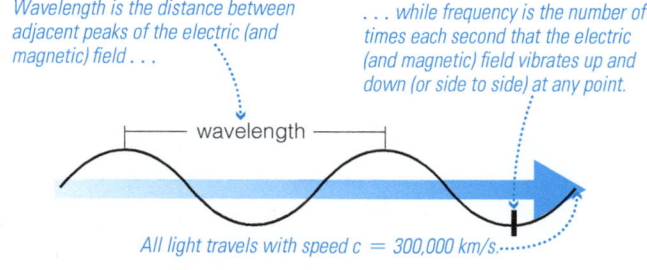

wavelength

All light travels with speed $c = 300,000$ km/s.

FIGURE 3.23 INTERACTIVE FIGURE
Light is an electromagnetic wave, but it also comes in individual pieces called *photons*, each characterized by a wavelength and a frequency.

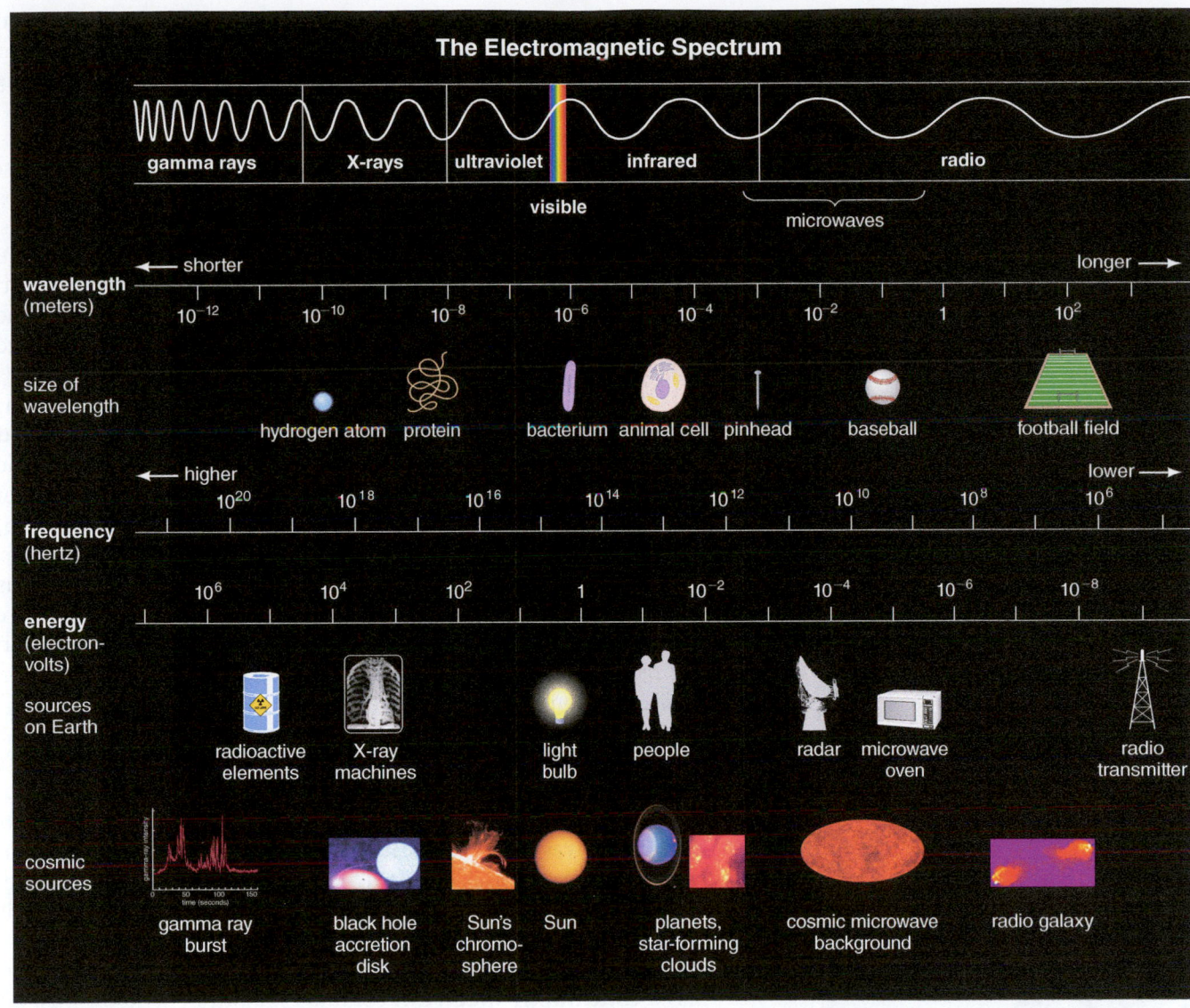

The Electromagnetic Spectrum

gamma rays | X-rays | ultraviolet | infrared | radio

visible

microwaves

wavelength (meters)

← shorter   longer →

10^{-12} 10^{-10} 10^{-8} 10^{-6} 10^{-4} 10^{-2} 1 10^2

size of wavelength

hydrogen atom   protein   bacterium   animal cell   pinhead   baseball   football field

frequency (hertz)

← higher   lower →

10^{20} 10^{18} 10^{16} 10^{14} 10^{12} 10^{10} 10^8 10^6

energy (electron-volts)

10^6 10^4 10^2 1 10^{-2} 10^{-4} 10^{-6} 10^{-8}

sources on Earth

radioactive elements   X-ray machines   light bulb   people   radar   microwave oven   radio transmitter

cosmic sources

gamma ray burst   black hole accretion disk   Sun's chromosphere   Sun   planets, star-forming clouds   cosmic microwave background   radio galaxy

FIGURE 3.24 INTERACTIVE FIGURE
The electromagnetic spectrum. Notice that wavelength increases as we go from gamma rays to radio waves, while frequency and energy increase in the opposite direction.

A simple formula relates the wavelength and frequency of a photon: wavelength × frequency = speed of light. Because all forms of light travel at the same speed, we find that *longer wavelength means lower frequency and shorter wavelength means higher frequency.* The energy of a photon is proportional to its frequency, so higher frequency light has higher energy.

THE ELECTROMAGNETIC SPECTRUM Light can in principle have any wavelength, frequency, or energy. The complete range of possibilities, shown in Figure 3.24, is called the **electromagnetic spectrum.** For convenience, we usually refer to different portions of the electromagnetic spectrum by different names. The **visible light** that we see with our eyes is only a tiny portion of the complete spectrum, with wavelengths from about 400 nm at the blue end of the rainbow to about 700 nm at the red end. (One nanometer [nm] is a billionth of a meter.)

Light with wavelengths somewhat longer than those of red light is called **infrared,** because it lies beyond the red end of the rainbow. **Radio waves** are the longest-wavelength light. (Be sure to note that radio waves are a form of light, *not* of sound.) The region near the border between infrared and radio waves, where wavelengths range from micrometers to centimeters, is often called **microwaves.** In astronomy, microwaves are sometimes divided further: Wavelengths from about one to a few millimeters are called *millimeter waves*, while wavelengths of tenths of a millimeter are called *submillimeter waves*.

On the other side of the spectrum, light with wavelengths somewhat shorter than those of blue light is called **ultraviolet,** because it lies beyond the blue (or violet) end of the rainbow. Light with even shorter wavelengths is called **X rays,** and the shortest-wavelength light is called **gamma rays.** Notice that visible light is an extremely small part of the entire electromagnetic spectrum: The reddest red that our eyes can see has only about twice the wavelength of the bluest blue, but the radio waves from your favorite FM radio station are a billion times longer than the X rays used in a doctor's office.

The various energies of light explain many familiar effects in everyday life. Radio waves carry so little energy that they have no noticeable effect on our bodies, but they can make electrons move up and down in an antenna, making them useful for radio communication. Molecules moving in a warm object emit infrared light, which is why we sometimes associate infrared light with heat. Receptors in our eyes respond to visible-light photons, making vision possible. Ultraviolet photons carry enough energy to damage skin cells, causing sunburn or skin cancer. X-ray photons have enough energy to penetrate skin and muscle but can be blocked by bones or teeth, which is why they can be used to make images of bone or tooth structures.

LEARNING FROM LIGHT The most obvious way of learning about a distant object from its light is to use a telescope to take a picture of it. But there are other ways to learn from light. For our purposes in this book, one particular way of learning from light is especially important: **spectroscopy,** which involves collecting light through a telescope, then dispersing it into a *spectrum* in much the same way a prism disperses white light into a rainbow of color (Figure 3.25).

Figure 3.26 shows the three basic types of spectra that we observe: (1) a **continuous spectrum** contains smooth light across a broad range of wavelengths; (2) an **emission line spectrum** has bright lines on a dark background; and (3) an **absorption line spectrum** has dark lines on a continuous background.

As Figure 3.26 shows, a hot object like a light bulb tends to produce a continuous spectrum. In fact, almost any dense object—including stars, planets, and people—emits light with a continuous spectrum that allows us to determine the object's surface temperature, which is why we refer to this type of light as **thermal radiation** (sometimes known as *blackbody* radiation). Figure 3.27 shows thermal radiation spectra for objects of different temperatures. Notice that a star like the Sun emits more strongly in visible light than at any other wavelength, while a typical planet emits infrared light but no visible light at all. This fact allows us to learn a distant object's temperature just by measuring where its thermal radiation spectrum peaks. Moreover, for objects like planets that reflect sunlight, we can learn even more by studying which wavelengths of light are reflected most strongly. For example, the planet Mars not only emits its own infrared light, from which we learn its temperature, but also reflects visible light from the Sun. The fact that Mars reflects red light more strongly than blue light (hence its red color) helps us identify minerals and ices on its surface.

FIGURE 3.25

When we pass white light through a prism, it disperses into a rainbow of color that we call a *spectrum*.

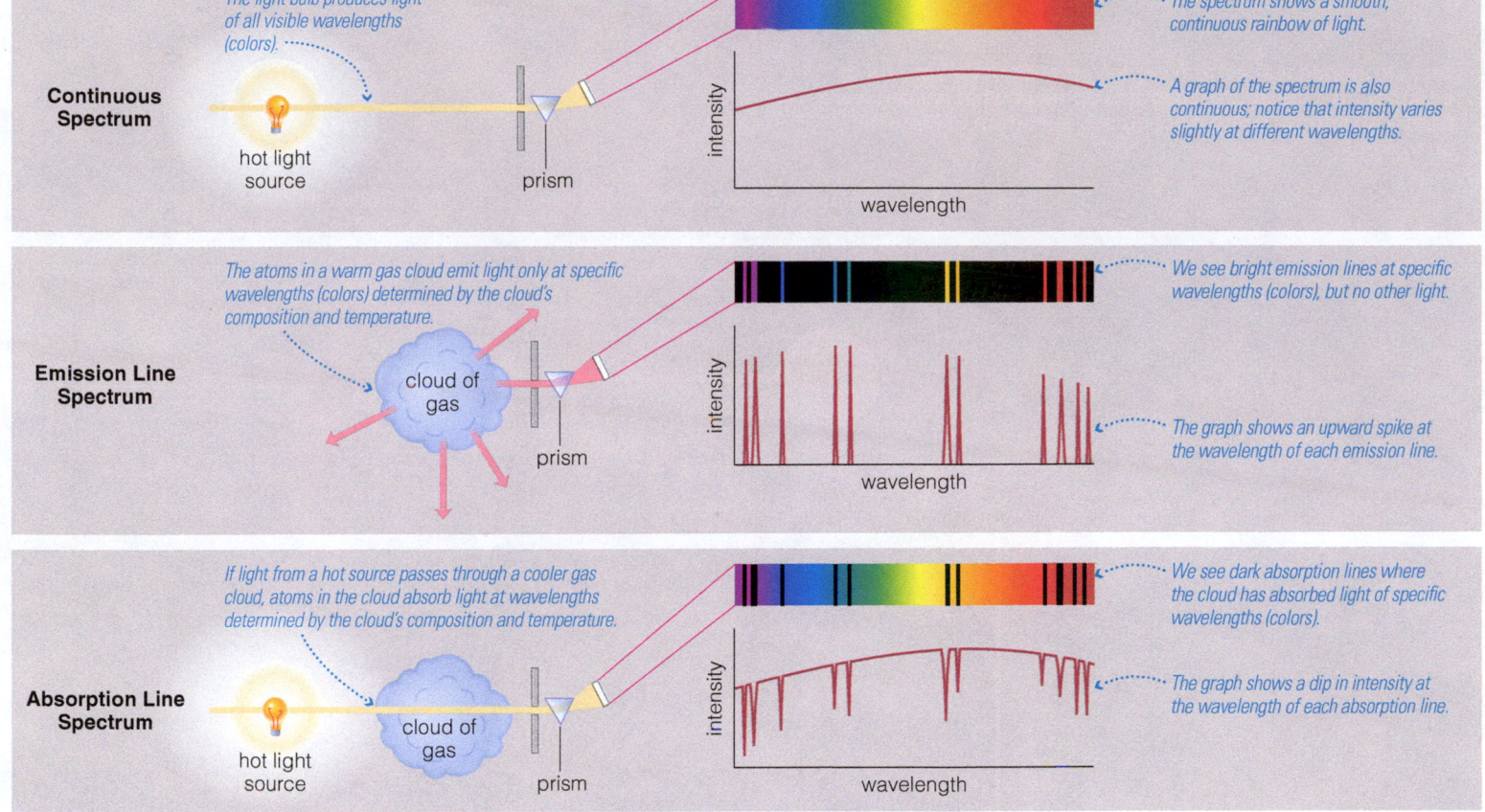

FIGURE 3.26 INTERACTIVE FIGURE
Examples of conditions under which we see the three basic types of spectra.

Spectral lines can provide even more information. Every chemical element, every ion of each element, and every molecule produces its own unique pattern of spectral lines; in essence, this pattern represents a "chemical fingerprint" that allows us to identify what produced it. Therefore, careful study of a spectrum can allow us to determine the chemical composition of distant objects. That is how we learn the chemical compositions of stars, gas clouds, and planetary atmospheres. Different isotopes of an element also have slightly different spectra, so we can sometimes even determine isotopic ratios in distant worlds.

By studying spectral lines in detail—for example, how bright or dark they are, how wide they are, and what precise set of atoms and ions is represented in a spectrum—scientists can infer even more information about distant objects. Perhaps most importantly, the *Doppler effect* (which we'll discuss further in Chapter 11) causes the precise positions of spectral lines to shift with an object's motion relative to us. We can use this effect to determine the speed of any distant object, a fact that has allowed us to discover and measure the masses of many planets around other stars. In addition, careful study of spectral lines can sometimes tell us such things as an object's temperature, rotation rate, pressure, density, and magnetic field strength.

Figure 3.28 (pages 78–79) shows a schematic spectrum of Mars, along with a summary of some of the many things we can learn from the spectrum. Although we won't do a lot with spectroscopy in this book,

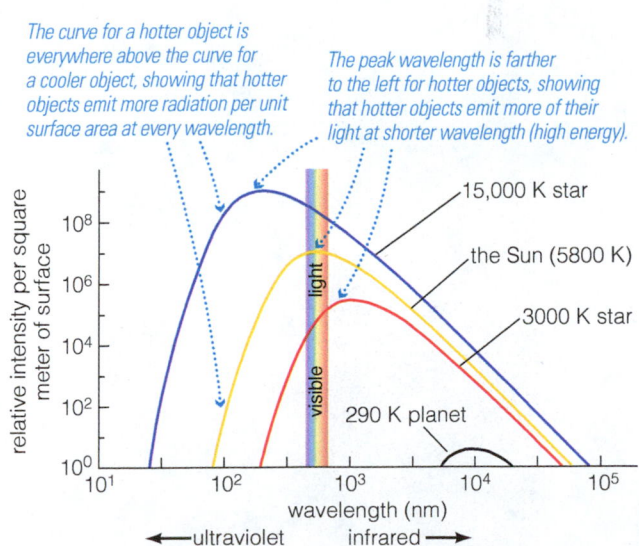

FIGURE 3.27 INTERACTIVE FIGURE
Graphs of idealized thermal radiation spectra demonstrate two laws of thermal radiation: (1) Each square meter of a hotter object's surface emits more light at all wavelengths; and (2) hotter objects emit photons with a higher average energy. Note that both axes of the graph use power-of-10 scales, which allow us to see all the curves even though the differences among them are quite large.

An astronomical spectrum carries an enormous amount of information. This figure illustrates some of what we can learn from a spectrum, using a schematic spectrum of Mars as an example.

① **Continuous Spectrum:** The visible light we see from Mars is actually reflected sunlight. The Sun produces a nearly continuous spectrum of light, which includes the full rainbow of color.

hot light source

prism

Like the Sun, a light bulb produces light of all visible wavelengths (colors).

② **Scattered/Reflected Light:** Mars is red because it absorbs most of the blue light from the Sun but reflects (scatters) most of the red light. This pattern of absorption and reflection helps us learn the chemical composition of the surface.

Like Mars, a red chair looks red because it absorbs blue light and scatters red light.

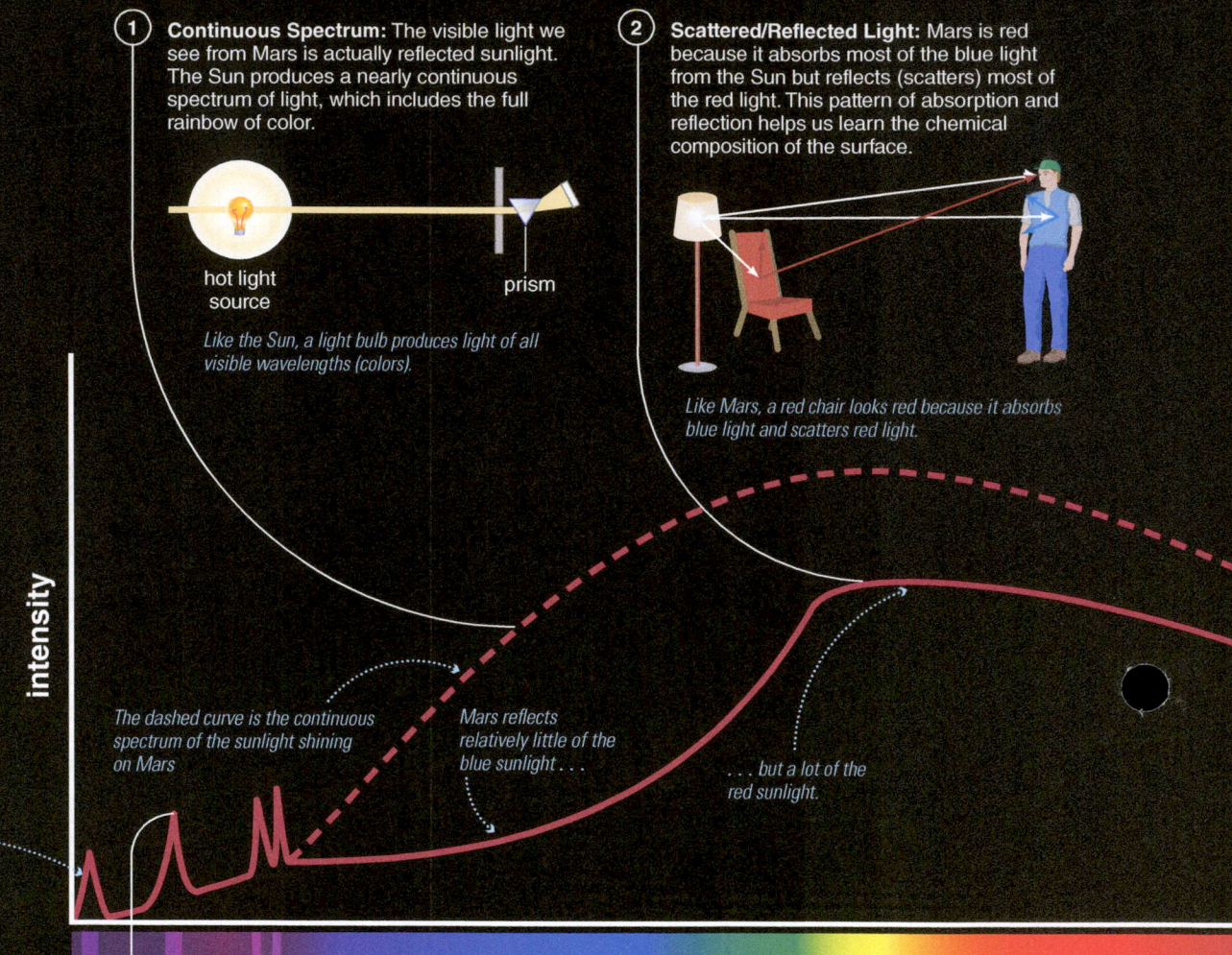

The dashed curve is the continuous spectrum of the sunlight shining on Mars

Mars reflects relatively little of the blue sunlight . . .

. . . but a lot of the red sunlight.

The graph and the "rainbow" contain the same information. The graph makes it easier to read the intensity at each wavelength of light . . .

intensity

. . . while the "rainbow" shows how the spectrum appears to the eye (for visible light) or instruments (for non-visible light).

ultraviolet   blue   green   red

wavelength

④ **Emission Lines:** Ultraviolet emission lines in the spectrum of Mars tell us that the atmosphere of Mars contains hot gas at high altitudes.

cloud of gas

prism

We see bright emission lines from gases in which collisions raise electrons in atoms to higher energy levels. The atoms emit photons at specific wavelengths as the electrons drop to lower energy levels.

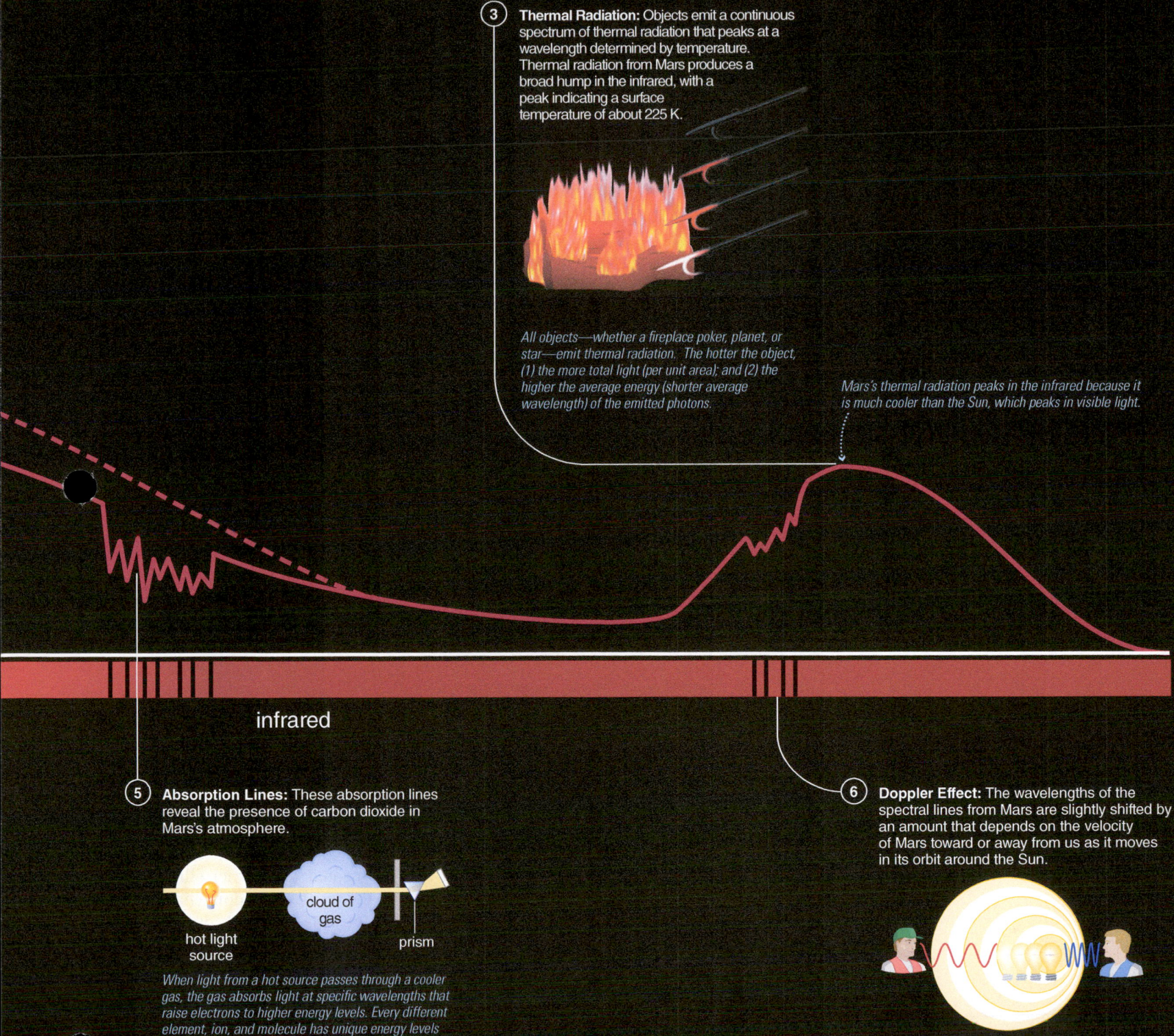

③ Thermal Radiation: Objects emit a continuous spectrum of thermal radiation that peaks at a wavelength determined by temperature. Thermal radiation from Mars produces a broad hump in the infrared, with a peak indicating a surface temperature of about 225 K.

All objects—whether a fireplace poker, planet, or star—emit thermal radiation. The hotter the object, (1) the more total light (per unit area); and (2) the higher the average energy (shorter average wavelength) of the emitted photons.

Mars's thermal radiation peaks in the infrared because it is much cooler than the Sun, which peaks in visible light.

infrared

⑤ Absorption Lines: These absorption lines reveal the presence of carbon dioxide in Mars's atmosphere.

hot light source

cloud of gas

prism

When light from a hot source passes through a cooler gas, the gas absorbs light at specific wavelengths that raise electrons to higher energy levels. Every different element, ion, and molecule has unique energy levels

⑥ Doppler Effect: The wavelengths of the spectral lines from Mars are slightly shifted by an amount that depends on the velocity of Mars toward or away from us as it moves in its orbit around the Sun.

you may find it useful to refer back to this figure whenever you need a review of how we learn from light.

3.4 Our Solar System

So far in this chapter, we have discussed general background about the universe and its basic constituents, but we have not yet talked much about the places where we might hope to find life beyond Earth. Those places are *worlds*, like planets or moons. After all, there does not seem to be any plausible way in which life could get its start in the near-vacuum of interstellar space or in the extremely high temperatures of a star. You might think of this idea as *even aliens need a world to call home*. In this section, we'll focus primarily on understanding our own solar system, though of course this understanding should also apply to planetary systems throughout the universe. Before we begin, you should briefly study Table 3.1, which lists the major features of the planets.

What are the major features of our solar system?

Imagine viewing the solar system from beyond the orbits of the planets. What would we see? Without a telescope, the answer would be "not much." Remember that the Sun and planets are all quite small compared to the distances between them—so small that if we viewed them from the outskirts of our solar system, the planets would be visible only as pinpoints of light, and even the Sun would be just a small bright dot in the sky. But if we magnify the sizes of the planets by about a thousand times compared to their distances from the Sun, and show their orbital paths, we get the central picture in Figure 3.29 (pages 82–83). The surrounding portions of the figure then summarize four major features of our solar system. One of the key questions in the search for life beyond Earth is whether other planetary systems should be expected to be similar to ours, and the answer depends on how the major features of our solar system arose. Therefore, as we discuss the four features in a little more detail, be sure to think about what they might be telling us about the formation of our solar system.

FEATURE 1: PATTERNS OF MOTION The first major feature of our solar system is that it exhibits several clear patterns of motion. For example, all the planets orbit the Sun in nearly the same plane and in the same direction—counterclockwise as viewed from far above Earth's North Pole. This orbital direction is also the same as the direction of the Sun's rotation, the direction of most planet rotations, and the direction in which most large moons orbit their planets. In addition, notice that all the planets have nearly circular orbits, even though Kepler's laws [Section 2.2] could in principle allow for much more elongated (eccentric) ellipses. These orderly patterns are far too clear to be the result of random chance, which means that they must be a result of the way in which our solar system formed.

Think About It You already know that modern science indicates that our solar system formed from the gravitational contraction of an interstellar cloud of gas. Based on the patterns of motion, which of the following must have been the shape of the cloud at the times the planets formed: (a) a sphere; (b) a circular disk; (c) a cigar-shaped bar? Explain your answer.

TABLE 3.1 Planetary Data*

Photo	Planet	Relative Size	Average Distance from Sun (AU)	Average Equatorial Radius (km)	Mass (Earth = 1)	Average Density (g/m3)	Orbital Period	Rotation Period	Axis Tilt	Average Surface (or Cloud-Top) Temperature†	Composition	Known Moons (2015)	Rings?
	Mercury	·	0.387	2440	0.055	5.43	87.9 days	58.6 days	0.0°	700 K (day) 100 K (night)	Rocks, metals	0	No
	Venus	•	0.723	6051	0.82	5.24	225 days	243 days	177.3°	740 K	Rocks, metals	0	No
	Earth	•	1.00	6378	1.00	5.52	1.00 year	23.93 hours	23.5°	290 K	Rocks, metals	1	No
	Mars	·	1.52	3397	0.11	3.93	1.88 years	24.6 hours	25.2°	220 K	Rocks, metals	2	No
	Jupiter	⬤	5.20	71,492	318	1.33	11.9 years	9.93 hours	3.1°	125 K	H, He, hydrogen compounds§	67	Yes
	Saturn	⬤	9.54	60,268	95.2	0.70	29.4 years	10.6 hours	26.7°	95 K	H, He, hydrogen compounds§	62	Yes
	Uranus	●	19.2	25,559	14.5	1.32	83.8 years	17.2 hours	97.9°	60 K	H, He, hydrogen compounds§	27	Yes
	Neptune	●	30.1	24,764	17.1	1.64	165 years	16.1 hours	29.6°	60 K	H, He, hydrogen compounds§	14	Yes

*Appendix E gives additional data, including data for moons.

†Surface temperatures for all objects except Jupiter, Saturn, Uranus, and Neptune, for which cloud-top temperatures are listed.

§Includes water (H_2O), methane (CH_4), and ammonia (NH_3).

The solar system's layout and composition offer four major clues to how it formed. The main illustration below shows the orbits of planets in the solar system from a perspective beyond Neptune, with the planets themselves magnified by about a thousand times relative to their orbits. (The Sun is not shown on the same scale as the planets; it would fill the page if it were.)

① **Large bodies in the solar system have orderly motions.** All planets have nearly circular orbits going in the same direction in nearly the same plane. Most large moons orbit their planets in this same direction, which is also the direction of the Sun's rotation.

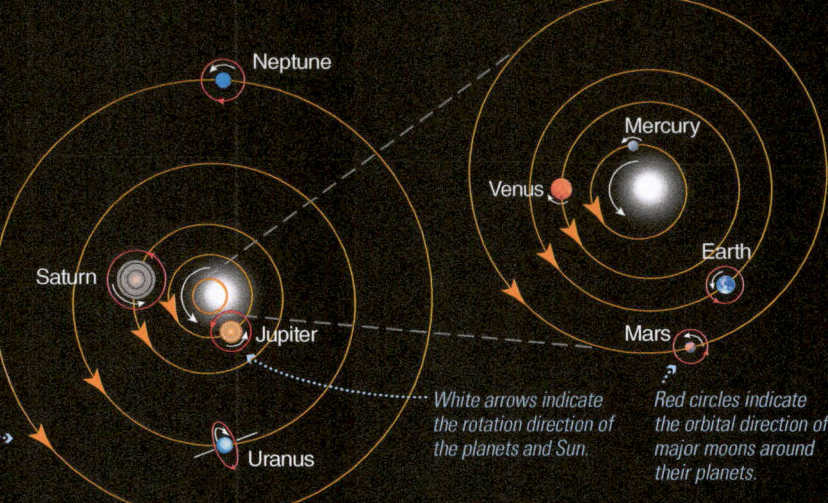

Seen from above, planetary orbits are nearly circular.

White arrows indicate the rotation direction of the planets and Sun.

Red circles indicate the orbital direction of major moons around their planets.

Each planet's axis tilt is shown, with small circling arrows to indicate the direction of the planet's rotation.

Orbits are shown to scale, but planet sizes are exaggerated about 1000 times relative to orbits. The Sun is not shown to scale. (Its size is exaggerated only about 50 times relative to the orbits.)

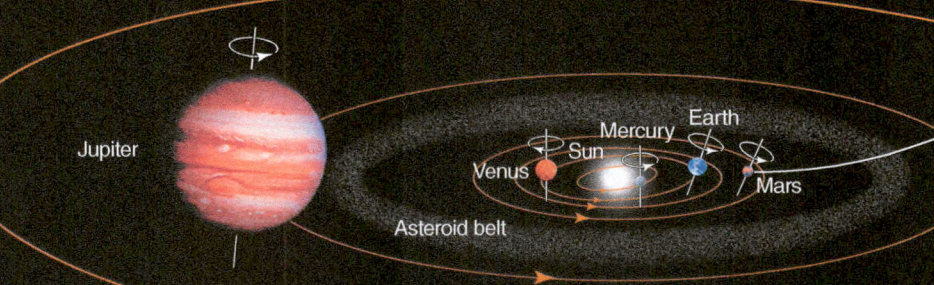

Orange arrows indicate the direction of orbital motion.

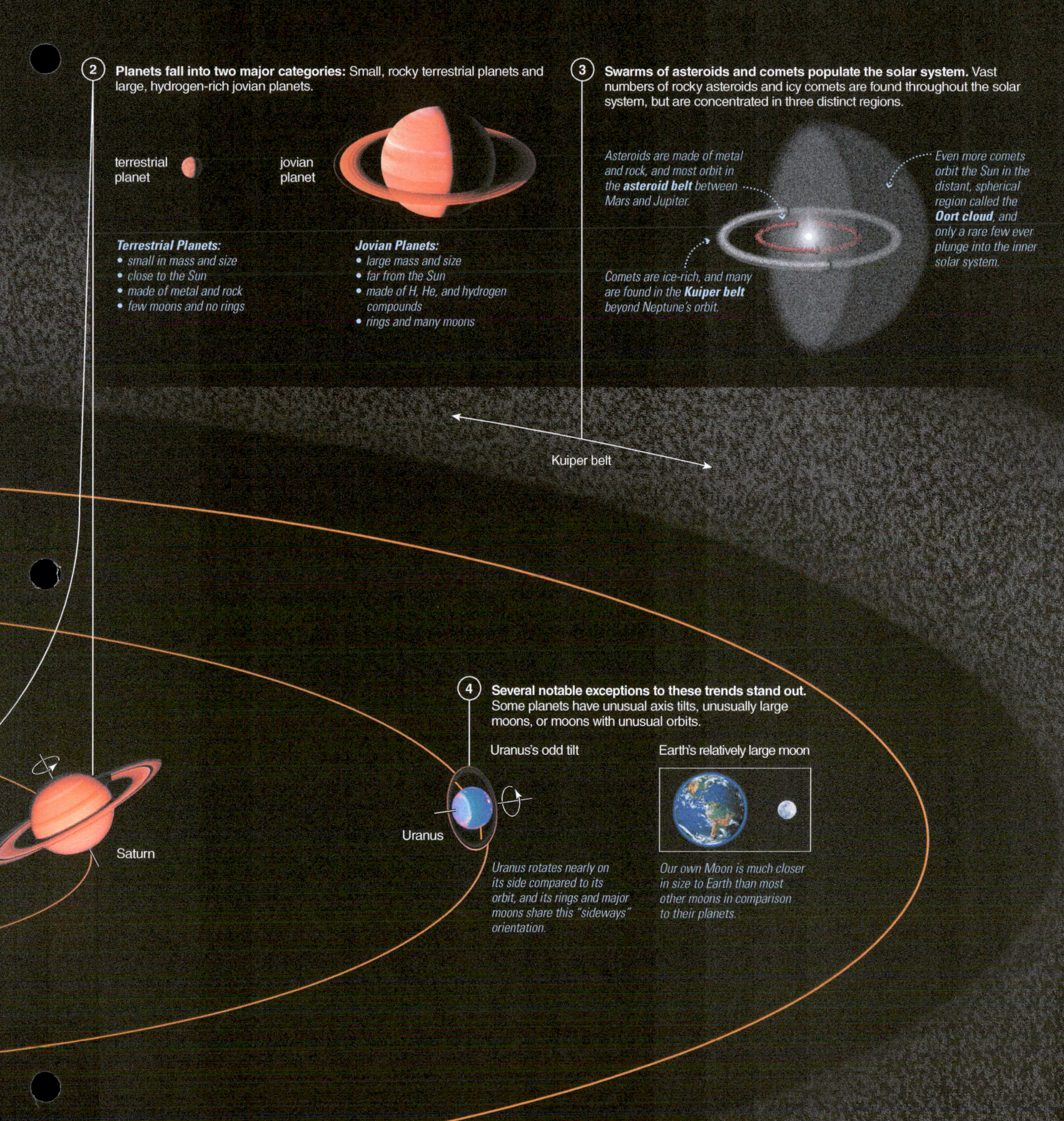

2 **Planets fall into two major categories:** Small, rocky terrestrial planets and large, hydrogen-rich jovian planets.

terrestrial
planet

jovian
planet

Terrestrial Planets:
- *small in mass and size*
- *close to the Sun*
- *made of metal and rock*
- *few moons and no rings*

Jovian Planets:
- *large mass and size*
- *far from the Sun*
- *made of H, He, and hydrogen compounds*
- *rings and many moons*

3 **Swarms of asteroids and comets populate the solar system.** Vast numbers of rocky asteroids and icy comets are found throughout the solar system, but are concentrated in three distinct regions.

*Asteroids are made of metal and rock, and most orbit in the **asteroid belt** between Mars and Jupiter.*

*Comets are ice-rich, and many are found in the **Kuiper belt** beyond Neptune's orbit.*

*Even more comets orbit the Sun in the distant, spherical region called the **Oort cloud**, and only a rare few ever plunge into the inner solar system.*

Kuiper belt

4 **Several notable exceptions to these trends stand out.** Some planets have unusual axis tilts, unusually large moons, or moons with unusual orbits.

Uranus's odd tilt

Earth's relatively large moon

Uranus

Uranus rotates nearly on its side compared to its orbit, and its rings and major moons share this "sideways" orientation.

Our own Moon is much closer in size to Earth than most other moons in comparison to their planets.

Saturn

FEATURE 2: TWO TYPES OF PLANETS The next major feature of our solar system is that the eight planets split clearly into two main groups, one group consisting of the four inner planets and the other consisting of the four outer planets.

The four inner planets—Mercury, Venus, Earth, and Mars—are all relatively small, close to the Sun, and close together. They also share similar compositions, in that all are made almost entirely of metal and rock. This composition gives them fairly high densities (average densities several times that of water) and solid surfaces. Their atmospheres are minor constituents of these planets by mass, though they have important effects on their surfaces (such as creating winds and weather). Because Earth is a member of this group, we refer generally to these rocky worlds as **terrestrial planets** (*terrestrial* means "Earth-like").

The four outer planets—Jupiter, Saturn, Uranus, and Neptune—are much larger, farther from the Sun, and farther apart than the terrestrial planets. We will refer to these worlds as **jovian planets** (*jovian* means "Jupiter-like"), because Jupiter is the largest of the group. The jovian planets are sometimes called "gas giants" because they are made largely of materials that are usually gaseous under earthly conditions: hydrogen, helium, and **hydrogen compounds** such as water (H_2O), methane (CH_4), and ammonia (NH_3). However, except in their outermost layers, the pressure inside these planets is so high that these "gases" are not actually in the gas phase; instead, they are compressed into liquid or other high-density forms that behave somewhat differently than any of the usual phases (gas, liquid, solid) that we experience in our daily lives. As a result, the jovian planets lack solid surfaces; if you plunged into one of them, you would continue your descent until you were crushed by the growing pressure.

The jovian planets differ from the terrestrial planets in another way that is important to the search for life: While the terrestrial planets have only three moons among them (one for Earth and two very small moons for Mars), the jovian planets each have many moons. Most of these moons contain substantial amounts of water ice in addition to rock and metal, and some of the larger moons exhibit active geology suggesting that they may have subsurface regions of liquid water. One moon, Titan (a moon of Saturn), even has a thick atmosphere and surface lakes of liquid methane and ethane. Therefore, even if the jovian planets themselves seem unlikely to harbor life, it is possible that some of their moons might be habitable. For this reason, we'll devote Chapter 9 to discussing the jovian moons in much more depth.

In addition to moons, all four jovian planets are also orbited by vast numbers of small particles that make up their **rings,** although only Saturn's rings are easily visible from Earth. The rings are fascinating in their own right, but are unlikely to play a substantial role in determining whether these planets or any of their moons have life.

Before we move on to the next solar system feature, it's worth noting that while the four jovian planets have much in common, they can also be split into two subgroups to reflect the general differences found when Jupiter and Saturn are compared to Uranus and Neptune. The most obvious difference is in size and mass; as you can see in Table 3.1, Jupiter and Saturn are substantially larger and more massive than Uranus and Neptune (though these are still much larger than any of the terrestrial worlds). In addition, while Jupiter and Saturn resemble the Sun in composition, being made mostly of hydrogen and helium, Uranus and

Neptune contain much higher proportions of hydrogen compounds (and also higher proportions of rock and metal).* These differences suggest that there might really be more than just two basic planet categories (terrestrial and jovian); as we'll discuss in Chapter 11, recent studies of extrasolar planets bear out this idea.

FEATURE 3: ASTEROIDS AND COMETS In addition to the eight planets, the solar system contains vast numbers of smaller bodies that we generally categorize into two groups: *asteroids* and *comets*.

Asteroids are essentially chunks of rock and metal that orbit the Sun but are much smaller in size than the planets. Asteroids come in many different shapes, a fact you can understand by thinking about two facts: (1) Objects become round when their own gravity is strong enough to pull all their material toward a common center, and (2) gravity is stronger for more massive objects. Most asteroids are so small in mass that gravity has been unable to make them round, which means they can have almost any shape (Figure 3.30); some asteroids appear to be little more than piles of rubble weakly held together by gravity. Only a handful of the largest asteroids have enough mass for gravity to have made them at least somewhat round. One asteroid, Ceres (Figure 3.31), is spherical enough to have been designated a *dwarf planet*, which essentially means that it is considered too small to count as one of the official planets but large enough for its own gravity to have made it round. (More technically, the current definition of "planet" requires not only that an object be round but also that it be large enough to have cleared most other objects away from its orbital path. Objects that are round but orbit among many other similar objects—like Ceres, Pluto, and Eris—are designated dwarf planets.)

Astronomers have identified hundreds of thousands of asteroids, but their mass does not add up to much; the combined mass of all asteroids is less than the mass of our Moon. Moreover, despite their large numbers, asteroids are spread out over such a large region of space that they are thousands to millions of kilometers apart on average.

Most asteroids orbit within the **asteroid belt** between the orbits of Mars and Jupiter (see Figure 3.29). However, some have orbits that move them through the inner solar system, which means they can potentially crash into one of the planets. Indeed, such crashes explain the many impact craters visible on planets and moons, and, as we will discuss in Chapter 6, some of these impacts have had profound effects on life on Earth.

Comets are much like asteroids except that they contain large amounts of ice (in addition to rock and metal) and are generally found much farther from the Sun. On the rare occasions when a comet enters the inner solar system, we may see it in the night sky as a fuzzy object with long tails (Figure 3.32); the "fuzziness" and tails arise because some of the comet's ice vaporizes when it is near the Sun, causing particles of dust and ice to be ejected into space. By studying the orbits of the rare comets that we see in the sky, scientists have concluded that comets come

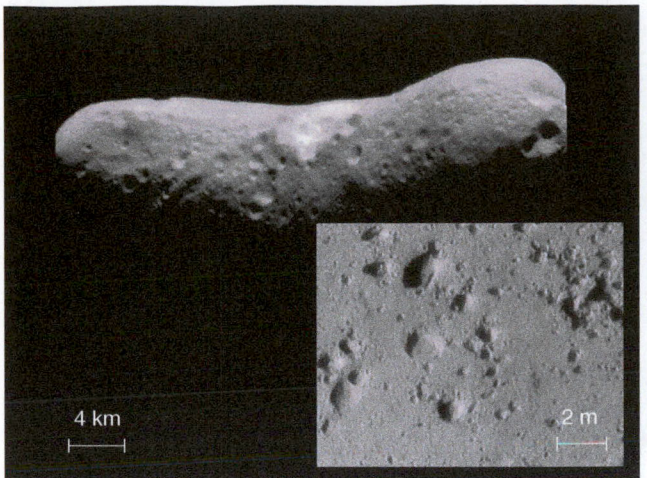

FIGURE 3.30
The asteroid Eros, photographed from the *NEAR* spacecraft, is typical of small asteroids in appearance. The inset shows its surface, on which *NEAR* landed at the end of its mission.

FIGURE 3.31
Ceres, the largest asteroid, photographed by the *Dawn* spacecraft. Ceres is the only asteroid that is fully round; this shape means that, in addition to being an asteroid, Ceres is also designated a *dwarf planet*.

---

*Because the most common hydrogen compounds are water, methane, and ammonia, and these all freeze to make ices at low temperatures, you may hear some planetary scientists refer to Uranus and Neptune as "ice giants." However, under the conditions found in these planets, the hydrogen compounds are not found in the form of ice; except for a few ice particles in clouds, all of these compounds are gaseous in the outer layers and in unfamiliar high-pressure phases deeper down in the interiors.

FIGURE 3.32
Comet McNaught and the Milky Way over Patagonia, Argentina, in 2007. (The fuzzy patches above the comet tail are the Small and Large Magellanic Clouds, which are satellite galaxies of the Milky Way.)

FIGURE 3.33
Pluto, photographed by the *New Horizons* spacecraft, which reached Pluto in 2015 after a 9-year journey. Pluto orbits the Sun within the Kuiper belt, and it is one of the two largest known objects in this region (the other is Eris). Pluto's composition identifies it as an unusually large comet, though it is still much smaller than any of the planets, with a mass 0.2% that of Earth.

from two vast regions (see Figure 3.29): the **Kuiper belt** (Kuiper rhymes with "piper"), a donut-shaped region beyond the orbit of Neptune, and the **Oort cloud** (Oort rhymes with "court"), a roughly spherical region that extends out to tens of thousands of times Earth's distance from the Sun.

Vast numbers of comets populate these two regions. Scientists estimate that there are at least 100,000 comets in the Kuiper belt and perhaps a trillion or more comets in the Oort cloud. A few dozen of the Kuiper belt comets (sometimes called "Kuiper belt objects" or "trans-Neptunian objects") are large enough to qualify as dwarf planets; the two largest of these are Pluto (Figure 3.33) and Eris.

FEATURE 4: EXCEPTIONS TO THE TRENDS The fourth key feature of our solar system is that there are a few notable exceptions to the general trends. For example, while most of the planets rotate in the same direction as they orbit, Uranus rotates nearly on its side and Venus rotates "backward" (opposite the direction of planetary orbits). Similarly, while most large moons orbit in their planet's equatorial plane in the same direction as their planet rotates, many small moons have inclined or backward orbits.

One of the most interesting exceptions is our own Moon. While the other terrestrial planets have either no moons (Mercury and Venus) or very tiny moons (Mars has two small moons), Earth has one of the largest moons in the solar system. As we'll see shortly, the various exceptions actually tell us something very important about the early history of our solar system, and about why we don't expect all planetary systems to be exactly like ours.

How does modern science explain the features of our solar system?

In science, we assume that the major features of our solar system must have arisen as a result of natural processes that occurred as the solar system formed. Today, we have a very successful scientific theory for the formation of the solar system, commonly called the **nebular theory** because it starts with the idea that our solar system was born from the gravitational collapse of an interstellar cloud, or *nebula* (*nebula* is Latin for "cloud"). In the rest of this section, we'll outline the nebular theory and how it explains the four features. We'll then discuss how the nebular model gained its status as a theory, in Section 3.5.

GRAVITATIONAL COLLAPSE OF THE SOLAR NEBULA The particular cloud that gave birth to our own solar system about 4½ billion years ago is usually called the **solar nebula.** Observations of interstellar clouds in which stars are being born today suggest that early in its history the solar nebula must have been a large and roughly spherical cloud of very cold, low-density gas. Initially, this gas was probably so spread out—perhaps over a region a few light-years in diameter—that gravity alone may not have been able to pull it together to start its collapse. Instead, the collapse may have been triggered by a cataclysmic event, such as the impact of a shock wave from the explosion of a nearby star (a supernova).

Once the collapse started, the law of gravity ensured that it would continue. Remember that the strength of gravity follows an inverse square law with distance [Section 2.4]. Because the mass of the cloud remained the same as it shrank, the strength of gravity increased as the diameter of

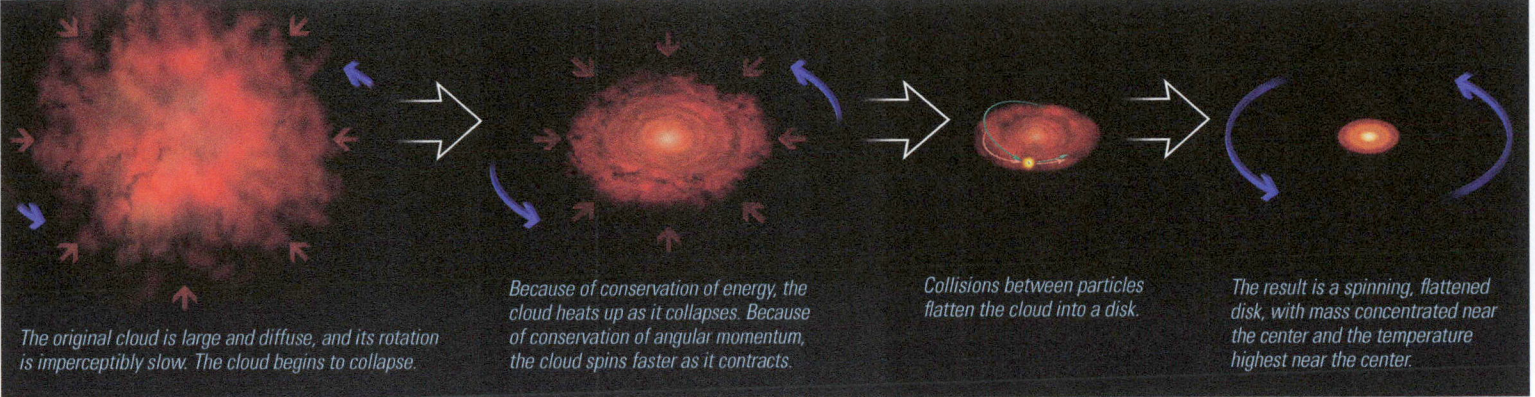

The original cloud is large and diffuse, and its rotation is imperceptibly slow. The cloud begins to collapse.

Because of conservation of energy, the cloud heats up as it collapses. Because of conservation of angular momentum, the cloud spins faster as it contracts.

Collisions between particles flatten the cloud into a disk.

The result is a spinning, flattened disk, with mass concentrated near the center and the temperature highest near the center.

FIGURE 3.34 INTERACTIVE FIGURE
This sequence of paintings shows how the gravitational collapse of a large cloud of gas causes it to become a spinning disk of matter. The hot, dense central bulge becomes a star, while planets can form in the surrounding disk.

the cloud decreased. For example, when the diameter decreased by half, the force of gravity increased by a factor of four.

Gravity pulls inward in all directions, so you might at first guess that the solar nebula would have remained spherical as it shrank. Indeed, the idea that gravity pulls in all directions explains why the Sun and the planets (and dwarf planets) are spherical. However, gravity is not the only physical law that affects a collapsing cloud of gas. The following three processes altered the solar nebula's density, temperature, and shape, changing it from a large spread-out cloud to a much smaller spinning disk; Figure 3.34 summarizes the processes, and Figure 3.35 shows an example of a real disk around a young star.

- *Heating.* The temperature of the solar nebula increased as it collapsed. Such heating represents energy conservation in action. As the cloud shrank, its gravitational potential energy was converted to the kinetic energy of individual gas particles falling inward. These particles crashed into one another, converting the kinetic energy of their inward fall to the random motions of thermal energy. The Sun formed in the center of the cloud, where temperatures and densities were highest.

- *Spinning.* The solar nebula rotated faster and faster as its own gravity made it shrink in radius. This increase in rotation rate represents a law called the **conservation of angular momentum,** which essentially states that the total amount of "circling motion" of an object (or set of objects) must be conserved. We won't go into the details here, but you've probably seen this law in action with ice skating: It explains why a spinning skater's rate of spin increases when she pulls her arms. In much the same way, a shrinking cloud of gas must spin faster as it contracts, as long as it had at least some small rate of spin to begin with. In the case of interstellar clouds, random motions ensure that they almost inevitably have some small overall rotation, though it is often imperceptible when the cloud is large. As the cloud shrinks, its spin becomes very clear.

- *Flattening.* The solar nebula flattened into a disk. This flattening is a natural consequence of collisions between particles in a spinning cloud. A cloud may start with any size or shape, and different clumps of gas and dust particles within the cloud may be moving in random directions at random speeds. These clumps collide and merge as the cloud collapses, and each new clump has the average velocity of the clumps that formed it. Much as in a spinning ball of pizza dough, this averaging of random motions tends to force all

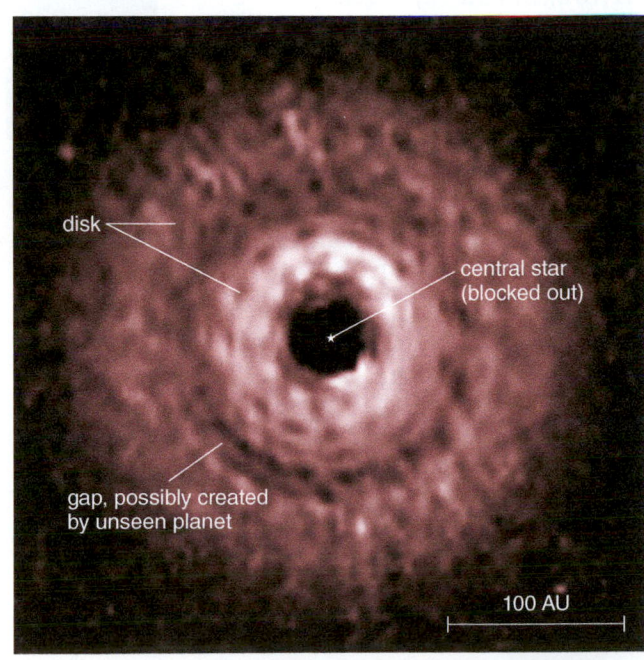

disk

central star (blocked out)

gap, possibly created by unseen planet

100 AU

FIGURE 3.35
This Hubble Space Telescope image shows a flattened, spinning disk around the star TW Hydrae. This particular disk also shows at least one circular "gap" in which material seems to have been cleared away, probably by a planet forming in the disk, which would have a gravitational attraction that would tend to sweep up material along its path. Also see Figure 3.38.

the particles to end up in a flattened, spinning disk in which all the particles have nearly circular orbits.

Note that these ideas about what happened in our solar system some 4½ billion years ago are not just theoretical; we observe the same processes occurring around newly forming stars today. Such stars are almost always surrounded by spinning disks of gas, just as the nebular theory predicts.

EXPLAINING THE PATTERNS OF MOTION We can now see how the nebular theory explains our first major feature of the solar system, which is its orderly motions. The planets all orbit the Sun in nearly the same plane because they formed in a flat disk. The direction in which the disk was spinning became the direction of the Sun's rotation and the orbits of the planets. Computer models show that planets would have tended to rotate in this same direction as they formed—which is why most planets rotate the same way—though the small sizes of planets compared to the entire disk allowed some exceptions to arise. The fact that collisions in the disk tended to make orbits more circular explains why most planets have nearly circular orbits.

Think About It Both the rotating solar nebula and a spinning ball of pizza dough flatten into disks, but the solar nebula shrank in size as it flattened, while a pizza grows in radius as it flattens. Explain why.

EXPLAINING THE TWO TYPES OF PLANETS In the center of the disk, gravity drew together enough material to form the Sun. In the surrounding disk, the gaseous material was too spread out for gravity alone to clump it up. Instead, material had to begin clumping in some other way and to grow in size until gravity could start pulling it together into planets. In essence, planet formation required the presence of "seeds"—solid bits of matter around which gravity could ultimately build planets.

The basic process of seed formation was probably much like that of the formation of snowflakes in clouds on Earth: When the temperature is low enough, some atoms or molecules in a gas may bond together and solidify. The general process in which solid (or liquid) particles form in a gas is called **condensation**—we say that the particles *condense* out of the gas. Different materials condense at different temperatures. Table 3.2 shows that the ingredients of the solar nebula fell into four major categories.

Figure 3.36 shows how the different condensation temperatures of the different materials led directly to the formation of terrestrial planets in the inner solar system and jovian planets in the outer solar system. Follow along with this figure as you read.

Notice that the hydrogen and helium gas that made up 98% of the solar nebula's mass was unable to condense (because these materials do not condense under low-pressure conditions), which means the vast majority of the nebula remained gaseous. However, the other three categories of material could condense wherever the temperature allowed. In the inner solar system, which means closer to the young Sun, temperatures were too high for hydrogen compounds to condense into ices; therefore, the only solid particles that condensed in the inner solar system were particles of rock and metal. Farther out, temperatures were lower, and that allowed hydrogen compounds to condense along with rock and metal; because hydrogen compounds were several times as abundant as rock and metal, these particles tended to be mostly ice.

TABLE 3.2 Materials in the Solar Nebula

A summary of the four types of materials present in the solar nebula. The squares in the final column represent the relative proportions of each type (by mass).

	Examples	Typical Condensation Temperature	Relative Abundance (by mass)
Hydrogen and Helium Gas	hydrogen, helium	do not condense in nebula	98%
Hydrogen Compounds	water (H_2O) methane (CH_4) ammonia (NH_3)	<150 K	1.4%
Rock	various minerals	500–1300 K	0.4%
Metals	iron, nickel, aluminum	1000–1600 K	0.2%

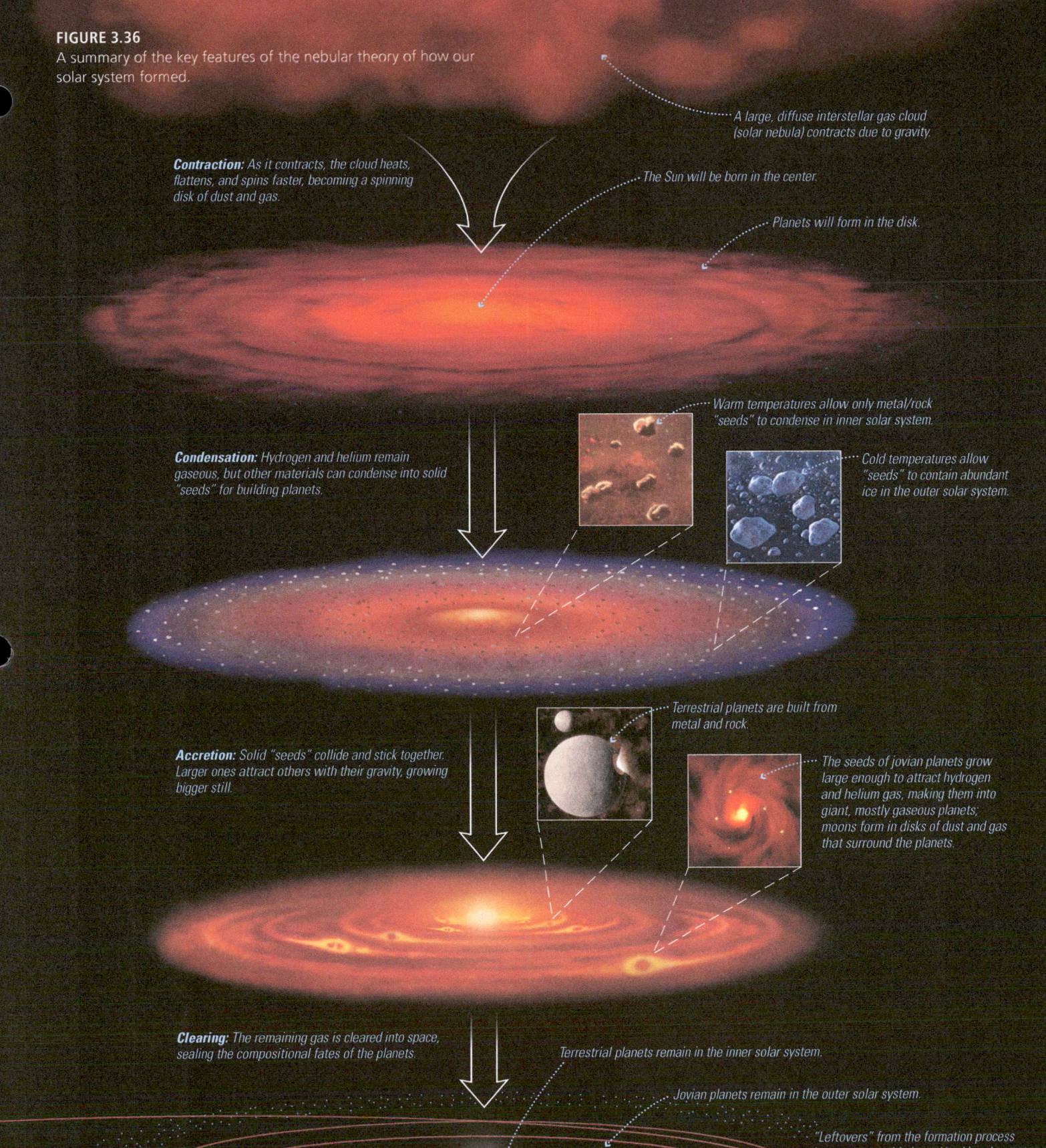

FIGURE 3.36
A summary of the key features of the nebular theory of how our solar system formed.

A large, diffuse interstellar gas cloud (solar nebula) contracts due to gravity.

Contraction: As it contracts, the cloud heats, flattens, and spins faster, becoming a spinning disk of dust and gas.

The Sun will be born in the center.

Planets will form in the disk.

Warm temperatures allow only metal/rock "seeds" to condense in inner solar system.

Condensation: Hydrogen and helium remain gaseous, but other materials can condense into solid "seeds" for building planets.

Cold temperatures allow "seeds" to contain abundant ice in the outer solar system.

Terrestrial planets are built from metal and rock.

Accretion: Solid "seeds" collide and stick together. Larger ones attract others with their gravity, growing bigger still.

The seeds of jovian planets grow large enough to attract hydrogen and helium gas, making them into giant, mostly gaseous planets; moons form in disks of dust and gas that surround the planets.

Clearing: The remaining gas is cleared into space, sealing the compositional fates of the planets.

Terrestrial planets remain in the inner solar system.

Jovian planets remain in the outer solar system.

"Leftovers" from the formation process become asteroids (metal/rock) and comets or Kuiper belt objects (ice/rock).

Not to scale

The first particles to condense were microscopic in size and orbited the Sun with the same orderly, circular paths as the gas from which they condensed. Individual particles therefore moved at nearly the same speed as neighboring particles, so "collisions" were more like gentle touches. Under these circumstances, particles could stick together through electrostatic forces—the same static electricity that makes hair stick to a comb. Small particles thereby began to combine into larger ones. As the particles grew in mass, they began to attract each other through gravity, accelerating their growth. The general process by which particles stick together and grow larger is called **accretion,** and the growing objects are often called **planetesimals,** which means "pieces of planets."

Planetesimals grew rapidly at first, with some probably reaching hundreds of kilometers in size in only a few million years—a long time in human terms, but only about 1/1000 the present age of the solar system. At this point, gravitational encounters between planetesimals began to play an important role. These encounters tended to alter orbits, particularly those of the smaller planetesimals. With different orbits crossing each other, collisions occurred at higher speeds, often shattering the planetesimals. Only the largest planetesimals could avoid shattering. In our solar system, four rocky planetesimals ultimately survived, becoming the four terrestrial planets (Mercury, Venus, Earth, and Mars).

The planet formation process probably began similarly in the outer solar system, except that the condensation of ice (much more abundant than rock and metal) meant that the planetesimals had enough material to grow much larger. A few of the icy planetesimals were able to grow to masses many times that of Earth. With these large masses, their gravity became strong enough to capture and hold some of the hydrogen and helium gas that made up the vast majority of the surrounding solar nebula. The more they grew, the stronger their gravity became. In this way, the nebular theory explains the existence of the four jovian planets: Each began its formation with the accretion of icy planetesimals (made from hydrogen compounds, rock, and metal), but ultimately collected so much gas that it also ended up with a substantial amount of hydrogen and helium. The differences between Jupiter/Saturn and Uranus/Neptune also make sense: Uranus and Neptune formed at a greater distance from the Sun, where the hydrogen and helium gas was more spread out and therefore more difficult to collect. As a result, the overall compositions of Uranus and Neptune remain more similar to those of the icy planetesimals around which they grew—which means compositions dominated by hydrogen compounds (and rock and metal)—while Jupiter and Saturn ended up mostly hydrogen and helium.

The jovian planet formation process also explains why these planets tend to have many moons. The same processes of heating, spinning, and flattening that made the disk of the solar nebula should have also affected the gas drawn by gravity to the young jovian planets. Each jovian planet came to be surrounded by its own disk of gas, spinning in the same direction as the planet rotated. Moons could accrete from icy planetesimals within these disks, and that probably explains the formation of most of the large moons of the jovian planets. The smaller moons likely were captured asteroids or comets. Because objects can be captured only if they are slowed enough to enter into an orbit around a planet, models predict that nearly all of the captures would have happened early in the solar system's history, when the jovian planets were still surrounded by disks of gas that could exert friction to slow down passing asteroids or comets.

The general lack of moons among the terrestrial planets also makes sense: Captures were far less likely since the terrestrial planets were not surrounded by large disks of gas, and there was no place for large moons to accrete. Of course, this leaves a key problem: explaining the existence of Earth's relatively large Moon. As we'll discuss in Chapter 4, the leading hypothesis for our Moon's formation invokes a gigantic collision between Earth and one of the other large planetesimals that must have roamed the solar system early in its history.

EXPLAINING ASTEROIDS AND COMETS You can probably see how the nebular theory accounts for the existence of so many asteroids and comets: They are simply "leftover" planetesimals from the era of planet formation. Asteroids are the rocky leftover planetesimals of the inner solar system, while comets are the icy leftover planetesimals of the outer solar system.

Asteroids tend to reside in the asteroid belt because the influence of Jupiter's gravity "herds" them in a way that makes them less likely to suffer collisions than asteroids in other regions of the solar system. Therefore, while most asteroids in other regions of the inner solar system long ago crashed into one of the planets, those in the asteroid belt had a decent chance of surviving to the present day.

The division of comets into two regions—the Kuiper belt and the Oort cloud—is slightly more difficult to explain, but scientists think they have a good handle on it. The Kuiper belt comets probably reside in the same general region in which they formed. This region, which lies beyond the orbit of Neptune, was relatively low in density. So while none of the planetesimals grew large enough to become a fifth jovian planet, some grew to the size of Pluto and Eris, while hundreds of thousands of smaller comets also survived to the present day.

The Oort cloud comets are thought to have originated in regions where they crossed the orbits of the jovian planets. When one of these comets passed near a jovian planet, it was likely to be flung out to a great distance by the planet's gravity, in much the same way that scientists have taken advantage of Jupiter's gravity to accelerate spacecraft to planets beyond. While it may sound strange for gravity to fling an object away, it's a direct consequence of the law of conservation of energy: When two objects interact through their gravity, their combined energy must remain unchanged, which means that one will lose energy and the other will gain it. In an interaction between Jupiter and a comet, Jupiter's loss (or gain) of energy would be unnoticeable because it is so much larger, while the comet's gain (or loss) would completely change its orbit.

EXPLAINING THE EXCEPTIONS As you look back through our discussion and review Figure 3.36, you'll probably see why we should *expect* at least some exceptions to the general trends. The formation of the solar system involved a lot of big things crashing into each other, and that inevitably means there would be some major accidents that might, for example, tilt a planet onto its side (like Uranus), blast out material that could form a large moon (like our Moon), or lead to a moon being captured with an orbit in the "wrong" direction (as is the case for a few of the jovian moons, including Neptune's moon Triton). In other words, the exceptions we observe actually help support the nebular theory; if there were no exceptions, it would mean something was wrong with our model, because random collisions and close gravitational encounters should inevitably lead to at least a few exceptions.

In addition, when scientists make computer models of the process of planet formation, they find that "near-miss" encounters between two objects will often speed one of them up to a velocity that's great enough to kick it out of the solar system altogether. It could be that our solar system, in its youngest days, had more planets than it has today, but that these were ejected into space to wander, cold and lonely, as planetary orphans.

CLEARING THE NEBULA One key question remains for us to answer: Given that the vast majority of the hydrogen and helium gas in the solar nebula never became part of any planet, what happened to it? Models and observations of other star systems suggest that it was cleared away by a combination of energetic light from the young Sun and the **solar wind**—a stream of charged particles continually blown outward in all directions from the Sun. The solar wind was almost certainly much stronger when the Sun was young than it is today.

Once the gas cleared, the compositional fate of the planets was sealed. If the gas had remained longer, it might have continued to cool until hydrogen compounds condensed into ices even in the inner solar system. In that case, the terrestrial planets might have accreted abundant ice, and perhaps some hydrogen and helium gas as well, changing their basic nature. At the other extreme, if the gas had been blown out much earlier, the raw materials of the planets might have been swept away before the planets could fully form. Although these extreme scenarios did not occur in our solar system, they may sometimes occur around other stars. And that leads us to our final topic for this chapter, which is the ongoing development of the nebular theory as we seek to explain recent discoveries about planetary systems beyond our own.

THE PROCESS OF SCIENCE IN ACTION

3.5 Ongoing Development of the Nebular Theory

We saw in the previous section that the nebular theory successfully accounts for all the major features of *our* solar system. But if this theory is correct, it should apply equally well to other planetary systems, and today we know of thousands of such systems. Because the nebular theory is critically important to the study of life in the universe—after all, it explains how a planet like ours came to exist—we will use its history and ongoing development as this chapter's case study in the process of science in action.

How did the nebular model gain acceptance?

The first step in understanding the ongoing development of the nebular theory of the solar system's formation is to understand how and why this theory gained its current acceptance. Recall from Chapter 2 that scientific progress generally involves someone proposing a model to explain a variety of observations and then putting that model to the test. More than one model may be proposed, leading to competition among models, with one model ultimately winning out. The development of the nebular model occurred in precisely this way.

TWO COMPETING MODELS After the Copernican revolution established Earth as just one planet in our solar system, scientists began to speculate about how our solar system came to be. Around 1755, German philosopher Immanuel Kant proposed that our solar system formed from the gravitational collapse of an interstellar cloud of gas. About 40 years later, French mathematician Pierre-Simon Laplace put forth the same idea independently (he apparently was unaware of Kant's proposal). Because an interstellar cloud is usually called a *nebula*, this idea became known as the *nebular hypothesis*.

Other scientists put forth other ideas. For example, in 1745, or 10 years before Kant's publication, French scientist Georges Buffon suggested that the planets had been born when a massive object (which he guessed to be a comet) collided with the Sun and splashed out debris that coalesced into the planets. This basic idea came to be the leading competitor to the nebular hypothesis. It took almost 200 years for science to collect enough data to allow us to choose between these two general ideas about how our solar system was born. To understand how one model eventually gained acceptance, we must first understand a bit more about how we decide what types of observations a theory must explain.

OBSERVATIONS THAT A THEORY MUST ADDRESS The primary goal of a scientific theory is to explain a broad range of diverse observations in terms of just a few fundamental principles. But what observations should we focus on? You might at first guess that we'd want a theory to be able to explain *everything* about a particular topic, but that is neither realistic nor even useful. For example, we expect any theory of gravity—whether it is Newton's theory, Einstein's theory, or a future theory that is even broader—to explain planetary orbits and Galileo's discovery that mass does not affect an object's rate of fall, but we do *not* expect the theory of gravity to explain why a sheet of paper falls more slowly when it is flat than when it is crumpled up into a small ball, because we know that the paper is also affected by other forces, such as air resistance.

In the case of the solar system, an enormous number of observations might seem potentially relevant, from general characteristics of planetary orbits to the shapes of individual asteroids. Historically, as we've learned more about the solar system, scientists at different times have focused on different sets of observations. At the time of Buffon, Kant, and Laplace, which was before asteroids had been discovered and before we recognized differences between terrestrial and jovian worlds, the focus was almost entirely on explaining the mere existence of planets. Moreover, at that time (and until just a couple of decades ago), we did not have evidence of planets around any other star, so the model only had to explain the planets of our own solar system. As a result, the choice between the competing models was not obvious.

THE FALL AND RISE OF THE NEBULAR MODEL Although Buffon's idea of planets forming from a collision with the Sun always had some supporters, by and large the Kant/Laplace nebular hypothesis was more popular throughout the nineteenth century. By the early twentieth century, however, scientists had found a few aspects of our solar system that the nebular hypothesis did not seem to explain, at least in its original form. In particular, Laplace had proposed a specific mechanism by which he thought the planets formed from rings of gas around the Sun, but as scientists investigated his mechanism in more detail, they found that it did not work.

Some scientists sought to modify the nebular hypothesis by looking for alternative ways to build planets, while others looked for entirely different ideas about how the solar system might have formed. Before too long, a new version of Buffon's old idea began to gain favor. In this new version, instead of a direct collision with the Sun, scientists imagined a *near*-collision between the Sun and another star. According to this *close encounter* hypothesis,* the planets formed from blobs of gas that had been gravitationally pulled out of the Sun during the near-collision.

For several decades, the two models battled almost to a draw. Each had at least some features that seemed to agree well with observations, and there was no conclusive evidence that favored one over the other. However, as scientists studied the models in greater depth, they learned to calculate the consequences of each model more precisely. By the mid-twentieth century, these calculations showed that the close encounter hypothesis could *not* account for either the observed orbital motions of the planets or the neat division of the planets into two categories (terrestrial and jovian).

With this clear failure, the close encounter model rapidly lost favor. Moreover, this failure made scientists take more seriously a second problem with the model: It required a highly improbable event. Given the vast separation between star systems in our region of the galaxy, the chance that any two stars would pass close enough to cause a substantial gravitational disruption is so small that it would be difficult to imagine it happening even in the one case needed to make our own solar system. While low probability alone could not rule out the close encounter hypothesis, it certainly did not help the case for a hypothesis that also failed on other grounds.

Think About It Explain why the close encounter hypothesis predicts that planetary systems should be extremely rare. How do recent observations demonstrate that this prediction does not agree with reality, and hence that we can definitively rule out the close encounter hypothesis?

At the same time that the close encounter hypothesis was losing favor, new ideas about the physics of planet formation led to modifications of the nebular hypothesis. Laplace's mechanism was discarded and replaced by the idea of condensation and accretion, and scientists soon realized that this important modification could indeed allow the nebular model to explain the major features of our solar system. Perhaps even more important, new discoveries about our solar system—such as learning of the existence of the Kuiper belt and learning more about the differing compositions of planets and moons—fit quite well into the nebular model. By the latter decades of the twentieth century, so much evidence had accumulated in favor of the nebular hypothesis that it achieved the status of a scientific theory—the nebular *theory* of our solar system's birth.

IMPACT ON ASTROBIOLOGY The historical competition between the nebular and close encounter models may sound like scientific trivia, but it had a profound effect on attitudes toward astrobiology and the search for life in the universe. The reason is probably clear: If the close encounter

---

*The close-encounter hypothesis is often called the Moulton-Chamberlin hypothesis, after two scientists who proposed it, but variations of their idea were proposed by others, including James Jeans.

hypothesis had been correct, then it would have been likely that no other habitable worlds would exist in our galaxy, and perhaps even in the universe. In that case, any chance of finding life beyond Earth would have been limited to the other worlds in our own solar system. The fact that this model seemed quite plausible for much of the first half of the twentieth century partially explains why scientific interest in extraterrestrial life waned during that period. Once the close encounter model was discarded and the nebular theory gained acceptance, it became immediately clear that other planetary systems were to be *expected*, making the possibility of life on other worlds seem far more reasonable.

How have other planetary systems altered our understanding of the nebular theory?

The nebular theory clearly predicts that other stars should form in the same basic way as our Sun, and therefore that other stars also should be accompanied by planets in many or most cases. In that sense, the discovery of other planetary systems represents a fantastic success for the nebular theory, since it predicted the existence of these systems before they were discovered.

The existence of other planetary systems not only confirms the basic idea of the nebular theory, but also allows us to put the details of the theory to the test. For the most part, the theory has passed these new observational tests with flying colors. However, in a few cases, other planetary systems display characteristics that differ from those of our own solar system, and these discoveries led scientists to conclude that our twentieth-century version of the nebular theory was incomplete. In other words, much as Einstein showed that Newton's theory of gravity was not the entire story of gravity [Section 2.4], we now know that the "old" version of the nebular theory was only part of the full story of how planetary systems form.

NEW CONFIRMATIONS OF THE NEBULAR THEORY Over the past three decades, new and more powerful telescopes have revolutionized our ability to study the universe, and observation of the formation of stars and planets is one of the areas that has advanced the most. Overall, these new observations provide strong confirmation of the nebular theory. For example:

- The nebular theory predicts that new stars should be born from clouds of gas contracting under gravity. We now have high-resolution images of many such clouds, clearly showing new stars being born; one is shown in the photo that opens this chapter (page 49) and another is the Orion Nebula (Figure 3.37).
- The nebular theory also predicts that these forming stars ought to be surrounded by spinning disks of material. Much higher resolution is needed to see these disks than to see the star-forming clouds, but we nevertheless have numerous examples. We've already seen one image of a *protoplanetary disk*—a disk in which planets are likely forming (see Figure 3.35)—and the close-ups in Figure 3.37 show several others among more than 30 that have been identified in the Orion Nebula.
- In some cases, these disks offer evidence that planets are already forming and clearing their orbital paths of debris as they do so. Again, Figure 3.35 showed a hint of this, and Figure 3.38 shows even stronger evidence of this process under way.

FIGURE 3.37
This image shows a portion of the Orion Nebula (see Figure 3.6 to see the full nebula) with close-ups of six protoplanetary disks (nicknamed "proplyds"), each surrounding a young star that is being born.

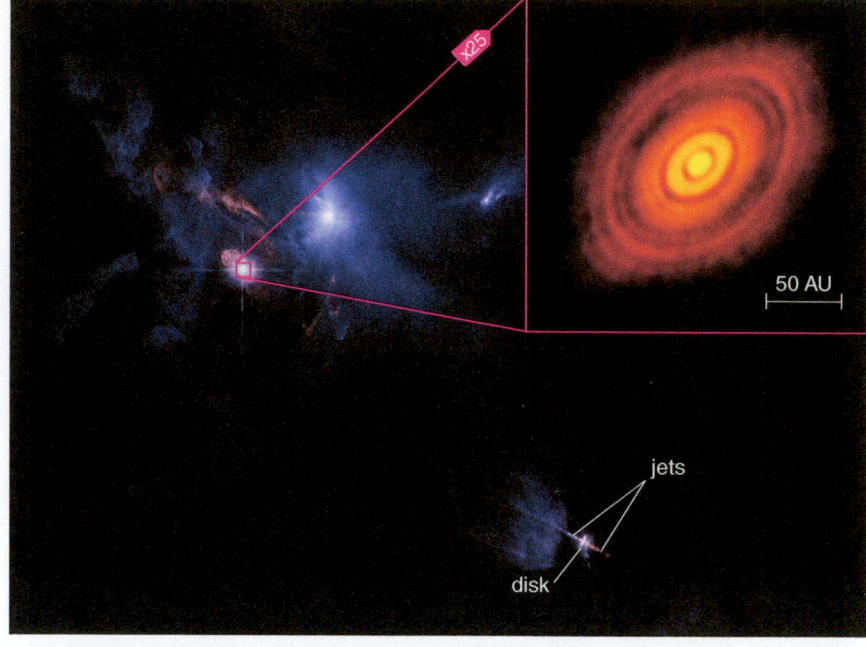

FIGURE 3.38

This is not an artist's conception! The inset shows a real image taken by the Atacama Large Millimeter/submillimeter Array (ALMA). ALMA observes light with wavelengths in the radio portion of the spectrum, which happens to be very useful for peering deep into star-forming clouds. The disk surrounds a star named HL Tauri, and the concentric gaps in the disk are almost certainly regions being cleared as planets form. The background image, from the Hubble Space Telescope, shows the star-forming region in which this disk is located. Another disk, seen edge-on with jets extending outward, appears at lower right.

- And, of course, the actual detection of planets around other stars provides clear confirmation of the nebular theory. In Chapter 11, we'll discuss detection methods and the nature of planets in the context of the search for life. For now, note that current data suggest that a large fraction of all stars are indeed orbited by planets, just as the nebular theory would lead us to expect.

MODIFICATIONS TO THE NEBULAR THEORY In addition to confirming the basic idea of the nebular theory, new observations have also turned up a few surprises. For example, observations of young stars show that even as gas falls inward to make a central star and a rotating disk, huge "jets" of matter may be shot outward along the disk's rotation axis (Figure 3.39). Astronomers still don't fully understand what causes these powerful outflows of matter, though they are not completely unexpected, since similar outflows are found in other astronomical cases in which material swirls in disks (such as from the *accretion disks* that swirl around black holes).

Perhaps a more important surprise is that some planetary systems do not appear to share our solar system's "feature 2"—the clear division into a set of terrestrial planets close to the Sun and jovian planets far from the Sun. Instead, we find some planets that seem to fall in between the terrestrial and jovian categories, and some planetary systems in which jovian planets are found very close to their stars.

With hindsight, the existence of planets with "in between" types is not too surprising. After all, even in our own solar system, we noted some clear differences when we compared Uranus and Neptune to Jupiter and Saturn, even though we lumped both pairs into the same jovian category. We also know that the final composition of the planets depended on when the remaining gas of the solar nebula was cleared into space by the solar wind, so we might expect different types of planets in systems in which the gas clears at different stages of development. (Look ahead to Figure 11.20 to see some of the different planetary types that have now been identified.)

Another surprise was the discovery that some planetary systems have Jupiter-size planets orbiting very close to the stars. This is surprising

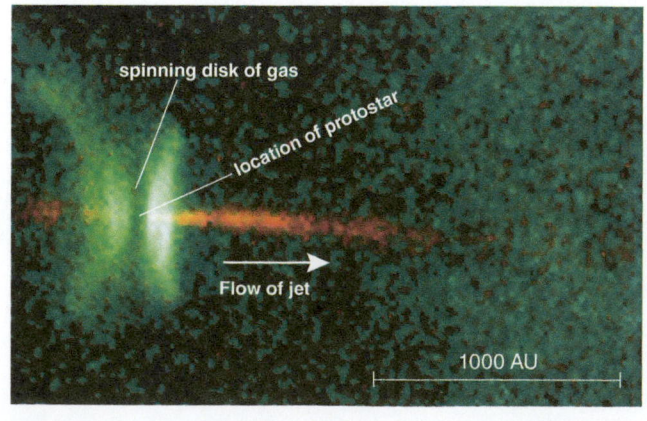

FIGURE 3.39

This image shows jets (red) being shot out along the axis of the disk of gas (green) that surrounds a *protostar*—a star that is still in the process of forming. (The disk is not really split in two as it appears; rather, the central region of the disk is darker and does not show up in this photo.)

because the nebular theory clearly predicts that jovian planets should form only far from their stars, where temperatures are low enough for hydrogen compounds to condense into solid bits of ice. Many of these close-in jovian planets also have fairly eccentric (stretched-out) elliptical orbits, which also seems contrary to the expectation, because the nebular theory predicts that planets should form with nearly circular orbits.

At first, scientists wondered if the surprising orbits of these close-in jovian planets meant that there was something wrong with the nebular model of how planets form. However, after more than a decade of careful study, scientists have concluded that these planets almost certainly formed on circular orbits far from their stars—just as the nebular theory predicts for jovian worlds—but that they then "migrated" inward to their current orbits.

The idea of migration is not as strange as it may sound, and computer models of solar system formation suggest at least one likely mechanism: Migration may be caused by waves passing through a gaseous disk (Figure 3.40). The gravity of a planet moving through a disk can create waves that propagate through the disk, causing material to bunch up as the waves pass by. This "bunched up" matter (in the wave peaks) then exerts a gravitational pull on the planet that tends to reduce its orbital energy, causing the planet to migrate inward toward its star.

This type of migration is not thought to have played a significant role in our own solar system, because the nebular gas was cleared out before it could have much effect. However, planets may form earlier in some other planetary systems, or the nebular gas may be cleared out later, allowing time for jovian planets to migrate substantially inward. In a few cases, the planets may form so early that they end up spiraling all the way into their stars. Indeed, astronomers have noted that some stars have an unusual assortment of elements in their outer layers, suggesting that they may have swallowed planets (including migrating jovian planets and possibly terrestrial planets shepherded inward along with the jovian planets), and at least one recently discovered planet appears to be on a million-year death spiral into its star.

Migration may also occur after the nebula has cleared, as long as small planetesimals are still abundant. Astronomers suspect that this type of migration affected the jovian planets in our own solar system. Recall that the Oort cloud is thought to consist of comets that were ejected outward by gravitational encounters with the jovian planets, especially Jupiter. In that case, the law of conservation of energy demands that Jupiter must have migrated inward, losing the same amount of orbital energy that the comets gained. Models further suggest that Jupiter's inward migration may, in turn, have caused outward migration for Saturn, Uranus, and Neptune.

Related mechanisms may explain the surprisingly high orbital eccentricities of many extrasolar planets. For example, planetary migration increases the chances that planets will influence each other gravitationally. In some cases, planets may pass close enough for a gravitational encounter in which one planet gains enough energy to escape from the star system entirely while the other is flung inward into a highly elliptical orbit. In other cases, continuing gravitational tugs may lead to "orbital resonances" that can cause orbits to become more elliptical than they would be otherwise; we'll discuss this idea further when we discuss an orbital resonance among Jupiter's moons Io, Europa, and Ganymede (see Figure 9.11).

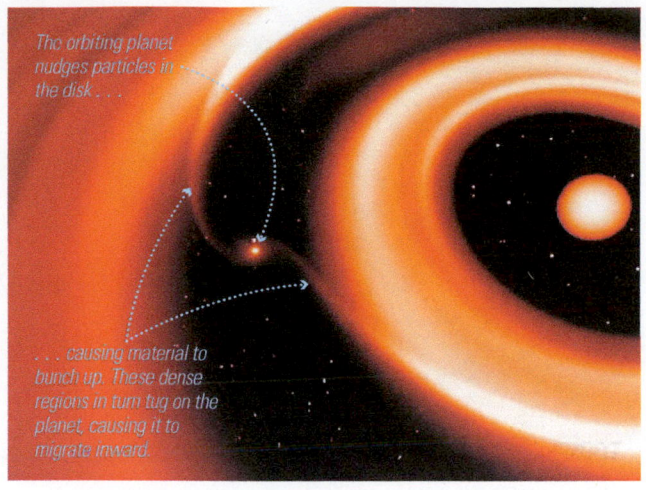

The orbiting planet nudges particles in the disk . . .

. . . causing material to bunch up. These dense regions in turn tug on the planet, causing it to migrate inward.

FIGURE 3.40
This figure shows a simulation of waves created by a planet embedded in a disk of material surrounding its star.

In summary, discoveries of other planetary systems have shown us that the nebular theory was incomplete. The original theory explained the formation of planets and the simple layout of a solar system such as ours, but we've needed to add new features—such as planetary migration—to explain the differing layouts of other solar systems.

IMPLICATIONS FOR LIFE IN THE UNIVERSE Current understanding of the nebular theory offers both good news and bad news for the search for life. On the good news front, confirmation of the prediction that planetary systems are common means that there must be an enormous number of planets. On the bad news front, the fact that many planetary systems have layouts that differ significantly from that of our own solar system may make those systems less likely to have habitable worlds (which we generally expect to be terrestrial in nature). The reason is disruption caused by migration: The same computer models that explain how jovian planets can migrate inward also indicate that this migration can disrupt the orbits of terrestrial planets, causing them either to be ejected into interstellar space or to end up with orbits that may not be conducive to life.

Of course, we must interpret this "good news, bad news" with an eye toward overall numbers. Even if planetary systems with layouts like ours are required for habitability, such systems may be very large in number even if they are relatively rare. For example, suppose that only 1 in 100 other planetary systems is like ours. With our Milky Way Galaxy likely to have 100 billion planetary systems, this would still mean 1 billion systems like ours. The bottom line is that, based on the strong evidence for the nebular theory, it seems almost inevitable that our galaxy must contain many worlds that should be at least potentially habitable.

The Big Picture

PUTTING CHAPTER 3 IN PERSPECTIVE

In this chapter, we have explored the universal context in which we conduct the search for life in the universe. As you continue in your studies, keep in mind the following "big picture" ideas:

- We are not the center of the universe, and we have no reason to think that any special circumstances contributed to the making of our solar system, Earth, or life. This simple idea is one of the major reasons why it seems reasonable to imagine life beyond Earth.

- We are "star stuff" in that we are made of elements that were manufactured in stars. The same elements are available to make planets and life throughout the universe, and while we do not yet know if life exists elsewhere, we know that the necessary raw materials are available.

- Today, we seem to understand the formation of our solar system quite well, and the theory that explains this formation successfully applies to other planetary systems too, giving us reason to think that an enormous number of habitable worlds should exist.

- Galaxies, stars, planets, and life all are ultimately a result of interactions between matter and energy. It is therefore important to have at least some understanding of matter and energy if we wish to understand the processes that make life possible, whether on Earth or beyond.

Summary of Key Concepts

3.1 The Universe and Life

What major lessons does modern astronomy teach us about our place in the universe?
Three major lessons of modern astronomy are (1) the universe is vast and old; (2) the elements of life are widespread; and (3) the same physical laws that operate on Earth operate throughout the universe.

3.2 The Structure, Scale, and History of the Universe

What does modern science tell us about the structure of the universe?

We now know that Earth is a planet orbiting a rather ordinary star among the more than 100 billion stars in the **Milky Way Galaxy,** which in turn is just one among billions of galaxies. The scale of the universe is truly astronomical: If we imagine the Sun the size of a grapefruit, Earth is the ball point from a pen about 15 meters away, while the nearest stars are thousands of kilometers away, and this is still just a tiny part of the cosmic distance scale.

What does modern science tell us about the history of the universe?

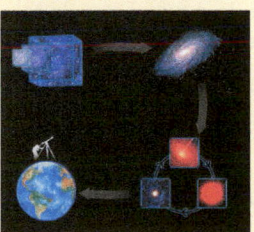

The universe began about 14 billion years ago in the **Big Bang.** It has been expanding ever since, except in localized regions where gravity has caused matter to collapse into galaxies and stars. The Big Bang essentially produced only two chemical elements: hydrogen and helium. The rest have been produced by stars and recycled within galaxies from one generation of stars to the next, which is why we are "star stuff." The universe is extremely old: On a cosmic calendar that compresses the history of the universe into 1 year, human civilization is just a few seconds old, and a human lifetime lasts only a fraction of a second.

How big is the universe?
Because light takes time to travel the vast distances across space, the age of the universe limits how far we can see. For a universe that is 14 billion years old, our **observable universe** extends to a distance of 14 billion light-years. This is extremely large: The total number of stars in all the galaxies of the observable universe is comparable to the number of grains of dry sand on all Earth's beaches combined.

3.3 A Universe of Matter and Energy

What are the building blocks of matter?
Ordinary matter is made of **atoms,** which are made of **protons, neutrons,** and **electrons.** Atoms of different **chemical elements** have different numbers of protons. **Isotopes** of a particular chemical element all have the same number of protons but different numbers of neutrons. **Molecules** are made from two or more atoms. The appearance of matter depends on its phase: **solid, liquid,** or **gas.**

What is energy?
Energy makes matter move, and while it comes in many different forms, we can group these forms into three basic categories: energy of motion, or **kinetic energy;** stored energy, or **potential energy;** and energy of light, or **radiative energy.** The law of **conservation of energy** tells us that energy can change its form but can never be created or destroyed.

What is light?

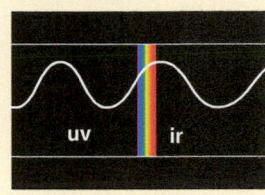

Light is an **electromagnetic wave,** but it also comes in individual "pieces" called **photons.** Each photon has a precise wavelength, frequency, and energy. In order of decreasing wavelength (or increasing frequency or energy), the forms of light are **radio waves, microwaves, infrared, visible light, ultraviolet, X rays,** and **gamma rays.** Light carries a great deal of information about the objects it comes from, and we can learn most of that through **spectroscopy,** in which we carefully study the makeup of the light.

3.4 Our Solar System

What are the major features of our solar system?
Our solar system has four major features: (1) The Sun, planets, and large moons generally rotate and orbit in a very organized way. (2) The planets divide clearly into two groups: **terrestrial planets** and **jovian planets.** (3) The solar system contains huge numbers of asteroids and comets. (4) There are some notable exceptions to these general patterns.

How does modern science explain the features of our solar system?

Planets formed around solid "seeds" of matter that condensed from the gas of the **solar nebula** and then grew through **accretion.** In the inner solar system, high temperatures allowed only metal and rock to condense, which explains why terrestrial worlds are made of metal and rock. In the outer solar system, cold temperatures allowed more abundant ices to condense along with metal and rock, making some **planetesimals** that grew large enough for their gravity to draw in hydrogen and helium gas, building massive jovian planets. **Asteroids** are the rocky leftover planetesimals of the inner solar system, and **comets** are the icy leftover planetesimals of the outer solar system. Most of the exceptions to the general trends probably arose from collisions or close encounters with leftover planetesimals.

3.5 Ongoing Development of the Nebular Theory

How did the nebular model gain acceptance?

 For almost 200 years, the nebular hypothesis competed with another idea that proposed a collision or near-collision between a massive object and the Sun as the mechanism for planet formation. The nebular model won out only after it was tested in great detail and the competing ideas failed the test of explaining the observed features of our solar system.

How have other planetary systems altered our understanding of the nebular theory?

Discoveries of other planetary systems led scientists to realize that the "old" version of the nebular theory was incomplete. It explained our solar system well, but needed additional features—most notably, planetary migration—to account for our observations of other planetary systems.

Exercises and Problems

MasteringAstronomy® *For instructor-assigned homework and other learning materials, go to MasteringAstronomy®.*

REVIEW QUESTIONS

Short-Answer Questions Based on the Reading

1. List three major ideas of astronomy that help frame the context of the search for life in the universe. Describe each one, along with its importance to astrobiology.

2. Briefly define and describe each of the various levels of structure illustrated in Figure 3.1.

3. Describe the solar system as it looks on the 1-to-10-billion scale used in the text. How far away are other stars on this same scale? How does this model show the difficulty of detecting planets around other stars? What does it tell us about the challenge of interstellar travel?

4. What is a *light-year*? Is it a unit of distance or time? Explain clearly.

5. Briefly describe the scale of the galaxy. How long would it take to count 100 billion stars?

6. What evidence makes scientists think the universe is made mostly of *dark matter* and *dark energy*, and why are these things so mysterious? Are these mysteries likely to have an impact on the question of life in the universe? Explain.

7. What do we mean when we say that the universe is expanding? How does expansion lead to the idea of the Big Bang? Briefly describe the evidence supporting the idea that our universe began with the Big Bang.

8. What do we mean when we say that Earth and life are made from "star stuff"? Explain how this star stuff was made, and briefly describe the evidence supporting this idea.

9. Imagine describing the cosmic calendar to a friend. In your own words, give your friend a feel for how the human race fits into the scale of time.

10. What do we mean by the *observable universe*? How big is it? Answer both in absolute terms (that is, a size in light-years) and by describing a way of putting its vast size into perspective.

11. What do we mean when we say that the universe appears to be "fine-tuned" for life? Briefly describe the possible implications of this idea.

12. Briefly describe the structure of an atom. What determines the atom's *atomic number*? What determines its *atomic mass number*? Under what conditions are two atoms different *isotopes* of the same element? What is a *molecule*?

13. What is the difference between matter in the phases of *solid*, *liquid*, and *gas*? What is *vaporization*? Distinguish between vaporization by *sublimation* and by *evaporation*.

14. Define and give examples of *kinetic energy, radiative energy,* and *potential energy*. What is the law of *conservation of energy*?

15. What are the characteristics of a *photon* of light? List the different forms of light in order from lowest to highest energy, lowest to highest frequency, and shortest to longest wavelength.

16. What is *spectroscopy*, and what can we learn from it?

17. Briefly describe the four major features of our solar system that provide clues to how it formed.

18. Briefly describe the *nebular theory* and how it accounts for each of the four major features.

19. What was the *close encounter hypothesis* for our solar system's formation, and why was it ultimately rejected in favor of the nebular theory? How did this rejection affect our understanding of possibilities for extraterrestrial life?

20. How have recent discoveries led scientists to modify the nebular theory? What implications do these modifications have for the search for life beyond Earth?

TEST YOUR UNDERSTANDING

Does It Seem Reasonable?

Suppose that, some day in the future, you heard the following announcements. (These are *not* real discoveries.) In each case, use what you've learned in this chapter to decide whether the announcement seems reasonable or difficult to believe. Explain clearly; because not all of these have definitive answers, your explanation is more important than your chosen answer.

21. The *Voyager 2* spaceship, launched in 1977, has just crash-landed on a planet orbiting another star.

22. At a middle school talent show, 14-year-old Sam Smally read off the names he had given to each of the 100 billion stars in the Milky Way Galaxy.

23. SETI researchers announced today that if they receive a message from a civilization located on the other side of the Milky Way Galaxy, they plan to respond with a message asking the aliens 20 questions about current mysteries in science.

24. A noted physicist today announced that he has found evidence that gravity operates only on Earth and nowhere else in the universe.

25. Astronomers have discovered a young solar system located in the Orion Nebula, with seven planets orbiting a central star.

26. Astronomers have discovered a galaxy in the far reaches of the observable universe that is moving *toward* us, rather than away from us.

27. Inventor John Johnson has patented a device that he is calling a "perpetual motion machine," which makes energy from nothing.

28. Astronomers announced that they had just found the largest extrasolar planet yet discovered, and it is made of solid gold.

29. Astronomers have discovered another star system that is virtually identical to ours except that it has no asteroids or comets.

30. Using new, powerful telescopes, biologists today announced that they had discovered evidence of complex organic molecules in the atmosphere of an extrasolar planet.

Quick Quiz

Choose the best answer to each of the following. Explain your reasoning with one or more complete sentences.

31. The *Milky Way Galaxy* is (a) another name for our solar system; (b) a small group of stars visible in our night sky; (c) a collection of more than 100 billion stars, of which our Sun is one.

32. If we represent the solar system on a scale that allows you to walk from the Sun to Pluto in a few minutes, then (a) the planets are the size of basketballs and the nearest stars are a few miles away; (b) the planets are marble-size or smaller and the nearest stars are thousands of miles away; (c) the planets are microscopic and the stars are light-years away.

33. A television advertisement claiming that a product is "light-years ahead of its time" does not make sense because (a) it doesn't specify the number of light-years; (b) it uses "light-years" to talk about time, but a light-year is a unit of distance; (c) light-years can only be used to talk about light.

34. When we say the universe is *expanding*, we mean that (a) everything in the universe is growing in size; (b) the average distance between galaxies is growing with time; (c) the universe is getting older.

35. According to observations, the overall chemical composition of our solar system and other similar star systems is approximately (a) 98% hydrogen and helium, 2% all other elements combined; (b) 98% ice, 2% metal and rock; (c) 100% hydrogen and helium.

36. The age of our solar system is about (a) one-third of the age of the universe; (b) three-fourths of the age of the universe; (c) 2 billion years less than the age of the universe.

37. The total number of stars in the observable universe is roughly equivalent to (a) the number of grains of dry sand on all the beaches on Earth; (b) the number of grains of dry sand on Miami Beach; (c) infinity.

38. How many of the planets orbit the Sun in the same direction that Earth does? (a) a few (b) most (c) all

39. Which of the following is *not* a general difference between terrestrial planets and jovian planets? (a) Terrestrial planets are much smaller and less massive than jovian planets. (b) Terrestrial planets are made largely of metal and rock while jovian planets also contain abundant hydrogen compounds such as methane, ammonia, and water. (c) Terrestrial planets have oceans of liquid water and jovian planets do not.

40. Some nitrogen atoms have seven neutrons and some have eight neutrons. These two forms of nitrogen are (a) ions of each other; (b) phases of each other; (c) isotopes of each other.

PROCESS OF SCIENCE

41. *Explaining the Past.* Is it really possible for science to inform us about things that may have happened billions of years ago? To address this question, test the nebular theory against each of the three hallmarks of science discussed in Chapter 2. Be as detailed as possible in explaining whether the theory does or does not exhibit these hallmarks. Use your explanations to decide whether the theory can really tell us about how our solar system formed. Defend your opinion.

42. *A Strange Star System.* Suppose that we discovered a star system with ten planets, in which nine orbit the star in the same direction but one travels in the opposite direction. Would this observation be consistent with what we would expect according to the nebular theory? Do you think this one observation would be enough to make us discard the nebular theory, or would we just seek to revise it? Defend your opinion.

GROUP WORK EXERCISE

43. *A Cold Solar Nebula.* **Roles:** *Scribe* (takes notes on the group's activities), *Proposer* (proposes explanations to the group), *Skeptic* (points out weaknesses in proposed explanations), *Moderator* (leads group discussion and makes sure everyone contributes). **Activity:** In our solar system, the point beyond which hydrogen compounds could condense into ices was located between Mars and Jupiter, but study of other solar systems suggests that our solar system could have turned out differently. Consider a hypothetical scenario in which the solar nebula was not cleared away by the solar wind until the entire disk of gas had cooled to 50 K.
    a. Make a list of materials that will condense at 50 K.
    b. Make a list of ways in which the terrestrial planets might have turned out differently under this alternative formation scenario.
    c. Repeat part b for the jovian planets.
    d. Discuss the likelihood that your predicted changes would match the actual characteristics of this alternative solar system.

e. Come up with additional "what if" scenarios, discussing various other ways in which the planets might have turned out differently.

INVESTIGATE FURTHER

In-Depth Questions to Increase Your Understanding

Short-Answer/Essay Questions

44. *Our Cosmic Origins.* Write one to three paragraphs summarizing why we could not be here if the universe did not contain both stars and galaxies.

45. *Perspective on Space and Time.* Come up with your own idea, different from any given in this chapter, to give perspective to some aspect of space or time, such as the size of our solar system, the Earth–Sun distance, the age of Earth, or the time scale of civilization. Your goal should be to explain the size or time you have chosen, perhaps using an analogy, in a way that will make sense to people who have not studied astronomy. Write up your explanation in the form of a short essay.

46. *Alien Technology.* Some people believe that Earth is regularly visited by aliens who travel here from other star systems. For this to be true, how much more advanced than our own would the space travel technology of the aliens have to be? Write one to two paragraphs to give a sense of the technological difference. (*Hint:* The ideas of scale in this chapter can help you contrast the distance the aliens travel easily with the distances we are now capable of traveling.)

47. *Common Levels of Technology.* In *Star Wars,* aliens from many worlds share approximately the same level of technological development. Does this seem plausible? Explain clearly. (*Hint:* Consider the scale of time and the amount of time for which our own civilization has so far existed.)

48. *Atomic Terminology Practice.*
    a. The most common form of iron has 26 protons and 30 neutrons in its nucleus. State its atomic number, atomic mass number, and number of electrons if it is electrically neutral.
    b. Consider the following three atoms: Atom 1 has seven protons and eight neutrons; atom 2 has eight protons and seven neutrons; atom 3 has eight protons and eight neutrons. Which two are *isotopes* of the same element?
    c. Consider a fluorine atom with nine protons and ten neutrons. What are the atomic number and atomic mass number of this fluorine? Suppose we could add a proton to this fluorine nucleus. Would the result still be fluorine? Explain. What if we added a neutron to the fluorine nucleus?
    d. The most common isotope of uranium is ^{238}U, but the form used in nuclear bombs and nuclear power plants is ^{235}U. Given that uranium has atomic number 92, how many neutrons are in each of these two isotopes of uranium?

49. *Origin of Your Energy.* Suppose you have just thrown a ball, and it is now in mid-flight so that it has energy of motion. Trace back the origin of that energy in as much detail as you can; for example, the ball got its energy from the throwing motion of your arm, but where did your arm get this energy? If possible, trace the energy all the way back to the Big Bang.

50. *Your Microwave Oven.* A *microwave oven* emits microwaves that have just the right wavelength needed to cause energy level changes in water molecules. Use this fact to explain how a microwave oven cooks your food. Why doesn't a microwave oven make a plastic dish get hot? Why do some clay dishes get hot in the microwave? Why do dishes that aren't themselves heated by the microwave oven sometimes still get hot when you heat food on them?

51. *Patterns of Motion.* In one or two paragraphs, explain why the existence of orderly patterns of motion in our solar system should suggest that the Sun and the planets all formed at one time from one cloud of gas, rather than as individual objects at different times.

52. *Two Kinds of Planets.* The jovian planets differ from the terrestrial planets in a variety of ways. Using phrases or sentences that members of your family would understand, explain why the jovian planets differ from the terrestrial planets in each of the following: composition, size, density, distance from the Sun, and number of satellites.

53. *Pluto and Eris.* How does the nebular theory explain the origin of objects like Pluto and Eris? How was their formation similar to that of jovian and terrestrial planets, and how was it different?

54. *Rocks from Other Solar Systems.* Many "leftovers" from planetary formation were likely ejected from our solar system, and the same has presumably happened in other star systems. Given that fact, should we expect to find meteorites, or even entire planets, that come from other star systems? How rare or common would you expect them to be? (Be sure to consider the distances between stars.) Suppose that we *did* find a meteorite identified as a leftover from another stellar system. What could we learn from it?

Quantitative Problems

Be sure to show all calculations clearly and state your final answers in complete sentences.

55. *Distances by Light.* Just as a light-year is the distance that light can travel in 1 year, we define a light-second as the distance that light can travel in 1 second, a light-minute as the distance that light can travel in 1 minute, and so on. Calculate the distance in both kilometers and miles represented by each of the following: (a) 1 light-second; (b) 1 light-minute; (c) 1 light-hour; (d) 1 light-day.

56. *Communication with Mars.* We use radio waves, which travel at the speed of light, to communicate with robotic spacecraft. How long does it take a message to travel from Earth to a spacecraft on Mars when (a) Mars is at its closest distance to Earth; (b) Mars is at its farthest distance from Earth? (*Data:* The distance from Earth to Mars ranges between about 56 and 400 million kilometers.)

57. *Scale of the Solar System.* The real diameters of the Sun and Earth are approximately 1.4 million kilometers and 12,800 kilometers, respectively. The Earth–Sun distance is approximately 150 million kilometers. Calculate the sizes of Earth and the Sun, and the distance between them, on a scale of 1 to 10 billion. Show your work clearly.

58. *Moon to Stars.* How many times greater is the distance to Alpha Centauri (4.4 light-years) than the distance to the Moon? What

does this tell you about the relative difficulty of sending astronauts to other stars compared to sending them to the Moon?

59. *Galaxy Scale.* Consider the 1 to 10^{19} scale on which the disk of the Milky Way Galaxy fits on a football field. On this scale, how far is it from the Sun to Alpha Centauri (real distance: 4.4 light-years)? How big is the Sun itself on this scale? Compare the Sun's size on this scale to the size of a typical atom (real diameter: about 10^{-10} meter).

60. *Counting Stars.* Suppose there are 400 billion stars in the Milky Way Galaxy. How long would it take to count them if you could count continuously at a rate of one per second? Show your work clearly.

61. *Interstellar Travel.* Our fastest current spacecraft travel away from Earth at a speed of roughly 50,000 km/hr. At this speed, how long would it take to travel the 4.4-light-year distance to Alpha Centauri (the nearest star system to our own)? Show your work clearly. (*Hint:* Recall that a light-year is approximately 9.5×10^{12} km.)

62. *Faster Trip.* Suppose you wanted to reach Alpha Centauri in 100 years. (a) How fast would you have to go, in km/hr? (b) How many times faster is the speed you found in (a) than the speeds of our fastest current spacecraft (around 50,000 km/hr)?

63. *Planet Probabilities.* Suppose that only one in one million stars is orbited by an Earth-like planet. If there are 100 billion stars in the Milky Way Galaxy, how many Earth-like planets are there in the galaxy? If there are 100 billion galaxies in the observable universe, how many Earth-like planets are there in the observable universe?

64. *What Are the Odds?* The fact that all the planets orbit the Sun in the same direction is cited as support for the nebular hypothesis. Imagine that there's a different hypothesis in which planets can be created orbiting the Sun in either direction. Under this hypothesis, what is the probability that ten planets would end up traveling in the same direction? (*Hint:* It's the same probability as that of flipping a coin ten times and getting ten heads or ten tails.)

Discussion Questions

65. *The Changing Limitations of Science.* In 1835, French philosopher Auguste Comte stated that science would never allow us to learn the composition of stars. Although spectral lines had been seen in the Sun's spectrum by that time, not until the mid-nineteenth century (primarily through the work of Robert Bunsen and Gustav Kirchhoff) did scientists recognize that spectral lines give clear information about chemical composition. Why might our present knowledge have seemed unattainable in 1835? Discuss how new discoveries can change the apparent limitations of science. Today, other questions seem beyond the reach of science, such as the question of *why* there was a Big Bang. Do you think such questions will ever be answerable through science? Defend your opinion.

66. *Lucky to Be Here?* Considering the overall process of solar system formation, do you think it was likely for a planet like Earth to have formed? Could random events in the early history of the solar system have prevented our being here today? What implications do your answers have for the possibility of Earth-like planets around other stars? Defend your opinions.

WEB PROJECTS

67. *Dark Matter and Dark Energy.* Look for recent discoveries that might shed light on the possible nature of dark matter or dark energy. Choose one such discovery, and write a short report on its implications for our understanding of the universe.

68. *Tour of the Solar System.* Visit one of the many websites that give virtual tours of the planets of our solar system. Write a few paragraphs about which planet is your personal favorite, and why.

69. *Star Birth.* Search the Internet for recent images from the Hubble Space Telescope and other telescopes that show young star systems in the process of formation. Choose five to ten favorite images, and create a photojournal with a page for each picture, along with a short description of the picture and what it may tell us about the process of star and planet formation.

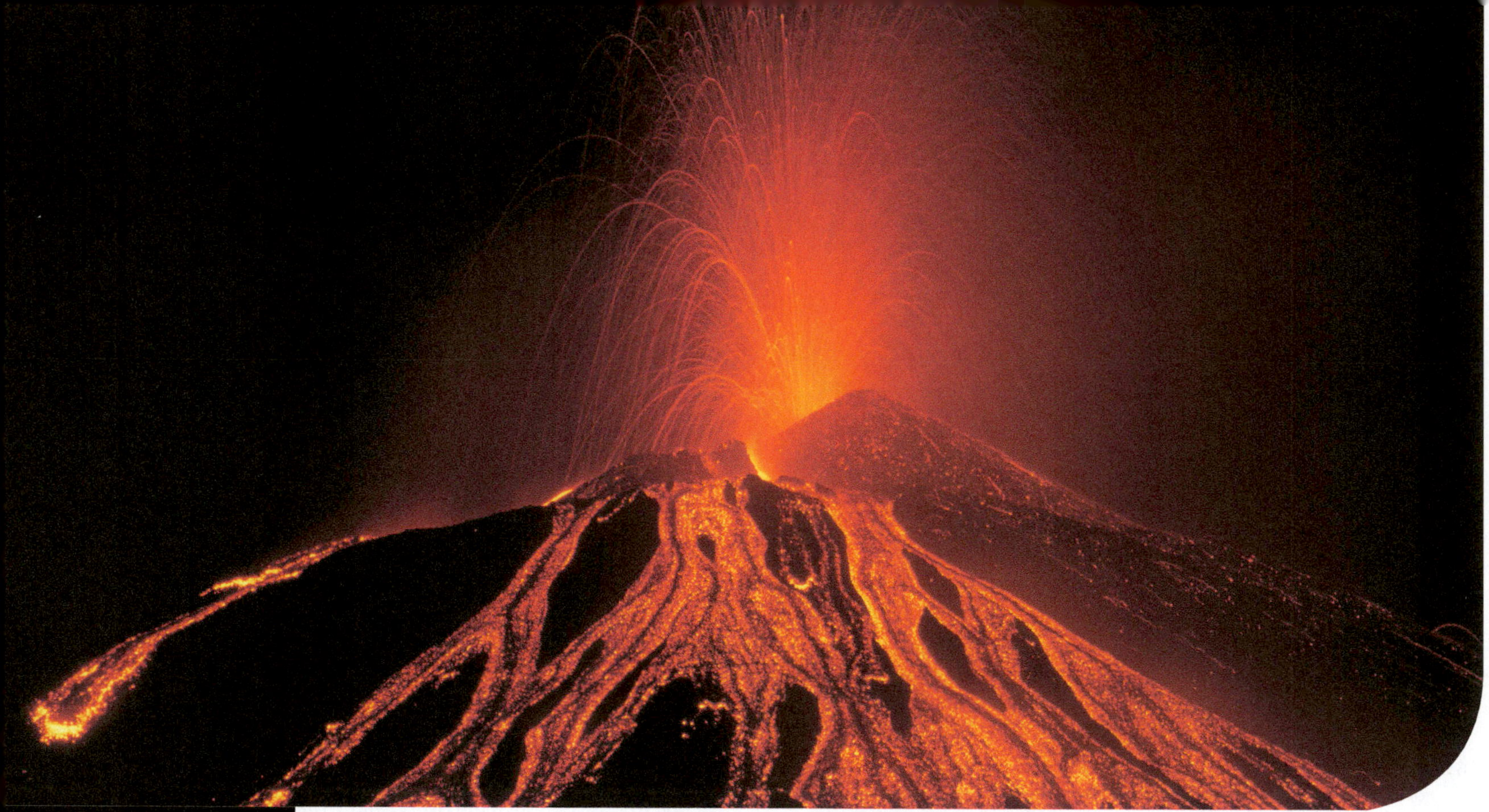

4 The Habitability of Earth

LEARNING GOALS

4.1 GEOLOGY AND LIFE
- How is geology crucial to our existence?

4.2 RECONSTRUCTING THE HISTORY OF EARTH AND LIFE
- What can we learn from rocks and fossils?
- How do we learn the age of a rock or fossil?
- What does the geological record show?

4.3 THE HADEAN EARTH AND THE DAWN OF LIFE
- How did Earth get an atmosphere and oceans?
- Could life have existed during Earth's early history?

4.4 GEOLOGY AND HABITABILITY
- What is Earth like on the inside?
- How does plate tectonics shape Earth's surface?
- Why does Earth have a protective magnetic field?

4.5 CLIMATE REGULATION AND CHANGE
- How does the greenhouse effect make Earth habitable?
- What regulates Earth's climate?
- How does Earth's climate change over long periods of time?

THE PROCESS OF SCIENCE IN ACTION

4.6 FORMATION OF THE MOON
- How did the giant impact model win out over competing models?
- Does the giant impact model count as science?

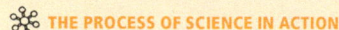

▲ **About the photo:** Volcanic eruptions like this one (Mount Etna in Sicily) play a major role in shaping the surfaces of Earth and other terrestrial worlds.

Some people think that our galaxy's vast number of worlds means that it must be full of life. Others think that life beyond Earth will prove to be rare or nonexistent. But no matter what you think, one fact seems indisputable: So far, we have no convincing evidence for the existence of life anyplace except right here on Earth.

This simple fact means that the scientific search for life in the universe must begin with the study of life on Earth. After all, unless we can understand why life exists here, we have little hope of understanding the prospects for finding life elsewhere. Our goal in this and the next two chapters is to understand why life thrives on our planet.

In this chapter, we will focus on understanding the physical conditions that make our planet habitable. We'll explore the role that geology plays in Earth's habitability, and discuss how, when, and why Earth became a suitable home for life. With this understanding we'll then be ready to turn our attention in Chapters 5 and 6 to life itself.

> In every outthrust headland, in every curving beach, in every grain of sand there is a story of the earth.
>
> *Rachel Carson*

4.1 Geology and Life

It's easy to take for granted the qualities that make Earth so suitable for human life: a moderate temperature, abundant water, a protective atmosphere, and a relatively stable environment. But we need look only as far as the neighboring worlds of the inner solar system—the Moon and the three other terrestrial planets—to see how fortunate we are (Figure 4.1). The Moon and Mercury are airless and barren, with surfaces covered by the craters of past impacts. Venus is a searing hothouse, with surface temperatures higher than that of a pizza oven and surface pressure nearly as great as we would measure a kilometer deep in Earth's oceans. Mars has many features that look almost Earth-like and might indicate a hospitable past, but today Mars has an atmosphere so thin that a visiting astronaut would require a full space suit at all times, and the only surface water is frozen as ice.

Why is Earth so different? Our distance from the Sun is clearly an important factor, but that cannot be the whole story. After all, the Moon is the same distance from the Sun, yet lacks all of the qualities that make Earth habitable. Comparing Earth and the Moon in Figure 4.1 suggests

FIGURE 4.1
The four inner planets and our Moon, together known as the "terrestrial worlds," shown to scale, along with sample close-ups from orbiting spacecraft.

Heavily cratered Mercury has long steep cliffs (arrow).

Cloud-penetrating radar revealed this twin-peaked volcano on Venus.

A portion of Earth's surface as it appears without clouds.

The Moon's surface is heavily cratered in most places.

Mars has features that look like dry riverbeds; note the impact craters.

that Earth's much larger size is important, but that can't be the whole story either, since hothouse Venus is only slightly smaller than Earth.

Scientists suspect that Earth owes its habitability primarily to a combination of its size and distance from the Sun, which together have shaped its geology and atmosphere. Because our atmosphere originated through geological processes, let's begin by looking at the surprising ways in which Earth's geology has made our existence possible.

How is geology crucial to our existence?

Geology is a word with multiple meanings. Taken literally, it is the study of Earth (because *geo* means "Earth"), but scientists commonly extend the meaning to encompass the study of any world with a solid surface. We also use the word *geology* to describe the processes and features that shape worlds; for example, when we speak of Earth's geology, we can mean anything from the composition of our planet to the volcanoes and other processes that rework the surface.

Although geology helps shape local environments, we can generally ignore geological interactions with life for relatively short time scales, such as decades or centuries. After all, only rarely is life affected directly by something like a volcanic eruption. However, over longer time scales, geology and life are deeply intertwined. In fact, it is Earth's geology that has made our planet habitable, ultimately allowing not only the existence of life but also the long-term evolution of life into complex forms that include us.

Geology is important to life on Earth in many ways, but three aspects of Earth's geology stand out as being especially important:

- **Volcanism.** A volcanic eruption can be a spectacular sight, but volcanoes are important to our existence on a much deeper level: Volcanic activity releases gases that became trapped in Earth's interior as our planet formed, and these gases were the original source of Earth's atmosphere and oceans. In addition, volcanism releases heat and creates chemical environments that, we suspect, helped lead to the origin of life on our planet.
- **Plate tectonics.** Earth's surface has been shaped largely by the movement and recycling of rock between the surface and the interior. This process, called *plate tectonics,* is best known for gradually rearranging the continents, but its most profound relevance to life involves Earth's climate: According to modern understanding, plate tectonics is largely responsible for the long-term climate stability that has allowed life to evolve and thrive for some 4 billion years.
- **Earth's magnetic field.** Compass needles point north because our planet has a global magnetic field generated deep in its interior. You may know that the magnetic field has at least a few biological effects—for example, some birds use the magnetic field to help guide their migrations—but its deeper significance is to our atmosphere. The magnetic field shields Earth's atmosphere from the energetic particles of the solar wind [Section 3.4], and without this shielding, it's likely that a significant portion of our planet's atmosphere would by now have been stripped away into space.

Because these factors seem so important to life on Earth, we'd like to understand the likelihood of finding them on other worlds. To do so, we must understand how they work and how they came about, a task to which we'll devote most of this chapter. First, however, it's important to understand how we learn about these and other processes that shape our planet, which ultimately comes down to methods for reconstructing Earth's history from the clues we find in rocks and fossils.

4.2 Reconstructing the History of Earth and Life

Human recorded history dates back only a few thousand years on a planet that has existed for about $4\frac{1}{2}$ billion years. To put this fact in perspective, imagine making a time line to represent Earth's history. On a time line the length of a football field, human civilization would occupy a sliver no thicker than a piece of paper at the end. How, then, can we possibly know anything about the long history that preceded human civilization?

The answer is that this history is recorded in rocks and **fossils**—relics of organisms that lived and died long ago—that preserve clues we can unravel to learn about the past. Reading this history is not as easy as reading words on a page, but with proper scientific tools it can be reliably deciphered. Our task in this section is to explore a few of the key scientific ideas that have helped us put together a chronology of geology and life on our planet.

What can we learn from rocks and fossils?

Recall that matter is found in three basic phases: solid, liquid, and gas. The atoms and molecules of liquids or gases are in constant motion, remixing so rapidly that we can't possibly learn much about how they were arranged in the past. But solid objects offer a different story: Because atoms and molecules are essentially locked in place in a solid, they preserve information about the time at which they first became locked together—that is, at the time the object solidified.

Many solid objects preserve past records, including ancient bones and archaeological artifacts such as pieces of pottery or cloth. But when we seek to learn about Earth's history, we must look to the rocks and fossils that make up what we call the **geological record.** (Some people use the term *fossil record* synonymously, but the latter technically refers only to relics of life.) To see how we read this record, let's begin by discussing how we classify the rocks that preserve past history.

TYPES OF ROCKS Geologists classify rocks into three basic types according to how they are made (Figure 4.2):

- **Igneous rock** is made from molten rock that cools and solidifies.
- **Metamorphic rock** is rock that has been structurally or chemically transformed by high pressure or heat that was not quite high enough to melt it.
- **Sedimentary rock** is made by the gradual compression of *sediments,* such as sand and silt at the bottoms of seas and swamps.

Note that rock can change from one type to another. For example, an igneous rock may be transformed by high pressure or heat into a metamorphic rock, and both igneous and metamorphic rock may be eroded into sediments and become part of a sedimentary rock. Sedimentary rock may then be carried deep underground, where it can melt and then resolidify as igneous rock. In fact, each of the three rock types can be transformed into the others, an idea often described as the *rock cycle* (Figure 4.3).

a This solidified lava is an example of igneous rock, found at the UNESCO World Heritage Site Giant's Causeway (County Antrim, United Kingdom).

b Metamorphic rock has gone through transformations that often give it a contorted appearance. This formation lies on the crest of Big Black Balsam Knob, in the Blue Ridge Mountains of North Carolina.

c Sedimentary rock tends to build up in layers like those visible in "The Wave" at Vermilion Cliffs National Monument in Arizona.

FIGURE 4.2

Samples of the three basic rock types. (All three photos are on approximately the same scale.)

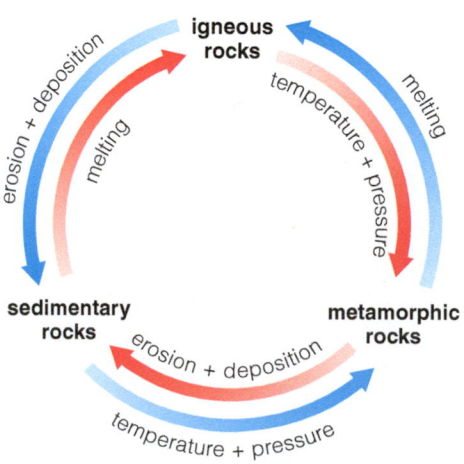

FIGURE 4.3

The rock cycle describes how rocks can change from one to another of the three basic types. Not shown are loops that transform rock within the individual types; for example, igneous rocks can melt and then reform as new igneous rocks.

Because rock can be recycled among the three types, a rock's type does not necessarily tell us much about its composition. Individual rocks of any of the three types usually contain a mixture of different crystals in close contact. Each individual crystal represents a **mineral,** which is the word we use to describe a crystal of a particular chemical composition and structure. Geologists have identified more than 4300 distinct mineral types, but we often group them by their primary constituents. For example, all minerals that contain substantial amounts of silicon and oxygen are called *silicates;* familiar silicates include quartz and feldspar. Similarly, *carbonates,* such as limestone, are minerals containing large amounts of carbon and oxygen.

Overall, a rock's type—igneous, metamorphic, or sedimentary—tells us *how* it was made, while its mineral composition tells us *what* it is made of. However, because rocks of any type may contain different mixes of minerals, geologists use many more names to subclassify rocks. For example, two subtypes of igneous rock form much of our planet's crust. **Basalt** is a dark, dense igneous rock that is commonly produced by undersea volcanoes and that is rich in iron and magnesium-based silicate minerals. **Granite,** which is much lighter in color and less dense than basalt, is an igneous rock common in mountain ranges; it gets its name from its grainy appearance and it is composed largely of quartz and feldspar minerals.

SEDIMENTARY STRATA Sedimentary rock is particularly important to our study of Earth's history for two reasons. First, most fossils are found in sedimentary rock. Second, sedimentary rock forms in a way that tends to produce a record of time. The sediments that make sedimentary rock are produced primarily by erosion on land. Wind, water, and ice can all help break up solid rock into small pieces, some smaller than a millimeter across, and these small pieces (or *grains*) comprise sediments. Sediments can be carried away by rivers and deposited on floodplains or in the oceans. Over millions of years, sediments pile up on the seafloor, and the weight of the upper layers compresses underlying layers into rock. Fossils can be made when remains of living organisms are buried along with the sediments. Remains of aquatic organisms may be buried in sediments simply because they settle to the bottom of the sea. Some land organisms form fossils when their remains are swept into bodies of

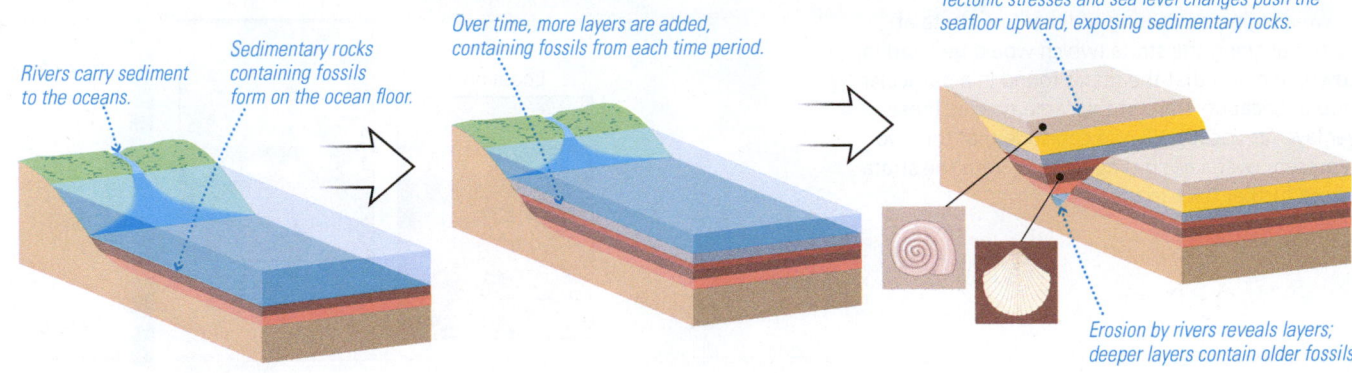

Rivers carry sediment to the oceans.

Sedimentary rocks containing fossils form on the ocean floor.

Over time, more layers are added, containing fossils from each time period.

Tectonic stresses and sea level changes push the seafloor upward, exposing sedimentary rocks.

Erosion by rivers reveals layers; deeper layers contain older fossils.

FIGURE 4.4
These diagrams depict one example of the formation of sedimentary rock. Note that each stratum, or layer, represents a particular time and place in Earth's history and is characterized by fossils of organisms that lived in that place and time. (Adapted from Campbell, Reece, *Biology*.)

water. In other cases, remains of land organisms may be buried in place by windblown silt and later compressed by sediments deposited on top of them when sea levels rise.

Sediments deposited at different times tend to look different as a result of changes in the rate of sedimentation, in the composition or grain size of sediments settling to the bottom, or in the type of organisms leaving fossils. As a result, sedimentary rock tends to be marked by distinct layers, or **strata** (singular, *stratum*). We can view these strata in sedimentary rocks that have been exposed by, for example, the gradual action of a river carving through the rock over millions of years or by a cut made through a mountain to make way for a road. Figure 4.4 summarizes one process by which sedimentary rock forms. Figure 4.5 shows part of the Grand Canyon, which was carved by the Colorado River. While the river itself took "only" millions of years to carve the Grand Canyon, the sedimentary rock walls it exposed contain layers dating from recent history (near the top) to hundreds of millions of years ago (near the bottom).

Because sedimentary rock builds up gradually over time, at any particular location the more deeply buried layers generally are older. This allows geologists to determine the *relative* ages of rocks and fossils buried in sediments. For example, because fossils of dinosaurs appear only in layers older than those in which we find fossils of primates, we conclude that dinosaurs lived before primates.

Sedimentary strata record most (but not all) of Earth's history, but no single location contains a full record. Nevertheless, geologists have put together a fairly detailed geological record by comparing sedimentary strata from many sites around the world. Scientists correlate the strata from different sites by looking for layers with similar fossils. For example, suppose an upper stratum at one location contains fossils of the same type as those found in a lower stratum at another location. In that case, the first location must contain more ancient strata than the second location (Figure 4.6).

ROCK ANALYSIS You can often tell a rock's type from its appearance. For example, a piece of recently solidified lava is obviously igneous, while a rock with an embedded seashell is probably sedimentary. Scientists, too, often start by studying a rock's appearance and considering where it was found, as both offer clues to the rock's origin and history. But if you really want to learn about a rock, you need to examine it in detail in the laboratory.

FIGURE 4.5
The walls of the Grand Canyon are exposed sedimentary rock in which the strata record more than 500 million years of Earth's history.

FIGURE 4.6

In this diagram, we imagine comparing sedimentary strata at two locations. After aligning the strata (which would be tilted in the hillsides shown), we find that the fossils found in a particular layer near the top at Location 1 are of the same type as those found in a lower layer at Location 2. We conclude that the two sets of strata represent overlapping time periods, with the strata at Location 1 going further back in time.

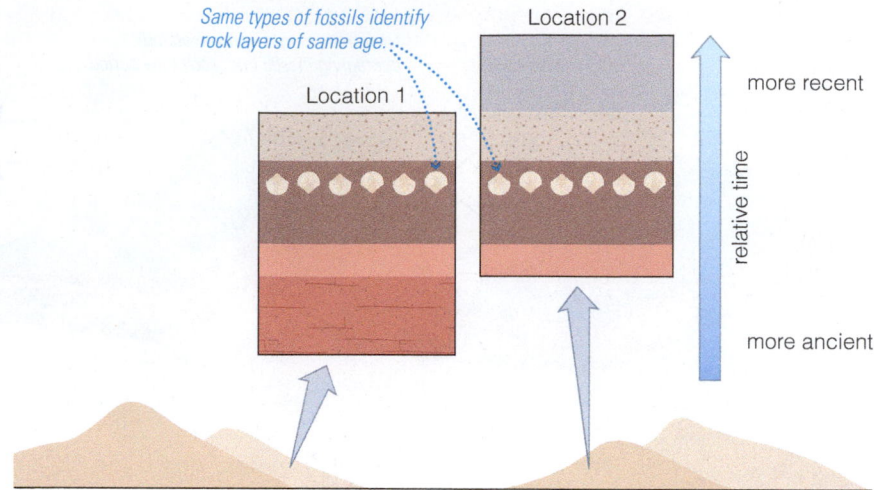

Scientists can analyze rocks in a variety of ways, but three types of analysis are particularly important in reconstructing a rock's history and hence the history of our planet.

- *Mineralogical analysis* generally means identifying the minerals present in a rock.
- *Chemical analysis* generally means determining the elemental or molecular composition of a rock or mineral. For example, chemical analysis will tell you the percentages of a rock that consist of iron, silicon, carbon, or other elements.
- *Isotopic analysis* generally means determining the ratio of different *isotopes* [Section 3.3] of elements in a rock. For example, oxygen has three stable isotopes: Oxygen-16 (eight protons and eight neutrons) is by far the most common, but it is always mixed with small amounts of oxygen-17 (eight protons and nine neutrons) and oxygen-18 (eight protons and ten neutrons); isotopic analysis can tell us the relative amounts of the three oxygen isotopes in a rock.

These three types of analysis are often used in tandem and each can provide clues about a rock's history; these techniques can also be applied to fossils and other solids. The importance of chemical analysis is probably obvious, since it is always useful to know what an object is made of. Mineralogical analysis can tell us about the temperature and pressure conditions under which a rock formed. For example, while graphite and diamond are both minerals made of nearly pure carbon (they differ in their crystal structures), diamond forms only under much higher pressure conditions than graphite.

Isotopic analysis can be particularly illuminating, because measurements show that isotopes tend to exist in particular overall ratios in nature. For example, the overall ratio of oxygen-16 to oxygen-18 in nature is about 2000 to 1. If we find a rock that has more than 1 in 2000 of its oxygen atoms in the form of oxygen-18, we know that something must have happened to cause the rock to become enriched with this heavier oxygen isotope. Even more important, some isotopes turn out to be **radioactive,** meaning that their nuclei are unstable and tend to change over time into other isotopes in a predictable way. This fact means that the ratios of certain radioactive isotopes and their products serve as natural clocks that can allow us to learn precisely *when* a rock (or other solid object) formed.

How do we learn the age of a rock or fossil?

The most reliable method for measuring the age of a rock, fossil, or other solid object is known as **radiometric dating.** This method relies on careful measurement of an object's proportions of various atoms and isotopes.

A **radioactive isotope** has a nucleus that can undergo spontaneous change, or **radioactive decay,** such as breaking apart or having one of its protons turn into a neutron. For example, carbon comes in two stable isotopes: carbon-12 (98.9% of carbon atoms), with six protons and six neutrons, and carbon-13 (1.1% of carbon atoms), with six protons and seven neutrons. This stable carbon is sometimes found mixed with much smaller amounts of the radioactive isotope carbon-14, which has six protons and eight neutrons. Other elements, such as uranium, are always unstable—and therefore radioactive—no matter which isotope we are dealing with. When a radioactive nucleus undergoes decay, we say that the original nucleus (before decay) is the *parent* nucleus (or parent isotope) and the changed nucleus (after decay) is the *daughter* nucleus (or daughter isotope).

Radioactive decay can occur in a variety of ways. Sometimes a large atomic nucleus ejects a helium nucleus, which consists of two protons and two neutrons. (This process is called *alpha decay.*) In that case, the remaining daughter nucleus has a lower atomic mass than its parent nucleus. For example, uranium-238 decays by ejecting a helium nucleus, leaving thorium-234 as its daughter. However, this isotope is not the final daughter of uranium-238, because thorium-234 is also radioactive. Through a chain of individual decays, eight of which involve the emission of a helium nucleus, uranium-238 ultimately decays into lead-206, which is stable and decays no further (Figure 4.7a).

In other cases, radioactive decay occurs when a nucleus spontaneously emits or absorbs an electron, causing one of its neutrons to turn into a proton, or a proton to turn into a neutron. (The processes are called *beta decay* and *electron capture,* respectively; an absorbed electron is one of the atom's own electrons that gets captured by the nucleus.) In these cases, the parent and daughter nuclei will both have the same atomic mass number (the same total number of protons and neutrons), but they will represent different elements because they have different numbers of protons. For example, carbon-14 decays by emitting an electron as one of its neutrons becomes a proton, so its daughter isotope has seven protons and seven neutrons and therefore is nitrogen-14 (Figure 4.7b).

Regardless of the decay process, radioactive decay always occurs at a specific and measurable rate that is different for every radioactive parent–daughter pair. The basic idea behind radiometric dating is to determine the age of a rock (or other solid object) from the ratio of parent and daughter atoms within it, which depends only on the decay rate and the length of time over which the decay has been occurring. To fully understand how this important technique works, we must explore the nature of radioactive decay in a bit more detail.

THE PROBABILISTIC NATURE OF RADIOACTIVE DECAY Radioactive decay is governed by the laws of quantum physics, which means it is a probabilistic process, much like coin tossing. If you toss a single coin, you cannot determine for sure whether it will land heads or tails. All you can say is that it has a 50% chance of landing heads and a 50% chance of landing tails. However, while the probabilistic nature of coin tossing means you

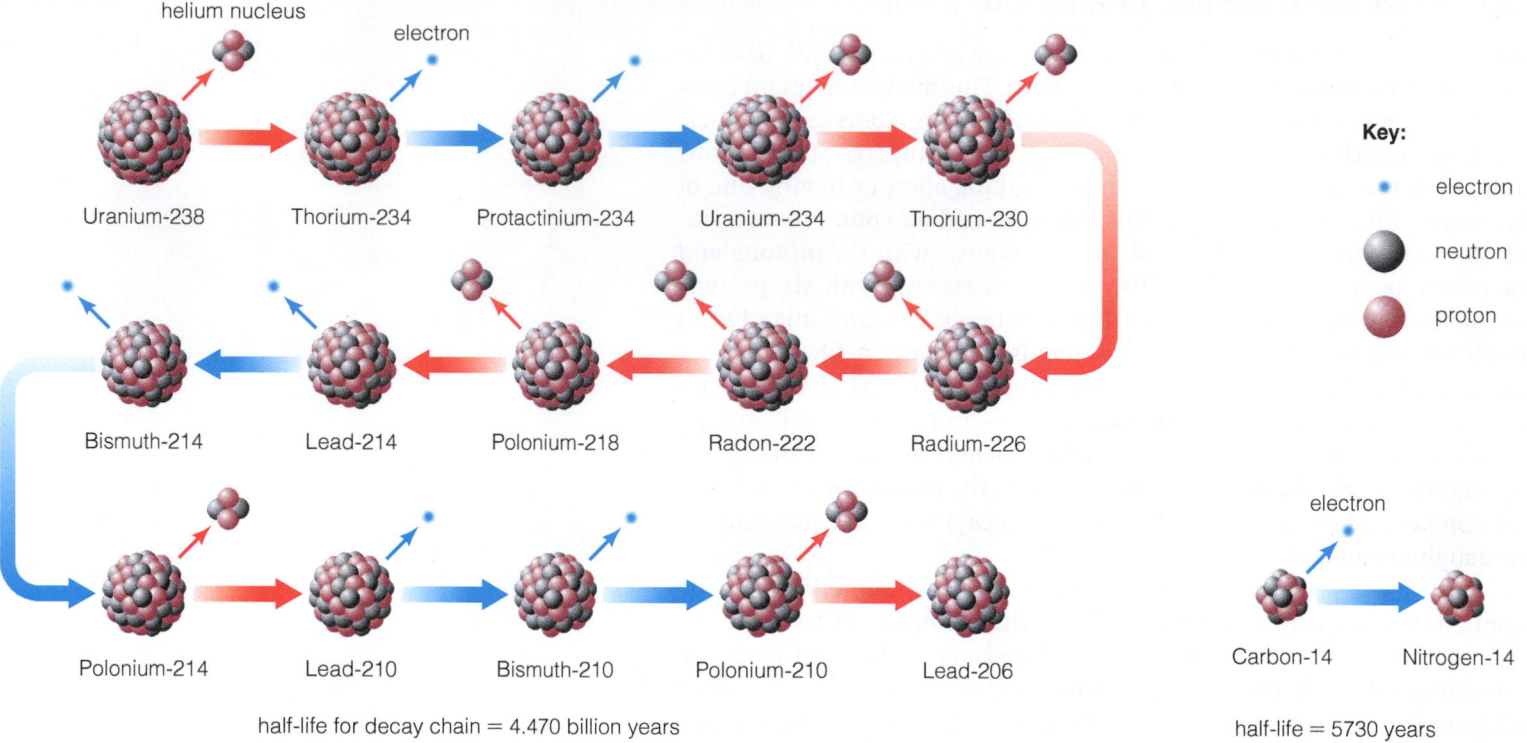

half-life for decay chain = 4.470 billion years

half-life = 5730 years

a Uranium-238 decays through a chain of individual decay processes, eight of which involve the emission of a helium nucleus, ultimately leaving lead-206 as its stable daughter isotope. The half-life for the decay chain as a whole is 4.47 billion years.

b Carbon-14 decays by emitting an electron from its nucleus, which changes one neutron into a proton to make nitrogen-14 as its stable daughter. The half-life of carbon-14 is 5730 years.

FIGURE 4.7
Examples of radioactive decay.

cannot predict the outcome for a single coin toss, it allows you to predict the outcome accurately for many coin tosses. For example, if you toss a fair coin one million times, you can be quite certain that heads will come up close to 500,000 times (but you would not expect the number to be *exactly* 500,000).

Think About It To convince yourself that probability can be used to make predictions involving large numbers of atoms or coins, toss a coin 100 times (or toss 100 coins all at once) and record your result. Compare your result with those of other students; did anyone get a result closer to 0 or 100 than to 50? Discuss what the results tell you about the nature of probability.

In the case of radioactive decay, the relevant probability describes the likelihood of a single nucleus decaying within a specific amount of time (such as within a year). However, because atoms are so tiny, we nearly always deal with such huge numbers of them that we can ignore the individual probabilities and focus instead on the rate at which large numbers of the radioactive nuclei decay.

For example, suppose we study a sample that contains 1 microgram of a radioactive substance, which despite the small weight (about one-millionth the weight of a paper clip) represents trillions of individual atoms. If we find that 1% of the radioactive atoms in the sample undergo radioactive decay within 1 year—meaning that at the end of the year we have 0.99 microgram of the parent substance remaining—we can conclude that any individual atom has a 1% probability of decaying in a 1-year period.

Note that, as with coin tosses, results from the past do not affect results for the future. Just as a coin landing heads on one toss does not change the 50% probability of its landing heads on the next toss, the past history of the radioactive sample does not affect the future probability of decay. In our example, we would find that 1% of the remaining radioactive atoms in the sample decay in any 1-year period, regardless of how we obtained the sample, how many atoms it contains, or how long we have been studying it. We can therefore reliably and reproducibly measure the decay rate for any radioactive substance by studying a sample of the substance in a laboratory.

THE CONCEPT OF A HALF-LIFE Although the probability of decay in a 1-year period is a perfectly good way to describe a substance's decay rate, we usually describe the decay rate by the substance's **half-life**—the time it would take for half the atoms in a sample of the substance to decay. In other words, the amount of radioactive substance in any size sample drops by half with each half-life.

Consider the radioactive decay of potassium-40, which undergoes spontaneous change into argon-40 when its nucleus absorbs an electron to change one of its protons into a neutron. (Potassium-40 also decays by other paths, but we focus only on decay into argon-40 to keep the discussion simple.) Laboratory measurements show that this decay process has a half-life of 1.25 billion years. Suppose a small piece of rock contained 1 microgram of potassium-40 and no argon-40 when it formed (solidified) long ago. The half-life of 1.25 billion years means that half the original potassium-40 had decayed into argon-40 by the time the rock was 1.25 billion years old, so at that time the rock contained $\frac{1}{2}$ microgram of potassium-40 and $\frac{1}{2}$ microgram of argon-40. Half of this remaining potassium-40 had then decayed by the end of the next 1.25 billion years, so after 2×1.25 billion $= 2.5$ billion years, the rock contained $\frac{1}{4}$ microgram of potassium-40 and $\frac{3}{4}$ microgram of argon-40. After three half-lives, or 3×1.25 billion $= 3.75$ billion years, only $\frac{1}{8}$ microgram of potassium-40 remained, while $\frac{7}{8}$ microgram had become argon-40. Figure 4.8 summarizes the gradual decrease in the amount of potassium-40 and the corresponding rise in the amount of argon-40.

Think About It Briefly describe how it is possible for us to know that potassium-40 has a half-life of 1.25 billion years, even though we have been capable of studying potassium-40 in the laboratory for only a few decades.

THE ESSENCE OF RADIOMETRIC DATING The potassium-40 example shows the essence of radiometric dating. Suppose you find a rock (or a mineral within a rock) that contains equal amounts of potassium-40 and argon-40. As long as you can be confident that the rock contained no argon-40 when it formed (which you can, as we'll discuss shortly) and has lost none during its history, you can conclude that the rock is one half-life, or 1.25 billion years, old. If instead the rock contains seven times as much argon-40 as potassium-40—that is, only $\frac{1}{8}$ of the original potassium-40 remains—you can conclude that it is three half-lives, or 3.75 billion years, old. More generally, you can get the age of any such rock from the relative amounts of potassium-40 and argon-40 and the graphs in Figure 4.8.

Radiometric dating with other isotopes works the same way. By comparing the amounts of parent and daughter isotopes in the rock (or other

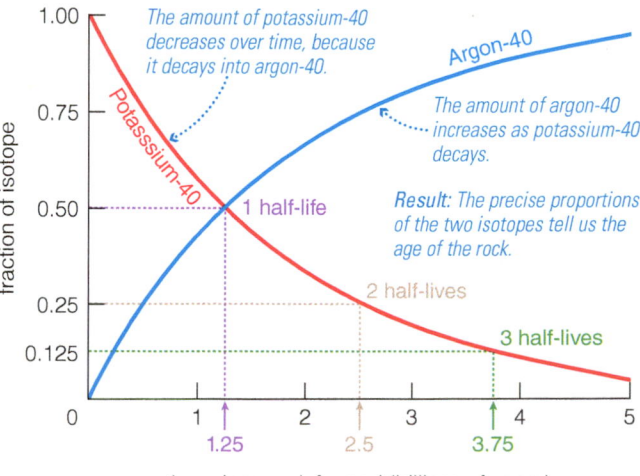

FIGURE 4.8
Potassium-40 is radioactive, decaying into argon-40 with a half-life of 1.25 billion years. The red line shows the decreasing amount of potassium-40, and the blue line shows the increasing amount of argon-40. The remaining amount of potassium-40 drops by half with each successive half-life.

TABLE 4.1 Selected Isotopes Used for Radiometric Dating of Rocks and Fossils

Some of the isotopes decay in several stages of parent–daughter pairs, and only the final daughter product is shown.

Parent Isotope	Daughter Isotope	Half-Life
Rubidium-87	Strontium-87	49.4 billion years
Lutetium-176	Hafnium-176	37.1 billion years
Thorium-232	Lead-208	14.0 billion years
Uranium-238	Lead-206	4.47 billion years
Potassium-40	Argon-40	1.25 billion years
Uranium-235	Lead-207	704 million years
Hafnium-182	Tungsten-182	8.9 million years
Aluminum-26	Magnesium-26	717,000 years
Carbon-14	Nitrogen-14	5730 years

Source: Berkeley Laboratory Isotopes Project

object), we can determine the fraction of the parent isotope that has decayed since the rock formed.* Then, based on laboratory measurements of the decay rate (half-life), we can determine the rock's age. Table 4.1 lists several of the parent–daughter isotope pairs commonly used in radiometric dating. Notice that the half-lives vary dramatically, so different pairs are useful for dating materials of different ages. For example, carbon-14 is useful only for dating objects less than about 50,000 years old, while uranium-238 is useful for dating the oldest rocks on Earth.

A major issue that sometimes arises in radiometric dating concerns a rock's original composition. If we're going to determine the rock's age by comparing the amounts of a radioactive isotope and its daughter product, we must have some way of knowing how much of the daughter was originally present, if any. In the case of potassium-40, this is quite easy. The daughter product, argon-40, is a gas that does not combine with other elements and can become trapped in rock only under special circumstances. Therefore, argon-40 generally cannot be part of a rock when it solidifies, so any argon-40 trapped inside a rock must have come from radioactive decay of potassium-40. Moreover, since the argon-40 atoms become trapped within the rock at the time they are produced by the decay of potassium-40, we can be similarly confident that no argon has escaped from the rock, unless the rock has undergone heating or some other transformation that could have released the trapped gas.

Radiometric dating with other substances is not always as easy, but detailed study of several isotope ratios or several different mineral grains within a single rock often allows geologists to determine a rock's original composition. Still, if a rock has undergone partial melting or a major shock during its history, or if water has removed some atoms from the rock, we may find somewhat different ages when we date the rock with different parent–daughter isotope pairs. Nevertheless, decades of experience in working with these complications have given geochemists a precise understanding of how they affect radiometric dating. As a result, it is nearly always possible to know whether an age is correct or uncertain.

Note that, while radiometric dating is an extremely powerful technique, it works only for dating solids that have not undergone significant change since they formed. For rocks, we get unambiguous ages only for igneous rocks. Metamorphic rocks are more challenging to date, because they may have undergone change that has altered their isotope ratios from what they would be due to decay alone. Still, we can often place fairly clear age constraints on metamorphic rocks, such as a minimum time since a particular rock formed, and learn much about the changes they have undergone. A sedimentary rock cannot be dated as a whole because it is a compressed mix of rock grains containing minerals of different ages; we can sometimes date mineral grains found within a sedimentary rock, but this tells us when the grains originally formed (in an igneous or metamorphic rock that was later broken down by weather) as opposed to when the sediments were deposited or compressed into a solid rock. To estimate ages of sedimentary rocks, we generally rely on dating igneous rocks buried above them or intruding into them. Similar ideas apply when we use radiometric dating for fossils and other artifacts: A radiometric age will be reliable if we can be confident that we know precisely how much

---

*The specific methodology varies in some cases. Carbon-14, for example, is continually produced in Earth's atmosphere by interactions between nitrogen-14 and high-energy particles coming from the Sun. An estimate of the original carbon-14 content of a fossil is based on the atmospheric production rate rather than on a parent–daughter isotope ratio.

of the original parent isotope has decayed, and any uncertainty in decay fraction will mean a corresponding uncertainty in the age.

THE RELIABILITY OF RADIOMETRIC AGES Scientists today have great confidence in the reliability of ages measured through radiometric dating. The underlying processes that govern radioactive decay are well understood, and the measurements needed to determine half-lives are straightforward when done with care. Perhaps most important, in many cases scientists can use several different radioactive isotopes to measure the ages of rocks or fossils. The ages from different parent–daughter isotope pairs almost always agree for a particular rock, fossil, or mineral grain, and when they don't it is usually possible to determine a cause for the difference. This agreement shows that the technique of radiometric dating is highly reliable. Ages of different rocks and fossils from particular strata of sedimentary rock also agree, further confirming the reliability of the technique.

Additional checks come from the fact that, at least in some cases, we can estimate ages in ways that are independent of radiometric dating. For example, if radiometric ages are correct, then they should confirm the ordering of the relative ages in sedimentary strata—and they do. A more precise check can be made for relatively young archaeological artifacts: Some ancient Egyptian artifacts have dates printed on them, and the dates agree with ages found with radiometric dating; in other cases, we can confirm radiometric ages by comparing them to ages that we can obtain from tree ring data. We can even confirm the $4\frac{1}{2}$-billion-year radiometric age for the solar system as a whole by comparing it to an age based on detailed study of the Sun. Theoretical models of the Sun, along with observations of other stars, show that stars slowly expand and brighten as they age. The model ages are not nearly as precise as radiometric ages, with uncertainties of up to a billion years or so, but the current state of the Sun agrees with what we expect for a star of its mass and age.

Taken together, our theoretical understanding of radioactive decay and the various ways of confirming ages found by radiometric dating leave no room for doubt about the reliability of the technique. Today, scientists routinely use radiometric dating to determine ages of rocks and fossils, and the overall uncertainty in radiometric age dates is usually less than 1–2%.

What does the geological record show?

Today, we have a detailed understanding of many of the events that mark the geological record, because geologists have carefully studied and dated tens of thousands of rocks and fossils. But before we get into the specifics, it's important to remember that while radiometric dates can be quite precise even very far back in time, our understanding of Earth's prevailing conditions becomes less certain as we look deeper into the past. Any individual rock or fossil may give us only limited information about the conditions under which it formed, so a fuller understanding requires studying many rocks and fossils from the same time period. However, the geological record is more sparse as we look further back. Earth's surface undergoes continual change through volcanism, plate tectonics, and erosion, making older rocks comparatively rarer than younger ones. For fossils left by life, the problem is compounded by other factors, best understood if we think about how fossils form.

FOSSIL FORMATION Although we tend to think of fossils as "remains" of living organisms, a *fossil* is any evidence of past life and most fossils

Cosmic Calculations 4.1

RADIOMETRIC DATING

The amount of a radioactive substance decays by half with each half-life, so we can express the decay process with a simple formula relating the current amount of a radioactive substance in a rock to the original amount:

$$\frac{\text{current amount}}{\text{original amount}} = \left(\frac{1}{2}\right)^{t/t_{\text{half}}}$$

where t is the time since the rock formed and t_{half} is the half-life of the radioactive material. We can solve this equation for t by taking the logarithm of both sides and rearranging the terms:

$$t = t_{\text{half}} \times \frac{\log_{10}\left(\dfrac{\text{current amount}}{\text{original amount}}\right)}{\log_{10}\left(\dfrac{1}{2}\right)}$$

Example: You chemically analyze a small sample of a meteorite. Potassium-40 and argon-40 are present in a ratio of approximately 0.85 unit of potassium-40 atoms to 9.15 units of gaseous argon-40 atoms. (The units are unimportant, because only the relative amounts of the parent and daughter materials matter.) How old is the meteorite?

Solution: Because no argon gas could have been present in the meteorite when it formed (see discussion in text), the 9.15 units of argon-40 must originally have been potassium-40 that has decayed with its half-life of 1.25 billion years. The sample must therefore have started with $0.85 + 9.15 = 10$ units of potassium-40 (the original amount), of which 0.85 unit remains (the current amount). The formula now reads

$$t = 1.25 \text{ billion yr} \times \frac{\log_{10}\left(\dfrac{0.85}{10}\right)}{\log_{10}\left(\dfrac{1}{2}\right)}$$

$$= 1.25 \text{ billion yr} \times \left(\frac{-1.07}{-0.301}\right)$$

$$= 4.45 \text{ billion yr}$$

This meteorite solidified about 4.45 billion years ago.

a Dinosaur bones preserved in sandstone in Dinosaur National Monument, which straddles Utah and Colorado.

b A more than 200-million-year-old petrified (stone) tree in Arizona's Petrified Forest National Park.

c These 200-million-year-old impressions are casts of snail-size, extinct organisms (called ammonites) made when minerals filled the empty space left after the organisms decayed.

d An insect preserved in hardened tree resin (often called *amber*), 45 million years old.

e These tusks belong to a whole 23,000-year-old mammoth discovered in Siberian ice in 1999.

f This man is looking at a 150-million-year-old dinosaur track in Colorado.

FIGURE 4.9
A gallery of fossils.

contain little or no organic matter. Figure 4.9 summarizes some of the ways in which fossils are made. In general, when an organism dies and gets buried in sediments, minerals dissolved in groundwater gradually replace organic material. Mineral-rich portions of organisms, such as bones, teeth, and shells, may be left behind, becoming fossils like those of the dinosaur bones displayed in many museums (Figure 4.9a). In some cases, the mineral replacement is complete and organisms literally turn to stone; the "stone trees" of Arizona's Petrified Forest formed in this way (Figure 4.9b). In many other cases, the organisms themselves decay, but in doing so they leave an empty mold that fills with minerals dissolved in water. The minerals may then make a cast in the shape of the dead organism (Figure 4.9c).

More rarely, some of the organic material from a dead organism may be preserved well enough to allow at least some study. Some fossil plant leaves are still green and well enough preserved for their cells to be studied with microscopes, even though they died millions of years ago. In other rare cases, whole organisms may be preserved in tree resin (Figure 4.9d) or frozen in ice (Figure 4.9e). One of the most interesting types of fossil is left not by a dead organism but by the activity of an organism while it was alive. For example, *coprolites* are rocks that consist of petrified excrement, which can allow us to learn about an animal's diet. In other cases, scientists have found fossilized dinosaur footprints, made when mineral processes preserved impressions left by a dinosaur as it walked through soft soil or mud (Figure 4.9f). Such fossil tracks provide clues about how dinosaurs walked and can help scientists learn something about dinosaur behavior.

Think About It Molecules of DNA tend to break down rapidly and easily after an organism dies, making it difficult or impossible to identify even fragments of DNA in most fossils. What types of fossils do you think are most likely to yield intact DNA? What types are least likely to yield intact DNA? Explain.

Although fossils can be made in numerous ways, only a tiny fraction of living organisms leave behind any kind of fossil at all. The vast majority decay—becoming food for living organisms in the soil or oceans—long before any mineral replacement can occur. If you've ever read about criminologists exhuming the skeletons of people who have been dead for just a few years, you know that even bones and teeth usually decay quite rapidly after death.

The tiny fraction of organisms that have left fossils in relatively recent times in Earth's history still amounts to a substantial fossil record. But fossils become rarer as we look deeper into the geological past, primarily for two reasons. First, fossils can suffer the same fates as rocks, so older fossils are more likely to have been destroyed over time by volcanism, erosion, or other geological processes. The second reason comes from the nature of past life itself. Studies of the geological record show that large plants and animals—which make the most easily discovered fossils—are relatively recent arrivals. For the first 90% or so of Earth's history, nearly all living organisms were microscopic, which means their fossilized remains must also be microscopic. Clearly, it is much more difficult to find and identify microscopic fossils than huge fossils of dinosaur bones. This fact is especially important in astrobiology, since we generally focus more on the origin and early evolution of life than on large plants and animals. Indeed, as we'll discuss in Chapter 6, the difficulty of finding and studying microfossils (and other potential traces of life in very old rocks) has led to significant scientific controversies about exactly when life took hold on Earth.

THE GEOLOGICAL TIME SCALE Although the geological record gives us far more information about recent events than the distant past, geologists have nevertheless been able to piece together a fairly detailed history of our planet going nearly all the way back to the time of its formation. To help organize this history, scientists divide it into a set of distinct intervals that make up what we call the **geological time scale.** Figure 4.10 shows the names of the various intervals on a time line, along with numerous important events that we will discuss later in this chapter and in Chapter 6. It is not necessary to memorize the names of the geological time intervals for the purposes of this book. However, the names are commonly used in books and articles about Earth or astrobiology, so it's useful to be familiar with them.

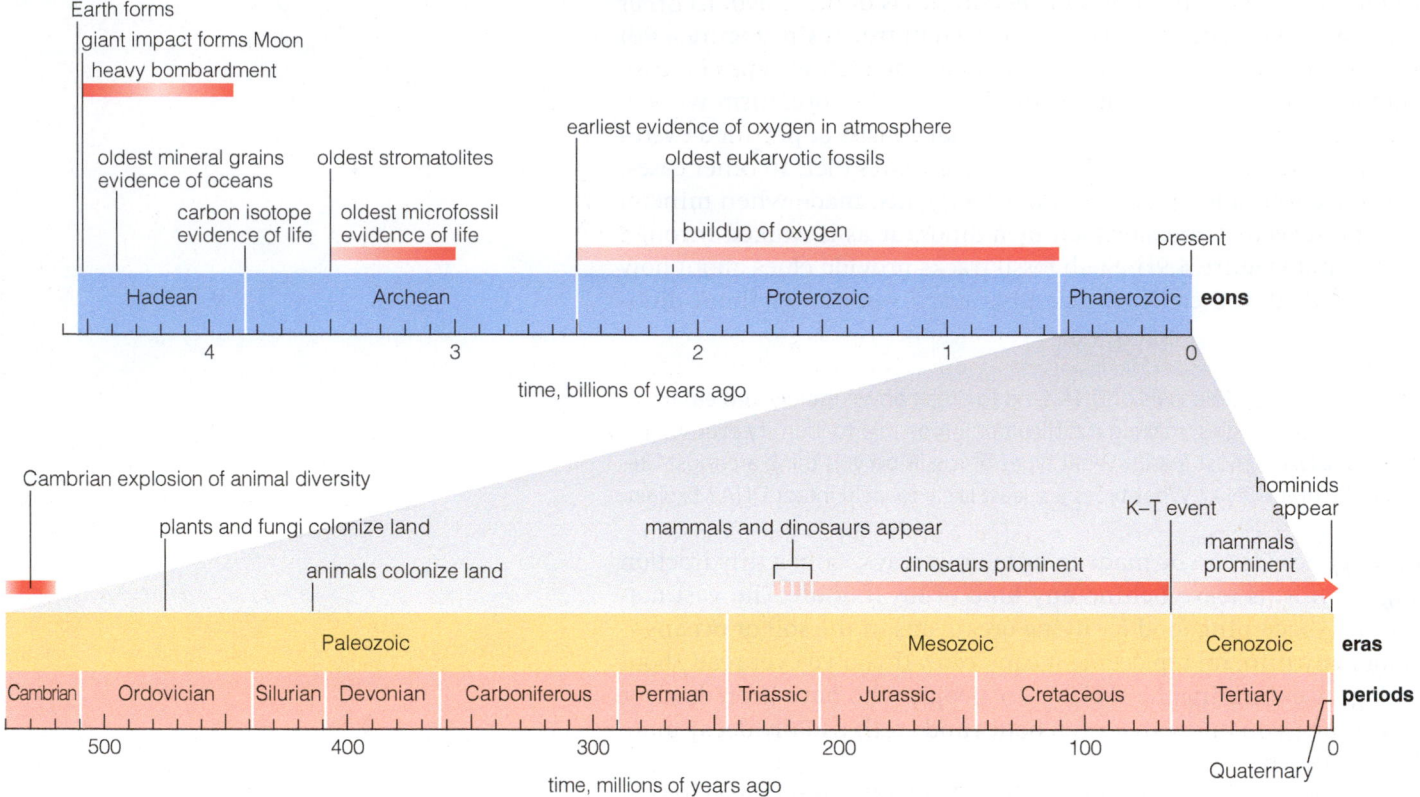

FIGURE 4.10

The geological time scale, along with key events discussed in this book. Notice that the lower time line is an expanded view of the last portion (Phanerozoic) of the upper time line. The eons, eras, and periods are defined by changes in the rocks and fossils present in the geological record. The absolute ages come from radiometric dating. (Although it is not shown, geologists now define about the last 100 million years of the Proterozoic eon as its own period, called the Vendian or Ediacaran period.)

The first major division of geological time is into a set of four **eons:** the Hadean, Archean, Proterozoic, and Phanerozoic. We can understand these names by looking at their Greek roots. The Phanerozoic eon extends from the present back to about 542 million years ago; its name comes from the Greek for "visible life" because it is marked by the presence of fossils visible to the naked eye. The Proterozoic eon, which extends from 542 million to about 2.5 billion years ago, takes its name from the Greek for "earlier life" because it predominantly shows fossils of single-celled organisms that lived before the Phanerozoic eon. The Archean eon extends from 2.5 to about 3.85 billion years ago and is named for "ancient life"; it got this name after the discovery of fossils from the first half of Earth's history. The Hadean eon got its name at a time when it was presumed that the early Earth would have had "hellish" conditions during this early time; Hades was the Greek mythological name for the underworld. However, as we'll discuss in the next section, recent evidence suggests that the Hadean may not have been quite that bad.

The fact that the geological record is much richer for more recent times is reflected in the more detailed naming system used for these times. The most recent, or Phanerozoic, eon is subdivided into three major **eras:** the Paleozoic, Mesozoic, and Cenozoic. These names also have Greek roots and mean, respectively, "old life," "middle life," and "recent life." The three eras are further subdivided into **periods.** The periods do not follow any consistent naming scheme. For example, the Cambrian period* gets its name from the Latin name for Wales (in Great Britain),

---

*The Cambrian is the earliest period in the Phanerozoic eon and was once thought to be the first period in which fossil organisms could be found. For that reason, the entire time before the Phanerozoic eon—that is, the Hadean, Archean, and Proterozoic eons—is often called the *Precambrian*.

the Jurassic period gets its name from rocks found in the Jura mountains of Europe, and the Tertiary period simply means "the third period." The recent geologic periods are even further subdivided into *epochs* and *ages,* but these are not shown in Figure 4.10.

Note that the eons, eras, and periods do not have uniform lengths. For example, the Proterozoic eon is about three times the length of time of the Phanerozoic eon, and the Paleozoic era is longer than the Mesozoic and Cenozoic eras combined. The divisions in the geological time scale are determined not by duration but by specific changes in the geological record, such as where certain species disappear and other new ones appear.

THE AGE OF EARTH Figure 4.10 starts with Earth's birth, but we cannot read the precise time at which that occurred from the geological record. The oldest known (as of 2015) intact Earth rocks date to about 4.02 billion years ago.* Apparently, rocks that formed prior to this time have been remelted or reshaped so much that we cannot obtain an age through radiometric dating. So how do we know the age of our planet?

We have a variety of ways of looking back to earlier times. The most direct way is based on studies of tiny mineral grains of zirconium silicate, or *zircons* for short (Figure 4.11). Although they are found embedded in much younger sedimentary rocks, radiometric dating (based on analysis of uranium and lead isotopes contained within them) shows that some zircons solidified as much as 4.38 billion years ago. Moreover, their oxygen isotopic content suggests that they formed at a time when liquid water was present and continents had already begun to form, suggesting that Earth's crust had separated from interior material before about 4.4 billion years ago.

Moon rocks also help us constrain Earth's age. Some Moon rocks brought back by the *Apollo* astronauts are considerably older than any intact Earth rocks, reflecting the fact that volcanism and other geological processes have done far less reshaping of the Moon's surface than they have of Earth's surface. Radiometric dating with isotopes of uranium and lead places the ages of the oldest Moon rocks at more than 4.4 billion years, telling us that Earth and the Moon must have formed before this time. In fact, the leading model for the Moon's formation holds that it formed as a result of a *giant impact* that blasted material out of the young Earth's outer layers; we'll discuss this model in detail in Section 4.6. Earth must therefore be at least slightly older than the Moon, which means older than 4.4 billion years.

While the zircons and lunar rocks set a minimum age for Earth, we can set a maximum age by dating the formation of the solar system as a whole. We do this by studying **meteorites,** which are rocks that have fallen to Earth from space. Some meteorites are younger than the solar system as a whole, because they are fragments of asteroids that formed and later shattered in collisions. However, a large number of meteorites have a chemical structure that suggests they were among the first pieces of solid material to condense in the early history of the solar system. These meteorites also all have about the same age, offering further evi-

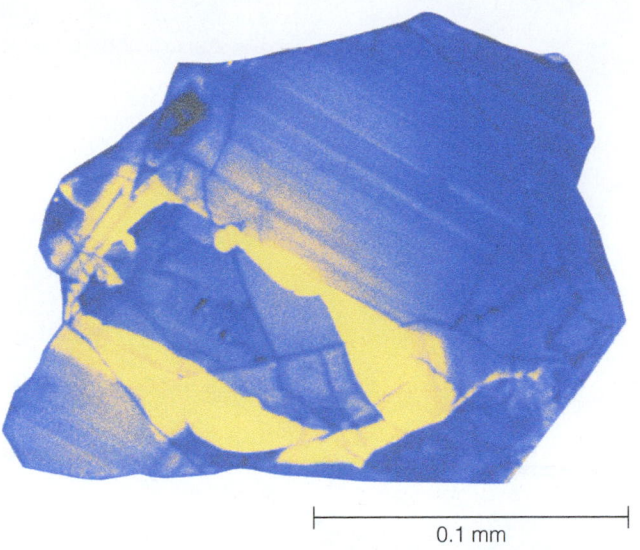

|———————————————————|
0.1 mm

FIGURE 4.11
This image shows a tiny zircon crystal—about 0.2 millimeter across—that was found embedded in a rock formation in Western Australia. Radiometric dating shows it to be nearly 4.4 billion years old. The image was made with a technique called cathodoluminescence, in which an electron beam is focused on the crystal, causing it to emit visible light; the colors in the image are not real, but instead correspond to differing intensities in the emitted light.

---

*Some researchers have claimed that the Nuvvuagittuq rock formation in Northern Canada contains intact rocks more than 4.3 billion years old, which would make them very valuable sources of information about the very early Earth. However, while other researchers agree that some of the material in the rocks is that old, they disagree that the material has been solid for that long, in which case the rocks are of limited value in telling us about conditions prior to the time they solidified.

dence that they represent material from the very beginning of the solar system; these meteorites date to 4.57 billion years ago (with an uncertainty of less than about 0.01 billion years).

The meteorite data allow scientists to use other techniques to further constrain the ages of both Earth and the Moon. These techniques are somewhat complex in their details, but they are based on comparisons of isotope ratios in meteorites, Earth rocks, and Moon rocks. The results show that both Earth and the Moon had formed within about 50 to 70 million years (0.05 to 0.07 billion years) after the oldest meteorites formed. We therefore conclude that our planet accreted quickly in the early solar system, and the giant impact that formed the Moon happened quite early in Earth's history as well. By about 4.5 billion years ago, our planet had its Moon and must have been essentially at its current mass and size, ready for its geology to begin shaping the features that would ultimately make it our home.

KEY GEOLOGICAL DEFINITIONS

ROCKS, MINERALS, FOSSILS

minerals: The basic pieces of solid rock; a particular mineral is distinguished from other minerals by its chemical composition or crystal structure (or both).

rocks: Intact solids that may contain a variety of minerals. We classify rocks into three basic types by their formation process: *Igneous rock* is of volcanic origin, made when molten rock cools and solidifies. *Sedimentary rock* is made by the gradual compression of sediments, which may contain bits of other rock types as well as fossils. *Metamorphic rock* was once either igneous or sedimentary but has since been transformed (but not melted) by high heat or pressure. Rocks of any of the three types may be subclassified by the minerals they contain.

fossil: Any relic left behind by living organisms that died long ago.

TERMS RELATED TO GEOLOGICAL TIME

geological record: The information about Earth's past that is recorded in rocks and fossils; the latter record is sometimes called the *fossil record*.

geological time scale: The time scale used to measure the history of Earth.

radiometric dating: The method of determining the age of a rock or fossil from study of radioactive isotopes contained within it.

half-life: The time it takes for half of the atoms to decay in a sample of a radioactive substance.

TERMS RELATED TO EARTH'S GEOLOGICAL HISTORY

differentiation: A process in which materials separate by density. In Earth, differentiation led to a dense *core* made mostly of iron and nickel, a rocky *mantle* made mostly of *silicates* (minerals rich in silicon and oxygen), and a *crust* made of the lowest-density rocks.

heavy bombardment: The period of time during which the planets were heavily bombarded by leftover planetesimals, starting from the time the planets first formed and likely ending some 3.8–4.0 billion years ago.

outgassing: The process of releasing gases trapped in a planetary interior into the atmosphere.

lithosphere: The layer of cooler, more rigid rock that sits above the warmer, softer mantle rock below. It encompasses both the crust and the uppermost portion of the mantle, extending to a depth of about 100 kilometers. On Earth, the lithosphere is broken into a set of large *plates*.

seafloor crust: The relatively dense, thin, young crust found on Earth's seafloors, composed largely of the igneous rock called *basalt*.

continental crust: The thicker, lower-density crust that makes up Earth's continents. It is made when remelting of seafloor crust allows lower-density rock to separate, and typically consists of granite. Continental crust ranges in age from very young to as old as about 4.0 billion years.

plate tectonics: The geological process in which lithospheric plates move around the surface of Earth. It acts like a conveyor belt, with new seafloor crust erupting and spreading outward from mid-ocean ridges and then being recycled back into the mantle by subduction at ocean trenches. It also explains *continental drift*, because plates carry the continents with them as they move.

TERMS RELATED TO CLIMATE AND CLIMATE REGULATION

greenhouse effect: The effect that makes a planet's surface warmer than it would be in the absence of an atmosphere. It is caused by the presence of *greenhouse gases*—such as carbon dioxide (CO_2), water vapor (H_2O), and methane (CH_4)—that can absorb infrared light emitted by the planetary surface (after the surface is heated by sunlight).

carbon dioxide cycle (CO_2 cycle): The cycle that keeps the amount of carbon dioxide in Earth's atmosphere small and nearly steady and hence keeps Earth habitable. Over time, this cycle has locked up most of Earth's carbon dioxide in *carbonate* rocks (rocks rich in carbon and oxygen) such as limestone.

ice ages: Periods of time during which Earth becomes unusually cold, so water from the oceans freezes out as ice and covers a substantial portion of the continents.

snowball Earth: Refers to periods of extreme ice ages that may have occurred several times before about 580 million years ago.

global warming: Usually refers to the current warming of Earth caused by human input of greenhouse gases into the atmosphere.

4.3 The Hadean Earth and the Dawn of Life

We know comparatively little about the Hadean eon, which constitutes a little more than the first half-billion years of Earth's history. Nevertheless, it was clearly an important time in the history of our planet, and evidence that we'll discuss in Chapter 6 shows that life arose during or not long after this period of time. So if we want to understand when and how our planet became habitable and gave birth to life, we need to start back in this earliest of Earth's time periods.

How did Earth get an atmosphere and oceans?

Before Earth could have life, it needed an atmosphere and liquid water oceans. Neither is likely to have existed when Earth first formed. Our planet was probably too small and too warm to capture significant amounts of hydrogen and helium gas as it accreted within the solar nebula [Section 3.4], and no other gases were present in large enough quantities to make a substantial atmosphere. In fact, the presence of Earth's atmosphere and oceans once posed a mystery to solar system formation theory, because the planetesimals that condensed at Earth's distance from the Sun should have been made only of rock and metal, with no gaseous or icy content at all. How, then, did Earth obtain the water and gases that make up our oceans and atmosphere?

Models of planetary formation now show that while Earth formed primarily from "local" planetesimals of rock and metal, it should also have incorporated some planetesimals from farther out in the solar system; these planetesimals were flung inward by gravitational interactions with other planetesimals and forming planets. Some of these planetesimals came from far enough away—probably from the region currently occupied by the asteroid belt (and perhaps also including some comets from farther out in the solar system)—that they contained ices or rock chemically bound with molecules of water or other common gases. As these planetesimals became part of the forming Earth, their gaseous content became trapped on or within our planet.

The young Earth therefore contained trapped gases, including water, held under pressure in the interior in much the same way that the gas in a carbonated beverage is trapped in a pressurized bottle. When molten rock erupts onto the surface as lava, the release of pressure violently expels the trapped gas in a process we call **outgassing.** Outgassing probably released most of the water vapor that condensed to form our oceans as well as most of the gas that formed our atmosphere. Some additional water and gas may also have been supplied directly to the surface by impacts after Earth formed, but recent studies suggest that this would have been a relatively minor contribution to the atmosphere and oceans.

Volcanism is the major source of outgassing, as you can probably guess from looking at any photo of a volcanic eruption (Figure 4.12). Some outgassing also occurs as a result of impacts—the heat of an impact can melt rock and allow gas to escape. Studies of present-day volcanoes show that the primary gases released by outgassing are water vapor (H_2O), carbon dioxide (CO_2), nitrogen (N_2), sulfur-bearing gases (H_2S or SO_2), and hydrogen (H_2). Because Earth's overall composition has not changed much since its formation, we expect that the same gases were released

FIGURE 4.12
This photo shows the eruption of Mount St. Helens (in Washington State) on May 18, 1980. Eruptions are accompanied by a tremendous amount of outgassing. (The gas itself is generally invisible, so what you see is dust and ash expelled along with the gas.)

in early times.* The water vapor condensed as rain to fill Earth's oceans (along with water that bubbled out of the ground at volcanic vents). The gases that remained airborne made up Earth's early atmosphere.

Isotopic analysis of the oldest zircon grains (see Figure 4.11) suggests that substantial amounts of water—and probably oceans—were already present on Earth at the time these grains solidified, some 4.4 billion years ago. If so, much of Earth's atmospheric gas must have been released quite early, perhaps as a side effect of Earth's interior melting or of the giant impact thought to have formed the Moon. Taken alongside the evidence noted earlier suggesting that continents had also begun to form when the zircons solidified, we are led to the intriguing idea that Earth may have had early continents, oceans, and an atmosphere within only about 100 million years after the planet first formed.

The composition of the early atmosphere was very different from that of the atmosphere today. The early atmosphere was probably dominated by carbon dioxide (though there are uncertainties that we will discuss in Section 6.2), while today's atmosphere is dominated by nitrogen (about 78% of the atmosphere) and contains only trace amounts of carbon dioxide (less than 0.1%). More important to us, the early atmosphere contained essentially no molecular oxygen (O_2), while the present atmosphere is about 21% oxygen. Therefore, we could not have breathed the atmosphere on the early Earth. Earth's present oxygen atmosphere is almost certainly a result of photosynthesis by living organisms, as we will discuss further in Chapter 6. Life therefore must have arisen in a nearly oxygen-free environment.

Could life have existed during Earth's early history?

The Hadean eon got its name because of its presumed hellish conditions, but the recent zircon evidence tells a different story. If the evidence is being interpreted correctly, it means that Earth may have been habitable within 100 million years after its formation. Many modern-day microbes survive just fine in the absence of oxygen, and it seems likely that such organisms could have thrived almost from the moment the oceans and atmosphere first formed.

However, while the Hadean might have been reasonably balmy at most times, those calm periods must have been frequently interrupted by great violence: Volcanic eruptions were probably more frequent and larger than those of today and, more significantly, the young Earth was bombarded by planetesimals left over from the birth of the solar system.

THE HEAVY BOMBARDMENT The accretion of the planets did not end suddenly. The young planets must have shared the solar system with vast numbers of "leftover" planetesimals. Some of these leftovers still survive as asteroids and comets, but many more must have crashed into the Sun and planets. The vast majority of these collisions occurred in the first few hundred million years of our solar system's history, during the period we call the **heavy bombardment.** On planets and moons with solid surfaces, such collisions leave **impact craters** as visible scars. A small

---

*Some present-day volcanoes release recycled gases, while others release "juvenile" gases (gases that have not previously been outgassed). The outgassing composition is determined from those releasing juvenile gases, since these should be indicative of the gases originally trapped in Earth's interior.

telescope reveals the presence of numerous impact craters on the Moon (see Figure 4.1).

Earth must have experienced many more impacts than the Moon, both because it presents a larger target and because its stronger gravity tends to attract more objects. However, while the Moon still shows scars of craters formed throughout much of its history, craters on Earth tend to get erased with time by erosion, volcanic eruptions, and plate tectonics. Therefore, if we wish to learn about the past cratering rate on Earth, we must look to the evidence recorded on the Moon.

Figure 4.13 shows a map of the Moon's entire surface. Notice that some regions are much more crowded with craters than others. The most heavily cratered regions are the **lunar highlands,** where craters are so abundant that we see overlapping crater boundaries and craters on top of other craters. Radiometric dating of Moon rocks from the lunar highlands shows them to be more ancient than rocks from other regions of the Moon; the highlands are the sources of the Moon rocks that date to more than 4.4 billion years ago, and even their youngest rocks are generally at least 4.0 billion years old. We conclude that most of the craters of the lunar highlands formed more than 4 billion years ago.

In contrast, we see relatively few craters in the regions known as the **lunar maria.** The maria are lava plains that fill huge, ancient impact craters. Scientists hypothesize that the impacts that made the original craters were so devastating that they fractured the lunar crust, creating cracks through which molten lava was able to escape at some later time. The lava then flooded the crater basins, covering the existing craters. The craters we now see within the maria must be the result of impacts that occurred *after* the lava flows, and radiometric dating of rocks from the maria shows that most of these lava flows occurred between about 3.9 and 3.0 billion years ago. The relatively small number of craters within the maria (about 3% as many per unit area as in the lunar highlands) tells us that the heavy bombardment must have essentially been over by the time the maria flooded with lava.

More detailed studies of the ages of lunar craters, as well as analysis of mineral grains in meteorites and old zircons in Earth rocks, can in principle tell us more precisely when the impact rate dropped. Although we might expect that the heavy bombardment would have tapered off gradually, there is some evidence for a spike in the number of impacts around 4 billion years ago. Assuming such a spike in the impact rate really occurred, models of planetary formation suggested that the reason might have been *planetary migration* [Section 3.5] in which gravitational interactions caused the orbits of the jovian planets to shift around this time. These shifts could have disturbed the orbits of asteroids and comets, causing more of them to end up on trajectories that ultimately sent them crashing into the planets.

Every object in the solar system must have been similarly pelted during the heavy bombardment, so all solid surfaces should originally have been as crowded with craters as the lunar highlands. So in addition to helping us understand the early history of Earth and the Moon, these ideas allow us to estimate surface ages for worlds throughout the solar system. Those surface regions that still have abundant craters, such as much of Mercury's surface (see Figure 4.1), must have undergone little change during the last 4 billion years or so. Surfaces with fewer craters must be correspondingly younger, indicating that their original craters

Lunar highlands are ancient and heavily cratered. Lunar maria are huge impact basins that were flooded by lava. Only a few small craters appear on the maria.

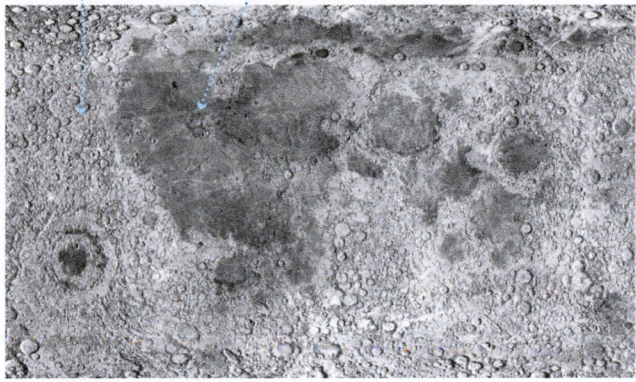

FIGURE 4.13
This flat map shows the entire surface of the Moon in the same way that a flat map of Earth represents the entire globe. Radiometric dating shows the heavily cratered lunar highlands to be a half-billion or more years older than the darkly colored lunar maria, telling us that the impact rate dropped dramatically after the end of the heavy bombardment.

have been erased in some way. In other words, thanks to the reconstruction of cratering rates made possible by studies of Moon rocks, planetary scientists can use simple counts of craters on other worlds to estimate the ages of the surfaces. These age estimates are not nearly as precise as radiometric dating—indeed, ages from crater counts are sometimes uncertain by as much as a billion years or so—but until we begin collecting rock samples from other worlds, they are the only age evidence we have to go by.

Think About It Crater abundance varies greatly in different regions of Mars. For example, the "southern highlands" are quite crowded with craters, while the "Tharsis Bulge" has volcanoes but few impact craters. Which region has the older surface? Does this imply any difference in the number of impacts that occurred in the two regions? Why or why not?

LARGE IMPACTS AND EARLY LIFE We have found that the Hadean Earth must have endured large impacts at a rate much higher than Earth has experienced since. What does this mean for the possibility of life during the Hadean? Remember that the Hadean lasted more than 600 million years, which means that even with "frequent" large impacts, individual events were typically separated by thousands or millions of years. There was never a time when impacts occurred so rapidly that they could have done the equivalent of hitting every living organism on the head. (Note that this is quite unlike the dramatic artist illustrations you frequently see in which the early Earth is being pummeled by a hail of rocks from space.) Given the evidence of continents and oceans all the way back to 4.4 to 4.5 billion years ago, there seems no reason why life could not have arisen during the Hadean. The question of life during the Hadean seems less a question of whether it could have *existed* than whether it could have *survived* the occasional large impacts.

Movie Madness ICE AGE: THE DAWN OF THE DINOSAURS

Let's face it: Everyone loves dinosaurs. Not because they're cuddly, but because they're not (Barney is an exception). Most of us have an innate fascination with predators—after all, those of our ancestors who didn't pay attention to the habits of creatures with big teeth were preferentially removed from the gene pool. Celebrity dinosaurs such as *Tyrannosaurus rex* have the dual attraction of being king of the carnivores while also being thoroughly extinct.

But in the third *Ice Age* film, *Dawn of the Dinosaurs*, a few hundred years of hard work by paleontologists is thrown out the window so that familiar mammals can confront their lizard-like predecessors on the silver screen. That's dramatically interesting, but geologically bonkers.

There's good evidence that ice ages have been frosting our planet for more than 2 billion years. But the *Ice Age* movies are set in *recent* history—a fact immediately obvious to nine out of ten moviegoers who've studied the Pleistocene epoch—because the films feature wooly mammoths. These hulking shag rugs with trunks first appeared less than 2 million years ago, and most of them faded from the tundras as the last ice age began to lose its cool around 10,000 B.C. Though not seen in the film, our human ancestors busily hunted these overgrown elephants, and may even have helped drive them to extinction. The dinosaurs, meanwhile, were long gone, having disappeared some 65 million years ago.

To make a film in which mammoths (or possibly unseen humans) share the landscape with Mesozoic monsters is like pairing Rambo with Julius Caesar. Hollywood seldom overestimates the intelligence of its customers, but even hard-nosed studio executives seem to have balked at the idea of mammoths and dinosaurs co-existing. So in *Dawn of the Dinosaurs*, the thunder lizards are all in the basement—in a kind of forgotten underground city.

Of course, you've got to wonder what they eat down there, other than one another. How can you grow a lot of lush vegetation where the sun doesn't shine? And meat eaters are merely the top of a food chain that begins with . . . plants. Then there's the fact that this subterranean sanctuary has got to be enormous—after all, there's an entire range of big critters, and they need plenty of room. Micro environments do not support mega fauna.

In truth, there really are oodles of dinosaurs lurking beneath the landscapes in which today's furry mammals caper and cavort. But they're all bones, and they don't move very much.

The effects of impacts depend primarily on their sizes. As we'll discuss in Chapter 6, the impact of an object about 10 to 20 kilometers across is thought to have precipitated a series of global changes that caused the extinction of the dinosaurs some 65 million years ago. Fortunately for us, impacts of that size are extremely rare today. But the lunar evidence tells us that far larger impacts occurred during the heavy bombardment.

To understand the effects of very large impacts, scientists calculate the amount of energy they would release and then model how the energy would heat the planet. These models suggest that the largest impacts might have completely vaporized the oceans, and in some cases they might have raised the global surface temperature high enough to melt the upper crust. Such events would have been **sterilizing impacts** that would have killed off any life on Earth at the time. However, the full extent of the devastation is still subject to debate, and some models suggest that while large impacts would have sterilized substantial portions of our planet, microscopic life living underground (and possibly in some deep ocean environments) could have survived.

All in all, it now seems likely that life could have arisen during the Hadean, and if it did, there is at least some chance that it may have survived the many impacts of the heavy bombardment. However, unless we someday find rocks old enough to contain fossil evidence of such life, we may never know whether it really existed during this ancient time.

4.4 Geology and Habitability

Although impacts can have significant short-term effects and leave behind craters as scars, they are of relatively minor importance in shaping Earth's geology. Instead, most features of Earth's surface have been shaped by one or more of the following three geological processes: volcanism, plate tectonics, and erosion. Why are these processes so important on Earth?

Remarkably, all three of these processes that continually reshape our planet's surface are directly attributable to internal heat. The connection is most obvious for volcanism, but it applies equally to plate tectonics. The connection to internal heat is more subtle for erosion by wind, water, and ice, but remember that these things exist only because our planet has oceans and an atmosphere, both of which were produced by volcanic outgassing.

Even more important, as we noted in Section 4.1, volcanism and plate tectonics have played major roles in Earth's long-term suitability for life. Therefore, if we hope to understand Earth's surface habitability, we must first understand what our planet is like on the inside and how the interior conditions drive these crucial geological processes. As we'll see, interior heat also plays a key role in explaining why our planet has a protective magnetic field.

What is Earth like on the inside?

Our deepest drills have barely pricked Earth's surface, reaching only a few kilometers down into the nearly 6400 kilometers to Earth's center. Nevertheless, we have managed to learn a lot about our planet's interior structure. One set of clues comes from the nature of surface rocks. For

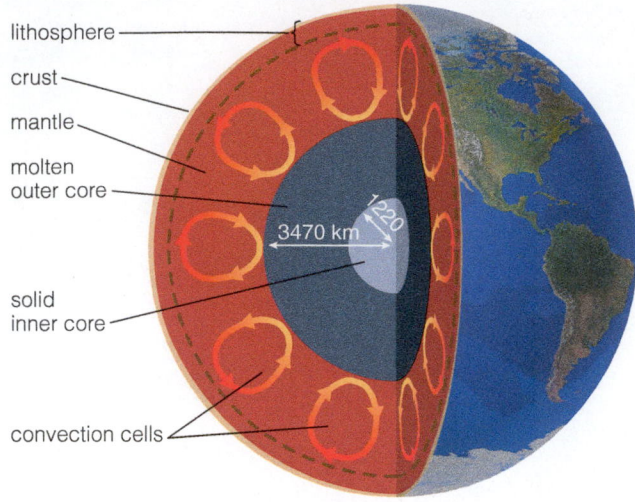

lithosphere
crust
mantle
molten
outer core

3470 km
1220

solid
inner core

convection cells

FIGURE 4.14

Earth's interior structure, determined from seismic studies. The layering by core-mantle-crust is based on density, while the identification of the lithosphere is based on rock strength. The circular arrows represent the general pattern of mantle convection.

example, the fact that the density of surface rocks is considerably less than Earth's overall average density tells us that much denser material must reside in the interior than on the surface. Other clues come from precise measurements of Earth's gravitational field strength in different locations and studies of Earth's magnetic field.

We learn much more about Earth's internal structure from the study of **seismic waves**—waves that propagate much like sound waves both through Earth's interior and along its surface after an earthquake. The precise speed and direction of seismic waves depend on the composition, density, pressure, temperature, and phase (solid or liquid) of the material they pass through. After an earthquake, geologists record the arrival times and the strengths of seismic waves at stations distributed all around the world. At each station, the arriving seismic waves tell us something about the average state of the material through which they have passed. By comparing data from many different stations, scientists have pieced together a fairly detailed picture of Earth's internal structure.

EARTH'S INTERIOR STRUCTURE Seismic studies reveal that Earth's interior is divided into three major layers by density (Figure 4.14):

- **Core:** The highest-density material, consisting primarily of the metals nickel and iron, resides in the central core. Earth's core has two distinct regions: a solid **inner core** surrounded by a molten (liquid) **outer core.**
- **Mantle:** Rocky material of moderate density—mostly silicate minerals rich in silicon and oxygen—forms a thick mantle that surrounds the core.
- **Crust:** The lowest-density rock, including igneous rocks such as granite and basalt, forms a thin crust that is essentially Earth's outer skin.

The molten lava from volcanoes gives many people the misconception that Earth is molten inside, but in fact the lava rises upward from a fairly narrow zone of rock in the upper mantle that is only partially molten. The vast majority of Earth's interior is solid rock, and only the outer core is fully molten. However, the interior rock is not equally strong throughout, and differences in rock strength are often more important than differences in density for understanding many geological processes.

The idea that rock can vary in strength may seem a bit surprising, since in our everyday lives we tend to think of rock as the ultimate in strength and stability (hence the phrase "solid as a rock"). However, the strength of rock depends on its temperature and the surrounding pressure, and under the conditions found in Earth's mantle, even "solid" rock can flow gradually. The popular toy Silly Putty provides a good analogy. The putty can feel pretty solid, especially when it is cold; you can even form it into a ball and bounce it. But if you put a pile of it on a table or inside its "egg" container, after a few days you'll see that it has flowed slowly outward.

In terms of rock strength, geologists define Earth's outer layer as the relatively cool and rigid rock called the **lithosphere** (*lithos* is Greek for "stone"), which "floats" on warmer, softer rock beneath. The lithosphere encompasses the crust and the upper part of the mantle (see Figure 4.14).

Beneath the lithosphere, the higher temperatures allow rock to deform and to flow much more easily. In fact, the mantle rock flows with a characteristic pattern called **convection,** in which hot material expands and rises while cooler material contracts and falls. You are probably familiar with convection in other situations, as it can occur any time a substance is strongly heated from below. For example, you can see convection in a pot of soup on a hot stove, and convection is important in weather because the warm air near the ground tends to rise while the cool air above tends to fall. In the mantle, convection is driven by heat from the core. This heat causes rock near the base of the mantle to expand, giving it a tendency to rise because it becomes lighter and less dense than the rock above it. Meanwhile, rock near the top of the mantle (just below the lithosphere) can cool as its heat flows to the surface (by conduction), causing it to contract and sink. The ongoing process of convection creates *convection cells,* indicated by the circular arrows in Figure 4.14. Keep in mind that mantle convection involves the flow of solid rock, so it is quite slow: Typically, mantle rock flows at a rate of only perhaps 10 centimeters per year, slow enough that it would take about 100 million years for a particular piece of rock to be carried from the base to the top of the mantle.

DIFFERENTIATION AND INTERNAL HEAT Earth's interior layering tells us that it underwent the process known as **differentiation,** in which materials separate according to their density. Earth must have undergone differentiation quite early in its history; as we'll discuss in Section 4.6, the giant impact that formed the Moon must have occurred *after* differentiation, and isotopic comparisons between meteorites and lead ore on Earth's surface have led scientists to conclude that most lead must have sunk to the core by the time Earth was about 30 million years old. For differentiation to have occurred so rapidly, our planet must have been molten or nearly molten throughout its interior. The melting allowed material to separate by density, much as oil separates from water when you mix them and let the mixture sit.

The heat that caused rock to melt came from three main sources. First, the impacts of accretion created heat that melted the outer layers of the young Earth. Second, as denser materials sank through the molten outer layers, their gravitational potential energy was converted into thermal energy that added further heat to the interior. Third, heat is continually released by the radioactive decay of elements within Earth. This heat source is still important today—in fact, it is the dominant heat source within present-day Earth—but it was even more important in early times, when there was more radioactive material to decay. Once the outer layers began to melt due to heat from the first source (accretion), the second and third heat sources (sinking of dense materials and radioactive decay) ensured that our planet would completely melt and differentiate fully.

All the terrestrial worlds in our solar system underwent similar melting and differentiation when they were very young. Since that time, they have never again been hot enough to melt fully, and they have all been slowly cooling with time. However, different worlds have cooled at different rates.

In general, two factors determine a world's cooling rate. The first is size: Large worlds tend to retain their internal heat much longer than do smaller worlds. You can see why by picturing a large world as a smaller world wrapped in extra layers of rock. The extra rock acts as insulation, making it take longer for interior heat to escape into space. If you now add the fact that the larger world contains more heat in the first place, it's clear that a large world will take much longer to cool than a small world. The Moon's relatively small size probably allowed it to cool substantially within a billion years or so after its formation, while Earth's interior still remains hot enough to keep iron molten in the outer core.

Think About It Give an example from everyday life of a small object cooling faster than a larger one and of a small object warming up more quickly than a larger one. How do these examples relate to the issue of geological activity on Earth and the Moon?

The second general factor in the cooling rate is ongoing heat deposition: If a world has a source of ongoing internal heat, it will tend to cool more slowly with time. For the terrestrial worlds, the only significant source of ongoing heat is radioactive decay. Because many radioactive materials have long half-lives, they can continue to add heat to the interior for billions of years. Over time, radioactive decay has contributed several times as much heat to Earth's interior as accretion and differentiation. As we'll discuss in Chapter 9, some moons of jovian planets have other sources of ongoing heat deposition (such as *tidal heating*) that can keep them much hotter than we would otherwise expect given their relatively small sizes.

How does plate tectonics shape Earth's surface?

Our discussion of internal heat explains most of the differences in geological activity that we see between Earth and the other terrestrial worlds. The Moon and Mercury have essentially no active, internally driven geology, because their small sizes allowed them to cool long ago. Mars is large enough to have had significant volcanism in the past, but its interior must now be much cooler than it was in its heyday. Venus is only slightly smaller than Earth, so it probably retains nearly as much internal heat; confirmation of this idea comes from the fact that Venus has few impact craters, indicating that lava flows or other processes that recycle surface material have erased craters from the more distant past. However, Earth appears to be geologically distinct from Venus and all the other terrestrial worlds in one crucial way: Earth is the only planet with ongoing plate tectonics.

The word *tectonics* comes from the Greek word *tekton*, meaning "builder"; notice the same root in the word *architect*, which means "master builder." In geology, tectonics refers to the "building" of surface features by stretching, compression, or other forces acting on the lithosphere. Tectonic processes have operated to some extent on all the terrestrial worlds, but they are most important on Earth, where they operate by the distinctive mechanism of plate tectonics.

THE MEANING OF PLATE TECTONICS The term *plate tectonics* refers to the scientific theory that explains much of Earth's surface geology as a result of the slow motion of *plates*—fractured pieces of the lithosphere—driven by the underlying convection of the mantle. According to the theory, the lithosphere fractured because of stresses generated by mantle convection, and the resulting plates essentially "float" over the mantle, gradually moving over, under, and around each other as convection moves Earth's interior rock. Because it refers to the theory, the term *plate tectonics* is generally considered to be singular rather than plural.

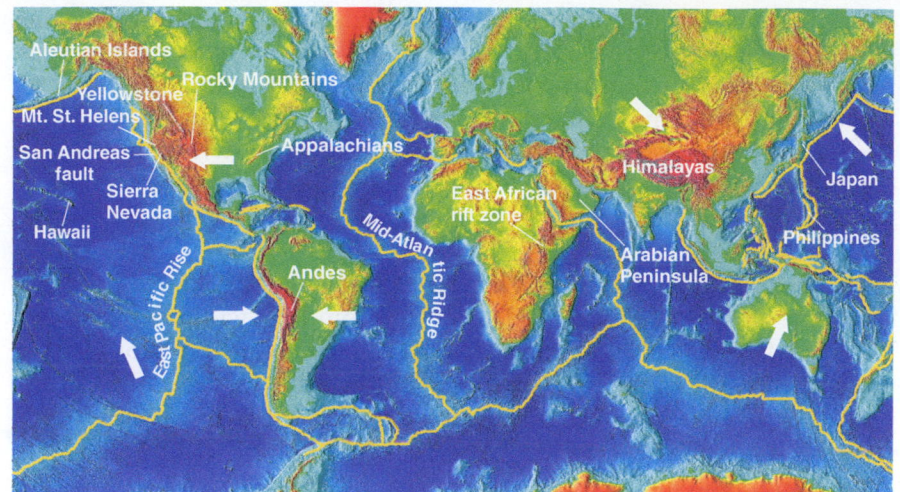

FIGURE 4.15
This relief map shows known plate boundaries (solid yellow lines), with arrows to represent directions of plate motion. Color represents elevation, progressing from blue (lowest) to red (highest).

Earth's lithosphere is broken into about a dozen plates (Figure 4.15). Except during earthquakes, the motions of the plates are barely noticeable on human time scales—a few centimeters per year, which is about the rate at which your fingernails grow. However, geologists can now measure plate motions by comparing readings taken with the global positioning system (GPS) on either side of plate boundaries.

EVIDENCE FOR PLATE TECTONICS The GPS measurements offer the most direct evidence of plate motion. However, the overall theory of plate tectonics rests on three additional significant lines of evidence: evidence of past continental arrangements, evidence that plates spread apart on seafloors, and a difference between the nature of Earth's crust on the seafloors and the continents.

We usually trace the origin of the theory of plate tectonics to a slightly different idea proposed in 1912 by German meteorologist and geologist Alfred Wegener: *continental drift,* the idea that continents gradually drift across the surface of Earth. Wegener got his idea in part from the puzzlelike fit of continents such as South America and Africa (Figure 4.16). This fit had been noted by earlier mapmakers, but Wegener took the idea further. He noted that similar types of distinctive rocks and rare fossils were found in eastern South America and western Africa, suggesting that these two regions once had been close together.

Despite these strong hints, no one at Wegener's time knew of a mechanism that could allow the continents to push their way through the solid rock beneath and around them. Wegener suggested that Earth's gravity and tidal forces from the Sun and Moon were responsible, but other scientists quickly showed that these forces were too weak to move continents around. As a result, Wegener's idea of continental drift was rejected by all but a few geologists for decades after he proposed it, even though his evidence of a "continental fit" for Africa and South America ultimately proved correct. Today, far more extensive fossil evidence makes it clear that the continents really were arranged differently in the past, and our understanding of how rock can flow in the mantle allows us to understand the real reasons that the continents move around.

Scientists began to recognize the mechanism of continental motion through the discovery of *mid-ocean ridges* in the mid-1950s. Mantle ma-

FIGURE 4.16
The puzzlelike fit of South America and Africa.

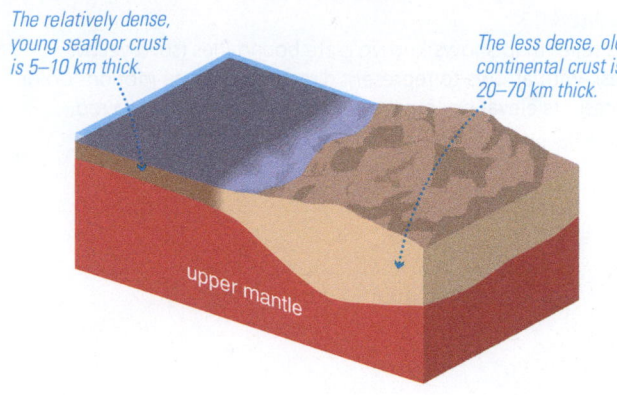

The relatively dense, young seafloor crust is 5–10 km thick.

The less dense, older continental crust is 20–70 km thick.

upper mantle

FIGURE 4.17
Earth today has two distinct kinds of crust.

terial erupts onto the ocean floor along these ridges, such as the mid-Atlantic Ridge shown in Figure 4.15, while the existing seafloor spreads outward to either side. This **seafloor spreading** helps explain how the continents could move apart with time. Because this idea is quite different from Wegener's original notion of continents plowing through the solid rock beneath them, geologists no longer use the term *continental drift* and instead consider continental motion within the context of plate tectonics.

The third line of evidence for plate tectonics (in addition to continental motion and seafloor spreading) comes from the fact that Earth's surface has two very different types of crust (Figure 4.17). **Seafloor crust** is made primarily of the relatively high-density igneous rock called *basalt*, which commonly erupts from volcanoes like those along mid-ocean ridges and in Hawaii. Seafloor crust is typically only 5–10 kilometers thick, and radiometric dating shows that it is quite young—the average age is about 70 million years, and even the oldest seafloor crust is less than about 200 million years old. **Continental crust** is made of lower-density rock, such as granite, and its rock spans a wide range of ages from the very young all the way back to the oldest rocks found on Earth. Continental crust is much thicker than seafloor crust—typically 20–70 kilometers thick—but it sticks up only slightly higher because its weight presses it down farther onto the mantle below.

The two types of crust make it clear that Earth's surface must undergo continual change. New seafloor crust continually emerges at sites of seafloor spreading, while the wide age range of continental crust tells us that the continents have gradually built up over time.

THE MECHANISM OF PLATE TECTONICS Today, we understand that plates move in concert with underlying mantle convection, driven by the heat released from Earth's interior. Over millions of years, the movements involved in plate tectonics act like a giant conveyor belt for Earth's lithosphere (Figure 4.18). This movement explains many of the major features of Earth's geology.

Seafloor spreading occurs at mid-ocean ridges because they are places where mantle material rises upward toward the surface. As it

FIGURE 4.18 INTERACTIVE FIGURE
Plate tectonics acts like a giant conveyor belt for Earth's lithosphere.

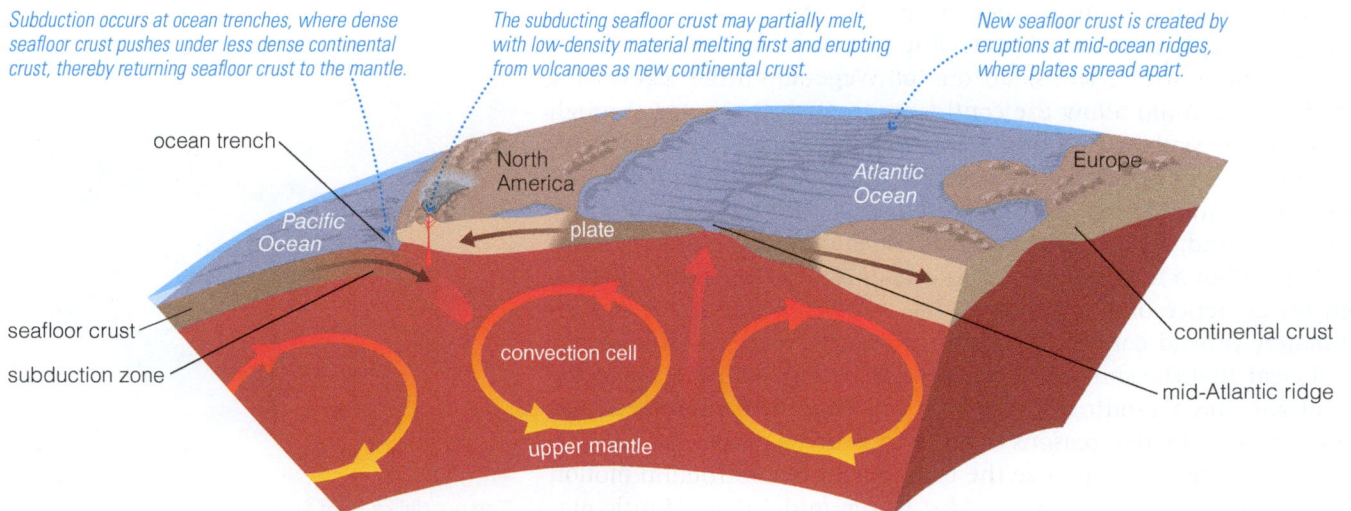

Subduction occurs at ocean trenches, where dense seafloor crust pushes under less dense continental crust, thereby returning seafloor crust to the mantle.

The subducting seafloor crust may partially melt, with low-density material melting first and erupting from volcanoes as new continental crust.

New seafloor crust is created by eruptions at mid-ocean ridges, where plates spread apart.

ocean trench

North America

Atlantic Ocean

Europe

Pacific Ocean

plate

seafloor crust

subduction zone

convection cell

upper mantle

continental crust

mid-Atlantic ridge

gets close to the surface, the lower pressure allows it to partly melt; the molten material then erupts to the surface, cooling and contracting as it spreads sideways. This explains both the formation of the basaltic sea-floor crust and the characteristic shapes of the ridges. Worldwide along the mid-ocean ridges, new crust covers an area of about 2 square kilometers every year, enough to replace the entire seafloor within about 200 million years—which explains the geologically young age of seafloor rocks.

Over tens of millions of years, any piece of seafloor crust gradually makes its way across the ocean bottom, then finally gets recycled into the mantle in the process we call **subduction.** Subduction occurs where a seafloor plate meets a continental plate, which is generally somewhat offshore at the edge of a sloping *continental shelf.* The continental rocks are less dense than those on the seafloor. As the dense seafloor crust of one plate pushes under the less dense continental crust of another plate, it can pull the entire surface downward to form a deep *ocean trench.* At some trenches, the ocean depth is more than 8 kilometers.

Beneath a subduction zone, the descending seafloor crust heats up and may begin to melt as it moves deeper into the mantle. If enough melting occurs, the molten rock may erupt upward. As you can see in Figure 4.18, the process of subduction tends to make the melting occur under the edges of the continents, which is why so many active volcanoes tend to be found along those edges. This volcanic activity explains the presence of many coastal mountain ranges (as well as many older mountain ranges that are no longer located along coasts). Moreover, the lowest-density material tends to melt first, which is why the continental crust emerging from these landlocked volcanoes is lower in density than seafloor crust.

Spreading and subduction are not the only ways in which plates interact. Two continental plates crashing into each other can push each other upward, providing a second way of creating mountain ranges. The Himalayas are still slowly growing in this way as the plate carrying India pushes into the plate carrying most of the rest of Eurasia (Figure 4.19). In places where continental plates are pulling apart, the crust thins and can create a large *rift valley.* The East African rift zone is an example (see Figure 4.15). This rift is slowly growing and will eventually tear the African continent apart. At that point, rock rising upward with mantle convection will begin to erupt from the valley floor, creating a new zone of seafloor spreading. A similar process tore the Arabian Peninsula from Africa, creating the Red Sea (Figure 4.20).

Places where plates slip sideways relative to each other are marked by what we call plate boundary **faults.** For example, the San Andreas Fault in California marks a line where the Pacific plate is moving northward relative to the continental plate of North America (Figure 4.21). At its current rate, this motion will bring Los Angeles (on the Pacific plate) and San Francisco Bay (on the North American plate) together in about 20 million years. The two plates do not slip smoothly against each other. Instead, their rough surfaces catch, and tension can build up until it is so great that it forces a rapid and violent shift, causing an earthquake. In contrast to the usual motion of plates, which proceeds at a few centimeters per year, an earthquake can move plates by several *meters* in a few seconds. The movement can level cities, set off destructive tsunamis, and make the whole planet vibrate with seismic waves (much like a ringing bell).

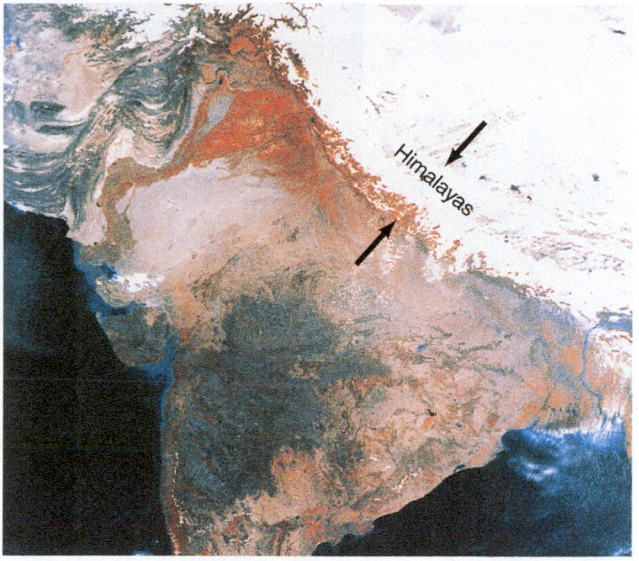

FIGURE 4.19
This satellite photo shows the Himalayas, which are still slowly growing as the plate carrying India pushes into the Eurasian plate. Arrows indicate the directions of plate motion. (See the plate boundaries in Figure 4.15.)

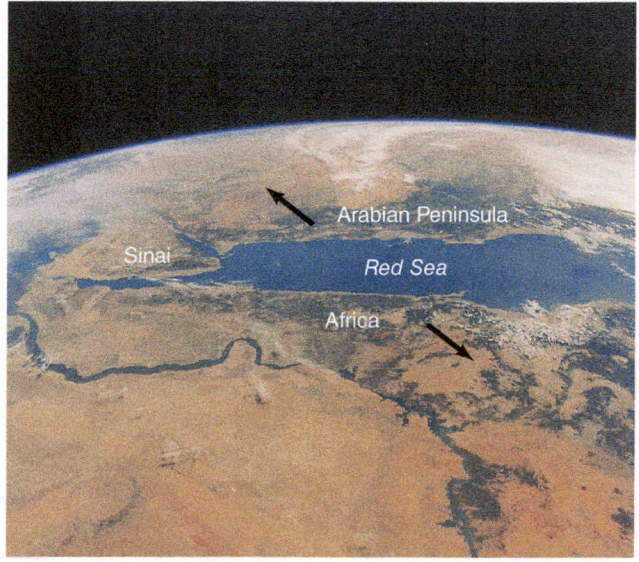

FIGURE 4.20
When continental plates pull apart, the crust thins and deep rift valleys form. This process tore the Arabian Peninsula from Africa, forming the Red Sea. Arrows indicate the directions of plate motion.

FIGURE 4.21

California's San Andreas Fault marks a boundary where plates are sliding sideways, as shown by the white arrows; asterisks indicate sites and years of major earthquakes. The inset shows an aerial photo of the San Andreas Fault (in the Coachella valley), which runs between the arrows; notice the offset of the land on the two sides of the fault.

FIGURE 4.22

The Hawaiian Islands are just the most recent of a long string of volcanic islands made by a mantle hot spot. The image of Loihi (lower right) was obtained by sonar, as it is still entirely underwater. The long chain records the past 60 million years of history of the oceanic crust in the region.

Think About It By studying the plate boundaries in Figure 4.15, explain why the west coast states of California, Oregon, and Washington are prone to more earthquakes and volcanoes than most other parts of the United States. Find the locations of recent earthquakes and volcanic eruptions worldwide. Do the locations fit the pattern you expect?

Not all earthquakes and volcanoes occur near plate boundaries. Earthquakes sometimes occur along old or buried faults that are now far from plate boundaries. For example, some of the biggest earthquakes in U.S. history occurred in 1811 and 1812 along the New Madrid fault zone, which runs through parts of Illinois, Kentucky, Missouri, Arkansas, and Tennessee.

Volcanic activity may occur far from plate boundaries when a plume of hot mantle material rises up to make what we call a **hot spot.** The Hawaiian Islands are the result of a hot spot that has been erupting basaltic lava for tens of millions of years. Plate tectonics gradually carries the Pacific plate over the hot spot, forming a chain of volcanic islands as different parts of the plate lie directly above the hot spot at different times (Figure 4.22). Today, most of the lava erupts on or near the Big Island of Hawaii, giving much of this island a young, rocky surface. About a million years ago, the Pacific plate lay farther to the southeast (relative to its current location), and the hot spot built the island of Maui. Before that, the hot spot created other islands, including Oahu (3 million years ago), Kauai (5 million years ago), and Midway (27 million years ago). The older islands are more heavily eroded. Midway has been eroded so much that it barely rises above sea level. The movement of the plate over the hot spot continues today, building underwater volcanoes that eventually will rise above sea level to become new Hawaiian Islands. The growth of a future island, named Loihi, is already well under way—prime beach real estate should be available there in a million years or so. Hot spots can also occur beneath continental crust. For example, a continental hot spot is responsible for the volcanism that supplies the heat for the geysers and hot springs of Yellowstone National Park.

PLATE TECTONICS THROUGH TIME We can use the current motions of the plates to project the arrangement of continents millions of years into the past or the future. For example, at a speed of 2 centimeters per year, a plate will travel 2000 kilometers in 100 million years. Figure 4.23 shows two past arrangements of the continents, along with the present and one future arrangement. Note that the present-day continents were once all stuck together in a single "supercontinent," sometimes called *Pangaea* (meaning "all lands"), which began to break up a little more than 200 million years ago.

Mapping the sizes and locations of continents at even earlier times is more difficult. However, studies of magnetized rocks (which can record the orientation of ancient magnetic fields) and comparisons of fossils found in different places around the world have allowed geologists to map the movement of the continents much further into the past. It seems that, over at least the past billion years or more, the continents have slammed together, pulled apart, spun around, and changed places on the globe. Central Africa once lay at Earth's South Pole, and Antarctica once was near the equator. The continents continue to move, and their current arrangement is no more permanent than any past arrangement.

Has Earth had plate tectonics throughout its history? We are not yet certain, but we have some evidence for plate tectonics going back quite far in time. The oldest fairly definitive evidence comes from seismic studies that suggest the presence of an ancient subduction zone (found in Canada) that formed some 2.7 billion years ago; subduction would be a sure sign of the conveyorlike action of plate tectonics. More controversial is the recent zircon evidence suggesting that continental crust had already begun to form more than 4.4 billion years ago. If the evidence is being properly interpreted, it suggests that plate tectonics began just shortly after the birth of Earth, since continental crust is generally formed as a direct result of the separation of rock by density along subduction zones.

CAUSE AND EFFECTS OF PLATE TECTONICS We have a fairly good understanding of how plate tectonics works and how it is driven by Earth's internal heat and mantle convection. However, we still face at least one significant mystery: Why does plate tectonics operate only on Earth among the terrestrial worlds?

The answer is probably simple for the Moon, Mercury, and Mars: Because their small size has allowed their interiors to cool much more than Earth's interior, their lithospheres have thickened. If they have any remaining internal convection at all, it is too weak to break their thick lithospheres into plates. Venus poses a greater mystery, since it is almost the same size as Earth and therefore should have retained a similar amount of internal heat.

We still do not know why Venus appears to lack plate tectonics today or whether it had plate tectonics in the past [Section 10.2]. However, we have at least one plausible hypothesis: As we'll discuss further in Chapter 10, Venus's high surface temperature has probably baked out water from its crust and upper mantle. This drying of the rock may have strengthened and thickened Venus's lithosphere so that it has resisted the fracturing that occurred on Earth; the high temperature may also make Venus's lithosphere less brittle than Earth's colder crust. If this hypothesis is correct, then we can ultimately trace the cause of plate tectonics to two factors: heat-driven mantle convection and a lithosphere thin and brittle enough to be fractured by the movement of the underlying mantle.

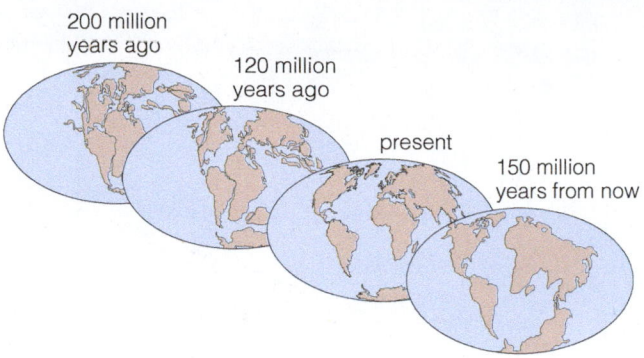

FIGURE 4.23
Selected past, present, and future arrangements of Earth's continents.

Whatever its cause, the effects of plate tectonics are profound. We've seen that plate tectonics is the most important geological mechanism on Earth, and it plays a key role in explaining nearly all of Earth's geological features. But its deeper significance to life lies in the fact that its recycling of rock turns out to play a crucial role in climate regulation, a topic we'll address in Section 4.5.

Why does Earth have a protective magnetic field?

As we noted briefly in Section 4.1, Earth's magnetic field plays a key role in our planet's habitability because it shields our planet from the solar wind. Let's investigate this protection—and why it exists—in a bit more detail.

IMPORTANCE OF THE MAGNETIC FIELD To understand the protective role of the magnetic field, we must first recognize that planetary atmospheres do not necessarily last forever. For Earth and other terrestrial worlds, atmospheres come primarily from outgassing, but the released gas can later be lost to space in three major ways.

First, gas molecules can sometimes move fast enough that they may exceed their world's escape velocity, in which case they may simply "take off" into space. This process is often called **thermal escape,** because gas particles move at higher speeds when their temperature is higher, and hence are more likely to escape when temperatures are high than low. Moreover, at any particular temperature, lightweight atoms or molecules, such as hydrogen and helium, tend to move faster than heavier ones such as oxygen, making it easier for them to escape. That is why the terrestrial worlds lack any significant amount of hydrogen or helium gas: Unlike the large, jovian planets, which can retain hydrogen and helium because their large masses give them high escape velocities, any hydrogen and helium that once surrounded the terrestrial worlds would have long ago escaped to space. More generally, smaller worlds have lower escape velocities, so gas escapes from them much more readily. That is why the Moon and Mercury, the two smallest of the terrestrial worlds, have become essentially airless, while Venus, Earth, and Mars retain significant atmospheres of relatively heavy molecules.

Second, impacts can also blast atmospheric gas into space. Again, smaller worlds are more prone to this type of loss because of their lower escape velocities: Equivalent impacts are more likely to blast material upward with escape velocity on a smaller world than a larger world.

Third, gas can be lost through a mechanism known as **solar wind stripping,** which occurs when particles from the solar wind in effect sweep atmospheric gas particles into space. This mechanism acts slowly, but calculations suggest that it can strip away significant amounts of gas over billions of years. The key role of the magnetic field lies in the way it protects our planet from solar wind stripping.

REQUIREMENTS FOR A GLOBAL MAGNETIC FIELD A **magnetic field** can affect charged particles or magnetized objects in its vicinity. For example, if you've ever used a compass, you know that Earth has a magnetic field that determines the direction in which the compass needle points. The global extent of Earth's magnetic field gives us a strong clue that the field is generated *inside* our planet.

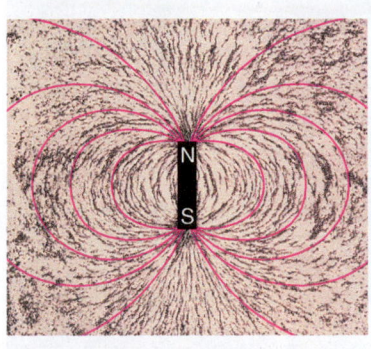

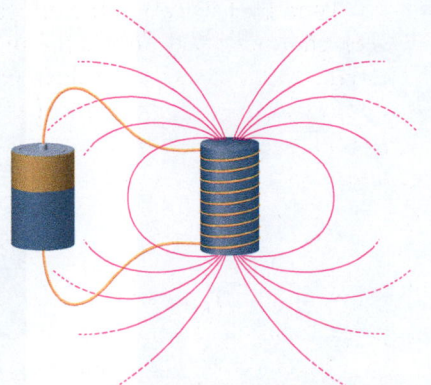

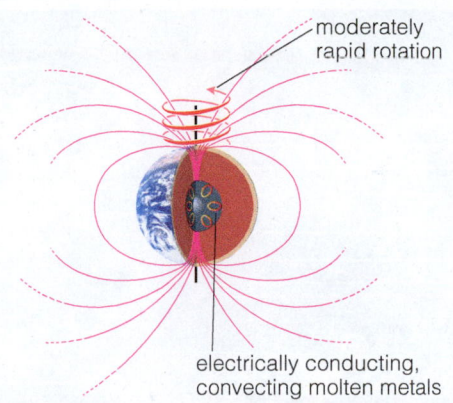

moderately rapid rotation

electrically conducting, convecting molten metals

a This photo shows how a bar magnet influences iron filings (small black specks) around it. The *magnetic field lines* (red) represent this influence graphically.

b A similar magnetic field is created by an electromagnet, which is essentially a wire wrapped around a bar and attached to a battery. The field is created by the battery-forced motion of charged particles (electrons) along the wire.

c Earth's magnetic field also arises from the motion of charged particles. The charged particles move within Earth's liquid outer core, which is made of electrically conducting, convecting molten metals.

FIGURE 4.24
Sources of magnetic fields.

You are probably familiar with the general pattern of the magnetic field created by an iron bar magnet (Figure 4.24a). Earth's magnetic field is generated by a process more similar to that of an *electromagnet,* in which the magnetic field arises as a battery forces charged particles (electrons) to move along a coiled wire (Figure 4.24b). Earth does not contain a battery, of course, but charged particles move with the molten metals in its liquid outer core (Figure 4.24c). Internal heat causes the liquid metals to rise and fall (convection), while Earth's rotation twists and distorts the convection pattern of these molten metals. The result is that electrons in the molten metals move within Earth's outer core in much the same way that they move in an electromagnet, generating Earth's magnetic field.

We can generalize what we know about Earth's magnetic field to other worlds. There are three basic requirements for a global magnetic field:

1. An interior region of electrically conducting fluid (liquid or gas), such as molten metal
2. Convection in that layer of fluid
3. At least moderately rapid rotation of the planet

Earth is unique among the terrestrial worlds in meeting all three requirements, which is why it is the only terrestrial world in our solar system with a strong magnetic field. The Moon has no magnetic field, presumably because its core has long since cooled and ceased convecting. Mars's core probably still retains some heat, but not enough to drive core convection, which is why it also lacks a magnetic field today. Venus probably has a molten core layer much like that of Earth, but either its convection or its 243-day rotation period is too slow to generate a magnetic field. Mercury poses a slight enigma: It possesses a measurable magnetic field despite its small size and slow, 59-day rotation. The reason the planet has a magnetic field may be tied to the fact that Mercury has a very large metal core, which may still be partly molten and convecting.

The same three requirements for a magnetic field also apply to jovian planets and stars. For example, Jupiter has a very strong magnetic field as a result of its rapid rotation (it rotates about once every 10 hours)

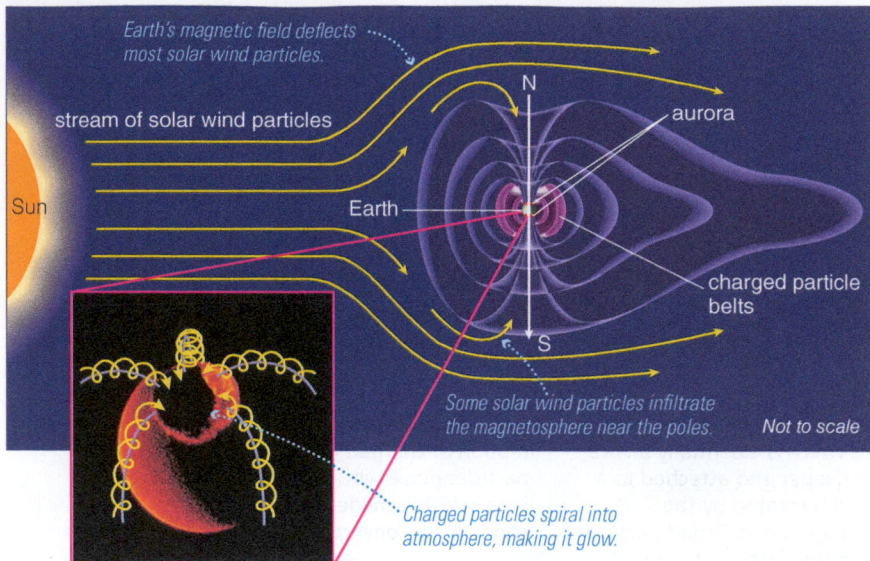

a This diagram shows how Earth's invisible magnetosphere (represented in purple) deflects solar wind particles. Some particles accumulate in charged particle belts encircling our planet. The inset is a photo of a ring of auroras around the North Pole; the bright crescent at its left is part of the day side of Earth.

Labels in diagram:
Earth's magnetic field deflects most solar wind particles.
stream of solar wind particles
N
aurora
Sun
Earth
charged particle belts
S
Some solar wind particles infiltrate the magnetosphere near the poles.
Not to scale
Charged particles spiral into atmosphere, making it glow.

b This photograph shows the aurora near Yellowknife, Northwest Territories, Canada. In a video, you would see these lights dancing about in the sky.

FIGURE 4.25
Earth's magnetosphere acts like a protective bubble, shielding our planet from the charged particles of the solar wind.

and an interior layer of convecting metallic hydrogen that conducts electricity. The Sun and other stars have magnetic fields generated by the combination of convection of ionized gas (plasma) in their interiors and rotation.

THE PROTECTIVE MAGNETOSPHERE The magnetic field protects Earth's surface and atmosphere from most of the energetic particles of the solar wind because it creates a **magnetosphere** that acts like a protective bubble surrounding our planet. The magnetosphere deflects most of the solar wind particles while channeling a few toward the poles, where they can cause auroras (Figure 4.25). The magnetosphere itself is invisible, but we can map its presence with devices that work much like compass needles and we can detect particles that become trapped within it (in the charged particle belts, also known as *Van Allen belts*).

The magnetosphere generally deflects particles while they are still high above our atmosphere, and it therefore prevents the solar wind from stripping Earth's atmospheric gas away. Evidence for this protective function comes from studying our neighboring planets. As we'll discuss in Chapter 8, Mars today apparently has much less atmospheric gas than it did in the distant past, and we suspect that Mars lost much of its gas when its interior cooled to the point that it no longer generated a strong magnetic field and protective magnetosphere. Careful studies of the isotopic composition of Venus's atmosphere suggest that it, too, has lost gas to solar wind stripping, as we would expect given its lack of magnetic field. However, Venus has such a thick atmosphere that its overall gas loss has been proportionally small.

For Earth, calculations indicate that without the protection of Earth's global magnetic field, our planet would by now have lost most of its atmosphere to solar wind stripping. In other words, without the magnetic field, Earth probably could not have retained the atmosphere upon which life depends.

4.5 Climate Regulation and Change

At the beginning of this chapter, we stated that three crucial ingredients in Earth's long-term habitability have been volcanism, plate tectonics, and the magnetic field. We have seen that volcanism's most important role has been in releasing the gases that formed our oceans and atmosphere, while the magnetic field has helped prevent atmospheric gas from being lost to space. As for plate tectonics, we have seen the way it is responsible for shaping our planet's surface; however, plate tectonics plays an even more important role, because it helps regulate the climate.

You are probably aware that Earth's climate has not been perfectly stable through time: Even during human history our planet has experienced what we call *ice ages*, and as we look back through geological time, we find other periods of far more severe cold or warmth. Nevertheless, the climate has been sufficiently stable for life to exist continually for some 4 billion years. Because life on Earth needs liquid water [Section 5.3], we infer that the oceans have remained at least partially liquid throughout this long period of time. Although the temperature range in which water can be liquid may seem wide to us humans, when we compare it to temperatures found on other worlds, we realize that Earth's climate has been remarkably stable through time.

At first, Earth's stable climate might not seem surprising. After all, the primary source of heat for the atmosphere and oceans is the Sun, and Earth's orbit about the Sun should not have changed much since its formation. However, theoretical models of the Sun and observations of other Sun-like stars reveal an important fact: Stars gradually brighten with age. Models suggest that the Sun today may be as much as 30% brighter than it was when Earth formed, which means the young Earth received a lot less solar warmth and light than it does today. How, then, could our planet have been warm enough for liquid water in the distant past, and why hasn't our planet overheated as the Sun brightened? To answer these and other questions about Earth's long-term climate, we must begin by investigating why Earth is warm enough for liquid water in the first place.

How does the greenhouse effect make Earth habitable?

Most people assume that Earth is warm enough for liquid water simply because it is at the "right" distance from the Sun, but this clearly is not the whole story. The Moon lies at the same distance, but its daytime temperatures rise to 125°C (257°F)—well above the normal boiling point of water—while its nighttime temperatures plummet to a frigid −175°C (−283°F). Moreover, the fact that the Sun has brightened with time means that even if we were at the "right" distance when Earth was young, that same distance would be too close to the now-brighter Sun.

It's fairly easy to calculate the Earth's expected temperature based solely on its distance from the Sun and the amount of incoming sunlight absorbed by its surface. Such a calculation shows that the **global average temperature**—that is, the average temperature for the entire planet—would be −16°C(3°F), well below the freezing point of water. But Earth is not frozen. The actual global average temperature today is about 15°C (59°F), and geological evidence shows that it has been warmer at various

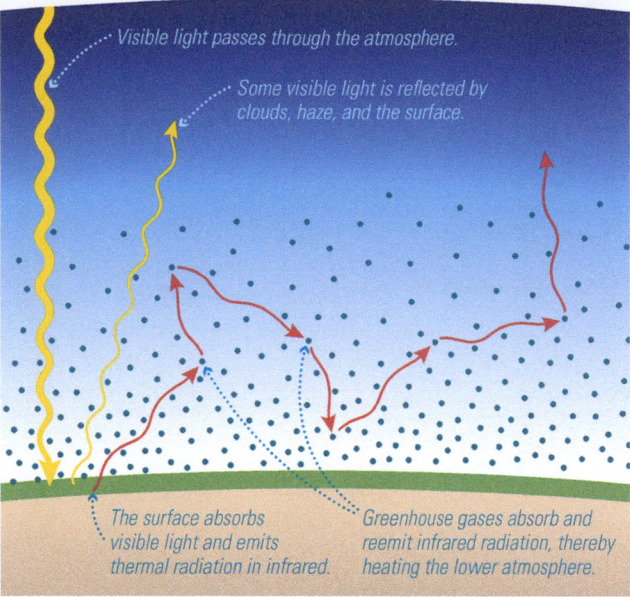

Visible light passes through the atmosphere.

Some visible light is reflected by clouds, haze, and the surface.

The surface absorbs visible light and emits thermal radiation in infrared.

Greenhouse gases absorb and reemit infrared radiation, thereby heating the lower atmosphere.

FIGURE 4.26
The greenhouse effect makes the surface and lower atmosphere much warmer than they would be without greenhouse gases such as water vapor, carbon dioxide, and methane.

times in the past. Something must be making our planet much warmer than we would expect based on its distance from the Sun alone, and that something is what we call the **greenhouse effect.**

Figure 4.26 shows the basic mechanism of the greenhouse effect. Sunlight consists mostly of visible light, which passes easily through most atmospheric gases. Some of this visible light gets absorbed by the ground, while the rest is reflected back to space (much of it by clouds). The ground must return the energy it absorbs back to space, because if it didn't the energy would make the ground heat up rapidly. However, the fact that the ground doesn't glow in the dark tells us that the ground does not return the energy in the same visible-light form in which it absorbs it. Instead, the ground returns the energy in the form of infrared light.

The greenhouse effect works by temporarily "trapping" some of the infrared light, slowing its return to space. This trapping occurs because some atmospheric gases can absorb the infrared light. Gases that are particularly good at absorbing infrared light are called **greenhouse gases,** and they include water vapor (H_2O), carbon dioxide (CO_2), and methane (CH_4). These gases absorb infrared light effectively because their molecular structures make them prone to begin rotating or vibrating when struck by an infrared photon (an individual "piece" of light); diatomic molecules such as nitrogen (N_2) and oxygen (O_2) generally cannot rotate or vibrate in these ways and hence do not absorb infrared light.

After a greenhouse gas molecule absorbs the energy of an infrared photon, it quickly releases the energy by emitting a new infrared photon. However, the new photon will be emitted in some random direction that is unlikely to be the same direction from which the original photon came. This photon can then be absorbed by another greenhouse gas molecule, which does the same thing. The net result is that greenhouse gases tend to slow the escape of infrared radiation from the lower atmosphere, while their molecular motions heat the surrounding air. In this way, the greenhouse effect makes the surface and the lower atmosphere warmer than they would be from sunlight alone. A blanket offers a good analogy: You stay warmer under a blanket not because the blanket itself provides any heat, but rather because it slows the escape of your body heat into the cold outside air. The more greenhouse gases that are present, the greater the degree of surface warming. On Earth, the naturally occurring greenhouse effect is strong enough to raise the global average temperature by about 31°C from what it would be without greenhouse gases. Without this warming, our planet would be frozen over.

Incidentally, you are probably aware that the greenhouse effect is often in the news, usually portrayed in a negative light as part of an environmental problem. But as we have just seen, the greenhouse effect is not a bad thing in and of itself, since life on Earth would not exist without it. Why, then, is the greenhouse effect discussed as an environmental problem? The reason is that human activity is adding more greenhouse gases to the atmosphere, thereby strengthening the greenhouse effect and further warming our planet ("global warming"). While the precise effects of this human-induced warming are difficult to predict [Section 10.5], we need only look to our hot neighbor Venus to see that changes in the greenhouse effect should not be taken lightly. While the greenhouse effect makes Earth livable, it is also responsible for the searing 470°C (878°F) temperature of Venus—proving that it's possible to have too much of a good thing.

Think About It Carbon dioxide makes up less than 1% of our atmosphere today, while nitrogen and oxygen together make up some 98% of our atmosphere. Why, then, do we focus on carbon dioxide when we talk about Earth's climate?

What regulates Earth's climate?

The case of Venus leads us to a crucial question about Earth's hospitable climate. As we'll discuss in Chapter 10, Venus's extreme greenhouse effect occurs because its atmosphere contains almost 200,000 times as much carbon dioxide as Earth's atmosphere. But Venus and Earth are nearly the same size and both were made from similar materials, so volcanic outgassing should have released similar amounts of carbon dioxide on both worlds. Moreover, outgassing from modern-day volcanoes on Earth shows that they do indeed release plenty of carbon dioxide, and over time, volcanoes must have outgassed nearly as much carbon dioxide into Earth's atmosphere as we find in the atmosphere of Venus. Where, then, did all of Earth's carbon dioxide end up?

Geological studies reveal the answer: Most of Earth's carbon dioxide is locked up in **carbonate rocks**—sedimentary rocks, such as limestone, that are rich in carbon and oxygen. By estimating the total amount of carbonate rock on Earth, we find that these rocks contain about 170,000 times as much carbon dioxide as our atmosphere, which means that Earth does indeed have almost as much total carbon dioxide as Venus. Venus lacks carbonate rock (for reasons we'll discuss in Chapter 10), so all of its carbon dioxide remains in its atmosphere. Keep in mind that this difference between the two planets in carbon dioxide location makes all the difference in the world: If Earth's carbon dioxide were in our atmosphere rather than in carbonate rocks, our planet would be nearly as hot as Venus and certainly uninhabitable.

Of course, the fact that Earth's carbon dioxide is locked up in rocks leads to a deeper question: How did it get there? The answer lies with a mechanism closely tied to plate tectonics.

THE CARBON DIOXIDE CYCLE The mechanism by which carbon dioxide has been removed from Earth's atmosphere and by which the current small amount of atmospheric carbon dioxide remains stable is called the inorganic **carbon dioxide cycle**, or the **CO_2 cycle** for short. Let's follow the cycle as illustrated in Figure 4.27, starting with the carbon dioxide in the atmosphere:

- Atmospheric carbon dioxide dissolves in rainwater, creating a mild acid.
- The mildly acidic rainfall erodes rocks on Earth's continents, and rivers carry the broken-down minerals to the oceans.
- In the oceans, calcium from the broken-down minerals combines with dissolved carbon dioxide and falls to the ocean floor, making carbonate rocks.*
- Over millions of years, the conveyor belt of plate tectonics carries the carbonate rocks to subduction zones, where they are carried downward.
- As they are pushed deeper into the mantle, some of the subducted carbonate rock heats up and releases its carbon dioxide, which then outgasses back into the atmosphere through volcanoes.

---

*During the past half-billion years or so, carbonate minerals have been made by shell-forming sea animals, falling to the bottom in the seashells left after the animals die. Without the presence of animals, chemical reactions would do the same thing—and apparently did for most of Earth's history.

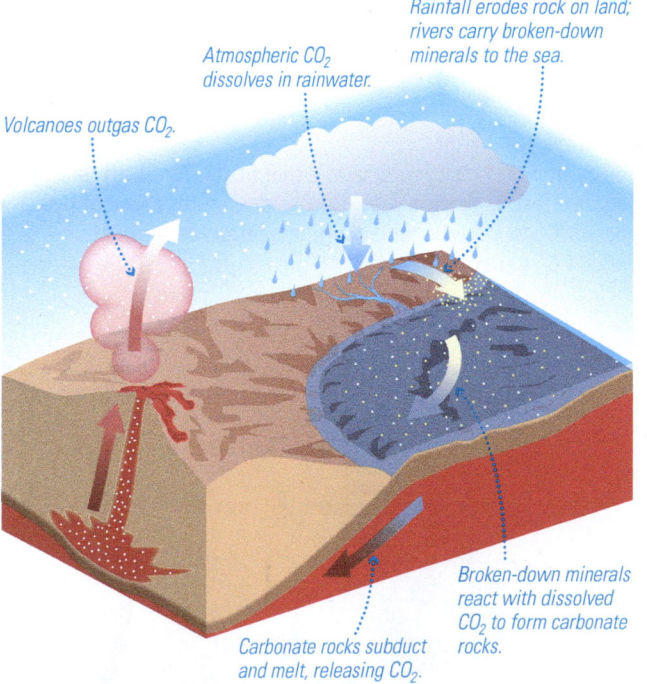

Volcanoes outgas CO_2.

Atmospheric CO_2 dissolves in rainwater.

Rainfall erodes rock on land; rivers carry broken-down minerals to the sea.

Carbonate rocks subduct and melt, releasing CO_2.

Broken-down minerals react with dissolved CO_2 to form carbonate rocks.

FIGURE 4.27
This diagram shows how the CO_2 cycle continually moves carbon dioxide from the atmosphere to the ocean to rock and back to the atmosphere. Note that plate tectonics (subduction in particular) plays a crucial role in the cycle.

In summary, the reason that Earth has so little carbon dioxide in its atmosphere is that most of the carbon dioxide was dissolved in the oceans, where chemical reactions converted it to carbonate minerals. In fact, about 60 times as much carbon dioxide is dissolved in the oceans as is present in our atmosphere, though this amount still pales in comparison to the 170,000 times as much that the CO_2 cycle has locked up in rock.

THE CO_2 CYCLE AS A THERMOSTAT The CO_2 cycle acts as a thermostat for Earth because of the way that changes in temperature feed back into the cycle. You are probably familiar with what we generally call **feedback processes**—processes in which a change in one property amplifies (positive feedback) or counteracts (negative feedback) the behavior of the rest of the system. For example, if someone brings a microphone too close to a loudspeaker, it picks up and amplifies small sounds from the speaker. These amplified sounds are again picked up by the microphone and further amplified, causing a loud screech. This sound feedback is an example of *positive feedback,* because it automatically amplifies itself. The screech usually leads to a form of *negative feedback:* The embarrassed person holding the microphone moves away from the loudspeaker, thereby stopping the positive sound feedback.

The CO_2 cycle has a built-in form of negative feedback that returns Earth's temperature toward "normal" whenever it warms up or cools down. This negative feedback occurs because the overall rate at which carbon dioxide is pulled from the atmosphere is extremely sensitive to temperature: the higher the temperature, the higher the rate at which carbon dioxide is removed. Figure 4.28 shows how it works. Consider first what happens if Earth warms up a bit. The warmer temperature means more evaporation and rainfall, pulling more CO_2 out of the atmosphere. The reduced atmospheric CO_2 concentration leads to a weakened greenhouse effect that counteracts the initial warming and cools the planet back down. Similarly, if Earth cools a bit, precipitation decreases and less CO_2 is dissolved in rainwater, allowing the CO_2 released by volcanism to build back up in the atmosphere. The increased CO_2 concentration strengthens the greenhouse effect and warms the planet back up.

FIGURE 4.28

The carbon dioxide cycle acts as a thermostat for Earth through negative feedback processes. Cool temperatures cause atmospheric CO_2 to increase, and warm temperatures cause atmospheric CO_2 to decline.

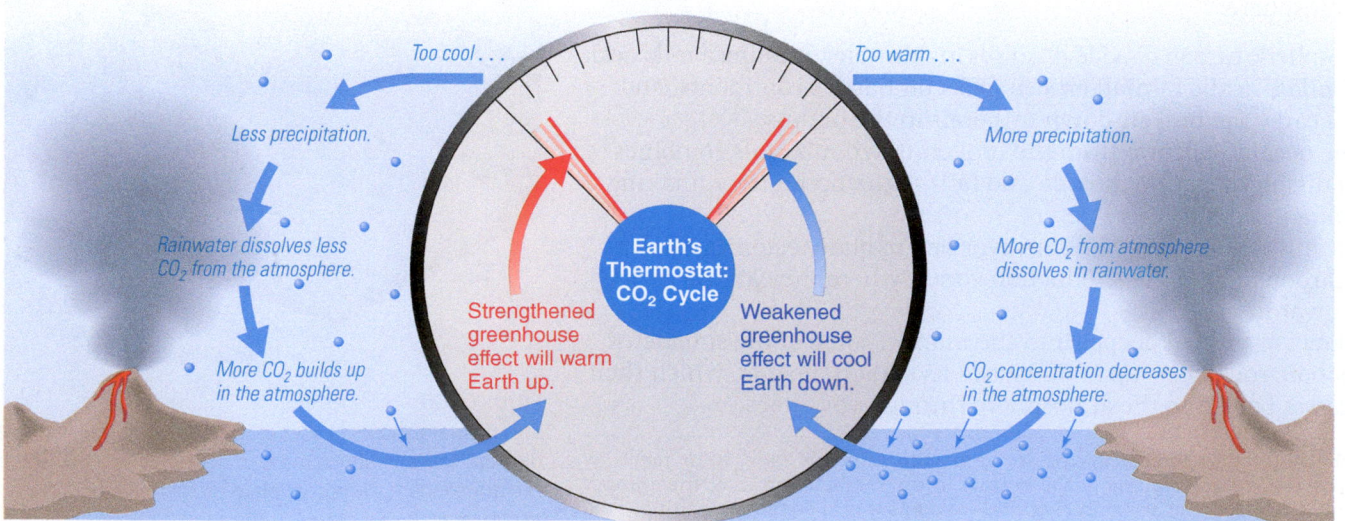

Overall, the natural thermostat of the CO_2 cycle has allowed the greenhouse effect to strengthen or weaken just enough to keep Earth's climate in a range that has allowed for liquid water, regardless of what other changes have occurred on our planet. And, because subduction plays a critical role in the CO_2 cycle, we now see the importance of plate tectonics to Earth's climate: Without plate tectonics, there would be no CO_2 cycle to regulate our planet's surface temperature.

How does Earth's climate change over long periods of time?

While Earth's climate has remained stable enough for the oceans to remain at least partly liquid throughout history, the climate has not been perfectly steady. Numerous warmer periods and numerous ice ages have occurred. Such variations are possible because the CO_2 cycle does not act instantly. When something begins to change the climate, it takes time for the feedback mechanisms of the CO_2 cycle to come into play, because these mechanisms depend on the gradual action of mineral formation in the oceans and of plate tectonics. Calculations show that the time required to stabilize atmospheric CO_2 through the CO_2 cycle is about 400,000 years. That is, if the amount of CO_2 in the atmosphere were to rise because of, say, increased volcanism, it would take some 400,000 years for the CO_2 cycle to restore temperatures to their current values.*

ICE AGES **Ice ages** occur when the global average temperature drops by a few degrees. The slightly lower temperatures lead to increased snowfall, which may cover continents with ice down to fairly low latitudes. For example, the northern United States was entirely covered with glaciers during the peak of the most recent ice age, which ended only about 10,000 years ago. In fact, relative to temperatures over at least the past 200 million years, we are still in an ice age. This ice age has persisted for the past 35 million years or so, with periods of deeper cold interspersed with periods of warmer temperatures, such as the present. Remarkably, recent evidence indicates that we can enter or leave a cold period very rapidly, within a time as short as a few decades.

The causes of ice ages are complex and not fully understood. Over periods of tens or hundreds of millions of years, the Sun's gradual brightening and the changing arrangement of the continents around the globe have at least in part influenced the climate. During the past few million years—a period too short for solar changes or continental motion to have a significant effect—the ice ages appear to have been strongly influenced by small, cyclical changes in Earth's rotation and orbit. These cyclical changes are often called *Milankovitch cycles*, after the Serbian scientist who suggested their role in climate change.

For example, while Earth's current axis tilt is about $23\frac{1}{2}°$, the tilt varies over time between about 22° and 25° (Figure 4.29). These small changes affect the climate by making seasons more or less extreme. Greater tilt means more extreme seasons, with warmer summers and colder winters. The extra summer warmth tends to prevent ice from building up, making the whole planet warmer on average. In contrast,

---

*This time scale applies to ocean/atmosphere equilibrium only. The time scale for crust recycling is much longer, while shorter-term climate variations in atmospheric CO_2 concentration can occur through factors besides the inorganic CO_2 cycle, such as cycling of carbon dioxide by life or the addition of CO_2 to the atmosphere through human activity.

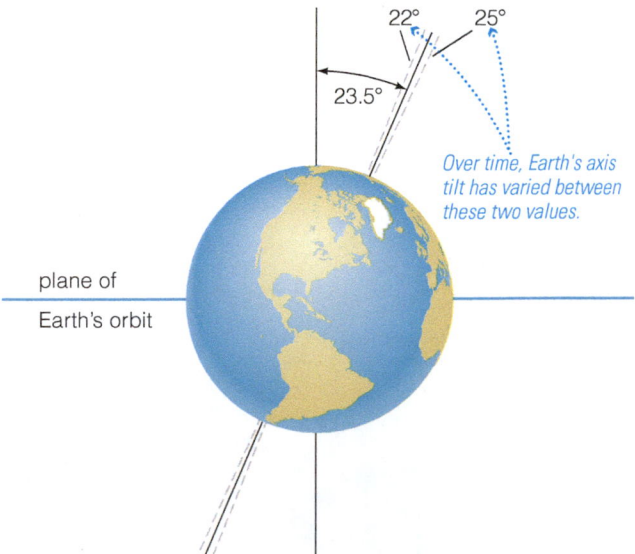

FIGURE 4.29
Small changes in Earth's axis tilt affect the climate: Greater tilt tends to mean a warmer climate.

smaller tilt means less extreme seasons that tend to keep polar regions colder and darker on average, allowing ice to build up. Earth's past periods of smaller axis tilt correlate well with colder climate and ice ages, especially when considered along with other cyclical changes in Earth's rotation and orbit.

SNOWBALL EARTH Geologists have discovered evidence of several particularly long and deep ice ages between about 750 and 580 million years ago, and another similar set between about 2.4 and 2.2 billion years ago. Although the limited evidence available from such distant times in the past can be difficult to interpret, it appears that at least some glaciers advanced all the way to the equator during these periods. There is significant scientific controversy over the full extent of the glaciation, but these periods of extreme cold have come to be called **snowball Earth.** We do not know what might have triggered these episodes or precisely how extreme the cold was, but a form of positive feedback probably played a major role: Once a significant increase in global ice began, the fact that ice can reflect up to about 90% of the sunlight hitting it means that much less sunlight would have been absorbed by the surface, causing Earth to cool even further. Some models suggest this positive feedback may have driven the global average temperature as low as −50°C (−58°F), causing the oceans to freeze to a depth of 1 kilometer or more. Other models suggest the oceans never froze completely, making Earth more of a "slushball" than a snowball. Either way, it seems that Earth became far colder during the snowball periods than in more recent ice ages.

How would Earth have recovered from a snowball phase? Figure 4.30 shows the current model. Even if Earth's surface got cold enough for the ocean surface to freeze completely, the interior would still have remained hot. As a result, volcanism would have continued to add CO_2 to the atmosphere. Oceans covered by ice would have been unable to absorb this CO_2 gas, and the CO_2 content of the atmosphere would have gradually built up and strengthened the greenhouse effect. Eventually, the strengthening greenhouse effect would have warmed Earth enough to start melting the ice. The feedback processes that started the snowball Earth episode then moved in reverse. As the ice melted, more sunlight

FIGURE 4.30
The CO_2 cycle rescues Earth from a snowball phase.

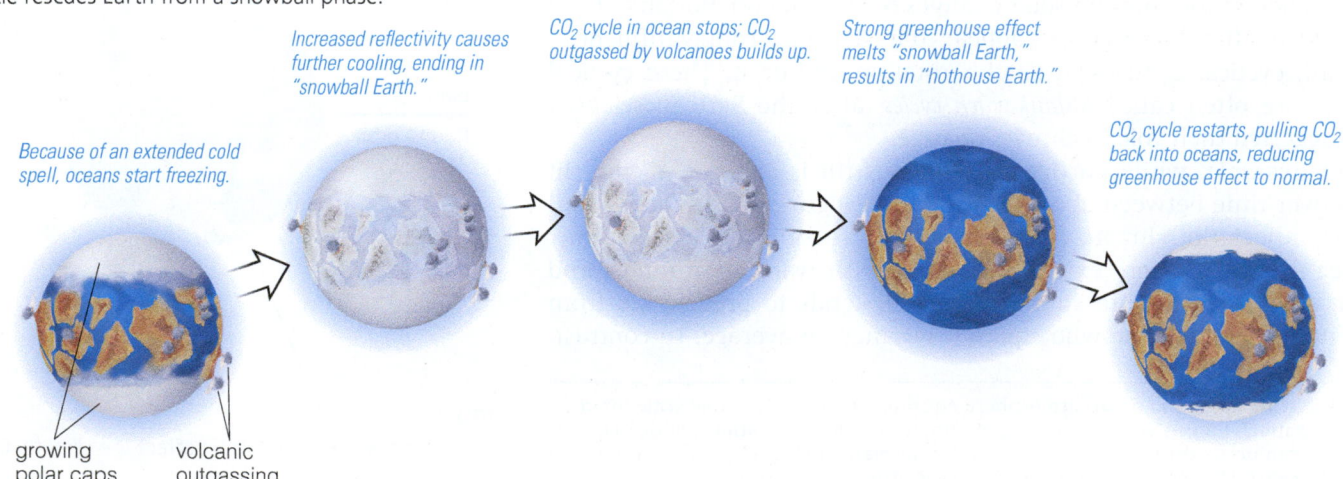

Because of an extended cold spell, oceans start freezing.

Increased reflectivity causes further cooling, ending in "snowball Earth."

CO_2 cycle in ocean stops; CO_2 outgassed by volcanoes builds up.

Strong greenhouse effect melts "snowball Earth," results in "hothouse Earth."

CO_2 cycle restarts, pulling CO_2 back into oceans, reducing greenhouse effect to normal.

growing polar caps   volcanic outgassing

would have been absorbed, warming the planet further. Moreover, because the CO_2 concentration was so high, the warming would have continued well past current temperatures, perhaps raising the global average temperature as high as 50°C (122°F). That is, in just a few centuries, Earth would have emerged from a snowball phase into what we might call a hothouse phase. Geological evidence supports the occurrence of dramatic increases in temperature at the end of each snowball Earth episode. Earth then slowly recovered from the hothouse phase as the CO_2 cycle removed carbon dioxide from the atmosphere.

Think About It Suppose Earth did not have plate tectonics. Could the planet ever recover from a snowball phase? Explain.

The snowball Earth episodes should have had severe consequences for any life on Earth at the time. Indeed, the end of the snowball Earth episodes roughly coincides with a dramatic increase in the diversity of life on Earth (the *Cambrian explosion* [Section 6.3]). Some scientists suspect that the environmental pressures caused by the snowball Earth periods may have led to a burst of evolution. If so, we might not be here today if not for the dramatic climate changes of the snowball Earth episodes.

EARTH'S LONG-TERM HABITABILITY We have covered a lot of ground in this chapter, but it has given us a clear picture of the major factors that have kept our planet habitable for the past 4 billion or more years. Let's briefly review a few of the key points:

- Volcanic outgassing released most of the gases that made the atmosphere and the water vapor that condensed to form the oceans.
- Earth has kept its atmosphere at least in part because the magnetic field has protected atmospheric gases from being stripped away by the solar wind.
- The greenhouse effect warms our planet enough for water to be liquid, but not so much that the water would boil away.
- This moderate greenhouse effect is maintained by the self-regulating mechanism of the CO_2 cycle, which depends on plate tectonics.
- Even with the regulation provided by the CO_2 cycle, the climate still goes through changes influenced by variations in Earth's axis tilt and other properties of its rotation and orbit.
- The regulatory mechanism sometimes breaks down, leading to periods such as the snowball Earth episodes, but the CO_2 cycle ultimately brings the climate back into balance.

These ideas should help us understand the prospects for finding other habitable worlds, especially those that might have the long-term habitability that could allow for the evolution of intelligent species and civilizations. Of course, these ideas also leave several questions unanswered. For example, should we expect to find plate tectonics and a CO_2 cycle on other worlds similar to Earth, or was some rare "luck" involved in Earth's getting these regulatory mechanisms? We might also wonder how long Earth's climate can continue to regulate itself as the Sun brightens with time, a topic we'll discuss in Chapter 10. Finally, and perhaps of the most immediate importance, we might wonder whether we humans could alter the regulatory mechanisms enough to cause serious consequences for our civilization. Unfortunately, the answer appears to be yes, but we'll save this discussion for Section 10.5.

4.6 Formation of the Moon

Earlier in this chapter, we stated that the Moon likely formed when a "giant impact" blasted away much of the material in the young Earth's outer layers. If you think about it, this is a rather astonishing idea, since it postulates a single event at a time so far in the distant past that we have little hope of finding rocks that survived the event and could tell its tale. For this chapter's case study in the process of science in action, we'll explore how and why this remarkable idea has gained widespread acceptance in the scientific community.

How did the giant impact model win out over competing models?

The existence of the Moon has long been puzzling. As we discussed in Chapter 3, the Moon counts as one of the "exceptions to the rules" when we consider the overall formation of the solar system, because it is unusually large compared to its planet (Earth). So how did the Moon form?

THREE MODELS, ALL FLAWED During the mid-twentieth century, three competing models were advanced to explain the Moon's existence. The first held that the Moon formed along with Earth through the same process of accretion; in essence, this idea suggested that the two worlds were born together. The second model suggested that the Moon had been an independent "planet" orbiting the Sun that was somehow captured into Earth's orbit. The third model suggested that a young, molten Earth had been spinning so rapidly that it split into two pieces, casting off the piece that became the Moon.

All three models had difficulties from the start. The joint formation model didn't work when scientists tried to calculate exactly what might have happened. If you try to build a planet and such a large moon in close proximity, gravitational interactions between them disrupt the process. Moreover, the Moon's average density is much lower than Earth's, which doesn't make sense if both worlds accreted from the same material.

The capture model seemed too improbable. It is difficult for a planet to gravitationally capture a passing object under any circumstances, because the passing object has its own orbital energy carrying it around the Sun. This energy cannot simply disappear, so an object can be captured only if it somehow loses some of its orbital energy. Captures therefore are most likely for small objects that lose orbital energy to friction with gas surrounding a planet. Mars probably captured its two small moons in this way, back at a time when it had a more extended atmosphere. Jupiter and the other jovian planets probably also captured many of their small moons in a similar fashion, back when they were still surrounded by gas from the solar nebula. But Earth never had an atmosphere thick enough to have slowed an object the size of the Moon. The only way that Earth could have captured the Moon would have been if another, similarly sized object had been passing by at precisely the same time as the Moon, and if the Moon and this other object had exchanged just enough energy for the Moon to end up in orbit of Earth. It's possible in principle, but highly unlikely.*

The splitting model also suffered from improbability. For example, it seemed unlikely that Earth could ever have been spinning fast enough to spin off the Moon. Still, like the other two models, it could not be ruled out completely at the time.

The *Apollo* missions to the Moon ended this debate, because study of rocks brought back by the astronauts ruled out all three models. The key finding was that the Moon rocks differed significantly in composition from Earth rocks. This immediately ruled out the joint formation model in which both worlds would have accreted from essentially identical material. The capture and splitting models were ruled out by chemical processing that had apparently occurred in the Moon rocks: The Moon rocks contained virtually no **volatiles,** or easily vaporized ingredients. In this context, volatile ingredients include not only things such as water but also elements such as lead and gold that vaporize at lower temperatures than other metals and rocks. Because these volatile elements should have been mixed in with other elements in any accreting object, the Moon could not have accreted first and been captured later. And because these volatiles are present on Earth, the Moon should also have had them if it split off from a spinning, molten planet.

THE GIANT IMPACT MODEL With all three models fatally flawed, it was back to the drawing board, which meant taking a closer look at the clues. Two key pieces of evidence soon began to stand out:

1. The Moon's average density is much smaller than Earth's, and in fact is about the same as the density of Earth's mantle. This suggests that the Moon lacks a large iron core like that of Earth and the other terrestrial planets, and instead is made almost entirely from material like that of Earth's mantle.
2. The overall composition of the Moon rocks looked quite similar to the composition of Earth's mantle material, except for the lack of volatile elements. Since heating could cause volatile elements to vaporize and escape into space, the rock composition suggested that the Moon was built from mantlelike material that had been strongly heated before it collected to form the Moon.

Taken together, the evidence had an almost obvious implication: The Moon was made from material that accreted in Earth orbit after first being violently blasted out of Earth's mantle. The idea that it was made from mantle material would explain the Moon's general resemblance to Earth's mantle, and the violence of being blasted out would explain the heat necessary to have allowed volatiles to vaporize and escape. But what could have blasted out a large portion of Earth's mantle?

Recall that models of planetary formation suggest that the late stages of accretion must have been extraordinarily violent [Section 3.4]. Rather than the four terrestrial planets that exist in the inner solar system today,

---

*There is one case in the solar system in which a fairly large moon apparently *was* captured: Neptune's moon Triton has orbital characteristics that make it almost certain to be a captured object. One model for Triton's capture assumes that it had a binary companion that served as the "other object" to carry off energy. However, because the ratio of Neptune's mass to Triton's mass is about 50 times that of Earth's mass to the Moon's mass, Triton's capture would have had a higher probability.

FIGURE 4.31

Artist's conception of the Moon's formation by a giant impact. As shown, the Moon formed quite close to a rapidly rotating Earth, but interactions related to tides over billions of years have slowed Earth's rotation and moved the Moon's orbit outward (see Figure 9.8).

A Mars-sized planetesimal crashes into the young Earth, shattering both the planetesimal and our planet.

Hours later, our planet is completely molten and rotating very rapidly. Debris splashed out from Earth's outer layers is now in Earth orbit. Some debris rains back down on Earth, while some will gradually accrete to become the Moon.

Less than a thousand years later, the Moon's accretion is rapidly nearing its end, and relatively little debris still remains in Earth orbit.

there may have been a dozen or more planet-size bodies. The current planets are the survivors both of close encounters of two bodies that would have sent one of them entirely out of the young solar system and of the shattering collisions that must have occurred. In other words, "giant impacts" in which one planet-size body struck another were not only possible but likely during this period. Such an impact on Earth would have had enough energy to blast much of the mantle into orbit, leading to the Moon's formation.

Today, scientists use sophisticated computer models to test the giant impact hypothesis, and it seems to work. The outcome of a giant impact depends on many factors, including the mass and speed of the incoming object and the precise place and angle at which it strikes Earth. By testing many scenarios, scientists have developed the model of the Moon's formation summarized in Figure 4.31. According to this model, a Mars-size object blasted into the young Earth. The impact must have occurred after Earth had differentiated but before the age of the oldest Moon rocks. Radiometric dating and other isotopic evidence tell us that the impact occurred within about 10 to 20 million years after Earth's iron sank to the core, which occurred more than 4.50 billion years ago.* The blast shattered and melted our planet, splashing out molten debris from the mantle. Much of this material fell back to Earth, but some remained in orbit. There, with its volatile content having vaporized and escaped, the material accreted to make the Moon. This model successfully reproduces the Moon's size, orbit, and composition, giving scientists confidence that it is on the right track.

*For example, the slightly different abundances on the Earth and Moon of tungsten-182, which is the daughter isotope of hafnium-182, indicates that hafnium-182 was still present when the Moon formed. The relatively short half-life of hafnium-182 (8.9 million years) means that no more than tens of millions of years could have passed since Earth had formed, because otherwise all the hafnium-182 would already have been gone. More precise measurements of the relative abundances can allow scientists to pin down the timing even more precisely.

Does the giant impact model count as science?

The giant impact model works so well and so successfully explains the Moon's characteristics that it is widely accepted among planetary scientists. But is it really "scientific" to invoke a single event for which we may never have more than indirect evidence?

One way to decide whether the giant impact model should count as science is to see how it stacks up against the hallmarks of science presented in Chapter 2. Looked at this way, the giant impact model certainly qualifies as science. It invokes a natural explanation for the origin of the Moon, one that even seems likely given what we know about the collisions that must have occurred in the solar system's early history. It also makes testable predictions. The most important of these predictions are the ones about composition—the idea that the Moon formed from mantle material leads to specific predictions about the composition of Moon rocks, and these predictions match the evidence. Note that the fact that the evidence was discovered before the model was proposed is not important here: Just as Kepler came up with his model of planetary motion to explain data that Tycho had already collected, the key point is that the model has been worked out in detail and it successfully matches the observations. The model will also be subject to ongoing tests in the future. For example, the *Apollo* astronauts collected Moon rocks from only a handful of sites on the Moon. In the future, we will presumably collect rocks from many other parts of the Moon. If these rocks were to turn up compositional surprises, it would cause us to reconsider and perhaps even discard the giant impact model. In summary, the giant impact model exhibits all the hallmarks of science: It is natural and testable, and we can imagine future discoveries that would cause us to call it into question.

Think About It Considering the evidence for the giant impact model, do you think it qualifies as a hypothesis or a theory or something in between? Note that even scientists disagree on this question, so be sure to defend your opinion.

The giant impact model has important consequences both for our understanding of the solar system and for the search for life in the universe. For our solar system, it may help explain other "exceptions to the rules." If a giant impact really was as likely as we have presumed, then Earth should not have been the only object to suffer one, and limited evidence points to several other giant impacts. Mercury's surprisingly large iron core is easily explained if a giant impact also blasted away much of its mantle, but without leaving a moon behind. Pluto's moon Charon also shows characteristics we'd expect if it formed in a giant impact. A similar event might also account for the huge axis tilt of Uranus, and perhaps for Venus's slow, backward rotation.

In terms of life, the consequences of the giant impact model lie with roles the Moon has played in shaping Earth's biological history. The Moon is the primary cause of Earth's tides (the Sun contributes less than half as much to Earth's tides as the Moon), and many living organisms have biological cycles tied to the tides. If there had been no giant impact and no Moon, these types of cycles might not have arisen. The Moon also plays a role in Earth's long-term climate stability. Recall that small changes in Earth's axis tilt can significantly change the climate, bringing on or ending ice ages. Models show that Earth's axis tilt would vary much more if we did not have the Moon; the Moon's gravity exerts a stabilizing influence on axis tilt. Indeed, as we'll see in Chapter 8, evidence suggests that Mars undergoes much more extreme changes in axis tilt, changes that are possible because it lacks a large moon. Some scientists therefore wonder if the existence of the Moon, along with the axis tilt stability that it brings, was necessary to our own evolution on Earth. If it was—and if the Moon really is the result of a random giant impact—then the possibility of finding intelligent life elsewhere may depend on the likelihood of giant impacts that result in the formation of large moons around terrestrial planets.

The Big Picture

PUTTING CHAPTER 4 IN PERSPECTIVE

In this chapter, we've explored the interconnections between geology and habitability, learning about the conditions that have made life on Earth possible. As you continue your studies, keep in mind the following "big picture" ideas:

- We can read Earth's geological history by studying the geological record. This history is not mere speculation. It is recorded in ways that we can verify independently in rocks and fossils from many places around the world and through the well-verified technique of radiometric dating for determining ages. While we may never know every detail of Earth's history, we already have a complete enough picture to understand the major processes and events that have shaped our planet.

- Earth's surface has been shaped by active geology driven by internal heat. This geology made life possible by causing the outgassing of the material that made our oceans and atmosphere, while also creating a magnetic field that has helped preserve the atmosphere. Moreover, through the action of plate tectonics and the carbon dioxide cycle, geology has kept our planet's climate stable enough for water to remain liquid for the past 4 billion years or more.

- Geology plays a role in making life possible, but life, once it takes hold, can also change the conditions on a planet. The oxygen in our atmosphere is a direct consequence of life. Today life affects our planet in another way: Our advanced civilization is capable of changing the way our climate functions, with consequences that we cannot fully predict.

- We do not yet know how common habitable planets may be in the universe, but geology clearly plays a crucial role in habitability. We'll return to geological considerations many more times throughout this book.

Summary of Key Concepts

4.1 Geology and Life

How is geology crucial to our existence?
Geology appears to be crucial to our existence in at least three ways: **Volcanism** released most of the gas that made the atmosphere and the water vapor that condensed to form the oceans; **plate tectonics** is crucial to the climate regulation that has kept Earth habitable over the long term; and Earth's **magnetic field** has helped protect the atmosphere from being stripped by the solar wind.

4.2 Reconstructing the History of Earth and Life

What can we learn from rocks and fossils?
Rocks and fossils preserve a record of the conditions under which they formed. By studying **mineral** structure and chemical and isotopic composition, we can learn such things as when the rock or fossil formed, how it formed, and what kinds of conditions prevailed on Earth when it formed.

How do we learn the age of a rock or fossil?
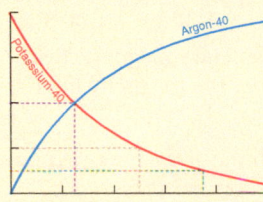
Radiometric dating is based on carefully measuring the proportions of radioactive isotopes and their decay products within rocks. The ratio of the two changes with time and provides a reliable measure of a rock's or fossil's age.

What does the geological record show?
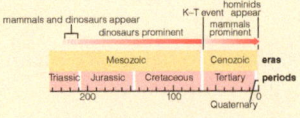
The geological record allows us to reconstruct Earth's history, which we summarize with the **geological time scale.** We divide this history into four **eons** (the Hadean, Archean, Proterozoic, and Phanerozoic), subdividing the last one into three **eras** and shorter **periods.** The time scale extends back to Earth's birth a little over 4.5 billion years ago.

4.3 The Hadean Earth and the Dawn of Life

How did Earth get an atmosphere and oceans?
The water and gases that made our atmosphere were originally trapped inside our planet. They were released by volcanic outgassing. Water vapor condensed to form the oceans, and the other gases made Earth's early atmosphere. Life has since transformed the atmosphere, most importantly by adding molecular oxygen.

Could life have existed during Earth's early history?
Earth probably had oceans and continents throughout much of the Hadean. However, it is possible that some of the largest impacts of the **heavy bombardment** might have killed off any life that existed at the time, so while life may have arisen during the Hadean, we do not know whether it could have survived continuously down to the present day.

4.4 Geology and Habitability

What is Earth like on the inside?
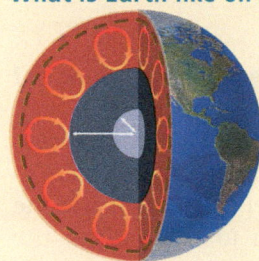
In order of decreasing density and depth, the interior structure consists of the **core, mantle,** and **crust.** The crust and part of the mantle together make up the rigid **lithosphere.** Internal heat allows the mantle rock to convect slowly.

How does plate tectonics shape Earth's surface?
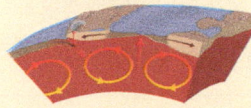
Plate tectonics has led to many unique features of our geology, especially **seafloor spreading** zones and the building of continents along **subduction zones.** The shifting of plates has completely changed Earth's geological appearance many times in the past billion years.

Why does Earth have a protective magnetic field?
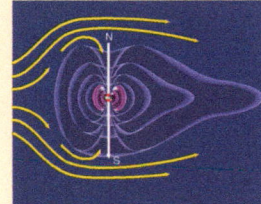
Earth's magnetic field is generated by the combination of its molten outer core, convection in that outer core, and a moderately rapid rotation rate. The magnetic field creates a **magnetosphere** that acts like a protective bubble surrounding our planet, shielding the atmosphere from loss through **solar wind stripping.**

4.5 Climate Regulation and Change

How does the greenhouse effect make Earth habitable?
Greenhouse gases such as carbon dioxide, methane, and water vapor absorb infrared light emitted from a planet's surface. The absorbed photons are quickly reemitted, but in random directions. The result acts much like a blanket, warming the planet's surface. Without the warming due to the greenhouse effect, Earth would be frozen.

What regulates Earth's climate?

Earth's long-term climate is remarkably stable because of feedback processes that tend to counter any warming or cooling that occurs. The most important feedback process is the **carbon dioxide cycle,** which naturally regulates the strength of Earth's greenhouse effect.

How does Earth's climate change over long periods of time?
Earth's temperature has remained in a range allowing liquid water at least since the end of the heavy bombardment, but there have been periods of unusual warmth or cold. Recent **ice ages** are tied to small changes in Earth's rotation and orbit. Ge-

ological evidence also suggests more extreme variations in the past, including a series of **snowball Earth** episodes that ended more than 500 million years ago.

THE PROCESS OF SCIENCE IN ACTION

4.6 Formation of the Moon

How did the giant impact model win out over competing models?

The giant impact model is the only model of the Moon's formation that successfully explains the differences in composition between Earth and the Moon. According to this model, the Moon formed from mantle material splashed out of Earth by the impact of a Mars-size object.

Does the giant impact model count as science?

The giant impact model shows all the hallmarks of science: It is natural and testable, and we can imagine future discoveries that would cause us to call it into question.

Exercises and Problems

MasteringAstronomy® *For instructor-assigned homework and other learning materials, go to MasteringAstronomy®.*

REVIEW QUESTIONS

Short-Answer Questions Based on the Reading

1. Briefly describe three aspects of geology that are especially important to our existence.
2. What do we mean by the *geological record*? Why is it important?
3. Describe the three basic types of rock and the *rock cycle*. What is a *mineral*? How do we study rocks in the laboratory?
4. How are sedimentary *strata* made, and how do they enable us to determine the relative ages of rocks and fossils?
5. Describe the technique of *radiometric dating*, and explain how we know it is reliable. Be sure to explain what we mean by a *radioactive isotope, parent and daughter isotopes,* and a *half-life.*
6. How do fossils form? Do most living organisms leave fossils? Do most fossils contain organic matter? Explain.
7. Summarize the *geological time scale.* What are *eons, eras,* and *periods*?
8. How old is Earth? How do we know?
9. Briefly describe how *outgassing* led to the origin of our oceans and atmosphere. How did Earth's early atmosphere differ from Earth's current atmosphere?
10. What was the *heavy bombardment,* and what effect might it have had on life?
11. Briefly describe Earth's *core-mantle-crust* structure and how it developed this structure. What is the *lithosphere*? What is *mantle convection*?
12. Briefly describe the conveyorlike action of *plate tectonics* and the evidence for this action. How does plate tectonics account for the observed differences in the *seafloor crust* and the *continental crust*?
13. Describe how plate tectonics shapes important geological features of Earth, including mid-ocean ridges, continents, mountain ranges, rift valleys, and earthquakes. How did Hawaii form?
14. What evidence do we have for the operation of plate tectonics in Earth's distant past? Why do we think Earth has plate tectonics?
15. What are the three requirements for a planetary magnetic field, and how does Earth meet them? How does the magnetic field protect our atmosphere?
16. Briefly describe the mechanism by which the *greenhouse effect* warms a planet. What are the most common *greenhouse gases*?
17. What has happened to most of the carbon dioxide outgassed through Earth's history? Describe the *carbon dioxide cycle* and how it helps regulate Earth's climate.
18. What are *ice ages,* and what may cause them? What do we mean by *snowball Earth* periods, and how might Earth recover from them?
19. Briefly summarize the key ways in which geology is important to Earth's long-term habitability.
20. How do we think the Moon formed, and what evidence supports this model? Why were other models ruled out? Should we be surprised that a giant impact could have affected our planet?

TEST YOUR UNDERSTANDING

Does It Make Sense?

Decide whether each statement makes sense (or is clearly true) or does not make sense (or is clearly false). Explain your reasoning; because not all of these have definitive answers, your explanation is more important than your chosen answer.

21. We can expect that if there are paleontologists a few million years from now, they will find the fossil remains of almost every human who ever lived.
22. Nearly all the rocks I found in the lava fields of Hawaii are igneous.
23. The most common rock type in the strata of the Grand Canyon is sedimentary rock.

24. Although Earth contains its densest material in its core, it's quite likely that terrestrial planets in other star systems would contain their lowest-density rock in their cores and their highest-density rock in their crusts.

25. If you had a time machine that dropped you off on Earth during the Hadean eon, you'd be quickly killed by a large impact.

26. If there were no plate tectonics on Earth, our planet would be far too hot to have liquid oceans.

27. Without the greenhouse effect, there probably would be no life on Earth.

28. If nitrogen were a greenhouse gas, our planet would be far hotter than it is.

29. We can learn a lot about Earth's early history by studying the Moon.

30. Science can never determine with confidence the times or sequence of events that occurred millions or billions of years ago.

Quick Quiz

Choose the best answer to each of the following. Explain your reasoning with one or more complete sentences.

31. A rock's type (igneous, metamorphic, or sedimentary) tells us (a) its age; (b) its chemical composition; (c) how it was made.

32. To learn a rock's age, we must (a) determine its chemical composition; (b) identify its mineral structure; (c) measure the ratios of different isotopes within it.

33. Radiometric dating now allows us to determine Earth's age to an accuracy of about (a) a billion years; (b) 10 million years; (c) less than 20 years.

34. Earth's oceans formed (a) during the late stages of accretion as water ice collected on the surface; (b) from water vapor outgassed by volcanoes; (c) when Earth underwent differentiation.

35. We learn about the heavy bombardment by studying (a) craters and rocks from the Moon; (b) zircon mineral grains; (c) Earth's oldest igneous rocks.

36. Earth has retained a lot of internal heat primarily because of its (a) distance from the Sun; (b) large iron core; (c) relatively large size.

37. Plate tectonics is best described as a process that (a) recycles rock between Earth's surface and upper mantle; (b) brings metal from Earth's core to the surface; (c) allows continents to plow through the crust.

38. Earth has far less atmospheric carbon dioxide than Venus because (a) Earth was born with less of this gas; (b) Earth's carbon dioxide was lost in the giant impact that formed the Moon; (c) Earth's carbon dioxide is locked up in carbonate rocks.

39. If Earth had more greenhouse gases in its atmosphere, it would (a) heat up; (b) cool off; (c) accelerate plate tectonics.

40. *Snowball Earth* refers to (a) one of a series of very deep ice ages that occurred more than 500 million years ago; (b) the idea that Earth would be frozen without the greenhouse effect; (c) any of the ice ages that have occurred in the past few million years.

PROCESS OF SCIENCE

41. *The Age of Earth.* Some people still question whether we have a reasonable knowledge of the age of Earth or the ability to date events in Earth's history. Based on what you have learned about both relative and absolute ages on the geological time scale, do you think it is reasonable for scientists to be confident of ages found by radiometric dating? Is there any scientific reason to doubt the reliability of our chronology of Earth? Explain.

42. *Unanswered Questions.* Choose one important but still not fully answered question about Earth's past history, and write two or three paragraphs in which you discuss how we might answer this question in the future and what the answer might mean for the study of life in the universe. Be as specific as possible.

GROUP WORK EXERCISE

43. *The Geology of Other Worlds.* **Roles:** *Scribe* (takes notes on the group's activities), *Proposer* (proposes explanations to the group), *Skeptic* (points out weaknesses in proposed explanations), *Moderator* (leads group discussion and makes sure everyone contributes). **Activity:** Imagine that scientists have discovered a planet orbiting a Sun-like star at the same distance that Earth orbits the Sun, and measurements show it is the same size and mass as Earth. However, no other details about it are known. Make a list of reasons why you would, or would not, expect it to have each of the following geological features: (a) a differentiated interior; (b) plate tectonics; (c) oceans; (d) continents; (e) a magnetic field. At the end, come to a group consensus about the likely geology of the planet.

INVESTIGATE FURTHER

In-Depth Questions to Increase Your Understanding

Short-Answer/Essay Questions

44. *Understanding Radiometric Dating.* Imagine you had the good fortune to find a meteorite in your backyard that appeared to be a piece of material from the early history of the solar system. How would you expect its ratio of potassium-40 and argon-40 to be different from that of other rocks in your yard? Explain why, in a few sentences.

45. *Dating Planetary Surfaces.* We have discussed two basic techniques for determining the age of a planetary surface: studying the abundance of impact craters and radiometric dating of surface rocks. Describe each technique briefly. Which technique seems more reliable? Which technique is more practical? Explain.

46. *Earth Without Differentiation.* Suppose Earth had never undergone differentiation. How would Earth be different? Write two or three paragraphs discussing likely differences. Explain your reasoning carefully.

47. *Earth Without Plate Tectonics.* Suppose plate tectonics had never begun on Earth. How would Earth be different? Write two or three paragraphs discussing likely differences. Explain your reasoning carefully.

48. *Earth Without the Moon.* Suppose the giant impact that formed the Moon had never occurred. How would you expect Earth to be different? Explain your reasoning carefully.

49. *Feedback Processes in the Atmosphere.* As the Sun gradually brightens in the future, how can the CO_2 cycle respond to reduce the warming effect? Which parts of the cycle will be affected? Is this an example of positive or negative feedback?

50. *Experiment: Geological Properties of Silly Putty.* Roll room-temperature Silly Putty into a ball and measure its diameter. Place the ball on a table and gently place one end of a heavy book on it. After 5 seconds, measure the height of the squashed ball. Repeat the experiment two more times, the first time warming the Silly Putty in hot water before you start and the second time cooling it in ice water before you start. How do the different temperatures affect the rate of "squashing"? How does the experiment relate to planetary geology? Explain.

51. *Experiment: Planetary Cooling in a Freezer.* To simulate the cooling of planetary bodies of different sizes, use a freezer and two small plastic containers of similar shape but different size. Fill each container with cold water and put both into the freezer at the same time. Checking every hour or so, record the time and your estimate of the thickness of the "lithosphere" (the frozen layer) in the two containers. How long does it take the water in each container to freeze completely? Describe in a few sentences the relevance of your experiment to planetary geology. Extra credit: Plot your results on a graph with time on the x-axis and lithospheric thickness on the y-axis. What is the ratio of the two freezing times?

Quantitative Problems

Be sure to show all calculations clearly and state your final answers in complete sentences.

52. *Geological Time.* Geological time scales are often written in ways that can mask their significance. For each of the following pairs of times, state which one is larger and by how much:
    a. 25,000 years, 0.1 million years
    b. 4 million years, 0.05 billion years
    c. 0.1 billion years, 1 million years

53. *Dating Lunar Rocks.* You are analyzing Moon rocks that contain small amounts of uranium-238, which decays into lead with a half-life of about 4.5 billion years.
    a. In one rock from the lunar highlands, you determine that 55% of the original uranium-238 remains; the other 45% decayed into lead. How old is the rock?
    b. In a rock from the lunar maria, you find that 63% of the original uranium-238 remains; the other 37% decayed into lead. Is this rock older or younger than the highlands rock? By how much?

54. *Carbon-14 Dating.* The half-life of carbon-14 is about 5700 years.
    a. You find a piece of cloth painted with organic dyes. By analyzing the dye in the cloth, you find that only 77% of the carbon-14 originally in the dye remains. When was the cloth painted?
    b. A well-preserved piece of wood found at an archaeological site has 6.2% of the carbon-14 that it must have had when it was living. Estimate when the wood was cut.
    c. Suppose a fossil is 570,000 years old, which is 100 half-lives of carbon-14. What fraction of its original carbon-14 would remain? Use your answer to explain why carbon-14 generally is not useful for dating fossils of this age.

55. *Martian Meteorite.* Some unusual meteorites thought to be chips from Mars contain small amounts of radioactive thorium-232 and its decay product, lead-208. The half-life for this decay process is 14 billion years. Analysis of one such meteorite shows that 94% of the original thorium remains. How old is this meteorite?

56. *Internal vs. External Heating.* In daylight, Earth's surface absorbs about 400 watts per square meter. All of Earth's internal radioactivity produces a total of 3 trillion watts, which leak out through the surface. Calculate the internal heat flow (watts per square meter) averaged over Earth's surface. Compare this internal heat flow quantitatively to solar heating, and comment on why internal heating drives geological activity.

57. *Plate Tectonics.* Typical motions of one plate relative to another are 2 centimeters per year. At this rate, how long would it take for two continents 3000 kilometers apart to collide? What are the global consequences of motions like this?

58. *More Plate Tectonics.* Consider a seafloor spreading zone creating 2 centimeters of new crust over its entire 2000-kilometer length every year. How many square kilometers of surface will this create in 100 million years? What fraction of Earth's surface does this constitute?

Discussion Questions

59. *Plate Tectonics and Us.* Based on what you learned in this chapter, can you imagine cases in which civilizations might arise on planets without plate tectonics? Defend your opinion.

60. *Implications for Other Worlds.* Overall, do you think that Earth's geological features are likely to be rare or common on other worlds? How does your answer affect your opinion of the prospects of discovering life or civilizations on other worlds?

61. *Evidence of Our Civilization.* Discuss how the geological processes will affect the evidence of our current civilization in the distant future. For example, what evidence of our current civilization will survive in 100,000 years? In 100 million years? Do you think that future archaeologists or alien visitors will be able to know that we existed here on Earth?

WEB PROJECTS

62. *Local Geology.* Learn as much as you can about how geological features in or near your hometown were formed. Write a one- to three-page summary of your local geology.

63. *Volcanoes and Earthquakes.* Learn about one major earthquake or volcanic eruption that occurred during the past decade. Report on the geological conditions that led to the event, as well as on its geological and biological consequences.

64. *Formation of the Moon.* Scientists continue to model and study the giant impact thought to have formed the Moon. Look for and report on one recent discovery that may shed more light on how or when the giant impact occurred.

5 The Nature of Life on Earth

▲ **About the photo:** Much life on Earth is very different from familiar plants and animals. These giant tube worms live so deep in the ocean that no sunlight penetrates, and most of their nutrients come from microbes that live off chemical reactions and the heat from a nearby volcanic vent.

> There is grandeur in this view
> of life, with its several powers,
> having been originally breathed
> by the Creator into a few forms
> or into one; and that, whilst this
> planet has gone cycling on ac-
> cording to the fixed law of grav-
> ity, from so simple a beginning
> endless forms most beautiful
> and most wonderful have been,
> and are being, evolved.
>
> *Charles Darwin,* The Origin of
> Species, *1859*

Having talked about the conditions that make life possible on Earth, we are now ready to begin talking about the nature of life and its interactions with a planetary environment. As with many aspects of astrobiology, we are limited by the fact that we have only one example to study: life on Earth. Life elsewhere might be quite different, but the great diversity of life on Earth still gives us plenty to study here. Moreover, anything we learn about life on this world can help us understand the possibilities for other worlds.

In this chapter, we'll explore the general nature of life on Earth. Along the way, we'll see that life elsewhere would almost certainly share at least a few characteristics with life on Earth, and we'll gain a few clues about where we might most profitably focus our search for life in the universe.

5.1 Defining Life

What is life? This seemingly simple question lies at the heart of research into life in the universe. After all, if we are interested in the possibility of life elsewhere, we must know what it is that we are looking for. Unfortunately, defining life is surprisingly difficult, even when we consider only life on Earth. Life on Earth is remarkably diverse; organisms range in size from tiny microbes to huge plants and animals, and can be found thriving in almost every conceivable place on and near our planet's surface. Defining life is all the more difficult when we consider life elsewhere, because we cannot be sure that life on other worlds would resemble life on Earth physically or chemically. Given the difficulty of defining life, the only sensible way to proceed is by studying the one example of life that we know, hoping it will yield fundamental insights into how life operates and into the environmental conditions required to support life. In this first section, we will explore general characteristics of life on Earth and attempt to come up with at least some reasonably useful definition of life.

What are the general properties of life on Earth?

A cat and a car have much in common. Both require energy to function—the cat gets energy from food, and the car gets energy from gasoline. Both can move at varying speeds and can turn corners. Both expel waste products. But a cat clearly is alive, while a car clearly is not. What's the difference?

In the case of a cat and a car, we can find many important differences without looking too far. For example, cats reproduce themselves, while cars must be built in factories. But as we look deeper into the nature of life, it becomes increasingly difficult to decide what characteristics separate living organisms from rocks and other nonliving materials. Indeed, the question can be so difficult to answer that we may be tempted to fall back on the famous words of U.S. Supreme Court Justice Potter Stewart, who, in avoiding the difficulty of defining pornography, wrote: "I shall

FIGURE 5.1
Six key properties of life.

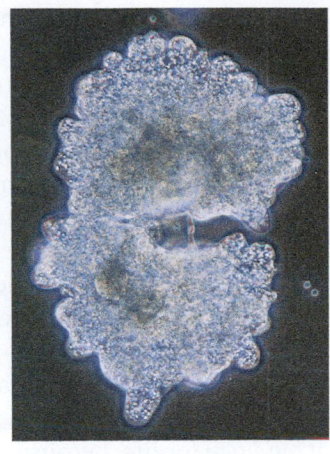

a Order: Living organisms exhibit order in their internal structure, as is apparent in this microscopic view of spiral patterns in two single-celled organisms.

b Reproduction: Organisms reproduce their own kind. Here, a single-celled organism (an amoeba) has already copied its genetic material (DNA) and is now dividing into two cells.

c Growth and development: Living organisms grow and develop in patterns determined at least in part by heredity. Here, we see a Nile crocodile hatching from an egg.

d Energy utilization: Living organisms use energy to fuel their many activities. These tube worms, which live near a deep-sea volcanic vent, obtain energy from chemical reactions made possible in part by the heat released from the vent.

e Response to the environment: Life actively responds to changes in its surroundings. Here, we see a Venus flytrap closing in response to being touched by an insect.

f Evolutionary adaptation: Life evolves in a way that leads to organisms that are adapted to their environments. Here, we see a katydid with camouflage that evolved to hide it among leaves.

not today attempt further to define [it]. . . But I know it when I see it."* If living organisms on other worlds turn out to be much like those on Earth, it may prove true that we'll know them when we see them. But if the organisms are fairly different from those on Earth, we'll need clearer guidelines to decide whether or not they are truly "living."

One way to seek distinguishing features of life is to study living organisms, looking for common characteristics. Given the difficulty of defining life, you probably won't be surprised to learn that there are exceptions to almost any "rule" we think of. Nevertheless, biologists have identified at least six key properties that appear to be shared by most or all living organisms on Earth, all of which are summarized in Figure 5.1. Let's briefly investigate each property.

---

*From Potter Stewart's concurring opinion in *Jacobellis v. Ohio*, 378 U.S. 184, 198 (1964).

ORDER The materials in living organisms always exhibit some type of order. For example, the molecules in living cells are not scattered randomly about but instead are arranged in patterns that make cell structures. These structures, in turn, make possible all the other properties of life that we will discuss. Note that order alone does not make something living: A book has order, because words are not scattered randomly on the pages, but it is not alive. The same is true for rock crystals, whose atoms are arranged in an orderly way, and even for the individual molecules of life such as proteins or DNA; these molecules clearly have order, but we consider them only to be building blocks of life, not life itself.

Nevertheless, it seems reasonable to expect that all living things will show order. In logical terms, we say that order is a *necessary condition* for life, because something cannot be alive without order. However, order is not a *sufficient condition* for life, because order alone does not make something alive.

Think About It The idea of necessary and sufficient conditions is important in science. To make sure you understand it, decide whether each of the following conditions is necessary or sufficient (or neither or both) for the given effect: (a) condition: oxygen in the atmosphere; effect: human survival; (b) condition: living in New York City; effect: living in the United States; (c) condition: meeting all requirements for a college degree; effect: receiving a college degree.

REPRODUCTION Living organisms reproduce or are products of reproduction. Simple life forms, such as bacteria, reproduce by dividing to make nearly exact copies of themselves. More complex organisms may reproduce in more sophisticated ways—including sexual reproduction, in which offspring inherit genetic material from two parents. Note that not all living organisms are capable of reproduction. For example, a mule is sterile and cannot reproduce. However, the mule still meets the reproduction criterion because it is the product of reproduction between two closely related animals (a horse and a donkey).

Reproduction seems necessary to any definition of life; without it, there would be no way for life as a whole to survive the death of individuals. However, it also exposes borderline cases about which even scientists disagree. For example, *viruses* are generally much smaller than bacteria and are incapable of reproducing on their own, but they can reproduce by infecting a living cell and commandeering the cell's reproductive machinery for their own purposes. The fact that viruses can reproduce when they infect other organisms but not when they are on their own seems to put them somewhere between the nonliving and the living. Another borderline case concerns the infectious proteins known as *prions*, which are thought to be the agents of *mad cow disease*. Prions appear to be abnormal forms of protein molecules that somehow cause normal protein molecules to change into the abnormal prion form. In other words, they make copies of themselves by causing other molecules to change rather than by actually replicating themselves. Most biologists therefore put them on the nonliving side of the gray region between nonlife and life, though they present at least some ambiguity.

GROWTH AND DEVELOPMENT Living organisms grow and develop in patterns directed at least in part by **heredity**—traits passed to an organism from its parent(s). The property of growth and development appears necessary to life in that all organisms grow or develop during at least some periods in their life cycles, but it is not sufficient to constitute life. For example, fire grows and develops as it spreads through a forest, but a fire is not alive. As we will discuss later in this chapter, all life on Earth passes on its heredity through the molecules known as DNA. (Some viruses use a related molecule called RNA, but we will leave viruses and prions out as we discuss "life" in the rest of this chapter.)

ENERGY UTILIZATION Living organisms use energy to create and maintain patterns of order within their cells, to reproduce, and to grow. Life without energy utilization is simply not possible (though some organisms can survive temporarily in dormant states). Of course, energy utilization is not sufficient to constitute life; any electrical or gas-powered appliance uses energy to function.

We can gain further insight into the importance of energy utilization by considering what is sometimes called the *thermodynamics* of life. Thermodynamics is a branch of science that deals with energy and the rules by which it operates. Recall the *law of conservation of energy* [Section 3.3], which tells us that energy can be neither created nor destroyed but only transformed from one form to another; this law is sometimes referred to as the "first law of thermodynamics."

The **second law of thermodynamics** states that, when left alone, the energy in a system undergoes conversions that lead to increasing disorder. Living organisms are a perfect example of this law's importance: If you place a living organism into a sealed box, it will eventually use up the available energy and therefore no longer be able to build new molecules or fuel any of the molecular processes needed for life. Its molecules will then become more disordered with time—for example, the molecules may decay or may lose the orderly relationships they maintain with other molecules when the organism is alive—causing the organism to die. To maintain order and survive, a living organism must have a continual source of energy that it can use to counter the tendency for disorder to take over. Living organisms get this energy from the environment, either through food or through chemical interactions with the environment. The environment, in turn, gets its energy either from an internal source, such as the heat of the planet itself, or from an external source, such as sunlight. Life is probably not possible on a world that lacks a long-term source of energy input to the environment.

RESPONSE TO THE ENVIRONMENT All living organisms interact with their surroundings and actively respond in at least some ways to environmental changes. For example, some simple organisms may move to a region where the temperature is more suited to their growth, and warm-blooded mammals may sweat, pant, or adjust blood flow to maintain a constant internal temperature. Like all the other properties on our list, response to the environment is a necessary but not sufficient condition for life. Many human-made devices also respond to changes in the environment; for example, a thermostat can respond to changes in temperature by turning on heating or cooling systems.

EVOLUTIONARY ADAPTATION Life has changed dramatically over time as the organisms that lived billions of years ago have gradually evolved into the great variety of organisms found on Earth today. Life evolves as a result of the interactions between organisms and their environments, leading over time to evolutionary adaptations that make species better suited to their environments. When the adaptations are significant enough, organisms carrying the adaptations may be so different from their ancestors that they constitute an entirely new species.

Before we continue, it's worth noting that, like life itself, the familiar term **species** is not so easy to define in a precise way. Traditionally, a species was defined as a group of organisms that share some set of common characteristics and are capable of interbreeding with one another to produce fertile offspring. For example, horses and donkeys represent different species because the result of their interbreeding is an infertile mule. However, while this definition of species works fairly well for animals and most plants, it does not work for organisms that reproduce asexually, including microorganisms that reproduce through cell division. As a result, biologists today recognize species as groups of organisms that are genetically distinct from other groups, though the precise border between one species and another is not always clear, especially with microorganisms.

Once a species is identified, it is given a scientific name that consists of two parts. The first part is the **genus,** which describes the "generic" category to which the organism belongs, while the second part distinguishes multiple species within the same genus. (You may recognize that the term *genus* is related to "generic" and that the term *species* is related to "specific.") The full name is written in italics, with the genus capitalized. For example, humans are scientifically classified as *Homo sapiens,* meaning that we are one specific species that has been identified within the genus *Homo* (the others are all extinct). Horses and donkeys are, respectively, *Equus caballus* and *Equus asinus,* names that show that both belong to the same genus (*Equus*).

What is the role of evolution in defining life?

All six properties of life that we have discussed are important, but biologists today regard evolutionary adaptation as the most fundamental and unifying of all these properties. It is the only property that can explain the great diversity of life on Earth, and an understanding of it allows us to understand how all the other properties of life came to be. Modern understanding of the capacity for evolutionary adaptation is described by the **theory of evolution.** Because this theory is so central to modern biology, let's briefly investigate the origin of the theory and the evidence that supports it.

AN ANCIENT IDEA The word *evolution* simply means "change with time," and the idea that life might evolve through time goes back more than 2500 years. The Greek scientist Anaximander (c. 610–547 B.C.) promoted the idea that life originally arose in water and gradually evolved from simpler to more complex forms. A century later, Empedocles (c. 492–432 B.C.) suggested that creatures poorly adapted to their environments would perish, foreshadowing the modern idea of evolutionary adaptation. Many of the early Greek atomists [Section 2.1] probably held similar beliefs, though the evidence is sparse. Aristotle, however, maintained that species are fixed and independent of one another and do not evolve. This Aristotelian view eventually became entrenched within the theology of Christianity, and evolution was not taken seriously again for some 2000 years. In the mid-eighteenth century, scientists began to suspect that many fossils represented extinct ancestors of living species. Then, in the early 1800s, French naturalist Jean Baptiste Lamarck suggested that the best explanation for the relationship between fossils and living organisms is that life-forms evolve by gradually adapting to perform successfully in their environments.

Lamarck's idea of evolution by adaptation represented the first clear attempt to explain what we now consider the "observed facts" of evolution. That is, observations of how fossils differ in different layers of the geological record and of relationships between living species make it quite clear that life has changed over time. However, Lamarck was unable to come up with a successful theory to explain *how* evolution occurs. His hypothesis concerning the mechanism of evolution, called "inheritance of acquired characteristics," suggested that organisms develop new characteristics during their lives and then pass these characteristics on to their offspring. For example, Lamarck would have imagined that weight lifting would enable a person to create an adaptation of great strength that could be genetically passed to his or her children. While this hypothesis may have seemed quite reasonable at the time, it has not stood

KEY BIOLOGICAL DEFINITIONS

TERMS RELATED TO EVOLUTION

evolution (biological): The gradual change in populations of living organisms that has transformed life on Earth from its primitive origins to the great diversity of life today.

evolutionary adaptation: An inherited trait that enhances an organism's ability to survive and reproduce in a particular environment.

theory of evolution: The theory, first advanced by Charles Darwin, that explains how and why living organisms evolve through time.

natural selection: The primary mechanism by which evolution proceeds. More specifically, natural selection refers to the process by which, over time, advantageous genetic traits naturally win out (are "selected") over less advantageous traits because they are more likely to be passed down through succeeding generations.

species: For our purposes, a population of organisms that is genetically distinct from other groups of organisms (precise definitions vary).

TERMS RELATED TO HEREDITY

heredity: The characteristics of an organism passed to it by its parent(s), which it can pass on to its offspring. The term can also apply to the transmission of these characteristics from one generation to the next. Hereditary information is encoded in DNA.

gene: The basic functional unit of an organism's heredity. A single gene consists of a sequence of DNA bases (or RNA bases, in some viruses) that provides the instructions for a single cell function (such as building a protein).

genome: The complete sequence of DNA bases in an organism, encompassing all of the organism's genes along with noncoding DNA in between.

genetic code: The specific set of rules by which the sequence of bases in DNA is "read" to provide the instructions that make up genes.

DNA (deoxyribonucleic acid): The basic hereditary molecule of life on Earth. A DNA molecule consists of two strands, twisted in the shape of a double helix, along each of which lies a long sequence of *DNA bases*. The four DNA bases are adenine (A), cytosine (C), guanine (G), and thymine (T), and they can be paired across the two DNA strands only so that A pairs with T and C pairs with G.

RNA (ribonucleic acid): A molecule closely related to DNA, but with only a single strand and a slightly different backbone and set of bases. RNA plays many crucial roles in cells.

TERMS RELATED TO THE MODERN CLASSIFICATION OF LIFE

cell: The basic structure of all life on Earth, in which the living matter inside is separated from the outside world by a barrier called a *membrane.*

domains of life: The three broad domains into which all known species of life fall: *bacteria, archaea,* and *eukarya.* The last includes all plants and animals, as well as fungi and many microbes.

tree of life: A representation of biochemical and genetic relationships between species. The three major branches of the tree are the three domains (bacteria, archaea, and eukarya).

TERMS RELATED TO CELLULAR CHEMISTRY

organic molecule: Generally, any molecule containing carbon and associated with life. Note that we do not generally consider molecules such as carbon dioxide (CO_2) and carbonate minerals to be organic, since they are commonly found independent of life.

organic chemistry: The chemistry of organic molecules.

biochemistry: The chemistry of life.

amino acids: The molecules that form the building blocks of proteins. Most organisms construct proteins from a particular set of 20 amino acids, although several dozen other amino acids can be found in nature. More technically, an amino acid is a molecule containing both an *amino group* (NH or NH_2) and a *carboxyl group* (COOH).

protein: A large molecule assembled from amino acids according to instructions encoded in DNA. Proteins play many roles in cells; a special category of proteins, called **enzymes,** catalyzes nearly all the important biochemical reactions that occur within cells.

catalysis: The process of causing or accelerating a chemical reaction by involving a substance or molecule that is not permanently changed by the reaction. The unchanged substance or molecule involved in catalysis is called a **catalyst.** In living cells, the most important catalysts are the proteins known as *enzymes.*

metabolism: The many chemical reactions that occur in living organisms to provide cellular energy and nutrients.

up to scientific scrutiny and therefore has been discarded as a model of how evolution occurs. Today, we understand that evolution proceeds primarily via a model first proposed by the British naturalist Charles Darwin.

THE MECHANISM OF EVOLUTION Charles Darwin described his theory of evolution in his book *The Origin of Species*, first published in 1859. In this book, Darwin laid out the case for evolution in two fundamental ways. First, he described his observations of living organisms (made during his voyages on the HMS *Beagle*) and showed how they supported the idea that evolutionary change really does occur. Second, he put forth a new model of *how* evolution occurs, backing up his model with a wealth of evidence. In essence, the geological record and the observed relationships between species together provide strong evidence that evolution *has* occurred, while Darwin's theory of evolution explains *how* it occurs.

As is the case with most scientific theories, the underlying logic of Darwin's model is really quite simple. As biologist Stephen Jay Gould (1941–2002) described, Darwin built his model from "two undeniable facts and an inescapable conclusion":

- *Fact 1: overproduction and competition for survival.* Any localized population of a species has the potential to produce far more offspring than the local environment can support with resources such as food and shelter. This overproduction leads to a competition for survival among the individuals of the population.
- *Fact 2: individual variation.* Individuals in a population of any species vary in many heritable traits (traits passed from parents to offspring). No two individuals are exactly alike, and some individuals possess traits that make them better able to compete for food and other vital resources.
- *The inescapable conclusion: unequal reproductive success.* In the struggle for survival, those individuals whose traits best enable them to survive and reproduce will, on average, leave the largest number of offspring that in turn survive to reproduce. Therefore, in any local environment, heritable traits that enhance survival and successful reproduction will become progressively more common in succeeding generations.

It is this unequal reproductive success that Darwin called **natural selection:** Over time, advantageous genetic traits will naturally win out (be "selected") over less advantageous traits because they are more likely to be passed down through many generations. This process explains how species can change in response to their environment—by favoring traits that improve adaptation—and it is the primary mechanism of evolution.

EVIDENCE FOR EVOLUTION BY NATURAL SELECTION Darwin backed up his logical claim that evolution proceeds through natural selection by documenting cases in which related organisms are adapted to different environments or lifestyles. He found a particularly striking example among the finches of the Galápagos Islands (Figure 5.2). The different islands have different species, with each species adapted to its particular environment. Darwin realized that natural selection could explain this situation. He presumed that an ancestral pair of finches reached the Galápagos from the mainland (perhaps by being blown off course by winds). Over time, local populations of island finches gradually adapted to become the distinct species that he observed.

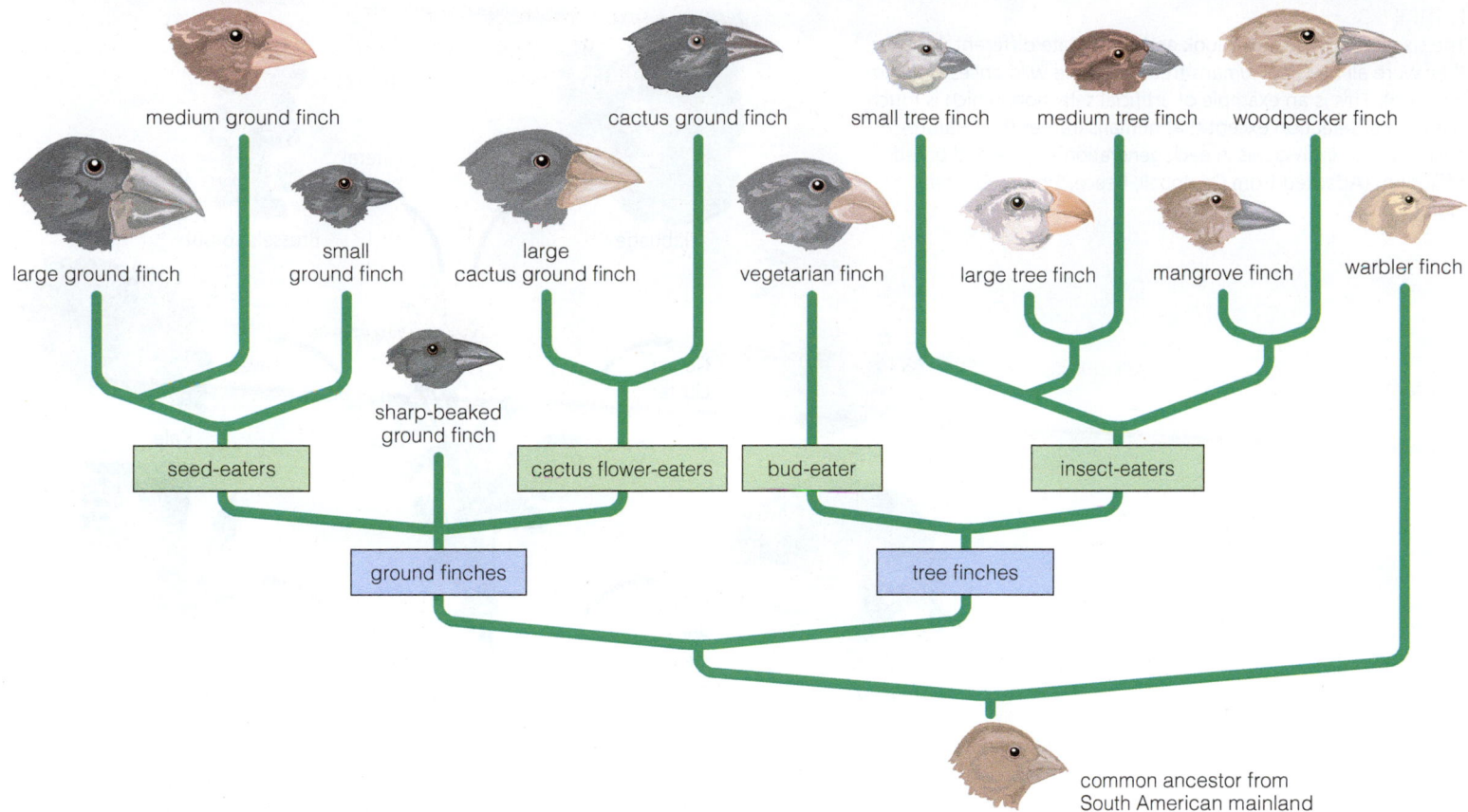

medium ground finch

cactus ground finch

small tree finch

medium tree finch

woodpecker finch

large ground finch

small ground finch

large cactus ground finch

vegetarian finch

large tree finch

mangrove finch

warbler finch

sharp-beaked ground finch

seed-eaters

cactus flower-eaters

bud-eater

insect-eaters

ground finches

tree finches

common ancestor from South American mainland

FIGURE 5.2

An evolutionary tree for the 13 species of Galápagos finches. These finch species are all closely related descendants of a common ancestor from the South American mainland. Note the diversity of beaks, which are adapted to certain food sources on the different islands. (Courtesy of Campbell, Reece, *Biology*.)

Darwin recognized similar patterns among many other species in the Galápagos and elsewhere in his round-the-world voyage. He also discovered fossils of extinct organisms that were clearly related to modern organisms, yet different in key respects. For example, in Brazil he found fossils of giant armadillos that he realized must be the ancestors of the modern armadillos found in the same region. The pieces of the puzzle gradually came together in Darwin's mind: He realized that natural selection not only explained the differences between closely related modern species like the finches, but also explained the fact that larger changes can occur over longer periods of time, with the result that entire species can become extinct and new ones can take their places.

Darwin also found strong support for his theory of evolution by looking at examples of *artificial selection*—the selective breeding of domesticated plants or animals by humans. Over the past few thousand years, humans have gradually bred many plants and animals into forms that bear little resemblance to their wild ancestors. Figure 5.3 shows how artificial selection has created a variety of vegetables from a single common ancestor. Similarly, dogs as different as Rottweilers and Chihuahuas were bred from a common ancestor within just a few thousand years. Darwin recognized that if artificial selection could cause such profound changes in just a few thousand years, natural selection could do far more over many millions of years.

Today, we can observe natural selection occurring right before our eyes. In many places on Earth, species have changed in time spans as short as a few decades in response to human-induced environmental changes. On a microbial level, natural selection is what allows a population of bacteria to become resistant to specific antibiotics; those few

FIGURE 5.3
The six vegetables shown look and taste quite different, but they were all bred by humans from the same wild ancestor (wild mustard). This is an example of artificial selection, which is much like natural selection except that humans (rather than nature) decide which individuals in each generation survive and breed offspring. (Adapted from Campbell, Reece, Simon, *Essential Biology*.)

Cabbage

Terminal bud

Lateral buds

Brussels sprouts

Cauliflower

Flower clusters

Leaves

Kale

Broccoli

Flowers and stems

Wild mustard

Stem

Kohlrabi

bacteria that acquire a genetic trait of resistance are the only ones that survive in the presence of the antibiotic. Indeed, bacterial cases of natural selection pose a difficult problem for modern medicine, because bacteria can quickly develop resistance to almost any new drug we produce. As a result, pharmaceutical companies are continually working to develop new antibiotics as bacteria become resistant to existing ones. Viruses can evolve even faster, which is one reason it has proved so difficult to fight viral diseases such as influenza and AIDS.

THE MOLECULAR BASIS OF EVOLUTION Darwin's theory of evolution by natural selection tells us that species adapt and change by passing hereditary traits from one generation to the next. However, Darwin did not know precisely how these traits were communicated across generations, nor did he know why there is always variation among individuals or how new traits can appear in a population. Today, thanks to discoveries in molecular biology made since Darwin's time, we know the answers to all these questions. In particular, we now know that organisms are built from instructions contained in a molecule called **DNA** (short for "*d*eoxyribo*n*ucleic *a*cid"), and biologists can now trace how evolutionary adaptations are related to changes that occur through time in DNA. We'll discuss how DNA makes evolutionary adaptations possible when we discuss its structure and function in Section 5.4, but for now the key point is that our understanding of DNA means that we understand the specific mechanisms by which natural selection occurs.

Our detailed understanding of how evolution proceeds on a molecular level, coupled with all the evidence for evolution collected by Darwin and others, puts the theory of evolution by natural selection on a solid foundation. That is, it is a *scientific theory* [Section 2.3] that has withstood countless tests and challenges. Like any scientific theory, the theory of evolution can never be proved beyond all doubt. However, as we'll discuss

Special Topic CHARLES DARWIN AND THE THEORY OF EVOLUTION

Charles Robert Darwin was born into a wealthy and educated family in England on February 12, 1809. His father was a physician, and he had two famous grandfathers. His paternal grandfather, Erasmus Darwin (1731–1802), was a renowned physician and scientist who was a strong proponent of the idea that life evolved gradually. His maternal grandfather, Josiah Wedgwood, started the famous Wedgwood Pottery and China company that still bears his name. Darwin's mother died when he was just 8 years old, but his father and his extended family provided him with a generally happy childhood.

At his father's urging, Darwin enrolled in medical school at age 16. However, he was so horrified by the sight of operations, then done without anesthesia, that he left after just 2 years. He next enrolled in Christ College at Cambridge University, intending to become a minister. While there, he began to indulge a childhood love for the study of nature. Shortly after graduating in 1831, Darwin was offered the opportunity to serve as the naturalist aboard a ship of exploration—the HMS *Beagle*. Darwin was 22 years old when the *Beagle* set sail on December 27, 1831. The voyage lasted nearly 5 years.

The *Beagle* spent much of its voyage exploring the coasts of South America and nearby islands. While the crew conducted surveys, Darwin went ashore to observe the geology and life, collecting numerous specimens that he took back to England. He also read extensively during the voyage, and one book proved particularly influential: Charles Lyell's *Principles of Geology*, published in 1830, presented the case for an ancient Earth sculpted by gradual geological processes. Darwin was given the book by a friend who expected Darwin to disagree with its conclusions; instead, Darwin found that his own observations of geology gave further credence to Lyell's theory.

Meanwhile, Darwin became intrigued by the many adaptations he observed among species in varied environments. He was particularly impressed by the animal life he observed during a 5-week stay on the Galápagos Islands, which lie approximately 1000 kilometers (600 miles) due west of the coast of Ecuador in South America. He focused special attention on the Galápagos finches, concluding that the different bird species must have evolved from a common mainland ancestor (see Figure 5.2). However, at the time he returned to England, he still did not understand how the evolutionary changes occurred.

In 1838, Darwin read Thomas Malthus's *An Essay on the Principle of Population*, in which Malthus famously argued that populations are capable of growing too fast for food supplies to support. The essay helped Darwin crystallize the idea of natural selection by making clear that individuals within a population must compete for survival (the idea embodied in Fact 1 on p. 160). He then began intensive study of how humans bred domestic plants and animals, which helped him understand the variation in populations (Fact 2 on p. 160). By 1842, Darwin was convinced that natural selection held the key to evolution, and he began to draft the text that would eventually be published as *The Origin of Species*.

Darwin is said to have been a pleasant man. He was an ardent opponent of slavery, and while he was not a feminist by contemporary standards, he believed that women should be treated with dignity and respect. He was deeply concerned with the impact his theory would have on those who believed in the biblical story of creation. That is probably why he did not publish his theory immediately—he wanted to take time building his case, in the hope that his theory would be so

Charles Darwin and his son William, photographed in 1842.

strong that it would be accepted by all without anyone taking offense. Indeed, Darwin might have delayed publication indefinitely if not for a manuscript he received from another scientist, Alfred Russel Wallace, on June 18, 1858.

Wallace had been observing geology and life in Indonesia and had independently come to the same conclusion as Darwin: that life evolves through natural selection. After reading Wallace's draft paper, Darwin worried that "all my originality will be smashed." Fortunately, both Darwin and Wallace were willing to share credit. Their first papers on the theory of evolution were read back to back at a scientific meeting in London on July 1, 1858. A little over a year later, Darwin finally published *The Origin of Species*. All 1250 copies in the first printing sold out on the first day. Within a decade, Darwin's theory was accepted by the vast majority of biologists, an acceptance that has grown stronger ever since.

Darwin never had a taste for arguing about evolution, leaving that to other scientists (especially Thomas Huxley, who called himself "Darwin's bulldog"). He continued his scientific work, publishing several more books on evolution and related topics. In his personal life, Darwin married a cousin, Emma. He and Emma had ten children, but two died in infancy and a third died at age 10. Darwin died on April 19, 1882, at the age of 73. A parliamentary petition won him burial in London's Westminster Abbey, where he lies next to Sir Isaac Newton.

further in Section 5.6, no credible scientific alternative to the theory of evolution has been proposed, and the evidence in the theory's favor is overwhelming. Indeed, it is difficult to imagine any aspect of biological science that can be understood without being examined in the context of the theory of evolution. That is why evolution has become the unifying theme of all modern biology.

Think About It The idea that life changes through time is quite ancient, and it was already well supported by observations of fossils before Darwin was even born. Moreover, Lamarck recognized that evolution occurs as a result of adaptations about a half-century before Darwin advanced the idea of natural selection. Given these facts, explain why we credit Darwin with the *theory* of evolution.

What is life?

Now that we have examined the fundamental properties of life on Earth, let's return to our original goal in this section: Can we come up with a definition of life?

Based on the central role of evolution, our simplest definition might be that *life is something that can reproduce and evolve through natural selection*. This definition is probably sufficient for most practical purposes, but some cases may still challenge this definition. For example, scientists can now write computer programs (lines of computer code) that can reproduce themselves (create additional sets of identical lines of code). By adding programming instructions that allow random changes, so that the programs can compete and change through a computer analog of natural selection, scientists can even make "artificial life" that evolves on a computer.

Think About It Do you think computer programs that can reproduce and evolve are alive? Why or why not? Would your opinion change if these programs evolved to the point where they could write their own computer code or exchange text messages with us? What if they wrote other programs that operated machinery to build other computers? Do you think it is possible to create true life on a computer?

Another issue with this definition concerns the origin of life. Darwin's theory of evolution does not tell us how the first life got started. For that, as we'll discuss in Chapter 6, we presume there must have been some type of molecular or *chemical evolution* (as opposed to biological evolution) that went on until the first living organism arose. The idea of chemical evolution is no more surprising than that of natural selection—certain chemical processes are energetically favored under certain circumstances, and laboratory experiments show that under the right conditions, chemicals can evolve in complexity much like life. However, it begs the question of whether we would recognize a clear distinction between, for example, the last case of chemical evolution and the first living organism capable of biological evolution. No one knows the answer; some scientists think there must have been a clear "first" living organism, while others think that the emergence of life may have been marked by a more gradual transition.

The fact that we have such difficulty distinguishing the living from the nonliving on Earth suggests that we should be cautious about constraining our search for life elsewhere. No matter what definition of life we choose, the possibility always remains that we'll someday encounter something that challenges our definition. Nevertheless, the ability to reproduce and evolve through natural selection seems likely to be shared by most, if not all, life in the universe.

5.2 Cells: The Basic Units of Life

Now that we have discussed general properties of life, we are ready to look more specifically at the nature of life on Earth. In this section, we will explore cells, the basic units of life. We will then be prepared to consider in Section 5.3 how cells make use of energy and in Section 5.4 how cells reproduce and evolve.

What are living cells?

All living organisms are made of **cells**—microscopic units in which the living matter inside is separated from the outside world by a barrier called a **membrane*** (Figures 5.4 and 5.5). Cells are the basic structures of life on Earth. Some organisms consist only of a single cell. Other organisms, like oak trees and people, are complex structures in which trillions of cells work cooperatively, dividing various tasks among specialized cells of different types. For reasons that should soon become clear, we generally presume that life elsewhere would also have to have some cell-like structure.

Despite the great diversity of life on Earth, all living cells on Earth share a great many similarities. For example, all pass on their hereditary information in the same basic way with DNA, and many other chemical processes are nearly the same in all cells. These similarities, which we'll discuss in more detail later, are profoundly important to our understanding of the origin of life. As far as we know, there is no reason why all living cells must share these characteristics. That is, while life elsewhere might also be composed of cells, we should not expect those cells to have the same biochemistry as cells on Earth. The many similarities among all cells on our planet therefore suggest a startling conclusion: *All life on Earth shares a common ancestor.* In other words, every living organism on Earth is related to every other one because all evolved over billions of years from the same origin of life.[†]

---

*Some organisms, such as some slime molds, do not perfectly fit this picture of discrete cells, because they consist of a large mass of protoplasm containing thousands of nuclei. Nevertheless, the basic idea that living tissue is contained in a package separated from the external environment still holds.

[†]We have not yet identified every type of living organism on Earth, so it is still conceivable that we'll someday discover organisms right here on Earth (perhaps deep underground or in other isolated ecosystems) that use a different biochemistry and hence seem to have come from a separate origin of life. If so, it would greatly expand our understanding of biology, since we'd have more than one form of life to study up close in our laboratories.

FIGURE 5.4
Microscopic views of four types of living cells. In each cell, a membrane separates the living matter inside the cell from the outside world.

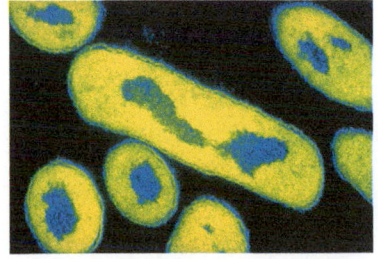
a Bacteria

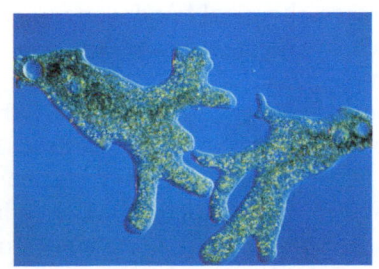
b Amoebas

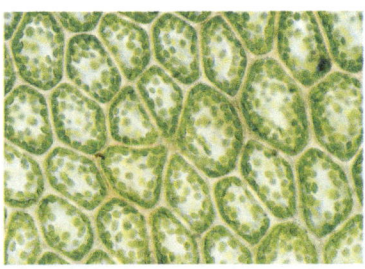
c Plant cells

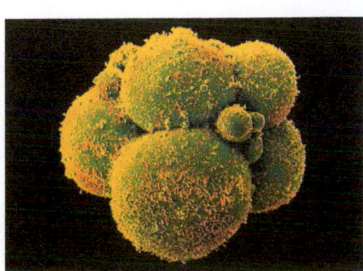
d Animal cells

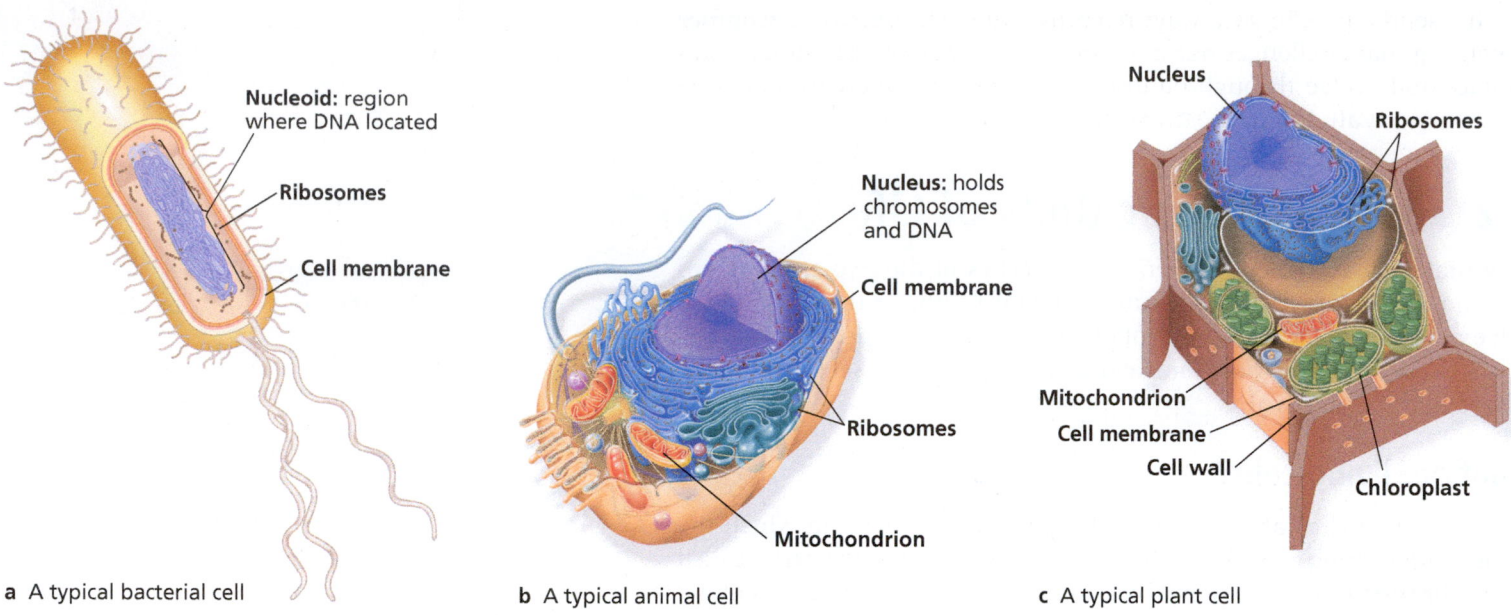

a A typical bacterial cell

Nucleoid: region where DNA located

Ribosomes

Cell membrane

b A typical animal cell

Nucleus: holds chromosomes and DNA

Cell membrane

Ribosomes

Mitochondrion

c A typical plant cell

Nucleus

Ribosomes

Mitochondrion

Cell membrane

Cell wall

Chloroplast

FIGURE 5.5

These diagrams shows the structures of typical cells with a few major components labeled.

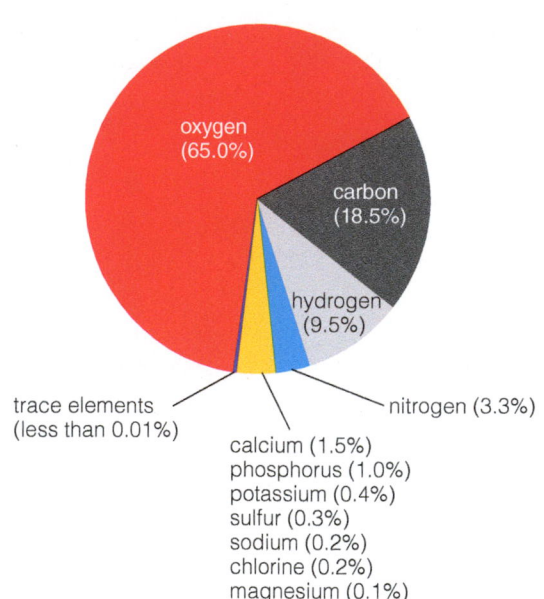

oxygen (65.0%)

carbon (18.5%)

hydrogen (9.5%)

trace elements (less than 0.01%)

nitrogen (3.3%)

calcium (1.5%)
phosphorus (1.0%)
potassium (0.4%)
sulfur (0.3%)
sodium (0.2%)
chlorine (0.2%)
magnesium (0.1%)

FIGURE 5.6

This pie chart shows the chemical composition of the human body by weight; this composition is fairly typical of all living matter on Earth. The trace elements include (in alphabetical order) boron, chromium, cobalt, copper, fluorine, iodine, iron, manganese, molybdenum, selenium, silicon, tin, vanadium, and zinc. (Adapted from Campbell, Reece, Simon, *Essential Biology*.)

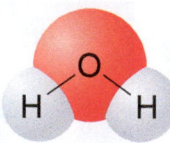

FIGURE 5.7

A water molecule has a single oxygen atom bonded to two hydrogen atoms. (Adapted from Campbell, Reece, Simon, *Essential Biology*.)

EARTH LIFE IS CARBON-BASED Life on Earth is made from more than 20 different chemical elements. However, just four of these elements—oxygen, carbon, hydrogen, and nitrogen—make up about 96% of the mass of typical living cells. Most of the remaining mass consists of just a few other elements, notably calcium, phosphorus, potassium, and sulfur (Figure 5.6).

Given that oxygen dominates Figure 5.6, you might be tempted to say that life on Earth is "oxygen-based." However, most of the oxygen in living cells is found in water molecules (H_2O). The molecules that account for a cell's structure and function owe their remarkable qualities to a different element: carbon. We therefore say that life on Earth is **carbon-based.**

Why is carbon so important to life on Earth? The primary answer to this question lies in how carbon can combine with other elements to make complex molecules. The atoms in any molecule are linked together by **chemical bonds,** which essentially involve sharing of electrons between the individual atoms of a molecule. Different elements can make chemical bonds in different ways. For example, hydrogen atoms generally can bond with only one other atom at a time, while oxygen atoms generally can bond with at most two other atoms at a time. We can see these properties in a water molecule (Figure 5.7): The oxygen atom has two chemical bonds, one to each of the two hydrogen atoms, while each hydrogen atom has only a single chemical bond to the oxygen atom.

Carbon is a particularly versatile chemical element because it can bond to as many as four atoms at a time. This allows carbon atoms to link together in an endless variety of carbon "skeletons" varying in size and branching patterns (Figure 5.8). The carbon atom sometimes uses two of its bonds to link with the same atom, forming a *double bond;* notice the double bonds in three of the four molecules at the bottom of Figure 5.8.

We generally refer to carbon molecules that are associated with life as **organic molecules.** The simplest organic molecules consist of carbon skeletons bonded only to hydrogen atoms; these simple organic molecules are often called *hydrocarbons* to reflect the fact that they contain only hydrogen and carbon. In more complex organic molecules, one or more carbon atoms are bonded to something besides hydrogen and other carbon atoms (Figure 5.9).

NON-CARBON-BASED LIFE When we consider the possibility of extraterrestrial life, it's natural to wonder whether it might be based on an element besides carbon. In truth, we cannot say for sure whether other elements would work. However, given the importance to life on Earth of carbon's ability to form four bonds at once, we might expect that any other elemental basis for life would have to have the same bonding capability. Among the elements common on Earth's surface—and likely to be common on other planets—silicon is the only element besides carbon that can have four bonds at once. As a result, science fiction writers have often speculated about finding silicon-based life on other worlds.

Unfortunately for science fiction, silicon has at least three strikes against it as a basis for life. First and most important, the bonds formed by silicon are significantly weaker than equivalent bonds formed by carbon. As a result, complex molecules based on silicon are more fragile than those based on carbon—probably too fragile to form the structural components of living cells. Second, unlike carbon, silicon does not normally form double bonds; instead, it forms only single bonds. This limits the range of chemical reactions that silicon-based molecules can engage in as well as the variety of molecular structures that can form. Third, carbon can be mobile in the environment in the form of gaseous carbon dioxide, but silicon dioxide is a solid (for example, quartz is made from silicon dioxide) that offers no similar mobility. Given the three strikes against silicon, most scientists consider it unlikely that biochemical life can be silicon-based ("artificial" life using silicon-based computer chips may be a different story). Moreover, observational evidence on Earth also argues against silicon: Silicon is about 1000 times as abundant as carbon in Earth's crust, so the fact that life here is carbon-based despite the greater abundance of silicon suggests that carbon will always win out over silicon as a basis for life.

A few other elements have also been suggested as possibilities for replacing carbon on other worlds, but most scientists believe carbon's natural advantages will still win out. We have found carbon-based (organic) molecules even in space (as identified in meteorites and interstellar clouds), suggesting that carbon chemistry is so easy and so common that even if life with another basis were possible, carbon-based life probably would arise first and then reproduce so successfully that it would crowd out the possibility of any other type of life. Nevertheless, we should not completely rule out the possibility of non-carbon-based life, and some scientists are therefore seeking to learn more about how we might recognize it, if it exists. As a recent report from the National Research Council stated, "Nothing would be more tragic in the . . . exploration of space than to encounter alien life without recognizing it."

What are the molecular components of cells?

All the major components of cells are made from complex organic molecules. Today, biologists know the precise chemical structure of a great

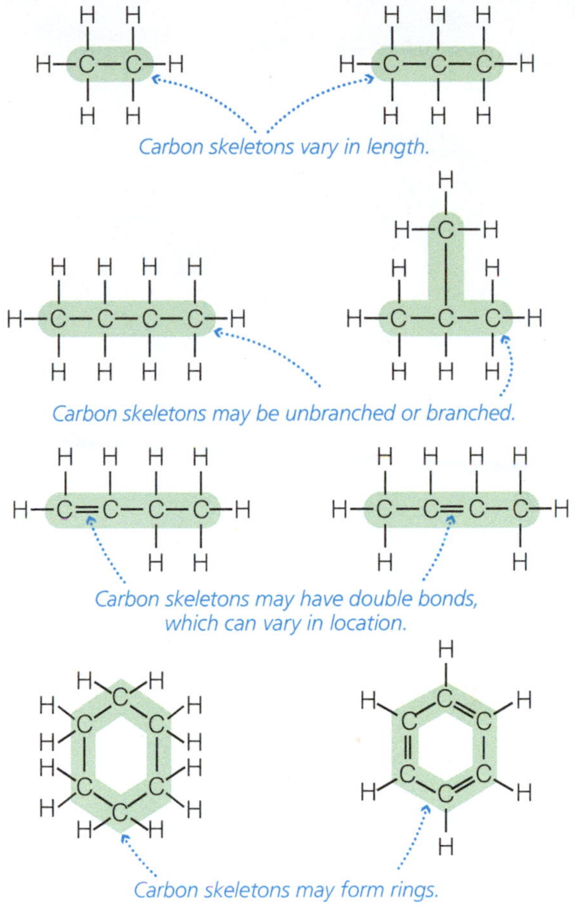

Carbon skeletons vary in length.

Carbon skeletons may be unbranched or branched.

Carbon skeletons may have double bonds, which can vary in location.

Carbon skeletons may form rings.

FIGURE 5.8
These diagrams represent several relatively simple hydrocarbons—organic molecules consisting of a carbon skeleton attached to hydrogen atoms. The carbon skeletons are highlighted in green. Each single line represents a single chemical bond; a double line represents a double bond. Note that every carbon atom has a total of four bonds (a double bond counts as two single bonds). (Adapted from Campbell, Reece, Simon, *Essential Biology*.)

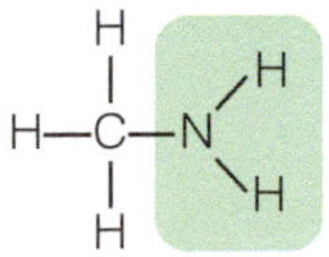

FIGURE 5.9
In a more complex organic molecule, at least one bond links a carbon atom to something besides hydrogen or another carbon atom. Here, one of the carbon atom's four bonds links it to an *amino group* (which consists of a nitrogen atom and two hydrogen atoms), highlighted in green. (Adapted from Campbell, Reece, Simon, *Essential Biology*.)

many of these molecules, and this knowledge has enabled them to gain a deep understanding of the biochemistry of life. If you take a course in biology, you will learn about much of this biochemistry; here we focus only on generalities about the molecules of life. The large molecular components of cells fall into four main classes: *carbohydrates, lipids, proteins,* and *nucleic acids.* Let's briefly investigate the properties of each class that are most important to life.

CARBOHYDRATES You're probably familiar with **carbohydrates** as a source of food energy—the sugars and starches known to athletes and dieters as "carbs." In addition to providing energy to cells, carbohydrates make important cellular structures. For example, a carbohydrate called *cellulose* forms the fibers of cotton and linen and is the main constituent of wood. Life on other worlds would presumably need molecules to play these same energy-storing and structural roles, but we do not yet know whether such molecules would have to resemble carbohydrates on Earth or if they could be very different in their chemistry. As a result, we will have little more to say about carbohydrates in this book.

LIPIDS Like carbohydrates, **lipids** can store energy for cells. The types of lipids that store energy are more commonly known as *fats.* That is, despite its bad reputation, fat is actually critical to living cells. Lipids also play a variety of other roles in cells on Earth, but from the standpoint of life in the universe, perhaps their most important role is as the major ingredients of the membranes that make it possible for intact cells to exist. Moreover, as we'll discuss in more detail in Chapter 6, the membrane-forming role of lipids is thought to have played a critical role in the origin of life: Lipids can spontaneously form membranes in water and probably did so on the early Earth. Other organic molecules would have been trapped inside the space formed by the membranes, making what were in essence tiny chemical factories. These tiny chemical factories may have facilitated the chemical reactions that ultimately led to the origin of life.

PROTEINS: KEY EVIDENCE FOR A COMMON ANCESTOR OF LIFE The molecules called **proteins** are often described as the workhorses of cells, because they participate in such a vast array of functions. Some proteins serve as structural elements in cells. Others, called **enzymes,** are crucial to nearly all the important biochemical reactions that occur within cells—including the copying of genetic material (DNA)—because they serve as *catalysts* for these reactions. A **catalyst** is any substance (not necessarily a single molecule) that facilitates or accelerates a chemical reaction that would otherwise occur much more slowly; the catalyst itself is not changed by the process. Enzymes are catalysts because they greatly accelerate the reactions in which they are involved, even though they enter and leave the reactions essentially unchanged. Moreover, because an enzyme is left unchanged after it catalyzes a reaction, a single enzyme can catalyze a specific chemical reaction many times without needing to be rebuilt.

All proteins, whether they serve as enzymes or in other roles, are large molecules built from long chains of smaller molecules called **amino acids.** The "amino" in amino acids refers to the *amino group* that they all share—a nitrogen atom bonded to two hydrogen atoms and a carbon atom (see Figure 5.9); amino acids also always contain what is called a *carboxyl group* (COOH). Different types of amino acids are distinguished by the different sets of atoms that are also bonded to the central carbon.

The nature of the amino acid chains that make proteins in living organisms provides important evidence supporting the idea that all life on Earth shares a common ancestor. Biochemists have identified more than 70 different amino acids, but most life on Earth builds proteins from only 20 of them. (Two additional amino acids are known to be used in rare cases by particular microorganisms, and scientists suspect that other cases of rare amino acids may yet be discovered.) If life on Earth had more than one common ancestor, we might expect that different organisms would use different sets of amino acids, but they don't. Moreover, naturally occurring amino acids come in two slightly different forms, distinguished by their **handedness** (or *chirality*): The "left-handed" and "right-handed" versions are mirror images of each other (Figure 5.10). Amino acids found in nonbiological circumstances generally consist of a mix of the left- and right-handed versions,* but living cells use only the left-handed versions of amino acids to build proteins. Again, the fact that all life on Earth makes use of the same versions of amino acids suggests a common ancestor. Carbohydrates provide some similar evidence, as life on Earth uses mainly the right-handed versions of sugars.

Think About It Large impacts can blast meteorites into space, allowing rocks from one world to travel to another. As we'll discuss in Chapter 6, some scientists hypothesize that microscopic life might survive such impacts and thereby might have migrated between the inner worlds of our solar system. Suppose we discover life on Mars and we find that, while it also has proteins, it builds them from a different set of amino acids than does life on Earth, and they are all the right-handed versions. Would this support or contradict the hypothesis that life migrated between Earth and Mars? Explain.

NUCLEIC ACIDS Perhaps no cellular molecule is more famous than DNA, which is the basic hereditary material of all life on Earth. A second important nucleic acid, RNA (short for "*ribo*nucleic *a*cid"), helps carry out the instructions contained in DNA. Together, the nucleic acids DNA and RNA are responsible for allowing cells to function according to precise, heritable instructions. Changing a cell's DNA changes the inherent nature of an organism; indeed, it is changes to DNA that allow species to evolve. We do not know whether other types of molecules could replace nucleic acids in life elsewhere, but it is difficult to imagine life existing in any form without a molecule or molecules to serve the hereditary functions of DNA and RNA. These molecules are so important that we'll devote Section 5.4 to discussing them in more detail.

What are the major groupings of life on Earth?

Until just a couple of decades ago, life was generally classified only by outward appearances. For thousands of years, these appearances suggested that life existed only in two basic forms: plants and animals. The first evidence of a different reality surfaced at about the time of the

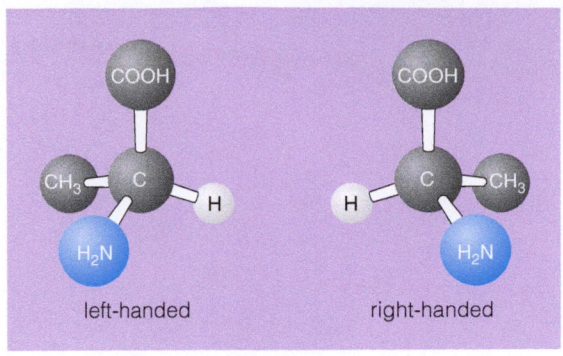

FIGURE 5.10
Any particular amino acid comes in two forms, distinguished by their *handedness*. These diagrams show the left- and right-handed versions of the amino acid *alanine*. Notice that the two versions are mirror images of each other.

---

*There is some evidence from studies of meteorites that left-handed versions of amino acids outnumber right-handed versions in interstellar space, in which case meteorites bringing amino acids to Earth might have tipped the scales to make left-handed versions more likely to be incorporated into life. Some scientists have suggested that this overabundance of left-handed versions may be a result of the way amino acids interact with polarized light in interstellar space.

Copernican revolution. While Galileo turned his telescopes to the heavens, other scientists began to employ similar lens technology to study the microscopic world. The precise origin of the microscope is not known, but the first practical microscopes used for scientific study were built by the Dutch scientist Anton Van Leeuwenhoek (1632–1723; his last name is pronounced "LAY-ven-huke"). During decades of observations beginning around 1674, Leeuwenhoek discovered the world of microscopic life. He was the first to realize that drops of pond water are teeming with microorganisms—a discovery now repeated by almost every elementary school student. He also discovered bacteria and studied the microscopic structure of many plant and animal cells.

With hindsight, it may seem surprising that anyone could have thought that all microorganisms might just be tiny plants or animals, since we now know that microorganisms are far more genetically diverse than larger organisms. But that is exactly what happened. If you look at an old-enough biology textbook, you'll see that life was classified into two "kingdoms," the plants and the animals, and microbes were generally just stuck into one of those two. In the 1960s, biologists expanded the list to five kingdoms, with two (*protista* and *monera*) reserved for microorganisms; the third new addition was *fungi* (which include mushrooms), by then recognized to be different from plants. However, as our understanding of biochemistry improved during the ensuing decades, biologists began to consider whether life could be classified by its cell structure or biochemistry (including genetics), rather than by its outward appearance. Today, we know that classification by the biochemistry of cells gives us much deeper insights into the relationships among different living species than does classification by appearances alone.

MICROSCOPIC LIFE Because we are more familiar with plants and animals than with microbes (meaning any single-celled organism), most people assume that microscopic life is a "minor" part of life on Earth. And because we tend to associate bacteria with disease, many people assume that microbes are generally harmful. Both assumptions are wrong.

Although we humans like to think of ourselves as the dominant form of life on Earth, measurements show that microbes are far more dominant in terms of mass and volume (see Cosmic Calculations 5.1). These microbes are remarkably diverse, varying substantially in size, cell structure, biochemistry, and genetics (Figure 5.11).

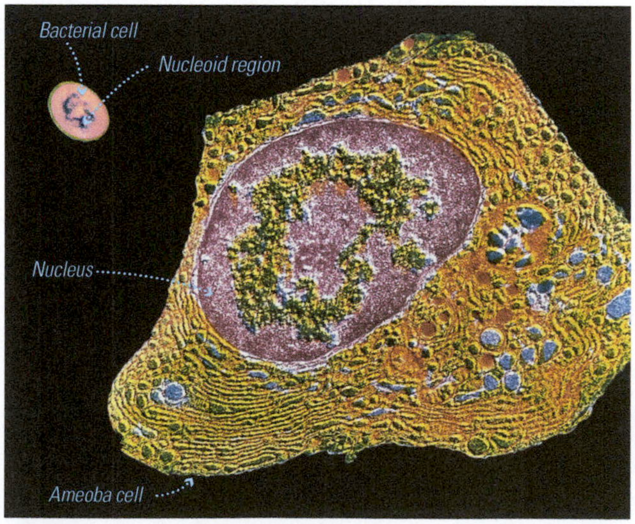

Moreover, most microbes are harmless to humans, and many are crucial to our survival. For example, bacteria in our intestines provide us with important vitamins, and bacteria living in our mouths prevent harmful fungi from growing there. Indeed, research into the human microbiome—the collection of bacteria and viruses that live in our bodies—suggests that it can influence our mood and many aspects of our health. Other microbes play crucial roles in cycling carbon and other vital chemical elements between organic matter and the soil and atmosphere; for example, microbes are responsible for decomposing dead plants and animals. Indeed, plant and animal life would be doomed if microbes somehow disappeared from Earth. In contrast, microbes could survive just fine without plants and animals, as they did during most of the history of life on Earth [Section 6.3].

THE THREE DOMAINS LIFE The classification of microbes long proved difficult. For decades during the twentieth century, biologists assumed

FIGURE 5.11

These microscope photographs contrast a typical amoeba (a single-celled organism of domain eukarya) and a typical bacterial cell. Notice the very different sizes and cell structures; differences in biochemistry and genetics are even greater.

that the presence or absence of a cell nucleus (such as the nucleus of the amoeba in Figure 5.11) represented a fundamental distinction. This visible distinction even led to different names for the two groups: Cells with nuclei were called *eukaryotes*, and those without were called *prokaryotes*. However, analysis of cellular biochemistry has shown that the latter are not a distinct group at all.

Today, biologists classify all life into three broad "superkingdoms," or **domains,** known as the **bacteria,** the **archaea,** and the **eukarya.** The domain eukarya includes not only thousands of known species of microbes, but also all complex plants, animals, and fungi. Cells of eukarya generally have cell nuclei, but this is no longer considered to be as fundamental as their biochemistry. The domains bacteria and archaea consist exclusively of microbes. While species within these two domains look similar under a microscope, study of their biochemistry—for example, the types of lipid structures in their cell membranes, the way in which they make cellular proteins, and most importantly their genetics—shows that they are not closely related. In fact, the archaea appear to be more closely related to eukarya than to bacteria.

THE TREE OF LIFE Biologists now routinely map relationships between species by comparing their DNA or the precise structures of molecules coded for by DNA. For reasons we'll discuss in more detail later, the greater the similarity in these molecules, the more closely the species are related. By studying these molecules in tens of thousands of species, both microbial and multicellular, biologists have mapped out what is usually called the **tree of life** (Figure 5.12).

Note that the tree of life gives us a very different picture of the diversity of life on Earth than the old idea of classifying life into "kingdoms" based on

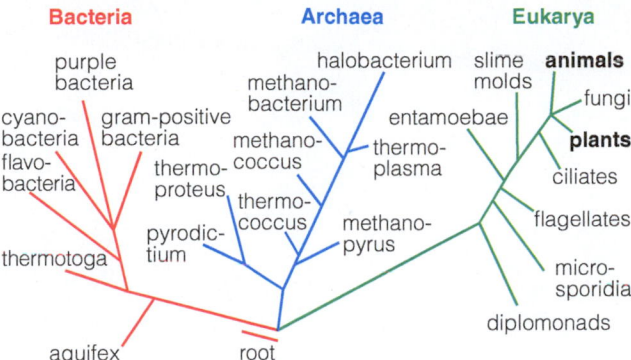

FIGURE 5.12
The tree of life has three major domains: bacteria, archaea, and eukarya; note that all plants and animals represent just two small branches of the domain eukarya. (Only a few of the many known branchings within each domain are shown in this diagram; it also remains possible that additional domains will be discovered.)

Cosmic Calculations 5.1 THE DOMINANT FORM OF LIFE ON EARTH

We can use estimation to show that microbes far outweigh human beings on Earth. We first estimate the total mass of the approximately 6 billion human beings on Earth. If an average person is 50 kilograms (110 pounds), the total human mass is about

$$6 \times 10^9 \text{ persons} \times 50 \frac{\text{kg}}{\text{person}} = 3 \times 10^{11} \text{ kg}$$

We next estimate the mass of microbes in the oceans. The density of microbes varies significantly with location and depth, but a rough average is *1 billion* (10^9) microbes per liter of water. Multiplying this value by the total volume of ocean water (a Web search reveals this to be about 1.4×10^9 km^3) gives us an estimate of the total number of microbes in the ocean:

$$\text{Total microbes} \approx \underbrace{10^9 \frac{\text{microbes}}{\text{liter}}}_{\substack{\text{density of microbes} \\ \text{in ocean water}}} \times \underbrace{(1.4 \times 10^9 \text{ km}^3)}_{\text{volume of ocean water}} \times \underbrace{(10^3 \tfrac{\text{m}}{\text{km}})^3 \times (10^3 \tfrac{\text{liter}}{\text{m}^3})}_{\text{convert from km}^3 \text{ to liters}}$$

$$= 10^9 \frac{\text{microbes}}{\text{liter}} \times (1.4 \times 10^9 \text{ km}^3) \times (10^9 \tfrac{\text{m}^3}{\text{km}^3}) \times (10^3 \tfrac{\text{liter}}{\text{m}^3})$$

$$= 1.4 \times 10^{30} \text{ microbes}$$

To find the total mass of microbes, we next need to know the typical microbe mass. A typical bacterium measures about 1 micrometer on a side, which means it has a volume of about 1 cubic micrometer.

There are 1 million (10^6) micrometers per meter, so the volume of a bacterium is

$$1 \text{ }\mu\text{m}^3 \times \left(10^{-6} \tfrac{\text{m}}{\mu\text{m}}\right)^3 = 10^{-18} \text{ m}^3$$

Because life is made mostly of water, we can use the density of water (1000 kg/m^3) as the density of a microbe. Multiplying the microbe volume by the density, we estimate that the typical microbe mass is

$$10^{-18} \text{ m}^3 \times 10^3 \tfrac{\text{kg}}{\text{m}} = 10^{-15} \text{ kg}$$

We combine our results to find the total mass of microbes in the oceans:

total mass of microbes $\approx$ (number of microbes) $\times$ (mass per microbe)

$$= 1.4 \times 10^{30} \text{ microbes} \times 10^{-15} \tfrac{\text{kg}}{\text{microbe}}$$

$$= 1.4 \times 10^{15} \text{ kg}$$

We can compare to the mass of human beings by dividing:

$$\frac{\text{total mass of microbes}}{\text{total mass of humans}} \approx \frac{1.4 \times 10^{15} \text{ kg}}{3 \times 10^{11} \text{ kg}} \approx 5000$$

The total mass of microbes in the oceans is roughly 5000 times that of all humans combined.

visible distinctions, and this new picture is thought to be far more accurate in depicting relationships among species. In particular, for our purposes in astrobiology, you should focus on three main features of the tree of life:

1. All large, multicellular organisms—meaning all plants, animals, and fungi—represent just three small branches of one domain (eukarya).
2. The true diversity of life on Earth is therefore found almost entirely within the microscopic realm. Biochemically and genetically, we humans (and all other animals) are much more closely related to mushrooms than most microbes are to one another.
3. The branch lengths in the tree of life represent the amount of genetic difference between species. Therefore, as we trace the branches back toward the "root," we are presumably looking back to species that split from the common ancestor at earlier times. The closer we get to the root, the closer we must be to finding an organism that resembles a common ancestor of all life on Earth.

Keep in mind that depictions of the tree of life are a work in progress. We have carefully studied only a tiny fraction of all the species that exist on Earth; indeed, we do not even know how many species there are, and we are only beginning to learn about many species that live in hard-to-reach environments such as the deep ocean and underground. We will certainly discover more branches within the three known domains, and it is even possible that we will discover entirely new domains in the future.

5.3 Metabolism: The Chemistry of Life

Why are cells so important to life on Earth? More to the point, is it possible that life elsewhere might exist without having a fundamental organizational unit like the cell? To answer these questions, we must understand the processes that take place inside living cells. These processes, which are all chemical in nature, make up what we call **metabolism.** More specifically, metabolism is a blanket term that refers to the many chemical reactions that occur in living organisms and are involved in providing energy or nutrients to cells.

What are the basic metabolic needs of life?

Most of the important chemical reactions that occur in cells share a common characteristic: Without the help provided by the cell itself, the reactions would occur too slowly to be useful for life. In this sense, a cell's primary purpose is to serve as a tiny chemical factory in which desired chemical reactions occur much more rapidly than they could otherwise, thereby making it possible to turn simple molecules into the great variety of complex organic molecules needed by living organisms. As is also the case in many factories, cellular work sometimes involves breaking down molecules as well as building them. Like any manufacturing process, this biochemical manufacturing process requires two basic things:

1. *A source of raw materials* with which to build new products. In the case of living cells, the key raw materials are molecules that provide the cell with carbon and other basic elements of life.
2. *A source of energy* to fuel the metabolic processes that break down old molecules and manufacture new ones.

Given the large variety of molecules involved in metabolic processes, you might think that cells would need an equally large variety of sources of raw materials and energy to survive. However, cells have the ability to build incredible variety from a limited set of starting materials. Part of this ability comes from the remarkable variety of enzymes in living cells. Each enzyme is specialized to catalyze particular chemical reactions needed in cellular manufacturing. The remarkable diversity of enzymes in living organisms today is a testament to the power of evolution. The instructions for enzyme creation are encoded in DNA and hence have been evolving for billions of years.

THE ROLE OF ATP Another reason why cells can produce so much variety from so little input is that, regardless of where they get their energy, all cells put the energy to work in the same basic way. Every living cell uses the same molecule, called **ATP** (short for "adenosine *tri*phosphate"), to store and release energy for nearly all its chemical manufacturing (Figure 5.13). Using ATP vastly simplifies the manufacturing process, because it means that a cell needs an outside energy source only for the purpose of producing ATP, rather than for producing the full variety of organic molecules in cells. Once ATP is produced, it can be used to provide energy for other cellular reactions. Moreover, the nature of ATP makes it completely recyclable. Each time a cell draws energy from a molecule of ATP, it leaves a closely related by-product, called ADP (short for "adenosine *di*phosphate"), that can be easily turned back into ATP.

The fact that all life on Earth uses the same molecule (ATP) for energy storage offers further evidence for a common origin of life. There's no known reason why other molecules could not fill the role of ATP. The fact that all living cells use ATP therefore suggests that they all evolved from a common ancestor that made use of this remarkable molecule.

CARBON SOURCES AND ENERGY SOURCES Because carbon compounds are the primary raw materials needed for life, the needs of metabolism essentially come down to the need for a carbon source and an energy source. We humans, like all animals, meet both of these needs with food. The food we eat gets digested and carried in molecular form by our bloodstreams to individual cells. There, the cells make use of the molecules from our food sources. Some of the molecules are used as the carbon source for cellular manufacturing, while others undergo chemical reactions that release energy the cell can use to fuel its ATP cycle. Of course, not all organisms get their carbon and their energy by eating other organisms. Plants get energy from sunlight, and some microorganisms get energy from chemical reactions that take place inside rocks or in hot springs.

How do we classify life by its metabolic sources?

Living cells on Earth get their carbon and energy from a surprisingly wide variety of sources. As we'll see, this wide variety makes life elsewhere seem more likely. For example, we already know of worlds within our own solar system, such as Mars and Europa, that may well have the necessary materials and energy source for metabolism, suggesting that life could at least in principle exist on those worlds.

Today, astrobiologists classify life into four major categories by its metabolic sources, summarized in Table 5.1. We can understand their rather long and technical names by looking first at their carbon sources and then at their energy sources.

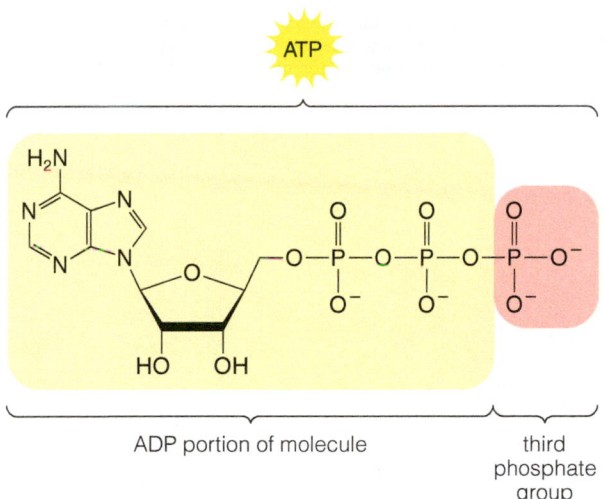

a The molecular structure of ATP. To understand the key parts of the structure, notice that the right side of the molecule shows three identical "phosphate groups," with the third one highlighted in pink. The portion of the molecule shown in yellow is ADP (adenosine diphosphate, because it has two of the phosphate groups), and the entire molecule, including the pink portion, is ATP (adenosine triphosphate, because it has three phosphate groups).

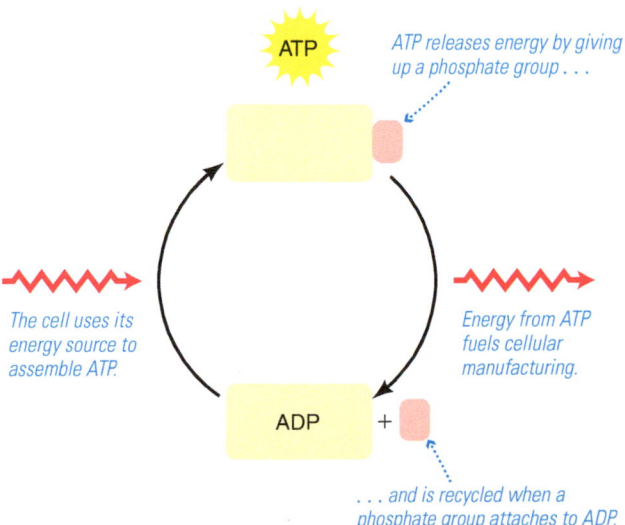

b Cells recycle ATP. The ATP molecule gives up energy when it splits into ADP (yellow) and a phosphate group (pink). Energy input puts the ATP molecule back together.

FIGURE 5.13
Every living cell on Earth uses the molecule ATP to store and release energy. (Adapted from Campbell, Reece, Simon, *Essential Biology*.)

TABLE 5.1 Metabolic Classifications of Living Organisms*

Metabolic Classification	Carbon Source	Energy Source	Examples
photoautotroph	carbon dioxide	sunlight	plants, photosynthetic bacteria
chemoautotroph	carbon dioxide	inorganic chemicals (e.g., iron, sulfur, ammonia)	some bacteria and archaea, especially in extreme environments
photoheterotroph	organic compounds	sunlight	some bacteria and archaea
chemoheterotroph	organic compounds	organic compounds	animals, many microbes

*You may see similar tables that add a third classification category based on the source of electrons for energy transfer reactions; organisms are designated "organo-" if the electrons come from an organic source and "litho-" if they come from an inorganic source. With this added distinction, the four classifications in the first column each branch into two (such as photolithoautotroph and photoorganoautotroph).

CARBON SOURCES: AUTOTROPHS AND HETEROTROPHS Cells need a source of carbon from which to build the skeletons of their organic molecules. In the broadest sense, cells can get their carbon in either of two ways:

1. Some cells get carbon by consuming preexisting organic compounds—that is, by eating. For example, we humans acquire carbon by eating plants or other animals. Any organism that gets its carbon by eating is called a **heterotroph;** the word comes from *hetero*, meaning "others," and *troph*, meaning "to feed." All animals are heterotrophs, as are many microscopic organisms.
2. Some cells get carbon directly from the environment by taking in carbon dioxide from the atmosphere or carbon dioxide dissolved in water. An organism that gets its carbon directly from the environment is called an **autotroph,** meaning "self-feeding." For example, trees and most other plants are autotrophs.

If you look at the first column of Table 5.1, you'll see that the first two entries are both autotrophs and the second two are both heterotrophs. However, in each case the entry carries a prefix of either *photo* (meaning "light") or *chemo* (meaning "chemicals"). These prefixes describe the energy source that goes with the carbon source of either eating (heterotroph) or taking in environmental carbon dioxide (autotroph).

ENERGY SOURCES: LIGHT OR CHEMICALS Broadly speaking, the energy that a living cell uses to make ATP can come from one of three sources (see Section 9.4 for more details about chemical energy for life):

1. Some cells get energy directly from sunlight, using the process we call *photosynthesis*. For example, plants acquire their energy from sunlight. Organisms that get energy from sunlight are given the prefix *photo*.
2. Some cells get energy from food; that is, they take chemical energy from organic compounds they've eaten and use it to make their own ATP. Since the energy comes from chemical reactions, these organisms get the prefix *chemo*.
3. Some cells get energy from *inorganic* chemicals—chemicals that do *not* contain carbon—in the environment. This type of energy source is different in character from organic food, and cells that get energy directly from the environment require neither sunlight nor other organisms to survive. However, because the energy still comes from chemical reactions, these organisms also get the prefix *chemo*.

THE FOUR METABOLIC CLASSIFICATIONS We can now put the carbon and energy sources together to understand the four metabolic classifications in Table 5.1. The first row of the table shows the **photoautotrophs,** which get their energy from sunlight (*photo*) and their carbon from carbon dioxide in the environment (making them autotrophs). This category therefore includes plants, as well as microorganisms that obtain their energy through photosynthesis.

The second row shows the **chemoautotrophs,** which obtain energy from chemical reactions (*chemo*) involving inorganic chemicals and carbon from environmental carbon dioxide (making them autotrophs). These are in some ways the most amazing organisms, because they need neither organic food nor sunlight to survive. For example, the archaea known as *Sulfolobus* (a genus that includes many distinct species) live in volcanic hot springs and obtain energy from chemical reactions involving sulfur compounds. As is the case with *Sulfolobus,* chemoautotrophs are often found in environments where most other organisms could not survive. For much the same reason, they may also be the organisms most likely to be found on other worlds, since a wider range of conditions seems suitable to them than to other forms of life.

The third row shows the **photoheterotrophs,** which get energy from sunlight (*photo*) but get their carbon by consuming other organisms or the remains of such organisms (making them heterotrophs). This category is much rarer, but some bacteria and archaea do indeed get their carbon by eating organic compounds while making ATP with energy from sunlight. Examples include bacteria known as *Chloroflexus,* which obtain their carbon from other bacteria but their energy from photosynthesis. These organisms live in lakes, rivers, hot springs, and some aquatic environments very high in salt content.

The fourth row shows **chemoheterotrophs,** which get both their energy and their carbon from food. This category includes us and all other animals, as well as many microorganisms. We are chemoheterotrophs because we extract chemical energy from food (*chemo*) and carbon from eating (making us heterotrophs).

Think About It Classify each of the following into one of the four metabolic categories listed in Table 5.1: (a) an organism that gets its energy from chemicals near an undersea volcano and gets its carbon from carbon dioxide dissolved in water; (b) a tomato plant; (c) a fly.

METABOLISM, WATER, AND THE SEARCH FOR LIFE The four metabolic classifications are quite general, so they ought to apply equally well to life elsewhere as to life on Earth. Moreover, any type of complex metabolism requires the existence of some kind of structure that allows carbon and energy to come together to manufacture (and break down) the molecules needed for life. Unless we are failing to imagine an entirely different potential mode of operation, it therefore seems likely that all living organisms must have a fundamental structure that functions much like cells on Earth. This crucial observation means that we can search for life in the universe by searching for cells rather than having to search for a much broader variety of possible structures.

This leaves us with one final ingredient to consider in metabolism: liquid water. On Earth, water plays three key roles in metabolism. First, metabolism requires that organic chemicals be readily available for reactions. Liquid water makes this possible by allowing organic chemicals essentially to float within the cell (because the chemicals dissolve in water). Second, metabolism requires a means of transporting chemicals to and within cells, and of transporting waste products away; water is the medium of this transport. Third, water plays a role in many of the metabolic reactions within cells; for example, water molecules are necessary for the reactions that store and release energy in ATP.

All living cells on Earth depend on liquid water to play these three roles, and this dependence limits the conditions under which we find life on our planet: We find life only in places where it is neither too cold nor too hot for liquid water to exist. Indeed, while we've seen that life on Earth can use a variety of different carbon and energy sources, liquid water is one thing that no organism on Earth can survive without. (Some organisms can become dormant in the absence of liquid water, but they cannot and grow or metabolize in such conditions.) Does this need for liquid water also apply to life on other worlds? Certainly, some kind of liquid seems necessary, but we'll save discussion of possibilities other than water for Chapter 7.

5.4 DNA and Heredity

In the previous two sections, we studied two key features of life on Earth that are likely to be crucial to life anywhere else as well: the structural units of cells and the metabolic processes that keep cells alive. A third feature that seems generally needed for all life is some means of storing information—that is, a set of "operating instructions" for the cell and a way of passing these instructions down through the generations. This information is what we generally call an organism's *heredity*.

All living things on Earth encode their hereditary information in the molecule known as DNA (although some viruses use RNA). That is, DNA holds the "operating instructions" for living organisms on Earth. DNA also allows organisms to reproduce, because it can be accurately copied. In this section, we will explore how DNA determines the nature of an organism and allows reproduction. We will also discuss how rare errors in the copying of DNA can lead to evolutionary adaptations, thus giving us the molecular-level understanding of natural selection that we first discussed in Section 5.1.

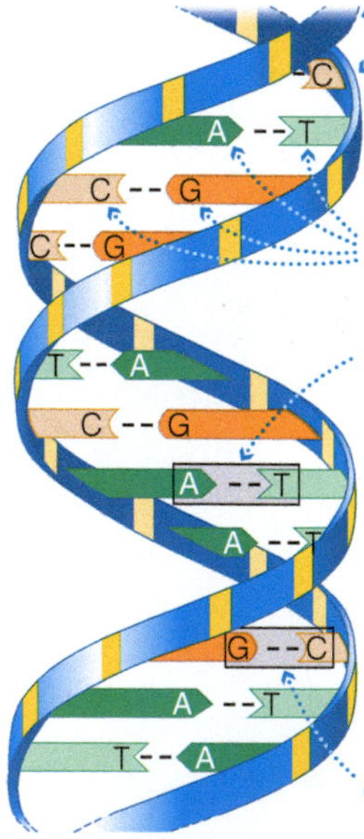

How does the structure of DNA allow for its replication?

The molecular structure of DNA, a *double helix,* is one of the most familiar scientific icons of our time (Figure 5.14). A helix is a three-dimensional spiral, such as you would make by extending a Slinky toy; a double helix has two intertwined strands, each in the shape of a helix. The structure looks much like a zipper twisted into a spiral. The fabric edges of the zipper represent the "backbone" of the DNA molecule, while the zipper teeth that link the two strands represent molecular components called **DNA bases.** The chemical structure of the backbone is interesting and important in its own right, but it is the DNA bases that hold the key to heredity. All known, naturally existing life on Earth makes use of only four DNA bases: adenine (abbreviated A), guanine (G), thymine (T), and cytosine (C). (However, in 2014 scientists engineered a living organism with two additional bases that don't occur naturally, indicating that additional DNA bases not only are possible but might be expected on other worlds.)

The two strands of the "backbone" are wound in the shape of a double helix.

The two strands are linked with four "bases": A, T, C, and G.

T attaches only to A.

C attaches only to G.

FIGURE 5.14
This diagram represents a DNA molecule, which looks much like a zipper twisted into a spiral. The important hereditary information is contained in the "teeth" linking the strands. These "teeth" are the DNA bases. Only four DNA bases are used, and they can link up between the two strands only in specific ways: T attaches only to A, and C attaches only to G. (The color coding is arbitrary and is used only to represent different types of chemical groups; in the backbone, blue and yellow represent sugar and phosphate groups, respectively.) (Adapted from Campbell, Reece, Simon, *Essential Biology.*)

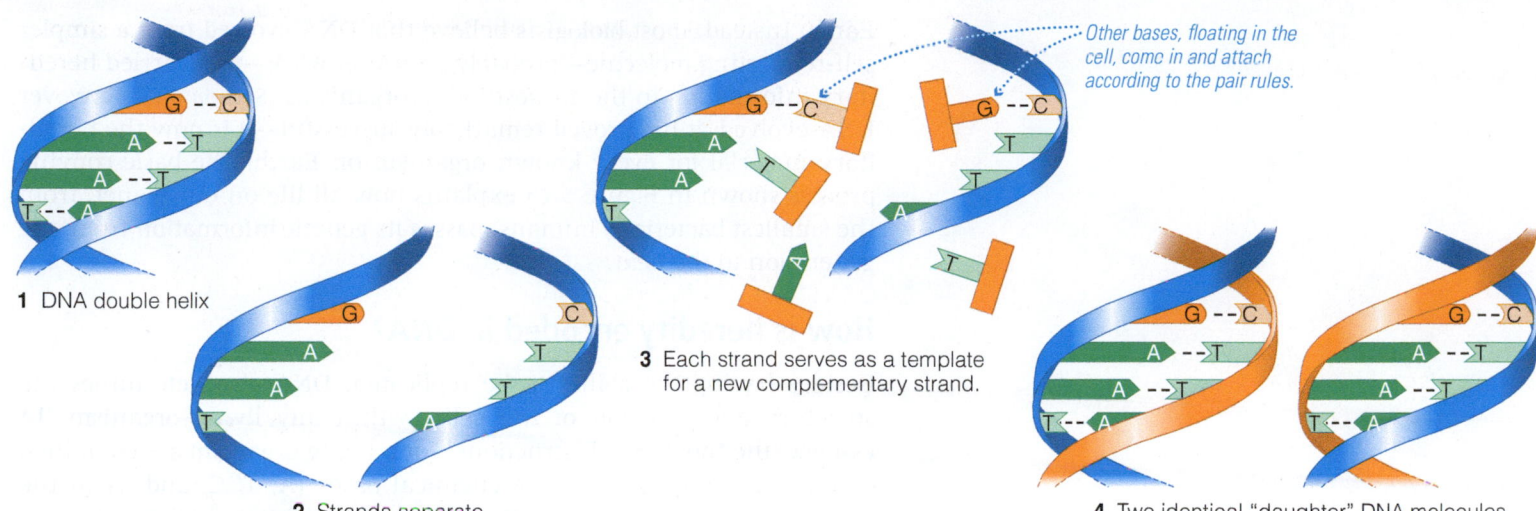

1 DNA double helix

2 Strands separate.

3 Each strand serves as a template for a new complementary strand.

Other bases, floating in the cell, come in and attach according to the pair rules.

4 Two identical "daughter" DNA molecules

FIGURE 5.15

DNA replication. DNA copies itself by "unzipping" its two strands, each of which then serves as a template for making a new, complementary strand built in accord with the base pairing rules: A goes only with T, and C goes only with G. The end result is two identical copies of the original DNA molecule. (Adapted from Campbell, Reece, Simon, *Essential Biology*.)

The key to DNA's ability to be duplicated by cellular machinery lies in the way the four DNA bases pair up to link the two strands: T can pair up only with A, while C can pair up only with G. Figure 5.14 shows this pairing by representing the different bases with different shapes. For example, the shape of A, which is depicted as ending with an open triangle, fits only into the notch in T. Similarly, what is shown as the curved end of G fits only into the curved notch in C. These diagrams are only schematic representations—there aren't literally notches and curves on the DNA bases—but the real chemical bases work much the same way. Their actual shapes and sizes determine how they pair up.

The process by which DNA is copied, called **DNA replication,** is illustrated in Figure 5.15. Step 1 begins with the complete double helix. In Step 2, the two strands separate, "unzipping" the links between the paired bases. Step 3 shows how, once the strands have been "unzipped," each strand can serve as a template for making a new strand. Because the "teeth" of each new strand must link to the existing strands according to the base pairing rules—T goes only with A, and C goes only with G—each new strand will be *complementary* to an existing one. (By saying that two strands are complementary, we mean that, while they are not identical, they contain the same information because knowing the base sequence on one strand automatically tells us the base sequence on the other strand.) The end result, shown in Step 4, is two identical copies of the original DNA molecule. When a cell divides, one copy goes to each daughter cell. Because cell division is the key to passing down genetic material from one generation to the next, DNA replication explains the basis of heredity.

Although the DNA copying process is easy in principle, the actual mechanics are fairly complex. More than a dozen special enzymes are involved in the various steps, performing tasks such as unzipping the double helix, making sure the correct bases pair up, checking for and correcting any errors in the copying process, and rezipping the new DNA molecules. This complexity is one reason why errors sometimes occur in DNA replication; as we'll discuss shortly, these errors are crucial to evolution. The complexity of replication also makes it extremely unlikely that DNA could have been the original hereditary molecule for life on

Earth. Instead, most biologists believe that DNA evolved from a simpler self-replicating molecule—probably a form of RNA—that carried hereditary information in the earliest living organisms [Section 6.2]. However DNA evolved, it has proved remarkably successful—it is now the hereditary material for every known organism on Earth. The basic copying process shown in Figure 5.15 explains how all life on our planet, from the smallest bacteria to humans, passes its genetic information from one generation to the next.

How is heredity encoded in DNA?

Besides having the ability to be replicated, DNA also determines the structure and function of the cells within any living organism. In essence, the "operating instructions" for a living organism are contained in the precise arrangement of chemical bases (A, T, C, and G) in the organism's DNA. Today, biologists have technology that allows them to rapidly determine the sequence of bases in almost any strand of DNA. This technology has been used to determine the DNA sequences that code for many cell functions, as well as to determine the complete DNA sequences of many living organisms. For example, the Human Genome Project, completed in 2003, was a 13-year effort in which scientists ultimately determined the order of all three billion bases that make up the DNA of a human being. (In humans, this DNA is spread among the 46 *chromosomes* found in normal human cells.)

GENES AND GENOMES Within a large DNA molecule, isolated sequences of DNA bases represent the instructions for a variety of cell functions. For example, a particular sequence of bases may contain the instructions for building a protein, for building a piece of RNA, or for carrying out or regulating one of these building processes. The instructions representing any individual function—such as the instructions for building a single protein—make up what we call a **gene.** A gene is the basic functional unit of an organism's heredity—a single gene consists of a sequence of DNA bases (or RNA bases, in some viruses) that provides the instructions for a single cell function.

Interestingly, among plants, animals, and other eukarya, most of the DNA is *not* part of any gene; that is, much of the DNA does not appear to carry the instructions for any particular cell function. For example, this so-called **noncoding DNA** (sometimes called "junk DNA") makes up more than 95% of the total DNA in human beings, and similarly large fractions of the DNA of many other eukaryotes. Biologists suspect that most of this noncoding DNA represents evolutionary artifacts—pieces of DNA that may once have had functions in ancestral cells but that no longer are important, much like the way the appendix is an organ that no longer plays an important role in our bodies. However, recent discoveries suggest that at least some of the noncoding DNA may function in ways that are not yet fully understood.

The complete sequence of DNA bases in an organism, encompassing all of the organism's genes as well as all its noncoding DNA, is called the organism's **genome.** Figure 5.16 summarizes the relationship between DNA, genes, chromosomes, and the full genome.

Different organisms have genomes that vary significantly both in total length (number of bases) and in their numbers of genes. For example, some simple microbes have DNA that extends only a few hundred

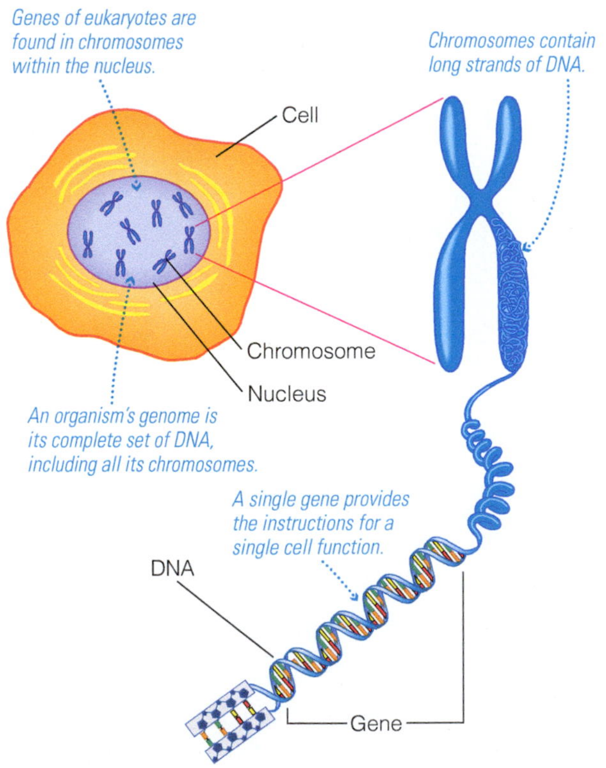

Genes of eukaryotes are found in chromosomes within the nucleus.

Chromosomes contain long strands of DNA.

Cell

Chromosome

Nucleus

An organism's genome is its complete set of DNA, including all its chromosomes.

A single gene provides the instructions for a single cell function.

DNA

Gene

FIGURE 5.16

An organism's genome is its complete set of DNA. This artwork summarizes the relationship between DNA, genes, chromosomes, and the full genome in a eukaryotic cell. (Adapted from Campbell, Reece, Simon, *Essential Biology.*)

thousand bases and contains only a few hundred genes.* We humans have a genome that contains an estimated 20,000 to 25,000 genes among its sequence of some three billion DNA bases. Note that, genetically speaking, we are by no means the most complex organisms on Earth. Rice, for example, has about 37,000 genes, though it has a shorter total DNA sequence than humans. Other organisms have far more DNA than people. For example, the simple plant known as the "whisk fern" (*Psilotum nudum*) has more than 70 times as many bases in its genome as humans, though most of this extra DNA is probably noncoding.

Every member of a particular species has the same basic genome. However, there is always some variation among individuals. For example, while all human beings have the same set of genes, the genes of different individuals may vary here or there in their precise sequence of DNA bases. These differences in the genes of individuals explain why we are not all identical, and they are also the source of the individual variation that underlies the theory of evolution (see Fact 2 on p. 160). Moreover, with a few exceptions, every cell in a living organism contains the same set of genes. Different cell types, such as muscle cells or brain cells, differ only because they *express,* or actually use, different portions of their full set of genes. That is, the DNA found in almost any cell in any organism contains the complete instructions for building an organism of that species. This fact underlies the science of *cloning,* in which a single cell from a living organism is used to grow an entirely new organism with an identical set of genes.

THE GENETIC CODE A strand of DNA contains a long, unbroken sequence of DNA bases; for example, a particular sequence might contain the bases ACTCAGCTTCAACGG.... For a sequence like this to be useful as the instructions for a cell function, there must be a set of rules for how to "read" the sequence. These rules must specify how to break the long sequence into individual "words," as well as where to start reading and where to stop reading the words that represent the instructions for a single gene. The set of rules for reading DNA is called the **genetic code** (Figure 5.17). More specifically, genetic "words" consist of three DNA bases in a row. For the purpose of protein building, each word represents either a particular amino acid or a "start reading" or "stop reading" instruction.

Because the genetic words consist of three DNA bases in a row and there are four DNA bases to choose from (A, C, T, G), the total number of words in the genetic code is $4^3 = 4 \times 4 \times 4 = 64$; all 64 words are spelled out in Figure 5.17. Notice that this is significantly more than the number of amino acids used to make proteins, which is 20 for most organisms. Therefore, the genetic code contains a fair amount of redundancy. For example, the genetic words ACC and ACA both represent the same amino acid. Moreover, a close examination of the genetic code offers a hint about the likely evolution of DNA: The codes for most amino acids really depend on just the first two bases in the three-base genetic words. For example, all four of the three-base words starting with AC (ACC, ACA, ACT, and ACG) code for the same amino acid (threonine). This suggests that the genetic code once depended only on two-base

---

*Many viruses are far simpler, with just a few thousand bases and a handful of genes. Mitochondria within eukaryotic cells, which are thought to have had free-living ancestors, are also much simpler than the simplest bacteria sequenced to date. For example, human mitochondria have fewer than 17,000 DNA base pairs, representing fewer than 40 genes.

FIGURE 5.17

The genetic code. This table shows how three-base-pair "words" of DNA code for particular amino acids or a start or stop instruction. For example, you can find the "word" CAG by looking along the left for C as the first base, along the top for A as the second base, and along the right for G as the third base; you'll then see that CAG codes for the amino acid glutamine. Notice that in most cases, the first two letters alone determine the amino acid; for example, if the first two letters are CT, the amino acid is always leucine regardless of the third letter. This suggests that the current genetic code evolved from an earlier version that used only two-letter "words" rather than three-letter "words."

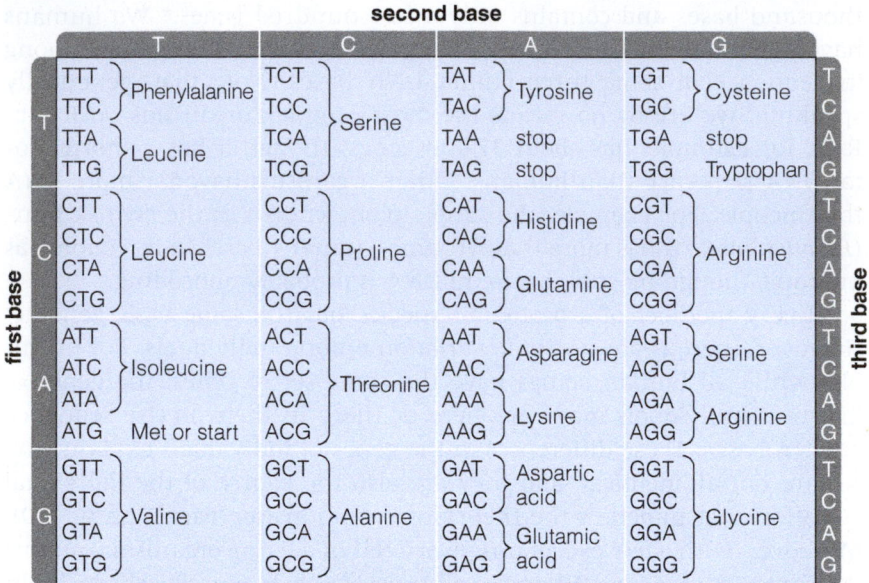

words rather than three-base words. Most biologists now believe that early life-forms used only a two-base language, which later evolved into the current three-base language of the genetic code.

Think About It Note that a two-base language would allow only $4 \times 4 = 16$ possible words—not enough for all the amino acids used by living organisms today. What does this imply about proteins in early life-forms? Explain.

Another important feature of the genetic code is that it is the same in nearly all living organisms on Earth. Only a few organisms show any variations at all on this code, and these variations are minor. (Variations in the genetic code are also found in mitochondria, structures within eukaryotic cells that contain their own DNA.) Nevertheless, the fact that some variations occur tells us that not all the specifics of the genetic code were inevitable. If we think of the genetic code as a language, the fact that nearly all organisms use the same genetic code is as if everyone on Earth spoke the same language, even though other languages are possible. This common language of the genetic code is further evidence for a common ancestor of all life on Earth.

THE ROLE OF RNA While the sequence of bases in a gene holds the instructions for its function, the actual implementation of these instructions is quite complex. As with DNA replication, many enzymes are involved in carrying out genetic instructions. In addition, the molecule RNA plays a particularly important role in these functions. A molecule of RNA is quite similar in structure to a *single* strand of DNA, except that it has a slightly different backbone and one of its four bases is different from one of the DNA bases. [RNA uses a base called *uracil* (U) in place of DNA's thymine (T).]

Several different types of RNA participate in carrying out genetic instructions in the cell. For example, in the process of building a protein, a molecule of *messenger* RNA (or mRNA) is first assembled along one strand of DNA, essentially transcribing the DNA instructions for use in another part of the cell. The messenger RNA then goes to a site in the cell known as a *ribosome*—made of *ribosomal* RNA (rRNA)—where amino acids are assembled into proteins. Assembling the proteins requires individual amino acids, which are collected from within the cell and brought to the ribosome by molecules of *transfer* RNA (tRNA). Working together, the different types of RNA attach the amino acids into the chains that make proteins. (This process is called *translation*, because it effectively *translates* the genetic instructions into an actual protein.) In recent years, biologists have learned that RNA can play many other vital roles in cells, but the roles we have discussed will be enough for our purposes in this book.

How does life evolve?

One of the most remarkable aspects of our current knowledge of DNA is that it has allowed us to further confirm Darwin's theory of evolution by natural selection. In particular, while Darwin had to base his theory on the variation in populations that he could directly observe, we now know precisely how such variation occurs at the molecular level. The key to this knowledge lies in understanding how DNA molecules gradually change through time. Based on what we've already said about DNA replication and protein building, we can see how and why changes in DNA occur.

MUTATIONS: THE MOLECULAR BASIS OF EVOLUTION Despite its complexity, DNA replication proceeds with remarkable speed and accuracy. Some microbes can copy their complete genomes in a matter of minutes, and copying the complete three-billion-base sequence in human DNA takes a human cell only a few hours. In terms of accuracy, the copying process generally occurs with less than one error *per billion* bases copied. Nevertheless, errors sometimes occur. For example, the wrong base may get attached in a base pair, such as linking C to A rather than to G. In other cases, an extra base may be inserted into a gene, a base may be deleted, or an entire sequence of bases might be duplicated or eliminated. Absorption of ultraviolet light or nuclear radiation or the action of certain chemicals (carcinogens) can also cause mistakes to occur. Any change in the base sequence of an organism's DNA is called a **mutation.**

Mutations can affect proteins in a variety of ways. Some mutations have no effect at all. For example, suppose a mutation causes the genetic word ACC to change to ACA in a gene that makes a protein. Because both of these words code for the same amino acid (threonine; see Figure 5.17), this mutation will not change the protein made by the gene. Other single-base mutations—such as changing ACC to CCC—will change a single amino acid in a protein. In some cases, such a change will alter a protein only slightly, hardly affecting its functionality. But in other cases, the change can be much more dramatic. For example, the cause of sickle-cell disease (Figure 5.18), which kills some 100,000 people each year worldwide, can be traced to a single mutation in the gene that makes hemoglobin, in which the base A changed to the base T in just one place within the gene.* Mutations that add or delete a base within a gene tend to have the most dramatic effects on protein structure. The reason is

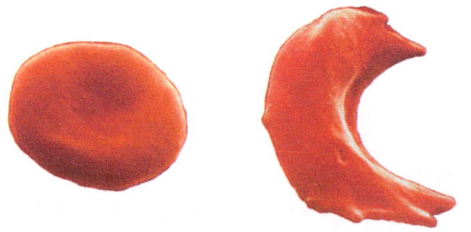

Normal red blood cell   Sickled red blood cell

FIGURE 5.18
These microscopic views contrast a normal human red blood cell with a blood cell found in patients with sickle-cell disease. The sickle shape makes it easier for the blood cells to clog tiny blood vessels, which can lead to debilitating disease. Sickle-cell disease occurs in people whose gene for hemoglobin differs from the "normal" gene in just a single DNA base.

---

*Humans have two copies of each gene, and sickle-cell disease generally occurs only in people who have the sickle-cell mutation in both copies of the gene. From an evolutionary standpoint, this mutation remains prevalent in the population because it actually confers an advantage—malaria resistance—to people with only one copy of the mutated gene.

that the genetic code has no "punctuation" or spacing between words; instead of saying something like "the fat cat ate the rat," for example, it says "thefatcatatetherat." If a letter (base) is added to or deleted from such a sequence, the result will be nonsense from that point on. For example, inserting an "a" so that the sequence becomes "theafatcatatetherat" would cause it to be read as "the afa tca tat eth era t."

Mutations that change proteins are often lethal, because the cell may not be able to survive without the correctly structured protein. However, if the cell survives, the mutation will be copied every time its DNA is replicated. In that case, the mutation represents a permanent change in the cell's hereditary information. If the cell happens to be one that gets passed to the organism's offspring—as is always the case for single-celled organisms and can be the case for animals if the mutation occurs in an egg or sperm cell—the offspring will have a gene that differs from that of the parent. It is this process of mutation that leads to variation among individuals in a species. Each of us differs slightly from all other humans because we each possess a unique set of genes with slightly different base sequences.

Think About It Ultraviolet radiation from the Sun can cause mutations in the DNA of skin cells. Based on what you've learned, explain why this is potentially dangerous (and, indeed, is the cause of skin cancer). How would sunscreen help prevent such mutations?

Mutations therefore provide the basis for evolution. Given that each individual of a species possesses slightly different genes, it is inevitable that some genes will provide advantageous adaptations to the environment. As we discussed in Section 5.1, the combination of individual variation and population pressure leads to natural selection, in which the advantageous adaptations will preferentially be passed down through the generations. Thus, what was once a random mutation in a single

Movie Madness WAR OF THE WORLDS

In 1898, when British novelist H. G. Wells wrote *War of the Worlds,* some astronomers were claiming that long, linear features could be seen criss-crossing the surface of Mars (see Chapter 8). They proposed that intelligent beings, stuck on a dying, drying planet, were lacing their landscape with irrigation canals.

It occurred to Wells that such thirsty Martians might choose to stop all the civil engineering, abandon their withered world, and invade Earth—a planet awash in water. He penned a classic alien invasion story that has since been reworked for radio, television, and two big-budget movies.

Today most people know that Mars is not home to a vast, canal-crazed society, so when director Steven Spielberg re-made *War of the Worlds* in 2005, he studiously avoided mentioning the Red Planet. The aliens in his film just come from *somewhere*. They look vaguely feline and definitely unattractive, and arrive in lightning bolts, a mode of transport that has not yet caught the attention of NASA.

As these bolts reach Earth, they punch right through the pavement to some previously buried military machinery. Despite being mothballed for a million years, these alien tanks-on-legs fire right up. It's hard to imagine anyone using such old weapons: Would today's Air Force be happy to mount an invasion with Neolithic stone axes? Nonetheless, the aliens and their machines quickly emerge and proceed to stomp across the landscape, happily zapping humanity en route. As usual in such movies, our own military wastes its time and a lot of ordnance in a vain effort to discourage them. One character has the bad form to note the obvious: "This is not a war; this is an extermination."

With only a few minutes of film time to go, it's looking bad for *Homo sapiens*, as cities get trampled and citizens get sucked for blood. (Perhaps hemoglobin is a delicacy on these aliens' world?) But then … a miracle occurs. The invaders get sick and keel over—done in not by us, but by earthly microbes. They have no immunity to our bacteria.

Frankly, it's a bit of a stretch to assume that alien biochemistry would be so similar to ours that the invaders would fall victim to terrestrial diseases. But the truly ironic thing about *War of the Worlds* is the idea that Martians (as they were identified in the original story) would invade us and be vanquished by microbes when—as we'll see later in this book— if there are any real Martians, they probably *are* microbes!

individual can eventually become the "normal" version of the gene for an entire species. In this way, species evolve through time. Notice that, while we often associate the word *mutation* with harm, evolution actually proceeds through the occasional beneficial mutation. Although such beneficial mutations may be relatively rare compared to other mutations, natural selection allows these beneficial mutations to propagate preferentially, so tremendous changes can accrue over time.

Evolution sometimes occurs in an even more dramatic way: In some cases, organisms can transfer entire genes to other organisms, a process called *lateral gene transfer*. This process is one of the primary ways that bacteria gain resistance to antibiotics. We humans have also learned to use this process for our benefit through what we call *genetic engineering*, in which we take a gene from one organism and insert it into another. For example, genetic engineering has allowed us to produce human insulin for diabetic patients: The human gene for insulin is inserted into bacteria, and these bacteria produce insulin that can be extracted and used as medicine. Lateral gene transfer can change a species more rapidly than individual mutations, but mutations are still the underlying basis, since they created the genes in the first place.

DNA AND LIFE ON OTHER WORLDS It is difficult to imagine life that does not have heredity, because it seems crucial for any form of life to have some means of storing its operating instructions and passing them on to its offspring. We've seen that DNA is the carrier of heredity for all life on Earth, though as we'll discuss in Chapter 6, we have good reason to believe that very early life on Earth used RNA for this role.* Should we expect DNA or RNA to also be the heredity molecule for life elsewhere? We do not yet know whether other, quite different molecules might be able to carry hereditary information in the same way as DNA. However, it seems a near certainty that any form of life anywhere else will have some molecule that plays the same functional role that DNA plays on Earth.

5.5 Life at the Extreme

We've discussed all the fundamental characteristics of life on Earth: the basic structure of cells, the metabolism of cells, and the means by which cells store and pass on their heredity. We've also discussed why these characteristics seem likely to be shared, at least in a general sense, by any life we find elsewhere. In essence, our bottom-line conclusion is that life elsewhere ought to share a lot of common features with life on Earth. But don't be tempted to think this means that life elsewhere will look like us—that is, like humans or any other animals. In fact, most life on Earth does not look much like "us." We've already noted that microbes are far more common than multicellular eukarya. Perhaps even more startling, in recent decades biologists have discovered that life can survive in an astonishing variety of environments that would be lethal to humans.

---

*Several other similar nucleic acids that can carry hereditary information have been synthesized in the laboratory, including molecules known as TNA and PNA. These molecules are in some ways even simpler than RNA, leading some astrobiologists to hypothesize that they may also have played a role in early life on Earth.

FIGURE 5.19

This photograph shows a volcanic vent on the ocean floor that spews out extremely hot, mineral-rich water. Organisms like *P. fumarii* and Strain 121 survive here in water above the normal boiling temperature.

FIGURE 5.20

A hot spring in Yellowstone National Park. To judge its size, notice the walkway winding along the lower right. The different colors in the water are from different bacteria that survive in water of different temperatures.

What kinds of conditions can life survive?

Deep on the ocean floor are places where volcanic vents release hot water and rock into the surrounding ocean (Figure 5.19). The water coming out of such vents often has temperatures above 350°C (660°F), far hotter than the normal boiling point of 100°C (212°F). However, the ocean pressure at these depths is so great that the water remains liquid despite its high temperature.

If you took any "ordinary" organism and placed it in the extremely hot water near a volcanic vent, it would die quickly because the high temperature would cause many of its critical cell structures to fall apart. Yet in recent decades scientists have discovered life—mostly microbes of the domain archaea*—thriving in the hot water around volcanic vents. For example, an organism called *Pyrolobus fumarii,* which was discovered *in* the walls of a volcanic vent (its name means "fire lobe of the chimney"), can grow in water heated to as high as 113°C (235°F). Another species of archaea, also found at volcanic vents, can grow in even hotter water: Nicknamed "Strain 121" (also called *Geogemma barossli*) because it can grow in water as hot as 121°C (250°F), it can also survive in the lab for up to 2 hours at temperatures of 130°C (266°F). Both *P. fumarii* and Strain 121 are chemoautotrophs that get their carbon from dissolved carbon dioxide and their energy from inorganic chemical reactions that occur in the hot water. Similar organisms thrive in hot springs on Earth's surface, such as in the springs around Yellowstone National Park (Figure 5.20).

Think About It Consider what you know so far about places where we might search for life in the solar system. Which of the following seems most likely to have habitats similar to the ones found near deep-sea volcanic vents: (a) the Moon; (b) Mars; (c) Europa; (d) Saturn? Explain why.

Organisms that survive in extremely hot water are sometimes called *thermophiles,* meaning "lovers of heat" (the suffix *phile* means "lover"), or, in the case of those living at the highest temperatures, *hyperthermophiles.* More generally, organisms that survive in extreme environments of any kind are called **extremophiles,** or "lovers of the extreme." Extremophiles are quite varied, though most of them are members of the domain bacteria or archaea. Some can live in "normal" as well as extreme conditions, while others can survive only in the extreme conditions. For example, many hyperthermophiles die when brought to "normal" temperatures because their enzymes have evolved to function only at the high temperatures in which they live. Many extremophiles are anaerobic (meaning they live without oxygen), and they are poisoned by the oxygen on which our own lives depend.

Hot environments are not the only extreme conditions favored by some organisms. Scientists have discovered numerous species of microbes and fungi living in some of the driest places on Earth, including the Atacama desert in Chile and the dry valleys of Antarctica. The most extreme Atacama organisms live on the slopes of volcanoes (Figure 5.21) where the average rainfall is less than 5 millimeters per year and

---

*Many known species of archaea live in extreme environments, but living in extreme conditions is not a general feature of domain archaea. Indeed, some of the most common organisms in "ordinary" environments on Earth are archaea. For example, some 20–50% of the living cells in cool ocean water typically represent various species of archaea.

temperatures during a single day can swing from a nighttime low of –10°C (14°F) to a daytime high of 56°C (133°F). Moreover, these organisms are found at very high altitudes—about 6000 meters (19,700 feet) above sea level—where they are subject to intense ultraviolet radiation. Researchers are still working to understand how these organisms manage to survive in such conditions, which seem as close to Mars-like as anything found on Earth.

The dry valleys of Antarctica also offer conditions that are somewhat Mars-like. Despite the presence of ice that flows into them, the dry valleys receive so little rain or snowfall that they are among the driest deserts on Earth, and temperatures rarely rise above freezing (Figure 5.22). Here, life manages to survive by living *inside* rocks. We often think of rocks as solid, but most rocks are composed of individual mineral grains packed together, leaving small spaces between the grains. Even in the dry valleys, these spaces within the rocks occasionally contain water from the rare snowfall. Sunlight can penetrate up to a few millimeters into the rock before being completely absorbed, so the layers just below the rock's surface can have temperatures slightly above freezing despite the freezing temperatures around them. Apparently, the microbes survive in these tiny pockets of liquid water inside rocks in these freezing cold valleys. Similar organisms have been found in many other cold environments on Earth, with some thriving in temperatures as low as –20°C (–4°F) as long as even a thin film of liquid water is available. (These cold-loving organisms are called *psychrophiles,* essentially the opposite of *thermophiles.*)

Another group of extremophiles with potential relevance to life on other worlds of our solar system consists of microbes called *endoliths* (meaning "within rocks"). Endoliths have been discovered living in rocks deep beneath Earth's surface in water that fills the pores within the rock. One community of endoliths consists of bacteria living deep beneath the surface of Oregon and Washington in a rock formation known as the Columbia River Basalt; others have been found living in rock as far as 3

FIGURE 5.21
This volcanic slope in the Atacama desert has conditions that are as Mars-like as anything found on Earth, as the desert is extremely dry and the high altitude allows for huge daily temperature swings and intense ultraviolet radiation. Yet numerous species of microbes and fungi appear to thrive here.

FIGURE 5.22
The main photo shows a cold, dry valley in Antarctica. These valleys are among the driest deserts on Earth (the ice is runoff from surrounding regions), yet they are still home to life. The inset shows a slice of rock (about 1.8 centimeters across) from a dry valley. The colored zones contain microbes that live inside the rock in the airspaces between tiny mineral grains. The organisms are dormant and frozen for most of the year, but can grow during the approximately 500 hours per year when sunlight warms the rock above the freezing point.

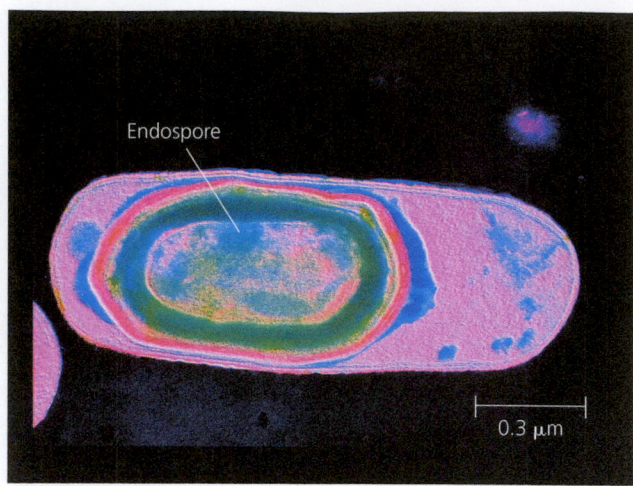

Endospore

FIGURE 5.23

This colorized microscopic photo shows an endospore created by the bacterium *Bacillus subtilis*. There are actually two cells here, one inside the other. The outer cell produced the specialized inner cell, which is the endospore; the red and green membrane layers give the endospore a thick, protective coat. Its interior is dehydrated, and no metabolism occurs. Under harsh conditions, the outer cell may disintegrate, but the endospore can survive. When the environment becomes more hospitable, the endospore absorbs water and resumes growth.

0.3 μm

kilometers underground. These organisms are chemoautotrophs that get their energy for metabolism from chemical reactions between the water and the surrounding rock, and they get their nutrients from chemicals within the rock itself and from carbon dioxide that has filtered down from the surface. The subsurface environments in which these endoliths live almost certainly exist in similar form on Mars and perhaps on other worlds in our solar system, even where the surface may be too cold for liquid water. Moreover, endoliths may be quite common. No one knows exactly how many of them exist here on Earth, but some estimates suggest that the total mass of subsurface organisms living in rock may exceed that of all the life on Earth's surface.

Other extremophiles live in conditions far too acidic, alkaline, or salty for "ordinary" life to survive, further widening the range of extraterrestrial environments that might be considered suitable for life. Some microbes can even survive high doses of radiation. A bacterial species known as *Deinococcus radiodurans* can survive radiation more than 1000 times stronger than what would be lethal to humans and other animals. These remarkable organisms actually thrive in radioactive waste dumps! They could survive the radiation exposure on a world without ozone and even in space, and they can survive extremely dry conditions as well.

The idea that life might not be killed even after some period of time in space raises the possibility that life could survive journeys aboard meteorites that are blasted off one planet and land on another [Section 8.5], and *D. radiodurans* is not the only species that shows this potential. Some bacteria have the ability to form what are known as **endospores**, which are essentially "resting" cells (Figure 5.23). Endospores allow the organisms that create them to become dormant, neither growing nor dying in extremely inhospitable conditions. (When not dormant, the organisms are not necessarily extremophiles; some live under more "normal" conditions.) For example, endospores of the bacterium *Bacillus anthracis*, which causes the deadly disease anthrax, can survive a complete lack of water, extreme heat or cold, and most poisons. Some endospores can survive even in the vacuum of space, which is why planetary scientists worry that our interplanetary spacecraft could potentially contaminate other worlds with life from Earth.

Finally, it's worth noting that while most organisms that can survive extreme conditions are microbes, there are a few multicellular examples, even including some animals. The most famous of these are the *tardigrades* (Figure 5.24), tiny animals found in droplets of water around the world. More than 1150 distinct species of tardigrades are known. Although not technically considered extremophiles because they are better able to survive in more ordinary environments, scientists have found tardigrades living under an enormous range of conditions. They apparently live (based on detection of traces of their DNA) even in Antarctica's Lake Vostok, a subterranean lake buried under 3.7 kilometers of ice, which has not been exposed to open air or sunlight for some 15 million years. Moreover, tardigrades can survive for a least short periods of time in temperatures ranging from close to absolute zero to well above the boiling point of water (100°C) and in pressures far greater than those at the bottoms of Earth's deepest oceans. They can become dormant and survive more than 10 years without food or water, and in this condition they can also survive the vacuum of space, making them another potential type of interplanetary traveler.

FIGURE 5.24
It almost doesn't look real, but the tiny animal in this photograph is a tardigrade (also called a "water bear"); it is about a millimeter long. Tardigrades can survive an incredible range of extreme conditions, including at least some time in the near-vacuum of space.

Are extremophiles really extreme?

From our human point of view, the environments in which extremophiles survive truly are extreme. But if an extremophile could think, it would probably claim that its environment is quite normal and that ours is the one that is extreme. So who's right, humans or the extremophiles?

In some sense, it's all just relative. Any species would naturally consider its own environment to be normal and others to be extreme. A more important question we might ask is which environment is more common. Surprisingly, if we look at the history of Earth, so-called extreme environments have been much more common than an environment suitable for humans. Earth's atmosphere may have contained oxygen at a level suitable for human life for only the past few hundred million years, or about 10% of Earth's history [Section 6.3]. Indeed, for the first couple of billion years after life first arose on Earth, extremophiles were the only organisms that could survive. Even today, it's an open question whether extremophiles are more or less common than organisms that live in conditions favorable to humans. All in all, extreme life appears to be much more the norm than is life that lives in an environment suitable for humans.

The study of extremophiles has several important implications for the search for extraterrestrial life, but two are particularly important. First, the fact that extremophiles apparently evolved earlier than other forms of life [Section 6.1] suggests that we should begin the search for life elsewhere by searching for similar extreme organisms. Second, the fact that extremophiles can survive such a broad range of conditions suggests that life may be possible in many more places than we would have guessed only a few decades ago: Any world containing an environment in which some type of extremophile might survive becomes a good candidate for the search for life.

Extremophiles also offer an important lesson about our role on our own planet. There's no doubt that human activity is doing great harm to many species of living organisms on Earth. However, if you ever hear someone claim that we might somehow wipe out life on our planet, you can reply that it's simply not true. Yes, we could lay waste to our own civilization with nuclear bombs, and there are numerous other ways in which we might ultimately cause our own demise as a species, perhaps taking millions of other animal and plant species along with us. But if any of this were to occur, the endoliths living miles beneath our cities in Oregon and Washington, the hyperthermophiles living near volcanic vents, and *D. radiodurans* and others wouldn't even bat a proverbial microscopic eye. We hold the future of the human race—and of many other animal and planet species—in our own hands, but life on Earth will continue with or without us. In this sense, astrobiology has provided the one more feather in the cap of the Copernican revolution by showing that we are not even the center of Earth's biological universe, let alone the center of the universe as a whole.

Think About It For most of human history, we humans generally assumed that we were the "dominant" form of life on Earth and that most other life—which was assumed to be plants and animals—was at least somewhat like us. Does the knowledge that much of life on Earth is very different from us change your perspective on our place in the universe? Explain.

THE PROCESS OF SCIENCE IN ACTION

5.6 Evolution as Science

In this chapter and throughout this book, we treat evolution as an established fact, a position consistent with official statements by virtually every scientific society, including the National Academy of Sciences, the American Association for the Advancement of Science, the National Association of Biology Teachers, and the American Astronomical Society. However, if you live in the United States, you've almost certainly heard about public battles in which the scientific idea of evolution is portrayed as controversial. How can an idea so well accepted in science be considered so differently by many among the general public? The answer lies in differences between the way science works and the way people often seek knowledge in their daily lives, and especially in the difference between the evidence-based approach of science and the faith-based approach of religion. Once this difference is understood, much of the supposed conflict between science and religion is not an issue, which explains why the vast majority of religious denominations see no inherent conflict between their faiths and the science of evolution. For this chapter's case study in the process of science, we'll explore why most scientists and most theologians agree that evolution and faith can coexist without difficulty.

Is evolution a fact or a theory?

By this point in the book, you should already recognize that the question of whether evolution is a fact or a theory offers a false choice, much like asking whether gravity is a fact or a theory [Section 2.4]. Gravity is a fact in that objects really do fall down and planets really do orbit the Sun, but we use the *theory* of gravity to explain exactly how and why these things occur. The theory of gravity is not presumed to be perfect and indeed has at least one known flaw (its inconsistency with quantum mechanics on very small scales). Moreover, Newton's original theory of gravity is now considered only an approximation to Einstein's improved theory of gravity, which itself will presumably be found to be an approximation to a more complete theory that has not yet been discovered.

The same idea holds for evolution. Nearly all scientists consider evolution to be a fact, because both the geological record and observations of modern species make clear that living organisms really do change with time. We use the *theory* of evolution to explain how and why these changes occur. For example, we use the theory of evolution to understand the changes in species recorded in the geological record, the genetic relationships among modern species, and the way that bacteria can rapidly acquire resistance to antibiotics.

The theory of evolution clearly explains the major features of life on Earth, but scientists still debate the details of the theory. For example, there is considerable debate about the rate at which evolution proceeds: Some scientists suspect that evolution is "punctuated" with periods of rapid change followed by long periods in which species remain quite stable, while others suspect that evolution proceeds at a steadier pace. This debate can be quite heated between individual scientists, but it does not change the overall idea that life evolves, and it will eventually be settled by additional evidence. Similarly, scientists often debate the precise relationships among species, especially those that are extinct. For example, we do not yet have enough evidence to put together a complete evolutionary tree for relationships among all dinosaur species, and the relationship between extinct dinosaurs and modern birds is not yet fully understood.

We can draw an analogy between Darwin's original theory of evolution by natural selection and Newton's original theory of gravity: Just as Newton's theory captured the main features of gravity but proved to be incomplete, Darwin's theory clearly captures the main features of evolution but is not complete. Moreover, like Newton's theory, Darwin's also has been modified and improved with time. Just as Einstein's general theory of relativity allowed us to understand gravity in more realms than Newton's original theory (by refining it, not refuting it), our modern understanding of DNA and mutations allows us to understand biological changes and relationships beyond those that Darwin was able to understand or was even aware of. And like the theory of gravity, the theory of evolution remains a work in progress. We expect to learn more about evolution as we continue to study relationships among species, how DNA works (especially the noncoding regions), and the biochemistry of life. Perhaps someday we'll even be able to broaden the theory through the study of *comparative evolution*, in which we'll explore the similarities and differences among living organisms on multiple worlds.

Does all this mean that you need to *believe* that evolution really occurred? From the standpoint of learning science, what counts is *understanding* evolution; belief is up to you. Remember, all the evidence in the world can never prove any scientific theory true beyond all doubt [Section 2.3]. Even if we had a complete geological record, with precise dating of every species that ever lived, you could still choose to believe, for example, that the fossils had been placed there intact at some single moment in the past, rather than having been deposited as the evidence suggests. All we can say from a scientific viewpoint is that a tremendous wealth of evidence points to the idea that life on Earth has evolved through time, that these evolutionary changes have been driven by natural selection, and that natural selection occurs on a molecular level as genes are modified through mutations of DNA.

FIGURE 5.25

The theory of evolution makes many predictions that have been verified. For example, it predicts that genetically similar species—such as humans and chimpanzees (which share more than 98% of the same genes)—should respond similarly to diseases and medications, and this is indeed the case.

Are there scientific alternatives to evolution?

In recent years, the public controversy over evolution has centered largely around the question of whether it should be the only idea about our origins taught in science classes or whether it should be taught alongside other, competing ideas. Other ideas certainly offer different visions of how we came to exist; for example, the idea that God created Earth and the universe a mere 6000 years ago is obviously quite different from the idea that life has evolved gradually over the past 4 billion years. The question is whether this idea or other competing ideas qualify as science.

The best way to determine whether any alternatives to evolution qualify as science is to consider them against the hallmarks of science that we discussed in Chapter 2. Let's start by showing that the theory of evolution *does* satisfy the standards of science. The first hallmark states that science seeks explanations for observed phenomena that rely solely on natural causes. The theory of evolution clearly does this, as it explains the geological record and observed relationships among species through the mechanism of natural selection and other natural causes.

The second hallmark states that science progresses through the creation and testing of models. Our understanding of evolution has indeed progressed in this way. The idea of evolution won out over Aristotle's competing idea of species that never changed. As the fact of evolutionary change gained acceptance, the first model proposed to explain these changes came from Lamarck (see Section 5.1); his model was later discarded because Darwin's alternative model explained the observations so much more successfully. Our current, molecular model of evolution is a refinement of Darwin's original model, and we can expect further refinements to the theory in the future as continued study turns up new evidence.

The theory of evolution also satisfies the third hallmark, which states that a scientific model must make testable predictions that would lead us to revise or abandon the model if the predictions did not agree with observations. Our modern, molecular theory of evolution clearly qualifies. For example, it predicts that diseases can and will evolve in response to medicines designed to combat them, a prediction borne out in the rapid way that many diseases acquire drug resistance. It also predicts that genetically similar species should respond to medicines in similar ways, a prediction confirmed by the fact that we can test many medicines in other primates and they do indeed have effects similar to those they have in humans (Figure 5.25). The theory of evolution also provides a road map that we can use to modify organisms through genetic engineering; in this sense, every genetically engineered grain of rice or kernel of corn represents a success of the predictive abilities of the theory of evolution.

In fact, even Darwin's original theory made testable predictions. For natural selection to be possible, living organisms must have some way of passing on their heritable traits from parent to offspring. So although Darwin did not predict the existence of DNA per se, his theory clearly predicted that some type of mechanism had to exist to carry the hereditary information. Similarly, now that we know about DNA and the genetic code, the theory of evolution predicts that closely related species should be genetically similar, a prediction that has been confirmed through genome sequencing. For example, in the ordering of their base sequences, the DNA of humans and chimpanzees is 98.5% identical. If we were not closely related to chimpanzees, we would not expect such similar genomes.

Now that we have established that evolution qualifies as science, we next turn to the question of whether any of the alternative models that have been suggested for inclusion in the classroom might also qualify as science. Since the time that Darwin first published his theory, the main alternatives have been religious ideas about creation. Here, we run into an immediate problem: There are so many different religious ideas about creation that we can't even define the potential alternatives clearly. For example, many Native American religious beliefs speak of creation in terms that bear little resemblance to the Judeo-Christian tradition found in Genesis. Even among people who believe in the literal truth of the Bible, there are differences in interpretation about creation. Some biblical literalists argue that the creation must have occurred in just 6 days, as the first chapter of Genesis seems to say, while others suggest that the term "day" in Genesis does not necessarily mean 24 hours and therefore that the story in Genesis is compatible with a much older Earth and with evolution.

Nevertheless, a few groups have tried to claim that scientific evidence supports some alternative to evolution. In the 1980s, an idea called "creation science" emerged, and its proponents tried to find scientific evidence to support the idea that Earth was created a mere 6000 years ago. However, to support this "young Earth" view, they not only had to reject evolution but also had to reject the tremendous weight of evidence that supports an old Earth and an old universe—evidence based on such things as radiometric dating of rocks, astronomical measurements of distances to other stars and galaxies (since their light has obviously had time to reach us), and even tree ring data that go back more than 6000 years. For all this evidence to be wrong, there would have to be fundamental errors in our basic understanding of the laws of nature, an idea that seems implausible, given the many successes of modern physics and chemistry.

More recently, some people have advanced an idea called "intelligent design," or ID; this idea holds that living organisms are too complex to be explained by natural selection, and so must have been designed by some transcendent entity or power. For example, proponents of this idea point to features of the human eye as suggesting design rather than natural processes, and some believe they see evidence of "digital code" in the arrangement of the bases in the genomes of living organisms.

For the vast majority of scientists, the primary problem with these claims of intelligent design is that they do not seem to stand up to scientific scrutiny. The features that the ID proponents cite as evidence for design are to most scientists well explained by natural selection. For example, the evolution of the eye seems well understood; indeed, biologists have traced the evolution of the eye from primitive beginnings all the way to the eyes found in different animal species today. Nevertheless, good scientists will always allow the possibility that evidence of design might someday be found. Moreover, even if no such evidence is found, absence of evidence would not preclude a role for a Designer.

The greater problems with intelligent design from a scientific perspective show up when we test it against the hallmarks of science. In particular, ID is clearly incompatible with the first hallmark—that science seeks explanations for observed phenomena that rely solely on natural causes. The very idea of a transcendent Designer implies something that natural processes cannot explain, no matter whether the Designer is or isn't explicitly named as God. As a result, some ID proponents have sought to redefine science to allow nonnatural explanations. The problem with such a redefinition is that it would render science impotent. As a simple analogy, consider the collapse of a bridge. You can choose to believe that the collapse was an act of God, and you might well be right—but this belief won't help you design a better bridge. We learn to build better bridges only by assuming that collapses happen through natural causes that we can understand and learn from. In precisely the same way, it is the scientific quest for a natural understanding of life that has led to the discovery of relationships among species, genetics, DNA, and much of modern agriculture and medicine. Many of the scientists who made these discoveries, including Charles Darwin himself (see the quotation at the beginning of this chapter), believed deeply that they could see God's hand in creation. But if they had let their belief stop them from seeking natural explanations, they would have discovered nothing.

Intelligent design also fails to be in accord with the second and third hallmarks of science, because it does not offer a predictive model that can be tested. The assumption of a Designer might or might not be correct, but it does not tell us how life would be different from what we'd see if there were no Designer. Moreover, as we've discussed, scientists continually modify the theory of evolution as new evidence requires. In the unlikely event that we found evidence that strongly contradicted the current theory—for example, fossil evidence proving that people and dinosaurs existed at the same time—scientists would willingly discard the theory and go back to the drawing board. In contrast, because most proponents of intelligent design are motivated by their religious faith, their belief in ID is unshakable.

The bottom line is that science and faith are different things, and the relative worth of one does not override the worth of the other. Whether you choose to believe the theory of evolution is up to you, and if you do believe it, you can choose whether to believe that it occurred through random chance or with the help of a guiding hand. But whatever your beliefs, the theory of evolution is a clear and crucial part of modern science, and it is integral to an understanding and appreciation of modern biology. And more important for the discussion in this book, the theory of evolution frames our ideas about how to search for life beyond Earth. No competing model offers any similar scientific benefits.

The Big Picture

PUTTING CHAPTER 5 IN PERSPECTIVE

In this chapter, we have surveyed the nature of life on Earth and explored some of the implications of this survey for the search for life elsewhere. As you continue in your studies, keep in mind the following "big picture" ideas:

- If we are going to search for life, it's useful to think about just what it is we are searching for. Defining life turns out to be surprisingly difficult, but at a minimum it seems that life must be capable of reproducing and evolving. Evolution therefore plays a central role in the definition of life as well as in our understanding of life on Earth.

- Life on Earth has at least three key features that are likely to be shared by any life we find elsewhere: (1) Life has a fundamental structural unit, which we call the cell; (2) living cells undergo metabolism, by which we mean chemical reactions that keep the cells alive; and (3) living cells have a heredity molecule, which is DNA for life on Earth, that allows them to store their operating instructions and to pass these instructions to their offspring.

- Life on Earth survives under a much wider range of conditions than we would have guessed a few decades ago, suggesting that life elsewhere might similarly be found in a fairly broad range of environments. This fact greatly increases the number of worlds on which we might hope to find life.

Summary of Key Concepts

5.1 Defining Life

What are the general properties of life on Earth?

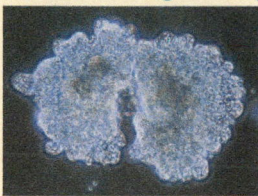

Six key properties of life on Earth are order, reproduction, growth and development, energy utilization, response to the environment, and evolutionary adaptation.

What is the role of evolution in defining life?

The **theory of evolution,** which holds that life changes over time through the mechanism of **natural selection,** is the unifying principle of modern biology. It holds this central role because it successfully explains all the other properties of life, the observations we make in the geological record, and the observations we make of relationships among living organisms.

What is life?

No known definition of life works in all circumstances, but for most purposes the following definition will suffice: Life is something that can reproduce and evolve through natural selection.

5.2 Cells: The Basic Units of Life

What are living cells?

Cells are the basic units of life on Earth, as they serve to separate the living matter inside them from the outside world. The barrier that marks this separation is called the **cell membrane.**

What are the molecular components of cells?

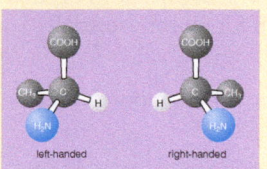

The major molecular components of cells fall into four main classes: **Carbohydrates** are sugars and starches that provide energy and build many cellular structures; **lipids,** which include fats, are the main ingredients of structures including cell membranes and also store cellular energy; **proteins** play a vast number of roles in cells, and the proteins known as **enzymes** act as **catalysts** to facilitate biochemical reactions; **nucleic acids,** which include **DNA** and **RNA,** are most important for the roles they play in heredity. Commonalities among the molecules used in different organisms, such as the fact that all life on Earth builds proteins from only left-handed versions of **amino acids,** provide strong evidence for the idea that all life evolved from a common ancestor.

What are the major groupings of life on Earth?

Modern biologists classify life into three **domains: bacteria, archaea,** and **eukarya.** The **tree of life** shows relationships among species within the three domains; note that all plants and animals are just two small branches of the eukarya.

5.3 Metabolism: The Chemistry of Life

What are the basic metabolic needs of life?

Life requires (1) a source of raw materials to build cellular structures, with carbon as the most important of these materials, and (2) a source of energy to fuel metabolic processes.

How do we classify life by its metabolic sources?

We classify life by its carbon source as either a **heterotroph,** which gets its carbon by eating, or an **autotroph,** which takes carbon directly from the environment in the form of atmospheric or dissolved carbon dioxide. We then subclassify these categories by energy source, using the prefix *photo* for life that gets energy from sunlight and the prefix *chemo* for life that gets energy either from eating or from inorganic chemical reactions.

5.4 DNA and Heredity

How does the structure of DNA allow for its replication?

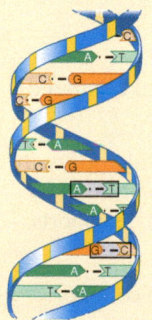

The double helix of DNA consists of two strands connected by the DNA bases. The bases connect according to precise pairing rules (T attaches only to A, and C attaches only to G), so that when the strands separate, each can serve as a template for making a new DNA molecule that is identical to the original one.

How is heredity encoded in DNA?

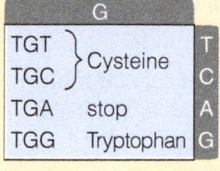

The precise sequence of the bases in a DNA molecule contains the instructions for assembling proteins and other cell functions. A segment of DNA that codes for a single cell function or protein is called a **gene,** and the complete base sequence of an organism represents its **genome.** The "language" used to translate from the base sequence into proteins is called the **genetic code.**

How does life evolve?

Life evolves because the copying of DNA is not perfect, although the error rate is quite small. The occasional, random copying errors, called **mutations,** can change the instructions in DNA. Most mutations either are lethal or have no effect at all, but a few carry benefits that can then be transmitted to offspring when the DNA replicates.

5.5 Life at the Extreme

What kinds of conditions can life survive?

Many living organisms can survive in a surprisingly wide range of conditions. These **extremophiles** include microbes that can survive in temperatures above the normal boiling point of water, in dry deserts and cold Antarctic valleys, deep underground in the tiny pores of rocks, and even under exposure to high levels of radiation.

Are extremophiles really extreme?

The conditions that we consider extreme are, overall, probably more typical of the conditions found on Earth during most of its history than the conditions we enjoy on the surface today. Many other worlds may have similar conditions, suggesting that extremophiles may in fact be more common in the universe than life similar to plants and animals.

✄ **THE PROCESS OF SCIENCE IN ACTION**

5.6 Evolution as Science

Is evolution a fact or a theory?

This question offers a false choice, because fact and theory are not considered to be opposites in science. Evolution is a well-confirmed theory based on a wide variety of observational and experimental evidence.

Are there scientific alternatives to evolution?

While there are many alternative explanations for our existence, including ideas such as creation science or intelligent design, none of these ideas qualifies as a *scientific* alternative to evolution.

Exercises and Problems

MasteringAstronomy® *For instructor-assigned homework and other learning materials, go to MasteringAstronomy®.*

REVIEW QUESTIONS

Short-Answer Questions Based on the Reading

1. Briefly describe the six key properties that appear to be shared by most living organisms on Earth.

2. What is *natural selection*? Summarize the logic by which Darwin came to the "inescapable conclusion" that evolution occurs by natural selection. Describe some of the evidence that supports Darwin's *theory of evolution*.

3. Briefly describe the evidence that points to a single common ancestor for all life on Earth.

4. Why do we say that living cells are *carbon-based*? Briefly discuss whether life elsewhere could be based on something besides carbon.

5. Briefly describe each of the four main classes of cellular molecules: *carbohydrates, lipids, proteins,* and *nucleic acids.* What are *enzymes,* and where do they fit into this picture?

6. What are *amino acids*? What do we mean by their *handedness*? How do amino acids offer further evidence for a common ancestor for all life on Earth?

7. What are the three *domains* of life? Which domain do we belong to?

8. What do we mean by the *tree of life*? List three important ideas that we learn from the tree and that differ from older ideas about biology.

9. What is *metabolism,* and what are the two basic metabolic needs of any organism? Explain the four metabolic classifications listed in Table 5.1.

10. Why is water so important to life on Earth? List the three major roles that water plays in metabolism.

11. Describe the double helix structure of DNA. How does a DNA molecule replicate?

12. What is a *gene*? A *genome*? The *genetic code*?

13. What are *mutations,* and what effects can they have? Briefly explain why mutations represent the molecular mechanism of natural selection.

14. What are *extremophiles*? Give several examples of organisms that live in extreme environments. What are the implications of the existence of extremophiles for the search for extraterrestrial life?

15. Describe several ways in which the theory of evolution is analogous to the theory of gravity.

16. Explain how evolution exhibits each of the three hallmarks of science, and discuss why alternatives such as creationism and intelligent design do not show these hallmarks.

TEST YOUR UNDERSTANDING

Surprising Discoveries?

Suppose we found an organism on Earth with the characteristics described. In light of our current understanding of life on Earth, should we be surprised to find such an organism existing? Why or why not? Explain clearly; because not all of these have definitive answers, your explanation is more important than your chosen answer.

17. A single-celled organism that builds proteins using 45 different amino acids.

18. A single-celled organism that lives deep in peat bogs, where no oxygen is available.

19. A multicellular organism that reproduces without passing copies of its DNA to its offspring.

20. A single-celled organism that can survive in a dormant state even in the complete absence of any liquid water.

21. A multicellular organism that can grow and reproduce even in the absence of water.

22. A bacterium with cells that lack the molecule ATP.

23. A species of archaea that lives in the 1000°C molten rock of a volcano.

24. A species of archaea that lives in the walls of a nuclear reactor.

25. Two different animal species whose genomes are more than 99% identical.

26. A species of bacteria that has a genome 99% identical to that of humans.

Quick Quiz

Choose the best answer to each of the following. Explain your reasoning with one or more complete sentences.

27. Which of the following is *not* a key property of life? (a) the maintenance of order in living cells; (b) the ability to evolve over time; (c) the ability to violate the second law of thermodynamics.

28. *Natural selection* is the name given to (a) the occasional mutations that occur in DNA; (b) the mechanism by which advantageous traits are preferentially passed on from parents to offspring; (c) the idea that organisms can develop new characteristics during their lives and then pass these to their offspring.

29. Which of the following is *not* considered a key piece of evidence supporting a common ancestor for all life on Earth? (a) the fact that all life on Earth is carbon-based; (b) the fact that all life on Earth uses the molecule ATP to store and release energy; (c) the fact that all life on Earth builds proteins from the same set of left-handed amino acids.

30. An organism's heredity is encoded in (a) DNA; (b) ATP; (c) lipids.

31. An enzyme consists of a chain of (a) carbohydrates; (b) amino acids; (c) nucleic acids.

32. Which of the following is *not* a source of energy for at least some forms of life on Earth? (a) inorganic chemical reactions; (b) energy release from plutonium; (c) consumption of preexisting organic compounds.

33. People belong to domain (a) eukarya; (b) archaea; (c) bacteria.

34. Which of the following mutations would you expect to have the greatest effect on a living cell? (a) a mutation that changes a single base in a region of noncoding DNA; (b) a mutation that changes the third letter of one of the three-base "words" in a particular gene; (c) a mutation that deletes one base in the middle of a gene.

35. Generally speaking, an *extremophile* is an organism that (a) thrives in conditions that would be lethal to humans and other animals; (b) could potentially survive in space; (c) is extremely small compared to most life on Earth.

36. Based on what you have learned in this chapter, it seems reasonable to think that life could survive in each of the following habitats *except* (a) rock beneath the martian surface; (b) a liquid ocean beneath the icy crust of Jupiter's moon Europa; (c) within ice that is perpetually frozen in a crater near the Moon's south pole.

PROCESS OF SCIENCE

37. *The History of Evolution.* Many people assume that Charles Darwin was the first person to recognize that life evolves, but this is not true. Write a few paragraphs summarizing the history of ideas about evolution and explaining why we give Darwin credit for the theory of evolution even though he was not the first person to realize that evolution occurs.

38. *A Separate Origin?* Suppose that we someday discover life on Mars. How might we be able to determine whether it shares a common origin with life on Earth (perhaps suggesting that life traveled on meteors between the two planets) or has a completely separate origin? Explain clearly.

GROUP WORK EXERCISE

39. *Searching for a Shadow Biosphere.* **Roles:** *Scribe* (takes notes on the group's activities), *Proposer* (proposes ideas to the group), *Skeptic* (points out weaknesses in proposed ideas), *Moderator* (leads group discussion and makes sure everyone contributes). **Activity:** Some scientists have suggested that life might have begun more than once on Earth, and that there could be an undiscovered biological realm based not on DNA but on some other molecule. How might we go about searching for such a "shadow biosphere," which would mean looking for microbe-size things that have the properties of life (as described in this chapter) but a different genetic molecule? Come up with a group plan and write a one-page bulleted summary of it.

INVESTIGATE FURTHER

In-Depth Questions to Increase Your Understanding

Short-Answer/Essay Questions

40. *Rock Life?* How do you know that a rock is not alive? In terms of the properties of life discussed in this chapter, clearly describe why a rock does not meet the criteria for being alive.

41. *Genetic Variation.* One of the underlying facts (Fact 2 on p. 160) that explains natural selection is that individuals in a population of any species vary in many heritable traits. Based on what you have learned about the molecular basis of evolution, explain why individuals of the same species are *not* expected to be genetically identical.

42. *Artificial Selection.* Suppose you lived hundreds of years ago (before we knew about genetic engineering) and wanted to breed a herd of cows that provided more milk than cows in your current herd. How would you have gone about it? Explain, and describe how your breeding would have worked in terms of the idea of artificial selection. How does this breeding offer evidence in favor of the idea of natural selection?

43. *Ingredients of Life.* Study the ingredients of life as shown in Figure 5.6, and consider them in light of what you've learned about the overall chemical composition of the universe. Would

you expect the ingredients to be rare or common on other worlds? Explain.

44. *Dominant Life.* While most of us tend to think of ourselves as the dominant form of life on Earth, biologists generally argue that the dominant life consists of microbes of the domains archaea and bacteria. In two to three paragraphs, explain why microbes seem more dominant than humans.

45. *The Human Power to Destroy.* We may have the ability to destroy ourselves today, perhaps as the result of nuclear war or perhaps through some type of environmental catastrophe. But is there anything we could do with our current abilities that would allow us to wipe out *all* life on Earth? Explain why or why not.

46. *The Search for Life.* Based on what you have learned about life on Earth, what are we searching for when we search for life elsewhere? For example, are we searching only for worlds with surface oceans and oxygen-rich atmospheres like Earth, or for something else? Write one to three paragraphs describing the types of worlds that we can consider as potential homes for life.

47. *Evolution and God.* Does the theory of evolution preclude the existence of God? Clearly explain your answer.

Quantitative Problems

Be sure to show all calculations clearly and state your final answers in complete sentences.

48. *Atomic Numbers in Life.* A typical bacterium has a volume of about 1 cubic micrometer. A typical atom has a diameter of about 0.1 nanometer. Approximately how many atoms are in a bacterium?

49. *Oxygen Atoms in People.* Figure 5.6 shows that oxygen makes up about 65% of the mass of a human being. A single oxygen atom has a mass of 2.66×10^{-26} kg. (a) Use this fact to estimate the number of oxygen atoms in *your* body. (*Hint:* If you know your weight in pounds, you can convert to kilograms by dividing by 2.2.) (b) Compare your answer to the number of stars in the observable universe (which is roughly 10^{22}).

50. *Cellular Energy.* A typical eukaryotic cell, such as a cell in the human body, uses about 2×10^{-17} joule of energy each second. The breakdown of a single molecule of ATP (in which a phosphate separates from ATP to make ADP; see Figure 5.13) releases about 5×10^{-20} joule of energy. (a) About how many molecules of ATP must be broken down and reassembled each second to keep a eukaryotic cell alive? (b) How many times does this ATP recycling occur each *day* in a typical cell? (c) The human body has roughly 10^{14} cells. Approximately how many cycles of the ATP reaction occur each day in your body?

51. *The Genetic Code.* Suppose that, as evidence suggests, very early life on Earth used a genetic code that consisted of only two-base "words" rather than three-base "words." Could such life have made use of the same set of 20 amino acids that life uses today? Explain, using quantitative arguments.

Discussion Questions

52. *Science and Religion.* Science and religion are often claimed to be in conflict. Do you believe this conflict is real and hence irreconcilable, or is it a result of misunderstanding the differing natures of science and religion? Defend your opinion.

53. *Computer Life.* Although scientists have already developed computer programs capable of reproducing themselves and evolving, few people consider such programs to be alive. But consider future developments in computing and robotic technology. Do you think we'll ever make something based on electronics that is truly alive? Could it also be intelligent? If so, what civil rights should we give to such "artificial" intelligent life?

54. *Genetic Engineering and Future Evolution.* For billions of years, evolution has proceeded through mutations and natural selection. Today, however, we have the ability to deliberately alter DNA in what we call "genetic engineering." How do you think this ability will affect the future evolution of life? How will it affect future human evolution on Earth? Based on your answers, should we expect extraterrestrial civilizations to have naturally evolved or to be products of their own genetic engineering? Discuss and defend your opinions.

55. *Gene Transfer and GMOs.* In some cases, organisms can transfer entire genes to other organisms. This fact causes some people to worry that organisms that we have genetically engineered— commonly referred to as GMOs, for "*g*enetically *m*odified *o*rganisms"—may transfer their genes to other organisms in unexpected ways. For example, a crop engineered with a gene that gives it resistance to some pest may transfer its gene to weeds, giving them the same resistance. Discuss how GMOs might affect other organisms. Overall, what, if any, controls do you think the government should put on the use of GMOs?

WEB PROJECTS

56. *The Dover Opinion.* In December 2005, a U.S. District Court issued its opinion on a case concerning the teaching of intelligent design (ID), deciding that ID does not belong alongside evolution in science classes. The full text of the opinion, commonly called the Dover opinion, is available online. Read the opinion, and discuss its implications for the ongoing public controversy about what belongs in science classes.

57. *Darwin on Evolution.* You can find online the entire text of Charles Darwin's *The Origin of Species.* Read the final chapter, in which Darwin addresses potential criticisms of his theory. Evaluate how well he presented his case. How much stronger does the theory seem today than at the time Darwin first described it in 1859? Summarize your conclusions in a one-page essay.

58. *Extreme Life.* Look for information about a recent discovery of a previously unknown type of extremophile. Describe the organism and the environment in which it lives, and discuss any implications of the finding for the search for life beyond Earth. Summarize your findings in a one-page report.

6 The Origin and Evolution of Life on Earth

LEARNING GOALS

6.1 SEARCHING FOR LIFE'S ORIGINS
- When did life begin?
- What did early life look like?
- Where did life begin?

6.2 THE ORIGIN OF LIFE
- How did life begin?
- Could life have migrated to Earth?

6.3 THE EVOLUTION OF LIFE
- What major events have marked evolutionary history?

- Why was the rise of oxygen so important to evolution?

6.4 IMPACTS AND EXTINCTIONS
- Did an impact kill the dinosaurs?
- What caused other mass extinctions?
- Is there a continuing impact threat?

6.5 HUMAN EVOLUTION
- How did we evolve?
- Are we still evolving?

 **THE PROCESS OF SCIENCE IN ACTION**

6.6 ARTIFICIAL LIFE
- How can we create artificial life?
- Should we create artificial life?

▲ **About the photo:** These large mats at Shark Bay, Western Australia, are colonies of microbes thought to closely resemble organisms that made some of the oldest known fossils of life on Earth.

We explored the habitability of our planet in Chapter 4, learning how and why Earth has remained a suitable home for life during the past 4 billion or more years. In Chapter 5, we saw how life has taken advantage of this habitability, as we explored the nature of current life and the wide variety of environments in which it lives. But we have yet to discuss the deepest questions of all: How did life arise in the first place, and what events shaped life's evolution to produce the current diversity of species?

The complete answers to these questions still elude us, but scientists have put together a fairly detailed outline of what is most likely to have occurred. The early Earth was a natural laboratory for organic chemistry, and we have at least some ideas about how this chemistry might have led to life. The geological record shows us how life evolved subsequently, and it is full of important surprises. For example, plants and animals appeared only relatively recently, and we've discovered that external forces, such as the impacts of asteroids or comets, can dramatically alter the course of evolution. In this chapter, we'll survey current ideas about the origin and subsequent evolution of life on Earth, in effect studying how we ourselves came to be.

6.1 Searching for Life's Origins

The geological record details much of the history of life on Earth, and the theory of evolution tells us how life has changed from the forms that existed long ago to those found on our planet today. However, neither the geological record nor ideas of biological evolution are likely to tell us precisely how life first arose on Earth. The geological record becomes increasingly incomplete as we look further back in time, and we know of no rocks that survive from Earth's first half-billion years. The theory of evolution explains how one species can evolve into others, but does not tell us how the first living organisms came to be. Attempts to understand the origin of life are therefore based on careful study of limited clues about what existed in Earth's early history, along with laboratory experiments that can help us reconstruct the processes that may have occurred on the young Earth. In this section we'll explore the clues that tell us about early life on Earth, and in the next section we'll explore how laboratory experiments shed light on a possible mechanism for the origin of the first life.

When did life begin?

Perhaps the first thing we'd like to know about the origin of life on Earth is when it occurred. The only way to approach this question is through the study of fossils. If we find a fossil of a particular age, then we know life already existed at that time. For example, a 3.5-billion-year-old fossil tells us that life on Earth arose *at least* 3.5 billion years ago. Because the geological record is incomplete and because we may not yet have discovered the oldest intact fossils, we do not know exactly how long life has existed on Earth. Nevertheless, three lines of fossil evidence all point to the idea that, geologically speaking, life arose quite early in Earth's history.

a These knee-high mats at Shark Bay, Western Australia, are colonies of microbes known as "living stromatolites."

b The banded structure in this section from one of the Shark Bay stromatolites is formed by layers of sediment attaching to the microbial mats.

5 cm

c This section of a 3.5-billion-year-old stromatolite (found in the Strelley Pool Formation in Western Australia) shows the same type of structure found in living stromatolites. The black layers are organic deposits that are the remains of ancient microbial mats. (The ruler is marked in centimeters.)

FIGURE 6.1

Rocks called *stromatolites* offer evidence of microbial life existing as early as 3.5 billion years ago.

STROMATOLITES The first line of evidence comes from **stromatolites** (from the Greek for "rock beds"), rocks that are characterized by a distinctive, layered structure. In size, shape, and interior structure, ancient stromatolites look virtually identical to sections of modern day, mat-shaped formations known as "living stromatolites" (Figure 6.1). Living stromatolites are formed by colonies of microbes, and they contain layers of sediment intermixed with microbes of different species. Microbes near the top generate energy through photosynthesis, and those beneath use organic compounds left as waste products by the photosynthetic microbes. The living stromatolites grow in size as sediments are deposited over them, forcing the microbes to migrate upward in order to remain at the depths to which they are adapted. This gradual migration creates the layered structures.

The similarity of structure between the ancient stromatolites and the living stromatolites suggests a similar origin, implying that stromatolites are fossil remnants of early life. There is some controversy about the biological origin of stromatolites, because geological processes of sedimentation can mimic their layering. However, the wide variety of structures seen in stromatolites and the results from chemical analysis make most scientists confident that they offer evidence for the existence of microbial colonies as far back as 3.5 billion years ago. Moreover, if the microbes that made the stromatolites are like the microbes in the living stromatolites today, then the implication is that at least some of these ancient microbes produced energy by photosynthesis. Because photosynthesis is a fairly sophisticated metabolic process, we presume that it must have taken at least a moderately long time for this process to evolve in living organisms. In other words, the oldest stromatolites suggest that photosynthetic life already existed some 3.5 billion years ago, in which case more primitive life must have existed even earlier.

MICROFOSSILS The second line of evidence comes from individual fossilized cells. Finding ancient microscopic fossils, or microfossils, is quite challenging, both because rocks become increasingly rare with age and because the oldest rocks have been altered by geological processes in ways that tend to destroy microfossils within them. Moreover, while

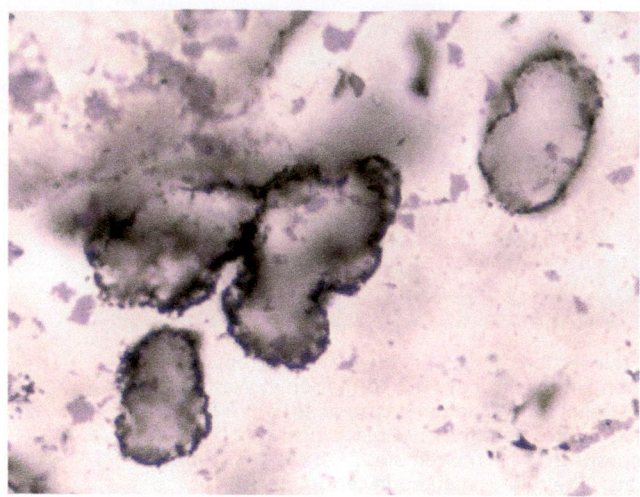

FIGURE 6.2
This microscopic photograph shows cell-like structures that appear to be fossils of microbes that lived about 3.4 billion years ago.

dinosaur bones are fairly "obvious" fossils, it can be quite difficult to determine whether an interesting-looking microscopic structure is biological or mineral in origin. As a result, claims of ancient microfossils often generate significant scientific controversy, with competing hypotheses attempting to explain their origin.

Not surprisingly, the greatest controversy surrounds the oldest claimed microfossils, which come from a nearly 3.5-billion-year-old rock formation in northwestern Australia. One team (headed by William Schopf of UCLA) claims to have found microfossils in rocks that are 3.465 billion years old; however, the opinions of other geologists are sharply divided as to whether these particular structures have a mineral or a biological origin. But in 2011, another team (led by David Wacey of the University of Western Australia) found a set of structures that seem more clearly to be microfossils in rocks from the same region that are only slightly younger—having a maximum age of 3.426 billion years. These microfossils show a much clearer cell structure (Figure 6.2), and chemical analysis shows that their makeup includes metabolic by-products that are consistent with what we would expect if they had once been alive. In fact, careful study suggests that these microbes used sulfur compounds as a source of nutrients and energy, which is also consistent with what we might expect of primitive chemoautotrophs (see Table 5.1) of the type likely to have existed on the early Earth. Overall, most geologists consider the evidence of biological origin for these 3.4-billion-year-old microfossils to be quite strong.

Think About It Consider the controversy over these microfossils in light of what you have learned about the process of science in this book. Do you think the controversy indicates strength or weakness in our methods of scientific inquiry? Defend your opinion.

Many other microfossils have also been discovered that are nearly as old, and likewise seem clearly to have been formed by living microbes. For example, fossilized cells dating to between 3.2 and 3.4 billion years ago have been found at two locations in southern Africa. Altogether, the microfossil evidence is generally taken to support the idea that life existed at least 3.4 billion years ago. The fact that this is quite consistent with the ages of the oldest stromatolites adds to the strength of the case, and again suggests that life not only existed but was well established on Earth by 3.4 to 3.5 billion years ago.

ISOTOPIC EVIDENCE The third line of evidence for an early origin of life comes from isotopic analysis of some of the most ancient rocks on Earth. Living organisms can change the ratios of isotopes from their background, nonliving values. For example, carbon has two stable isotopes: carbon-12, with six protons and six neutrons in its nucleus, and carbon-13, which has one extra neutron. Carbon-12 is far more common, but any inorganic carbon sample always contains a small proportion of carbon-13 atoms mixed in with the more numerous carbon-12 atoms. (Typically, carbon-13 accounts for about 1 out of every 89 carbon atoms.) When living organisms metabolize carbon, they incorporate carbon-12 atoms into cellular molecules slightly more easily than they do carbon-13 atoms. As a result, living organisms—and fossils of living organisms—always show a slightly lower fraction of carbon-13 atoms than that found in inorganic material.

On an island off the coast of Greenland, this lower carbon-13 ratio has been found in rocks that are more than 3.85 billion years old,* suggesting that the rocks contain remnants of life from that time (Figure 6.3). The claim of biological origin (made by a team led by the University of Colorado's Stephen Mojzsis) has been challenged by a few other scientists. The rocks are metamorphic, meaning they have been transformed substantially by high pressure or heat, which would explain why no intact microfossils remain within them. For these rocks to have contained life, we must presume that they were sedimentary before they underwent the metamorphic transformation; if they were volcanic (igneous) rocks, then it is much more difficult to see how they could have been home to or preserved evidence of living microbes.

While the controversy is by no means settled, recent evidence seems to be swaying the debate in favor of the hypothesis that these rocks really do indicate that life already existed by 3.85 billion years ago. The new evidence falls into two basic categories. First, if the Greenland rocks hold evidence of life, we would expect to find something similar in other rocks dating to the same general time. This is indeed the case, as other rocks dating to some 3.8 billion years ago have been found to show similar carbon isotope ratios. Second, life can also alter the isotopic ratios of other elements, such as iron, nitrogen, and sulfur. Recent studies show that these isotopes also are present in ratios characteristic of life within the ancient rocks, just as we would expect if the carbon isotope data are a result of biological origin.

IMPLICATIONS OF THE EARLY ORIGIN OF LIFE We have seen that the stromatolite evidence suggests the presence of fairly advanced life by nearly 3.5 billion years ago. The microfossil evidence is only slightly younger and therefore roughly consistent with the idea that life was present at that time. Carbon isotope evidence may push the existence of life back to at least 3.85 billion years ago. Given the fact that the geological record is so sparse for such early times, we would expect to find evidence of life in these ancient rocks only if life had already been widespread on Earth. Therefore, if we are interpreting all the data correctly, it is likely that life arose considerably earlier than 3.85 billion years ago. Geologically speaking, this would mean life arose quite early in Earth's history (see Figure 4.10).

By itself, this early origin of life proves nothing about life elsewhere, since it is always possible that Earth was the lucky beneficiary of a highly improbable event. However, if we assume that what happened here would be typical of what might happen elsewhere, then the early origin of life is profoundly important: It suggests that we could expect life to also arise rapidly on any other world with similar conditions. Because we expect many other worlds to have conditions similar to those that prevailed on the young Earth, this idea gives us reason to think that life might be quite common in the universe.

What did early life look like?

The earliest living organisms presumably went extinct long ago, replaced by others with evolutionary adaptations that allowed them to

FIGURE 6.3
This ancient rock formation on the island of Akilia (off the coast of southern West Greenland) may hold the oldest known evidence for life on Earth.

---

*We say that the rocks are *older* than 3.85 billion years because the rocks containing the isotopic evidence are sediments that cannot be dated. However, they are cut through by igneous rocks that date to 3.85 billion years, which means the sediments must be even older.

outcompete their ancestors. Nevertheless, just as we know that sharks and alligators are evolutionarily much older than primates (because the geological record shows them coexisting with dinosaurs while primates emerged much later), some modern-day microbes must be more closely related to the earliest living organisms than others. Study of these more primitive species should help us understand what early life looked like, which in turn should help us investigate the question of how life arose.

MAPPING EVOLUTIONARY RELATIONSHIPS The geological record tells us a great deal about how life evolved, but it is not the only way to study evolution. Because living species have evolved from common ancestors, the base sequence in the DNA of living organisms provides a sort of map of the genetic changes that have occurred through time. By comparing the genomes of different organisms, we should be able to reconstruct the evolutionary history of much of life on Earth.

To understand how the technique works, consider the DNA of an organism that long ago became the common ancestor of all life today. Mutations created variations on this DNA, and each new species therefore had slightly different DNA sequences than did the older species from which it evolved. Lateral gene transfer [Section 5.4] may also have been common among early living organisms, changing genomes even more rapidly. Over millions and billions of years, continuing evolution led to new species with DNA molecules increasingly different from the DNA of the common ancestor. But, always, the new molecules were built by changes to the older ones so that, in principle, the changes are traceable in the precise base sequences of living organisms.

Determining the sequence of bases in an organism's DNA is a difficult and time-consuming task (though the technology for doing so is rapidly improving), and to date only a small fraction of known species have had their complete genome sequences determined. In many more cases, biologists have compared smaller pieces of the DNA of many species.* By comparing the DNA sequences in similar genes among many different species, biologists can map the evolutionary history of the genes. For example, two species with very similar DNA sequences probably diverged relatively recently in evolutionary history, while two species with very different DNA sequences probably diverged much longer ago.

These types of DNA sequence comparisons are what has enabled biologists to map out the relationships shown in the tree of life (see Figure 5.12). That is, DNA studies tell us that life can be divided into the three domains that we discussed in Chapter 5—bacteria, archaea, and eukarya—and also tell us about the branching patterns within each domain. Despite the many uncertainties that remain in the tree of life, the branching patterns still reveal a lot about evolutionary history. For example, the fact that animals and plants represent two branches that split off in about the same place from other eukarya tells us that all animals and plants are genetically quite similar, at least in comparison to organisms on most other branches. Moreover, organisms on branches located closer to the "root" of the tree must contain DNA that is evolutionarily

---

*A particularly common technique relies on determining the sequence of bases in molecules of ribosomal RNA (rRNA), which tells us the sequence of the DNA that coded for it.

older, suggesting that they more closely resemble the organisms that lived early in Earth's history.

CONCLUSIONS ABOUT EARLY LIFE The genetic studies that have led to mapping the tree of life tell us that species near the root of the tree must be more similar than other species to the common ancestor of life on Earth. Unfortunately, aside from recognizing these species as more primitive than most others, we have not yet been able to draw clear conclusions about their nature. Initially, it was thought that most of the organisms closest to the root were extremophiles such as those living near deep-sea vents or underground [Section 5.5], but more recently scientists have found non-extreme living archaea that are genetically similar. Nevertheless, scientists remain hopeful that as we sequence the genomes of more organisms, we'll get a clearer picture of what the earliest life may have looked like.

Where did life begin?

We rely primarily on geological considerations to come up with ideas about *where* life first arose on Earth.

It seems unlikely that life could have arisen on land. The early atmosphere contained practically no molecular oxygen, so our planet could not have had a protective layer of ozone. Ozone (O_3) is a form of oxygen produced in the upper atmosphere by interactions between ordinary oxygen (O_2) and ultraviolet light from the Sun. Today, ozone shields Earth's surface from the Sun's dangerous ultraviolet radiation. Before the ozone layer existed, any surface life would have been exposed to high levels of this radiation. While we can't rule out the possibility that life might have arisen in such an environment—some organisms today (such as *D. radiodurans* [Section 5.5]) can survive high-radiation conditions—the environment would have been much more hospitable under water (because water also absorbs ultraviolet light) or in rocks beneath the surface.

One such possibility, first suggested by Darwin, is shallow ponds. As we'll discuss in the next section, organic compounds may have formed spontaneously in such ponds. Once the compounds formed, tides or cycles of wetting and evaporation could have increased their concentration near the pond edges, spurring reactions that might have led to life. (Note that tides would have been much stronger early in Earth's history, because the Moon was closer to Earth at that time.) Volcanic hot springs may also have offered energy to support an origin of life. However, while these factors suggest ponds could have been a good location for an origin of life, the shallow water would not have offered much protection against solar ultraviolet radiation.

A better possibility might be deep-sea or underground environments, which would have been protected from high-energy radiation. Deep-sea volcanic vents offer plenty of chemical energy to fuel reactions that might have led to life, and chemical energy is also available underground in reactions between water and minerals in rock. Moreover, even if life first arose in ponds at the surface, the larger impacts of the heavy bombardment probably would have allowed the survival only of life that had migrated to deep-sea or underground environments. For that reason, it now seems likely that the common ancestor of all life on Earth today evolved from organisms that lived near deep-sea vents or underground, even if the first origin of life occurred elsewhere.

6.2 The Origin of Life

Even the simplest living organisms today seem remarkably advanced. Metabolic processes involve many intricate molecules and enzymes working together. The complex chemistry of DNA and RNA is deeply intertwined with the proteins and enzymes that help in making them [Section 5.4]. Indeed, every cellular component and process depends on many other components and processes. Given the complexity and interdependency of these processes, it might at first seem difficult to conceive of ways in which they might have come to be. However, over the past few decades, laboratory experiments have given us insights into the chemical processes that likely occurred on the early Earth. While these experiments have not yet told us precisely how life first arose—and it's possible that they never will—we'll see in this section that they give us good reason to think that life may have started through natural, chemical processes.

Before we begin, it's worth noting two important caveats. First, the laboratory experiments generally try to re-create the chemical conditions that should have prevailed on the early Earth, an assumption that makes sense if life originated here. However, it is conceivable that life migrated to Earth from another world—a possibility we will also discuss in this section—in which case it might have arisen in a somewhat different chemical environment. Second, we have not said anything about the possibility that life arose through any kind of divine intervention; as we have discussed in prior chapters, that possibility falls outside the realm of science and instead is a matter of personal faith.

How did life begin?

Life today is based on the chemistry of organic molecules, making it logical to assume that the first life was somehow assembled from organic molecules produced by chemical reactions on the early Earth. Such reactions do not occur naturally today, because Earth's oxygen-rich atmosphere prevents complex organic molecules from forming readily outside living cells. Oxygen is such a highly reactive gas that it tends to attack chemical bonds, removing electrons and destroying organic molecules. However, the oxygen in our atmosphere today is a product of life, produced by photosynthesis, which means it could not have been present before life arose. It therefore seems reasonable to think that the young Earth might have been much like a giant laboratory for organic chemistry.

THE MILLER–UREY EXPERIMENT As early as the 1920s, some scientists recognized that Earth's early atmosphere should have been oxygen-free, and they hypothesized that sunlight-fueled chemical reactions could have led to the spontaneous creation of organic molecules. (This idea was proposed independently by Russian biochemist A. I. Oparin and British biologist J. B. S. Haldane.) This hypothesis was most famously put to the test in the 1950s in an experiment credited to Stanley Miller and Harold Urey, now known as the **Miller–Urey experiment** (Figure 6.4).

The original Miller–Urey experiment used small glass flasks to simulate chemical conditions that scientists thought represented those on the early Earth. One flask was partially filled with water to represent the sea and heated to produce water vapor. Gaseous methane and ammonia were added and mixed with the water vapor to represent the atmosphere.

FIGURE 6.4
Stanley Miller poses with a reproduction of the experimental setup he first used in the 1950s to study pathways to the origin of life. (He worked with Harold Urey, so the experiment is called the Miller–Urey experiment.)

These gases flowed into a second flask, where electric sparks provided energy for chemical reactions. Below this flask, the gas was cooled so that it could condense to represent rain and then was cycled back into the water flask. The water soon began to turn a murky brown, and a chemical analysis (performed after letting the experiment run for a week) showed that it contained many amino acids and other organic molecules.

We now know that the methane and ammonia mixture in the original Miller–Urey experiment was *not* representative of Earth's early atmosphere, so the experiment's specific results probably don't tell us a lot about what happened on the early Earth. Nevertheless, the experiment demonstrated that, at least under some conditions, the building blocks of life form naturally and abundantly.

Scientists have since tried different approaches to the Miller–Urey experiment, changing the ingredients to try to better represent conditions on the early Earth. Unfortunately, we still don't have a clear understanding of those conditions. For example, hydrogen can play a major role in facilitating the production of organic chemicals, but the hydrogen content of the early atmosphere is a topic of great debate. Scientists long assumed that any hydrogen in the atmosphere would have escaped quickly to space, but some models suggest that conditions in Earth's early upper atmosphere would have slowed its escape, allowing hydrogen to make up as much as 30% of the early atmosphere. Other aspects of the early environment are subject to similar debate. Indeed, the issue is so unsettled that one scientist (Gerald Joyce, Scripps Institute) recently quipped, "Just wait a few years, and conditions on the primitive Earth will change again."

OTHER SOURCES OF ORGANIC MOLECULES Although the jury is still out on precisely how much organic material might have been made through processes like those in the Miller–Urey experiment, we know of at least three other potential sources of organic molecules on the early Earth.

The first of these potential "other" sources is chemical reactions near deep-sea vents. As these undersea volcanoes heat the surrounding water, a variety of chemical reactions can occur between the water and the minerals. These chemical reactions would have occurred spontaneously in the conditions thought to have prevailed in the early oceans, and they should have resulted in the production of the same types of organic molecules thought to have been necessary for the origin of life.

The second additional source of organic molecules may have been material from space, including asteroids, comets, and interplanetary dust. Analysis of meteorites (which are fragments of asteroids) shows that they often contain organic molecules, including complex molecules such as amino acids. Telescopic and spacecraft study of comets, along with analysis of comet dust collected and returned to Earth by the *Stardust* mission, shows that they also contain organic molecules, sometimes even including some of the DNA bases [Section 5.4]. Interplanetary dust grains from the solar nebula may also play a role, because research has shown that ultraviolet light from the young Sun could have catalyzed reactions on the dust grains that would have produced organic molecules. The young Earth therefore could have been seeded with the building blocks of life from impacts of asteroids and comets and from a cosmic "rain" of interplanetary dust.

The third additional source of organic molecules may have been the heat and pressure generated by large impacts themselves during the heavy bombardment. In 2014, scientists in the Czech Republic reported results of an experiment that used a laser to reproduce the heat and pressure that would have occurred when a large impact hit a mixture of chemicals known to have been present on the early Earth. The experiment created all four DNA bases—the first time all four had been produced in a single experiment—and many other organic molecules. Given the prevalence of impacts during the heavy bombardment, this process could have generated enormous quantities of the chemical building blocks of life.

It's likely that all four sources of organic molecules—chemical reactions near the ocean surface, chemical reactions near deep-sea vents, material from space, and the heat and pressure of impacts—played a role in shaping the chemistry of the early Earth. More important, given at least four different ways of obtaining organic molecules, it seems likely that at least parts of the early Earth would have contained substantial amounts of the organic molecules needed for life.

THE TRANSITION FROM CHEMISTRY TO BIOLOGY Having found strong evidence that the young Earth should have had all the building blocks needed to make life, we now turn to the question of how these ingredients might have assembled themselves to make a living cell.

There seems little doubt that natural processes would have produced all the essential building blocks of life, but, to paraphrase the late Carl Sagan, these represent only the notes of the music of life, not the music itself. Viewed in terms of simple probability, the likelihood of a set of simple building blocks ramming themselves together to form a complete living organism is at least as small as that of letting monkeys loose in a roomful of musical instruments and hearing Beethoven's Ninth Symphony. It simply wouldn't happen, even if the experiment was repeated over and over again for billions of years. There must have been at least a few intermediate steps—each involving a chemical pathway with a relatively high probability of occurring—that eased the transition from chemistry to biology.

One way to explore the transition is to work backward from organisms living now. Heredity today is shaped by DNA, which serves this function primarily because of its ability to replicate. Early life must also have had a self-replicating molecule, but it probably was not DNA: Double-stranded DNA seems far too complex, and its replication far too intertwined with RNA and proteins, to have been the genetic material of the first living organisms. We are therefore looking for a molecule that is simpler than DNA but still capable of making fairly accurate copies of itself. The most obvious candidate is RNA (though, as noted briefly in Chapter 5, a few scientists hypothesize that RNA may itself have been predated by the nucleic acid TNA or PNA).

RNA WORLD RNA is much simpler than DNA because it has only one strand rather than two and the manufacture of its backbone structure requires fewer steps. But it still possesses hereditary information in the ordering of its bases, and in principle it can serve as a template for making copies of itself. For a while there seemed to be a problem with this idea. In modern organisms, neither DNA nor RNA can replicate itself. Both require the help of enzymes. These enzymes are proteins that are made from genetic instructions contained in DNA and carried out with the help of RNA. This fact seemed to present a "chicken and egg" dilemma: RNA cannot replicate without enzymes, and the enzymes cannot be made without RNA.

A way around this dilemma was discovered in the early 1980s by Thomas Cech and his colleagues at the University of Colorado, Boulder. They found that RNA can catalyze biochemical reactions in much the same way as enzymes (work for which Cech shared the Nobel Prize in 1989). We now know that RNA molecules play this type of catalytic role in many cellular functions, and we call such RNA catalysts *ribozymes* (by analogy to enzymes). Follow-up work has shown that some RNA molecules can at least partially catalyze their own replication. These discoveries have led biologists to envision that modern, DNA-based life may have arisen from an earlier **RNA world,** in which RNA molecules served both as genes and as chemical catalysts for copying and expressing those genes.

How might an RNA world have gotten started? The first requirement would have been the spontaneous production of self-replicating strands of RNA. Even under the most optimistic assumptions, the concentration of organic molecules on the early Earth would have been far too low to allow those building blocks to assemble spontaneously into full-fledged RNA molecules. RNA assembly almost certainly would have required some sort of catalytic reaction to facilitate it. Here, again, laboratory experiments offer evidence for such a process.

Experiments show that several types of inorganic minerals can facilitate the self-assembly of complex, organic molecules. Minerals of the type that geologists call *clay** may have been especially important. Clay is extremely common on Earth and in the oceans, where it forms through simple weathering of silicate minerals; indeed, the oldest zircon grains [Section 4.2] suggest the widespread abundance of clays by about 4.4 billion years ago, so we expect clay to have been common at the time of the origin of life. Moreover, clay minerals contain layers of molecules to which other molecules, including organic molecules, can adhere. When organic molecules stick to the clay in this way, the mineral surface structure can force them into such close proximity that they react with one another to form longer chains.

Laboratory experiments show that this natural process quickly and easily produces strands of RNA up to a few dozen bases in length. These strands are thought to be too short to have produced a self-replicating RNA; other experiments suggest a minimum length of at least 165 bases for a molecule capable of catalyzing self-replication. But the process would not have stopped with these short strands.

The RNA strands are only weakly bound to the clay on which they form, so they can easily peel away. At that point, some of them naturally fold in ways that make it much easier for other RNA strands to attach to them. Moreover, while the short RNA strands probably could not have catalyzed self-replication, they could have catalyzed other chemical reactions; in 2010, scientists discovered an RNA strand only 5 bases long that can act as a ribozyme. Given the countless grains of clay that could have facilitated chemical reactions, it seems reasonable to *expect* the natural formation of simple ribozymes that could have catalyzed the attachment of folded RNA molecules, making the strands longer and more complex. This would have dramatically increased the probability of getting an RNA molecule capable of self-replication. Figure 6.5 summarizes these ideas.

1. *Clay minerals catalyze the formation of RNA strands up to a few dozen bases long.*

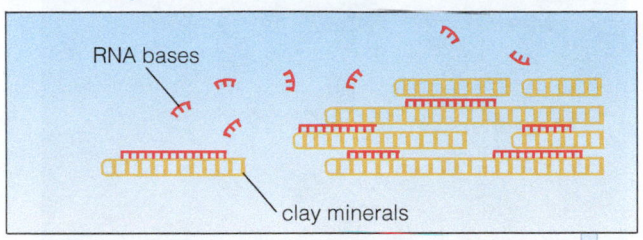

2. *RNA strands peel away from clay and fold; some are capable of catalyzing chemical reactions.*

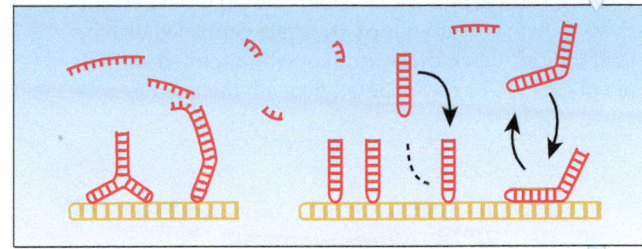

3. *Aided by catalysis, folded RNA molecules attach to each other to make longer RNA strands.*

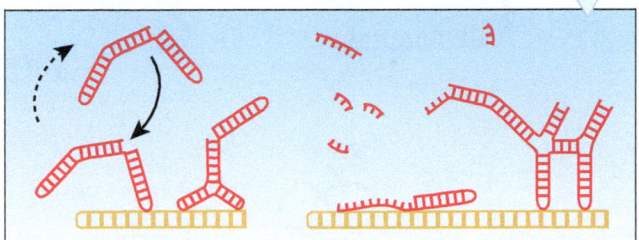

4. *Longer strands can perform more catalysis, eventually leading to self-replication.*

FIGURE 6.5
These diagrams show steps through which self-replicating RNA may have originated, as RNA bases (already present on Earth) interacted with clay minerals. (Adapted from Briones, Stich, and Manrubia, "The Dawn of RNA World," *RNA Journal,* May 2009.)

---

*In this context, *clay* refers to silicate minerals with a particular physical structure; this mineralogical definition is somewhat different from what you may think of as clay in the context of pottery or sculpture.

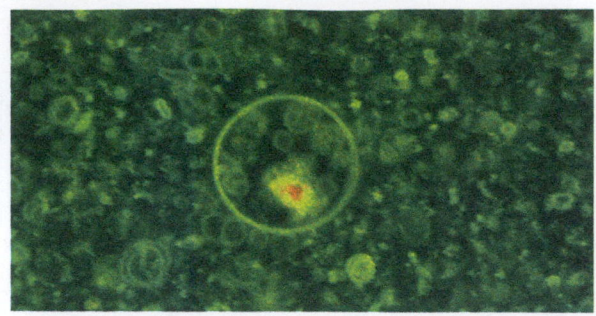

FIGURE 6.6
This microscopic photo (made with the aid of fluorescent dyes) shows short strands of RNA (red) contained within a lipid pre-cell (green circle), both of which formed with the aid of catalysis by clay minerals beneath them.

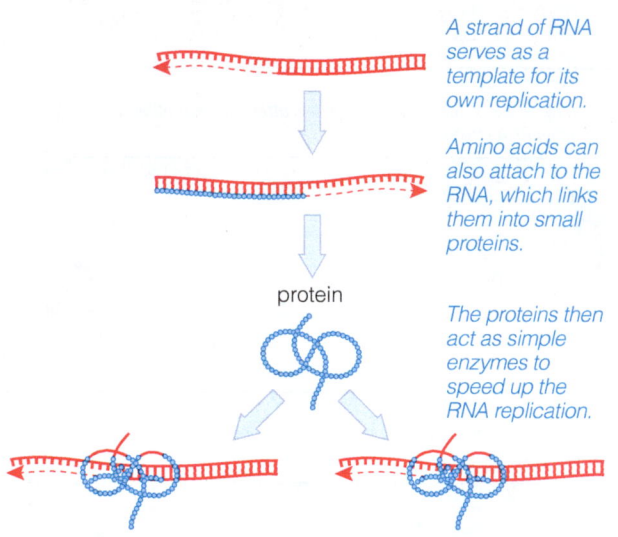

A strand of RNA serves as a template for its own replication.

Amino acids can also attach to the RNA, which links them into small proteins.

protein

The proteins then act as simple enzymes to speed up the RNA replication.

a This diagram shows a self-replicating RNA molecule that has evolved the capability to produce a primitive enzyme that helps its own replication.

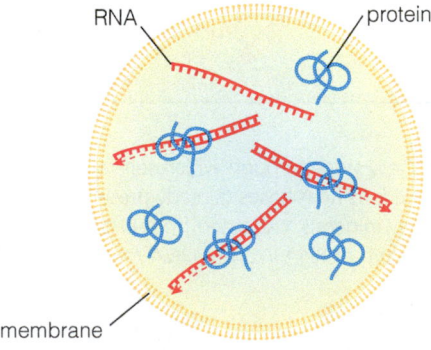

RNA   protein

membrane

b If the RNA and the enzyme are isolated from the outside environment inside a pre-cell, then only the molecules in this particular pre-cell will benefit from the new enzyme, a fact that can speed up the molecular evolution.

FIGURE 6.7
Self-replicating RNA could have rapidly evolved through a molecular analog to natural selection. (Adapted from Campbell, Reece, Simon, *Essential Biology*.)

Other experiments show that RNA, along with other organic molecules and even tiny bits of clay mineral, could easily have become confined within naturally forming microscopic enclosures often called "pre-cells" (or *vesicles*). Scientists have known for decades that such pre-cells can be formed naturally in at least two different ways: by cooling a warm-water solution of amino acids so that they form bonds among themselves to make an enclosed spherical structure or by mixing lipids with water. These structures can exhibit some of the most important properties of the membranes of living cells. For example, they can selectively allow some types of molecules to cross into or out of the enclosure, and some can store energy in the form of an electrical voltage across their surfaces, which can be discharged in a way that facilitates reactions inside them. In some cases, they can also grow until they reach an unstable size, at which point they split to form "daughter" spheres. Moreover, experiments show that lipid pre-cells can form on the surface of the same clay minerals that help assemble RNA molecules, sometimes with RNA inside them (Figure 6.6).

Confining RNA and other organic molecules within pre-cells could have facilitated an origin of life in two important ways. First, keeping molecules concentrated and close together should have increased the rate of reactions among them, making it far more likely that a self-replicating RNA would have arisen; the high rate of reactions would also have greatly increased the probability that cooperative relationships between RNA molecules and proteins could arise. Second, once self-replicating RNA molecules came to exist, pre-cells would have kept them isolated from the outside in a way that should have facilitated a molecular analog to natural selection, in which RNA molecules that replicated faster and more accurately would rapidly come to dominate the population. For example, suppose a particular self-replicating RNA molecule assembled amino acids into a primitive enzyme that sped up replication (Figure 6.7a). If the enzyme floated freely within the ocean water, it might just as easily have helped the replication of other RNA molecules as the one that made it. But inside a pre-cell, the enzyme would help only the RNA that made it, giving this RNA an advantage over less capable RNA molecules in other pre-cells (Figure 6.7b).

Experiments suggest that the mutation rate in simple, self-replicating RNA molecules would have been quite high, so molecular natural selection would have inevitably led RNA molecules to gain complexity and evolve more efficient replication pathways. At some point, the RNA pre-cells would likely have become sufficiently good at reproducing and evolving to be "alive." The process probably would have been gradual; there might never have been a particular moment when we would have been able to say that the "first living cell" had appeared on the scene.

Once the first living organisms of the RNA world arose, biological natural selection could take over. It then seems easy to understand why the RNA world would have given way to the present DNA world. The structural similarities between RNA and DNA make it likely that DNA molecules would eventually have evolved within living cells. Because DNA is a more flexible hereditary material and is less prone to copying errors than RNA, life that used DNA for its genome would quickly have outcompeted the remaining organisms that used RNA. But RNA served many other cell functions well, so those would have been retained and would have continued to evolve, explaining why RNA still plays so many important roles in cells, even though it no longer plays a hereditary role (except in some viruses).

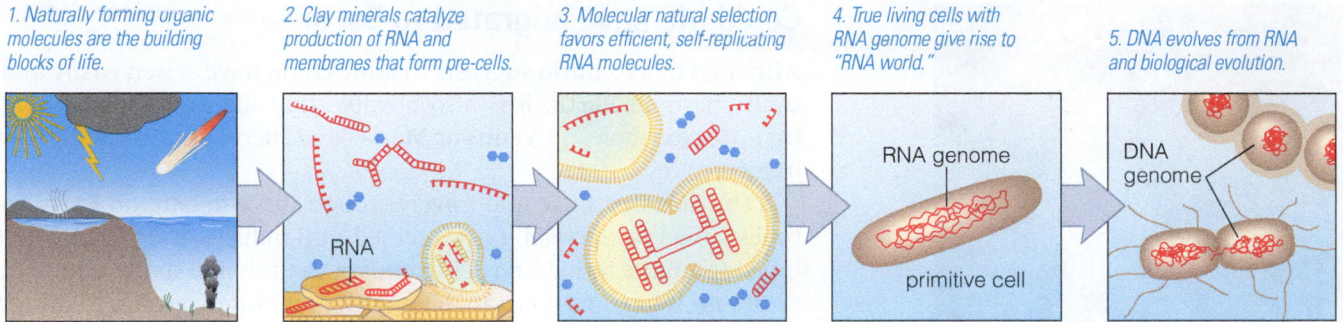

1. Naturally forming organic molecules are the building blocks of life.

2. Clay minerals catalyze production of RNA and membranes that form pre-cells.

RNA

3. Molecular natural selection favors efficient, self-replicating RNA molecules.

4. True living cells with RNA genome give rise to "RNA world."

RNA genome

primitive cell

5. DNA evolves from RNA and biological evolution.

DNA genome

FIGURE 6.8
A summary of the steps by which chemistry on the early Earth may have led to the origin of life. (Adapted from Campbell, Reece, Simon, *Essential Biology*.)

PUTTING IT ALL TOGETHER Let's review the sequence we've discussed here as the possible explanation for the origin of life (Figure 6.8).

1. Through some combination of atmospheric chemistry, chemistry near deep-sea vents, and molecules brought to Earth from space or formed in impacts, the early Earth had at least localized areas with significant amounts of organic molecules that could serve as building blocks for more complex organic molecules.

2. More complex molecules, including short strands of RNA, grew from the organic building blocks, probably with the aid of reactions catalyzed by clay minerals. The minerals also helped catalyze the production of microscopic pre-cells in which RNA and other organic chemicals became enclosed.

3. The concentration of RNA molecules within pre-cells facilitated reactions that eventually led to self-replicating RNA, at which point molecular natural selection favored the spread of those RNA molecules that replicated most accurately and efficiently.

4. Natural selection among the RNA molecules in pre-cells gradually led to an increase in complexity, until eventually some of these structures became true living organisms.

5. DNA evolved from RNA, and its advantages made it the preferred hereditary molecule. Natural selection continued, enabling organisms to adapt to a great many environmental niches on planet Earth.

We may never know for certain whether life actually originated in this way, in some similar way, or in some completely different way. Nevertheless, this scenario seems quite reasonable and perhaps even "easy," given geological time scales. It seems especially reasonable given that a number of different components of the scenario have been demonstrated in laboratory experiments. Even if life did not originate in this way, it seems that it could have—which suggests that the actual path to life must have been equally easy, or else life would have followed the path we've described. In summary, we have good reason to believe that the origin of life was a *likely* consequence of conditions on the early Earth, in which case it seems likely that life should also have arisen on many other worlds that once had similar conditions.

Think About It We've noted that the probability of life's arising through the random mixing of simple organic building blocks is so small as to seem impossible. Yet, in the scenario we've described, the likelihood of getting life seems quite good. In your own words, describe why these two probabilities are so different.

FIGURE 6.9
Chemical analysis of this meteorite, known as NWA 7034, indicates that it came from Mars. The small block shown for scale just below the meteorite is 1 cubic centimeter, about the size of a typical sugar cube. This particular martian meteorite was found in the Sahara desert in 2011.

Could life have migrated to Earth?

Although our scenario suggests that life could have arisen easily and naturally here on Earth, it is also possible that life arose somewhere else first—for example, on Venus or Mars—and then migrated to Earth within meteorites.

The idea that life could travel through space to land on Earth, sometimes called *panspermia,* once seemed outlandish. After all, it's hard to imagine a more forbidding environment than that of space, where there's no air, no water, and constant bombardment by dangerous radiation from the Sun and stars. However, the presence of organic molecules in meteorites and comets tells us that the building blocks of life can form and remain stable in the space environment, and we've already discussed some forms of Earth life that are capable of surviving at least moderate periods of time in space [Section 5.5]. It therefore seems possible that life could migrate from one planet to another, if it could hitch a suitable ride.

THE POSSIBILITY OF MIGRATION We know that meteorites can and do travel from one world to another. Among the tens of thousands of meteorites that scientists have identified and cataloged, careful chemical analysis has so far revealed dozens with compositions that clearly suggest that they came from Mars (Figure 6.9); even more have been found that come from the Moon. Apparently, these meteorites were blasted from their home worlds by large impacts, then followed orbital trajectories that eventually caused them to land on Earth. Examination of these meteorites, along with theoretical calculations based on the amount of material blasted into space by impacts, suggests that over time the inner planets have exchanged many tons of rock. In a sense, Earth, Venus, and Mars have been "sneezing" on each other for billions of years, offering the possibility of microscopic life hitchhiking between worlds on one of the meteorites.

For a living microbe to arrive intact on Earth after such a journey, it would have to survive at least three potentially lethal events: the impact that blasts it off the surface of its home world, the time it spends in the harsh environment of interplanetary space, and the fiery plunge through our atmosphere. Examination of martian meteorites suggests that neither the first nor the last of these events poses insurmountable obstacles. The interiors of martian meteorites show only minimal disruption, suggesting that microbes inside these rocks could survive both the initial impact and the later fall to Earth. The larger question is whether they could survive their time in space.

The chance of surviving the trip between planets probably depends on how long the meteorite spends in space. Once a rock is launched into space, it orbits the Sun until its orbit carries it directly into the path of another planet. Most meteorites will orbit for many millions of years before reaching Earth, even if they come from a world as nearby as Venus or Mars. It seems highly unlikely that living organisms could survive in space for millions of years. However, a few meteorites are likely to be launched into orbits that cause them to crash to Earth during one of their first few trips around the Sun. For example, calculations suggest that about 1 in 10,000 meteorites may travel from Mars to Earth in a decade or less. Because experiments in Earth orbit have already shown that some terrestrial organisms can survive at least 10 years exposed to the vacuum and cold of space, and microbes in a spore state could remain viable for far longer times in rocks, it seems quite reasonable to imagine microbes from Mars arriving safely on Earth.

While migration between planets seems possible, similar considerations almost certainly rule out the possibility of migration from other star systems. Under the best of circumstances, meteorites from planets around other stars would spend millions or billions of years in space before reaching Earth; any living organisms would almost surely be killed by exposure to cosmic rays during this time, or simply die because of desiccation—the lack of water. Moreover, calculations suggest that the probability of a rock from another star system hitting Earth is extremely low, which may also explain why we have never yet found a meteorite from beyond our own solar system.

REASONS TO CONSIDER MIGRATION Given the reality that the inner planets exchange rocks in the form of meteorites, the key question probably is not whether life *could* migrate through space but whether we have any reason to suppose it originated elsewhere rather than right here on Earth. Many scientists have debated this question, with the debate taking many different twists. Today, most ideas about migrating life fall into one of two broad categories.

The first broad category of reasons why some people favor a migration hypothesis suggests that life does not form as easily as we have imagined, at least under the conditions present on the early Earth. In this view, the only explanation for life on Earth (other than invoking the supernatural) would be migration from elsewhere. Although this idea in some sense only moves the problem of life's origin to another place, it at least allows for the possibility that another world had conditions that were more conducive to rapid development of life, or that life arose on a world that offered more time. The primary drawback to this idea is that, as we've discussed, it seems that Earth *did* have conditions that would have allowed an origin of life. Moreover, on the off chance that this idea is incorrect, we know of no compelling reason why any of the other worlds in our solar system would have offered either better conditions or substantially more time for an origin of life. In that case, we'd be left with the possibility that life migrated from another star system, but we have already explained why that seems highly unlikely.

The second broad category suggests that life forms so easily that we should expect to find life originating on any planet with suitable conditions. In that case, the origin of life in our solar system would have occurred on whichever planet got those conditions *first;* for example, if the very early Venus or very early Mars had suitable conditions for life before Earth did, life from one of those worlds might have migrated to Earth and taken hold on our planet as soon as conditions allowed. In essence, this idea suggests that life might never have gotten the chance to originate indigenously on Earth because life from another planet got here first.

IMPLICATIONS OF MIGRATION TO THE SEARCH FOR LIFE BEYOND EARTH
While ideas about microbes migrating *to* Earth are speculative, it seems a near-certainty that microbes *from* Earth have many times made the journey to Mercury, the Moon, Venus, and Mars. After all, Earth has

Cosmic Calculations 6.1

BACTERIA IN A BOTTLE I: LESSONS FOR EARLY LIFE

Once the first organisms took hold, how quickly could they have spread and evolved? A thought experiment* offers insight into this question. Suppose that you place a single bacterium in a nutrient-filled bottle at noon, and that this species is capable of replicating by cell division every minute. The original bacterium grows until it divides into two bacteria at 12:01. These two bacteria divide at 12:02 into 4 bacteria, which divide at 12:03 into 8 bacteria, and so on. Then the number of bacteria in the bottle at any time t minutes after noon is

$$\text{number of bacteria at } t \text{ minutes after } 12:00 = 2^t$$

We'll explore general characteristics of this **exponential growth** (t is in the exponent) in Cosmic Calculations 6.2. Here, to understand how rapidly early life could have spread, let's consider the volume of a bacterial colony. A typical bacterium is 10^{-7} m (0.1 micrometer) across, which means it has a volume of about $(10^{-7} \text{ m})^3 = 10^{-21} \text{ m}^3$. So the volume of bacteria at any time t minutes after noon is

$$\text{bacterial volume} = 2^t \times 10^{-21} \text{ m}^3$$

Our two formulas tell us that after 60 minutes the number of bacteria is $2^{60} \approx 1 \times 10^{18}$, or a *million trillion;* their volume is about $2^{60} \times 10^{-21} \text{ m}^3 \approx 0.001 \text{ m}^3$, or 1 liter (the volume of a typical bottle). But let's imagine they could somehow continue to multiply. By the end of the second hour, they would number an astonishing $2^{120} \approx 1.3 \times 10^{36}$, and their volume would be $2^{120} \times 10^{-21} \text{ m}^3 \approx 1.3 \times 10^{15} \text{ m}^3$—large enough to cover the surface of the Earth to a depth of about 2 meters (see Problem 53 at the end of the chapter). Continuing the calculations, you'd find that the bacteria would exceed the total volume of the world's oceans (about $1.3 \times 10^{18} \text{ m}^3$) at $t = 130$ minutes. Note that changing the doubling time from one minute to a year hardly matters; a time of $t = 130$ years rather than 130 minutes is still geologically insignificant.

Although the bacteria could not really continue this hypothetical growth, the implication should be clear: The first self-replicating organisms would have spread rapidly as far as conditions allowed, leaving the door wide open for biological evolution through natural selection.

---

*This thought experiment is adapted from one created by Professor of Physics Albert A. Bartlett of the University of Colorado.

suffered plenty of impacts large enough to blast rock into space during its long history, offering abundant opportunities for hitchhiking microbes. Therefore, if it were possible for Earth life to survive on any of these other worlds, we should *expect* to find it there. As we'll discuss in Chapter 7, we can almost certainly rule out the possibility of survival on the Moon and Mercury, and probably on Venus as well. Mars, however, may well have habitats that could provide at least temporary refuge to terrestrial microbes, and Mars may have been globally habitable in the distant past.

The likelihood of such interplanetary migration raises at least two important issues in astrobiology. First, if we someday find life on Mars, we will have to wonder if it is native or if it arrived there from Earth. The only way we may ever be confident that Mars life is not transplanted Earth life will be if its biochemistry is too different from that of terrestrial life to allow for a common ancestor.

Second, the possibility of life migrating among the planets raises the question of whether we could ever distinguish between an indigenous origin of life on Earth and an origin based on migration from elsewhere. To date, no one knows how we might choose between these possibilities. It's conceivable that a fossil record from Mars might suggest an earlier origin of life there, though even then we might not be certain that this life came to Earth. Venus poses a more intractable problem: As we'll discuss in Chapter 10, it is possible that Venus once had oceans and a habitable climate in which life might have arisen. However, Venus now is so hot that any fossil record would almost certainly have been destroyed by the heat and subsequent geological activity.

Despite these potential uncertainties, the major lessons of our study of life's origins still hold: One way or another, life arose on Earth quite soon after conditions first allowed it, and even if life migrated here from another world, we have good reason to think that it evolved naturally, through chemical processes that favor the creation of complex, organic molecules and the subsequent evolution of self-replicating molecules.

6.3 The Evolution of Life

Regardless of exactly how or where it originated, we've seen that there's little doubt that life on Earth was established by 3.5 billion years ago, and the origin may go back as far as 4 billion years or more. Life on Earth has been evolving ever since. Careful studies of the geological record provide the key data with which we attempt to re-create the evolutionary time scale, while genome comparisons offer data that help us map relationships among species. In this section, we'll briefly retrace the history of life on Earth as it is currently understood, which should in turn help us understand the possibilities for finding similarly complex life on other worlds.

What major events have marked evolutionary history?

Reconstructing 4 billion years of history from limited clues is an obviously difficult task. Nevertheless, we have identified at least a few of the key events that have marked the evolution of early life to its current diversity, and that help us understand our own origins. You may find it helpful to look back at the geological time scale in Figure 4.10 as you read this section, since it shows the timeline of the major events.

EARLY MICROBIAL EVOLUTION The earliest organisms must have been quite simple, but they undoubtedly had at least a few enzymes and a rudimentary metabolism. Their cells probably looked somewhat like those of the simplest modern bacteria or archaea, lacking cell nuclei and other complex structures that we find in eukarya. Moreover, because the atmosphere at that time was essentially oxygen-free, all early life must have been **anaerobic,** meaning that it did not require molecular oxygen; by contrast, we are **aerobic** organisms, because we cannot survive without molecular oxygen.

We expect that the first microorganisms were *chemoautotrophs* (see Table 5.1)—organisms that obtained their carbon from carbon dioxide dissolved in the oceans and their energy from chemical reactions involving inorganic chemicals. The reason for this expecation is that both photosynthesis and the ability to digest other organisms involve more complex metabolic pathways than those used by chemoautotrophs, and these more complex pathways presumably would have evolved some time later. Some modern archaea that appear to be fairly close to the root of the tree of life, such as those thriving in hot sulfur springs, are chemoautotrophs that obtain their energy through chemical reactions involving hydrogen, sulfur, and iron compounds. Because similar compounds were abundant on the early Earth, perhaps especially so in hot springs and near deep-sea vents, it seems reasonable to assume that early life used the available inorganic chemical energy in a similar way. (See Section 9.4 for discussion of possible chemical pathways.)

Natural selection probably caused rapid diversification among the early life-forms. Modern DNA replication involves a variety of enzymes that help keep the mutation rate low. Early organisms, with a much more limited set of enzymes, probably experienced many more errors in DNA copying. Because more errors mean a higher mutation rate, evolution would have been rapid among early microbes. As life diversified, many new metabolic processes evolved, making some of the new organisms biochemically quite different from their ancestors. Because of the rapid pace of early evolution, it may not have taken long to establish many of the major branches in the tree of life (see Figure 5.12).

Fossil evidence supports the idea of rapid diversification. Recall that stromatolites suggest the presence of organisms that obtained energy by photosynthesis some 3.5 billion years ago, and some of the oldest microfossils also resemble modern photosynthetic organisms. Because photosynthesis is a complex metabolic pathway, its relatively early emergence indicates that early evolution was rapid.

Photosynthesis probably evolved through multiple steps. At first, some organisms may have developed light-absorbing pigments that absorbed excess light energy—especially ultraviolet—that was harmful to life near the ocean surface. Over time, some of these pigments evolved to enable the cell to make use of the absorbed solar energy. Modern organisms known as purple sulfur bacteria and green sulfur bacteria may be much like the early photosynthetic microbes. These organisms use hydrogen sulfide (H_2S) rather than water (H_2O) in photosynthesis and therefore do not produce any oxygen. Photosynthesis using water, which produces oxygen as a by-product, probably came later and ultimately caused the buildup of oxygen in Earth's atmosphere. The timing of the oxygen buildup is still uncertain, but it probably did not get under way much before about 2.5 billion years ago.

The rise of oxygen created a crisis for life, because oxygen attacks the bonds of organic molecules. Many species of microbes probably went extinct, and those that survived had to somehow avoid the detrimental effects of oxygen. Some avoided these effects because they lived in (or migrated to) underground locations where the oxygen did not reach them. We still find many anaerobic microbes in such locales today, living in soil or deeper underground in rocks. Others survived because the oxygen content of the atmosphere rose gradually, allowing them time to evolve new metabolic processes and protective mechanisms that enabled them to thrive rather than die in the presence of oxygen. Plants and animals, including us, still use the metabolic processes that evolved in response to the "oxygen crisis" faced by living organisms some 2 billion or more years ago.

THE EVOLUTION OF EUKARYA The evolution of eukarya was the crucial first step in our own eventual evolution; we are, after all, members of this domain. Even single-celled eukaryotes exhibit much more diversity in cellular structure than exists among bacteria or archaea, and multi-celled eukaryotes enjoy diversity far beyond that. Because more variations are possible on complex structures than on simple ones, the complexity of eukaryotic cells allowed for the selection of many more adaptations than were possible in simpler cells. Indeed, multicellularity appears to have evolved independently in several different branches of eukarya, suggesting that the complex structure of eukaryotic cells opened the door for the evolution of more advanced organisms.

When did eukarya arise? Despite the fact that modern eukarya have more complex cellular structures than bacteria and archaea, genome studies do not suggest any substantial differences in the evolutionary ages of the three domains. That is, it is quite possible that members of all three domains—bacteria, archaea, and eukarya—split from a common ancestor early in Earth's history. Early eukarya could not yet have had cell nuclei and other complex intracellular structures. These must have come later, and the oldest known fossils that clearly show cell nuclei date to about 2.1 billion years ago. However, because cell nuclei do not fossilize well, it's possible that eukarya began to have such structures much earlier but we are unable to recognize them in the fossil record.

Modern, complex eukarya probably evolved through a combination of at least two major adaptations that arose in their simpler ancestors. First, some early species of eukarya may have developed specialized infoldings of their membranes that compartmentalized certain cell functions, ultimately leading to the creation of a cell nucleus (Figure 6.10a). Second, some relatively large ancestral host cells absorbed small bacteria within them, creating a **symbiotic relationship** in which both the invading organisms and the host organisms benefited from living together (Figure 6.10b).

The key evidence for symbiosis (the development of a symbiotic relationship) comes from two structures in eukarya that appear to be "cells within cells": **mitochondria,** the cellular organs in which oxygen helps produce energy (by making molecules of ATP), and **chloroplasts,** structures in plant cells that produce energy by photosynthesis. Besides the fact that mitochondria and chloroplasts look like tiny bacterial cells, both also have their own DNA and reproduce themselves within their eukaryotic homes. Moreover, sequencing of the DNA in mitochondria and chloroplasts clearly groups them with the domain bacteria, rather than

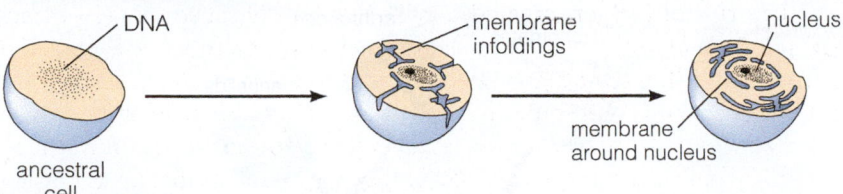

a Early eukarya probably lacked a cell nucleus, but some large cells may have developed membrane infoldings that compartmentalized certain cell functions, ultimately leading to the creation of a cell nucleus.

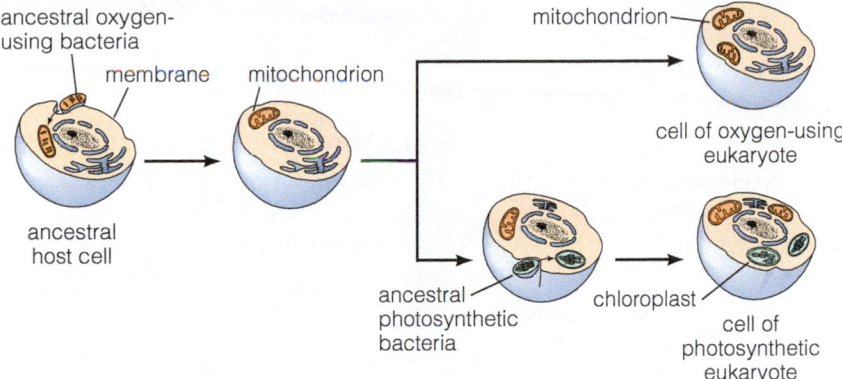

b Mitochondria and chloroplasts may have evolved as small bacteria invaded a larger host cell, forming a symbiotic relationship.

FIGURE 6.10

Hypotheses concerning the origin of eukaryotes. (Adapted from Campbell, Reece, Taylor, Simon, *Biology Concepts & Connections*.)

with eukarya, making it a near-certainty that they originated as free-living bacteria. Assuming that these bacteria had already evolved the ability to make efficient use of oxygen (in the case of mitochondria) or to carry out photosynthesis (in the case of chloroplasts), a symbiotic relationship might have developed easily. The host cell would have benefited from the energy produced by the incorporated bacteria, while the bacteria would have benefited from the protection offered by the host cell.

THE CAMBRIAN EXPLOSION We have seen that life on Earth existed at least 3.5 billion years ago—and perhaps hundreds of millions of years before that—and that all three domains of life were well established by at least 2.1 billion years ago. However, the fossil record tells us that all this life remained microscopic (aside from microbes organized into colonies) until much later. The earliest fossil evidence for complex, multicellular organisms—all of which are eukarya—dates to only about 1.2 billion years ago. In other words, microbes had our planet to themselves for more than 2 billion years after the origin of life. Even today, the total biomass of microbes far exceeds that of multicellular organisms like fungi, plants, and animals [Section 5.2]. But mass isn't everything, and we have a special interest in understanding the evolution of multicellular life, even if it is comparatively rare on our world.

In particular, we have a special interest in animal evolution: Animals may represent only one small branch on the tree of life, but it's *our* branch. Moreover, we generally assume that extraterrestrial intelligence, if it exists, will belong to animal-like beings from other worlds. The fossil record suggests that animal evolution progressed slowly at first, with relatively little change seen between fossils from 1.2 billion years ago and those from a half-billion years later. But then something quite dramatic happened.

FIGURE 6.11
This tree shows the major phyla of the animal kingdom.
Humans, along with all other mammals and reptiles, belong
to the chordates (phylum *Chordata*).

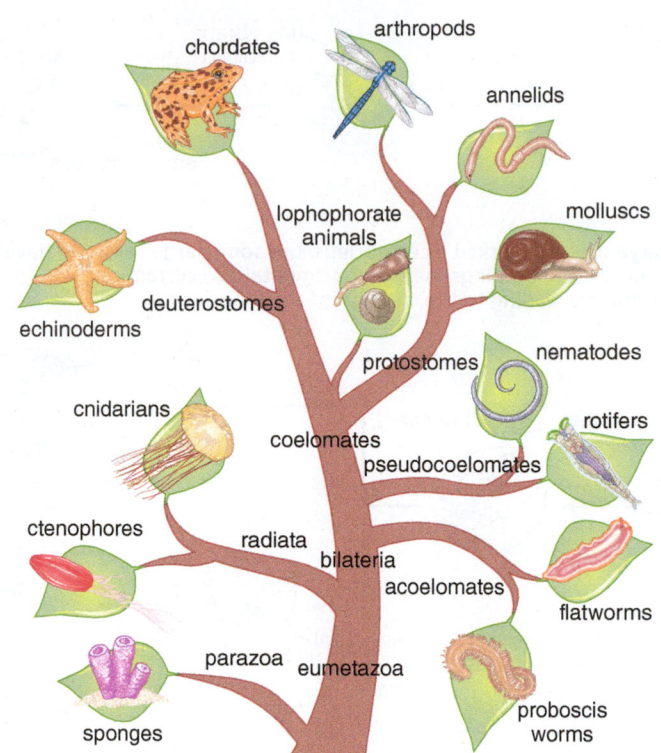

In the broadest sense, biologists classify animals according to their basic "body plans." For example, the basic body plan shared by mammals and reptiles is fundamentally different from that of insects. Animals are grouped by body plan into what biologists call **phyla** (singular, *phylum*), which is the next level of classification below kingdoms (such as the plant kingdom and the animal kingdom [Section 5.2]). Mammals and reptiles both belong to the phylum *Chordata*, which represents animals with internal skeletons. Insects, crabs, and spiders belong to the phylum *Arthropoda*, which represents animals with body features such as jointed legs, an external skeleton, and segmented body parts. Classifying animals into phyla is an ongoing effort by biologists, but modern animals appear to comprise about 30 different phyla, each representing a different body plan (Figure 6.11).

Remarkably, nearly all of these different body plans, plus a few others that have gone extinct, make their first known appearance in the geological record during a period spanning only about 40 million years—less than about 1% of Earth's history. This remarkable flowering of animal diversity appears to have begun about 542 million years ago, which corresponds to the start of the Cambrian period (see Figure 4.10).* Hence, it is called the **Cambrian explosion.**

Think About It One early development in the evolution of multicellular organisms was a trend toward larger size. Briefly discuss how larger size might have conferred an evolutionary advantage.

---

*Although the Cambrian explosion is among the most vivid events in the geological record, evidence indicates that the lineages we first see in the Cambrian explosion actually began evolving earlier. The Cambrian explosion didn't involve only animals; other groups, such as algae, also diversified.

The fact that the Cambrian explosion marks the only major diversification of body plans in the geological record presents us with two important and related questions: Why did the Cambrian explosion occur so suddenly, at least in geological terms, yet so long after the origin of eukaryotes, and why hasn't any similar diversification happened since?

No one knows the answers to these questions, but we can identify at least four possible contributing factors. First, the oxygen level in our atmosphere may have remained well below its present level until about the time of the Cambrian explosion. If so, the dramatic change in animal life may have occurred at least in part because oxygen reached a critical level for the survival of larger and more energy-intensive life-forms.

A second factor was the evolution of genetic complexity. As eukaryotes evolved, they developed more and more genetic variation in their DNA, which opened up ever more possibilities for further variation. Perhaps the Cambrian explosion marks a point in time when organisms had become sufficiently complex that a great diversity of forms could evolve over a short period.

The third factor may have been climate change. Recall that geological evidence points to a series of snowball Earth episodes [Section 4.5] ending around the time of the Cambrian explosion. The extreme climate conditions that marked these episodes may have exerted evolutionary pressure that aided the diversification of life and then fueled the Cambrian explosion when the environmental conditions eased.

The fourth factor may have been the absence of efficient predators. Early predatory animals were probably not very sophisticated, so some adaptations that later might have been snuffed out by predation were given a chance to survive if they arose early enough. The beginning of the Cambrian period may have marked a window of opportunity for many different adaptations to gain a foothold in the environment.

This last idea may partly explain why no similar explosion of diversity has taken place since the Cambrian. Once predators were efficient and widespread, it would have been much more difficult for entirely new body forms to find an available environmental niche. In addition, the fact that certain body forms were already selected during the Cambrian explosion may have limited other options. That is, while more body plans may have been possible than actually arose, once some were in existence there may not have been clear evolutionary pathways to others. Alternatively, perhaps the various body forms that arose during the Cambrian explosion represent the full range of forms possible, at least within the constraints of the genetic variability available on Earth (Figure 6.12). In any case, we and nearly all other animals living today can trace our ancestry to species that arose during the Cambrian explosion.

THE COLONIZATION OF LAND We do not know when life first migrated onto land, in part because most fossils form in sediments deposited in the oceans, regardless of whether the organisms making the fossils lived in the water or on the land. However, given the wide variety of environments in which microbes survive today and the fact that many different genetic lineages seem to have appeared quite early in evolutionary history, it's likely that microbial life quickly established itself wherever it could find liquid water and protection from ultraviolet radiation. Plenty of such locations are available on land—including underground and any place where water can pool under a shelter of overhanging rock—so it is hard to imagine reasons why microbial life would not have taken hold

FIGURE 6.12
This fossil is about 505 million years old and shows one of the many animal forms that arose during the Cambrian explosion. It is from the Burgess Shale, a rock formation in British Columbia (Canada) famous for its well-preserved fossils.

on land quite early. However, the situation is different for multicellular organisms. While microbes may have thrived on land, larger organisms, including all animals, remained confined to the oceans (and other bodies of water) even after the Cambrian explosion.

For larger organisms, surviving on land was more difficult than surviving in the oceans, primarily because it required evolving a means of obtaining water and mineral nutrients without simply absorbing them from their surroundings. The timing of the development of the ozone layer may also have played a role in the late colonization of land. Recall that ozone is a molecule (O_3) made from oxygen atoms, so a protective ozone layer could not exist until the atmosphere contained some threshold level of oxygen. Uncertainties regarding both the oxygen levels through time and the level needed for a substantial ozone layer make it difficult to know when the ozone layer first appeared. But until it did, life on any exposed land surface would have been difficult or impossible.

Fossil evidence shows that plants (and perhaps fungi as well) were the first large organisms to develop the means to live on the land. The colonization of land by plants appears to have begun about 475 million years ago. DNA evidence suggests that plants evolved from a type of alga. Some ancient algae might have survived in salty shallow-water ponds or along lake edges. Because such locales occasionally dry up, natural selection would have favored adaptations, such as thick cell walls, that allowed the algae to survive during periods of dryness. Cell walls would have given the organisms structure that would have helped them survive on land. The first fully land-based organisms would have had even more advantages, because there were no land animals around to eat them. Large plants gradually developed complex bodies with some parts specialized for energy collection above ground (where sunlight is available) and other parts specialized for collecting water and nutrients from the soil.

Once plants moved onto the land, it was only a matter of time until animals followed them out of the water. Within about 75 million years, amphibians and insects were eating land plants. By the beginning of the Carboniferous period, about 360 million years ago, vast forests and abundant insects thrived around the world (Figure 6.13). These Carboniferous forests were important not only as a major step in evolution, but also because they became an important part of our modern economy. Much of the land area of the continents was flooded by shallow seas during the Carboniferous period, hindering the decay of dead plants. Thick layers of dead organic matter piled up in the stagnant waters. Over millions of years, as these layers were buried, pressure and heat gradually converted the organic matter to coal. Nearly all the coal that has helped fuel our industrial age is, in fact, the remains of these forests of the Carboniferous period.

Think About It Based on the preceding discussion, explain why coal is called a "fossil fuel." What other fossil fuels do we use to generate energy?

Why was the rise of oxygen so important to evolution?

We are oxygen-breathing animals, so there's no question that the rise of oxygen was critical to our eventual emergence. More generally, the rise of oxygen was important to evolution because oxygen can react so strongly with organic molecules. While these reactions can kill organisms that are not adapted to oxygen's presence, they also offer the possibility of

FIGURE 6.13
This painting, based on fossil evidence, shows a forest of the Carboniferous period.

much more efficient cellular energy production (that is, making molecules of ATP [Section 5.3]) than is possible through anaerobic processes. That is, as aerobic organisms evolved, they were able to develop adaptations that demanded much more energy than would have been available to their anaerobic ancestors. The rise of oxygen ignited an explosion of eukaryotic diversification and, as we've discussed, may have helped fuel the Cambrian explosion.

THE ORIGIN OF ATMOSPHERIC OXYGEN Molecular oxygen is a highly reactive gas that would disappear from the atmosphere in just a few million years if it were not continually resupplied by life. Fire, rust, and the discoloration of freshly cut fruits and vegetables are everyday examples of **oxidation reactions**—chemical reactions that remove oxygen from the atmosphere. Many elements and molecules can participate in oxidation reactions. Today, most reactions that remove oxygen from the atmosphere occur in living organisms that use oxygen, including ourselves. Before oxygen-breathing organisms evolved, oxidation reactions involved primarily volcanic gases, dissolved iron in the oceans, and surface minerals (especially those containing iron) that could react with oxygen. Such reactions essentially "rust" the minerals, causing them to turn reddish in color. In the oceans, oxidation reactions with dissolved iron create minerals that precipitate to the bottom, forming "red beds" on the ocean floor. On land, the reddish color of much of Earth's rock and clay is a direct result of oxidation reactions involving surface minerals.

The fact that free oxygen would not last long without life tells us that our atmosphere must have been essentially oxygen-free before life existed. Moreover, while today we recognize plants as a major source of oxygen, we know that plants arrived relatively recently on the evolutionary scene. Where, then, did the oxygen come from? Remarkably, it seems that we owe our oxygen atmosphere to microscopic bacteria sometimes called "blue-green algae" but more technically known as **cyanobacteria** (Figure 6.14). Fossil evidence suggests that cyanobacteria were producing oxygen by at least 2.7 billion years ago, and perhaps for hundreds of millions of years before that. It took at least 2 billion years for oxygen to build up in the atmosphere to its present levels, but in the end the oxygen we breathe originally entered the air through the action of microscopic cyanobacteria.

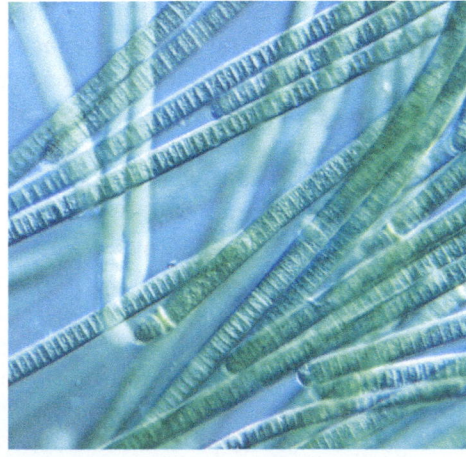

a The blue-green color of this lake (in Anhui Province, China) is the result of a population explosion, or "bloom," of cyanobacteria.

b This micrograph shows individual cyanobacteria.

FIGURE 6.14
Cyanobacteria split water and release oxygen in photosynthesis and are thought to have been responsible for the rise of oxygen in Earth's atmosphere.

FIGURE 6.15

This rock is an example of a banded iron formation (BIF) formed more than 2 billion years ago. Such rocks could have formed only before the atmosphere contained significant amounts of oxygen. The pen is included for scale.

TIMING OF THE OXYGEN RISE The precise timing of the rise of oxygen is difficult to study, because we have no direct way to sample air from hundreds of millions or billions of years ago. However, we can learn about oxygen content from a variety of other clues. For example, fossils of oxygen-breathing organisms indicate that at least a certain minimum amount of oxygen was present in the atmosphere in order for them to survive. Careful study of rock chemistry offers even more clues.

Studies of rocks that are between about 2 and 3 billion years old, especially rocks of a type called *banded iron formations* (Figure 6.15), show that the atmosphere during that time contained less than 1% of the amount of oxygen it contains today. The banded iron formations were made from iron-containing minerals dissolved in the oceans, and such iron minerals cannot dissolve if there is substantial oxygen in the atmosphere and ocean. Other mineral studies, based on sulfur isotope ratios in ancient rocks, constrain the timing of the rise of oxygen more tightly. Atmospheric oxygen alters the chemistry of sulfur compounds in the atmosphere (SO_2 and H_2S from volcanic outgassing) in ways that change the ratios of sulfur isotopes that end up in surface rock. The oldest rocks showing sulfur isotopes in a ratio that indicates the presence of atmospheric oxygen are about 2.35 billion years old. If these data are being properly interpreted, the abundance of atmospheric oxygen must have been less than 20 parts per million (0.002%) up until that time. At that point, sometimes called the "great oxidation event," oxygen began to build up in the atmosphere.

The timing of the great oxidation event poses a mystery, however: If we are correct in assuming that cyanobacteria began to produce oxygen at least 2.7 billion years ago—or at least 350 million years before the great oxidation event at 2.35 billion years ago—what took so long? Our best guess is that nonbiological processes, such as oxidation of surface rock and ocean minerals, were at first able to remove oxygen from the atmosphere as rapidly as the cyanobacteria could make it; only after the rock and ocean minerals were saturated with oxygen could the atmospheric buildup begin. Evidence for this possibility comes from study of shales dating to more than 100 million years before the great oxidation event. These shales, thought to have formed on the ocean bottom, show evidence that the ocean contained low levels of oxygen at least 2.5 billion years ago, even though the atmospheric oxygen level at that time was probably less than about 1/100,000 of its modern level.

An even greater mystery concerns what happened once the buildup began. Other isotopic evidence suggests that the "great oxidation event" wasn't really that great, and that oxygen levels remained only 1% or less of modern levels for at least the next billion years, and perhaps all the way up to nearly the time of the Cambrian explosion. Indeed, a related hypothesis ties a buildup of oxygen starting around 800 million years ago with the onset of the snowball Earth episodes. In this scenario, Earth before that time was warmed primarily by the greenhouse effect of atmospheric methane. The rise of oxygen then caused the methane to react with oxygen to form carbon dioxide, and because carbon dioxide is a weaker greenhouse gas (per molecule) than methane, the overall greenhouse effect weakened and the snowball Earth period started.

Regardless of the precise reasons, it now seems likely that oxygen levels remained far too low for complex animals until somewhere near the time of the Cambrian explosion, which occurred when Earth was already nearly 4 billion years old. The oxygen-breathing animals that evolved at that time probably needed oxygen levels of at least 10% of the modern value. It's possible that the oxygen level was higher than that, but the first clear evidence of an oxygen level near or above its current value appears in the geological record only about 200 million years ago. That is when we first find charcoal in the geological record, implying that enough oxygen was present in the atmosphere for fires to burn. In that case, if you had a time machine and could randomly spin the dial to take you back to any point in Earth's history, you'd have less than about a 1 in 10 chance—and perhaps only about a 1 in 20 chance—of appearing at a time recent enough that you could step out and breathe the air.

Think About It What does the late appearance of substantial atmospheric oxygen tell you about the difference between our planet's being habitable *in general* and being habitable *for us*?

IMPLICATIONS FOR LIFE ELSEWHERE The importance of oxygen to advanced life on Earth and the timing of its rise could potentially have important implications for life on other worlds. Our study of the origin of life gives us reason to think that life might be common on worlds with conditions like those of the early Earth. But the fact that it took so long for oxygen to build up in the atmosphere on Earth should make us wonder about the likelihood of getting oxygen-breathing life on other worlds. Could Earth have been "lucky" to get conditions that allowed the buildup of oxygen? If so, perhaps life on most other worlds would never evolve past microscopic forms; life might then be common, but advanced or intelligent life quite rare. Alternatively, maybe Earth was "unlucky" in having conditions that prevented the oxygen buildup for so long. In that case, other worlds might have complex plants and animals by the time they are just 1 to 2 billion years old, instead of having to wait until they are 4 billion years old. For the time being, we have no way to distinguish between these and other, intermediate possibilities. We will therefore continue our study with the assumption that Earth has been "typical," until and unless we learn otherwise.

6.4 Impacts and Extinctions

Once animals colonized the land, the evolutionary path that led to humans becomes much clearer. Reptiles evolved from amphibians; by about 245 million years ago, dinosaurs and mammals followed. But the fossil record shows that the path was not smooth. In particular, there is evidence for a number of striking transitions in the nature of living organisms. The most famous of these defines the boundary between the Cretaceous and Tertiary periods, which dates to about 65 million years ago. Dinosaur fossils exist below this boundary, but not above it. Somehow, after some 180 million years as Earth's dominant animals, the dinosaurs went extinct in a geological blink of the eye. Significant evidence now points to the idea that this extinction, and perhaps others, may have been caused by the impact of asteroids or comets crashing into Earth.

Did an impact kill the dinosaurs?

There's no doubt that major impacts have occurred on Earth in the past. Meteor Crater in Arizona (Figure 6.16) formed about 50,000 years ago

100 m

FIGURE 6.16
Meteor Crater in Arizona was created about 50,000 years ago by the impact of an asteroid about 50 meters across. Because the asteroid hit Earth at high speed, it left a crater more than 1 kilometer across and almost 200 meters deep (compare to the size of the parking lot and buildings at the crater's left). The K–T impact was about 200 times as large in size and tens of thousands of times more energetic.

when a metallic asteroid roughly 50 meters across crashed to Earth with the explosive power of a 20-megaton nuclear bomb. Although the crater is only a bit more than 1 kilometer across, the blast and ejecta probably battered an area covering hundreds of square kilometers. Meteor Crater is relatively small and recent, and it is a popular tourist stop because it is so obvious. But it is not alone: Geologists have identified more than 150 impact craters on our planet, and many others have presumably been destroyed by erosion and other geological processes.

THE K–T BOUNDARY LAYER In 1978, while analyzing geological samples collected in Italy, a scientific team led by Luis and Walter Alvarez (father and son) made a startling discovery. They found that the thin layer of sediments that marks the Cretaceous–Tertiary boundary, called the **K–T boundary** for short (the K comes from the German word for "Cretaceous," *Kreide*), is unusually rich in iridium, an element that is rare on Earth's surface (because Earth's iridium sank to the core when our planet underwent differentiation [Section 4.4]) but common in meteorites. Subsequent studies found the same iridium-rich sediment marking the K–T boundary at many other sites around the world (Figure 6.17). The Alvarez team proposed a stunning hypothesis: The extinction of the dinosaurs was caused by the impact of an asteroid or comet. They calculated that it would have taken an asteroid about 10–15 kilometers in diameter to deposit the iridium distributed worldwide in the K–T boundary layer.

In fact, the death of the dinosaurs was only a small part of the biological devastation that seems to have occurred 65 million years ago. The geological record suggests that up to 99% of all living plants and animals died around that time and that up to 75% of all existing plant and animal *species* were driven to extinction. This makes the event a clear example of a **mass extinction**—the rapid extinction of a large percentage of all living species. Could it really have been caused by an impact?

EVIDENCE FOR THE IMPACT There's still scientific debate about whether a period of active volcanism also contributed to the mass extinction, but there's little doubt that a major impact coincided with the death of the dinosaurs. Key evidence comes from further analysis of the K–T sediment layer. Besides being unusually rich in iridium, this layer contains four other unusual features: (1) high abundances of several other metals, including osmium, gold, and platinum; (2) grains of "shocked quartz," quartz crystals with a distinctive structure that indicates they experienced the high-temperature and high-pressure conditions of an impact; (3) spherical rock "droplets" of a type known to form when drops of molten rock cool and solidify in the air; and (4) soot (at some sites) that appears to have been produced by widespread forest fires.

All these features point to an impact. The metal abundances look much like what we commonly find in meteorites rather than what we find elsewhere on Earth's surface. Shocked quartz is also found at other known impact sites, such as Meteor Crater in Arizona. The rock "droplets" presumably were made from molten rock splashed into the air by the force and heat of the impact. Some debris would have been blasted so high that it rose above the atmosphere, spreading worldwide before falling back to Earth. On their downward plunge, friction would have heated the debris particles until they became a hot, glowing rain of rock.

A layer rich in iridium and soot tells us a huge impact occurred at this point in geological (and biological) history.

FIGURE 6.17
Around the world, sedimentary rock layers that mark the 65-million-year-old K–T boundary share evidence of the impact of a comet or asteroid. Fossils of dinosaurs and many other species appear only in rocks below the iridium-rich layer.

The soot probably came from vast forest fires ignited by radiation from this impact debris.

In addition to the evidence within the sediments, scientists have identified a large, buried impact crater that appears to match the age of the sediment layer. The crater, about 200 kilometers across, is located on the coast of Mexico's Yucatán Peninsula, about half on land and half underwater (Figure 6.18). Its size indicates that it was created by the impact of an asteroid or a comet measuring about 10 kilometers across, large enough to account for the iridium and other metals. (It is named the *Chicxulub crater*, after a nearby fishing village.)

THE MASS EXTINCTION If the impact was indeed the cause of the mass extinction, here's how it probably happened: On that fateful day some 65 million years ago, the asteroid or comet slammed into Mexico with the force of a hundred million hydrogen bombs (Figure 6.19). It apparently hit at an angle, sending a shower of red-hot debris across the continent of North America. A huge tsunami sloshed more than 1000 kilometers inland. Much of North American life may have been wiped out almost immediately. Not long after, the hot debris raining around the rest of the world ignited fires that killed many other living organisms. Indeed, the entire sky may have been bright enough to roast most life on land.

Dust and smoke remained in the atmosphere for weeks or months, blocking sunlight and causing temperatures to fall as if Earth were experiencing a global and extremely harsh winter. The reduced sunlight would have stopped photosynthesis for up to a year, killing large numbers of species throughout the food chain. This period of cold may have been followed by a period of unusual warmth: Some evidence suggests that the impact site was rich in carbonate rocks, so the impact may have released large amounts of carbon dioxide into the atmosphere. The added carbon dioxide would have strengthened the greenhouse effect, so the months of global winter immediately after the impact might have been followed by decades or longer of global summer.

The impact probably also caused chemical reactions in the atmosphere that produced large quantities of harmful compounds, such as nitrous oxides. These compounds dissolved in the oceans, where they probably were responsible for killing vast numbers of marine organisms. Acid rain may have been another by-product, killing vegetation and acidifying lakes around the world.

Perhaps the most astonishing fact is not that so many plant and animal species died but that some survived. Among the survivors were a few small mammals. These mammals may have survived in part because they lived in underground burrows and managed to store enough food to outlast the global winter that immediately followed the impact.

The evolutionary consequences of the extinctions were profound. For 180 million years, dinosaurs had diversified into a great many species large and small, while most mammals (which had arisen at almost the same time as the dinosaurs) had generally remained small and rodent-like. With the dinosaurs gone, mammals became the new animal kings of the planet. Over the next 65 million years, the small mammals rapidly evolved into an assortment of much larger mammals—ultimately including us. Had it not been for the K–T impact, dinosaurs might still rule Earth.

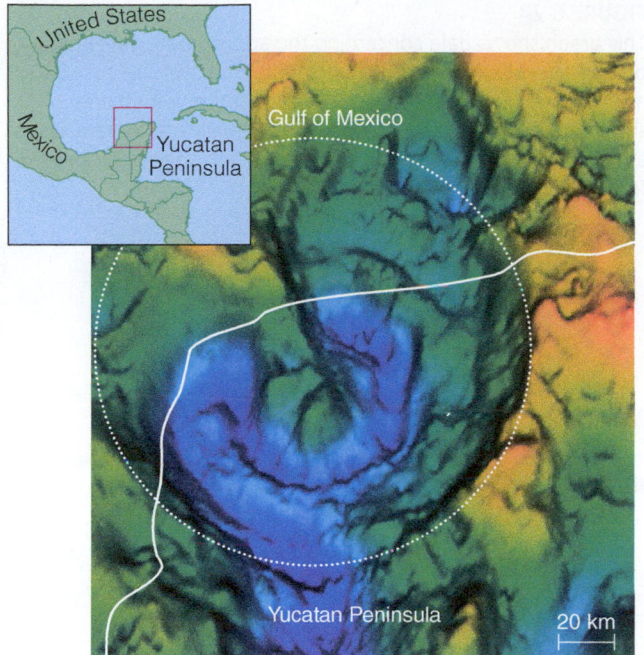

FIGURE 6.18
This computer-generated image, based on measurements of small local variations in the strength of gravity, reveals a buried impact crater about 200 kilometers across (dashed circle). The crater straddles the coast of Mexico's Yucatán Peninsula.

FIGURE 6.19
This painting shows an asteroid or comet moments before its impact on Earth, some 65 million years ago. The impact, known as the K–T impact, probably caused the extinction of the dinosaurs, and if it hadn't occurred, the dinosaurs might still rule Earth today.

FIGURE 6.20
This graph shows data concerning the approximate percentage of plants and animals to go extinct with time over the past 500 million years. Peaks represent mass extinctions, with names shown for five major events. (The data actually are for families, a higher level of classification than genus and species.) (Adapted from Campbell, Reece, *Biology*.)

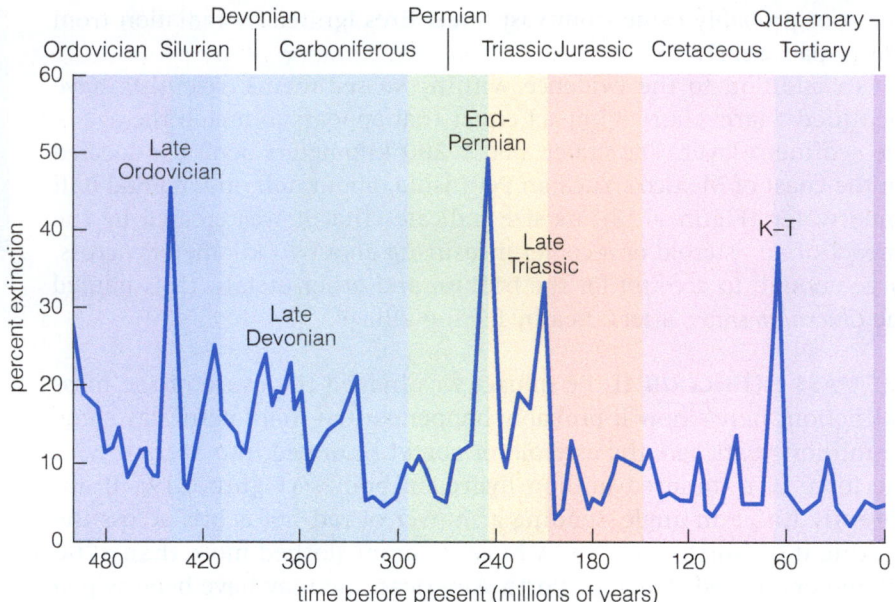

What caused other mass extinctions?

The K–T extinction seems quite clear, but it's generally difficult to measure past extinction rates precisely. The primary problem is that identifying the extinction of a species requires finding its *last* occurrence in the fossil record, which means we can be misled if we've yet to find a more recent occurrence of the species. Nevertheless, we have enough data to be sure that extinction rates vary considerably with time.

Figure 6.20 shows current data on the extinction rate for plants and animals over the past 500 million years. The data reveal at least five major mass extinctions, including the K–T extinction, and numerous smaller extinction events. Could some of these extinctions also have been caused by impacts?

Simple probability makes impacts a plausible hypothesis. On average, impacts the size of the K–T event should happen about every 100 million years or so, which is roughly the same as the average time between mass extinctions. However, none of the other major mass extinctions are as closely tied to impacts as the K–T extinction. No clear-cut evidence of iridium or shocked quartz has yet emerged in the rock layers dating to these extinctions, and no craters of just the right ages have been found. Of course, we should not necessarily expect to find a crater from such ancient events. Remember that most impacts occur in the oceans (because oceans cover nearly $\frac{3}{4}$ of Earth's surface), and seafloor crust is almost completely recycled in about 200 million years [Section 4.4]; this recycling would destroy any evidence of a crater on the seafloor.

EXTREME VOLCANISM The lack of clear evidence for impacts tied to other mass extinctions has led geologists to consider other possible causes. In at least one case, the end-Permian extinction, strong evidence points to a role for a form of extreme volcanism.

The end-Permian extinction, which occurred about 252 million years ago, appears to be by far the most devastating extinction in Earth's history; for that reason, it is sometimes called "the great dying." By some estimates, it may have killed off up to 90% of all species living at the time, and it is the only mass extinction in which insects were significantly affected.

The key evidence tying the end-Permian extinction to volcanism is a large formation of volcanic (igneous) rock in Siberia known as the *Siberian Traps*; the term *traps* comes from the Swedish word *trappa*, meaning "stairs," and refers to the structure of hills found in the region (Figure 6.21). The Siberian Traps cover an enormous area—about 2 million square kilometers—and contain enough basalt to cover Europe to a depth of a kilometer. They appear to have formed from a single set of enormous volcanic eruptions. Indeed, these eruptions are the largest known in Earth's history, and radiometric dating shows that they occurred at just the right time to have been a cause of the end-Permian extinction.

There is great debate about precisely how the volcanic eruptions might have led to the mass extinction, but the general scenario is somewhat similar to that thought to have followed the K–T impact. That is, the eruptions initially spewed enormous amounts of ash into the air, blocking sunlight and photosynthesis and cooling the planet. The ash may also have led to acid rain as it washed out. After a few years of cooling, the situation reversed as carbon dioxide and methane released by the eruptions greatly strengthened the greenhouse effect. Some studies (based on oxygen isotope ratios from the time period) suggest that Earth's global average temperature may have risen to as high as 40°C (104°F) during the post-eruption period, and high temperatures may have lasted hundreds of thousands to millions of years. Further evidence for this type of climate catastrophe comes from careful study of the Siberian Traps, which shows that they are made of a type of lava that was infused with unusually large amounts of carbon dioxide and that the eruptions likely occurred in shallow seas that also contained large amounts of methane. In other words, it seems likely that the eruptions would indeed have released enough greenhouse gas to account for catastrophic global warming. A 2014 study even added another methane source, citing evidence that a species of methane-releasing archaea (called *methanosarcina*) spread globally as other species died off.

If the Siberian Traps eruptions were indeed responsible for the end-Permian extinction, there remains the question of what caused the massive eruptions and whether similar eruptions played a role in other mass extinctions. No one knows the answers to either question. Some scientists speculate that impacts might have triggered the eruptions, but no clear evidence exists to make this linkage. Another suggestion is that the eruptions occurred because of a deep mantle plume—a column of molten, hot rock coming from far below the surface—that punched through Siberia when that land mass was drifting over the North Atlantic, as it was during the time of the Permian extinction.

MUTATION RATE CHANGES Another set of hypotheses envisions extinctions tied to changes in the mutation rate. While many mutations occur simply as the result of copying "errors" within cells, others are caused by external influences. For example, ultraviolet light can cause mutations, which is why sun exposure can lead to skin cancer. If the concentration of Earth's ozone layer varied with time, then the amount of ultraviolet light reaching the surface would also vary. Perhaps some of the mass extinctions occurred when the ozone layer thinned, allowing solar ultraviolet light to cause many more mutations.

Mutations can also be caused by high-energy particles that stream continuously from the Sun (the *solar wind*). Recall that Earth's magnetosphere deflects most of these particles, preventing them from reach-

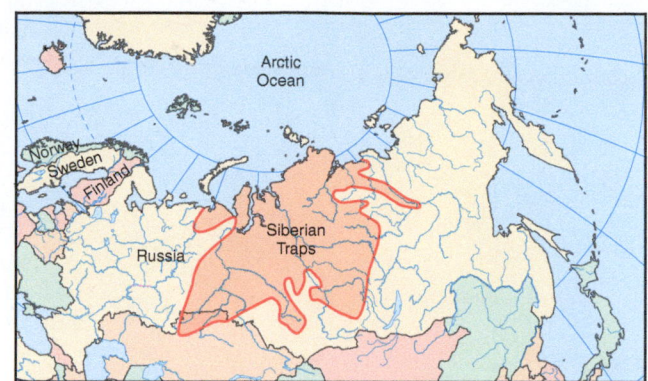

FIGURE 6.21

The map shows the vast extent of the Siberian Traps, formed by enormous volcanic eruptions that occurred at the time of the end-Permian mass extinction. The photo shows basalt formations in the region of the Traps known as the Putorana Plateau, a UNESCO World Heritage Site.

ing the surface [Section 4.4]. However, studies of magnetized rocks show that Earth's magnetic field varies significantly in strength with time and sometimes reverses itself entirely, with the north magnetic pole becoming the south magnetic pole and vice versa. These magnetic reversals occur every few million years on average, and the magnetic field may disappear altogether for thousands of years while a reversal is in progress. The mutation rate might spike upward during this time because of the absence of the normal protection from high-energy particles. Although magnetic field reversals happen much more frequently than mass extinctions, some reversals may have occurred at times when life was more susceptible to major change and thus might have played a role in extinction events.

SUPERNOVAE AND GAMMA-RAY BURSTS Some scientists hypothesize that mass extinctions on Earth could be triggered by more distant events, including *supernovae,* the explosions of massive stars [Section 3.2]. Supernovae are rare events. Out of the more than 100 billion stars in the Milky Way Galaxy, we expect only about one star per century to explode in a supernova. Most of these supernovae occur far from our solar system. Nevertheless, because the Sun orbits the center of the galaxy independently of other stars, different sets of stars make up our galactic neighborhood at different times. Simple probability calculations suggest that our planet must occasionally be located within a few tens of light-years of an exploding star. Supernovae generate prodigious numbers of very-high-energy particles called *cosmic rays.* When a supernova occurs near Earth, we might expect a big upward spike in the number of cosmic rays reaching Earth. These cosmic rays could in principle cause lethal mutations in many living organisms, leading to a mass extinction.

A related idea suggests that mass extinctions might be caused by *gamma-ray bursts*—bursts of gamma rays [Section 3.3] from space that last just minutes or less—most of which are produced by unusually powerful supernovae. Atmospheric models suggest that a gamma-ray burst occurring within a few thousand light-years of Earth could generate enough gamma rays to destroy half of Earth's ozone layer, thereby leading to a massive die-off through exposure to solar ultraviolet light. Probability arguments suggest that Earth should have been exposed to such a nearby gamma-ray burst at least once in the past billion years, and a few scientists have attempted to link a gamma-ray burst to the Ordovician mass extinction some 450 million years ago.

LESSONS FROM MASS EXTINCTIONS While we remain unsure of their causes, mass extinctions clearly have had tremendous effects on the evolution of life. With each mass extinction, many of the dominant species on the planet have disappeared, creating changes in environmental conditions and predator–prey relationships. These changes allow new species to evolve over the millions of years that follow. Just as the K–T event apparently paved the way for the rise of mammals, other extinctions may have caused similarly critical junctures in the evolutionary path that made our present existence possible.

The topic of mass extinctions also holds a cautionary lesson for our species today. Human activity is driving numerous species toward extinction. The best-known cases involve relatively large and wide-ranging animals, such as the passenger pigeon (extinct since the early 1900s) and the Siberian tiger (nearing extinction). But most of the estimated 10 million or more plant and animal species on our planet live in localized habitats, and most of these species have not even been cataloged. The destruction of just a few square kilometers of forest may mean the extinction of species that live only in that area. According to some estimates, human activity is driving species to extinction so rapidly that half of today's species could be gone within a few centuries or less. On the scale of geological time, the disappearance of half the world's species in just a few hundred years would qualify as another of Earth's mass extinctions, potentially changing the global environment in ways that we are unable to predict.

Think About It The geological record suggests that the dominant animal species are nearly always victims in a mass extinction. If we are causing a mass extinction, do you think we will be victims of it? Or will we be able to adapt to the changes so that we survive even though many other species go extinct? Defend your opinion.

Is there a continuing impact threat?

The discovery that at least one mass extinction is tied to an impact has spurred scientific concern over whether our civilization might be vulnerable to future impacts. How serious is this threat?

Small particles hit Earth almost continuously, burning up in our atmosphere as **meteors** (sometimes called "shooting stars," a misleading name that arose long before we knew their true source). Most meteors are caused by particles no bigger than a pea. The particle itself is too small for us to see, but it enters the atmosphere at such high speed (typically between about 45,000 and 250,000 km/hr) that it burns up and heats the surrounding air, producing the meteor flash. An estimated 25 million particles enter our atmosphere and burn up as meteors each day. Interestingly, these particles add a total of about 20,000 to 40,000 tons to Earth's mass each year. This sounds like a lot in human terms, but it is negligible compared to Earth's total mass of 6 *billion trillion* (6×10^{21}) tons.

Somewhat larger objects entering our atmosphere may be heated to the point where they explode, producing an extraordinarily bright flash called a *fireball*. (Some so-called UFO sightings are actually fireballs.) If debris from the explosion hits the ground, we may find some of it as the rocks we call *meteorites*. Larger impacts have also occurred in human history. In 1908, a tremendous explosion occurred over Tunguska, Siberia, flattening and setting fire to the surrounding forest (Figure 6.22). Seismic disturbances were recorded up to 1000 kilometers away, and atmospheric pressure fluctuations were detected at distances of almost 4000 kilometers. The explosion, now estimated to have released energy equivalent to that of nearly two hundred atomic bombs, is thought to have been caused by a small asteroid no more than about 40 meters across. Atmospheric friction caused it to explode completely before it hit the ground, so it left no impact crater. We have also witnessed a far larger impact on another world: In 1994, astronomers recorded the impact of Comet Shoemaker–Levy 9 (SL9) as it slammed into Jupiter; the comet had broken into pieces before the impact, and each piece crashed into Jupiter with energy equivalent to that of a million hydrogen bombs (Figure 6.23). We've since observed the aftermaths of at least two more impacts on Jupiter.

For anyone who still doubted the reality of an ongoing threat, February 15, 2013, was a date to remember. Astronomers were already aware that a 40-meter-long asteroid would pass just 28,000 kilometers

FIGURE 6.22
Damage from the 1908 impact over Tunguska, Siberia, shown in a photo taken many years later.

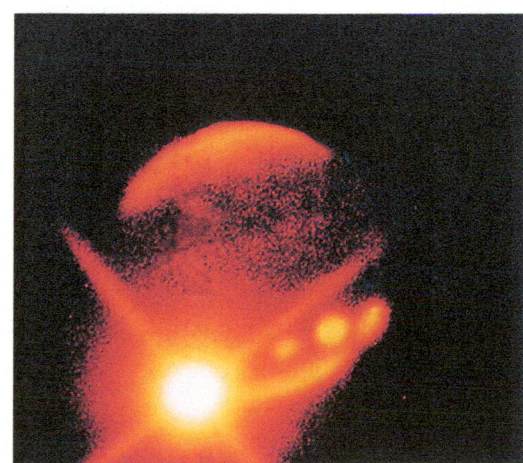

FIGURE 6.23
This infrared photo shows the brilliant glow of a rising fireball created when one of the pieces of Comet Shoemaker–Levy 9 (SL9) crashed into Jupiter (the round disk in the background) in 1994. Although overexposure exaggerates the size of the fireball, you can get a sense of scale by remembering that about ten Earths could fit side by side across Jupiter's diameter.

This photo shows the meteor trail of the previously unknown 10,000-ton asteroid that detonated in the sky above Chelyabinsk, Russia, on February 15, 2013. The explosion had the power of a 500-kiloton nuclear bomb and caused injuries to more than a thousand people.

above Earth's surface on that day. But as observers around the world prepared to watch the event, the people of Chelyabinsk, Russia, got a huge surprise: They suddenly saw a brilliant flash of light as a smaller, previously undiscovered asteroid entered the atmosphere above them at a speed of more than 60,000 kilometers per hour (Figure 6.24). Later estimated to have been about 20 meters long with a mass of 10,000 tons, the asteroid streaked across the sky as a giant meteor until friction with the atmosphere made it detonate with the power of a 500-kiloton nuclear bomb. More than a thousand people were injured, mostly by glass shattered by the shock wave. This was in some sense lucky: Had the asteroid's trajectory been directed more vertically toward the ground, the detonation would have occurred at lower altitude, causing much greater damage. The mid-air detonation left no large crater, but meteorites were found scattered across a wide area near Chelyabinsk. Note that, despite the close coincidence in timing, the 40-meter asteroid that passed safely and the Chelyabinsk asteroid were unrelated and approached on completely different orbits.

More impacts are virtually guaranteed to occur in the future. Figure 6.25 shows how often, on average, we expect Earth to be hit by objects of different sizes, based on geological data from past impacts. The good news is that we are highly unlikely to be hit by an asteroid as large as the one that killed the dinosaurs. Impacts of that size occur tens of millions of years apart on average, which means we face a far greater danger of doing ourselves in than of being done in by a large asteroid or comet. The bad news is that smaller impacts occur much more frequently. Objects a few meters across probably enter Earth's atmosphere about every week, and objects the size of the one that caused the Tunguska event probably strike our planet every few hundred years or so, making them a real threat. Until we more closely monitor the skies for potential impact threats, we cannot rule out the possibility of a very damaging impact.

Several efforts to search for potential impact threats are currently under way. As of 2015, astronomers have identified more than 12,000 asteroids with orbits that pass near Earth's orbit, including most or all of those larger than 1 kilometer in diameter, and none yet appear to pose an imminent threat. However, statistical estimates suggest that tens of thousands of smaller "potentially hazardous asteroids" still remain to be found. Comets pose a similarly unknown threat; while there are probably far fewer comets that might collide with our world, we are unlikely to see a comet plunging in from the outer solar system until it is well on its way, so our advance notice might be as short as a few years.

If we were to find an asteroid or a comet on a collision course with Earth, could we do anything about it? Many people have proposed schemes to save Earth by using nuclear weapons or other means to demolish or divert an incoming asteroid, but no one knows whether current technology is really up to the task. We can only hope that the threat doesn't become a reality before we're ready.

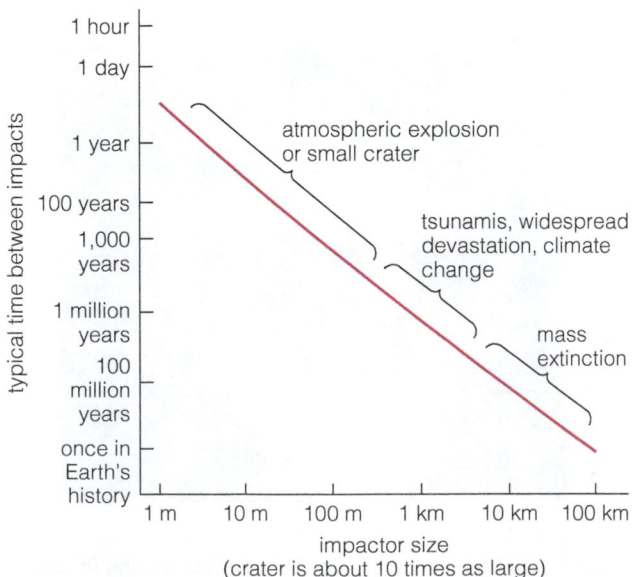

This graph shows that larger objects (asteroids or comets) hit Earth less frequently than smaller ones. The labels describe the effects of impacts of different sizes.

Think About It After the Chelyabinsk impact, the following statement circulated widely: "Meteors are nature's way of asking 'How's that space program coming?'" Comment on the meaning of this statement. How much time and money do *you* think we should be spending to counter the impact threat? Defend your opinion.

6.5 Human Evolution

We've traced the course of evolution from the origin of life through the extinction of the dinosaurs. We've seen that evolution took many surprising twists and turns to that point. The subsequent evolution of mammals and humans was just as interesting. In this section, we'll briefly investigate the pathway that led to our emergence as the first species on Earth capable of learning about its own origins.

How did we evolve?

We are primates, as are all the great apes, monkeys, and prosimians (such as lemurs). The ancestor of all of today's primates lived in trees, and many of the traits that make us so successful evolved as adaptations to tree life. For example, the limber arms that allow us to throw balls and work with tools evolved so that our ancestors could swing through trees, and our dexterous hands evolved to hang from branches and manipulate food. The eyes of primates are close together on the front of the face, providing overlapping fields of view that enhance depth perception—an obvious advantage when swinging from branch to branch. For the same reason, primates developed excellent eye–hand coordination.

Parental care is essential for young animals in trees, and primates evolved close parent–child bonds. These bonds, in turn, made it possible for primates to be born in a much more helpless state than the babies of most other types of animals. Although many primate species, including us, eventually moved down from the trees, most primates continue to nurture their young for a long time. This trait reaches its extreme in humans. Human babies are nearly helpless at birth and require parental care for more years than the offspring of any other species.

Movie Madness ARMAGEDDON

In 1994, a lot of people who believed that the dinosaurs went extinct thanks to encroaching mammals or simple lack of survival skills changed their minds. That was the year Comet Shoemaker–Levy 9 smacked into Jupiter, leaving entrance wounds the size of planet Earth. It was a graphic demonstration of cosmic catastrophe.

The public grasped that death by rock isn't all that improbable. If it happened to the dinos, it could happen to us. An errant asteroid a dozen miles across might someday careen into our planet and raise enough dust, and burn enough forests, to darken the world for years. We'd all slowly starve.

Alerted to this possibility for havoc and destruction, Hollywood lost little time in showing how ingenuity and some gutsy guys (with the emphasis on the latter) could save us even if Nature hurls a large space rock our way. Two theatrical films and a small torrent of TV specials soon appeared, showing Earth under mortal threat from ballistic boulders.

In the film *Armageddon*, the incoming object is as big as Texas, and a mere few weeks away. That's a real slap in the face for astronomers. Picture this: An asteroid as big as Ceres (the largest rock in the asteroid belt) is headed our way, and the astronomers only find the darn thing when it's as close as Mars?

The end of the world is nigh, but not to worry. NASA is in high gear to divert this king-sized clod and decides the best thing to do is to blow it into two large pieces with a nuclear bomb. Presumably, the two pieces will diverge slightly and sail harmlessly by on opposite sides of Earth. Needless to say, geeky NASA personnel aren't up to this kind of macho mission, so the space agency recruits a bunch of oil-rig roughnecks to plant and detonate the bomb. The NASA folk refer to this group of gritty misfits as "the wrong stuff."

In fact, it's a bad idea to try to blow up an incoming asteroid. The chances are that you'd only turn a single shell into buckshot. More practical schemes envision fastening some sort of rocket engine to the side of the rock and slowly nudging it out of the way. Another approach, at least for asteroids that we find when they are still far from Earth, is to paint the asteroid white and let the gentle pressure of sunlight do the job (light exerts a small force on anything it hits). Neither scheme involves roughnecks (or human pilots at all).

The threat, of course, is real. Astronomers estimate that rocks comparable to the one that obliterated the dinosaurs will slam into Earth roughly every 50–100 million years. But by carefully keeping tabs on those asteroids that cross Earth's orbit, we can see disaster coming. We'll have years to mount a defense.

The bottom line is that this is one kind of disaster we can probably avoid. After all, unlike the dinos, we've got a space program.

FIGURE 6.26

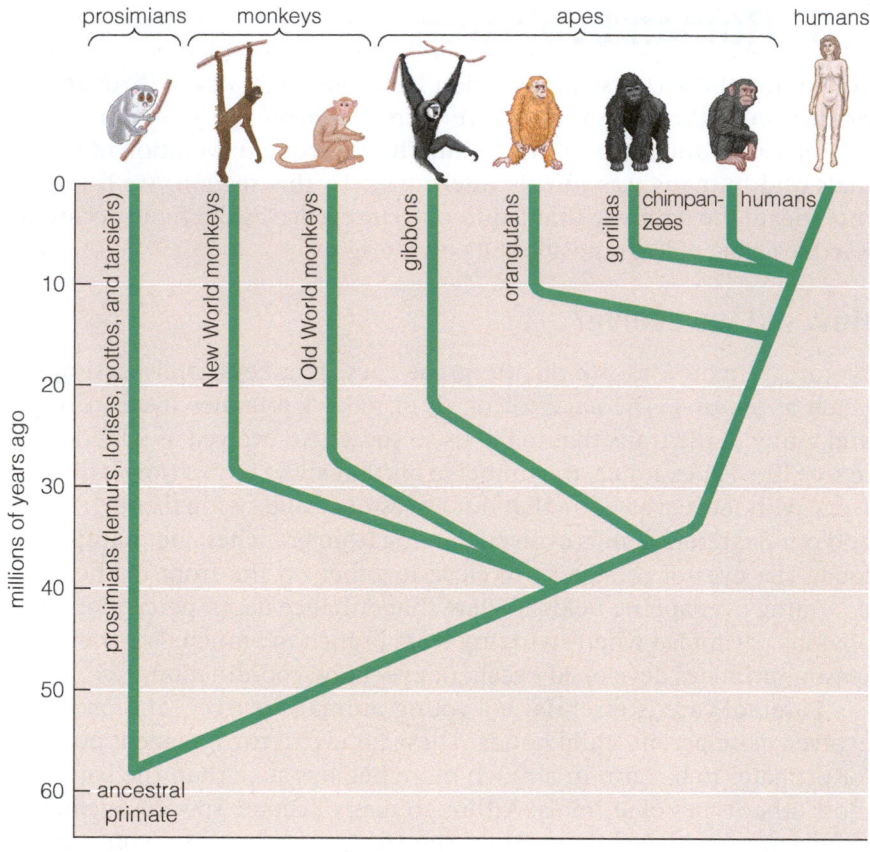

The evolutionary history of the major primate branches. Notice, for example, that the common ancestor of modern humans, chimpanzees, and gorillas lived between about 6 and 8 million years ago. (Adapted from Campbell, Reece, Simon, *Essential Biology*.)

FIGURE 6.27

This famous type of illustration suggests that humans evolved along a simple pathway from apes. However, it is almost completely wrong.

Contrary to a common myth, humans did not evolve *from* gorillas or other modern apes. Rather, modern apes and humans share a common ancestor that is now extinct. Figure 6.26 shows the evolutionary history of major primate branches. Our closest living relatives, chimpanzees and gorillas, shared a common ancestor with us just a few million years ago.

THE EMERGENCE OF HUMANKIND Even after hominids (human ancestors) diverged from the ancestors of chimpanzees and gorillas, human evolution followed a remarkably complex path. Indeed, one of the most pervasive but incorrect myths about human evolution is that it followed a simple pathway from stooped apes to upright humans (Figure 6.27).

The reality is that there have been numerous hominid species, some of which may be part of the lineage of modern humans and others that may have come to evolutionary dead ends. The oldest known fossil (as of 2015) that appears to be distinct from the lineage that led to chimpanzees and gorillas dates to between about 6 and 7 million years ago. This fossil, nicknamed Toumaï (officially called *Sahelanthropus tchadensis*), shows features intermediate between those of apes and humans (Figure 6.28). A stronger case for human ancestry is found in the fossils of the genus *Ardipithecus*, particularly fossils of *Ardipithecus ramidus*, nicknamed Ardi. Ardi lived about 4.4 million years ago, and reconstructions of Ardi fossils suggest at least partial upright walking. The case for upright walking is even stronger for the fossil known as Lucy (a female of the species *Australopithecus afarensis*), who lived about 3.2 million years ago.

The earliest fossil skulls that look essentially like those of modern humans are about 200,000 years old. However, even then our ancestors shared the planet with at least two other hominid species. The *Neandertals* were quite similar in appearance and brain size to *Homo sapiens*, and excavations of sites where they lived indicate they had culture, arts, and possibly religion and speech. The Neandertals disappeared for unknown reasons about 30,000 years ago, but their genes may still survive: Recent comparisons of human DNA with fossilized Neandertal DNA indicate that up to 4% of the modern human genome originated with the Neandertals, which means that *Homo sapiens* and Neandertals must have interbred. Another hominid species, called *Homo floresiensis* and discovered in 2004, lived on an Indonesian island as recently as about 12,000 years ago. These people apparently stood no more than about a meter tall, and for that reason have been nicknamed "hobbits." Figure 6.29 summarizes hominid lineages from our last common ancestor with other apes to modern *Homo sapiens*.

Deciphering the details of human ancestry is a rich field of research, and much remains subject to scientific debate. We will not discuss such details in this book, but before we leave the topic, it's worth dispelling two common myths. First, there is no "missing link" in human evolution. While a few mysteries may always remain, we now know enough from the geological record and genome comparisons to see a clear path from the earliest microbes to ourselves. Second, despite the many species of hominids that have come and gone, all modern humans are members of the same species. That is, while people often focus on outward differences between races, such as skin color or hair texture, all human genomes are nearly identical. Moreover, most of the small racial differences that might once have arisen have since been spread across races by the extensive interbreeding of our ancestors. The remaining genetic differences between human races are generally much smaller than the genetic variation among the individuals within each race.

FIGURE 6.28

This computer reconstruction, based on the actual fossils, shows the skull of *S. tchadensis*, or Toumaï, first excavated in Chad in 2001.

FIGURE 6.29

This time line shows some of the likely ancestors of modern humans (based on known fossils) over about the past 7 million years, since our lineage separated from that of modern gorillas and chimpanzees. To keep the diagram simple, not all known species are shown.

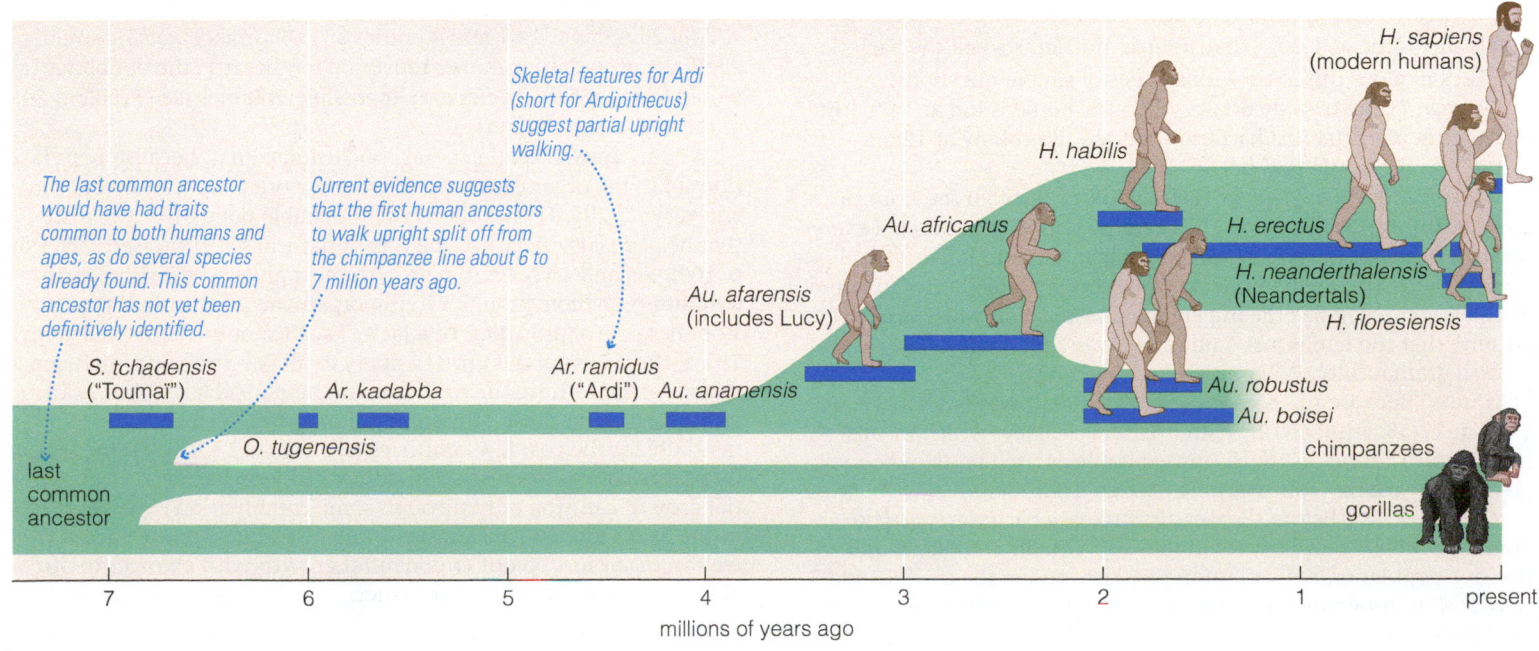

IMPLICATIONS FOR EXTRATERRESTRIAL INTELLIGENCE The fact that modern gorillas, chimps, and humans all evolved from the same ancestor has at least two important implications for understanding the possibility of extraterrestrial intelligence. First, it shows that relatively small genetic differences can make a big difference in species success. More than 98% of the DNA sequences that make up the human genome are identical to the sequences that make up the chimpanzee genome. That is, a relatively small genetic difference is all that separates our success on this planet from the current predicament of chimpanzees, which survive naturally in only a few isolated locations in Africa. Second, it suggests that the evolution of intelligence is a complex process. Gorillas and chimpanzees have been evolving from our common ancestor just as long as we have, but we are the only species building cities and radio telescopes. This fact raises the question of whether advanced intelligence is an inevitable outcome of evolution. We'll discuss this question further in Chapter 12.

Are we still evolving?

Given the very recent arrival of *Homo sapiens* on the evolutionary time scale, it's natural to wonder whether we are still evolving. Recent discoveries suggest that humans have continued evolving throughout our time on this planet. Nevertheless, the changes during the past 10,000 to 40,000 years have probably been relatively small (though a few scientists argue for more substantial changes). If we could sequence the genome of a human from 40,000 years ago, it would be difficult to distinguish it from the genome of a person living today. Nevertheless, we have clearly gone through dramatic changes as a species.

Cosmic Calculations 6.2   BACTERIA IN A BOTTLE II: LESSONS FOR THE HUMAN RACE

Recall the thought experiment of the bacteria in a bottle in Cosmic Calculations 6.1, in which we start with a single bacterium at 12:00 that replicates each minute. Suppose that bacterial growth completely fills the bottle at 1:00, exhausting all nutrients so all the bacteria die. Let's explore this issue with a series of simple questions.

Question 1: The tragedy occurs when the bottle is full at 1:00, just 60 minutes after the first bacterium started the colony at 12:00. When was the bottle *half*-full?

Answer: Most people guess 12:30, halfway through the hour of growth. But this is incorrect: The bacterial population doubled every minute, so the bottle went from half-full at 12:59 to full at 1:00.

Question 2: You are a mathematically sophisticated bacterium, and at 12:56 you recognize the impending disaster. You warn your fellow bacteria that the end is just 4 minutes away unless they slow their growth dramatically. Will anyone believe you?

Answer: Because the bottle was full at 1:00, it was $\frac{1}{2}$-full at 12:59, $\frac{1}{4}$-full at 12:58, $\frac{1}{8}$-full at 12:57, and $\frac{1}{16}$-full at 12:56—which means there's still 15 times as much unused bottle as used bottle when you give your warning. Your warning may go unheeded.

Question 3: Just before the disaster strikes, a bacterial space program discovers 3 more bottles in the lab. With an immediate and massive population redistribution program among the original and 3 new bottles, how much more time will the bacteria buy?

Answer: You may be tempted to think that 3 more bottles should give the colony 3 more hours, but in fact it gives them only 2 more minutes: Because the growth occurs through doubling, the colony will fill 2 bottles at 1:01 and 4 bottles at 1:02. In fact, *nothing* could allow the growth to continue much longer, because the doublings would soon lead the bacteria to impossible volumes (see Problem 54 at the end of the chapter).

We can draw several general conclusions. First, because populations of living organisms tend to grow exponentially, numbers can rise very rapidly. This explains the inevitable population pressure that helped Darwin realize the role of natural selection (see Fact 1 on p. 160). Second, exponential growth must always be a short-term, temporary phenomenon; for living organisms, the growth typically stops because of predation or a lack of sufficient nutrients or energy. Third, these laws about growth apply to all species—our intelligence cannot make us immune to simple mathematical laws. This is a critical lesson, because human population has been growing exponentially for the past few centuries (see Figure 13.14). Of course, our intelligence gives us one option not available to bacteria. Exponential growth can stop only through some combination of an increase in the death rate and a decrease in the birth rate. Unlike bacteria, we can *choose* to stop our exponential growth with changes to our birth rate before we "fill" our planet.

CULTURAL AND TECHNOLOGICAL EVOLUTION These dramatic changes are not due to biological evolution, but rather to what we might call **cultural evolution**—changes that arise from the transmission of knowledge accumulated over generations. In other words, we humans can transmit our history, using both spoken and written language, which allows us to learn from what has been done before. The know-how to build tractors, computers, and spaceships is stored not in our genes but in the cumulative product of hundreds of generations of human experience. Although some other species, including chimpanzees, demonstrate aspects of culture, humans appear to be unique (on Earth) in having reached the point where cultural evolution is far more important to our changing nature than is biological evolution.

Because biological evolution is driven by random mutations, it tends to proceed at a slow and relatively steady rate. Cultural evolution, in contrast, tends to accelerate over time. The development of agriculture and written language took tens of thousands of years, while less than two centuries separate the beginning of the industrial revolution from the first walk on the Moon. More recently, we have begun to develop another new type of evolution that is accelerating even more rapidly—**technological evolution.**

You are probably familiar with the way new computers get faster each year. These speed increases come from an interesting coupling of technology and science: Increased computing power enables scientists to make new discoveries, which in turn lead to further increases in computing power, and so on. Similar couplings of technology and science affect almost every field of human knowledge. One of the most striking cases involves the way technology is helping us understand biological evolution. As we've discussed, much of our current knowledge about the origin and evolution of life comes from studies of genomes that are made possible by technological advances such as machines that can read DNA sequences.

EVOLUTION AMONG HUMANS AND OTHER INTELLIGENT CIVILIZATIONS The same type of technology that allows us to read genomes also now allows us to reengineer living organisms (genetic engineering) in such a way that we may soon outpace nature in developing new species. It seems inevitable that we will also develop the capability to change human DNA, thereby opening the door to attempts to "improve" our species. The moral and ethical dimensions of this power are already profound, and you will undoubtedly have to confront these issues many times during your life. For our purposes in this book, perhaps the primary lesson is that advanced civilizations can alter the course of evolution through their choosing, rather than remaining subject to the random processes of natural selection. Indeed, modern medicine has already taken us out of the realm of Darwinian evolution, because we routinely save individuals who would have died earlier in generations past. The course of our future evolution is in our own hands. If other advanced civilizations exist, they must similarly control their own destinies.

 **THE PROCESS OF SCIENCE IN ACTION**

6.6 Artificial Life

In this chapter, we have discussed some of the laboratory experiments through which scientists seek to understand the origin of life. Some researchers are attempting to go even further, by trying to create life in the lab. Their efforts are rather different from those of the fictional Dr. Frankenstein, who sewed together dead body parts and jolted a living, human-like creature into existence with high voltage. Today's researchers are trying to put together novel organisms that can reproduce and grow, but on a microbial scale.

Success could have a variety of implications. Many of these could be practical applications, such as improving on the living things that evolution has produced: for example, plants that need less water to grow or require fewer insecticides. Another payoff from artificial life studies would be to help us understand how life got started on Earth. If the creation of life in the lab involves only straightforward chemical reactions, we might conclude that life is not a rare phenomenon and might spring up on any world where the conditions were favorable. This research might also aid us in our search for life in extraterrestrial environments, such as subsurface aquifers on Mars or a subterranean ocean of Europa. Moreover, we might be able to build organisms that could provide medical and other benefits. Of course, it might also be possible to create dangerous organisms, forcing us to confront the ethical dilemmas of our work. For this chapter's case study in the process of science in action, we turn our attention to current work in artificial life.

How can we create artificial life?

It's still far beyond our means to create a bacterium that's comparable to those found in nature from elementary chemicals. Nonetheless, hundreds of researchers are working to spawn A-life ("artificial life") by either rearranging bits and pieces of existing organisms (often called *genetic engineering*) or trying to build an extremely simple living cell in the lab (known as *synthetic biology*). The former scheme is referred to as a "top-down" approach and is similar to hot-rodding, where a car is stripped down and rebuilt with different components, producing a vehicle that differs from any existing auto. The latter is more akin to building a car from scratch.

ENGINEERING NEW SPECIES FROM EXISTING ORGANISMS Craig Venter, the man behind the private-sector effort that first sequenced the human genome, began a top-down program to make designer organisms in 1995. His goal is to create microbes that do useful things, such as fighting malaria or converting atmospheric carbon dioxide into methane, a trick that could be valuable for reducing our dependence on fossil fuels. His group succeeded in creating the first artificial organism in May 2010.

Venter's basic approach is to start with an existing species of bacteria that has a relatively small genome. Even then, some of the genes have functions that go beyond basic survival, so Venter strips out as many genes as he can while still leaving the cells viable and able to reproduce. He then tries to build up this "minimalist" genome using short sequences of DNA that he buys from a supply house. The length of these segments is typically 1000 base pairs. He assembles hundreds of these segments to build up a replica of the original bacteria's genomic structure, which had been sequenced earlier. To keep tabs on this artificially produced genetic material, he inserts "watermarks" into its base pairs, such as the coded names and email addresses of colleagues.

When the synthetic genome is complete, he inserts it into the cell body of another bacterium, *Mycoplasma mycoides,* whose own genome has been removed. The modified bacterium then "boots up," comes to life, and starts functioning as a naturally occurring cell. It even reproduces. In this way, Venter has created a new life-form, using factory-supplied DNA segments. He called his first success *Mycoplasma mycoides JCVI-syn1.0;* the *JCVI* stands for J. Craig Venter Institute. Although this new genome does not have any particularly useful functions, it is a proof of concept upon which Venter hopes to build in the future.

Note that this is not really creating life from scratch, but merely streamlining and modifying an existing microbe. It's a more efficient method for producing desirable organisms than simply breeding them and selecting those that have the desired properties. That scheme depends on mutation to produce a range of characteristics, and you simply cull those that are, by chance, closer to what you want. Venter's approach would permit the deliberate introduction of genes that will result in the desired behavior. Top-down engineering promises to give us entirely new species of great practical value, and it will also undoubtedly teach us much about how cells work. However, it's less likely to clarify how life got started on Earth, or to help us understand whether terrestrial biology is a fortunate accident or a virtually inevitable development. The bottom-up approach to A-life addresses those questions more directly.

MAKING LIFE FROM RAW INGREDIENTS As we discussed in Section 6.2, it seems likely that the young Earth had plenty of building blocks for life. Indeed, the many possible ways in which the building blocks could have formed or arrived has led some people to believe that the creation of life in the laboratory should be just around the corner.

Few researchers think it's that easy, but several have dared to take a bottom-up approach to making A-life in the lab. One major effort is being pursued by Jack Szostak, a Harvard University geneticist who initially studied the genes of yeast. He decided to switch gears when he read of the work of Thomas Cech, who discovered that RNA could serve as both a blueprint for reproduction and a catalyst for making proteins. Cech suggested that RNA may have been the basis of life before DNA arrived on the scene, an idea we discussed earlier as "RNA world." This idea encouraged Szostak to leave his yeast cells behind and try building synthetic organisms out of strands of this older nucleotide.

There were already research hints as to how RNA life might have arisen, since prior lab experiments had shown that short strands of RNA can form when a dilute organic soup washes over the surface of clay or rock. If, by chance, one of those RNA strands could reproduce, it would have staying power. Those that could do this most quickly, and with the fewest number of copying errors, would soon dominate their environment. In this way, a robust form of RNA life could have evolved. But did it really happen that way?

Szostak seeks to gain insight into this question by attempting to produce simple RNA-based cells that can replicate. If he succeeds, he will have created a new life-form after beginning with only nonliving materials—a rather different accomplishment than the A-life efforts described above, which mimicked an existing genome. He begins with some material containing thousands of trillions of short RNA fragments, chosen from simple RNA strands known to have special talents. They can, for example, grab onto a particular type of molecule or duplicate parts of it. He then puts his fragment collection into a test tube, and uses a technique called "in vitro selection" to fast-forward chemical evolution. He gives his collection the opportunity to chemically react, and then screens the samples to find, for example, those that have managed to replicate a short sequence of RNA. Keep in mind that these are merely organic molecules: They're not life. Szostak then filters out the winners in this test, and makes trillions of copies of those (some copy errors, or mutations, are allowed in this process). He runs the experiment again with the imperfectly cloned winners. Those that are even better, or at least faster, are retained for the next round—and so forth through dozens of cycles. In essence, Szostak is trying to emulate what might have happened on the early Earth by compressing time with laboratory evolution. It is somewhat analogous to producing large kernels of corn by repeatedly using seeds from only the largest ears of the crop.

So far, Szostak's lab has made strands that can replicate other RNA sequences, although they are short and the replication isn't always very accurate. The big prize is an RNA strand that can really sift through the material of an organic soup and build up a copy of itself. There seems little doubt that this will take place eventually. Then you could watch this self-replicating molecule evolve in the laboratory, because the sample would soon be dominated by the type of RNA that made copies of itself, rather than of other RNA. Presumably, it might soon evolve the ability to produce enzymes or other components that would help it function better.

We've noted that freely floating organic molecules aren't cells. The molecules Szostak hopes to build need to be confined by a wall to keep aggressive compounds in the cruel outside world from dismantling them or diluting their concentration, while at the same time allowing the building blocks necessary for reproduction to enter their mini-habitat. Consequently, Szostak also seeks natural processes that could make precells to enclose the RNA. He has already had some success: When he mixes his RNA strands with fatty acids (waxy substances such as the oleomargarine you spread on your toast), some of the RNA gets trapped in tiny bubble-like membranes (vesicles). While far less complex than a modern cell wall, these simple vesicles provide a protective space for the RNA inside (Figure 6.30). Szostak has found a fatty acid that would have been present in the ancient seas of Earth and is just porous enough to permit the molecular building blocks necessary for RNA reproduction to enter, while keeping the larger RNA molecules caged inside.

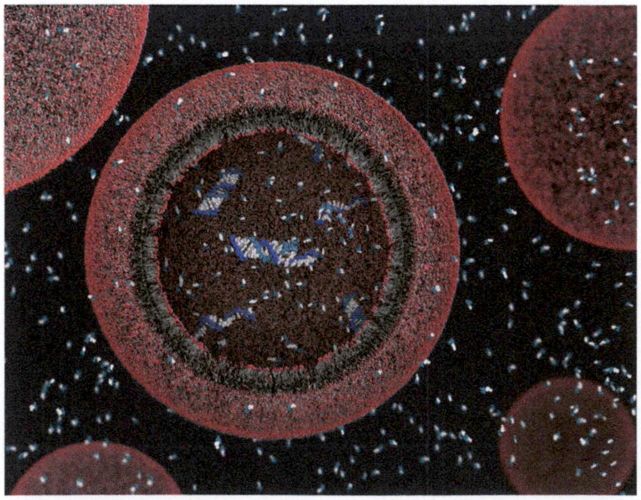

Szostak (who won the 2009 Nobel Prize in physiology) is wrestling with the problem of producing self-replicating DNA without having to introduce enzymes—specialized protein molecules that greatly speed up chemical reactions. So far, he hasn't yet produced RNA that will fully reproduce itself. But he reckons that creation of a replicating strand is not only within reach, but possibly just a few years away. If so, we may soon know whether the origin of life could have occurred as easily as we have imagined under the conditions that existed on the young Earth.

Another potentially important development in building artificial life is the recent engineering of two new DNA bases. As we saw in Section 5.4, life on Earth uses the four bases A, T, C, and G to build genes, but researchers have formulated two additional bases that can be incorporated into DNA and will replicate. They are also recognized by the enzymes that facilitate the process of translation by RNA. While this may seem like no more than an interesting chemical trick, having six bases rather than four would allow an organism to use more than the normal number of amino acids in constructing proteins, opening up more possibilities for

FIGURE 6.30

This illustration shows the idea behind Jack Szostak's approach to building artificial life from raw ingredients. The rendering represents short strands of RNA encapsulated in a pre-cell, where they can combine with free-floating nucleotide building blocks and, perhaps, eventually give rise to a self-replicating RNA.

biological function. It also suggests that life elsewhere might not be limited to the four base pairs that are found in terrestrial genomes.

Should we create artificial life?

The payoff in Venter's work to produce what are essentially self-replicating nanobots for specific tasks is obvious. We've mentioned the possibility of generating hydrocarbons for fuel, but A-life could also be used to target cancer cells or clean up toxic waste. Still, there's the danger that someone might eventually use this technology to engineer deadly organisms that have no natural enemies, a potent agent for biowarfare. Although Szostak's work is less intent on creating useful life and more focused on learning how biology began, it too poses a potential risk for misuse.

These dangers are recognized by the researchers themselves. While they're not overly worried about the threat A-life might pose (they point out that it would be extremely fragile and would have a difficult time living outside the laboratory environment), the scientists have occasionally convened panels to consider the ethical implications of their work. Do humans have the moral right to create new types of life? We don't seem to mind the development of hybrid corn or improved cattle (although many consumers are resistant to genetically modified foodstuffs), but the production of an entirely new species is likely to be more controversial. In 2015, the World Economic Forum recognized the potential of synthetic biology not only to solve many pressing problems but also to introduce hard-to-foresee risks.

One possible way to reduce threats from A-life is to engineer it to be vulnerable in any environment that is not artificial. As an example, scientists have altered the DNA in samples of *Escherichia coli* (*E. coli*), a bacterium that resides in your gut and is one of the best studied of all organisms. They rebuilt the *E. coli* genetic code so that the bacterium makes proteins that include an amino acid that's not found in naturally occurring life. In the lab, the modified bacteria can be "fed" this unusual amino acid, but in the wild they would be unable to find it, and therefore wouldn't survive. Constructing A-life that has such slightly altered biochemistry is a way to keep it from escaping from research environments and causing either accidental or deliberate harm.

There's an alternative to the A-life scenario that avoids most of these ethical dilemmas and risks to the public, and that is to build virtual life by modeling the functions of living cells using computer software. (Some people have claimed that computer viruses are a kind of synthetic life—"vandalware" that can manipulate its host environment and reproduce.) Describing the behavior of cells with software might not seem particularly interesting, but it would allow us to do biological experiments at the keyboard, without the difficulty and potential danger of using real microbes. It's somewhat analogous to testing aircraft with computer simulations of the air flow over their wings and fuselage, as opposed to building scale models and putting them in wind tunnels.

Software life could be very useful for determining the effects of new drugs, or even for predicting the consequences of removing or adding genes (genetic engineering). *E. coli* is frequently used in this research. But it has more than 4000 genes, of which approximately 1000 have functions that are still not understood. Consequently, creating a complete computer model of a real *E. coli* is still beyond us. However, borrowing a leaf from the top-down researchers, computer scientists are now programming a stripped-down model of the bacterium, with only about 1000 genes. This effort might bear fruit in the very near future, and even though the model is only an approximation to the actual microbe, it might still be useful in helping us understand how life works. Building a complete computer model of *E. coli,* while not on the immediate horizon, would surely encourage scientists to modify it to more closely resemble human cells. Suppose that could be done. Then, even aside from the insight it would give into how biology functions, it would permit us to, for example, quickly evaluate drugs for fighting cancer without the necessity of testing them on laboratory animals or people.

Of course, even if the life is only programmed on a computer, once we know the necessary DNA sequences it would be possible to put it together for real. One way or other, we are likely to be forced to confront the ethical dilemmas of creating artificial life. Indeed, with all the research that is already under way, it may be too late to put the genie back in the bottle, even if we wanted to. The future of biological science—and perhaps of our species—will depend largely on the ethical choices that we make as modern biotechnology continues its technological evolution.

The Big Picture

PUTTING CHAPTER 6 IN PERSPECTIVE

In this chapter, we have completed our overview of life on Earth. We have built on our understanding of Earth's habitability (Chapter 4) and of the nature of life (Chapter 5) to develop a modern picture of its origin and evolution. As you continue in your studies, keep in mind the following "big picture" ideas:

- We may never know precisely how life arose on Earth. However, we have found *plausible* scenarios for the origin of life based on natural, chemical processes. These scenarios are based on solid evidence found in the geological record, in comparisons of the genomes of different species, and in laboratory experiments.

- Earth has supported life for most or all of the past 4 billion or more years, but life remained microbial for most of this period. Animal life rapidly diversified only about 542 million years ago (with the Cambrian explosion), and plants colonized the land only about 475 million

years ago. If aliens had observed Earth during about the first 90% of its history, they would have found a planet that was home to nothing more than microscopic species.

- The course of evolution has been drastically changed at least several times by mass extinctions, and at least one of these mass extinctions is clearly linked to an asteroid or comet impact. As we'll discuss in more detail later in the book, the likelihood of impacts must be an important consideration in assessing the habitability of other planets.

- From the time of the first living organism to today, the evolution of life on our planet has been shaped by natural selection. However, we have developed or are on the verge of developing the capability to engineer existing species, including ourselves, and perhaps to create entirely new species. This power must be available to any advanced civilization and therefore is an important consideration in the search for extraterrestrial intelligence.

Summary of Key Concepts

6.1 Searching for Life's Origins

When did life begin?

Stromatolite and sulfur isotope evidence tells us that life existed by about 3.5 billion years ago; microfossil evidence is consistent with this view. Carbon and other isotopic evidence may push the time back to more than 3.85 billion years ago. Life must have existed even earlier than the oldest fossil evidence of it, though we do not know exactly how much earlier.

What did early life look like?

Although early life is long gone, genetic comparisons allow us to determine which modern organisms are evolutionarily oldest, suggesting that they are most similar to early life. These studies suggest that early life looked like some species of bacteria and archaea, though we cannot yet be more specific.

Where did life begin?

The origin of life would have required a source of chemical energy, leading scientists to suggest warm ponds, volcanic hot springs, or deep-sea vents. No one knows which is most likely, although impacts make it probable that a common ancestor lived in the deep ocean or underground.

6.2 The Origin of Life

How did life begin?

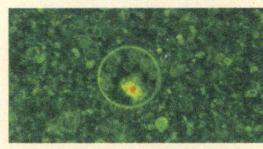

According to laboratory studies, the most likely scenario holds that organic molecules, either produced chemically or brought here from space, were found in ocean locations where clay and other minerals were common. Clay helped catalyze the building of RNA strands that became enclosed in lipid pre-cells. Some RNA molecules were able to partially or completely self-replicate, allowing natural selection among them to improve their replication until true life emerged.

Could life have migrated to Earth?

If life originated first on Venus or Mars, it may have migrated to Earth when impacts blasted rocks from one world to another. Meteorites from Mars show that life could in principle make the journey, and the durability of some microbes might allow them to take hold near deep-sea vents or elsewhere. Longer migrations, such as from planets around other stars, are highly unlikely.

6.3 The Evolution of Life

What major events have marked evolutionary history?

Life probably diversified rapidly after its origin, but remained microscopic for more than 2 billion years. Keys to the eventual transition included the origin of oxygen-producing photosynthesis, which released the oxygen now in our atmosphere, and the evolution of cell nuclei and other complex structures in eukaryotes. Multicellular animals diversified in the **Cambrian explosion,** starting about 542 million years ago. Plants and animals migrated onto land not long after.

Why was the rise of oxygen so important to evolution?

Aerobic processes, using oxygen, offer the possibility of much more efficient cellular energy production than anaerobic processes, and thus can lead to much greater evolutionary diversification. The precise timing of the rise of oxygen is not well known, but it began before about 2.5 billion years ago and probably did not reach levels near those of the present until the time of the Cambrian explosion or later.

6.4 Impacts and Extinctions

Did an impact kill the dinosaurs?

It may not have been the sole cause, but a major impact clearly coincided with the **mass extinction** that killed the dinosaurs about 65 million years ago. Sediments from the time contain iridium and other evidence of an impact, and a crater of the right age lies buried beneath the Yucatán coast of Mexico.

What caused other mass extinctions?

At least five mass extinctions have occurred in the past 500 million years. Although only the K–T extinction is clearly linked to an impact, it's plausible to imagine that impacts have played a role in other extinctions. The end-Permian extinction has been linked to extreme volcanism that was accompanied by severe climate change. Other possible causes of mass extinctions include changes in mutation rates, possibly influenced by changes in Earth's ozone layer or magnetic field, and radiation from distant supernovae or gamma-ray bursts.

Is there a continuing impact threat?

Impacts certainly pose a threat, though the probability of a major impact in our lifetimes is small. Impacts like the Tunguska event occur more frequently (about once a century) and would be catastrophic if they occurred over cities.

6.5 Human Evolution

How did we evolve?

Humans share a common ancestor with modern gorillas and chimpanzees, an ancestor that lived between about 6 and 8 million years ago. The earliest *Homo sapiens* emerged about 200,000 years ago.

Are we still evolving?

Genetically, humans have probably changed little in at least the past 40,000 years. However, we are now changing in new ways, through **cultural evolution** and **technological evolution.**

 THE PROCESS OF SCIENCE IN ACTION

6.6 Artificial Life

How can we create artificial life?

Scientists seek to make artificial life in two basic ways. A "top-down" approach starts with existing organisms and genetically strips them down, then transplants a synthetic version of this genome into a new species. A "bottom-up" approach starts in the laboratory with the raw ingredients of life and seeks to reproduce life in much the same way that it presumably originated billions of years ago.

Should we create artificial life?

The desirability of artificial life is a subject of great debate, even though it could offer important practical benefits. In addition to having moral concerns, some people worry that artificial life might be the source of new diseases or toxins to which terrestrial life would have little resistance.

Exercises and Problems

MasteringAstronomy® *For instructor-assigned homework and other learning materials, go to MasteringAstronomy®.*

REVIEW QUESTIONS

Short-Answer Questions Based on the Reading

1. What are the three lines of fossil evidence that point to an early origin of life on Earth? Discuss each line and what it tells us about when life arose. What are the implications of an early origin for the possibility of life elsewhere?

2. How do studies of DNA sequences allow us to reconstruct the evolutionary history of life? What living organisms appear to be most closely related to the common ancestor of all present life?

3. Based on current evidence, what locations on Earth seem likely for the origin of life? What locations can we rule out?

4. What was the *Miller–Urey experiment*, and how did it work? Why is its relevance now subject to scientific debate? How else might Earth have obtained the organic building blocks of life?

5. What do we mean by an "RNA world," and why do scientists suggest that such a world preceded the current "DNA world"?

6. Briefly summarize current ideas about the sequence of events through which life may have originated on Earth. What role(s) might clay or other inorganic materials have played?

7. Briefly discuss the possibility that life migrated to Earth. Also discuss the possibility that Earth life might have migrated to other worlds, and the implications of migration to the search for life elsewhere.

8. Why do we think that evolution would have proceeded rapidly at first, and what fossil evidence supports this conclusion?

9. Briefly discuss the early evolution of life, from the first organisms to the development of photosynthesis and oxygen production.

10. How do we think that eukaryotes evolved? What time constraints can we place on when eukaryotes first got cell nuclei?

11. What was the *Cambrian explosion*? Briefly discuss ideas about what might have caused it and why no similar event has happened since.

12. How and when did life colonize land? Why did it take so long after the origin of life in the oceans?

13. How do we know that the early Earth could not have had an oxygen atmosphere? Where did the oxygen in our atmosphere come from? How did the introduction of oxygen affect early life?

14. Summarize the history of the oxygen buildup as it is understood today, and describe key mysteries that still remain. When did oxygen reach current levels?

15. What was the *K–T impact*, and how is it thought to have led to the demise of the dinosaurs? What evidence supports this scenario? How did this event pave the way for our existence?

16. Briefly discuss the evidence for other mass extinctions, and list a few of their possible causes.

17. Discuss the threat that future impacts may pose to us and our planet, and how we know that the threat is real.

18. Describe several adaptations that evolved so primates could live in trees and that have proved useful to us as humans.

19. When did hominids arise, and when did modern humans arise?

20. Briefly describe and clarify a few common misconceptions about human evolution.

21. What do we mean by *cultural and technological evolution*? What implications do they have for extraterrestrial intelligence?

22. Briefly describe two main approaches to creating artificial life. Then describe the possibility and potential benefit of constructing computer programs that can mimic the biological functions of cells, and the ethical aspects of making artificial life.

TEST YOUR UNDERSTANDING

Would You Believe It?

Each of the following statements describes a hypothetical future discovery. In light of our current understanding of Earth and evolution, briefly discuss whether each discovery seems plausible or surprising. Explain clearly; because not all of these have definitive answers, your explanation is more important than your chosen answer.

23. We discover evidence of life, in the form of a particular ratio of carbon-12 to carbon-13, in rock that was originally formed in sediments and is 3.9 billion years old.

24. We discover an intact fossil of a eukaryotic cell, with a cell nucleus, that is 3.0 billion years old.

25. We discover a preserved, 3.5-billion-year-old microfossil that apparently had a genome identical to that of many modern animals.

26. We discover clear evidence that life arose on a high mountain-top, not in the oceans.

27. We discover a fossil of a large dinosaur that lived approximately 750 million years ago.

28. We discover that, contrary to present belief, oxygen was abundant in Earth's atmosphere at the time when life arose.

29. We discover a crater from the impact of a 10-kilometer asteroid that dates to about 2500 years ago.

30. We discover an asteroid about 300 meters across that is on a collision course with Earth.

31. We find fossil remains of an early primate that lived about 50 million years ago and was, from all appearances, identical to a modern gorilla.

32. The first life created in the laboratory has an RNA genome, rather than a DNA genome.

Quick Quiz

Choose the best answer to each of the following. Explain your reasoning with one or more complete sentences.

33. The origin of life on Earth most likely occurred (a) before 4.5 billion years ago; (b) between about 4.5 and 3.5 billion years ago; (c) between about 3.0 and 2.5 billion years ago.

34. The earliest living organisms probably were (a) cells without nuclei that used RNA as their genetic material; (b) cells with nuclei that used RNA as their genetic material; (c) cells with nuclei that used DNA as their genetic material.

35. The importance of the Miller–Urey experiment is that (a) it proved beyond doubt that life could have arisen naturally on the young Earth; (b) it showed that natural chemical reactions can produce building blocks of life; (c) it showed that clay can catalyze the production of RNA.

36. "RNA world" refers to (a) the possibility that life migrated from Mars; (b) the idea that RNA was life's genetic material before DNA; (c) the idea that early life was made exclusively from RNA, needing no other organic chemicals.

37. Early life arose in an oxygen-free environment, but if any of these microbes had somehow come in contact with oxygen, the most likely effect would have been (a) nothing at all; (b) to increase their metabolic rates; (c) to kill them.

38. The oxygen in Earth's atmosphere was originally released by (a) outgassing from volcanoes; (b) plants; (c) cyanobacteria.

39. The *Cambrian explosion* refers to (a) a dramatic increase in animal diversity beginning about 542 million years ago; (b) the impact that killed the dinosaurs; (c) the sudden emergence of eukaryotic life in the fossil record dating to about 2.1 billion years ago.

40. Which statement about Earth's ozone layer is true? (a) It formed only after the atmosphere became rich in oxygen. (b) It has existed since life first arose on Earth. (c) It first formed a few hundred million years after life colonized the land.

41. The hypothesis that an impact killed the dinosaurs seems (a) well supported by geological evidence; (b) an idea that once made sense but now can be ruled out; (c) just one of dozens of clear examples of impacts causing mass extinctions.

42. According to the fossil evidence, modern humans (a) evolved from chimpanzees; (b) evolved on a lineage that split from other apes 6 million years ago or more; (c) lack any known ancestors during the past few million years.

PROCESS OF SCIENCE

43. *Origin of Life Studies.* We cannot go back in time to see exactly how life first originated on Earth, which means that no matter how much we learn about ways in which life *might* have originated, we may never know for sure the path that was actually followed. Given this fact, can we still address the question of the origin of life scientifically? Defend your opinion.

44. *Unanswered Questions.* Choose one important but still not fully answered question about the history of life on Earth, and write two or three paragraphs in which you discuss how we might answer this question in the future and what the answer might mean for the study of life in the universe. Be as specific as possible.

GROUP WORK EXERCISE

45. *Looking for Life on Mars.* **Roles:** *Scribe* (takes notes on the group's activities), *Proposer* (proposes ideas to the group), *Skeptic* (points out weaknesses in proposed ideas), *Moderator* (leads group discussion and makes sure everyone contributes). **Activity:** Your group is a science team designing a lander that will drill beneath the surface of Mars looking for evidence of microbes, dead or alive. What sorts of evidence would you look for, and what experiment(s) would you propose that could provide "smoking gun" (i.e., highly convincing) evidence that Mars has life? How would you decide whether the life was indigenous to Mars or possibly migrated from Earth? Write a one-page summary of your mission plan.

INVESTIGATE FURTHER

In-Depth Questions to Increase Your Understanding

Short-Answer/Essay Questions

46. *A Brief History of Life on Earth.* Take all the ideas about the origin and evolution of life on Earth and try to condense them into a one- to three-page essay on the history of life on Earth. Or, if you prefer, try to capture the ideas in a poem.

47. *Geology and Life.* In Chapter 4, we discussed the role of plate tectonics and the CO_2 cycle in climate regulation on Earth. Suppose neither of these processes had ever operated. Could life still have arisen on Earth as discussed in this chapter? If so, how far could evolution have progressed before the lack of climate regulation would have blocked further major developments? Write a one-page essay summarizing and explaining your answers.

48. *Keys to Our Existence.* Identify and describe four crucial events in evolutionary history without which our current existence would have been highly unlikely. Explain your reasoning clearly.

49. *Extinction and Oxygen.* Suppose we somehow killed off a large fraction of the photosynthetic life on Earth. What consequences would this have for the oxygen content of our atmosphere? Explain your reasoning.

50. *Impact Movie Review.* Watch one of the Hollywood movies concerning the threat of an impact to our civilization, such as *Deep Impact* or *Armageddon*. Based on what you have learned in this chapter, write a one- to two-page critical review in which you include discussion of whether the impact scenario is realistic.

51. *Artificial Life Review.* Numerous science fiction stories and movies involve the creation of artificial life. Review one such story or movie, and identify at least three ideas in it that either do or do not meet the standard of being testable by science. Describe each in detail.

Quantitative Problems

Be sure to show all calculations clearly and state your final answers in complete sentences.

52. *Bacterial Evolution.* Suppose that a mutation occurs in about 1 out of every 1 million bacterial cells, and suppose that you have a bacterial colony in a bottle like that described in Cosmic Calculations 6.1 (in which the bacteria divide each minute). Given the number of bacteria in the bottle after 1 hour, calculate approximately how many bacteria would have some type of mutation. What does this tell you about why bacteria often evolve resistance to new drugs?

53. *Deep in Bacteria.* In Cosmic Calculations 6.1, we calculated that the volume of bacteria after 120 doublings would be 1.3×10^{15} m^3. The total surface area of Earth is about 5.1×10^{14} m^2. Use these facts to calculate the average depth of the bacteria at that time, if we spread them evenly over Earth's entire surface.

54. *Bacterial Universe.* Suppose the bacteria described in Cosmic Calculations 6.1 and 6.2 could continue to multiply and spread out.
    a. Recall that the observable universe extends about 14 billion light-years in all directions from us. Calculate its volume in cubic light-years, then convert your answer to cubic meters.
    b. How long would it take for the bacteria to reach this volume? (*Hint:* You can proceed by trial and error, testing different values of t in the formula.)
    c. Even assuming that nutrients and energy were available, why couldn't the bacteria really grow this fast?

55. *Human Population Growth.* During the twentieth century, human population grew with a doubling time of about 40 years, reaching about six billion in 2000. Suppose this growth rate continued. What would human population be in 2200? In 2600? Do these populations seem possible on Earth? Explain.

56. *Impact Energy.* Consider a comet about 2 kilometers across with a mass of 4×10^{12} kg. Assume that it crashes into Earth at a speed of 30,000 meters per second (about 67,000 miles per hour).
    a. What is the total energy of the impact, in joules? (*Hint:* The "kinetic energy" formula tells us that the impact energy in joules will be $\frac{1}{2} \times m \times v^2$, where m is the comet's mass in kilograms and v is its speed in meters per second.)
    b. A 1-megaton nuclear explosion releases about 4×10^{15} joules of energy. How many such nuclear bombs would it take to release as much energy as the comet impact?

c. Based on your answers, comment on the degree of devastation the comet might cause.

57. *Impact Probability.* Impacts the size of the Tunguska event occur about once every century or two on average. Estimate the probability that the next impact will occur over a major city, killing hundreds of thousands or millions of people. Be sure to explain all the numbers in your estimate clearly.

Discussion Questions

58. *Our Bacterial Ancestry.* Some of Darwin's early detractors complained that evolution implied we were descended from monkeys or apes. In fact, as we saw in this chapter, our evolution is built on far more primitive organisms. The oxygen we breathe was produced by bacteria and is processed in our cells by mitochondria that probably represent bacteria living symbiotically within us. Does our relationship to bacteria affect the way we should view ourselves as a species? Defend your opinion.

59. *The Missing Link.* As we discussed in this chapter, there is no critical "missing link" in human evolution, and the theory of evolution has never suggested that humans evolved *from* present-day apes. Nevertheless, huge numbers of Americans profess belief in both of these myths about evolution. Why do you think these erroneous claims continue to be popular? What can or should be done to better educate the public?

60. *Evolution by Choice.* Consider the technology we are likely to have in the near future that would enable us to genetically engineer our own species, allowing us to choose the path of our future evolution. How do you think society can or should regulate the use of this awesome power? Do you think its potential benefits outweigh its risks, or vice versa? Overall, do you think it likely that advanced civilizations, if they exist, have engineered their own evolution? Defend your opinions.

WEB PROJECTS

61. *The Origin of Life.* NASA's Astrobiology home page frequently covers new discoveries about the origin and evolution of life. Learn about one recent important discovery, and write a short essay summarizing the discovery and how it affects our understanding of how life might have evolved on Earth.

62. *Impact Programs.* The discovery that impacts could pose a threat to our civilization has led to calls for new programs to help alleviate the threat. In a few cases, legislation has even been proposed to implement such programs. Learn about one proposal, such as that of the B612 Foundation, for dealing with the impact threat. Write a short essay explaining the proposal and discussing your opinion of its merits.

63. *Extinction.* Learn more about how biologists estimate the rate at which human activity is driving species to extinction. What conclusions can we draw about the present rate of extinction? Based on this rate, are we in danger of causing a mass extinction comparable to past mass extinctions on Earth? Summarize your findings in a one-page essay.

64. *Artificial Life.* Find the websites of three research groups that are working to produce artificial life, and write a one-page summary of each group's goals and methods.

7 Searching for Life in Our Solar System

LEARNING GOALS

7.1 ENVIRONMENTAL REQUIREMENTS FOR LIFE

- Where can we expect to find building blocks of life?
- Where can we expect to find energy for life?
- Does life need liquid water?
- What are the environmental requirements for habitability?

7.2 A BIOLOGICAL TOUR OF THE SOLAR SYSTEM: THE INNER SOLAR SYSTEM

- Does life seem plausible on the Moon or Mercury?
- Could life exist on Venus or Mars?

7.3 A BIOLOGICAL TOUR OF THE SOLAR SYSTEM: THE OUTER SOLAR SYSTEM

- What are the prospects for life on jovian planets?
- Could there be life on moons or other small bodies?

✻ **THE PROCESS OF SCIENCE IN ACTION**

7.4 SPACECRAFT EXPLORATION OF THE SOLAR SYSTEM

- How do robotic spacecraft work?

▲ **About the photo:** Saturn's moon Enceladus spews geysers of frozen water into space, suggesting the presence of underground reservoirs or a global ocean of liquid water. Future spacecraft might sample the geyser spray in search of evidence of microbial life in the subsurface aquifers.

... for what can more concern us than to know how this world which we inhabit is made; and whether there be any other worlds like it, which are also inhabited as this is?

Bernard le Bovier de Fontenelle,
Conversations on the Plurality of Worlds, *1686*

Having studied Earth's habitability and life, we now turn to the search for life elsewhere in our solar system. Because there are many places to look—including other planets and their moons, and thousands of known asteroids and comets—we need a strategy to help focus our efforts on the worlds most likely to be habitable.

The first step in such a strategy is to determine where it makes sense to look, so we begin this chapter by discussing the environmental requirements that we expect to be necessary for life on any world. We'll then take a biological tour of the solar system, seeking to determine where the requirements might be met. This will enable us to decide which worlds deserve the greatest attention, both in research and in our studies in this book.

Keep in mind that, in our solar system at least, we are looking primarily for microbes or other simple life. We have already learned enough to be confident that no other advanced civilization has ever arisen in our solar system. Still, the discovery of life of any kind would be profound, both to our understanding of biology and to philosophical considerations of our place in the universe. Even if we don't find life in our own solar system, the search itself will teach us much about the characteristics that can make a planet habitable and will thereby help us when we extend the search to other planetary systems.

7.1 Environmental Requirements for Life

There's no place like home (Figure 7.1)—at least, not within our own solar system. No world besides Earth has an atmosphere that we could breathe or abundant surface water that we could drink. No other world has a combination of surface temperature and pressure under which we could survive outside without a space suit. Few worlds have atmospheres that offer any protection from dangerous ultraviolet radiation from the Sun or from high-energy particles from space. Indeed, without undertaking major engineering projects to build self-contained environments (or greatly altering the basic conditions of a planet through "terraforming" [Section 8.4]), we have no hope of long-term survival on any other world in our solar system.

However, when we discuss habitability, we generally mean an environment in which life of *some* kind might survive, not necessarily human life. This greatly broadens the possibilities. After all, we could not have survived even on our own planet for much of its history, yet life flourished just the same. Past and present life on Earth has managed to thrive in a far greater variety of environments than we ourselves can endure [Section 5.5]. If we are going to identify potentially habitable worlds in our solar system, we must specify the range of environments that we can consider acceptable for life. We've touched on some of these ideas in previous chapters. Here, we'll try to tie them all together into a clear list of environmental requirements for life.

Where can we expect to find building blocks of life?

Perhaps the most obvious requirement for life is a set of chemical elements with which to make the components of cells. Life on Earth uses about 25 of the 92 naturally occurring chemical elements, although just

FIGURE 7.1
Although it is possible that some life may exist on other worlds, Earth is the *only* world in our solar system on which humans can survive without space suits or self-contained environments.

four of these elements—oxygen, carbon, hydrogen, and nitrogen—make up about 96% of the mass of living organisms (see Figure 5.6). Therefore, the presence of most or all of the elements used by life on Earth might be a reasonable first requirement for life in general.

In fact, this requirement is not very limiting and probably can be met by almost any world. Recall that essentially all chemical elements besides hydrogen and helium were produced by stars [Section 3.2]. Although all of these "heavy elements" are quite rare compared to hydrogen and helium, they are found just about everywhere. Moreover, the elements oxygen, carbon, and nitrogen—arguably the most crucial elements for life—are the third-, fourth-, and sixth-most-abundant elements in the universe, respectively. The proportions of heavy elements vary: While they make up about 2% of the chemical content (by mass) of our solar system, they make up less than 0.1% of the mass in some very old star systems. Nevertheless, every star system we've studied has at least some amount of all the elements used by life.

The nature of solar system formation gives us additional reason to expect the elements of life to be common on other worlds. Recall that, according to the nebular theory of solar system formation, the planets were built when solid particles condensed from gas in the solar nebula, and these particles then accreted into planetesimals and ultimately into planets, moons, asteroids, and comets. The first step in this process—condensation—affects only the heavier elements or hydrogen compounds containing heavy elements (such as carbon, nitrogen, and oxygen), because pure hydrogen and helium always remain gaseous [Section 3.4]. As long as condensation and accretion can occur,* we expect the resulting worlds to contain the elements needed for life.

Note that this basic argument doesn't change even if we allow for life quite different from life on Earth. Life on Earth is carbon-based, and, as we discussed in Chapter 5, we have good reason to think that life elsewhere would also be carbon-based. However, we can't absolutely rule out the possibility of life with another chemical basis. The set of elements (or their relative proportions) used by life based on some other element might be somewhat different from that used by carbon-based life on Earth. But the elements are still products of stars that should be found everywhere. No matter what kind of life we are looking for, we are likely to find the necessary elements on almost every planet, moon, asteroid, and comet in the universe.

A somewhat stricter requirement is the presence of these elements in molecules that can be used as ready-made building blocks for life, just as the early Earth probably had at least moderate abundances of amino acids and other complex molecules [Section 6.2]. Recall that Earth's organic molecules likely came from some combination of chemical reactions on the surface or near deep-sea vents in the oceans, and molecules brought to Earth from space or created by the heat and pressure of impacts. Chemical reactions would likely occur only on worlds with atmospheres or oceans, but sources from space should have brought similar molecules to worlds throughout our solar system.

Studies of meteorites and comets suggest that organic molecules are widespread among both asteroids and comets. Because every world was pelted by asteroids and comets during its early history [Section 4.3], every world should have received at least some organic molecules;

---

*Observations indicate that condensation and accretion can occur in star systems with as little as one-quarter the proportion of heavy elements of our own solar system [Section 11.1]. We do not yet know if these processes also occur in systems with much smaller abundances of heavy elements.

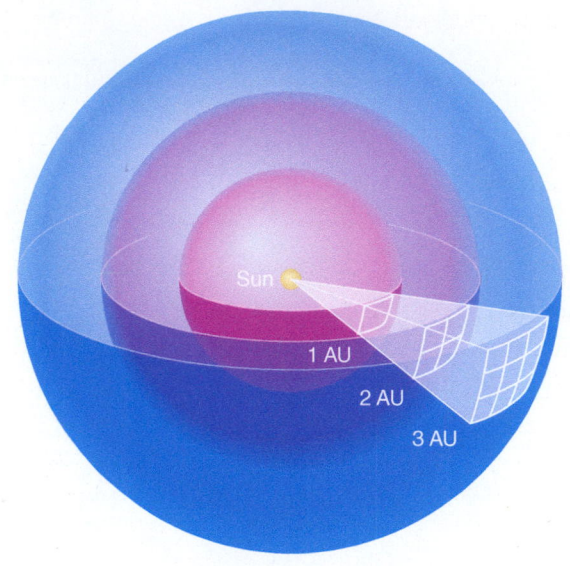

FIGURE 7.2

Any given amount of sunlight is spread over a larger area with increasing distance from the Sun. As shown in this diagram, the area over which the sunlight is spread increases with the square of the distance: At 2 AU the sunlight is spread over an area $2^2 = 4$ times as large as at 1 AU, and at 3 AU the sunlight is spread over an area $3^2 = 9$ times as large as at 1 AU. (Recall that 1 AU is the average Earth–Sun distance, or about 150 million kilometers.) Thus, the energy contained in sunlight (per unit area) decreases with the square of the distance from the Sun.

interplanetary dust may also have contained organic molecules that rained down on young worlds. However, organic molecules tend to be destroyed by solar radiation on surfaces unprotected by atmospheres. Moreover, while these molecules might stay intact beneath the surface—as they evidently do on asteroids and comets—they probably cannot react with each other unless some kind of liquid or gas is available to move them about. Given that it makes sense to start our search with worlds on which organic molecules are likely to be involved in chemical reactions, we should concentrate on worlds that have either an atmosphere or a surface or subsurface liquid medium, such as water, or both.

Where can we expect to find energy for life?

In addition to a source of molecular building blocks, life requires an energy source to fuel metabolism [Section 5.3]. Recall that life on Earth uses a wide variety of energy sources. Some organisms get energy directly from sunlight through photosynthesis. Others get energy by consuming organic molecules (for example, by eating photosynthetic organisms) or through chemical reactions with inorganic compounds of iron, sulfur, or hydrogen.

Sunlight is available everywhere in our solar system, though it becomes much weaker with increasing distance from the Sun. The energy available in sunlight decreases with the *square* of the distance from the Sun (Figure 7.2). For example, if we could put a leaf of a particular size on a world twice as far from the Sun as Earth, the leaf would receive only one-fourth as much energy as the same leaf on Earth (over the same period of time). At 10 times Earth's distance from the Sun—roughly the distance of Saturn—it would receive only $\frac{1}{10^2} = \frac{1}{100}$ of the energy it would receive on Earth. Photosynthetic life on such a world would have to be either much larger than life on Earth (giving it a larger surface area for collecting light), much more efficient at collecting solar energy, or much slower in its metabolism and reproduction. In the far outer solar system, sunlight is almost certainly too weak to support life.*

Chemical energy sources also place constraints on life. Chemical reactions can occur under a wide variety of circumstances, but only if the potential reactants are brought into contact with each other. This means that the ongoing reactions needed to provide energy for life can occur only on worlds where materials are being continually mixed. On a practical level, this probably requires either an atmosphere to mix gases or a liquid medium to mix materials on or below a world's surface—the same requirements we found for obtaining the building blocks of life.

Does life need liquid water?

In addition to organic building blocks and energy, one more ingredient is essential to all life on Earth: liquid water. Recall that water plays at least three vital roles for life on Earth [Section 5.3]: (1) It dissolves organic molecules, making them available for chemical reactions within cells; (2) it allows for the transport of chemicals into and out of cells; and (3) it is involved directly in many of the metabolic reactions that occur in cells. It is difficult to imagine life in the absence of a liquid substance to play these roles. But could these roles be fulfilled by some liquid other than water?

---

*Interestingly, some deep-sea bacteria on Earth appear to get energy from photosynthesis in which the light source is the weak infrared and visible light emitted by molten volcanic rock. The same energy could in principle be tapped for photosynthesis near volcanic sites on other worlds, but the total amount of energy available in this way is small.

POTENTIAL LIQUIDS FOR LIFE No one knows whether other liquids could support life in the absence of liquid water, but there are a number of constraints to consider. For example, a substance that might fulfill the roles of water must, like water, be fairly common. No other liquid seems to meet this condition on Earth, except perhaps the molten rock associated with volcanoes, which is so hot that it's difficult to imagine life surviving within it. However, several other common substances might take liquid form on colder worlds.

Table 7.1 lists the temperature ranges over which water and four other potential candidates—ammonia, methanol, methane, and ethane—remain liquid. The given temperature ranges are those that apply under the atmospheric pressure on Earth. At different pressures, or with the presence of dissolved minerals, the ranges can be different. For example, salt water can remain liquid at temperatures slightly below 0°C—sea water freezes at about −2°C and water fully saturated with table salt (NaCl) freezes at −21°C—and under sufficient pressure water can remain liquid at temperatures well above 100°C. (Near deep-sea vents, the pressure keeps water liquid at temperatures as high as about 375°C.) Despite such variation in melting and boiling temperatures, the ranges in Table 7.1 provide a useful comparison.

Think About It Oil (petroleum) is also found in liquid form on Earth. Why doesn't oil seem likely as a liquid medium for the origin of life? (*Hint*: Remember that oil is a fossil fuel.)

ADVANTAGES OF WATER Although we cannot rule out the other liquids in Table 7.1, water has at least three advantages that make it seem far more suitable as a liquid medium of life. First, as you can see in the table, water remains liquid over a relatively wide range of temperatures, and the top of its range is significantly higher than that of any of the other liquids. (Methanol, sometimes called wood alcohol and the principal ingredient of antifreeze and windshield wiper fluid, has an even wider range and stays liquid to a moderately high 65°C. However, there are no known reservoirs of it in the solar system, although massive interstellar clouds of methanol have been found by radio astronomers.) A wide range makes it more likely that the substance can stay liquid through changes in the weather or climate. A higher temperature range facilitates chemical reactions. As a general rule, chemical reactions proceed more rapidly at higher temperatures because the molecules themselves move more rapidly. Typically, the rate of a given chemical reaction doubles with each 10°C increase in temperature. As a result, chemical reactions in water should generally proceed much more rapidly than similar reactions in liquid ammonia, methane, or ethane. Therefore, life using colder liquids would probably have a much slower metabolism than life on Earth. The slower rate of reactions may also make it less likely that life would arise in the first place in these colder liquids, because the origin of life probably requires many complex chemical reactions.

The second advantage of liquid water involves an oddity in the way water freezes. Most substances are denser as solids than as liquids, but water is a rare exception (Figure 7.3): Ice is *less* dense than liquid water, which is why ice floats. No other liquid in Table 7.1 shares this property with water. This property helps life survive on Earth. In the winter, when surface temperatures are low enough for water to freeze, floating

TABLE 7.1 Potential Liquids for Life

Freezing and boiling points (at 1 atmosphere of pressure) for common substances that may be found in liquid form in our solar system. The last column gives the width of the liquid range, found by subtracting the freezing point from the boiling point.

Substance	Freezing Temperature	Boiling Temperature	Width of Liquid Range
Water (H_2O)	0°C	100°C	100°C
Ammonia (NH_3)	−78°C	−33°C	45°C
Methanol (CH_3OH)	−98°C	65°C	163°C
Methane (CH_4)	−182°C	−164°C	18°C
Ethane (C_2H_6)	−183°C	−89°C	94°C

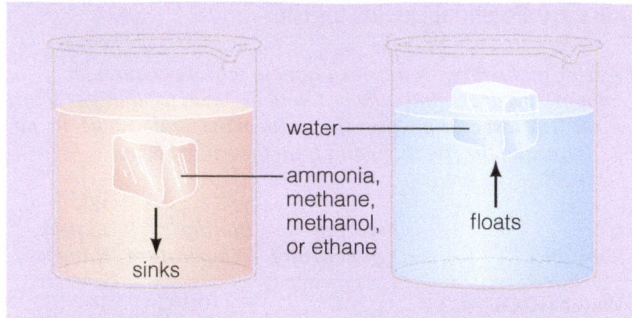

a Of the substances in the table of potential liquids for life, water is the only one whose solid form (ice) floats.

b The ability of ice to float is important both to life and to long-term climate stability, because floating ice provides an insulating layer that allows water to remain liquid beneath it.

FIGURE 7.3

Most substances are denser as solids than as liquids, so when solid and liquid forms exist together, the solid form sinks. Water is an exception; solid water in the form of ice is less dense than liquid water, and so ice floats in liquid water.

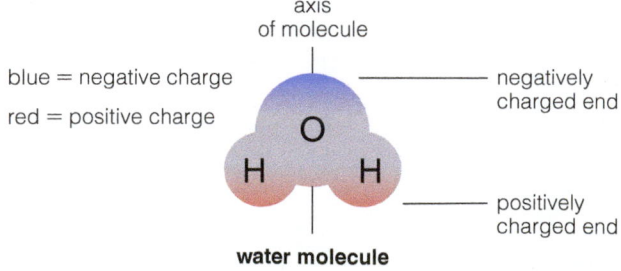

water molecule

FIGURE 7.4

Within individual water molecules, the electrons tend to be distributed in a way that makes one side have a net positive charge and the other side have a net negative charge.

ice forms a layer on the tops of lakes and seas. This layer of ice insulates the water beneath, allowing it to remain liquid—which allows life to survive within it.

The flotation of solid ice may be even more important to long-term climate stability. During periods when Earth cools a bit, such as the ice ages [Section 4.5], lower temperatures allow more ice to form on the surface. If this ice sank, the surface would still be covered with liquid water, which in turn would freeze and sink. This process would continue until no liquid water was left at the surface—that is, until lakes and oceans were completely frozen. Instead, because ice floats, a cool period thickens the insulating layer of ice on the surfaces of lakes and oceans, making it less likely that they will freeze completely.

The third advantage of liquid water over the other liquids in Table 7.1 comes from the way that electrical charge is distributed within water molecules (Figure 7.4). Within individual water molecules, the electrons tend to be distributed in a way that makes one side have a net positive charge and the other side have a net negative charge. (Molecules with such charge separation are called *polar* molecules; the term *polar* comes from the positive and negative charges being concentrated at opposite ends, or "poles," of the molecule's axis.) This charge separation affects the way in which water dissolves other substances. Molecules and salts that also have charge separations dissolve in water easily. Molecules that do not have any charge separation—such as molecules of oil—do not dissolve in water.

On Earth, the charge separation property of water is critical to life. Living cells have membranes that do not dissolve in water, so the membranes effectively protect the interior contents of cells. If we place living cells in liquid ethane, methane, methanol, or ammonia— molecules with less charge separation than water—their membranes tend to come apart. Charge separation also makes possible the formation of a special type of chemical bond, called a *hydrogen bond*, that is important to the biochemistry of life on Earth. (The formation of hydrogen bonds as water freezes also explains why ice is less dense, and hence why it floats—because these bonds force the molecules into a slightly expanded structure.)

THE BOTTOM LINE ON WATER We've identified three advantages of water over other liquid candidates for life: (1) a high and wide range of temperatures over which it remains liquid; (2) the fact that solid water floats; and (3) the fact that the charge separation of water molecules allows for types of chemical bonds that are not possible with the other liquids. These advantages make a strong case for the need for liquid water as a basis for life, but the case is not definitive. For example, it's possible to imagine circumstances under which the third advantage might be turned around. If life elsewhere had cell membranes made of molecules with different charge separation properties, they might dissolve in water but not in other liquids. We do not know if this is possible, but we cannot rule it out. The bottom line is that we do not yet know enough to draw a conclusion about the possibility of life using liquid mediums besides water. Nevertheless, because of the known advantages of water and the fact that liquid water is more common in the solar system than any of the other liquids, a search for liquid water seems like a good way to start a search for life.

What are the environmental requirements for habitability?

We began this section with the goal of making a list of environmental requirements for life that can help us decide which worlds to focus on as we begin the search for possible abodes of life in our solar system. In the broadest terms, we've found that the environment must satisfy three major requirements:

1. It must have a source of molecules from which to build living cells.
2. It must have a source of energy to fuel metabolism.
3. It must have a liquid medium—most likely liquid water—for transporting the molecules of life.

The first requirement is probably met by most if not all worlds. The second requirement is somewhat more limiting, but there are still plenty of worlds that should have sufficient sunlight or chemical energy for life. The third requirement—the need for a liquid—is the most stringent. Moreover, any world that meets this third requirement stands a good chance of meeting the first two as well. A liquid like water can facilitate chemical reactions with inorganic planetary materials, offering at least a potential source of energy for life—and, indeed, a source tapped by many microbes on Earth. To summarize, based on our current understanding of life, we can consolidate the requirements into a single "litmus test" for habitability: *A world can be habitable only if it has a liquid medium, probably meaning liquid water but possibly including one of the other liquids listed in Table 7.1.*

This requirement for a liquid certainly narrows the possibilities for life in our solar system, but not as much as you might at first guess. The wide variety of habitats in which we find both liquid water and life on Earth, including the deep ocean and rocks buried far underground, tells us that habitability requires only the presence of a liquid *somewhere*, not necessarily on the surface. As we'll discuss in the rest of this chapter, a large fraction of the worlds in our solar system have probably met this condition at least at some point in the past, and many still do.

Keep in mind that everything we know about life comes from the study of life on only a single world—our own. It is therefore possible that our discussions of habitability are based on too narrow a view of life, in which case our litmus test may be too narrow as well. Indeed, science fiction writers have imagined all sorts of bizarre life-forms existing under conditions far outside those we've considered here. Nature might be even more inventive. Nevertheless, a search for life must start somewhere, and it makes sense to begin by looking for the conditions that might support life "as we know it." If it turns out that our initial search is too narrow, we can always expand it in the future.

7.2 A Biological Tour of the Solar System: The Inner Solar System

Imagine living a century from now, when we have sent orbiters and landers to every major world in our solar system and when humans may even be living and working on other worlds. At that time, a "life in the universe" course might begin with a true biological tour of the solar sys-

a The Moon   b Mercury

FIGURE 7.5

Similar views of the Moon and Mercury, shown to scale. See Figure 4.1 for a size comparison to Earth.

tem, in which we could discuss with certainty which worlds have life, and why.

We cannot undertake such a complete biological tour today. However, we already know enough to make educated guesses about which worlds are most likely to be habitable. In this section, we'll focus on our neighbors in the inner solar system: our Moon and the terrestrial planets Mercury, Venus, and Mars.

Does life seem plausible on the Moon or Mercury?

Although Mercury is a planet and the Moon orbits Earth, the two worlds share many characteristics. Both are pockmarked with craters (Figure 7.5). Both are much smaller than Earth (see Figure 4.1)—so small that by now they have lost most of their internal heat, leaving them with little if any ongoing volcanism and without significant tectonic activity. The lack of volcanism means no outgassing to release gases into an atmosphere, and their small sizes mean weak gravity that has long since allowed any past atmospheric gases to escape to space. That is why both worlds are essentially airless. Mercury and the Moon also share the distinction of being among the places least likely to be habitable in the solar system, primarily because neither world is likely to have any liquids anywhere.

THE MOON The Moon contains very little water in any form, which makes sense if it was formed by a giant impact [Section 4.6]. However, scientists long suspected that water ice might be hidden at the bottoms of polar craters, where it would have accumulated from eons of comet impacts and remained frozen by being in perpetual shadow. This suspicion was confirmed in 2009, when the rocket from NASA's *LCROSS* spacecraft crashed into a crater near the south pole and splashed up ice-bearing debris. Shortly thereafter, a radar sensor aboard India's *Chandrayaan-1* spacecraft detected evidence for at least 600 million tons of water ice in craters near the Moon's north pole (Figure 7.6a). (More surprising, spectra show evidence of small numbers of water molecules mixed into lunar soil over much of the lunar surface; the origin of these molecules is unknown.) Nevertheless, while this water ice could prove valuable to future human colonists, it doesn't offer a liquid environment for life.

MERCURY Mercury may contain some water chemically bound in surface rock from the time of its formation, but it is unlikely that any of this water ever takes liquid form. The combination of Mercury's 58.6-day rotation period and 87.9-day orbital period gives Mercury days and nights that last about three Earth months each. Daytime temperatures reach 425°C, far too hot for liquid water. The lack of atmosphere means nighttime temperatures plummet to −150°C, far too cold for liquid water. Like the Moon, Mercury also contains ice in crater bottoms near its poles, where perpetual shadow keeps the temperatures perpetually low (Figure 7.6b). However, as on the Moon, this perpetually frozen ice seems unlikely to be a potential abode for life.

Note that while the Moon and Mercury seem to be lost causes when it comes to finding life, they can still teach us a great deal about the origin and history of our solar system and help us learn why some worlds are habitable and others are not. That is why both worlds are targets of ongoing exploration. Scientists hope that human exploration of the

Moon will resume in the not-too-distant future. As for Mercury, scientists are still analyzing data from the *MESSENGER* mission, which orbited Mercury from 2011 to 2015 (when it crashed to the surface), and look forward to learning more from the European/Japanese *BepiColombo* mission, scheduled to launch in 2016 and follow a complex path on the way to entering Mercury orbit in 2024.

Could life exist on Venus or Mars?

Prospects for past or present life look far better when we turn to our nearest planetary neighbors, Venus and Mars. Venus has been called our "sister planet," because it is the nearest planet to Earth in distance and is nearly identical to Earth in size. Mars is considerably smaller than Earth (see Figure 4.1), but spacecraft photos of its surface reveal an eerily Earth-like landscape (see Figure 1.7). Both planets have at least some potential for supporting past and present life, although Mars looks far more promising.

VENUS Venus is completely enshrouded in thick clouds (Figure 7.7), and past generations of scientists could only guess about surface conditions. Of course, that didn't stop speculation. Simple calculations based on Venus's distance from the Sun (about two-thirds of Earth's distance) show that *if* Venus had an Earth-like atmosphere, its global average temperature would be about 35°C (95°F). This fact led past generations of science fiction writers to imagine Venus as a lush, tropical paradise.

The reality is far different. Venus's surface temperature is an incredible 470°C (about 880°F)—easily hot enough to melt lead. This extreme temperature persists planetwide, both day and night. All the while, a thick atmosphere bears down on the surface with a pressure 90 times that on Earth's surface—equivalent to the pressure at a depth of nearly 1 kilometer (0.6 mile) in the oceans. In addition to the crushing pressure and searing temperature, the atmosphere of Venus contains sulfuric acid and other chemicals that are toxic to us. Indeed, landers sent to Venus by the Soviet Union provided data for only a brief period before being disabled by the high temperatures (Figure 7.8). Far from being a beautiful sister planet to Earth, Venus resembles a traditional view of hell.

What causes such extreme conditions on Venus? The answer is the *greenhouse effect*—the same effect that makes our own planet so comfortable [Section 4.5]. Recall that, in the absence of the greenhouse effect, Earth would be frozen over. Our planet is habitable because a moderate greenhouse effect traps enough heat to raise temperatures above the freezing point of water. The greenhouse effect operates the same way on Venus, but it is much stronger.

The strong greenhouse effect on Venus (often called a *runaway greenhouse effect*, for reasons we will discuss in Chapter 10) comes primarily from carbon dioxide in its atmosphere. Earth has a modest greenhouse effect because carbon dioxide makes up less than 1% of our atmosphere. In contrast, carbon dioxide makes up more than 96% of Venus's far thicker atmosphere. Interestingly, we can explain this difference in atmospheric carbon dioxide by contrasting the fate of carbon dioxide gas on each planet. Both planets have outgassed similar total amounts of carbon dioxide over the course of their histories. However, while almost all of the carbon dioxide outgassed on Earth is now locked up in carbonate rocks or dissolved in the oceans (170,000 times as much as is

a This radar map shows a region near the Moon's north pole, imaged by a NASA instrument on India's *Chandrayaan-1* spacecraft. The green circles represent craters in which water ice was detected.

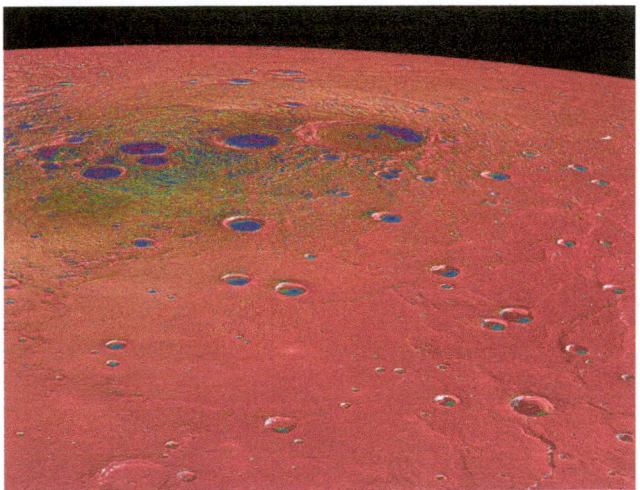

b This color-coded image, based on data from the *MESSENGER* spacecraft, represents surface temperatures in a region near Mercury's north pole. The purple regions are perpetually shadowed, with temperatures as low as −220° C, and water ice has been detected in these locations. The red regions receive sunshine and have daytime temperatures far too high for frozen or liquid water.

FIGURE 7.6
Both the Moon and Mercury have water ice at the bottoms of craters near their poles that are in perpetual shadow.

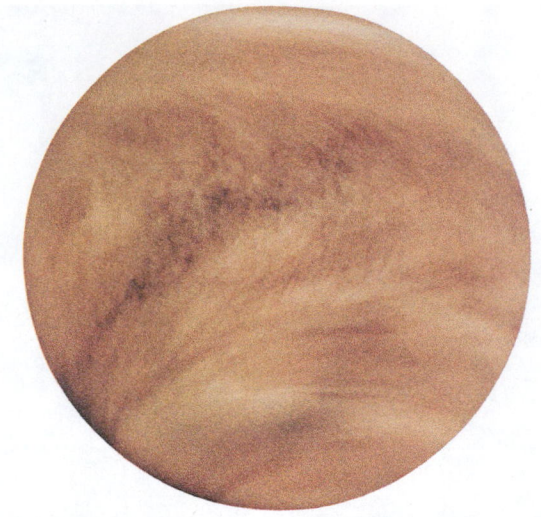

FIGURE 7.7
Clouds are all that can be seen in this ultraviolet image of Venus from the *Pioneer Venus* orbiter; no surface features can be seen at all.

FIGURE 7.8
The Soviet Union sent several landers to Venus during the 1970s and early 1980s. This photo from one of the landers shows volcanic rocks on the surface; part of the lander is visible at the bottom of the image, and sky is visible near the top.

in our atmosphere [Section 4.5]), all of Venus's outgassed carbon dioxide remains in its atmosphere. Venus's high temperature is a result of the extremely strong greenhouse effect produced by this atmospheric carbon dioxide, and the high pressure results from the sheer amount of this gas.*

The high surface temperature all but rules out the possibility of life on the surface of Venus today. It is far too hot for liquid water, let alone any of the other potential liquids for life from Table 7.1. However, the surface may have been habitable in the past. Recall that carbon dioxide gets locked up in Earth's carbonate rocks through the mechanism of the carbon dioxide cycle (see Figure 4.27). This cycle depends on both plate tectonics and the presence of oceans: Carbon dioxide dissolves in the oceans, where it reacts chemically to form carbonate minerals. Venus has no similar cycle because it lacks oceans (and apparently lacks plate tectonics as well) and therefore has no mechanism for removing carbon dioxide from the atmosphere. But the situation may have been different in the distant past. As we'll discuss further in Chapter 10, the fact that the Sun should have been much dimmer earlier in its history may have allowed Venus to have oceans prior to about 4 billion years ago. In fact, if you consider "Earth-like" to mean conditions found on Earth today, it's conceivable that ancient Venus may have been more "Earth-like" than Earth itself was at that time.

If Venus once had oceans—a big "if"—it's possible that life arose. If so, there's one place on Venus where microbes might still survive: in the clouds. At altitudes of about 50 kilometers above the surface, the greenhouse effect is far weaker and droplets containing liquid water can and do exist. The clouds are extremely acidic, but their sulfur content could allow chemical reactions that might provide sufficient energy for extremophiles adapted to survive in this environment.

Unless we find life in the venusian clouds (which seems unlikely), we may never know whether Venus had life in the distant past. Crater counts suggest that Venus's entire surface is less than about a billion years old, meaning that volcanism or tectonic processes have reshaped or paved over rocks from earlier times. If life arose and survived until the runaway greenhouse effect made Venus uninhabitable, any fossil evidence would almost certainly have been destroyed long ago.

MARS While Venus overheated early on, Mars went in the other direction. There's little doubt that Mars once had flowing water with a thicker and warmer atmosphere, but some 3 billion years ago, climate change caused the planet in effect to freeze over. Any remaining surface water froze, and Mars has since lost so much atmospheric gas that the surface pressure is too low for liquid water. Nevertheless, the evidence of past water makes Mars a prime candidate for past life. Moreover, Mars retains enough internal heat that liquid water may be possible underground, in which case life might still survive there.

Indeed, Mars seems such a good candidate for past or present life that, if it turns out never to have had it, we might have to revisit our assumptions about the likelihood of life arising under the "right" conditions. Because Mars is such a good candidate for habitability and life, we'll defer discussion of it to Chapter 8.

---

*Calculations show that carbon dioxide alone accounts for most but not all of Venus's high temperature. Other greenhouse gases—notably sulfur dioxide (SO_2), carbon monoxide (CO), and hydrochloric acid (HCl)—also make significant contributions.

7.3 A Biological Tour of the Solar System: The Outer Solar System

Beyond Mars, the weakening strength of sunlight makes surface life increasingly less likely. But there are numerous worlds with internal heat that could keep some water liquid and provide energy for life. In this section, we'll take our biological tour to the outer reaches of our solar system.

What are the prospects for life on jovian planets?

The four jovian planets—Jupiter, Saturn, Uranus, and Neptune—are very different from the terrestrial worlds. They are far more massive, lower in density, and composed largely of hydrogen, helium, and hydrogen compounds like water, methane, and ammonia. Figure 7.9 shows the bulk properties of the jovian planets, with Earth shown for scale.

The jovian planets are also different from the terrestrial planets in their interior structures, which scientists have deduced from observations and theoretical models of how their materials behave, given the planetary masses and sizes. These planets lack anything resembling Earth's solid surface. Instead, as shown in Figure 7.10, their outer layers contain visible clouds surrounding extended layers of gaseous hydrogen, mixed with helium and hydrogen compounds. Deeper in their interiors, the high pressure compresses the hydrogen into liquid, and within Jupiter and Saturn the pressure becomes so high that the hydrogen takes on a metallic form. Near their centers, each has a core of rock, metal, and hydrogen compounds, but the pressure is so great that these materials would be in a phase different from anything we ever see on Earth. If you plunged into any one of these worlds, you would just continue downward until you were crushed by the increasing pressure. Ultimately, your remains would sink into a hot sea of strange liquids.

Because these planets are so far from the Sun, temperatures in their upper atmospheres are extremely cold. However, observations show that all must be quite hot in their deep interiors. Because these worlds lack

FIGURE 7.9

The four jovian planets of our solar system, shown to scale. The values for distance, mass, and radius are given in terms of Earth units: M_{Earth} is Earth's mass and R_{Earth} is Earth's radius; 1 AU is Earth's distance from the Sun.

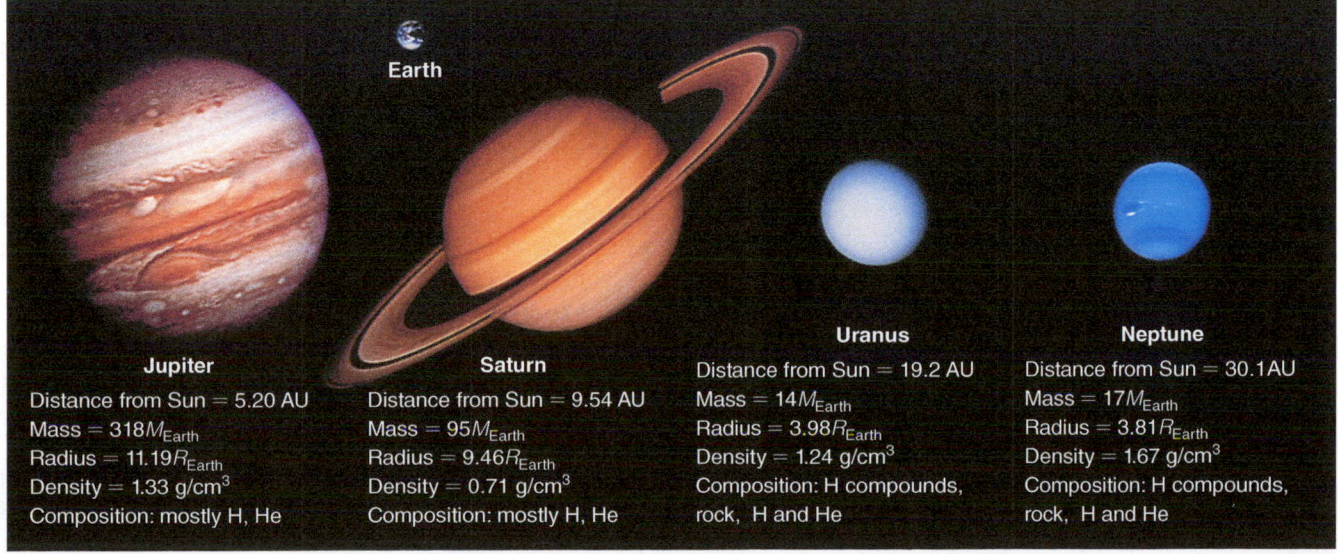

Earth

Jupiter
Distance from Sun = 5.20 AU
Mass = $318 M_{Earth}$
Radius = $11.19 R_{Earth}$
Density = 1.33 g/cm³
Composition: mostly H, He

Saturn
Distance from Sun = 9.54 AU
Mass = $95 M_{Earth}$
Radius = $9.46 R_{Earth}$
Density = 0.71 g/cm³
Composition: mostly H, He

Uranus
Distance from Sun = 19.2 AU
Mass = $14 M_{Earth}$
Radius = $3.98 R_{Earth}$
Density = 1.24 g/cm³
Composition: H compounds, rock, H and He

Neptune
Distance from Sun = 30.1 AU
Mass = $17 M_{Earth}$
Radius = $3.81 R_{Earth}$
Density = 1.67 g/cm³
Composition: H compounds, rock, H and He

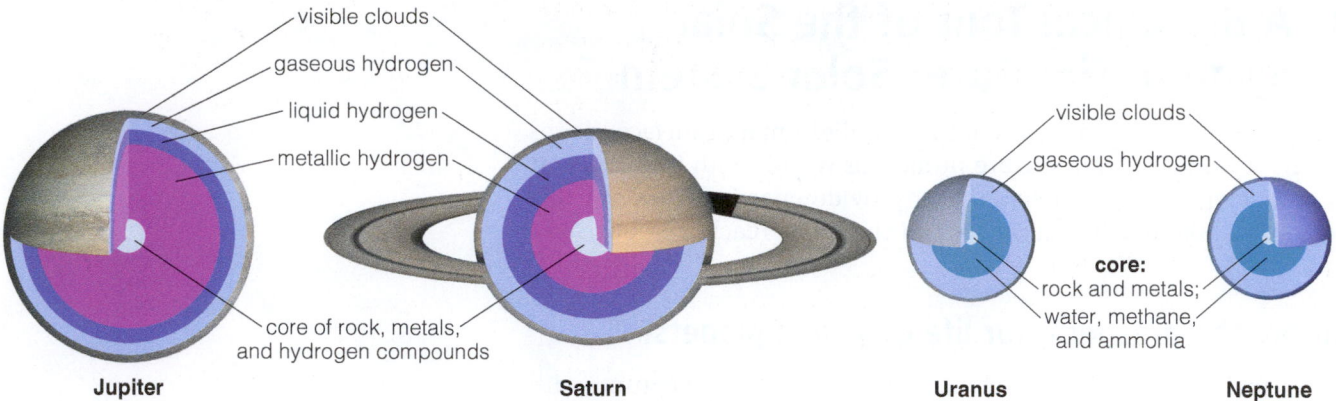

visible clouds
gaseous hydrogen
liquid hydrogen
metallic hydrogen
core of rock, metals, and hydrogen compounds

Jupiter

Saturn

visible clouds
gaseous hydrogen
core:
rock and metals;
water, methane, and ammonia

Uranus

Neptune

FIGURE 7.10

These diagrams compare the interior structures of the jovian planets, shown approximately to scale. All four have cores of rock, metal, and hydrogen compounds; the cores of Uranus and Neptune are differentiated into separate layers of rock/metal and hydrogen compounds. All four cores have about the same mass (ten Earth masses), so the planets differ primarily in the amount of material and the depth of the layers around the cores.

solid surfaces like those of the terrestrial worlds, this heat cannot fuel any geological activity. However, the heat ensures that some altitudes in their atmospheres have temperatures that should allow for liquid water. Moreover, chemical reactions powered by frequent lightning that has been observed in their atmospheres could potentially provide energy for life. So is it reasonable to imagine life in jovian atmospheres?

JUPITER AND SATURN Let's start by considering Jupiter, the largest of the four jovian planets. As Figure 7.10 shows, Saturn is so similar in its interior structure that the same considerations should apply to it.

Jupiter's temperature is far below freezing at its cloud tops, but the temperature rises rapidly with depth. Jupiter has several cloud layers, each formed as different types of gases condense (Figure 7.11). Clouds containing droplets of liquid water can form at a depth of about 100 kilometers into Jupiter's atmosphere, which is just over 1% of the way from the highest clouds to the center. Given the comfortable temperature and the presence of liquid water, we might therefore wonder if Jupiter could be habitable and host life at this depth in its atmosphere.

Unfortunately, Jupiter's atmosphere appears to present a fatal difficulty for life: It has strong, vertical winds with speeds that would make a hurricane seem like a gentle breeze in comparison. Any complex organic molecules that might form would quickly be carried to depths at which the heat would destroy them, so it is difficult to see how life could arise. We might imagine microbes reaching Jupiter on meteorites from elsewhere, but again the vertical winds make their survival seem impossible. Such microbes would be thrown quickly onto a nonstop elevator ride between cloud layers that are unbearably cold and others that are insufferably hot.

The only way that anyone has imagined life surviving in Jupiter's atmosphere is by supposing that it might have some sort of buoyancy that allowed it to stay at the right altitude while the vertical winds rushed by it. However, such buoyancy would require large gas-filled sacs, making the organisms themselves enormous. Given the difficulty of envisioning a way for microbes to survive, it's even more difficult to imagine a way in which large, buoyant organisms could evolve in the first place. They might survive if they arrived on Jupiter from elsewhere, but we know of no way that such large organisms could manage a journey through space, even if they existed elsewhere. As a result, most scientists do not consider Jupiter to be habitable. For essentially the same reasons, Saturn is unlikely to be habitable.

altitude above cloud tops (km)

0
−50
−100

cold enough for ammonia to condense to form clouds

0
−50
−100

cold enough for ammonium hydrosulfide to condense to form clouds

0
−50
−100

cold enough for water to condense to form clouds

0
−50
−100

FIGURE 7.11

This illustration shows how temperature changes within Jupiter's upper atmosphere, leading to at least three distinct cloud layers, with water droplets possible in the lowest clouds. But strong vertical winds mean that any microbes would be quickly killed either by the cold at high altitudes or by the heat far below.

URANUS AND NEPTUNE Uranus and Neptune also seem unlikely candidates for habitability. Their atmospheres are much colder than those of Jupiter and Saturn, mainly because of their greater distance from the Sun. If they have clouds in which liquid water droplets can form, the clouds must be deep in their atmospheres, and vertical winds similar to those of Jupiter and Saturn (though slower) would probably be fatal to any life.

However, Uranus and Neptune have one potential zone of habitability that Jupiter and Saturn lack: their outer cores of water, methane, and ammonia. Theoretical models suggest that these materials may be in liquid form, making for very odd "oceans" in the deep interiors of Uranus and Neptune. The high pressures, strange mix of liquids, and lack of any obvious way to extract energy from these "oceans" make life seem unlikely, but we cannot rule it out. If such life exists, we probably won't know about it for a long time, because no one has thought of a viable way to explore planetary cores.

Could there be life on moons or other small bodies?

The outer solar system contains vast numbers of small bodies, including the moons of the jovian planets, asteroids and comets, and dwarf planets like Ceres and Pluto. Could any of these worlds be habitable?

LARGE MOONS The best candidates for habitability are a few of the large moons of the jovian planets. This might seem surprising when you consider their sizes: Even the largest of these moons (Jupiter's moon Ganymede) is only slightly larger than Mercury, so you might expect all of them to be as geologically dead as Mercury and the Moon. However,

---

Movie Madness 2001—A SPACE ODYSSEY

It's the year 2001, and five clean-cut astronauts and a conniving computer named HAL are on their way to Jupiter. Yes, we know, this didn't really happen, and 2001 is now ancient history. But this Stanley Kubrick epic is a cinema classic, with music and scenes that remain a part of our popular culture and special effects that were revolutionary when the movie came out in 1968. If you haven't seen it, you should.

So why are astronauts and a wayward computer going to this behemoth world? Is it to study Jupiter's complex, churning atmosphere? Umm, no, that's not it. Are they on a mission to examine the giant planet's imposing magnetic field? Er, negative on that. Perhaps they're searching for signs of simple life on the moons Europa, Callisto, and Ganymede, all of which could have unseen oceans beneath their crusty exteriors? Actually, no, that's not it either.

In fact, these jovian rocket jockeys are headed to the king of the planets because some unknown race of aliens has been messing with our solar system. Four million years ago, some kindly extraterrestrials took a look at Earth and decided our planet's simian population had potential. They planted a dark gray monolith that, when touched by the local apes, converted them from a bunch of howling half-wits into tool-using primates.

This is an amazing tale (and one that, if it were true, would confound all paleontologists), but the aliens weren't content merely to introduce intelligence on Earth. They also buried a second monolith on the Moon, figuring that a brainy species might eventually develop enough technology to leave its planet and find this weird artifact. The lunar monolith, in turn, directed us to Jupiter.

So that's why the interplanetary spacecraft *Discovery One* is on its way to the biggest world of the solar system. The hope, it seems, is that by journeying to this massive planet, *Homo sapiens* can finally learn whatever profound secrets these altruistic aliens wish to share.

2001 is as much a space oddity as a space odyssey. Forget the fact that the message found at Jupiter is mostly a puzzling psychedelic light show. Ignore the fact that the ship's smooth-talking, onboard computer is more scheming than Machiavelli (though in the sequel, *2010*, we learn it was the humans' fault). No, in the end the most distressing thing about *2001* (other than the dismaying truth that our real space program still isn't even close to sending manned missions to Jupiter) is that our past and our destiny are simply some extraterrestrials' science-fair experiment. We're little more than a bunch of choreographed puppets.

It's likely that someday we will travel to many other worlds of the solar system, not just in our mind's eye or in the virtual worlds of our theaters and planetaria, but in honest-to-goodness spacecraft. However, we will make these voyages of discovery based on our own curiosity and quest for knowledge—not because some inscrutable aliens pointed to the gas-giant worlds of the outer solar system with their monolith fingers and said, "Go there."

a Ceres, photographed by the *Dawn* mission, which has been orbiting it since March 2015. Notice the mysterious bright spots in a crater near the upper left.

b Pluto, photographed by *New Horizons*, which flew past it in July 2015 and now continues on a trajectory that will eventually take it beyond the solar system.

FIGURE 7.12
Views of the two dwarf planets visited by spacecraft to date.

the moons of the jovian planets differ from the terrestrial planets in an important way: They were born containing a great deal of ice, because they formed in parts of the solar system where it was cool enough for ices to condense from the solar nebula (see Figure 3.36). Because ices melt at much lower temperatures than do metal or rock, these moons can have internally driven "ice geology" with much less internal heat than is needed for the "rock geology" of the terrestrial planets.

In addition, some of the jovian moons—most notably Jupiter's moons Io and Europa—have an ongoing source of internal heat quite different from any heat source on the terrestrial planets. We'll discuss this source, called tidal heating, in Chapter 9. The available heat can in principle melt subsurface ice into liquid water. Strong evidence for subsurface seas or oceans of liquid water has been found for Europa, Ganymede, and Enceladus, and some evidence suggests possible layers of liquid water on several other moons, including Callisto and Titan. Titan also has surface lakes of liquid methane and ethane. Numerous other moderate- to large-size moons in the outer solar system also show evidence of past or present geological activity that could have allowed for some liquid medium in the past.

The prospects for habitability are so good for some of these moons that we'll defer their discussion, in order to give them our full attention in Chapter 9. The prospects are much poorer for the more numerous but much smaller moons, so we can consider them along with other small bodies of the solar system.

DWARF PLANETS By definition, dwarf planets are worlds that are large enough to be round but too small to be considered "full" planets. The best known dwarf planets are the asteroid Ceres and comet-like Pluto, but three other solar system bodies also make the list (as of 2015): Eris, Makemake, and Haumea. More may be added to the list later, if other large asteroids or comets prove to be round.

The dwarf planets are very similar to the large jovian moons, which are also round. Like those moons, the dwarf planets tend to contain relatively large proportions of ice, thereby allowing for low-temperature "ice geology"; they also likely have differentiated interiors, so that their cores are rockier and their outer layers ice-rich. These facts have made scientists wonder if dwarf planets could harbor subsurface oceans, or might have harbored them in the past. On the plus side, they clearly have the right ingredients for such liquid layers. On the minus side, their isolation from other worlds means that they are not expected to have strong tidal heating, making it seem much less likely that they could retain enough internal heat to keep a subsurface layer melted.

Scientists got their first chance to explore dwarf planets in detail during 2015, when the *Dawn* spacecraft entered orbit of Ceres and the *New Horizons* spacecraft flew past Pluto (Figure 7.12). Both missions have provided a wealth of information. Ceres is covered with craters much as expected, but as this book goes to press, scientists are still analyzing mysterious bright spots that could potentially indicate a low level of ongoing geological activity. If such activity is indeed present, then there is at least a slim possibility of subsurface liquids. Pluto has proven even more surprising. Its surface shows relatively few craters and abundant evidence of recent (within millions of years, and likely ongoing) geological activity, including glacial movement of nitrogen ice. Data from the *New Horizons* flyby are still being received as this book goes to press, but already scientists are speculating about the possibility of subsurface liquids on Pluto. An underground ocean might have been created as two large bodies collided to create both Pluto and its five known moons, billions of years ago. Nevertheless, prospects for life on worlds like Pluto still seem dim, given the extremely cold temperatures.

Think About It Find the latest results concerning Ceres and Pluto. What conclusions have been drawn about the possibility of ongoing geological activity? How do these results affect prospects of habitability on these worlds? Explain.

SMALL BODIES Our biological tour now brings us to the numerous smaller bodies of the outer solar system. We'll include small moons in this group, such as Mars's moons Phobos and Deimos (Figure 7.13), since they probably were once asteroids orbiting the Sun independently. Scientists suspect the two moons were captured by Mars early in its history, when it had an extended atmosphere that created the friction necessary to slow and capture passing asteroids. The many small moons of the jovian planets are probably also bodies that once orbited the Sun independently and were captured at a time when those planets still had significant amounts of gas and dust around them.

These small bodies seem very unlikely to be habitable today. They are too small to have any leftover internal heat that might melt ices contained within them, and most of them—particularly the comets of the Kuiper belt and Oort cloud [Section 3.4]—orbit so far from the Sun that they are in a perpetual state of deep freeze. The only time any melting might occur

   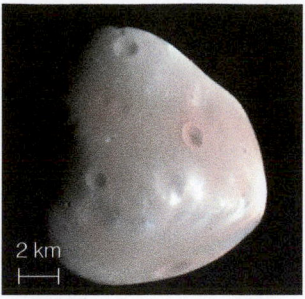

4 km   2 km

a Phobos   **b** Deimos

FIGURE 7.13
Mars has two small moons, Phobos (13 kilometers across) and Deimos (8 kilometers across), that are probably captured asteroids and much like many of the small moons of the outer solar system. Their small sizes make them unlikely to have any liquid water or life. (Colors are exaggerated in these photos from the *Mars Reconnaissance Orbiter*.)

Cosmic Calculations 7.1   NEWTON'S VERSION OF KEPLER'S THIRD LAW

How do we know the masses of distant objects? In many cases, we can use a modified version of Kepler's third law ($p^2 = a^3$). Recall that this law applies only to objects orbiting the Sun (see Cosmic Calculations 2.1). However, Newton found that Kepler's original law was just a specific case of a more general law, usually called *Newton's version of Kepler's third law*:

$$p^2 = \frac{4\pi^2}{G(M_1 + M_2)}a^3$$

where M_1 and M_2 are the masses of the two objects, p is their orbital period, and a is the distance between their centers. The term $4\pi^2$ is simply a number ($4\pi^2 \approx 4 \times 3.14^2 \approx 39.44$), and $G = 6.67 \times 10^{-11} \frac{m^3}{kg \times s^2}$ is the gravitational constant.

This law provides us with a way to measure the masses of distant objects. Any time we measure an orbiting object's period (p) and orbital distance (a), Newton's equation allows us to calculate the sum $M_1 + M_2$ of the two objects involved in the orbit. If one object is much more massive than the other, we essentially learn the mass of the massive object, as the following example shows.

Example: Use the fact that Earth orbits the Sun in 1 year at an average distance of 150 million kilometers (1 AU) to calculate the mass of the Sun.

Solution: Newton's version of Kepler's third law becomes

$$p_{\text{Earth}}^2 = \frac{4\pi^2}{G(M_{\text{Sun}} + M_{\text{Earth}})}a_{\text{Earth}}^3$$

Because the Sun is much more massive than Earth, the sum of their masses is nearly the mass of the Sun alone: $M_{\text{Sun}} + M_{\text{Earth}} \approx M_{\text{Sun}}$. Using this approximation, we find

$$p_{\text{Earth}}^2 \approx \frac{4\pi^2}{GM_{\text{Sun}}}a_{\text{Earth}}^3$$

To find an expression for the mass of the Sun, we multiply both sides by M_{Sun} and divide both sides by p_{Earth}^2 :

$$M_{\text{Sun}} \approx \frac{4\pi^2 a_{\text{Earth}}^3}{G p_{\text{Earth}}^2}$$

Because G is given above with units of seconds and meters, we must use the same units for Earth's orbital period ($p_{\text{Earth}} = 1$ year $\approx 3.15 \times 10^7$ seconds) and average orbital distance ($a_{\text{Earth}} = 1$ AU $\approx 1.5 \times 10^{11}$ m). We find

$$M_{\text{Sun}} \approx \frac{4\pi^2(1.5 \times 10^{11}\,\text{m})^3}{\left(6.67 \times 10^{-11}\frac{m^3}{kg \times s^2}\right)(3.15 \times 10^7\,\text{s})^2} = 2.0 \times 10^{30}\,\text{kg}$$

Simply by substituting in Earth's orbital period and distance from the Sun and the gravitational constant G, we have used Newton's version of Kepler's third law to find that the Sun's mass is about 2×10^{30} kilograms.

is following rare impacts or in the interiors of those comets that have had their original orbits perturbed enough to send them plunging into the inner solar system. Even in those cases, however, any melting would last for time periods that seem far too short to allow life to arise.

The prospects for life on small bodies may have been better in the past, though it still seems unlikely. Recall that we have substantial evidence of complex organic molecules in both asteroids and comets, and studies of meteorites show that many of them must have contained liquid water during the earliest history of the solar system. The liquid water may have persisted over time periods as long as a few tens of millions of years. Could life have arisen then? We cannot rule it out, but we have found no evidence of past life in meteorites from the asteroid belt, and most scientists consider it unlikely that life could have originated on any of the countless small bodies in the solar system.

TOUR RECAP We have completed our brief biological tour of the solar system. Although every world, large and small, has the raw chemical elements needed for life, the possibilities for life are much more limited when we focus on the environmental requirements we have found from the study of life on Earth. In particular, the need for liquid water, or possibly some other liquid medium, seems to rule out life on the numerous small worlds of our solar system. We have similarly ruled out life on Mercury and the Moon, and found life to be unlikely on the jovian planets. That leaves us with the slim chances for life in the atmosphere of Venus, and the much better chances for life on Mars and a few of the moons of the jovian planets. We will discuss these cases in the coming chapters.

✳ **THE PROCESS OF SCIENCE IN ACTION**

7.4 Spacecraft Exploration of the Solar System

We have discussed a lot of details about the different worlds in our solar system during our biological tour, and we'll describe some of them in even more detail in the next three chapters. How have we learned so much about these worlds?

Much of our knowledge comes from telescopic observations, using both ground-based telescopes and telescopes in Earth orbit such as the Hubble Space Telescope. In one case—our Moon—we have learned a lot by sending astronauts to explore the terrain and bring back rocks for laboratory study. In a few other cases, we have studied samples of distant worlds that have come to us as meteorites. But most of the data fueling the recent revolution in our understanding of the solar system have come from robotic spacecraft. To date, we have sent robotic spacecraft to all of the eight planets as well as to many moons, asteroids, comets, and dwarf planets. For this chapter's case study in the process of science in action, we'll briefly investigate how we use robotic spacecraft to explore the solar system.

How do robotic spacecraft work?

The spacecraft we send to explore the planets are robots suited for long space journeys and jam-packed with specialized equipment for scientific study (Figure 7.14). All spacecraft have computers used to control their major components: power sources such as solar cells, propulsion systems, and scientific instruments to study their targets. Robotic spacecraft

FIGURE 7.14

The *Cassini* spacecraft before launch. It is now nearly a billion miles away as it orbits Saturn. Notice major components, such as rocket thrusters at the bottom, the communications dish at the top, and various scientific instruments arrayed all around the main skeleton of the spacecraft. The *Huygens* probe (which landed on Titan) was not yet attached when this photo was taken. (For details on what you are seeing, go to the *Cassini* website.)

operate primarily with preprogrammed instructions, but they also carry radio transmitters and receivers that allow them to communicate with controllers on Earth. Most robotic spacecraft make one-way trips, never physically returning to Earth but sending their data back from space in the same way we send radio and television signals.

Broadly speaking, the robotic missions we send to explore other worlds fall into four major categories:

- **Flyby:** A spacecraft on a flyby goes past a world just once and then continues on its way.
- **Orbiter:** An orbiter is a spacecraft that orbits the world it is studying, allowing longer-term observation during its repeated orbits.
- **Lander or probe:** These spacecraft are designed to land on a planet's surface or probe a planet's atmosphere by flying through it. Some landers have carried rovers to explore wider regions.
- **Sample return mission:** A sample return mission requires a spacecraft designed to return to Earth carrying a sample of the world it has studied.

The choice of spacecraft type depends on both scientific objectives and cost.

FLYBYS Flybys tend to be cheaper than other missions because they are generally less expensive to launch into space. Launch costs depend largely on weight, and onboard fuel is a significant part of the weight of a spacecraft heading to another world. Once a spacecraft is on its way, the lack of friction or air drag in space means that it can maintain its orbital trajectory through the solar system without using any fuel at all. Fuel is needed only when the spacecraft needs to change from one trajectory (orbit) to another.

Moreover, some flybys gain more "bang for the buck" by visiting multiple planets. For example, *Voyager 2* flew past Jupiter, Saturn, Uranus, and Neptune before continuing on its way out of our solar system (Figure 7.15). This trajectory allowed additional fuel savings by using the gravity of each planet along the spacecraft's path to help boost it onward to the next planet. This technique, known as a *gravitational slingshot*, can not only bend the spacecraft's path but also speed the spacecraft up by essentially stealing a tiny bit of the planet's orbital energy, though the effect on the planet is unnoticeable. The boost in speed can be quite dramatic: During its flyby of Jupiter, the speed of the *New Horizons* spacecraft increased about 20%, shaving more than 3 years off the time it would otherwise have taken to reach Pluto.

Think About It Study the *Voyager 2* trajectory in Figure 7.15. Given that Saturn orbits the Sun every 29 years, Uranus orbits the Sun every 84 years, and Neptune orbits the Sun every 165 years, would it be possible to send another flyby mission to all four jovian planets if we launched it now? Explain.

Although a flyby offers only a relatively short period of close-up study, it can provide valuable scientific information. Spacecraft on flybys generally carry small telescopes, cameras, and spectrographs. Because these instruments are brought relatively close to other worlds (typically within a few thousand kilometers or less), they can obtain much higher-resolution images and spectra than even the largest telescopes on Earth or in Earth orbit. In addition, flybys sometimes give us information that would be difficult to obtain from Earth. For example, *Voyager 2*

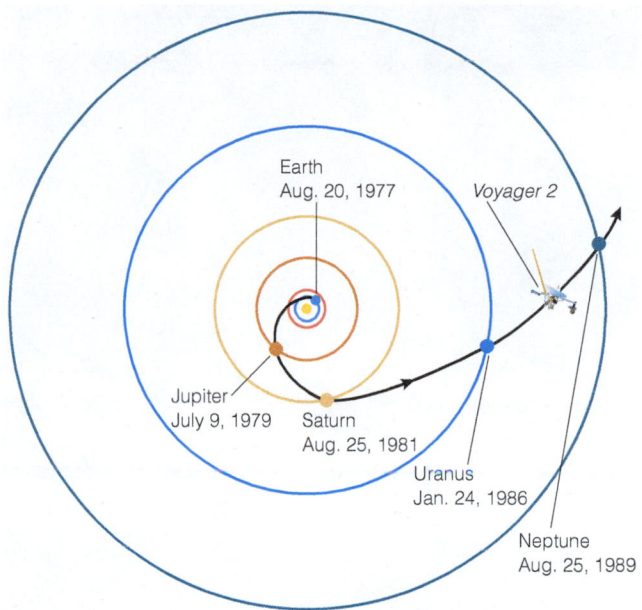

FIGURE 7.15
The trajectory of *Voyager 2*, which made a flyby of each of the four jovian planets in our solar system.

FIGURE 7.16

Comet 67P/Churyumov-Gerasimenko, photographed by the *Rosetta* spacecraft, which orbited the comet as it followed it on its course toward the Sun. Notice the jets of gas and dust spraying out from the comet.

helped us discover Jupiter's rings and learn about the rings of Saturn, Uranus, and Neptune through views in which the rings were backlit by the Sun. Such views are possible only from beyond each planet's orbit.

Spacecraft on flybys may also carry instruments to measure local magnetic field strength or to sample interplanetary dust. The gravitational effects of the planets and their moons on the spacecraft itself provide information about object masses and densities. Like the backlit views of the rings, these types of data cannot be gathered from Earth. Indeed, most of what we know about the masses and compositions of moons comes from data obtained by spacecraft that have flown past them.

ORBITERS An orbiter can study another world for a much longer period of time than a flyby. Like the spacecraft used for flybys, orbiters often carry cameras, spectrographs, and instruments for measuring the strength of magnetic fields. Some orbiters also carry radar instruments. Radar works by sending radio waves from the spacecraft to bounce off the surface: The time it takes for the bounced signals to return to the spacecraft tells how far they traveled, allowing precise measurements of surface altitude. Radar can even "see" through thick cloud cover (such as on Venus and Titan), revealing the nature of otherwise hidden terrain.

An orbiter is generally more expensive than a flyby for an equivalent weight of scientific instruments, primarily because it must carry added fuel to change from an interplanetary trajectory to a path that puts it into orbit around another world. Careful planning can minimize the added expense. For example, recent Mars orbiters have saved on fuel costs by carrying only enough fuel to enter highly elliptical orbits around Mars. The spacecraft then settled into the smaller, more circular orbits needed for scientific observations by skimming the martian atmosphere at the low point of every elliptical orbit. Atmospheric drag slowed the spacecraft with each orbit and, over several months, circularized the spacecraft orbit. (This technique is sometimes called *aerobraking*.) We have sent orbiters to all the planets except Uranus and Neptune, and to two asteroids and a comet (Figure 7.16).

LANDERS AND PROBES The most "up close and personal" study of other worlds comes from spacecraft that send probes into the atmospheres or landers to the surfaces. For example, in 1995, the *Galileo* spacecraft—which orbited Jupiter for more than 5 years—dropped a probe into Jupiter's atmosphere. The probe collected temperature, pressure, composition, and radiation measurements for about an hour as it descended, teaching us a great deal about Jupiter's winds and atmospheric conditions before it was destroyed by Jupiter's high interior pressures and temperatures. The *Cassini* spacecraft, currently in orbit of Saturn, also carried a probe, called *Huygens*, that descended to the surface of Saturn's moon Titan, studying the atmosphere on its way down. We'll discuss the findings from Huygens in Chapter 9, when we treat Titan in more detail.

On planets with solid surfaces, a lander can offer close-up surface views, local weather monitoring, and the ability to carry out automated experiments. Some landers carry robotic rovers able to venture across the surface, including the *Spirit* and *Opportunity* rovers that landed on Mars in 2004, and the *Curiosity* rover that landed in August 2012. Because of its weight, *Curiosity*'s landing required a spectacular feat of engineering (Figure 7.17). The spacecraft carrying the lander first used a parachute

1 Friction slows spacecraft as it enters Mars atmosphere.

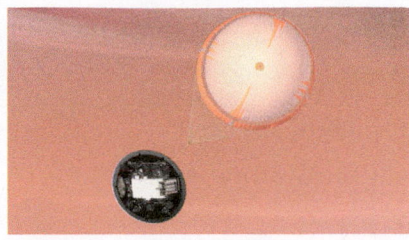

2 Parachute slows spacecraft to about 350 km/hr.

3 Rockets slow spacecraft to halt; "sky crane" tether lowers rover to surface.

4 Tether released, the rocket heads off to crash a safe distance away.

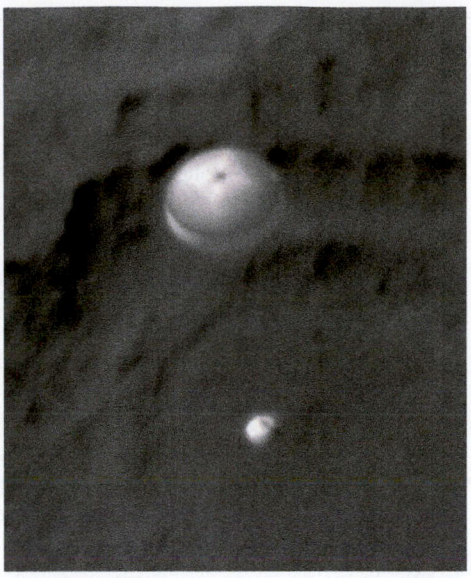

As it flew overhead, the *Mars Reconnaissance Orbiter* took this photo of the spacecraft with its parachute deployed.

FIGURE 7.17
An artist's conception of the landing sequence that brought the *Curiosity* rover to Mars, along with a photo of its descent taken from orbit.

to slow it down in the martian atmosphere, and then fired rockets that brought it to a halt while hovering about 7 meters above the surface. Finally, a "sky crane" lowered the rover to the surface.

SAMPLE RETURN MISSIONS Although probes and landers can carry out experiments on surface rock or atmospheric samples, the experiments must be designed in advance and the instrumentation must fit inside the spacecraft. One way around these limitations is to design missions in which samples from other worlds can be scooped up and returned to Earth for more detailed study. To date, the only sample return missions have been to the Moon (samples of which were brought back by the *Apollo* astronauts and by robotic spacecraft sent in the 1970s by the then–Soviet Union) and to an asteroid (Japan's *Hayabusa* mission). Many scientists are working toward a sample return mission to Mars, and they hope to launch such a mission within the next decade or so. A slight variation on the theme of a sample return mission is the *Stardust* mission, which collected comet dust on a flyby and returned to Earth in 2006.

ROBOTIC MISSIONS AND ASTROBIOLOGY Over the past few decades, many dozens of robotic spacecraft have explored various worlds in our solar system. Table 7.2 lists a few of the most significant missions. While not all of these missions were designed with astrobiology in mind, everything we learn about the solar system helps us understand more about the possibilities for life in our universe.

TABLE 7.2 Selected Robotic Missions to Other Worlds (as of 2015)

Destination	Mission	Arrival Year	Agency*
Mercury	*MESSENGER* orbiter studies surface, atmosphere, and interior	2011	NASA
Venus	*Venera* probes (eight of them) successfully landed on the surface	1970–81	USSR
	Magellan orbiter mapped surface with radar	1990	NASA
	Venus Express focused on atmosphere studies	2006	ESA
Moon	The United States, China, Japan, India, and Russia all have current or planned robotic missions to explore the Moon	—	—
Mars	*Viking 1* and *2* landers were the first landers on Mars and carried out the first search for extraterrestrial life on that planet (there were also *Viking* orbiters)	1976	NASA
	Spirit and *Opportunity* rovers found evidence for water on ancient Mars	2004	NASA
	Mars Express orbiter studies Mars's climate, geology, and polar caps	2004	ESA
	Mars Reconnaissance Orbiter takes very high-resolution photos	2006	NASA
	Phoenix lander studied soil near the north polar cap	2008	NASA
	Curiosity rover explores Gale Crater to understand prospects for life	2012	NASA
	MAVEN orbiter studies how Mars has lost atmospheric gas over time	2014	NASA
Asteroids	*Hayabusa* orbited and landed on asteroid Itokawa; returned sample to Earth in 2010	2005	JAXA
	Dawn visited asteroid Vesta and the dwarf planet Ceres	2011/2015	NASA
Jovian planets	*Voyagers 1* and *2* visited all the jovian planets and left the solar system	1979	NASA
	Galileo's orbiter studied Jupiter and its moons; probe entered Jupiter's atmosphere	1995	NASA
	Cassini orbits Saturn; its *Huygens* probe (built by ESA) landed on Titan	2004	NASA
	Juno orbiter to study Jupiter's deep interior	2016	NASA
Pluto and comets	*New Horizons* flew past Pluto in 2015; now en route to another Kuiper belt comet	2015	NASA
	Stardust flew through the tail of Comet Wild 2; returned comet dust in 2006	2004	NASA
	Deep Impact observed its "lander" impacting Comet Tempel 1 at 10 km/s	2005	NASA
	Rosetta orbits Comet 67P/Churyumov-Gerasimenko and sent a lander to its surface	2014	ESA

*ESA = European Space Agency. JAXA = Japan Aerospace Exploration Agency.

The Big Picture

PUTTING CHAPTER 7 IN PERSPECTIVE

In this chapter, we have discussed the general requirements for habitability and taken a biological tour of the solar system as we know it today. As you continue in your studies, keep in mind the following "big picture" ideas:

- The general environmental requirements for life are much broader than the requirements for complex beings such as humans. We would count a world as habitable if any form of life could survive on it, even if the life were microscopic.

- Our solar system contains a vast number of worlds, but most of them are unlikely to have life because they lack a liquid medium of any kind. Nevertheless, a few worlds may meet the criteria for habitability in at least some regions of their surfaces, subsurfaces, or atmospheres. More worlds may have had liquid water or other liquids in the distant past.

- Much of our current knowledge about the solar system and the potential for life comes from studies conducted by robotic spacecraft. We are living during a time when many spacecraft are simultaneously exploring different worlds in our solar system, and more missions are being planned for the future.

Summary of Key Concepts

7.1 Environmental Requirements for Life

Where can we expect to find building blocks of life?
The chemical elements needed for life should be present on almost any world, but a smaller number of worlds will contain more complex organic molecules that can serve as building blocks for life. Still, the fact that these building blocks are present in asteroids and comets suggests that we'll find them in many places.

Where can we expect to find energy for life?
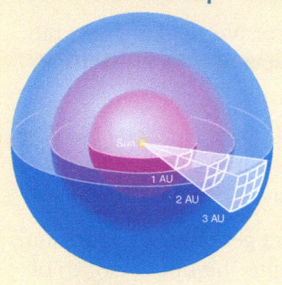
Energy for life can come from sunlight or chemical reactions. Sunlight weakens with distance from the Sun and is unlikely to be sufficient to sustain life at large distances. Chemical energy is probably available in many more places and is likely on any world with a substantial atmosphere or a liquid medium that can mix and support chemical reactions.

Does life need liquid water?
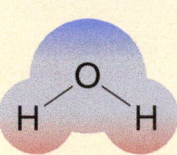
Life almost certainly requires some liquid, and water has at least three advantages over other liquids: a wider and higher range of temperatures at which it is liquid, the fact that ice floats, and the type of chemical bonding made possible by charge separation within water molecules. Nevertheless, we can't completely rule out other liquids, such as liquid ammonia, methane, methanol, or ethane.

What are the environmental requirements for habitability?
Life requires a source of molecules from which to build living cells, a source of energy for metabolism, and a liquid medium for transporting chemicals. In practice, these requirements probably come down to a need for liquid water, so the possibility of liquid water is the main requirement we search for in looking for habitable worlds.

7.2 A Biological Tour of the Solar System: The Inner Solar System

Does life seem plausible on the Moon or Mercury?
The Moon and Mercury are probably not habitable, since neither has liquid water or any other liquid medium for life.

Could life exist on Venus or Mars?

Venus is far too hot for liquid water to exist on or under its surface, making life seem unlikely. However, life might be possible high in Venus's atmosphere, where clouds contain droplets of water. Mars almost certainly had habitable conditions in the distant past and might still have habitable regions underground.

7.3 A Biological Tour of the Solar System: The Outer Solar System

What are the prospects for life on jovian planets?
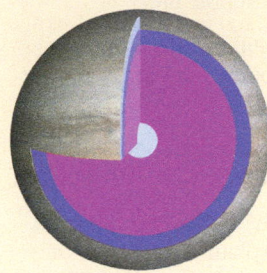
Liquid water could exist at certain depths in the atmospheres of the jovian planets, but strong vertical winds make life seem unlikely. Uranus and Neptune may have "oceans" of water and other liquids in their deep interiors, but at present we have no way to search such depths for life.

Could there be life on moons or other small bodies?
A few large moons may contain liquid water or other liquids, and therefore seem like potential candidates for life. Smaller moons and other small bodies probably do not have any liquids at present, though many may have had liquid water in the distant past.

❀ THE PROCESS OF SCIENCE IN ACTION

7.4 Spacecraft Exploration of the Solar System

How do robotic spacecraft work?
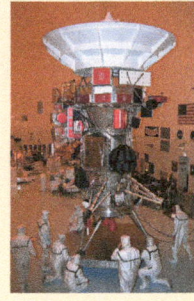
Spacecraft can be categorized as flybys, orbiters, landers and probes, or sample return missions. In all cases, robotic spacecraft carry their own propulsion, power, and communication systems and can operate under preprogrammed control or with updated instructions from ground controllers.

Exercises and Problems

MasteringAstronomy® *For instructor-assigned homework and other learning materials, go to MasteringAstronomy®.*

REVIEW QUESTIONS

Short-Answer Questions Based on the Reading

1. Why do we expect the elements of life to be widely available on other worlds? How does the requirement of organic building blocks further constrain the prospects of habitability?

2. How does the strength of sunlight vary with distance from the Sun? Discuss the implications for photosynthetic life.

3. Under what conditions does it seem reasonable to imagine a chemical energy source for life?

4. Why is a liquid medium important for life? Why does water seem the most likely liquid medium for life? Briefly discuss a few other liquids that could potentially support life.

5. Summarize the three major environmental requirements for life. Overall, what "litmus test" seems appropriate for constraining our search for habitable worlds, and why?

6. Why do the Moon and Mercury seem unlikely to be habitable? Does evidence for ice in polar craters affect the answer? Explain.

7. Why is Venus so much hotter than Earth? How does this heat affect the possibility of life on Venus? Explain why Venus may nonetheless have been habitable in the past and might still be habitable in some parts of its atmosphere.

8. Why does Mars seem such a good candidate for life?

9. Briefly discuss the possibility of life on Jupiter and Saturn.

10. With regard to habitability, how do the cases of Uranus and Neptune differ from those of Jupiter and Saturn? Explain.

11. What characteristics make some of the large moons of jovian planets seem like potential candidates for habitability?

12. Briefly describe the prospects for habitability of dwarf planets and of the many small bodies in the solar system.

13. Describe and distinguish between space missions that are *flybys, orbiters, landers* or *probes,* and *sample return missions.* What are the advantages and disadvantages of each mission type?

14. For a few of the most important past, present, or future robotic missions within the solar system, describe the targets, types, and mission highlights.

TEST YOUR UNDERSTANDING

Would You Believe It?

Each of the following gives a statement that a future explorer might someday make. In each case, decide whether the claim seems plausible in light of current knowledge. Explain clearly; because not all of these have definitive answers, your explanation is more important than your chosen answer.

15. On the smallest moon of Uranus, my team discovered a vast, subsurface ocean of liquid water.

16. New spacecraft images show lakes of liquid water on the dwarf planet Eris.

17. We are pumping water for our new Moon colony from a well we found by drilling about a kilometer down into the Moon's surface.

18. I was part of the first group of people to land on Venus, where we found huge, ancient cities that had been hidden from view by cloud cover.

19. We sent a robotic airplane into the atmosphere of Jupiter, but we could not keep it at a steady altitude and it was quickly ripped apart.

20. On a moon of Neptune, we discovered photosynthetic life with a metabolism that operates nearly a hundred times as fast as that of any photosynthetic organism on Earth.

21. We deposited bacteria (from Earth) that get energy from chemical reactions with sulfur compounds into the upper clouds of Venus, and they are surviving.

22. The drilled sample showed no signs of life on asteroid B612, but we found many complex organic molecules.

23. We cut holes in the frozen surface of a methane lake on Titan, so that we could search for swimming organisms in the liquid methane underneath it.

24. The drilled sample from Mars brought up rock that contained microscopic droplets of liquid water.

Quick Quiz

Choose the best answer to each of the following. Explain your reasoning with one or more complete sentences.

25. Oxygen and carbon are (a) rarer than almost all other elements; (b) found only on worlds close to a star; (c) the third- and fourth-most-abundant elements in the universe.

26. On an asteroid that is twice as far as Earth from the Sun, the strength of sunlight would be (a) twice as great as on Earth; (b) $\frac{1}{2}$ as great as on Earth; (c) $\frac{1}{4}$ as great as on Earth.

27. Compared to liquid water, liquid methane is (a) colder; (b) hotter; (c) denser.

28. Frozen lakes often have liquid water beneath their icy surfaces primarily because (a) Earth's internal heat keeps the water liquid; (b) ice floats and provides insulation to the water below; (c) sunlight penetrates the ice and warms the water below.

29. Temperatures on Mercury are (a) always very hot; (b) very hot in the day and very cold at night; (c) about the same as those on Venus.

30. On Venus, liquid water (a) does not exist anywhere; (b) exists only deep underground; (c) exists only high in the atmosphere.

31. The reason Venus is so much hotter than Earth is that (a) it has many more volcanoes; (b) its closer distance to the Sun makes sunlight dozens of times stronger; (c) its thick,

carbon dioxide atmosphere creates a far stronger greenhouse effect.

32. Life is probably not possible in Jupiter's atmosphere because (a) it is too cold there; (b) there is no liquid water at all; (c) winds are too strong.

33. Which of the following are you most likely to find if you randomly choose a small moon of one of the jovian planets to examine? (a) water ice (b) organic molecules (c) an abundance of heavy metals, such as gold

34. The *Cassini* spacecraft (a) flew past Pluto; (b) landed on Mars; (c) is orbiting Saturn.

PROCESS OF SCIENCE

35. *Bizarre Forms of Life.* Discuss some forms of life that have appeared in science fiction and that fall outside the general types of life that we've discussed in this chapter. Which forms seem most plausible, and why? Overall, do you think the definition of life we've used in this chapter is too constraining? Defend your opinion.

36. *Making a Living.* Consider various methods by which life "makes a living" (the ways in which it acquires energy for metabolism). For example, some microbes use direct chemical reactions to obtain energy, plants most commonly rely on sunlight, and animals generally depend on consuming plants or other animals. Discuss the relative merits of the schemes generally used by plants and animals. Do you think it's plausible that an intelligent species could obtain its energy from sunlight? For example, would a chlorophyll-skinned mammal be possible? Defend your opinion by drawing on your understanding of basic concepts of energy.

GROUP WORK EXERCISE

37. *Mission Plan.* **Roles:** *Scribe* (takes notes on the group's activities), *Proposer* (proposes ideas to the group), *Skeptic* (points out weaknesses in proposed ideas), *Moderator* (leads group discussion and makes sure everyone contributes). **Activity:** Suppose you could send a robotic mission of any type (flyby, orbiter, probe/lander) to any one of the places said in this chapter to be *unlikely* to be habitable. Which place and type of mission would you choose, and why? Write up a one-page mission plan and a one-page scientific justification for the plan.

INVESTIGATE FURTHER

In-Depth Questions to Increase Your Understanding

Short-Answer/Essay Questions

38. *Solar System Tour.* Based on the brief tour in this chapter, which world in our solar system (besides Earth) do you think is most likely to have life? Explain why.

39. *Galileo Spacecraft.* In 2003, scientists deliberately ended the *Galileo* mission to Jupiter by causing the spacecraft to plunge into Jupiter's atmosphere. They did this to avoid any possibility that the spacecraft might someday crash into Europa, potentially "contaminating" this moon with microbes from Earth. Do you think that the scientists should also have been worried about contaminating Jupiter itself? Why or why not?

40. *Greenhouse Effect.* The text (in Chapter 4) makes the statement that the greenhouse effect on Venus proves "that it is possible to have too much of a good thing." Explain this statement in two or three paragraphs.

41. *Transplanting Life.* Suppose you could genetically engineer organisms on Earth in any way that you chose. What, if any, features could you give them that would enable them to survive on (a) the Moon or Mercury; (b) Venus; (c) Jupiter; (d) a comet?

42. *Science Fiction Life.* Choose a science fiction book or movie that describes some form of alien life that falls outside the bounds of the type of life we have considered in this chapter. Write a one-to two-page critical review in which you discuss the plausibility of the life-form.

Quantitative Problems

Be sure to show all calculations clearly and state your final answers in complete sentences.

43. *Understanding Newton's Version of Kepler's Third Law I.* Imagine another solar system, with a star of the same mass as the Sun. Suppose there is a planet in that solar system with a mass twice that of Earth orbiting at a distance of 1 AU from the star. What is the orbital period of this planet? Explain. (*Hint:* The calculations for this problem are so simple that you will not need a calculator.)

44. *Understanding Newton's Version of Kepler's Third Law II.* Suppose a solar system has a star that is four times as massive as our Sun. If that solar system has a planet the same size as Earth orbiting at a distance of 1 AU, what is the orbital period of the planet? Explain. (*Hint:* The calculations for this problem are so simple that you will not need a calculator.)

45. *Earth Mass.* The Moon orbits Earth in an average time of 27.3 days at an average distance of 384,000 kilometers. Use these facts to determine the mass of Earth. (*Hint:* You may neglect the mass of the Moon, since its mass is only about $\frac{1}{80}$ of Earth's.)

46. *Jupiter Mass.* Jupiter's moon Io orbits Jupiter every 42.5 hours at an average distance of 422,000 kilometers from the center of Jupiter. Calculate the mass of Jupiter. (*Hint:* Io's mass is very small compared to Jupiter's.)

47. *Pluto/Charon Mass.* Pluto's moon Charon orbits Pluto every 6.4 days with a semimajor axis of 19,700 kilometers. Calculate the *combined* mass of Pluto and Charon. Compare this combined mass to the mass of Earth, which is about 6×10^{24} kg. Can you determine the individual masses of Pluto and Charon from the given data? Explain.

48. *Mission to Pluto.* The *New Horizons* spacecraft took about 9 years to travel from Earth orbit to Pluto. About how fast was it traveling on average? Assume that its trajectory was close to a straight line. Give your answer in AU/year and in km/hr. (*Hint:* You can find needed data in Appendix D.)

Discussion Questions

49. *Artificial Life.* Imagine that future humans decide to breed new organisms tailored to as many different environments as

possible. Discuss some of the places in our solar system where we could potentially plant such artificially created species, even if life probably would not arise naturally in those places. Do you think it likely that we will someday develop life-forms for other worlds? What are the philosophical ramifications of being able to custom-tailor life for worlds that don't have any natural life?

50. *Future Astrobiology Course.* Imagine that you are living a century from now and are taking a course about life in our solar system. Based on the current rate of exploration and reasonable rates for the future, how much more do you think we will know then about life in our solar system than we know now? Speculate about some of the discoveries you think may occur in the next century.

WEB PROJECTS

51. *Project Apollo.* Learn more about NASA's *Apollo* project, the only set of missions that has ever sent humans to another world. Describe the goals and objectives of each of the *Apollo* missions. Which were successful, and which were not? What lessons does *Apollo* offer for future attempts to send humans to the Moon and beyond? Summarize your findings and your opinions about lessons for the future in a one- to two-page essay.

52. *Planetary Missions.* Visit the web page for one of the missions listed in Table 7.2. Write a one- to two-page summary of the mission's basic design, goals, and current status.

8 Mars

▲ **About the photo:** Sunset on Mars (photographed by the *Curiosity* rover).

The idea of a civilization on Mars was once taken so seriously that the term *Martians* became nearly synonymous with alien life. Spacecraft sent to Mars have since shattered this fictional image of a world of cities and sophisticated beings, but the possibility of past or present microbial life on Mars remains a subject of intense scientific investigation.

Substantial evidence suggests that water once flowed on Mars, and it seems likely that Mars once had surface or subsurface environments similar to those in which life thrived on the early Earth. If life arose on Mars (or was transported there on meteorites from the early Earth), we may be able to find its fossil remains. It's even possible that life still survives somewhere on Mars, perhaps underground where volcanic heat can keep some water liquid.

We have not yet reached the point where we can undertake a definitive search for life on Mars, but we are rapidly learning about Mars and its history. In this chapter, we'll explore what we've learned to date and what this implies about the possibility of life on the red planet.

8.1 Fantasies of Martian Civilization

Shining brightly and noticeably red in the nighttime sky, Mars has long captured the human imagination. Most of our modern understanding of Mars comes from observations by robotic spacecraft (Figure 8.1). But interest in life on Mars began much earlier, and for decades was a mainstay of popular culture.

How did Mars invade popular culture?

The story begins with the noted English astronomer William Herschel (1738–1822). Though best known for discovering the planet Uranus, Herschel made numerous other astronomical discoveries, usually with help from his sister and fellow astronomer Caroline Herschel (1750–1848). The Herschels often observed Mars through their telescopes, noting its polar ice caps and discovering that the length of its day (24 hours 37 minutes) is similar to that of an Earth day. In a talk presented to Britain's Royal Society in 1784, William Herschel claimed that Mars possessed an atmosphere and that, consequently, "its inhabitants probably enjoy a situation in many respects similar to our own." With the mention of "inhabitants," the possibility of living beings on the red planet had been broached by a respected scientist in an academic setting, and Martians were assumed to exist. (It should be noted that Herschel was not overly particular when it came to populating the cosmos. As far as he was concerned, everything in the solar system was inhabited, including the Moon and the Sun.)

During the following century, Mars rose to the top of the astronomical charts. In 1877, Italian astronomer Giovanni Schiaparelli claimed to have seen a network of 79 linear features that he called *canali*, by which he meant the Italian word for "channels." However, it was often translated as "canals." Coming amid the excitement that followed the 1869

FIGURE 8.1

Mars, photographed by the *Viking* orbiter. The horizontal "gash" across the center is the giant canyon Valles Marineris.

opening of the Suez Canal, Schiaparelli's discovery soon inspired visions of artificial waterways built by a martian civilization. Schiaparelli himself remained skeptical of such claims, and it's not clear whether he even thought the *canali* contained water. But his work caught the imagination of a young Harvard graduate, Percival Lowell (1855–1916).

Lowell, whose degree was in mathematics, came from a wealthy and distinguished New England family. His brother Abbott became famous as a president of Harvard, and his sister Amy gained fame as a poet. After spending a few years as a businessman and as a traveler in the Far East, Percival Lowell turned to astronomy. Impassioned by his belief in the martian canals and enabled by his wealth, Lowell commissioned the building of an observatory in Flagstaff, Arizona. He chose Flagstaff because he thought its dry air and high altitude would limit the blurring caused by Earth's atmosphere, making it easier for him to map the martian canals. The Lowell Observatory opened in 1894 and is still operating today.

Over the next two decades, Lowell mapped close to 200 canals that he claimed to see on Mars, publishing his first book about them in 1895. He assumed Mars's polar caps were made of water ice like Earth's, so he imagined that the canals were built to carry water from the poles to agricultural areas and thirsty cities nearer the equator. From there it was a short leap to imagine the Martians as an old civilization on a dying planet. The global network of canals convinced Lowell that the Martians were citizens of a single, global nation. Such ideas provided the "scientific" basis for H. G. Wells's *The War of the Worlds*, published in 1898. Public belief in Martians became so widespread that, decades later, a radio broadcast of *The War of the Worlds* created a famous panic as many people thought an invasion was actually under way (Figure 8.2).

Think About It Think of as many popular references to a civilization of "Martians" as you can; be sure to consider novels, movies, television shows, advertisements, and music. Do these references tell us anything about the influence of science on the public imagination? Defend your opinion.

Lowell was an effective advocate for the canals, but they do not really exist. Even in his own time, other scientists shot holes through most of Lowell's claims. One notable problem was that most other astronomers did not see any canals either by eye or in photographs—even when using telescopes larger than Lowell's. Lowell's basic assumptions and interpretations also came under fire. Writing in 1907, Alfred Russel Wallace (the co-discoverer with Darwin of evolution by natural selection [Section 5.1]) used physical arguments to suggest that Mars is too cold for liquid water to flow. He also pointed out that Lowell's canals followed straight-line paths for hundreds or thousands of miles, while real canals would be built to follow natural contours of topography (for example, to go around mountains). In summarizing this argument, Wallace wrote that "[a network of canals,] as Mr. Lowell describes, would be the work of a body of madmen rather than of intelligent beings."*

What was Lowell seeing? In a few cases, his canals correspond to real features on Mars. For example, the canal he claimed to see most often (which he called *Agathodaemon*) coincides with the location of the

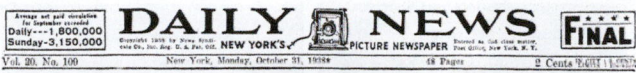

FIGURE 8.2

This front-page story from the New York *Daily News* described the panic caused by a 1938 radio broadcast of *The War of the Worlds*. (The radio voice was that of Orson Welles, no relation to the novelist H. G. Wells.)

---

*Excerpted in K. Zahnle, "Decline and Fall of the Martian Empire," *Nature*, vol. 412, July 12, 2001.

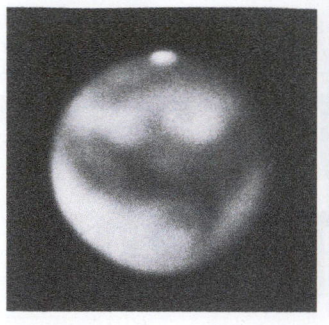

FIGURE 8.3

Can you see how the markings on Mars in the telescopic photo on the left might have resembled the geometrical features in the drawing by Percival Lowell on the right? Try squinting your eyes.

huge canyon network now known as Valles Marineris (see Figure 8.1). A few other canals also roughly follow the contours of real features on Mars, but most of the canals were pure fantasy. Figure 8.3 compares a telescopic photo of Mars with one of Lowell's maps of the same regions. You can probably see how the dark and light regions match up in the photo and the drawing, but seeing any canals requires a vivid imagination. (Some scholars speculate that the particular optics of his telescope caused Lowell to map images of blood vessels on his own retina as some of the canals.)

Lowell's story illustrates both the pitfalls and the triumphs of modern science. The pitfall is that individual scientists, no matter how upstanding and dedicated, may still bring personal biases to bear on their scientific work. In Lowell's case, he was so convinced of the existence of canals and Martians that he simply ignored all evidence to the contrary. But the story's ending shows why modern science ultimately is so successful. Despite Lowell's stature, other scientists did not accept his claims on faith. Instead, they sought to confirm his observations and to test his underlying assumptions. They found that Lowell's claims fell short on all counts. As a result, Lowell became an increasingly isolated voice as he continued to advocate a viewpoint that was clearly wrong.

8.2 A Modern Portrait of Mars

The public debate about martian canals and cities was not entirely put to rest until NASA began sending spacecraft to Mars. In 1965, NASA's *Mariner 4* spacecraft flew to within 6000 miles of the martian surface, transmitting a few dozen low-resolution images of the landscape below. Mars's surface was littered with craters, not canals, and measurements of the atmospheric pressure and temperature made from the spacecraft indicated a cold, dry planet seemingly incapable of supporting life.

Nevertheless, all was not lost when it came to the potential for life on Mars. There was no evidence of any intelligent beings, but the thin atmosphere and the polar caps left open the possibility of the existence of microbes or perhaps even some primitive plants or animals. On July 20, 1976, seven years to the day after Neil Armstrong's history-making walk on the Moon and nearly a century since Schiaparelli's description of *canali*, the thin skies above Mars were pierced by a NASA space probe. The *Viking 1* lander touched down on the Chryse Planitia, a sprawling, rock-strewn plain about 1300 kilometers north of the martian equator. Two months later, *Viking 2* landed on the other side of the planet. Meanwhile, two *Viking* orbiters began studying the planet from above.

When the *Viking* landers' cameras opened their eyes in the frigid martian air, they found a bleak landscape with red dust and scattered rocks. No creatures stared back at the cameras, and no plants were huddled in the weak sunlight. For months the images continued to come in, but the view scarcely changed. Nothing grew other than some occasional patches of frost, and nothing moved other than windblown dust (Figure 8.4). Though neither lander could move from the spot where it had settled, each had a robotic arm with which it collected soil for some onboard experiments designed to look for microbes (see Section 8.4).

The *Viking* orbiters and landers provided a wealth of scientific data about Mars. But they also left many questions unanswered, and the scientific community was itching for follow-up missions. Unfortunately,

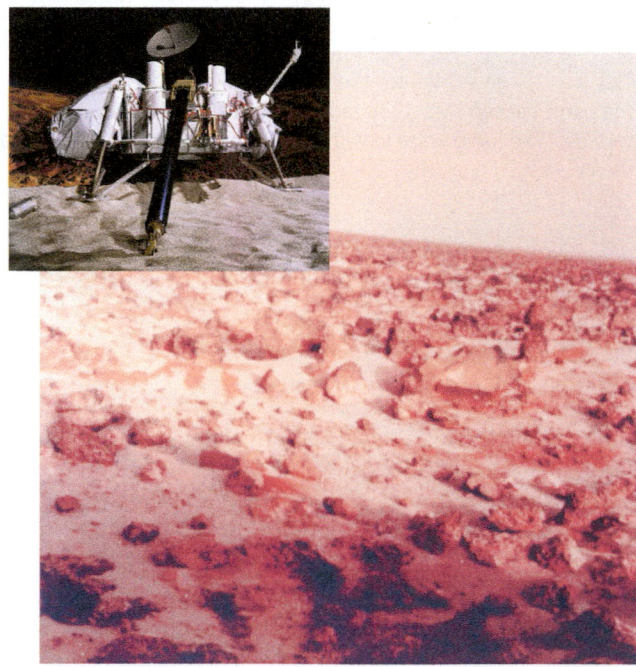

FIGURE 8.4

The surface of Mars photographed by the *Viking 2* lander in 1979, showing a thin coating of ice on the rocks and soil. The inset shows a working model (actually, a spare spacecraft) of the *Viking* landers, identical to those that landed on Mars, on display at the National Air and Space Museum in Washington, D.C.

budgetary and political considerations, along with the failure of two Russian missions to Mars (*Phobos 1* and *2*) and one American mission (*Mars Observer*), all conspired to stop spacecraft exploration of Mars for some 20 years. The long mission drought did not end until July 4, 1997, with the landing of *Pathfinder* and its little rover, *Sojourner* (Figure 8.5). Named for Sojourner Truth, an African American heroine of the Civil War era, the rover could travel only a few tens of meters—just enough for it to check the chemical compositions of nearby rocks.

As of 2015, nine more robotic spacecraft had reached Mars successfully (five orbiters, three rovers, and one stationary lander), the most recent being the *Curiosity* rover (Figure 8.6) and the orbiters from NASA's *MAVEN* mission and India's *Mars Orbiter Mission*. Additional missions may have reached Mars by the time you read this book (see Section 8.4). By combining data from these missions with past data, we are beginning to put together a realistic portrait of the past and present habitability of Mars.

What is Mars like today?

The present-day surface of Mars may look much like some deserts or volcanic plains on Earth, but its thin atmosphere makes the conditions quite different. Table 8.1 summarizes basic Mars data. The low atmospheric pressure—less than 1% of that on Earth's surface—means the air is so thin that a visiting astronaut could not survive outside for more than a few minutes without a pressurized spacesuit. The atmosphere is made mostly of carbon dioxide, but the total amount of gas is so small that it creates only a weak greenhouse effect. As a result, the temperature is usually well below freezing, with a global average of about −50°C (−58°F). The lack of oxygen means we could not breathe the thin air, and it also explains why Mars lacks an ozone layer (recall that ozone [O_3] is made of oxygen). The lack of ozone allows much of the Sun's damaging ultraviolet radiation to pass unhindered to the surface.

Nevertheless, martian conditions are much less extreme than those on the Moon (mainly because of the moderating effects of the atmosphere), and it's easy to imagine future astronauts living and working in airtight research stations while occasionally donning space suits for outdoor excursions. Martian surface gravity is about 38% of that on Earth, so everyone and everything would weigh about 38% of Earth weight. Astronauts could walk around easily even while wearing space suits with heavy backpacks. It would also probably be easy to adapt to patterns of day and night, since the martian day is only about 40 minutes longer than an Earth day.

THE LACK OF SURFACE LIQUID WATER The low atmospheric pressure explains one of the key facts relevant to the search for life on Mars: There are no lakes, rivers, or even puddles of liquid water on the surface of Mars today. We know this not only because we've studied most of the surface in reasonable detail but also because the surface conditions do not allow it. In most places and at most times, Mars is so cold that any liquid water would immediately freeze into ice. Even when the temperature rises above freezing, as it often does at midday near the equator, the air pressure is so low that liquid water would quickly evaporate. In other words, liquid water is *unstable* on Mars today: If you put on a space suit and took a cup of water outside your pressurized spaceship, the wa-

FIGURE 8.5 The view from the *Pathfinder* lander (partially visible in the foreground); the scattered rocks were probably carried to the site by an ancient flood. The little rover, *Sojourner,* is at the upper right, studying a rock that scientists named Yogi.

FIGURE 8.6 This self-portrait, assembled from dozens of individual images, shows the *Curiosity* rover on Mars.

TABLE 8.1 Basic Mars Data

Average Distance from Sun	1.52 AU = 227.9 million km
perihelion distance	206 million km
aphelion distance	249 million km
Orbital Period	1.881 Earth yr
Equatorial Radius	3397 km
Mass (Earth = 1)	0.107
Rotation Period	24 hr 37 min
Axis Tilt	25°
Surface Gravity (Earth = 1)	0.38
Atmospheric Composition	95% CO_2, 2.7% N_2, 1.6% argon
Average Surface Temperature	−50°C
Average Surface Pressure	0.007 bar*

*1 bar ≈ sea level pressure on Earth

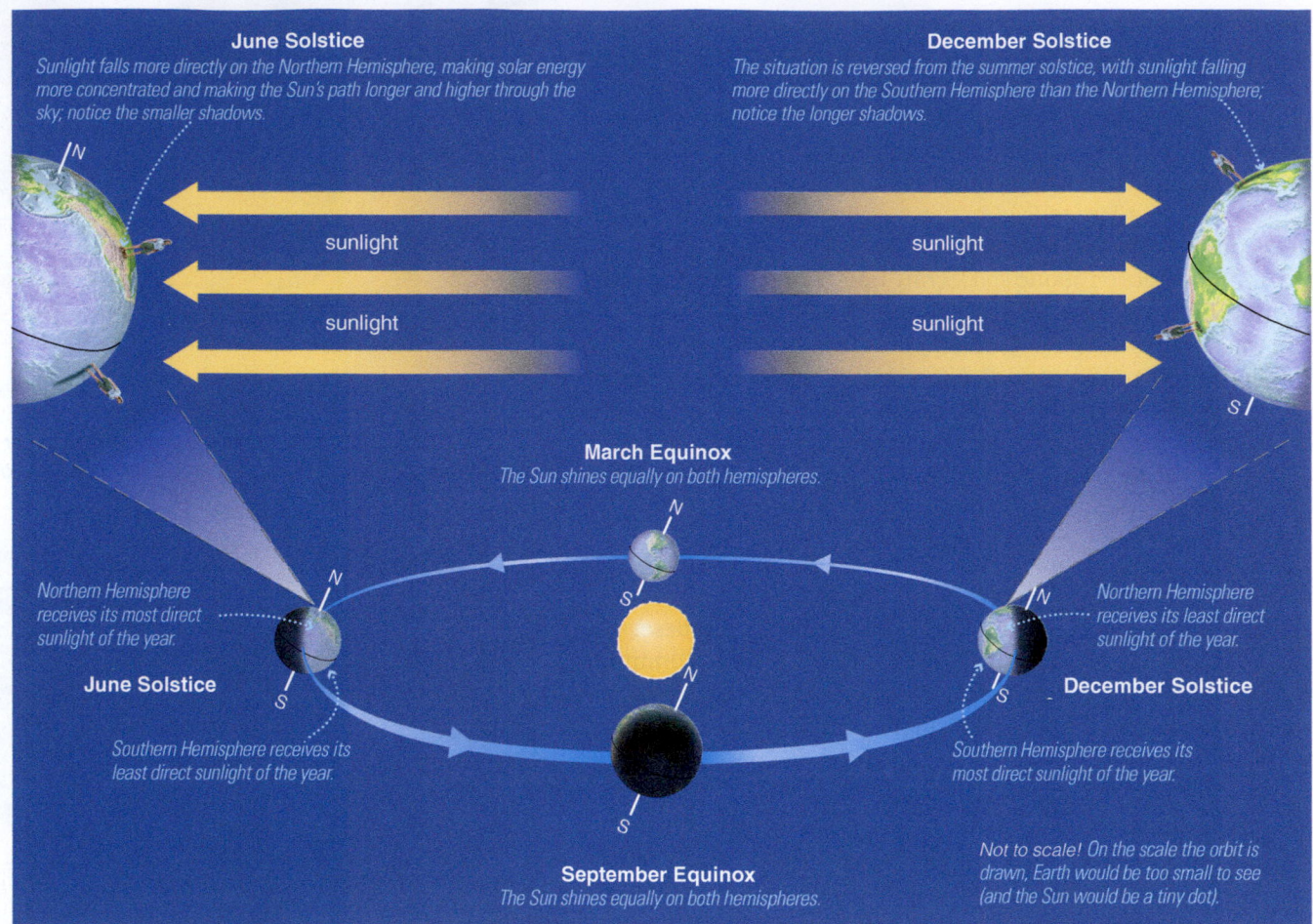

June Solstice
Sunlight falls more directly on the Northern Hemisphere, making solar energy more concentrated and making the Sun's path longer and higher through the sky; notice the smaller shadows.

December Solstice
The situation is reversed from the summer solstice, with sunlight falling more directly on the Southern Hemisphere than the Northern Hemisphere; notice the longer shadows.

sunlight

sunlight

sunlight

sunlight

March Equinox
The Sun shines equally on both hemispheres.

Northern Hemisphere receives its most direct sunlight of the year.

June Solstice

Southern Hemisphere receives its least direct sunlight of the year.

Northern Hemisphere receives its least direct sunlight of the year.

December Solstice

Southern Hemisphere receives its most direct sunlight of the year.

Not to scale! On the scale the orbit is drawn, Earth would be too small to see (and the Sun would be a tiny dot).

September Equinox
The Sun shines equally on both hemispheres.

FIGURE 8.7   INTERACTIVE FIGURE
Earth's seasons are caused by the tilt of the axis. Notice that the axis points in the same direction (toward Polaris) throughout the year, which means the Northern Hemisphere is tipped toward the Sun on one side of the orbit and away from the Sun on the other side. The same is true for the Southern Hemisphere, but on opposite sides of the orbit.

Seasons on Mars

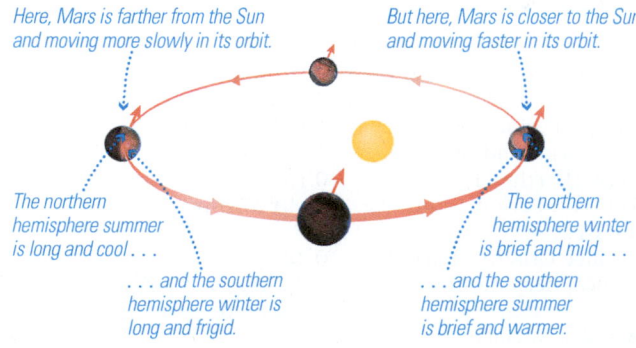

Here, Mars is farther from the Sun and moving more slowly in its orbit.

But here, Mars is closer to the Sun and moving faster in its orbit.

The northern hemisphere summer is long and cool . . .

The northern hemisphere winter is brief and mild . . .

. . . and the southern hemisphere winter is long and frigid.

. . . and the southern hemisphere summer is brief and warmer.

FIGURE 8.8
The ellipticity of Mars's orbit makes seasons more extreme in the southern hemisphere than in the northern hemisphere.

ter would almost immediately either freeze or evaporate away (or some combination of both). However, as we'll discuss shortly, Mars almost certainly had liquid water in the past, and still has ample water ice and some water vapor and perhaps even pockets of liquid water underground.

MARTIAN SEASONS AND WINDS Recall that Earth has seasons because of the tilt of our planet's axis (Figure 8.7). Earth's axis remains pointed in the same direction (toward the north star, Polaris) throughout the year, which means the Northern and Southern Hemispheres are angled toward the Sun on opposite sides of Earth's orbit. It is summer when your hemisphere is angled toward the Sun, and winter when it is angled away.

Mars's axis tilt today is only slightly greater than Earth's (25° versus 23.5°), so Mars has seasons for the same basic reason. However, the martian seasons differ from Earth seasons in two important ways. First, because the martian year is nearly twice as long as an Earth year, each season lasts nearly twice as long on Mars. Second, while Earth's nearly circular orbit means that tilt is the only significant factor in our seasons, Mars's more elliptical orbit puts Mars significantly closer to the Sun during southern hemisphere summer and farther from the Sun during southern hemisphere winter (Figure 8.8). Mars's southern hemisphere therefore has more extreme seasons (shorter and warmer summers, longer and colder winters) than its northern hemisphere. (The seasons differ in length because planets move faster near perihelion and slower near aphelion, in accord with Kepler's second law [Section 2.2].)

Seasonal changes lead to several major features of martian weather. Temperatures at the winter pole drop so low (about −130°C) that carbon dioxide condenses into "dry ice" at the polar cap; that is why the polar caps are so much larger in winter than in summer (Figure 8.9). Meanwhile, frozen carbon dioxide at the summer pole vaporizes into carbon dioxide gas,* and by the peak of summer only a residual cap of water ice remains (Figure 8.10). The atmospheric pressure therefore increases at the summer pole and decreases at the winter pole. Overall, as much as one-third of the total carbon dioxide of the martian atmosphere moves seasonally between the north and south polar caps.

The strong winds associated with the cycling of carbon dioxide gas can initiate huge dust storms, particularly when the more extreme summer approaches in the southern hemisphere (Figure 8.11). At times, the martian surface becomes almost completely hidden by airborne dust. As the dust settles out, it can change the surface appearance over vast areas (for example, by covering dark regions with brighter dust); such changes fooled astronomers of the past into thinking they were seeing seasonal changes in vegetation.

Martian winds can also spawn *dust devils*, swirling winds that you may have seen over desert sands or dry dirt on Earth. Dust devils look much like miniature tornadoes, but they rise up from the ground rather than coming down from the sky. The air in a dust devil is heated from below by the sunlight-warmed ground; it swirls because of the way the rising air interacts with prevailing winds. Dust devils on Mars are especially common during summer in either hemisphere. While many are quite small (Figure 8.12), some can be far larger than their counterparts on Earth.

COLOR OF THE MARTIAN SKY Martian winds and dust storms leave Mars with perpetually dusty air, which helps explain the colors of the martian sky. The air on Mars is so thin that, without the suspended dust, the sky would be essentially black even in daytime. However, light scattered by the suspended dust tends to give the sky a yellow-brown color. Different hues can occur as the amount of suspended dust varies, and in the mornings and evenings. For example, the martian sunset photo that opens this chapter shows the scene approximately as it would look to the human eye (but with slightly exaggerated colors).

What are the major geological features of Mars?

The surface of Mars may be desolate and barren today, but it was not always so. Many surface features appear to have been shaped by liquid water, leading scientists to conclude that Mars must once have had a much more hospitable climate. Before we discuss the evidence for

---

*Note that dry ice vaporizes directly from solid phase into gas phase, a process more technically called *sublimation*. On Mars, the low atmospheric pressure means water ice sublimates directly into gas in the same way. If you are not familiar with dry ice sublimation, it is easy to obtain some dry ice and watch this phenomenon for yourself, or you can find videos.

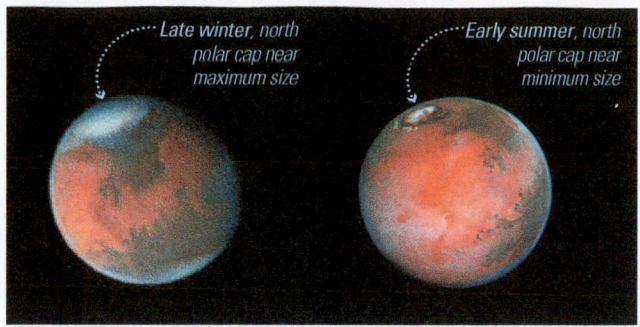

FIGURE 8.9
These images from the Hubble Space Telescope show the dramatic change in the size of the north polar ice cap with the martian seasons. (Mars is oriented slightly differently in the two photos.)

FIGURE 8.10
This image, from *Mars Global Surveyor*, shows the residual south polar cap during summer. A layer of frozen carbon dioxide around 8 meters thick overlies a much thicker cap of water ice. In winter, the whole area shown in the image is covered in CO_2 frost.

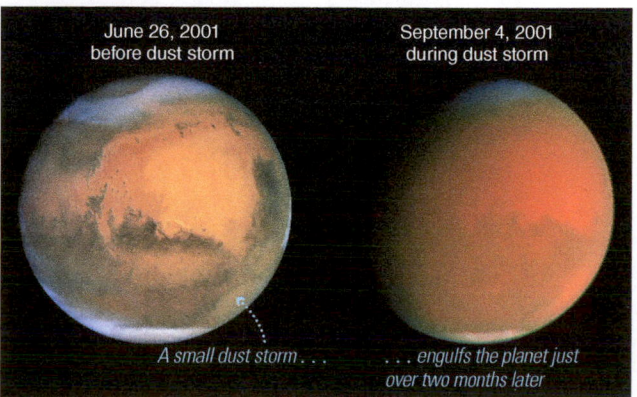

FIGURE 8.11
These two Hubble Space Telescope photos contrast the appearance of the same face of Mars in the absence (left) and presence (right) of a global dust storm.

FIGURE 8.12   INTERACTIVE PHOTO
This photograph shows a dust devil on Mars, photographed by the *Spirit* rover.

FIGURE 8.13   INTERACTIVE PHOTO
This image showing the full surface of Mars was made by combining more than 1000 images with more than 200 million altitude measurements from the *Mars Global Surveyor* mission. Several key geological features are labeled, and landing sites of Mars missions are marked. On the same scale, a map of Earth would be about twice as tall and twice as wide.

surface water and ideas about the climate history of Mars, it's useful to get our bearings by looking at the large-scale geographic features of the planet.

A MAP OF MARS Figure 8.13 shows the full surface of Mars, with the poles at the top and bottom and the equator running horizontally across the middle (in much the same way that an atlas shows the full globe of Earth). Study the map briefly, and familiarize yourself with major features such as the polar caps, the Tharsis bulge, Valles Marineris, and the large impact crater known as Hellas Basin. You should also recognize many smaller impact craters, particularly in the southern hemisphere, and numerous large volcanoes—including Olympus Mons—which you can identify by their dome shapes and central calderas (the "craters" in the tops of volcanoes). To understand the scale of these features, recall that Mars is about half as large in diameter as Earth, so its surface area is about one-fourth that of Earth (surface area is proportional to the square of the radius). Note that, because water covers about three-fourths of Earth's surface, the total land area of Mars is about the same as the total land area of Earth.

Think About It Any flat map of a round world will distort some features; the map projection in Figure 8.13 (a *Mollweide* projection) allows *areas* of different features to be accurately compared. Find a similar projection of Earth and print it so that it is scaled correctly relative to Figure 8.13 (that is, twice as long and wide). Compare the sizes of various Mars features to the sizes of familiar features on Earth.

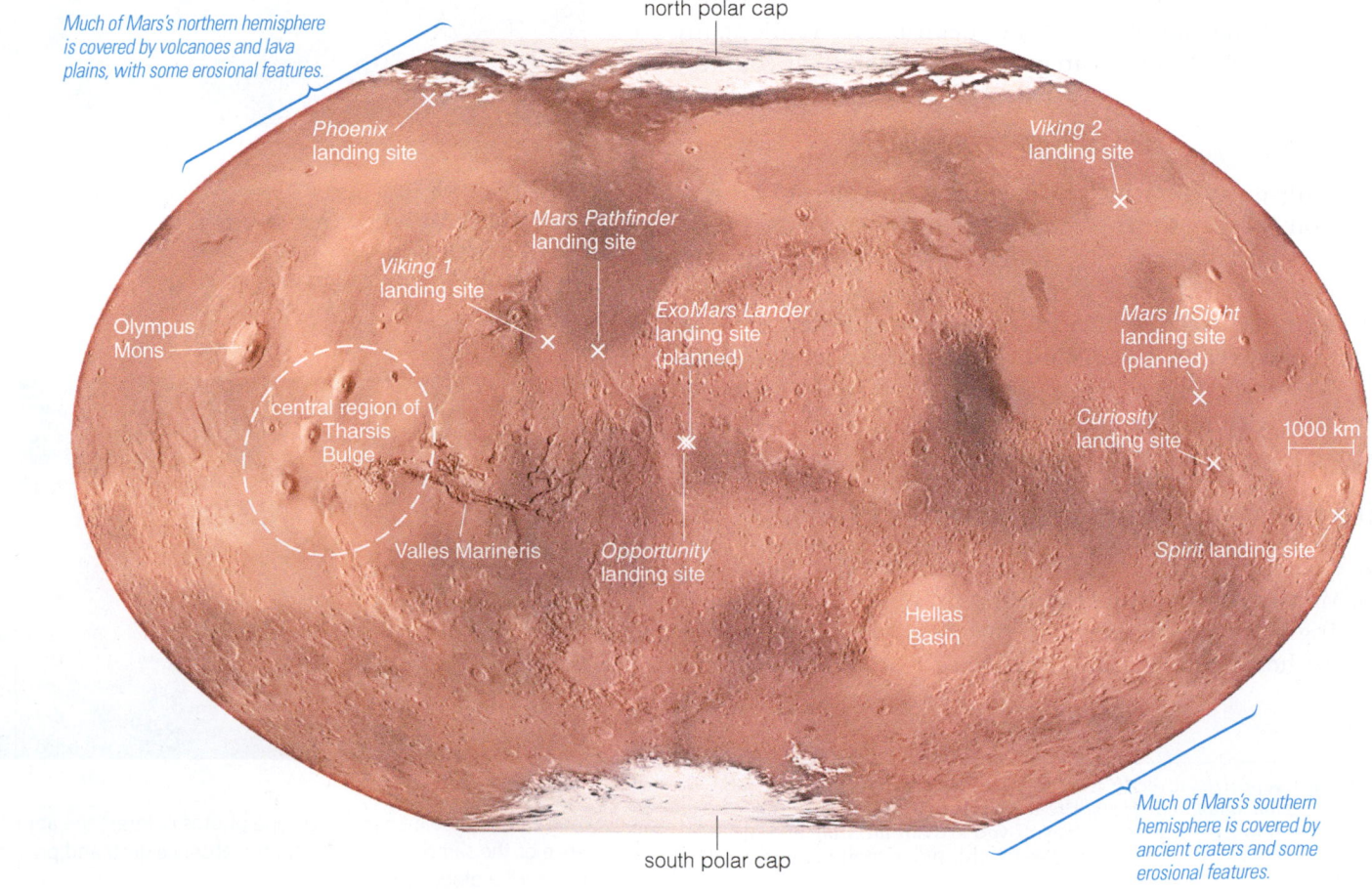

Much of Mars's northern hemisphere is covered by volcanoes and lava plains, with some erosional features.

north polar cap

Phoenix landing site

Viking 2 landing site

Mars Pathfinder landing site

Viking 1 landing site

ExoMars Lander landing site (planned)

Mars InSight landing site (planned)

Olympus Mons

central region of Tharsis Bulge

Curiosity landing site

1000 km

Valles Marineris

Opportunity landing site

Spirit landing site

Hellas Basin

south polar cap

Much of Mars's southern hemisphere is covered by ancient craters and some erosional features.

DIFFERING SURFACE REGIONS One of the most striking features in Figure 8.13 is the dramatic difference in terrain between the northern and southern hemispheres. The southern hemisphere is heavily scarred by impact craters, while craters are few and far between in the northern plains. The visible difference in cratering corresponds to general differences in elevation (Figure 8.14) and crustal thickness. Most of the northern hemisphere is well below the average martian surface level, and measurements indicate a relatively thin crust (about 40 kilometers thick) in these regions. In contrast, most of the southern hemisphere is high-altitude highlands residing on a thicker crust (about 80 kilometers thick). No one knows the reason for this "martian north-south dichotomy," though one hypothesis suggests a giant impact blasting away crust from the northern hemisphere.

Recall that differences in cratering tell us something about surface ages [Section 4.3]: Older surfaces are more heavily cratered than younger ones. We therefore conclude that the heavily cratered southern highlands are generally an older surface than the northern plains. More detailed crater counts have led planetary scientists to divide the surface of Mars into regions of *three* different ages (Table 8.2). The most heavily cratered regions must still look much as they did about 3.7 billion years ago, shortly after the end of the heavy bombardment; these regions therefore represent the "early" era in the history of Mars (more formally called the Noachian ["no-AH-ki-an"] era). Regions that are more moderately crowded with craters represent the "middle" (or Hesperian) era, which apparently ended about 3 billion years ago. The most lightly cratered regions, which include much of the northern plains and the lava-covered terrain around the Tharsis volcanoes, represent the "recent" (or Amazonian) era on Mars. Keep in mind that there is a great deal of uncertainty in ages based on crater counts, so the given age ranges for the three eras are only approximate. The timing of the end of the middle era (and beginning of the recent era) is especially uncertain and might be anywhere from about 3.5 to 2 billion years ago. More precise ages will be known only after we collect rocks from the different eras and measure their ages through radiometric dating [Section 4.2].*

Think About It Much like Mars, Earth's surface also has regions of different ages. Which regions of Earth's surface are generally the youngest? Explain.

VOLCANISM AND TECTONICS ON MARS Mars shows abundant evidence of volcanism. The northern plains show features that are characteristic of lava flows, suggesting that eruptions of an extremely fluid lava covered up the older impact craters. Interestingly, we can see faint "ghost" craters

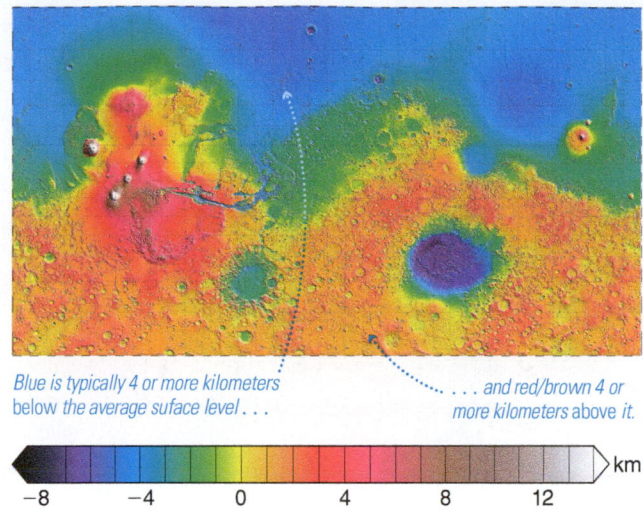

Blue is typically 4 or more kilometers below *the average suface level . . .*   *. . . and red/brown 4 or more kilometers above it.*

km
−8   −4   0   4   8   12

FIGURE 8.14
This map uses the same data as Figure 8.13 but is color-coded to show changes in elevation (and uses a different projection [equirectangular]). Notice the striking difference in elevation between the northern and southern hemispheres. In general, the crust is also thinner in regions with lower elevations than in regions with higher elevations.

TABLE 8.2 Eras of Martian History

Era	Time Period	Representative Surface Region
Early (Noachian)	Before 3.7 billion years ago	Heavily cratered southern highlands
Middle (Hesperian)	About 3.7 to 3 billion years ago	Moderately cratered terrain south of Tharsis
Recent (Amazonian)	Less than 3 billion years ago	Lightly cratered northern plains; volcanic slopes

---

*Scientists have successfully used *Curiosity*'s on-board Sample Analysis on Mars (SAM) instrument—which was designed for other purposes—to date a rock by looking at potassium-argon decay. The measured age (3.86 to 4.56 billion years old) had a fairly large uncertainty, but agreed with what had been expected from orbital data based on cratering rates. More important, it demonstrated that future rovers could be designed to do more such analysis.

in some of these regions, suggesting that the lava flows were not thick enough to completely erase the underlying features and confirming that the entire planet was once densely cratered. Plenty of mysteries remain, however. For example, no one knows why volcanism should have affected the northern plains so much more than the southern highlands, though perhaps it is a consequence of the thinner crust in northern regions. It's also possible that in some places the craters were erased by sedimentary rather than volcanic processes. In this case, the craters may have been submerged in sand and other martian material that was transported by wind and water, eventually building up layers of sediment that cover the landscape.

Regardless of their role in reshaping the northern plains, Mars boasts impressive volcanoes, many of which are concentrated on or near the continent-size Tharsis Bulge (some 4000 kilometers across). Most of Tharsis lies at least several kilometers above the average martian surface level, suggesting that it was created by a long-lived plume of rising mantle material that bulged the surface upward and provided the molten rock for the eruptions that built the giant volcanoes.

The Tharsis volcanoes dwarf any found on Earth. The largest of them, Olympus Mons (Figure 8.15), is the tallest known mountain in the solar system. Its peak rises about 26 kilometers above the average martian surface level, or about three times as high as Mount Everest stands above sea level on Earth. Its base is some 600 kilometers across, large enough to cover an area the size of Arizona, and is rimmed by a cliff that in places is 6 kilometers high. Two factors probably explain why the martian volcanoes are so much larger than volcanoes on Earth. First, Mars's weaker gravity makes it easier for tall structures to be built up. Second, the lack

FIGURE 8.15
Olympus Mons is the tallest known mountain in the solar system.

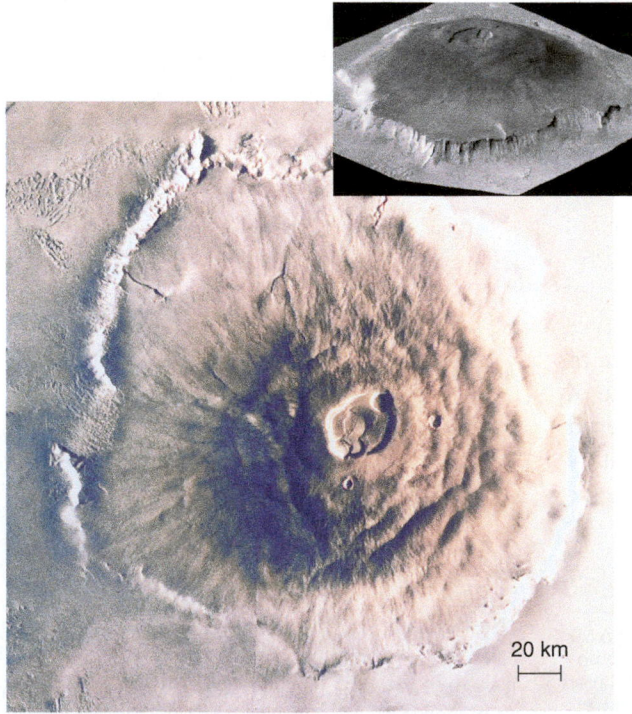

a Olympus Mons, photographed from orbit. The inset shows a 3-D perspective on this immense volcano.

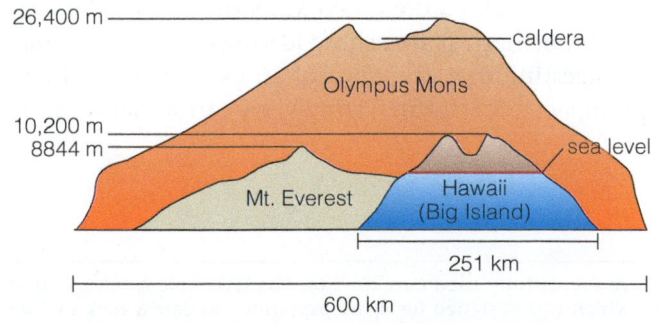

b This diagram compares the size of Olympus Mons to those of Mount Everest and the Big Island of Hawaii. The latter is shown as it appears starting from the bottom of the ocean, with the blue region indicating the part that lies below sea level.

of plate tectonics on Mars means that mantle plumes remain stationary relative to the surface, building up huge, single mountains. In contrast, the gradual motion of Earth's crust due to plate tectonics means that a single mantle plume tends to build a chain of volcanic islands (see Figure 4.22).

East of Tharsis and just south of the equator is the long, deep system of valleys called *Valles Marineris* (Figure 8.16). Named for the *Mariner 9* spacecraft that first imaged it, Valles Marineris is as long as the United States is wide and almost four times as deep as the Grand Canyon. No one knows exactly how Valles Marineris formed, but its location (see Figure 8.13) suggests a link to the Tharsis Bulge. Perhaps it formed through tectonic stresses accompanying the uplift of material that created Tharsis, cracking the surface and leaving the tall cliff walls of the valleys. A few features of the valley network appear to have been shaped by flowing water, and spectra from orbit show the presence of minerals likely to have formed in water. Some of the canyon walls also show evidence of layering that may have been caused by deposits of sediments, though the layering could also be due to repeated lava flows. In any event, the canyon is so deep that, if we are correct in assuming it was created by uplift, some of its walls must once have been several kilometers underground, where they may have been exposed to liquid water. For all these reasons, Valles Marineris may be one of the best places to look for fossil evidence of past martian life.

By examining the types of geological features that appear on surfaces of different ages on Mars, we can get an idea both of what processes helped shape the surface and of when they operated. For example, we can look at features that indicate volcanic eruptions, such as lava flows or volcanoes, and deduce the history of volcanism. Such studies suggest that the frequency of volcanic eruptions on Mars has decreased over time, just as we would expect for a planet small enough to have lost much of its internal heat by now.

We have not witnessed any ongoing volcanic or tectonic activity on Mars, and we expect Mars to be much less volcanically active than Earth, because its smaller size has allowed its interior to cool much more. However, the era of martian volcanism may not be completely over. Crater counts on the slopes of martian volcanoes suggest that some lava flows may have occurred as recently as tens of millions of years ago (and perhaps more recently than that), which is not so long ago in geological terms. In addition, radiometric dating of meteorites that appear to have come from Mars (so-called *martian meteorites* [Section 6.2]) shows some of them to be made of volcanic rock that solidified from molten lava as little as 180 million years ago—also quite recent in the $4\frac{1}{2}$-billion-year history of the solar system. This suggests that Mars still retains some internal heat. No one knows if it is enough to cause the volcanoes to erupt again in the future, but it is almost certainly enough to melt some underground ice into liquid water.

What evidence tells us that water once flowed on Mars?

We now turn our attention to the evidence that makes scientists confident that Mars once had substantial amounts of flowing water. It is this evidence that makes Mars a prime candidate in the search for past or present life beyond Earth.

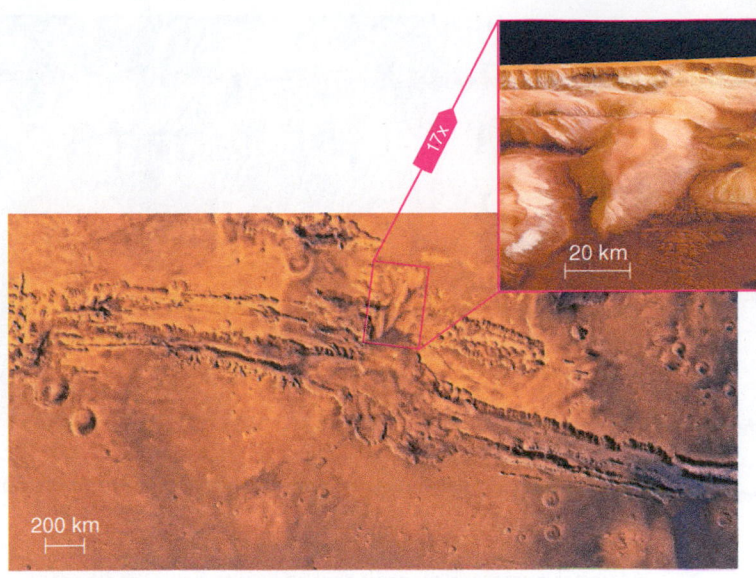

FIGURE 8.16

Valles Marineris is a huge system of valleys on Mars created in part by tectonic stresses. It extends nearly a fifth of the way around the planet (see Figure 8.13), and in some places is 10 kilometers deep. The inset shows a perspective view looking north across the center of the canyon.

a This photo from a *Viking* orbiter shows what appears to be a network of tributaries flowing from the upper left into the larger "river" near the lower right.

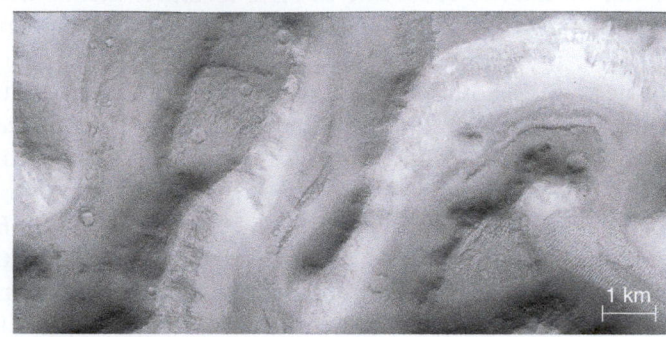

b This photo, taken by the *Mars Reconnaissance Orbiter*, shows what appears to be a meandering riverbed, now filled with dunes of windblown dust.

FIGURE 8.17

Mars has numerous channels that appear to be dry riverbeds. Notice the many small craters in the photos, which tell us that the riverbeds dried out by at least about 3 billion years ago.

ORBITAL EVIDENCE The first evidence of past water came from photos taken by *Mariner 9* and the *Viking* orbiters, some of which showed features that look much like dry riverbeds on Earth seen from above (Figure 8.17a). More recent orbiters have photographed these channels with much higher resolution (Figure 8.17b). Careful study indicates that the channels were almost certainly carved by running water, though no one knows whether the water came from runoff after rainfall, from erosion by water-rich debris flows, or from an underground source. Crater counts in and near the channels, along with study of the local terrain, indicate that the channels are in general at least about 3 billion years old, meaning that water has not flowed through most of them since that time. Nevertheless, they tell us an important story about the martian past: Their nature suggests they were carved over a long enough period of time that liquid water must have been stable at or just below the surface. Because the low temperature and atmospheric pressure make liquid water unstable today, we conclude that Mars must have had a warmer and thicker atmosphere during at least some times in its distant past.

Careful examination of impact craters also provides evidence that Mars had surface water long ago. Figure 8.18 shows just three among thousands of orbital images that provide evidence of water erosion. Figure 8.18a shows a broad region of the ancient, heavily cratered southern highlands. Notice the indistinct rims of many large craters and the relative lack of small craters. Both facts argue for ancient rainfall, which would have eroded crater rims and erased small craters altogether. Figure 8.18b shows a three-dimensional perspective of the surface that suggests water once flowed between two ancient crater lakes. Figure 8.18c shows what appears to be a river delta where water flowed into an ancient crater.

Even more convincing evidence comes from images and spectra that tell us about the mineral composition of the martian surface. Three general types of **hydrated minerals**—minerals containing water or hydroxide (OH), indicating that they formed in the presence of liquid water—have been found at numerous locations on Mars: clay minerals, hydrated sulfates, and hydrated silica, more commonly known as opal. For example, the green color coding in Figure 8.18c indicates the pres-

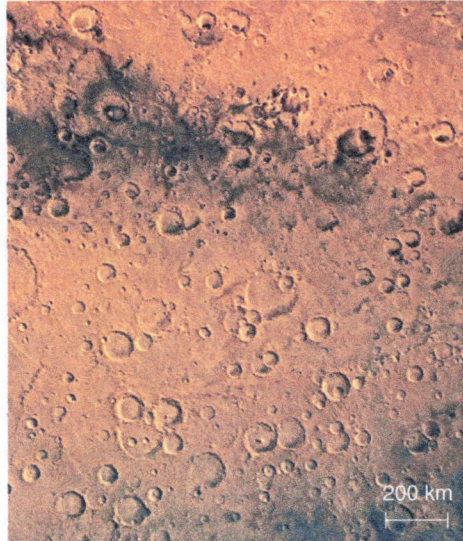

a This photo shows a broad region of the southern highlands. The eroded rims of large craters and the lack of many small craters suggest erosion by rain, wind, or glaciers.

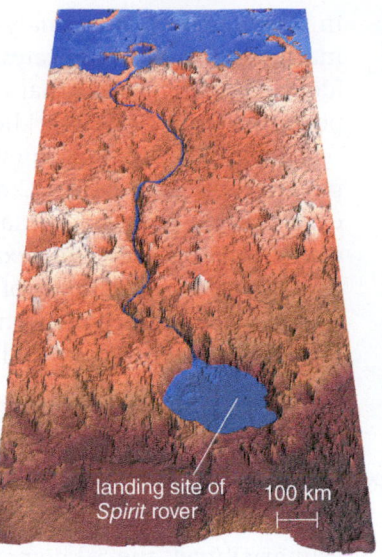

b This computer-generated perspective shows how a martian valley forms a natural passage between two possible ancient lakes (shaded blue). Vertical relief is exaggerated 14 times to reveal the topography.

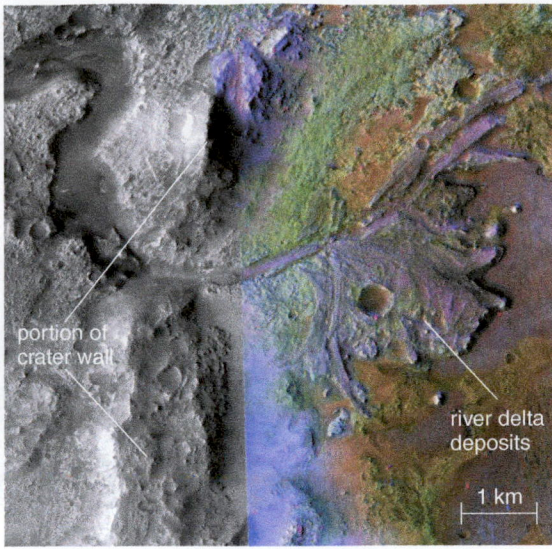

c Combined visible/infrared image of an ancient river delta that formed where water flowing down a valley emptied into a lake filling a large crater (portions of the crater wall are identified). Clay minerals are identified in green.

FIGURE 8.18
More evidence of past surface water on Mars.

ence of clay minerals that may have been deposited by sediments flowing down the river. The opaline minerals are particularly significant, for two reasons. First, they are thought to form in hot springs or hydrothermal environments—and recall that such environments may have been important to the origin of life on Earth [Section 6.1]. Second, some of the regions in which they are found appear to have formed as much as a billion years later than the thick, ancient clay deposits. If this interpretation is correct, the timing suggests that Mars remained wet for an extended period in its ancient history, giving more time for life to arise and evolve. Figure 8.19 shows a region where the *Mars Reconnaissance Orbiter* detected opal near Valles Marineris.

ROVER MINERAL EVIDENCE Surface studies further strengthen the case for past water. In 2004, the robotic rovers *Spirit* and *Opportunity* landed on opposite sides of Mars (see Figure 8.13). The twin rovers long outlasted their design lifetime of 3 months, with *Spirit* lasting more than 6 years and *Opportunity* still going as this book was being written, more than 11 years (and 42 kilometers of travel) after arrival. The more recent *Curiosity* rover (see Figure 8.6) carries the most powerful set of scientific instruments ever landed on another world, including cameras, drills, microscopes, rock and soil analyzers, a laser to vaporize rock, and a spectrograph to analyze the vaporized material.

All three rovers have found abundant mineral evidence of past liquid water. For example, rocks at the *Opportunity* landing site show a layered structure and odd indentations indicative of formation in water, and their composition reveals hematite and other minerals (such as the sulfur-rich mineral jarosite) that often form in water. Perhaps most significantly, the rock contains tiny hematite spheres (nicknamed "blueberries") that are strikingly similar in both appearance and composition to hematite spheres found on Earth (Figure 8.20). The ones on Earth clearly formed

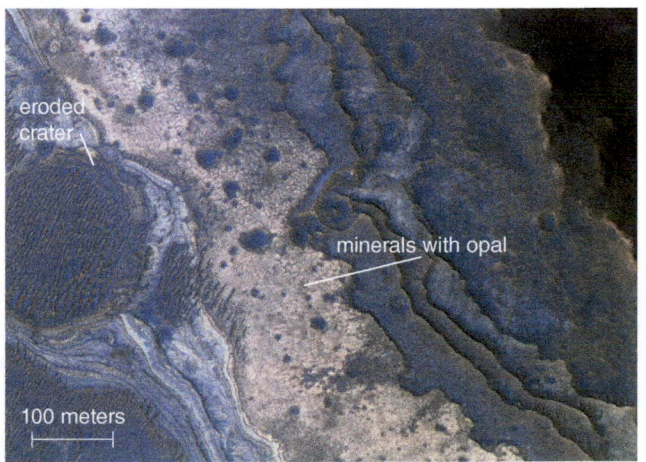

FIGURE 8.19
This color-coded image from the *Mars Reconnaissance Orbiter* shows one of many places where spectral data indicate the presence of opal, possibly formed in hot springs or similar environments.

Mars (Endurance Crater)   **Earth (Utah)**

FIGURE 8.20
Small, round hematite "blueberries" (foreground) are found on both Earth and Mars. Those on Earth formed from sedimentary rock (like that in the background) in water, then later eroded out and rolled downhill. The martian "blueberries" probably formed similarly. (The background rocks are about twice as far away from the camera in the Earth photo as in the Mars photo [taken by the *Opportunity* rover]; the Mars photo combines infrared and visible light.)

in water, and detailed analysis of the structure and composition of the martian blueberries indicates that they formed similarly. Overall, the evidence clearly suggests that *Opportunity* landed at a site that was once a pond or shallow lake of acidic and salty water.

Even more impressive evidence has been found as *Curiosity* has explored a site called Gale Crater, which scientists had already suspected of having once held an ancient lake. The rover's observations have confirmed this view. For example, *Curiosity* drove through an ancient streambed, where clumps of pebbles with rounded surfaces clearly indicate formation in flowing water (Figure 8.21a), and found numerous rock formations that are sediments laid down in water (Figure 8.21b). Moreover, chemical analysis of the region studied by *Curiosity* indicates that the lake that once covered it contained relatively pure ("drinkable") water. In other words, the Gale Crater region was almost certainly habitable in the distant past.

Curiosity is continuing its journey through Gale Crater to study its primary target, the 5-kilometer-tall Mount Sharp (more formally called Aeolis Mons) that is the crater's central peak. Mount Sharp was chosen as a destination because images from orbit indicate that it contains sedimentary rock layers dating to many different times over the past several billion years, an assumption that early images from *Curiosity* appear to confirm (Figure 8.22). Scientists hope that careful study of these layers by *Curiosity* will help us learn much more about Mars's geological and climatological history, and perhaps even shed light on whether Mars has ever been home to life.

Think About It Find the current status of the *Curiosity* and *Opportunity* rovers and images they have taken recently. Have they made any significant discoveries since this book went to press in 2015?

THE EXTENT AND TIMING OF ANCIENT WATER The case for past liquid water on Mars now seems very strong. But was the water shallow and localized, or widespread and deep? Did it exist in liquid form only intermittently,

FIGURE 8.21
Evidence of past water in Gale Crater, from the *Curiosity* rover.

Mars   **Earth**

a Clumps of rounded pebbles found by *Curiosity* in an ancient streambed on Mars show structure nearly identical to that found in a typical streambed on Earth, providing strong evidence that the pebbles were rounded by flowing water.

b The even layers of the foreground rocks in this image from *Curiosity* are characteristic of sediment deposited over time in the delta of a river that emptied into a lake.

FIGURE 8.22
A view of the base of Mount Sharp showing the tilted sedimentary rock layers that scientists hope will reveal much about martian history. This image was taken by *Curiosity* when it was still more than 20 kilometers away.

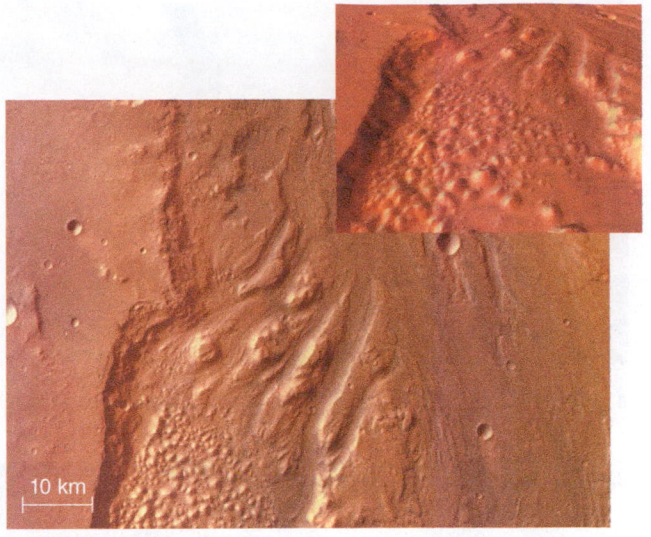

10 km

FIGURE 8.23
This image from Europe's *Mars Express* orbiter shows outflow channels likely carved by floodwaters. The inset is a perspective view of the region.

or were there lakes (or even oceans) that lasted for millions of years? Vigorous debate still surrounds these questions, and the martian surface seems to yield conflicting clues.

In many places, Mars shows evidence of having suffered catastrophic floods. For example, Figure 8.23 shows a region near the top of a long valley (called Ares Vallis) marked by outflow channels that look like channels carved by floodwaters on Earth. Tracing the channels upstream to their source reveals a landscape lacking in anything that looks like a past lake or reservoir, suggesting that the floodwaters emerged from underground. The *Pathfinder* mission, which landed downstream of this region in 1997, provided support for this hypothesis by showing a surface that appears to be a vast floodplain, with rocks scattered and stacked against each other in the same way that we see them after floods on Earth (see Figure 8.5).

The timing and source of past floods are uncertain, but orbital images suggest a link between temporary heating events and some of the floods. Some channels may have been created after ice was melted by impacts. Others may be tied to ice melted by volcanic heating. Figure 8.24 shows a volcano with numerous downhill channels flowing outward in all directions from its central caldera. Toward the lower right, we see a much wider channel that was probably carved by floodwaters released during one or more eruptions. Such events might produce liquid water flows only for short times, but if enough water is released at once, it could result in a large flood. Moreover, the evidence of flooding near volcanoes suggests that volcanic heat can create subsurface pockets of liquid water, offering potential habitats for life.

One way to estimate how much water might once have flowed on Mars is to look at water ice that still exists today. Water ice is clearly present in and around the polar caps, and orbital studies suggest that a significant amount of water ice is present in the martian subsoil at middle and high latitudes around much of the planet. The full extent of the water ice is only beginning to become clear. For example, scientists sent the *Phoenix* lander to explore the martian arctic in 2008. Although the spacecraft (which did not have roving capability) landed several hundred kilometers from the polar cap itself, its landing rockets exposed a patch

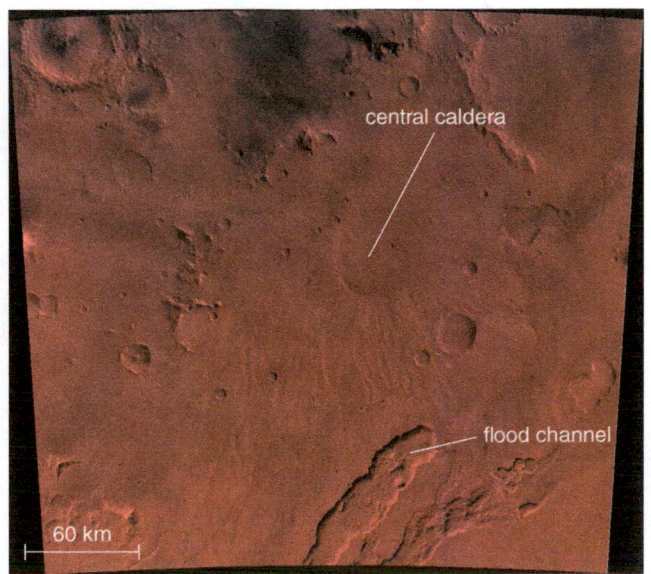

central caldera

flood channel

60 km

FIGURE 8.24
This photo from a *Viking* orbiter shows the volcano Hadriaca Patera; its central caldera is marked. Note the many channels flowing downhill from the caldera and the wide flood channel (called Dao Vallis) toward the lower right.

a The view from the lander, showing part of its robotic arm.

b The robotic arm camera found a bright patch of water ice right under the lander; the lander's rockets (visible at top) had blasted away an overlying layer of dust.

FIGURE 8.25
The *Phoenix* lander operated in the martian arctic in 2008.

FIGURE 8.26
Mars's northern hemisphere may once have held a vast ocean; this artist's conception shows what it might have looked like some 4 billion years ago.

of water ice (Figure 8.25). This surprising finding added to evidence suggesting that water ice is widespread in the arctic region, mixed in with the surface soil or hidden just beneath a layer of dust. The total amount of ice now known to be on Mars represents (if melted) enough water to make an ocean averaging about 10 meters deep over the entire planet.

Far more water may have been present on Mars in the distant past. Scientists can estimate how much water Mars has lost over time by measuring the ratio of deuterium (hydrogen with a neutron) to ordinary hydrogen. As we'll discuss in more detail shortly, Mars may have lost water because ultraviolet light from the Sun split atmospheric water vapor molecules apart, allowing some of the lightweight hydrogen atoms to escape to space. Deuterium does not escape as easily as ordinary hydrogen because of its greater weight, so an elevated ratio of deuterium to hydrogen suggests that a lot of ordinary hydrogen has been lost to space. One recent study found that the ratio of deuterium to hydrogen suggests that Mars may have once had enough water to cover the entire surface to a depth of more than 130 meters. Rather than having water uniformly distributed over its surface, however, some scientists hypothesize that the low-elevation northern plains once held a vast, deep ocean (Figure 8.26). Additional evidence for this hypothesis comes both from the fact that many of the largest flood channels appear to have drained into the northern plains and from geographic features that look like an ancient shoreline. Moreover, radar data suggest that the rock along the proposed shoreline is sedimentary rather than volcanic, just as we would expect if it had once been the edge of an ocean.

Think About It Do a web search for "ocean on Mars" and look for the latest news about the hypothesis. Overall, do you think an ancient ocean seems likely? Why or why not?

The overall question of the persistence of liquid water on Mars remains unsettled, though the recent discoveries of *Curiosity* are beginning to tilt scientific opinion toward the idea that surface liquid water persisted for millions of years. Learning whether this is the case will have important implications for the possibility of past or present life on Mars. If lakes and rivers were present only intermittently, then these periods of surface habitability may have been too short for life to arise. But if habitable lakes or oceans were present for millions of years, life might have had a chance to take hold and perhaps to evolve sufficiently so that it could survive to this day in any available pockets of subsurface liquid water. Either way, it is important to remember that the period of widespread surface liquid water ended at least about 3 billion years ago, which will shortly bring us to the question of why the martian climate changed.

RECENT WATER FLOWS? Although large-scale water flows on Mars seem to be confined to the distant past, orbital photographs offer tantalizing hints of smaller-scale water flows in much more recent times. The strongest evidence comes from photos showing dark streaks on crater walls (Figure 8.27). The streaks appear to grow during the warmest times of year—when ice would tend to melt—suggesting that they are created by flowing water. The water would have to be very salty to stay liquid under the expected conditions near the surface. However, no one has yet explained how water or ice might have gotten trapped in the rock layers from which the streaks appear to flow.

If any liquid water does still occasionally flow, the amount must be a tiny fraction of the water that flowed when riverbeds and lakes were formed long ago. Mars clearly was warmer and wetter at times in the past than it is today. Ironically, Percival Lowell's supposition that Mars was drying up has turned out to be basically correct, although in a very different way than he imagined.

8.3 The Climate History of Mars

While we have much left to learn about water in Mars's past, the evidence we've discussed makes it seem clear that liquid water was stable or nearly stable during at least some time periods prior to about 3 billion years ago. For that to have been possible, both the atmospheric pressure and the temperature must have been significantly higher than they are today. Mars in the past offered a much more hospitable climate than it does now, and perhaps one in which life could have arisen and taken hold.

Why was Mars warmer and wetter in the past?

It's easy to conclude that Mars must have been warmer and wetter in the past, but more challenging to explain why. The basic answer presumably lies with the greenhouse effect. Recall that the greenhouse effect can make a planet's surface much warmer than it would be otherwise. A moderate greenhouse effect keeps our own planet Earth from freezing over [Section 4.5], while an extremely strong greenhouse effect is responsible for the blistering temperatures on Venus [Section 7.2].

Today, Mars has such a thin atmosphere that it has only a weak greenhouse effect, despite the fact that 95% of its atmosphere is composed of the greenhouse gas carbon dioxide (see Table 8.1). However, Mars almost certainly had a much stronger greenhouse effect in the past. Calculations suggest that martian volcanoes should have outgassed enough carbon dioxide to make the atmosphere about 400 times as dense as it is today (and enough water to fill oceans tens to hundreds of meters deep).

If Mars had this much carbon dioxide today, it would have a surface pressure about three times that of Earth and a temperature above freezing—in other words, a climate in which liquid water could flow. However, because we think that the Sun was dimmer in the distant past [Section 4.5], even more greenhouse warming would have been needed to allow for liquid water when Mars was young. Current models are unable to account for the necessary additional warming with carbon dioxide gas alone. Many scientists hypothesize that additional warming was provided by a greenhouse effect due to carbon dioxide ice clouds or methane and/or sulfur gasses. Alternatively, perhaps Mars never had an extended period of warmth, but instead had only intermittent wet periods, possibly triggered by the heat of large impacts or volcanic action. But even in this case, the evidence we've found for extensive water flows means that Mars's atmosphere must have been much thicker and warmer in the distant past than it is today.

Why did Mars change?

Given that Mars must once have had a much denser atmosphere with a much stronger greenhouse effect, we can explain the current extremely

FIGURE 8.27

This image shows a crater wall with dark streaks that may be created by flowing water. The streaks, which originate below the rock outcrops visible across the center of the image, change from season to season, becoming more prominent in spring and summer. The image, constructed from data from the *Mars Reconnaissance Orbiter*, is shown in a perspective view with enhanced color.

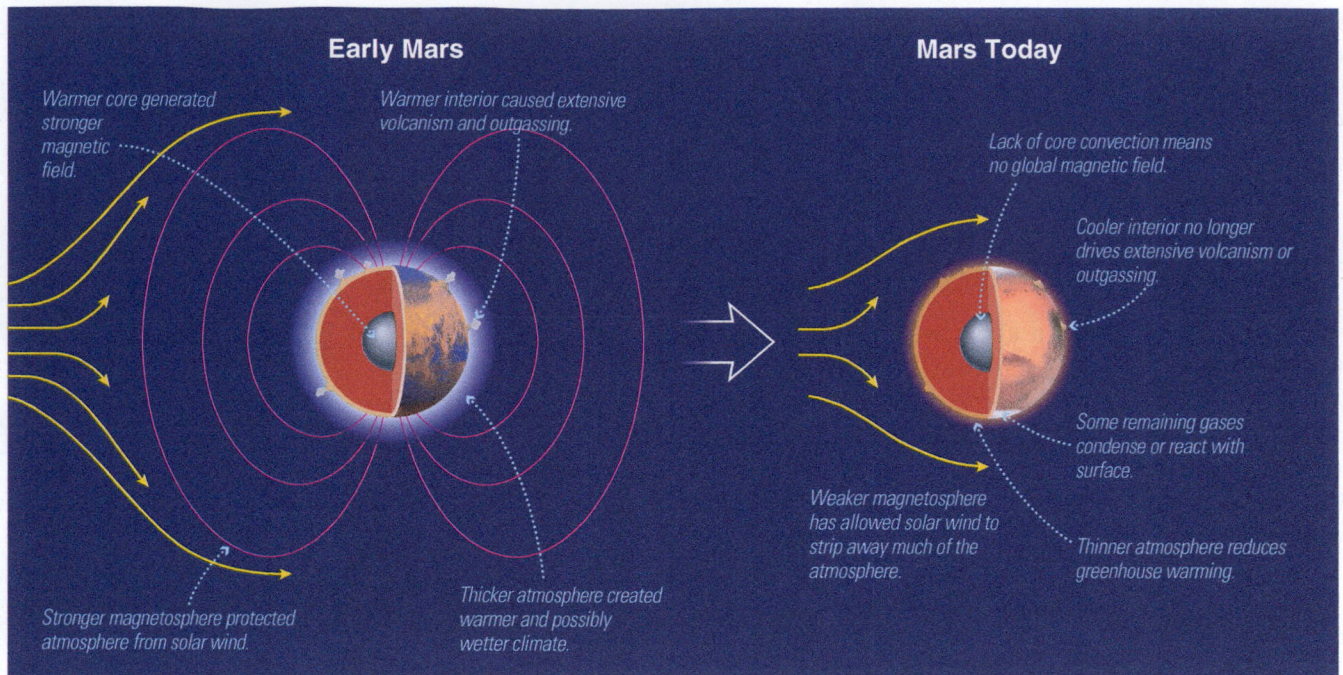

FIGURE 8.28

Some 3 billion years ago, Mars underwent dramatic climate change, probably because it lost its global magnetic field, leaving its atmosphere vulnerable to the solar wind.

different conditions only if Mars somehow lost a vast quantity of carbon dioxide gas. This loss would have weakened the greenhouse effect until the surface of the planet essentially froze over. Where did all this gas go? Some of the carbon dioxide condensed to make the polar caps, some may be chemically bound to surface rock, and some still makes up the martian atmosphere today. However, the bulk of the gas was probably lost to space.

LOSS OF CARBON DIOXIDE AND WATER The precise way in which Mars lost its carbon dioxide gas is not clear, although some gas was almost certainly blasted away by large impacts. However, the leading hypothesis suggests that an even more important loss mechanism was linked to a change in Mars's magnetic field (Figure 8.28). Early in its history, Mars probably had molten, convecting metals in its core, much like Earth today [Section 4.4]. The combination of this convection with Mars's rotation should have produced a magnetic field and a protective magnetosphere. The magnetic field would have weakened as the small planet cooled and core convection ceased, leaving the atmosphere vulnerable to solar wind particles. More specifically, the hypothesis suggests that carbon dioxide molecules were dissociated into carbon and oxygen atoms by sunlight or chemical processes, and the resulting atoms were then stripped away by the solar wind.

This hypothesis is being put to the test by the *MAVEN* mission, which has been orbiting Mars since 2014 and is measuring the escape of gases from Mars's atmosphere today. The first two panels of Figure 8.29 show results demonstrating that Mars is indeed losing the carbon and oxygen atoms that once made up carbon dioxide molecules, supporting the idea that Mars has lost substantial amounts of carbon dioxide gas through this process over the past few billion years.

Much of the water once present on Mars is also probably gone for good. Like the carbon dioxide, some water vapor may have been lost through stripping by the solar wind. In addition, Mars lost water in

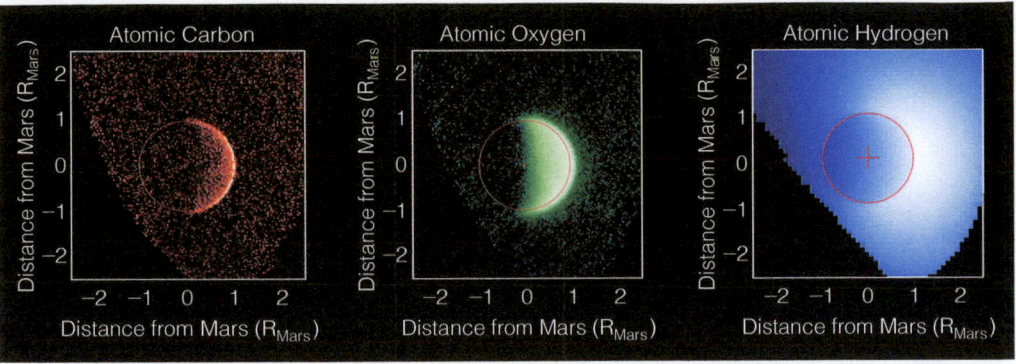

FIGURE 8.29
These ultraviolet images from NASA's *MAVEN* spacecraft show carbon, oxygen, and hydrogen atoms—which came from dissociated carbon dioxide and water molecules—in the martian atmosphere; the red circle shows the size of Mars, and black areas lack data. Notice that the atomic gases extend high above the surface, where some of the carbon and oxygen is escaping through solar wind stripping and hydrogen is being lost to thermal escape. Hydrogen extends the highest because light gases move fastest.

another way. Because Mars lacks an ultraviolet-absorbing stratosphere, atmospheric water molecules would have been easily broken apart by ultraviolet light from the Sun. The lightweight hydrogen atoms that broke away from the water molecules would have been lost rapidly to space through thermal escape—the process in which low-mass gas atoms can reach escape velocity and escape into space [Section 4.4]. The third panel of Figure 8.29 confirms the loss of hydrogen atoms in this way, and as we discussed earlier, the deuterium-to-hydrogen ratio lends further support to the idea that a lot of hydrogen was lost to space. Once the hydrogen atoms were lost, the water molecules could not be made whole again. Initially, oxygen from the water molecules would have remained in the atmosphere, but over time this oxygen was lost, too. Some was probably stripped away by the solar wind, and the rest was drawn out of the atmosphere through chemical reactions with surface rock. This process literally rusted the martian rocks, giving the "red planet" its distinctive tint.

In summary, the hypothesis we have described suggests that Mars changed primarily because of its relatively small size. It was big enough for volcanism and outgassing to release plenty of water and atmospheric gas early in its history, but too small to maintain the internal heat needed to create a strong magnetic field and magnetosphere that could prevent this loss of water and gas. As its interior cooled, its volcanoes quieted and released far less gas into the atmosphere, while its relatively weak gravity and the loss of its magnetic field allowed existing gas to be stripped away to space. If Mars had been as large as Earth, so that it could still have outgassing and a global magnetic field, it might still have a pleasant climate today. Mars's distance from the Sun also helped seal its fate: Even with its small size, Mars might still have some flowing water if it were significantly closer to the Sun, where the extra warmth could melt the water that remains frozen underground and at the polar caps.

MARS CLIMATE AND AXIS TILT With the gas that once warmed the planet now gone, there is little hope that Mars will ever again have a warm, wet climate, and the planet's weather varies little from year to year. However, both theoretical and observational studies suggest that Mars undergoes longer-term cycles of climate change on time scales of hundreds of thou-

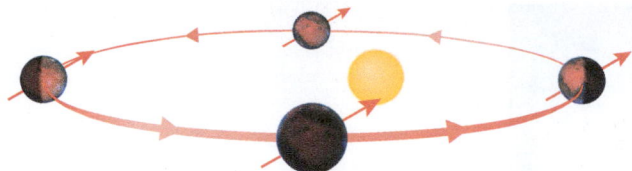

a When the axis is highly tilted, the summer pole receives fairly direct sunlight and becomes quite warm.

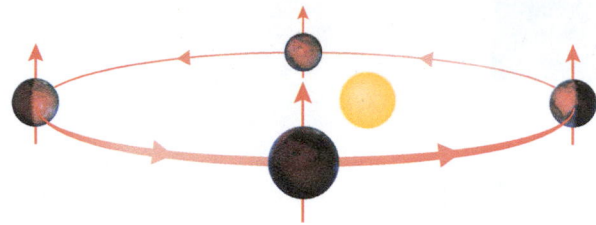

b When the axis tilt is small, the poles receive little sunlight at any time of year.

FIGURE 8.30
Mars's axis tilt probably varies dramatically, causing climate change because of the effect on the seasons.

FIGURE 8.31
The *Mars Reconnaissance Orbiter* captured this image of a land-slide in layered terrain in the north polar region. Despite the dark appearance, water makes up the bulk of the material. Layers of dusty ice more than 700 meters thick built up over many cycles of climate change. During the northern spring of 2010, warming conditions apparently weakened the cliff walls and triggered landslides.

sands to millions of years. This climate change arises from changes in axis tilt, and it may have significant implications for the potential habitability of Mars.

Recall that Earth experiences long-term climate cycles, such as ice ages, due to small changes in its rotation and orbit, including small changes in axis tilt. Earth's axis tilt doesn't change much—varying only between about 22° and 25°—because our large Moon exerts a gravitational pull that helps stabilize it. Mars lacks a large moon, and its two tiny moons (Phobos and Deimos) are far too small to offer any stabilizing influence on its rotation axis. In addition, because Mars is closer to Jupiter than Earth is, Jupiter's gravity more strongly perturbs Mars as it orbits the Sun, an effect that may be further magnified by the fact that Mars is less perfectly spherical than any other planet (as evidenced by measurements of small differences in gravity over different regions).

Calculations suggest that the lack of a stabilizing moon and the effects of Jupiter together may cause the martian axis tilt to vary over time from 0° to as much as 60°, which means that the current 25° is near the middle of the range. These changes in tilt would have dramatic effects on the climate (Figure 8.30). When the tilt is small, the martian poles may stay in a perpetual deep freeze for tens of thousands of years, which means much of the carbon dioxide now in the atmosphere is instead frozen at the poles. The atmosphere therefore becomes thinner, which lowers the pressure and weakens the greenhouse effect. When the tilt is large, the summer pole becomes warm enough to allow both the carbon dioxide and substantial amounts of water ice to vaporize into gas, thereby raising the atmospheric pressure and warming the planet through the stronger greenhouse effect. The martian polar regions show layering of dust and ice that probably reflects changes in climate due to the changing axis tilt (Figure 8.31).

Even at the greatest tilts, the atmospheric pressure probably does not become high enough to allow liquid water to pool in surface lakes or ponds. Nevertheless, models suggest that liquid water might form just beneath the surface or at rock/ice boundaries on the surface whenever the tilt is greater than about 40°, implying that Mars could have zones of liquid water during those epochs.

Is Mars habitable?

Mars clearly has the elements needed for life, and energy is available for life in the form of sunlight (on the surface) and chemical energy (underground). The question of whether Mars is habitable therefore hinges on the availability of liquid water.

The geological evidence strongly suggests that Mars once had abundant liquid water at its surface, meaning that the surface *was* habitable some time before about 3 billion years ago. In particular, findings from the *Curiosity* rover clearly indicate a habitable environment in Gale Crater around 3.5 billion years ago. The only question is whether it was habitable for a long period of time—making an indigenous origin of life seem plausible—or only for shorter, intermittent periods. As we have discussed, there is still debate on this question, though recent findings make it seem increasingly likely that liquid water was present for millions or tens of millions of years. In that case, the young Mars may have been quite similar to the young Earth, with a habitable environment during the same time period in which life arose on Earth [Section 6.1].

The *surface* of Mars is no longer habitable, because of the lack of liquid water (as well the lack of ozone or other gases to absorb solar ultraviolet light). However, we've noted that pockets of liquid water could still exist underground, as a result of remaining internal heat, so it is possible that Mars still has underground zones of habitability. Moreover, the climate changes tied to the changing martian axis tilt imply that these zones may reach to rock/ice boundaries at the surface during some epochs. On Earth, we find microbes that live in thin films of liquid water at such boundaries, which opens the intriguing possibility that similar life might be found on Mars.

Unless we are drastically misinterpreting the evidence, the conclusion seems clear. The surface of Mars was habitable during some periods of its early history, and it might still sometimes be habitable when the axis tilt is greater. Moreover, the subsurface probably has been habitable throughout the planet's history and may still contain habitable zones even today. Given the apparent habitability of Mars, it is time for us to turn our attention to the search for actual life.

8.4 Searching for Life on Mars

While we have some confidence in the past and present habitability of Mars, we do not yet know whether Mars has ever actually had life. The only way to learn whether life existed in the past is to search for fossil evidence in martian rocks, and the only way to learn whether life exists today is to find it. To date, only very limited searches for life have been carried out on Mars, but many more are planned for the future.

Is there any evidence of life on Mars?

The discovery of life on Mars would forever alter our view of life in the universe, so it should be no surprise that many scientists are working hard in hopes of being the first to discover evidence for life on Mars. As we'll discuss, some scientists already claim to have found such evidence. But are they interpreting data correctly, or are they engaged in the same type of wishful thinking that led Percival Lowell astray?

The answer is the subject of heated scientific debate, but at the moment the vast majority of scientists are skeptical of claimed evidence for martian life. Nevertheless, it is worth examining the claims, both to illustrate why there is scientific controversy and because they may point us toward ways of resolving the question in the future.

The claims of evidence of life fall into three main categories: claims based on results from the *Viking* landers, claims based on evidence of methane in the martian atmosphere, and claims based on studies of martian meteorites found on Earth. We'll examine the first two categories here; we'll save discussion of the martian meteorites for Section 8.5.

THE VIKING EXPERIMENTS One obvious way to search for life on Mars is to study the soil to see whether it contains living microbes. This type of search was first carried out by the two *Viking* landers in 1976. Each of the landers was equipped with materials for four on-board, robotically controlled experiments, along with a robotic arm for scooping up soil samples (Figure 8.32) to test in the experiments. The arm could push aside rocks to get at shaded soil that was less likely to have been sterilized by ultraviolet light from the Sun.

FIGURE 8.32
This pair of before (left) and after (right) photos from the *Viking 2* lander shows how the robotic arm pushed away a small rock on the martian surface. (You can see the entire robotic arm in the inset of Figure 8.4.)

Cosmic Calculations 8.1

THE SURFACE AREA–TO–VOLUME RATIO

The total amount of heat contained in Mars or any other planet depends on the planet's *volume*, but this heat can escape to space only from the planet's *surface*. As heat escapes, more heat flows upward from the interior to replace it, until the interior is no hotter than the surface. Therefore, the time it takes for a planet to lose its internal heat is related to the ratio of the *surface area* through which it loses heat to the *volume* that contains heat:

$$\text{surface area–to–volume ratio} = \frac{\text{surface area}}{\text{volume}}$$

A spherical planet (radius r) has surface area $4\pi r^2$ and volume $\frac{4}{3}\pi r^3$, so the ratio becomes

$$\underset{\text{(for a sphere)}}{\text{surface area–to–volume ratio}} = \frac{4\pi r^2}{\frac{4}{3}\pi r^3} = \frac{3}{r}$$

Because r appears in the denominator, we conclude that *larger objects have smaller surface area–to–volume ratios.* (Although we've considered a sphere, this idea holds for objects of any shape.)

Example: Compare the surface area–to–volume ratios of the Moon and Earth.

Solution: Dividing the surface area–to–volume ratios for the Moon and Earth, we find

$$\frac{\text{surface area–to–volume ratio (Moon)}}{\text{surface area–to–volume ratio (Earth)}} = \frac{3/r_{\text{Moon}}}{3/r_{\text{Earth}}} = \frac{r_{\text{Earth}}}{r_{\text{Moon}}}$$

The radii of the Moon and Earth are $r_{\text{Moon}} = 1738$ km and $r_{\text{Earth}} = 6378$ km:

$$\frac{\text{surface area–to–volume ratio (Moon)}}{\text{surface area–to–volume ratio (Earth)}} = \frac{6378 \text{ km}}{1738 \text{ km}} = 3.7$$

The Moon's surface area–to–volume ratio is nearly four times as large as Earth's, which means the Moon would cool four times faster if all else were equal. In fact, Earth has retained heat much longer, because its larger size gave it more heat to begin with and because Earth has a higher proportion of radioactive elements. (See Problem 53 for a similar analysis of Mars.)

The first experiment (the *carbon assimilation experiment*) mixed a sample of martian soil with carbon dioxide (CO_2) and carbon monoxide (CO) gas brought from Earth. (In some runs of the experiment, the soil was also mixed with water.) The carbon dioxide and carbon monoxide from Earth could be distinguished from the same gases in the martian atmosphere because they had been "tagged" with radioactive carbon-14. Results showed that the carbon-14 became incorporated into the soil, which at first seemed to suggest that life was present and was using the carbon for metabolism. However, when the experiment was repeated with soil heated to 175°C (347°F), the tagged carbon still became incorporated into the soil. Because 175°C is hot enough to break chemical bonds between carbon and other atoms—and presumably to kill any carbon-based organisms—most scientists concluded that a chemical rather than a biological process was responsible for all the experiment's results.

The second experiment (the *gas exchange experiment*) mixed martian soil with a "broth" containing organic nutrients from Earth. As soon as the soil was exposed to the nutrients, oxygen was released into the chamber, a result that seemed to suggest a process of photosynthesis. However, the process occurred in the dark rather than in sunlight, making photosynthesis seem implausible. Moreover, oxygen was released even when the soil was exposed only to water vapor (rather than to the nutrients), and as in the first experiment, the reactions continued even when the soil was heated to temperatures that should have killed any organisms present. Again, most scientists concluded that the results were due to chemical and not biological processes.

The third experiment (the *labeled release experiment*) provided the most intriguing results. This experiment also mixed martian soil with organic nutrients from Earth, which were tagged with radioactive carbon-14 and sulfur-35. It then looked for changes in the level of radioactivity in the chamber gas that might occur if living organisms consumed the nutrients and released the tagged, radioactive gases. Just as would be expected if life were present, the radioactivity rose at first and then leveled off as the nutrients were used up. Moreover, in contrast to the cases of the first two experiments, heating of the soil in this experiment produced results consistent with life: Heating to 50°C (122°F) substantially reduced the amount of radioactivity, and heating to 160°C (320°F) eliminated any sign of the tagged isotopes in the chamber gas.

The seemingly contradictory results of the first three experiments were further exacerbated by the fourth (the *gas chromatograph/mass spectrometer experiment*), which sought to measure the abundance of organic molecules in the martian soil. This experiment found no sign of organic molecules in the martian soil, seemingly ruling out the possibility of carbon-based life in the samples studied.

Recent results from the *Phoenix* lander and *Curiosity* rover may have finally solved the mystery of the *Viking* results. The martian soil contains a salt known as *perchlorate*, which, when heated, releases oxygen and chlorine that can destroy organic molecules. *Curiosity* has found strong evidence that the martian soil *does* contain at least some organic molecules—though, rather than having a biological source, these might have been brought by meteorites—suggesting that the reason *Viking* failed to find organics was because they were destroyed by heating the perchlorate. The chemistry of perchlorate also appears to explain the conflicting results of the other *Viking* experiments.

The bottom line is that scientists now generally regard the *Viking* results as inconclusive and recognize that different experiment designs will be needed for more definitive searches for life on Mars.

METHANE ON MARS Atmospheric studies can also provide clues about potential life, and scientists are particularly intrigued by the apparent detection of methane in the martian atmosphere. Methane gas cannot last more than a few centuries in the martian atmosphere before chemical reactions transform it into other gases.* Therefore, if methane is present, Mars must have an active source of methane gas.

Claims of methane gas detection were first announced in 2003 and 2004, based on telescopic observations from Earth and measurements from the *Mars Express* orbiter at Mars. Subsequent telescopic observations also suggested that the methane level varies with time. However, the signals were weak enough that many scientists doubted the claims. Scientists hoped that the *Curiosity* rover could put the controversy to rest with measurements on the ground, and results announced in 2014 and 2015 appear to do just that: *Curiosity* has detected measurable and variable levels of methane gas in its travels through Gale Crater.

Assuming the detections hold up to further scrutiny, the key question is the source of the methane release. We know of at least three possible sources: comet impacts, geological activity, or life. The first possibility is highly unlikely, since impacts are such rare events and no recent ones have been identified. That leaves us with geological activity or life, both of which might release methane in a way that varies with time. For

---

*The methane is oxidized to form water and carbon dioxide; the oxidation occurs because the martian atmosphere always contains some amount of free oxygen made by the breakup of atmospheric carbon dioxide molecules.

Movie Madness MISSIONS TO MARS

There was a time, not so long ago, when the term *Martian* was just about synonymous with "space alien." You could frequently meet the Martians at the local cinema, where they were busy invading Earth and ruining everyone's whole day.

The classic example of this type of smooth move was in *War of the Worlds,* which has so far spawned one scary radio play and two moderately scary films. In H. G. Wells's story, sophisticated Martians abandon their turf to grab ours. Mars, you see, was drying up and dying. Earth, on the other hand, was a world with abundant water: a sanctuary for our desperate neighbors.

But a varied assortment of landers and rovers sent to Mars during the last four decades has shown us a landscape that's as sterile as a mule. There are simply no indications of technologically sophisticated inhabitants, either dead or alive. So Hollywood, ever flexible, switched gears. Earthlings now go to Mars—but with the exception of the fairly realistic 2015 blockbuster *The Martian*, it is usually to find hidden signs of habitation that would startle astrobiologists.

In the film *Red Planet,* our descendants try to rebuild Mars into a kinder, gentler world in order to escape environmental disaster on Earth. Robotic craft are sent to melt the polar caps and sow the planet with blue-green algae in a barely plausible bid to produce a warm, breathable atmosphere. In the course of this terraforming project, visiting humans stumble across some complicated life-forms—indigenous Martians—who look like economy-size lice. The lice eat everything from space suits to spacemen, and frankly it's a puzzle why they haven't eaten Mars itself.

As implausible as this may be, astronauts in another space opera, *Mission to Mars,* find something even less reasonable. While checking out an odd mountain in Mars's Cydonia region, they discover that it's really a massive alien "face," disguised by dust and rock. Venturing inside, the astronauts eventually learn that the ancient martian civilization that built the face was wiped out by an asteroid a half-billion years ago. Just before abandoning their planet, the Martians launched a rocket to seed Earth with their DNA. This molecular emigration supposedly produced the Cambrian explosion that began the reign of multicellular life on Earth. One has to wonder how the Martians could be confident that the jellyfish and trilobites that resulted from their seeding would eventually evolve into humans who would look pretty much like … the long-gone Martians!

Life may, indeed, have once existed on the red planet, and perhaps it still does. But our current understanding of conditions on Mars strongly suggests that this life would never have resembled either voracious lice or us. It's probable that any real Martians would be visible only in a microscope.

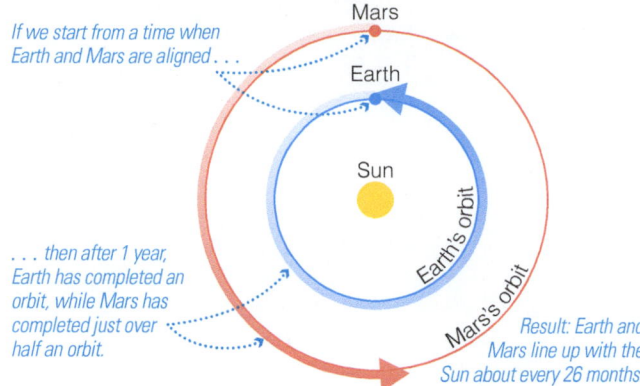

If we start from a time when Earth and Mars are aligned . . .

Mars

Earth

Sun

Earth's orbit

Mars's orbit

. . . then after 1 year, Earth has completed an orbit, while Mars has completed just over half an orbit.

Result: Earth and Mars line up with the Sun about every 26 months.

FIGURE 8.33

Mars takes almost 2 years to orbit the Sun, compared to Earth's 1 year. If you start from an alignment as shown here, after 1 year Earth will be back in the same place, but Mars will be only halfway around. It therefore takes a second year before the two planets line up again. Using Mars's more precise orbital period of 1.88 years, we find the alignments occur every 780 days, or about 26 months.

example, life might emit methane in seasonal or other cyclical patterns, while geological sources may release differing amounts of gas depending either on seasonal changes or on winds that can plug or unplug holes from which gases can escape from underground.

Either way, the presence of methane has important implications for the possibility of life. Even if the source is geological rather than biological, the amount of volcanic heat necessary for methane release would probably also be sufficient to maintain pockets of liquid water underground. That would make subsurface life seem more likely. And, of course, if the source is biological, it should be only a matter of time until we are able to identify the organisms responsible.

How do we plan to search for life on Mars?

The world scientific community has ambitious plans for continued exploration of Mars. The orbits of Earth and Mars bring the two planets to closest approach about every 26 months (Figure 8.33), and scientists hope to take advantage of upcoming alignments to send new and ever more sophisticated spacecraft to Mars.

UPCOMING MISSION PLANS Several new missions in various stages of development are designed to build upon the successes of recent and ongoing missions. Many of you will be reading this book after the next missions are on their way to or operating on Mars, as NASA's *Mars Insight* lander and the European Space Agency's *ExoMars 2016* mission, which includes both an orbiter and a lander, are scheduled to launch in early 2016 and arrive later that year. *Mars Insight* includes a seismic station that should help us understand the interior structure of Mars. In 2018, the European Space Agency plans a second *ExoMars* mission, this time with a rover that will explore the surface and search for signs of past or present life. Then, in 2020, NASA plans to launch another rover that will be essentially an upgraded version of *Curiosity*. Other nations are also getting involved. In 2014, India's *Mars Orbiter Mission* (also called *Mangalyaan*) successfully reached Mars, and the Indian Space Research Organization hopes to follow up with additional missions. Russia is participating in the European Space Agency's projects, and Japan, China, and the United Arab Emirates also are working on future missions to Mars.

Think About It Find the current status of some of the missions described in the above paragraph. Have any of them yet made any significant discoveries?

SEARCHING FOR LIFE There is always a possibility that upcoming missions will find evidence of life, but a more definitive search will probably have to wait until further in the future. The reason is at least two-fold: First, scientists are still not in full agreement on the best experiments to send to Mars in search of life; the results from *Viking* showed that it's easy to get inconclusive results, so experiments must be designed with great care. Second, even once scientists agree on the best experiments, the difficulty and expense of designing them to fit on a robotic spacecraft present a great challenge.

One of the key limitations of the *Viking* experiments, which was not known at the time, was that they in essence attempted to "culture" life (get it to metabolize or grow in an experimental chamber), and even most forms of life on Earth are very difficult to culture. Therefore, a better approach to searching for life is to look for biochemical markers. The difficulty lies in deciding what markers to look for. For example, on Earth we can now search for life in almost any sample by looking for DNA, but martian life may not use DNA, and even if it does, its DNA might be sufficiently different from that of Earth life that detectors that work on Earth may not work on Mars. The same is true for experiments that might search for various proteins.

For this reason, scientists are particularly interested in a "sample return mission" in which a robotic spacecraft would collect rocks on Mars and bring them back to Earth, where they could be analyzed in many more ways than would be possible with pre-designed robotic experiments sent to Mars. The primary difficult with such a mission is cost, because it requires not only sending a sophisticated spacecraft to Mars but also sending it with a rocket and enough fuel for the return trip. Still, scientists are optimistic that such a mission might be possible within a decade or so. Indeed, NASA's 2020 rover may include a system for collecting and storing samples that might eventually be picked up and returned to Earth by future missions.

PREVENTING CONTAMINATION—IN BOTH DIRECTIONS Given the likelihood that some Earth organisms could survive in at least a few locations on Mars, it's important to make sure that our robotic missions don't accidentally contaminate Mars with life from Earth. Otherwise, microbes that hitched a ride from Earth aboard a spacecraft might fool us into thinking we'd found evidence of martian life. The possibility of contamination also poses an ethical issue: It's at least conceivable that terrestrial life could outcompete any indigenous martian life, driving the martian life to extinction. Do we have a right to do something that could endanger native life on another planet? Clearly, the best way to avoid these problems is to prevent contamination in the first place. An international treaty, signed in 1967, requires that any spacecraft sent to Mars must have a less than 1 in 1000 chance of causing contamination. Today, scientists strive for even lower contamination probabilities by sterilizing spacecraft before they are launched.

Similarly, but in the other direction, the prospect of a sample return mission has caused some people to fear we might unleash dangerous martian microbes on Earth. Could such microbes cause disease for which we are unprepared or outcompete terrestrial organisms on our own turf? We cannot completely rule out any danger, but it is quite unlikely, because disease-causing microbes are highly adapted to the species they infect. For example, diseases that infect plants generally do not infect animals. Indeed, "species jumping" by diseases is quite rare and generally occurs only between species that are evolutionarily close. HIV (the virus that causes AIDS), for example, is thought to have jumped from chimpanzees to humans, but on an evolutionary level this is a fairly small jump between different species of primates. Therefore, even if martian microbes were accidentally released and subsequently survived on Earth, it's unlikely that they would cause disease. In addition, because martian meteorites must frequently land on Earth, any life that hides in martian rocks would almost certainly have reached Earth already. The fact that we do not see any harmful effects from this "natural contamination" makes it unlikely that any martian life can harm Earth life.

Nevertheless, it pays to be cautious, given the high stakes involved, and samples brought back from Mars will surely be transported in sealed containers that would not break open even if they were to crash on Earth. Once here, they will be quarantined and subjected to biological tests such as exposing terrestrial microbes to them. Biologists already know how to deal with dangerous terrestrial microbes, such as the Ebola virus, and scientists are developing protocols to ensure safe handling of any harmful martian organisms, if they actually exist and are actually harmful to terrestrial life.

Think About It Should we allow samples from Mars to be brought to Earth, or should they be studied only in space, such as on the Space Station or at a Moon colony? Defend your opinion.

Should we send humans to Mars?

Many people hope that we will soon be able to send humans to Mars. Sending people is far more difficult than sending robots. Even with the most advanced rockets that anyone has on the drawing board for the next couple of decades, the trip to Mars would take at least 3 to 4 months in each direction. A human mission would have to carry not only the weight of the astronauts and their living quarters, but also that of enough food, air, and water to last the trip. Shielding against dangerous radiation would also be necessary, which means having an on-board "storm cellar" in case a violent flare erupted on the Sun. Moreover, because the rockets could travel between Earth and Mars only when the two planets were nearly aligned every 26 months, the astronauts would likely have to spend almost 2 years on Mars before they could return home. Although they might conceivably get water from the subsurface ice and chemically extract oxygen from martian water or rock, they would still need food, which would have to be either taken along with them or sent separately aboard other spacecraft. They'd also need fuel for the return journey, which would add far more weight to the mission, unless they could manufacture the fuel for the return mission on Mars (an idea that is being actively explored). No matter how you look at it, the enormous amount of stuff required for a human mission ensures it would cost at least as much as dozens of robotic missions, and it would pose many dangers to the crew.

Think About It Some people have proposed reducing the cost of human missions to Mars by getting volunteers who would make only a one-way trip, living out the rest of their lives on Mars. A few people have already volunteered for such a mission, and many say they would do it if they were terminally ill. Do you support or oppose such ideas? Defend your opinion.

SCIENTIFIC PROS AND CONS While the cost and inherent danger of sending humans to Mars would be very high, the scientific payoff could potentially be even higher. We humans are far more capable than any robot, and a team of scientists with vehicles for traveling around the planet and equipment for drilling into the crust might well answer our questions about martian life long before they would be answered by robotic explorers. However, sending humans to Mars also has at least one significant scientific drawback: It vastly complicates the issue of avoiding contamination by terrestrial organisms. People are veritable warehouses of microbes: The number of bacteria in the average person's mouth, for example, is far greater than the number of people who have ever lived. We harbor microbes on our skin, in our breath, in our food, and in our excrement. Preventing all these microbes from escaping into the martian environment during an extended stay on the planet would be nearly impossible.

The scientific pros and cons of sending humans to Mars are fairly clear, but the history of the space program shows that human exploration has rarely been driven by science. The human space program began for political reasons, largely as part of a "race" between the United States and the Soviet Union. For example, while the *Apollo* program provided valuable scientific data about the Moon, its primary purpose was to prove to the world that the Americans could get there before the Soviets. If we decide to send humans to Mars, the decision will also probably be based more on social and political considerations than on scientific ones.

TERRAFORMING MARS Some people dream of establishing permanent colonies on Mars. For the near future, any such colonies would have to be self-contained environments, and no one would dare venture outside without a space suit. For the more distant future, a few scientists have suggested the possibility of altering the martian environment to make it more hospitable to us. This process goes by the name **terraforming,** because the changes would tend to make the planet more Earth-like.

Proposals to terraform Mars envision raising the atmospheric pressure and temperature. The temperature might be raised by adding a greenhouse gas to the atmosphere, while increasing the pressure simply requires more gas of any type. One suggestion involves manufacturing chlorofluorocarbons (CFCs), which are strong greenhouse gases, and releasing them into the martian atmosphere. If we could strengthen the greenhouse effect enough, the warmer temperatures might begin to release frozen carbon dioxide from the polar caps and elsewhere beneath the surface, which would further increase the atmospheric pressure and strengthen the greenhouse effect. There still wouldn't be oxygen to breathe, but if the pressure rose enough, we might be able to walk around on Mars carrying only an oxygen tank (and some protection from ultraviolet radiation) rather than having to wear a full, pressurized space suit. Such conditions might also allow plants to survive outdoors, making it much easier to grow food and eventually increasing the concentration of atmospheric oxygen.

The idea might just work, but putting it into practice wouldn't be easy. Because CFCs tend to be broken apart by sunlight, we would have to manufacture them continually and in great abundance in order to start the greenhouse warming. Calculations suggest that we would need a manufacturing capability about a million times greater than our recent CFC-manufacturing capability on Earth and would need to keep it up for *a few hundred thousand years* before the surface warmed enough to drive substantial quantities of carbon dioxide into the atmosphere. If it is possible at all, we have plenty of time to consider the ethical issues of terraforming, which could be quite significant if Mars turns out to have life: Do we have a right to alter a planet in a way that could harm its native life?

Think About It A similar ethical issue surrounds endangered species on Earth. Some people say that we have no right to drive any species to extinction—an idea that was embodied in the U.S. Endangered Species Act. Others say that potential extinctions must be weighed against the human and economic costs of preventing them. Where do you stand on this issue? Does your answer affect your opinion of whether it would be ethical to terraform Mars? Explain.

Interestingly, some of the ethical issues involved in Mars colonization were explored by science fiction writers well before the idea of terraforming ever arose. In particular, back in the days when people believed in canals and a dead or dying martian civilization, many stories dealt with the conditions under which humans might colonize Mars. So for our last word on the topic of human colonization, we turn to a science fiction story called "The Million-Year Picnic," written in 1946 by Ray Bradbury and included in his book *The Martian Chronicles*. It tells the story of a human family who escape to Mars just as people on Earth are finishing off our civilization through hatred and war. On Mars, the family finds plenty of water and the vacant cities left by extinct Martians. The story ends with the family on the bank of a canal, where one of the children asks his father about a promise made earlier:

> *"I've always wanted to see a Martian," said Michael. "Where are they, Dad? You promised."*
> *"There they are," said Dad, and he shifted Michael on his shoulder and pointed straight down.*
> *The Martians were there. Timothy began to shiver.*
> *The Martians were there—in the canal—reflected in the water. Timothy and Michael and Robert and Mom and Dad.*
> *The Martians stared back up at them for a long, long silent time from the rippling water. . . .*

 **THE PROCESS OF SCIENCE IN ACTION**

8.5 Martian Meteorites

As we briefly noted earlier, one claim of evidence for life on Mars comes from the study of rocks from Mars that have fallen to Earth—the so-called *martian meteorites* [Section 6.2]. The story begins in 1984, when a team of American scientists scooped up a 1.9-kilogram meteorite from the Allan Hills region of Antarctica. It was cataloged as ALH84001: "ALH" for Allan Hills, "84" for the year in which it was found, and "001" to indicate that it was the first meteorite found on the expedition (Figure 8.34). It did not immediately draw special attention, but an analysis a decade later showed that it was one of those rare meteorites to have come from Mars. It then proved itself special even among this small group of rocks, and was subject to intense study. In 1996, a team of researchers (led by David McKay at NASA) made an astonishing claim: They said that ALH84001 might contain fossil evidence of past life on Mars. Because this claim would be so important if true, and because it has proved so controversial, we use it as this chapter's case study of the process of science in action.

Is there evidence of life in martian meteorites?

To evaluate the claims about ALH84001, we must begin by understanding the rock. Scientists are fairly confident (though with some doubts) that ALH84001 really is a meteorite from Mars. It is definitely not an Earth rock, because its relative abundances of the isotopes oxygen-16, oxygen-17, and oxygen-18 are significantly different from those found in terrestrial rocks. But neither does it match what we'd expect from a piece of an asteroid or a rock from the Moon. Most important, gas trapped within ALH84001 appears very similar in its chemical and isotopic

FIGURE 8.34
Chemical analysis of this meteorite, known as ALH84001, indicates that it came from Mars. The small block shown for scale to the lower right is 1 cubic centimeter, about the size of a typical sugar cube.

composition to that of the martian atmosphere—and distinctly different from any other known source of gas in our solar system—leading to the suspicion that it came from Mars.

ALH84001 was singled out to be studied more intensely than other martian meteorites for a simple reason: While other known martian meteorites are geologically young, radiometric dating showed ALH84001 to be a piece of igneous rock that solidified about 4.1 billion years ago. That is, it formed about 400 million years after Mars was born, which means that it resided on Mars at times when liquid water flowed on the surface. Scientists therefore wondered if it might tell us something about the past habitability of Mars.

HISTORY OF THE METEORITE Careful study tells us quite a lot about the history of ALH84001. Radiometric dating tells us its age, while study of its structure reveals evidence of later shocks, probably due to the effects of impacts that occurred long before the one that ultimately launched it into space. The meteorite also contains carbonate grains (about 0.1 to 0.2 millimeter in diameter) that date to about 3.9 billion years ago and tell us that the rock must have been infiltrated by liquid water from which the carbonate minerals precipitated out—evidence that is at least consistent with the idea that Mars once had flowing water.

We can determine the timing of the impact that blasted ALH84001 into space by looking for effects of exposure to *cosmic rays,* high-energy particles that leave telltale chemical signatures on anything unprotected by an atmosphere. The results tell us that ALH84001 spent about 16 million years in space, which means the impact that started its journey occurred on Mars about 16 million years ago. By studying decay products from radioactive isotopes produced by the cosmic rays, we can learn when cosmic rays stopped disturbing the meteorite—which must be when it fell to Earth and gained the protection of Earth's atmosphere. Such analysis shows that ALH84001 landed in Antarctica about 13,000 years ago. Table 8.3 summarizes the history of ALH84001.

EVIDENCE OF LIFE The claimed evidence of life in ALH84001 came from detailed studies of its carbonate grains and the surrounding rock. In brief, four types of evidence were cited as pointing to the existence of biology on Mars:

- The carbonate grains have a layered structure, with alternating layers of magnesium-rich, iron-rich, and calcium-rich carbonates. On Earth, this type of layering generally occurs only as a result of biological activity.
- The carbonate grains contain complex organic molecules known as *polycyclic aromatic hydrocarbons,* or *PAHs.* These molecules can be produced by both biological and nonbiological processes, and they have indeed been found in many meteorites that are not from Mars. However, they are much more abundant in ALH84001 than in other meteorites, and on Earth these molecules are most commonly produced by the decay of dead organisms or by reactions between such decay products and the environment (for example, in the burning of fossil fuels).
- Electronic microscopy techniques reveal crystals of the mineral magnetite within the iron-rich layers of the carbonate grains. The sizes, shapes, and arrangements of these crystals were claimed to match those of magnetite grains that on Earth occur only when made by bacteria.

TABLE 8.3 The History of Meteorite ALH84001

Time	Event
4.1 billion years ago	Solidifies from molten rock in the southern highlands of Mars
4.0–4.1 billion years ago	Is affected by nearby impacts, but not launched into space
3.9 billion years ago	Is infiltrated by water, leading to the formation of carbonate grains within the rock
16 million years ago	Is blasted into space by an impact on Mars
13,000 years ago	Falls to Earth in Antarctica
December 27, 1984	Is found by scientists
October 1993	Is recognized as a martian meteorite
August 1996	Is cited as containing possible evidence of martian life

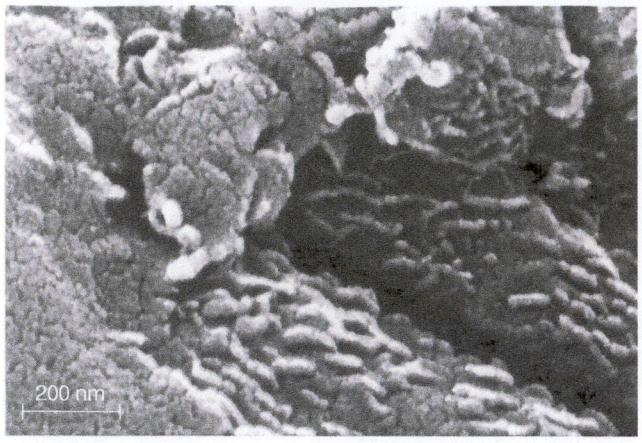

FIGURE 8.35

The tiny rod-shaped structures in this microscopic photo (of a slice of ALH84001) look much like fossilized bacteria, except they are smaller.

• Highly magnified images of the carbonate grains revealed rod-shaped structures that look much like fossilized bacteria, except they are much smaller in size (Figure 8.35).

While none of these lines of evidence alone would prove biological activity, the original investigators argued that, on the whole, biology seemed a much more likely explanation than nonbiological processes. They felt that it would be a "simpler and thus better" scientific explanation (in effect invoking Occam's razor [Section 2.3]) if only a single process—biology—could account for each observation than if a different process was required to explain each result.

ALTERNATIVE EXPLANATIONS The four lines of evidence for fossil life in ALH84001 might seem to make a strong case for past life on Mars. However, other scientists proposed alternative, nonbiological mechanisms that could have produced each observed phenomenon. Let's look at these alternatives in the same order that we presented the evidence:

• There are nonbiological ways to get layered carbonate. For example, several pulses of hot water with different dissolved elements might have passed through the rock and laid down the different mineral layers.
• Other meteorites prove that PAHs can be produced by chemical rather than biological processes, and their high abundance might also be explained by terrestrial contamination during the time the rock resided in Antarctica.
• The resemblance between the magnetite crystals in the meteorite and those made by bacteria on Earth may be coincidental, and some scientists have proposed nonbiological ways in which the crystals might have been formed.
• The rod-shaped structures may look like bacteria, but some are about 100 times smaller than typical terrestrial bacteria. Indeed, they are so small (only 10 to 20 nanometers in width) that it is difficult to see how the complex molecules presumably needed for life (such as RNA- or DNA-like molecules) could fit inside them. Furthermore, similar-looking structures have been found in meteorites that haven't come from Mars, suggesting that these are not reliable evidence of life.

Perhaps most significantly, subsequent study of the meteorite found modern, terrestrial bacteria living inside it, which means the meteorite has been contaminated by Earth life. While this is not too surprising in retrospect—after all, the meteorite spent 13,000 years sitting in Antarctica before scientists found it—it clearly complicates the issue of distinguishing organic materials from Mars from those that could have been made on Earth.

Think About It Given the fact that ALH84001 has apparently been contaminated by terrestrial bacteria, do you think we could ever be sure that a martian meteorite holds evidence of life on Mars? Defend your opinion.

OTHER MARTIAN METEORITES In 1996, when ALH84001 became a big news story, there were only six known meteorites from Mars. That tally has since grown significantly, and several of these meteorites are also old enough to date from a time when liquid water, and possibly life, may have existed on the martian surface. One of these, known as the *Tissint* *meteor*, fell to Earth 40 miles from the Moroccan town of the same name in 2011. It too has been said to contain evidence of martian life, on the basis of organic carbon lining fissures in its interior. These carbon deposits are postulated to have come from organic-rich fluids that seeped into the rock a half-billion years ago, lining the rock with evidence of past life. However, many experts remain skeptical of these claims, and point out that such organic-rich liquids can also be the result of volcanic action.

SUMMARY OF THE CONTROVERSY The debate over possible evidence of life in ALH84001 and other martian meteorites continues, though most scientists now lean toward nonbiological explanations of the evidence. Nevertheless, the debate has taught us at least two crucial facts relevant to the search for life on Mars. First, it now seems unlikely that a meteorite found on Earth could make a conclusive case about life on Mars; instead, we'll need to study rocks on Mars itself (or bring rocks back from Mars to Earth for study). Second, while meteorites are unlikely to tell us about life, they can tell us a great deal about past conditions on Mars. The geological history revealed by martian meteorites strongly supports the idea that Mars once had water, heat sources, and perhaps organic molecules, all of which strengthen the case for the planet's past habitability.

The Big Picture

PUTTING CHAPTER 8 IN PERSPECTIVE

In this chapter, we have discussed past fantasies about martian civilization, our current understanding of the habitability of Mars, and the search for life on the red planet. As you continue your studies, keep in mind the following "big picture" ideas:

- Mars holds a special allure not only because of legitimate scientific questions, but also because past fantasies led many people to imagine a martian civilization. Mars and Martians became deeply embedded in modern popular culture, helping generate great public interest in Mars exploration both by robotic spacecraft and by future human explorers.

- Different regions of the martian surface appear to be almost frozen in time, representing different eras in the planet's history. As a result, we can piece together at least a partial story of Mars from its earliest times to the present. We find a planet that has gone through dramatic change. Its surface, once warm and wet, is now dry and frozen.

- According to present understanding, Mars almost certainly was a habitable planet in the past and may still have habitable zones underground. This makes Mars a prime target in the search for life beyond Earth.

Summary of Key Concepts

8.1 Fantasies of Martian Civilization

How did Mars invade popular culture?

Superficial similarities between Mars and Earth led to speculation about martian civilization. Astronomer Percival Lowell thought he saw canals built by an advanced society, but the canals do not really exist.

8.2 A Modern Portrait of Mars

What is Mars like today?

Mars is cold and dry, with an atmospheric pressure so low that water is unstable. Martian weather is driven largely by seasonal changes that cause carbon dioxide alternately to condense and to sublime at the poles, creating winds that sometimes generate huge dust storms.

What are the major geological features of Mars?

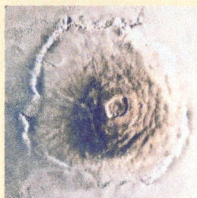
Mars has regions that are densely cratered and must be very old, and other regions with fewer craters that must be much younger. The different regions also vary greatly in elevation, with younger terrain generally in low-lying northern plains and older terrain in southern highlands. Giant volcanoes dot certain regions of Mars, and we also see evidence of past tectonics, which probably created Valles Marineris.

What evidence tells us that water once flowed on Mars?

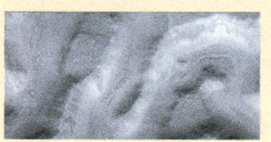
Orbital images of eroded craters, dry river channels, and floodplains all point to past water flows, and supporting evidence is found in chemical analysis of martian rocks studied by landers and rovers. The era of lakes (or possibly oceans) seems to have ended at least 3 billion years ago, but some flooding may have occurred later. Mars today still has water ice underground and in its polar caps, and could possibly have pockets of underground liquid water.

8.3 The Climate History of Mars

Why was Mars warmer and wetter in the past?

Mars's atmosphere must once have been much thicker with a much stronger greenhouse effect, though we do not yet know for certain whether this made Mars warm and wet for an extended period of time or only intermittently.

Why did Mars change?

Change must have occurred due to loss of atmospheric gas, which weakened the greenhouse effect. Some gas was proba-

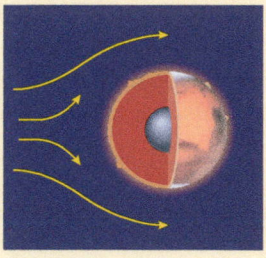
bly blasted away by impacts, but more likely was stripped away by the solar wind as Mars cooled and lost its magnetic field and protective magnetosphere. Water was probably also lost because ultraviolet light could break apart water molecules in the atmosphere, and the lightweight hydrogen then escaped to space.

Is Mars habitable?

Mars almost certainly had a habitable surface during its wet period(s) more than 3 billion years ago. Its surface or near-surface might still sometimes be habitable when its axis tilt is greater than it is now, and the subsurface may still have habitable regions today.

8.4 Searching for Life on Mars

Is there any evidence of life on Mars?

The *Viking* experiments produced results that were in some ways suggestive of life but are now deemed to have been inconclusive. The apparent detection of methane gas that varies in concentration in the atmosphere suggests a source that is either biological or due to geological activity, and the latter would still indicate enough internal heat to raise hopes for the existence of life on Mars.

How do we plan to search for life on Mars?

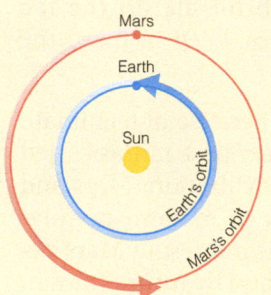
Space scientists plan an ongoing series of Mars missions, timed for the close approaches of Mars to Earth that occur about every 26 months.

Should we send humans to Mars?

Human missions to Mars could probably answer scientific questions about life much more quickly than robotic missions, but humans also pose a risk of contamination. Ultimately, the question will probably be decided by considerations beyond science alone.

❋ THE PROCESS OF SCIENCE IN ACTION

8.5 Martian Meteorites

Is there evidence of life in martian meteorites?

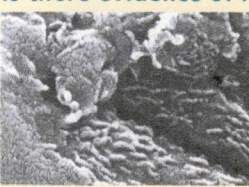
Several lines of evidence have been put forth as suggesting the presence of past life in martian meteorites, but each also has a potential nonbiological explanation. Overall, most scientists doubt current claims for evidence of life in these meteorites, but research continues.

Exercises and Problems

REVIEW QUESTIONS

Short-Answer Questions Based on the Reading

1. Briefly summarize the evidence, both real and imagined, that had led to widespread belief in a martian civilization by the end of the nineteenth century.

2. What would it be like to walk on Mars today? Briefly discuss the conditions you would experience.

3. Why isn't liquid water stable at the martian surface today? What happens to water ice that melts on Mars?

4. How do martian seasons differ from Earth seasons? Describe major seasonal changes that occur on Mars.

5. Give a brief overview of the geography and major features of Mars.

6. How do we know that different regions of the martian surface date to different eras in the past? What have we learned about changes in martian volcanism during the past eras?

7. Summarize the evidence suggesting that Mars must have been warm and wet, possibly with rainfall, in its distant past.

8. What evidence suggests that water might still flow at or beneath the martian surface today? Why do we think that Mars might still have subsurface liquid water today?

9. Why do we conclude that Mars must once have had a thicker atmosphere with a stronger greenhouse effect, and what gases could have made such an atmosphere possible?

10. What is the leading hypothesis concerning how Mars lost its once-thick atmosphere? What role does Mars's size play in this hypothesis?

11. How and why does Mars's axis tilt change with time, and how do these changes affect the climate?

12. Based on all the geographic and geological evidence, summarize the current view about the past and present habitability of Mars.

13. Briefly summarize the *Viking* experiments and their results. Do the results constitute evidence of life? Explain.

14. What is the potential significance of atmospheric methane to the search for life on Mars?

15. Briefly summarize plans for Mars exploration over the next few years. Why do we send missions to Mars only about every 26 months?

16. Discuss the issue of biological contamination in either direction between Earth and Mars. How serious is each problem? What steps can we take to prevent contamination in each direction?

17. Summarize the scientific pros and cons of sending humans to Mars. What other considerations are likely to play a role in decisions about such missions?

18. What do we mean by *terraforming* Mars? Is it something we could do within our lifetimes?

19. How do we know that ALH84001 really came from Mars, and how have we learned its history?

20. Briefly summarize the possible evidence of past life discovered in studies of martian meteorites and why this evidence generates controversy.

TEST YOUR UNDERSTANDING

Surprising Discoveries?

Suppose we were to make the following discoveries. (These are *not* real discoveries.) In light of your understanding of Mars, decide whether the discovery would be considered plausible or surprising. Explain clearly; because not all of these have definitive answers, your explanation is more important than your chosen answer.

21. The first human explorers on Mars discover that the surface is littered with the ruins of an ancient civilization, including remnants of tall buildings.

22. We discover a string of active volcanoes in the heavily cratered southern highlands.

23. We find underground pools of water on the slopes of one of the Tharsis volcanoes.

24. We discover that Mars was subjected to global, heavy rainfall less than 1 billion years ago.

25. A future orbiter finds a plume of volcanic gas emerging from Olympus Mons.

26. We find a lake of liquid water filling a small crater close to one of the dry river channels.

27. The first fossils discovered on Mars come from the canyon walls of Valles Marineris.

28. A sample return mission finds fossil evidence not only of martian microbes, but also of photosynthetic plants that lived on the exposed surfaces of martian rocks.

29. We discover that the martian polar caps have in the past extended more than twice as far toward the equator as they do now.

30. We find rocks on Mars showing clearly that the planet once had a global magnetic field nearly as strong as Earth's magnetic field.

Quick Quiz

Choose the best answer to each of the following. Explain your reasoning with one or more complete sentences.

31. When we say that liquid water is *unstable* on Mars, we mean that (a) a cup of water would shake uncontrollably; (b) it is impossible for liquid water to exist on the surface; (c) any liquid water on the surface would quickly either freeze or evaporate.

32. Mars's seasonal winds are driven primarily by (a) dust; (b) vaporization of carbon dioxide ice; (c) vaporization of water ice.

33. Olympus Mons is (a) a giant volcano; (b) a huge canyon network; (c) a continent-size plateau.

34. We can recognize the oldest surface regions of Mars by the fact that they have (a) the most impact craters; (b) the most volcanoes; (c) the most evidence of past water flows.

35. Minerals in surface rock studied by the martian rovers seem to tell us that (a) they formed in water; (b) they were formed by impacts; (c) they hold fossil evidence of life.

36. Rivers on Mars (a) have never existed; (b) existed in the past but are dry today; (c) continue to have flowing water today.

37. Which must be true if Mars was warmer and wetter in the past? (a) Mars was once closer to the Sun. (b) Mars once had a much thicker atmosphere. (c) Mars must somehow have avoided the effects of the heavy bombardment.

38. Which of the following fundamental properties of Mars could explain why it once had a global magnetic field but later lost it? (a) its small size (b) its larger distance than Earth from the Sun (c) a rotation rate that is slightly slower than Earth's

39. According to the leading hypothesis, if Mars once had much more carbon dioxide in its atmosphere, most of this carbon dioxide is now (a) gone, because it was lost to space; (b) frozen at the polar caps; (c) locked up in the form of carbonate rocks, just like on Earth.

40. The *Viking* experiments found (a) no evidence of life on Mars; (b) clear evidence of life on Mars; (c) results that were inconclusive about the possibility of life.

PROCESS OF SCIENCE

41. *The Role of the Martians.* Percival Lowell may have been sadly mistaken in his beliefs about Martians, but he succeeded in generating intense public interest in Mars. If he had never made his wild claims about canals and civilization, do you think we would be exploring Mars with the same fervor today? Defend your opinion.

42. *Learning from Past Mistakes.* The *Viking* missions landed on Mars about 40 years ago, with ambitious experiments designed to search for life. Today, however, we know that the experimental designs were not sufficient to determine whether or not the missions were finding signs of life. Do you think the scientists should have known better at the time? How can the lessons from *Viking* inform future searches for life on other worlds? Write a short summary of your answers to these questions, and defend your opinions clearly.

GROUP WORK EXERCISE

43. *Human Mission to Mars.* **Roles:** *Scribe* (takes notes on the group's activities), *Proposer* (proposes ideas to the group), *Skeptic* (points out weaknesses in proposed ideas), *Moderator* (leads group discussion and makes sure everyone contributes). **Activity:** You are the planning team for a mission that will carry humans to Mars. The journey will take a few months in each direction, and the explorers will spend about 2 years on the martian surface. Make a list of key provisions needed for the mission, explaining the purpose of each item. In addition, briefly discuss whether any of these provisions could be found or manufactured on Mars rather than having to be brought from Earth.

INVESTIGATE FURTHER

In-Depth Questions to Increase Your Understanding

Short-Answer/Essay Questions

44. *Hold Your Breath.* If you held your breath, would it be safe to walk outside on Mars? Why or why not?

45. *Miniature Mars.* Suppose Mars were significantly smaller than its current size—say, the size of our Moon. How would this have affected its potential habitability? Explain.

46. *Larger Mars.* Suppose Mars were significantly larger than its current size—say, the same size as Earth. How would this have affected its potential habitability? Explain.

47. *Civilization on Mars.* Based on what we can see on the surface of Mars, does it seem possible that Mars once had a civilization with cities on the surface but that the evidence has now been erased or buried underground? Explain.

48. *Martian Fossil Hunting.* On Earth, we cannot find fossil evidence of life dating to times prior to about 3.8 billion years ago. If life ever existed on Mars, is it possible that we would find older fossils than we find on Earth? Explain.

49. *Future Landing Site.* Suppose you were in charge of a mission designed to land on Mars. Assume the mission carries a rover that can venture up to about 50 kilometers from the landing site. What landing site would you choose? Write a one-page summary of why you think your site is a good target for a future mission.

50. *Terraforming Mars.* Make a list of the pros and cons of terraforming Mars, assuming that it is possible. Overall, do you think it would be a good idea? Write a short defense of your opinion.

51. *Mars Movie Review.* Watch one of the many science fiction movies that concern trips to Mars. In light of what you now know about Mars, does the movie give a realistic view of the planet? Are the plot lines that concern Mars plausible? Write a critical review of the movie, focusing on these issues.

52. *Martian Literature.* Read a book of science fiction about Mars, such as H. G. Wells's *The War of the Worlds,* Ray Bradbury's *The Martian Chronicles,* or any of the Edgar Rice Burroughs books about Martians. Write a critical review of the book, being sure to consider whether it still merits interest in light of current scientific understanding of Mars.

Quantitative Problems

Be sure to show all calculations clearly and state your final answers in complete sentences.

53. *Interior Heat.* Compare the surface area–to–volume ratios (that is, total surface area divided by total volume) of the Moon, Earth, and Mars. What does your answer tell you about how quickly each world should have cooled with time? What does your answer tell you about the implications of planetary size for habitability?

54. *Mars's Elliptical Orbit.* Mars's distance from the Sun varies from 1.38 AU to 1.66 AU. How much does this change the globally

averaged strength of sunlight over the course of the martian year? Give your answer as a percentage by which sunlight at perihelion (the orbital point closest to the Sun) is stronger than that at aphelion (the farthest orbital point). Comment on how this affects the martian seasons. (*Hint:* Remember that light follows an inverse square law; see Figure 7.2.)

55. *Atmospheric Mass of Earth.* What is the total mass of Earth's atmosphere? Use the fact that, under Earth's gravity, the sea level pressure of 1 bar is equivalent to 10,000 kilograms pushing down on each square meter of the surface. Also remember that the surface area of a sphere of radius r is $4\pi r^2$.

56. *Atmospheric Mass of Mars.* The weaker gravity of Mars means that 1 bar of pressure on Mars would be that exerted by about 25,000 kilograms pushing down on each square meter of the martian surface. Based on this approximation, the atmospheric pressure on Mars (see Table 8.1), and the size of Mars, estimate the total mass of Mars's atmosphere. Compare to Earth's atmospheric mass from Problem 55.

57. *Past Gas on Mars.* Models suggest that Mars today could have liquid water on its surface if the atmosphere were about 400 times as dense as it actually is. What would the atmospheric mass be in that case? How does this compare to the present mass of Earth's atmosphere? Does it seem plausible that Mars might once have had this much gas? Explain why or why not.

Discussion Questions

58. *Lessons from Mars.* Discuss the nature of the climate change that occurred on Mars some 3 billion years ago. Do you think this

climate change holds any important lessons for us as we consider potential climate changes that humans are causing on Earth? Explain.

59. *Human Exploration of Mars.* Should we send humans to Mars? If so, when? How much would you be willing to see spent on such a mission? Would you volunteer to go yourself? Discuss these questions with your classmates, and try to form a class consensus regarding the desirability and nature of a human mission to Mars.

WEB PROJECTS

60. *Martian Photo Journal.* By now, we have many thousands of photos of Mars taken both on the surface and from orbit, and virtually all of them can be found on the Web. Make your own photo journal of "Mars's Greatest Photo Hits" by choosing ten of your favorite photos. For each one, write a short descriptive caption and explain why you chose it.

61. *Current Mars Missions.* Pick one of the Mars missions that is currently operating and visit its website. Write a short report about the mission's history, goals, and accomplishments to date.

62. *Future Mars Missions.* Pick one of the Mars missions that is being planned or considered for the future (for example, the Mars One initiative) and visit its website. Write a short report about the purpose of the mission and its current status.

9 Life on Jovian Moons

▲ **About the photo:** Sunlight reflecting off a lake on Saturn's moon Titan.

Jupiter orbits the Sun at more than five times Earth's distance, and the other jovian planets (Saturn, Uranus, and Neptune) are even farther away. Sunlight in this distant realm is faint—too weak to provide much warmth. Nevertheless, several of the moons in these frigid outer reaches of our solar system are now considered to be possible places to find life beyond Earth. In this chapter, we will investigate these distant worlds.

We'll begin by examining the general characteristics of these moons and how and why it might be possible for some of them to have liquid water (or other liquids). We'll then turn our attention to the most promising potential abodes of life, including Jupiter's moons Europa, Ganymede, and Callisto, and Saturn's moons Enceladus and Titan, which is blanketed by an atmosphere thicker and denser than Earth's.

Aside from addressing our general curiosity about the habitability of jovian moons, our discussion in this chapter will lead to an intriguing possibility: It's conceivable that these cold and distant moons could be the most numerous homes to life in our solar system. If any of these moons does prove to harbor living things, the possibilities for finding biology elsewhere in the universe will be greatly broadened.

I distinctly recall the dreamy feeling of being in one universe one moment and in another universe the next. But it was no dream. We had, without doubt, journeyed to Titan, 10 times farther from the Sun than the Earth, and touched it. The solar system suddenly seemed a very much smaller place.

Carolyn Porco, Cassini Imaging Team Leader, recalling the 2005 landing of a probe on Titan (full text at http://www.ciclops.org/index/79861)

9.1 The Moons of the Outer Solar System

As we discussed in Chapter 7, the jovian planets themselves seem unlikely to be habitable. However, these planets are orbited by many moons, which we call **jovian moons** because they orbit jovian planets.

The idea of finding life in the cold outer solar system once seemed farfetched, but several of the jovian moons now seem potentially habitable. In this section, we'll introduce the major moons and explore the mechanisms thought to make it possible for at least some of them to have liquid water (or other liquids) on or within them, creating possible habitats for life.

What are the general characteristics of the jovian moons?

As of mid-2015, the four jovian planets were known to have at least 170 moons among them. A few of the larger of these moons have been known to astronomers for nearly four centuries, but many of the smaller ones have been found only with the improved telescopes and space probes of the last few decades. Robotic probes in particular have given us the chance to study these worlds from close by. While previous generations knew jovian moons only as points of light, today we have photographs of many of them that show enough detail that we can name individual craters and other features. One of the biggest surprises to emerge from our modern reconnaissance of these worlds is learning that some of them might be potential homes to life.

DISCOVERING THE MOONS Among the most notable of Galileo's many discoveries [Section 2.2] was finding the four large moons of Jupiter. Having heard of the telescope's recent invention, Galileo built his own homemade versions beginning in 1609. At the time, the telescope was thought to be

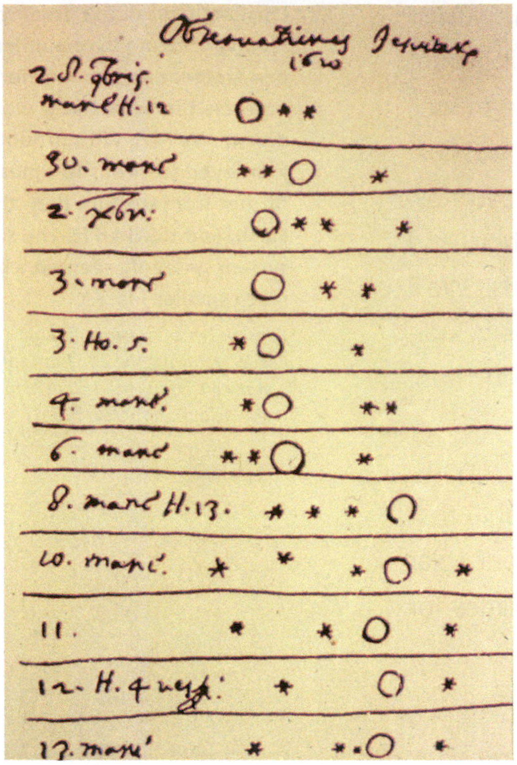

FIGURE 9.1

A page from Galileo's notebook written in 1610. His sketches show four "stars" near Jupiter (the circle), but in different positions at different times (and sometimes hidden from view). Galileo soon realized that the "stars" were moons orbiting Jupiter.

primarily a toy, or possibly useful for defense. Galileo, however, did something different with his telescopes: He turned them toward the sky. On January 7, 1610, while gazing at Jupiter, he saw what at first seemed to be three small stars in its vicinity. What intrigued Galileo was that the three were close to one another and in a line. The following night, he looked again and was surprised to note that the three stars had moved relative to Jupiter, but not in the direction or to the extent expected of background stars. A few days later he noted a fourth point of light, and within a week he realized that these four "stars" always stayed close to Jupiter and were clearly in orbit around it (Figure 9.1). In March 1610, Galileo published his results in a pamphlet he called *The Starry Messenger,* claiming to have found four bodies moving around the giant planet "as Venus and Mercury around the Sun." These four bodies are what we now call the **Galilean moons** of Jupiter; proceeding outward from that planet, we know them individually as Io, Europa, Ganymede, and Callisto (Figure 9.2).

Other scientists soon discovered additional moons in the outer solar system. The accomplished Dutch scientist Christiaan Huygens (1629–1695) discovered Titan, the largest of Saturn's moons, in 1656. Before the close of the seventeenth century, Giovanni Domenico Cassini (1625–1712), an Italian astronomer who became director of the Paris Observatory, had discovered four more moons around the ringed planet.*

Even today, astronomers sometimes make new discoveries of moons orbiting the jovian planets, though all of the larger moons have surely been discovered by now. The new discoveries involve small moons—usually no more than a few kilometers across—detected with the aid of new telescopes or spacecraft.

SIZES AND ORBITS OF THE MOONS The jovian moons come in a wide range of sizes. While many small ones are not much bigger than a single mountain on Earth, others are much larger. The two largest—Jupiter's moon Ganymede and Saturn's moon Titan—are bigger than the planet

---

*Saturn's rings were first sighted by Galileo, but the resolution of his telescope was too low for him to make out what they were. Huygens was the first to realize that the rings do not touch Saturn's surface, and Cassini showed that the rings were not solid but instead were marked by a dark division, which we still call the *Cassini division.*

FIGURE 9.2

This set of photos, taken by the *Galileo* spacecraft, shows global views of the four Galilean moons as we know them today. Sizes are shown to scale.

1000 km   Io   Europa   Ganymede   Callisto

Mercury. The three other Galilean moons (Io, Europa, and Callisto) and Neptune's moon Triton are bigger than Pluto. Figure 9.3 shows a montage of jovian moons larger than about 350 kilometers in diameter.

Almost all of the moderate-size and large moons orbit their planets in much the same way that planets orbit the Sun: They orbit nearly in the equatorial plane of their host world, moving in the same direction as their planet's spin. In this sense they resemble miniature solar systems, which suggests that they were formed in a smaller-scale version of the same processes that gave birth to the planets [Section 3.4]. As shown in Figure 9.4, the gravity of the jovian planets drew in gas and dust from the surrounding solar nebula. Like the solar nebula as a whole, this gas and dust formed a swirling disk-shaped cloud around each jovian planet. Condensation and accretion then built moons that shared the orbital properties of the original disks of gas and dust.

The small jovian moons differ from their larger brethren in both appearance and the properties of their orbits. Most have an irregular shape and often resemble peanuts, potatoes, or other snack foods (Figure 9.5). This is hardly surprising: The lesser gravity of these small objects is too weak to force the rigid material of which they're composed into spheres.

Some small moons may have formed "in place" like most of the larger moons, and others come in groups that share common orbital characteristics, suggesting that they may be fragments of larger moons that broke apart. But many of the smaller moons also have orbits that are highly elliptical or inclined to the equator of their host planet; in some cases, the moons orbit backward, or *retrograde*, relative to their host planet's spin. These orbital characteristics are telltale signs of moons that are captured asteroids or comets, meaning that they once orbited the Sun independently. As we discussed briefly in Section 4.6, it's not easy for a planet's gravity to grab a wayward asteroid or comet: Because energy must be conserved, the passing object must somehow lose some of its orbital energy or else it would simply fly on by and continue in its path around the Sun. The leading hypothesis holds that captures occurred when the jovian planets were young, when they had large, extended atmospheres that could have served to slow down small bodies as they passed nearby, thereby removing orbital energy and allowing the objects to be captured.

Surprisingly, one large moon also has orbital characteristics suggesting it is a captured object: Neptune's moon Triton, which orbits backward relative to its planet's rotation. This orbit makes it a near-certainty that Triton once orbited the Sun as an independent object, although capturing such a large object poses a trickier problem than capturing smaller moons. Recent research suggests that Triton could have been captured if it had once been orbited by its own satellite. In that case, it could have been snagged while passing close to Neptune, because the excess orbital energy that Triton had to lose could have been carried off by the satellite as it was ejected from its orbit. Regardless of the specific mechanism of the capture, for several decades scientists have assumed that Triton was once one of the *Kuiper belt objects* (KBOs) of which we now consider Pluto to be a member [Section 3.4]. Indeed, the fact that Triton is larger than Pluto is one reason some astronomers suspected that Pluto might not be the largest such object in the outer solar system, a suspicion confirmed with the 2005 discovery of Eris—an ice-rich world that is more massive than Pluto (but slightly smaller in diameter) and nearly a hundred times as far from the Sun as Earth.

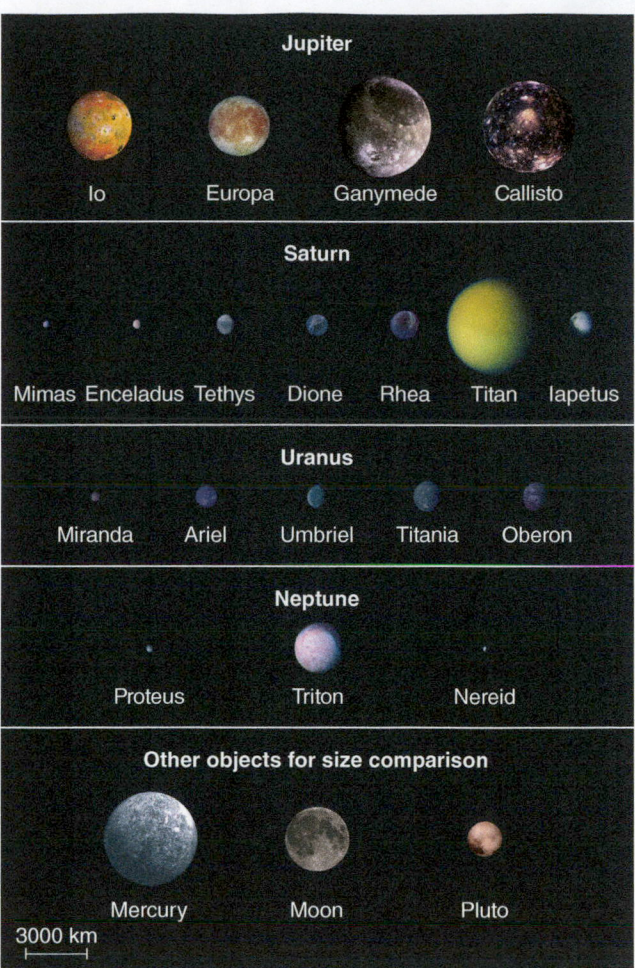

FIGURE 9.3

The larger moons of the jovian planets, with sizes (but not distances) shown to scale. Mercury, our own Moon, and Pluto are included for comparison.

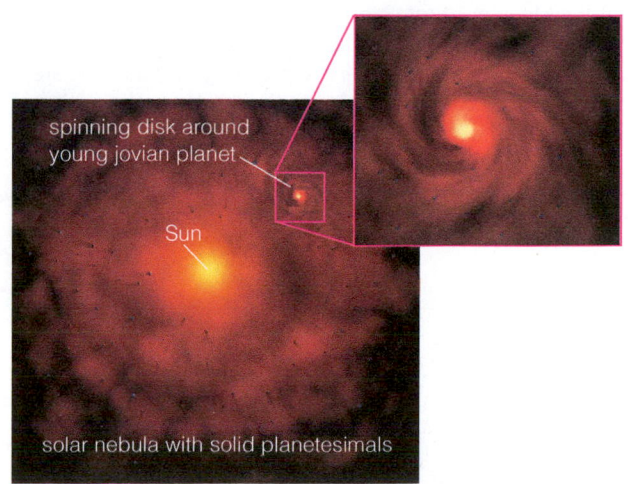

FIGURE 9.4

The young jovian planets are thought to have been surrounded by disks of gas and dust much like the solar nebula as a whole but smaller in size.

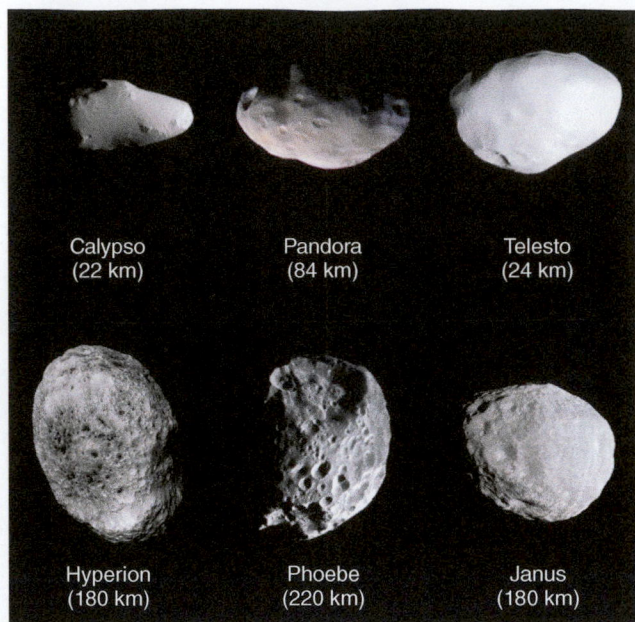

FIGURE 9.5

These photos from the *Cassini* spacecraft show six of Saturn's smaller moons (not to scale). All are much smaller than the smallest moons shown in Figure 9.3. Because they are not spherical, the sizes in parentheses represent approximate lengths along their longest axes.

FIGURE 9.6

The fact that we always see the same face of the Moon means that the Moon must rotate once in the same amount of time that it takes to orbit Earth once, an idea you can understand by walking around a model of Earth while imagining that you are the Moon. The same idea applies to the synchronous rotation of jovian moons.

COMPOSITION OF THE MOONS As we discussed in Chapter 3, the outer solar system was cold enough to allow ices to condense along with metal and rock. We therefore would expect the jovian moons to be made of a mixture of ice and rock, and that is indeed the case for most of them. The average densities of most jovian moons are significantly lower than that of Earth, reflecting the fact that they contain substantial quantities of ice, which is low in density. Within individual moon systems, we see variations in composition reflecting the fact that the moons formed at different distances from a hotter, central planet. For example, the Galilean moons show a decrease in density with distance from Jupiter, just as we would expect if they formed from a swirling cloud of gas that was hotter in the center than in its outer regions: Io's density indicates that it has virtually no water in any phase, Europa is mostly rock with water ice (and a liquid ocean) only in its outer layers, and Ganymede and Callisto have more significant amounts of water ice relative to their amounts of rock (and may also have liquid water below their surfaces).

In addition to the compositional variation within individual moon systems, there is variation in the composition of moons as we move from one planet to the next. These differences came about because of temperature differences in the overall solar nebula. Water ice condensed easily at the temperatures found near Jupiter's orbit, but methane and other ices condensed only at the colder temperatures found at significantly greater distances from the Sun. As a result, Jupiter's moons contain significant quantities of water ice but only a smattering of other ices. Moons of the more distant planets contain higher overall proportions of ice than of rock, and contain not only water ice but also methane and other ices.

SYNCHRONOUS ROTATION OF THE MOONS Nearly all jovian moons share a common characteristic: Like our own Moon, they always keep the same face turned toward their planet. This behavior, called **synchronous rotation,** means that each moon completes exactly one rotation around its axis while it makes one orbit around the planet. You can see how this works with a simple demonstration (Figure 9.6). Place a ball on a table to represent a planet like Earth, and walk around the ball so that your head represents an orbiting moon. If you do not rotate as you walk around the

a If you do not rotate while walking around the ball representing Earth, you will not always face it.

b You will face Earth at all times only if you rotate exactly once during each orbit.

ball, you'll be facing away from it by the time you are halfway around your orbit. The only way you can face the ball at all times is by completing exactly one rotation while you complete one orbit.

Synchronous rotation is not a coincidence: Rather, it develops naturally as a consequence of the same gravitational effects that lead to tides on Earth. Recall that the strength of gravity declines with distance, following an inverse square law [Section 2.4]. As a result, the gravitational attraction of each part of Earth to the Moon becomes weaker as we go from the side of Earth facing the Moon to the side facing away from the Moon. This *difference* in attraction creates a "stretching force," or **tidal force**, that stretches the entire Earth to create two tidal bulges, one facing the Moon and one opposite the Moon (Figure 9.7). If you are still unclear about why there are *two* tidal bulges, think about a rubber band: If you pull on a rubber band, it will stretch in both directions relative to its center, even if you pull on only one side (while holding the other side still). In the same way, Earth stretches on both sides even though the Moon is tugging harder on only one side. Tides affect both land and ocean, but we generally notice only the ocean tides because water flows much more readily than land. (Recall that solids *can* flow, though slowly, which is why Earth's solid mantle can undergo convection [Section 4.4].)

Think About It The Sun exerts a stronger gravitational force on Earth than does the Moon—after all, Earth orbits the Sun, not the Moon. So why is the Moon rather than the Sun primarily responsible for Earth's tides? (The Sun's tidal effect on Earth is a little less than half as strong as the Moon's.) Do you think other planets have any significant effect on tides on Earth? Explain.

Because tidal forces stretch Earth itself, the process necessarily creates some friction, called **tidal friction.** As shown in Figure 9.8, this tidal friction has important consequences for Earth and the Moon, because it allows Earth's rotation to pull the bulges slightly ahead of the Earth–Moon line. This slight misalignment of the tidal bulges with the Earth–Moon line means the Moon's gravity is always pulling back on the bulges, causing Earth's rotation to slow gradually. At the same time, the gravity of the bulges pulls the Moon slightly ahead in its orbit, causing the Moon to move gradually farther from Earth. These effects are barely noticeable on human time scales; the Moon is moving farther from Earth at only about 4 centimeters per year (as measured by laser beams bounced off the lunar surface), and tidal friction increases the length of Earth's day by only about 1 second every 50,000 years. (On short time scales, this effect is overwhelmed by other effects on Earth's rotation.) But the effects add up over billions of years. Early in Earth's history, a

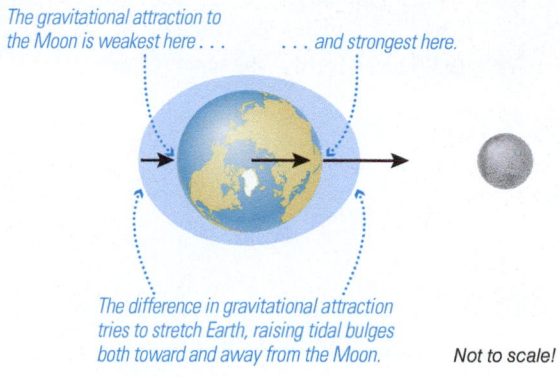

The gravitational attraction to the Moon is weakest here . . .   . . . and strongest here.

The difference in gravitational attraction tries to stretch Earth, raising tidal bulges both toward and away from the Moon.   *Not to scale!*

FIGURE 9.7
Tides on Earth are created by the difference in the force of attraction between the Moon and different parts of Earth. The two daily high tides occur as a location on Earth rotates through the two tidal bulges. The diagram greatly exaggerates the bulges, which raise the oceans only about 2 meters and the land only about a centimeter.

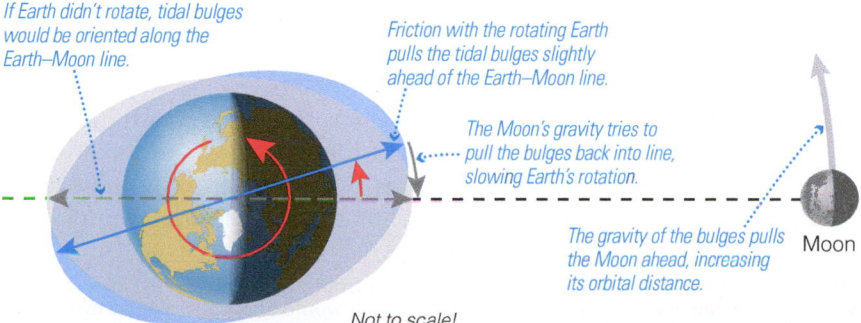

If Earth didn't rotate, tidal bulges would be oriented along the Earth–Moon line.

Friction with the rotating Earth pulls the tidal bulges slightly ahead of the Earth–Moon line.

The Moon's gravity tries to pull the bulges back into line, slowing Earth's rotation.

The gravity of the bulges pulls the Moon ahead, increasing its orbital distance.   Moon

Not to scale!

FIGURE 9.8
Earth's rotation pulls its tidal bulges slightly ahead of the Earth–Moon line, leading to gravitational effects that very gradually slow Earth's rotation and increase the Moon's orbital distance.

day may have been only 5 or 6 hours long and the Moon may have been one-tenth or less its current distance from Earth.

We can understand the Moon's synchronous rotation by turning the situation around. Just as the Moon raises tides on Earth, Earth must raise tides on the Moon—in fact, much stronger tides, because Earth's gravity is much stronger than the Moon's. If the Moon rotated rapidly, these tides would generate substantial tidal friction that would tend to slow the Moon's rotation. The tidal friction would cause the Moon's rotation to slow until the Moon kept the same face to Earth at all times. In other words, no matter how fast the Moon may have been rotating at its birth, it was inevitable that tidal friction would slow this rotation until it was synchronous. At that point, the synchronous rotation was permanently "locked in," because there was no more tidal friction to further slow the Moon's rotation.

Tidal friction similarly explains the synchronous rotation of the jovian moons.* From the time of its birth, when a moon was rotating faster than it orbited, tidal friction slowed its rotation until the rotation and the orbit were synchronized. Given the large sizes of the jovian planets compared with their moons, synchronous rotation likely set in for the close-in moons within no more than a few million years.

Why do we think that some moons could harbor life?

Most of the jovian moons are almost certainly lifeless. As we discussed in Chapter 7, they are too small and too far from the Sun to have any reasonable likelihood of having liquid water or other liquids. However, the situation may be different for a few of the bigger moons. One large moon, Titan (which we'll discuss in Section 9.3), has a temperature and atmospheric pressure that allow liquid methane and ethane to form lakes and rivers on its surface. In addition, several moons have enough internal heat to keep water liquid in their interiors—making their interiors potentially habitable.

TIDAL HEATING Based on what we've learned about internal heat on the terrestrial planets, it might seem surprising to find much internal heat in any of the jovian moons. After all, only Ganymede and Titan are even as large as Mercury, and we know that Mercury's interior has cooled enough over time that it no longer supports active volcanism. But when the *Voyager 1* spacecraft passed Jupiter in 1979, we got definitive proof that at least some distant moons have plenty of internal heat. Confirming a theoretical prediction made just weeks before *Voyager*'s arrival, the spacecraft photographed active volcanic eruptions on Io, the innermost of the four Galilean moons. The *Galileo* orbiter then provided much more detailed views (Figure 9.9). How did scientists predict this surprising volcanism? They realized that jovian moons can have a type of internal heating different from that found on the terrestrial worlds, and that Io would be the extreme example of this heating.

Recall that Earth and the other terrestrial worlds were all hot inside at their births, with heat from accretion, differentiation, and radioactive decay [Section 4.4]. Over time, heat has escaped, and only radioactive decay continues to supply new heat. For small terrestrial worlds, like

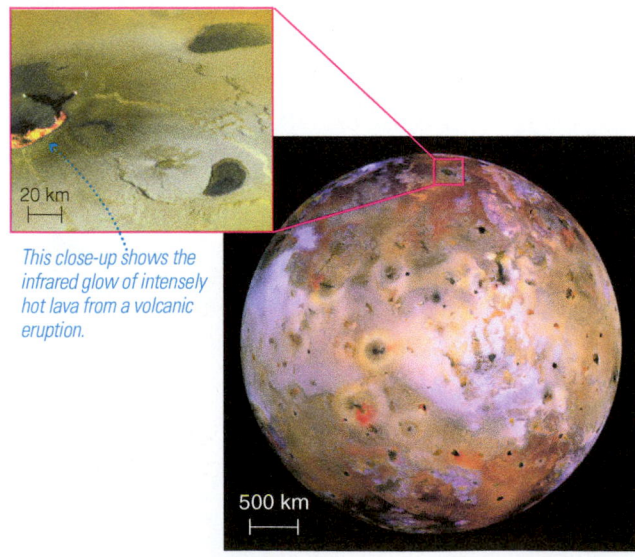

This close-up shows the infrared glow of intensely hot lava from a volcanic eruption.

20 km

500 km

FIGURE 9.9 INTERACTIVE FIGURE
Io is the most volcanically active body in the solar system. Most of the black, brown, and red spots on the surface are related to recently active volcanoes.

---

*Europa shows evidence of a very slight deviation from perfectly synchronous rotation, which may be an effect of a subsurface ocean, which would allow a free-floating ice shell above to avoid being synchronously locked.

the Moon and Mercury, the supply of new heat has not been enough to match the heat lost by escape, which is why these worlds have so little geological activity today. If jovian moons had only the same three heat sources as the terrestrial worlds, they probably would be geologically inactive by now too. Their additional heat source is called **tidal heating,** because it arises from effects of tidal forces.

Io's tidal heating arises from a combination of two factors: (1) Its proximity to Jupiter means it experiences a strong tidal force from the massive planet; and (2) Io has a slightly elliptical orbit, which causes the strength and direction of the tidal force to change slightly as Io moves through each orbit. Figure 9.10 shows the result: The constantly changing orientation of Io's tidal bulges means that Io is continuously being flexed in different directions, which generates friction inside it in much the same way that flexing warms Silly Putty.* Tidal heating generates tremendous heat on Io—more than 200 times as much heat (per gram of mass) as the radioactive heat driving much of Earth's geology. This heat makes Io the most volcanically active place in the solar system. Materi-

---

*The ultimate source of the energy that drives tidal heating is Jupiter's rotation, which is gradually slowing, much like Earth's rotation, although at a rate too small to be observed.

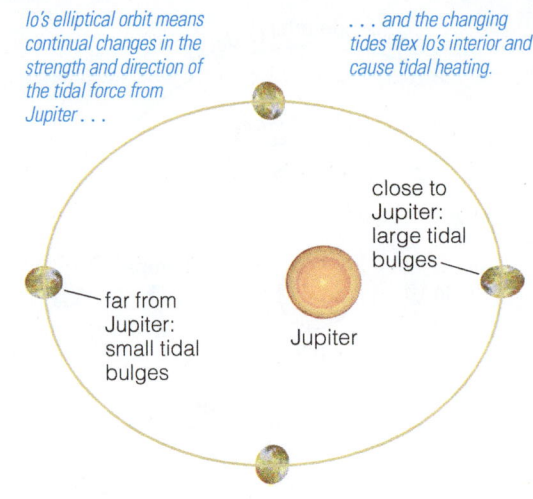

Io's elliptical orbit means continual changes in the strength and direction of the tidal force from Jupiter . . .

. . . and the changing tides flex Io's interior and cause tidal heating.

close to Jupiter: large tidal bulges

far from Jupiter: small tidal bulges

Jupiter

FIGURE 9.10 INTERACTIVE FIGURE
Tidal heating arises on Io from the combination of its elliptical orbit and the strong tidal force exerted on it by Jupiter. The bulges and orbital eccentricity are exaggerated.

Cosmic Calculations 9.1 THE STRENGTH OF THE TIDAL FORCE

Recall that the force of gravity acting between two objects is

$$F_g = G\frac{M_1 M_2}{d^2}$$

The tidal force that a planet exerts on a satellite is the *difference* between the gravitational force on the near and far sides of that satellite. One way to get a sense of this difference is to consider a "test mass"—say, a 1-kg rock—on the satellite's surface. If d is the distance between the center of the planet and the center of a satellite of radius r_{sat}, then on the near side of the satellite the distance from the planet's center to the 1-kg rock is $d - r_{sat}$ and on the far side it is $d + r_{sat}$. We then calculate the gravitational force in both positions and subtract to get the tidal force acting on the satellite per kilogram of mass.

Example: Calculate the tidal force that Earth exerts on the Moon (per kilogram) and compare it to the force exerted by the Moon's own gravity. Useful data: $M_{Earth} = 5.97 \times 10^{24}$ kg, $r_{Moon} = 1.74 \times 10^6$ m, the Earth–Moon distance is $d = 3.84 \times 10^8$ m, and $M_{Moon} = 7.35 \times 10^{22}$ kg. As long as you use G in standard units and the masses and distances in kilograms and meters, your answer will come out in *newtons* (1 N $\approx$ 1 kg $\times$ m/s^2 $\approx$ 0.225 lb).

Solution: We find the tidal force F_{tidal} that Earth exerts on the rock by subtracting the gravitational force F_g when the rock is on the far side of the Moon (where the rock's gravitational attraction to Earth is weaker) from the gravitational force when the rock is on the near side (where the gravitational attraction is stronger):

$$F_{tidal} = F_g(\text{on near side}) - F_g(\text{on far side})$$

Let's first consider the near-side term:

$$F_g(\text{near side}) = G\frac{M_{Earth}M_{rock}}{(d_{Earth\text{-}Moon} - r_{Moon})^2}$$

$$= \left(6.67 \times 10^{-11}\frac{m^3}{kg \times s^2}\right)\frac{(5.97 \times 10^{24}\text{ kg})(1\text{ kg})}{(3.84 \times 10^8\text{ m} - 1.74 \times 10^6\text{ m})^2}$$

$$= 0.00273\text{ N}$$

For the far-side term, we find

$$F_g(\text{far side}) = G\frac{M_{Earth}M_{rock}}{(d_{Earth\text{-}Moon} + r_{Moon})^2}$$

$$= \left(6.67 \times 10^{-11}\frac{m^3}{kg \times s^2}\right)\frac{(5.97 \times 10^{24}\text{ kg})(1\text{ kg})}{(3.84 \times 10^8\text{ m} + 1.74 \times 10^6\text{ m})^2}$$

$$= 0.00267\text{ N}$$

The difference gives us the tidal force (per kilogram):

$$F_{tidal} = F_g(\text{on near side}) - F_g(\text{on far side})$$
$$= 0.00273\text{ N} - 0.00267\text{ N} = 0.00006\text{ N}$$

For comparison, the Moon's gravitational force acting on the rock is

$$F_g = G\frac{M_{Moon}M_{rock}}{r_{Moon}^2}$$

$$= \left(6.67 \times 10^{-11}\frac{m^3}{kg \times s^2}\right)\frac{(7.35 \times 10^{22}\text{ kg})(1\text{ kg})}{(1.74 \times 10^6\text{ m})^2}$$

$$= 1.6\text{ N}$$

Note that this 1.6 N gravitational force (which is the weight of the rock on the Moon and is equivalent to 1.6 N $\times$ 0.225 lb/N = 0.36 lb) is more than 15,000 times greater than the 0.00006 N tidal force that Earth exerts on the rock. Clearly, the tidal force is quite small in comparison to the gravitational force, though over time it has been large enough to bring the Moon into synchronous rotation. (The tidal force was stronger when the Moon was closer to Earth.)

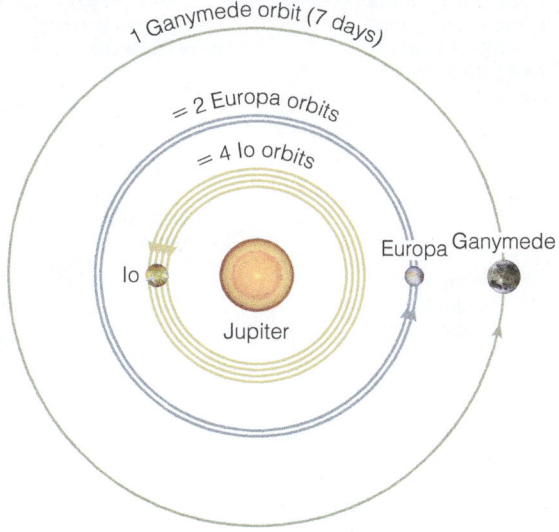

1 Ganymede orbit (7 days)

= 2 Europa orbits

= 4 Io orbits

Io

Jupiter

Europa Ganymede

FIGURE 9.11 INTERACTIVE FIGURE

Io, Europa, and Ganymede share an orbital resonance that returns them to the positions shown every 7 days, and the recurring gravitational tugs explain why all three orbits are slightly elliptical. The effect on Io is greatest, since it is closest to Jupiter.

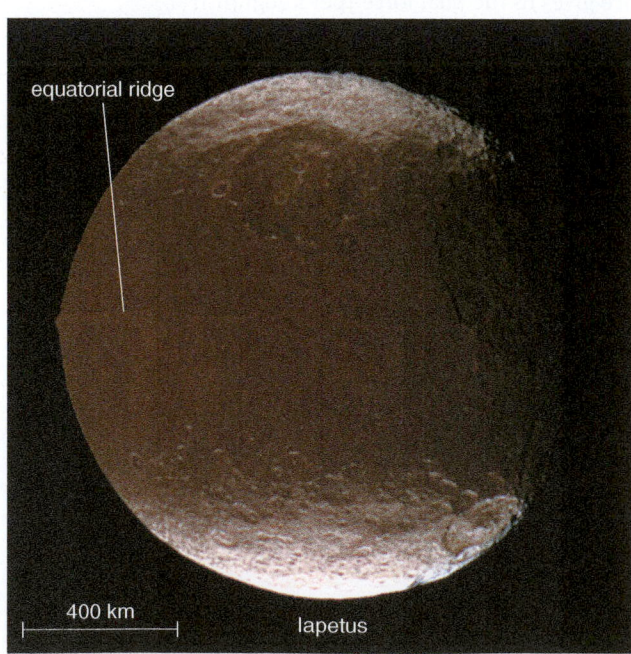

equatorial ridge

400 km

Iapetus

FIGURE 9.12

Saturn's moon Iapetus (about 1440 kilometers in diameter) has a steep ridge running along its equator, suggesting that it must once have been surprisingly warm inside.

al from Io's volcanic vents reaches temperatures of more than 1000°C (1800°F), and eruptions often spew plumes of sulfur and other gases to heights of hundreds of kilometers.

Although the combination of the elliptical orbit and Jupiter's tidal force seems to explain tidal heating, there is still another question to ask: Why is Io's orbit slightly elliptical, when most large satellites have nearly circular orbits? The answer lies with an "orbital resonance" that occurs among Io, Europa, and Ganymede: The three moons periodically line up, because during the time Ganymede takes to complete one orbit of Jupiter (about 7 days), Europa completes two orbits and Io completes four orbits (Figure 9.11). The effect is much like that of pushing a child on a swing. If timed properly, a series of small pushes can add up to a *resonance* that causes the child to swing quite high. For the three moons, the periodic alignments mean they experience repeated gravitational tugs that act much like the repeated pushing of the child on the swing (hence the term *orbital resonance*). In this case, these gravitational tugs cause the orbits to be more elliptical than they would be otherwise.

Like synchronous rotation, orbital resonances arise naturally. The orbital resonances of Io, Europa, and Ganymede probably came about because of feedback with tides that the moons raise on Jupiter. Just as our Moon raises tides on Earth that are gradually causing the Moon to move farther from Earth (and slowing Earth's rotation), the Galilean moons raise tides in Jupiter's atmosphere that cause the moons to move farther from Jupiter. The effect is greatest on the closest-in moon, Io, so we expect that its orbit would have moved outward until it achieved a resonance with the next moon out, Europa. Then Io and Europa would have continued to move outward in lockstep until they achieved a resonance with the third moon, Ganymede. This may be how Io, Europa, and Ganymede came to have orbital resonances among themselves. Indeed, these moons may now be moving slowly outward in lockstep toward Callisto, though it will be many billions of years before they could reach a resonance with this fourth moon.

OTHER HEAT SOURCES Tidal heating certainly explains Io's volcanic activity and probably explains Europa's internal heat. As we'll see in the next two sections, however, it may not be enough to explain the heating that apparently occurs or has occurred in the past on a few other jovian moons. Saturn's moon Iapetus, for example, has a striking equatorial bulge that indicates it must once have been warm enough to be soft inside (Figure 9.12), but tidal heating does not seem to explain how it became hot enough for its interior to become sufficiently "plastic" to flow. One possible explanation lies with radioactive decay, if Iapetus and other moons could have incorporated enough short-lived radioactive material during their formation to explain the level of heating that they have apparently experienced.

9.2 Life on Jupiter's Galilean Moons

Tidal heating has turned Io into a veritable hell, and its lack of water and extreme volcanic activity essentially rule it out as a home for life. But tidal heating has lesser effects on Europa and Ganymede (and no effect on Callisto, because it does not participate in the orbital resonance). In the case of Europa, in particular, it is possible that the level of tidal heating is "just right" for life. Without tidal heating, Europa would be wrapped in solid ice. However, its appearance and other characteristics give hints that this moon has a subsurface ocean that makes it a possible home to life. In this section, we'll discuss the prospects of habitability and life on Europa, as well as the prospects for the outer two Galilean moons, Ganymede and Callisto.

Does Europa have an ocean?

Even before the *Voyager* spacecraft snapped the first detailed photos of Europa in 1979, scientists already knew that Europa is covered with an icy shell. Spectroscopic observations made from Earth indicated the presence of water ice on the surface, although the amount was unknown. Theoretical studies, based on our ideas about the formation of moons, suggested that Europa was likely built of a mixture of rock and water. If heat had caused the moon to differentiate sometime in the past, then the less dense water would have migrated to the surface. The *Voyager* spacecraft confirmed at least part of this speculation: Europa's exterior is bright white and ice-covered. Also, its surface is remarkably smooth, with few features rising higher than about a kilometer. Faint ridge lines—giant cracks in the ice—crisscross the surface, looking almost like a spider web of highways.

The *Galileo* orbiter revealed Europa in much more detail (Figure 9.13). Notice the small number of recognizable impact craters, which tells us that Europa's surface has been repaved in recent geologic times. However, it is difficult to tell whether the resurfacing has been caused by the action of liquid water or of ice that is relatively warm and therefore soft enough to flow (as glaciers flow on Earth).

By using the *Galileo* spacecraft to measure subtle variations in Europa's gravitational field, scientists were able to determine its internal layering. Crudely speaking, Europa seems to consist of a central metallic (probably iron) core, overlaid with a thick mantle of silicate rock and an 80–170-kilometer-thick outer skin of water or water ice. From the gravity measurements alone, the water layer could contain just about any combination of solid ice, liquid water, or slush (partially melted ice), because all of these have about the same density (1 g/cm^3). As a result, over the past couple of decades scientists have debated which form of water is most likely. The debate is not yet fully settled, but as we'll discuss shortly, careful analysis of data from the *Galileo* spacecraft suggests that the most likely answer is the model shown in Figure 9.14: Europa probably has a brittle, icy crust underlaid by a layer of warmer ice that can flow easily and undergo convection, and a liquid water ocean beneath that. Lakes may sometimes form within the ice layer. Surface temperatures on Europa are brutally cold (typically $-150°C$), so the top portion of its icy crust must be as stiff as granite.

EVIDENCE FOR A LIQUID OCEAN The spacecraft pictures provide strong evidence suggesting that liquid water, not just more ice, underlies Europa's frozen skin. One key piece of evidence comes from the surprising lack of impact craters. Europa has only a few dozen large craters (10 kilometers or more across), a number that should accumulate in only a few tens of millions of years, not the billions of years since the solar system's formation. Clearly, something has erased the craters on Europa. An obvious candidate is resurfacing by an occasional breakthrough of subsurface water or slushy ice.

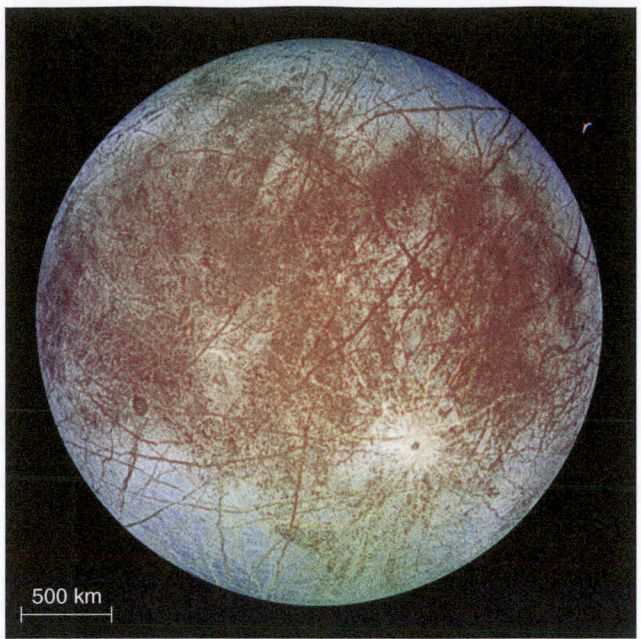

FIGURE 9.13

A global view of Europa, as seen from the *Galileo* spacecraft. Colors are enhanced to bring out subtle details.

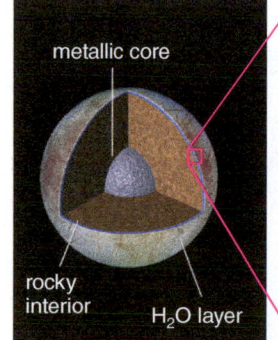
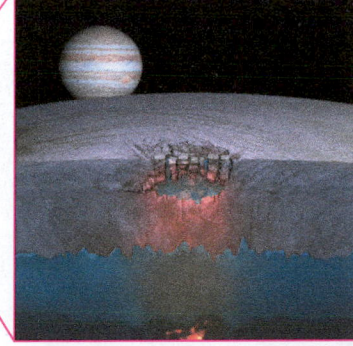

Europa may have a 100-km-thick ocean under an icy crust.

Rising plumes of warm water may sometimes create lakes within the ice, causing the crust above to crack.

FIGURE 9.14

This diagram shows one model of Europa's interior structure. There is little doubt that the H_2O layer is real, but questions remain about whether the material beneath the icy crust is liquid water, relatively warm convecting ice, or some of each.

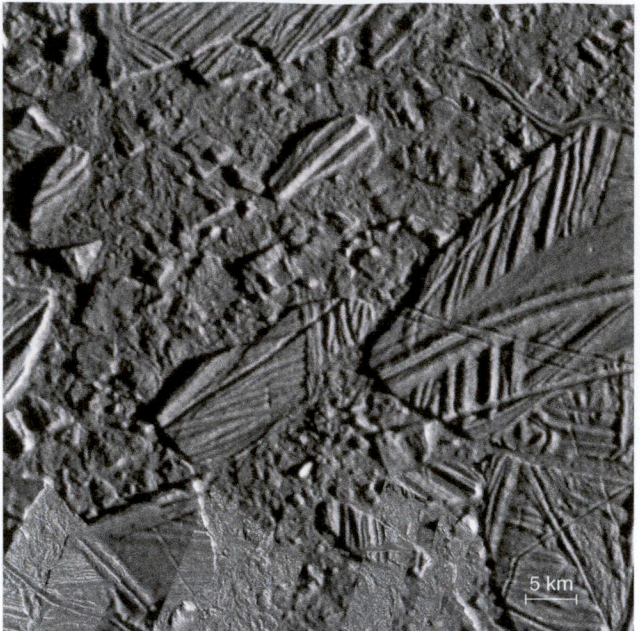

FIGURE 9.15

Chaotic terrain on Europa. This landscape suggests that liquid water (or slushy ice) has welled up from below, breaking apart the surface and then freezing in place. The pieces can be mentally reassembled by matching up the details of the ice blocks. However, such a reassembly indicates that some of the pieces are missing; they presumably sank, melted, or crumbled during the reshuffle.

FIGURE 9.16

More evidence of upwelling water on Europa. The smooth, dark terrain just to the right of the center of this photo mosaic could represent an area where liquid water broke through the surface and covered over some of the ridged landscape.

Another suggestive piece of evidence is the appearance of Europa's so-called *chaotic terrain,* which resembles photos of an arctic ice pack (Figure 9.15). The surface is clogged with iceberglike blocks—some as small as football fields and others as large as a city—crisscrossed by ridge lines and suspended in what appears to have been a slushy ocean that froze. These blocks are often separated, as if they have rafted away from their original positions. Imagine putting your palms down on an assembled jigsaw puzzle and then spreading them slightly. Gaps will form between the pieces, and the picture printed on them will become disjointed. Europa's ridge lines similarly form a telltale picture on the ice, a picture that in places has fractured into small, jostled pieces. Individual ridges can be traced by the eye from one block to another, clearly showing the motion that must have occurred. Some of the ridges could be analogous to mid-ocean ridges on Earth, such as we find in the mid-Atlantic Ocean (see Figure 4.15), in which case they may have been caused by upwelling of relatively warm slush from beneath the frozen crust. Further evidence for an ocean comes from features suggesting that liquid water has gushed from below and frozen in spaces between some of the ridges (Figure 9.16).

While the photographic evidence suggests that Europa's ice-pack exterior has been churned by an underlying ocean, it still leaves a shadow of a doubt. Could it be that the ocean that produced the tortured and broken surface existed millions of years ago and is now frozen, which would mean that Europa's intriguing face is merely the frozen remnant of an earlier time? Perhaps the icy crust sits atop relatively warm soft or slushy ice, and the surface features are the result of this warm ice convecting upward from below and melting pockets of ice within the crust, so the only liquid water resides in subterranean lakes similar to those found deep under the ice in Antarctica. In that case, Europa's frigid skin might extend all the way to the moon's rocky interior. But there's another line of evidence, arguably even more convincing than the photos, that makes the case for a liquid ocean.

In 1996, a magnetometer aboard the *Galileo* spacecraft detected a magnetic field near Europa. Because moons seldom have a magnetic field (though, as we'll discuss shortly, Ganymede does), researchers asked themselves what was causing magnetism on Europa. Jupiter has a strong magnetic field (which causes it to be a relatively powerful source of natural radio static), and Europa orbits within this field. Because Jupiter's magnetic equator is tilted with respect to Europa's orbit, its field at Europa's position is constantly changing as the giant planet spins on its axis. Just as a moving magnet produces an electric current in a coil of wire, so too could Jupiter's changing field induce currents in an electrical conductor within Europa. Of course, there's no giant coil of wire in Europa, but a salty ocean would conduct electricity in much the same way. The currents in such an electrically conducting ocean would act to set up a magnetic field that opposed Jupiter's—and therefore would change as Jupiter's field changed. Additional measurements with *Galileo*'s magnetometer showed that Europa's magnetic field does indeed change as Jupiter spins. Most researchers consider this the strongest evidence that a liquid, salty ocean exists under Europa's icy surface. The magnetometer data require that Europa's subsurface water be global in extent, not limited to just a few isolated liquid pockets. They also imply that this ocean could be as salty as Earth's seas. *Galileo*'s infrared spectrometer also found evidence for what appear to be salts on Europa's surface—possible seepage from a briny deep.

If we tally up the evidence for an ocean on Europa, we can list the following:

- Calculations show that tidal heating can supply enough heat to keep most of Europa's ice melted beneath a solid ice crust.
- The relatively small number of craters implies that the moon's surface is young, perhaps only a few tens of millions of years old, indicating that it has been recently repaved.
- Various features on the surface (chaotic and flooded terrain) suggest that liquid water sometimes wells up from below.
- Europa has a magnetic field that is likely caused by currents produced in something that conducts electricity—like a salty ocean—as Jupiter's magnetic field changes.

While no single piece of evidence would make an overwhelmingly strong case by itself, together the evidence gives us good reason to suspect that the model in Figure 9.14 is correct, and that a liquid ocean really does exist beneath the surface of Europa. Still, the case is not definitive. For that, we'll need new space missions.

PROVING THE CASE FOR AN OCEAN Scientists are hoping to confirm the case for Europa's ocean with a mission that NASA is developing for launch sometime between 2022 and 2025. The mission consists of a spacecraft that will orbit Jupiter while making dozens of flybys of Europa. One method for confirming an ocean relies on the fact that the gravitational tug-of-war that Europa undergoes in its elliptical orbit of Jupiter causes distortions in Europa's shape. If a deep liquid ocean exists under the ice, Europa's surface should bulge in and out by up to 30 meters with each orbit. In contrast, if there is only solid ice beneath the surface, the bulging will be by only about 1 meter. The Europa mission will include a radar altimeter to measure this bulging as well as instruments to measure the composition of Europa's surface and thin atmosphere, cameras to study and map surface features, and magnetometers to determine the thickness and salinity of the ocean. The radar can also probe the icy crust, looking for radio reflections from a subsurface water interface. This same technique was used to discover Antarctica's Lake Vostok, which lies hidden beneath 4 kilometers of ice.

Think About It Antarctica's buried Lake Vostok probably offers the closest analogy on Earth to the possible ocean on Europa. By the beginning of 2015, a bore hole drilled by researchers had reached the top of the lake, 3759 meters below the surface, and plans were being made to lower a capsule to start recovering samples of the lake's water. Search for news about Lake Vostok. Has sampling begun? Have scientists found anything that might help us understand the possibilities for life on Europa? Explain.

If there is an ocean on Europa, how big might it be? The current model (shown in Figure 9.14) suggests that the frozen crust (including the region of warmer, convecting ice) is about 20–25 kilometers thick. Given that our models of Europa's interior indicate that the total thickness of the water/ice layer is between about 80 and 170 kilometers, a 20–25-kilometer-thick crust leaves plenty of room for a deep ocean. In fact, if a global, subsurface ocean really exists, it could easily be 100 kilometers or more deep—some ten times as deep as the deepest ocean trenches on Earth. A 100-kilometer-deep europan ocean would contain roughly twice as much water as all of Earth's oceans combined.

Could Europa have life?

The oceans on Earth teem with life, and in Chapter 6 we discussed reasons why many scientists suspect that life on Earth first arose in the oceans, most likely near deep-sea volcanic vents. If Europa really does have a deep water ocean, tidal heating has probably kept it liquid for billions of years. Might life have arisen in the europan ocean as it arose on Earth, and could Europa be home to life today?

In Section 7.1, we identified the three key environmental requirements for life: (1) a source of elements and molecules from which to build living organisms, (2) a source of energy for metabolism and growth, and (3) a liquid medium for transporting the molecules of life. If Europa has an ocean, then it clearly meets the third requirement. As we discussed in Chapter 7, the first requirement is likely met by nearly all worlds, and Europa should be no exception, especially given its rock/water composition. In fact, the elements needed for life would almost certainly be present at the rock/water boundary of the ocean floor, which is where we might expect any life to have arisen. That leaves open the question of whether Europa satisfies the second requirement: Does it have an energy source for life?

You might at first guess that the existence of a liquid water ocean would automatically mean there is energy for life. After all, if there is liquid water, there must be enough energy from tidal heating (and radioactive decay) to keep the water temperature above freezing. However, the existence of energy in the form of heated water is not by itself enough to allow the energy to be put to use. That is why fish can't live off ocean heat alone.

In general, extracting energy from a reservoir of warm water is possible only if there is also an adjacent "sink" of much colder water and a substantial difference in temperature between the two. And while we can build machines to take direct advantage of such differences in temperature, life has an additional requirement: There must be chemicals present that can react to generate energy for biological use. In practice, this additional requirement is probably met whenever there are temperature differences at a rock/water boundary, because molecules in the rock and water can react together in a variety of useful ways. (We'll discuss some of the specific types of reactions that can supply energy to life in Section 9.4.)

Overall, then, the possibility of life in a europan ocean probably rests on the question of whether there is *enough* energy in a useful form to support biology. We can separate this question into two parts. First, is there enough energy to support an origin of life? Second, if so, is there enough energy to support a reasonable total biomass of ongoing life?

ENERGY FOR AN ORIGIN OF LIFE From an energy standpoint, the possibility that life on Earth might have begun near deep-sea vents makes sense. The volcanic vents heat the water near them to very high temperatures, creating a large temperature difference between this water and the surrounding cooler water. This temperature difference would have facilitated chemical reactions between the water and the rock erupted from the vents, leading to the formation of complex organic molecules and perhaps to the origin of life. We might therefore wonder whether the same type of vents exist on Europa, providing the energy needed for an origin of life.

We cannot see through Europa's icy crust, which is why we don't even know for certain that a liquid ocean lies beneath it. Clearly, then, we have no way at present to know whether Europa has volcanic activity on an ocean floor. Nevertheless, the possibility of volcanic vents seems reasonable. Our models of Europa's interior suggest that it should have a rocky ocean floor, and tidal heating and the decay of radioactive elements may provide enough energy to melt pockets of interior rock that could erupt into the ocean. Europa might even have large undersea volcanoes. In that case, it certainly seems plausible to imagine an independent origin of life on Europa.

ENERGY TO SUPPORT ONGOING LIFE If life did arise in a europan ocean, how widespread could it be today? While deep-sea vents might lead to enough energy for an origin of life, they could not by themselves support more than a small total biomass because they simply don't make enough energy available to organisms. This fact might surprise you if you think about the great communities of life that live near deep-sea vents on Earth today (Figure 9.17), but most of this life actually gets its energy from materials that filter down from above, such as dead organisms and oxygen produced by photosynthetic life near the surface. Only a small fraction of the life near Earth's deep-sea vents lives solely off energy from the vents themselves. For life to be similarly abundant or widespread on Europa, it would need some other energy source in addition to the chemical reactions near deep-sea vents.

On Earth, sunlight is the best-known source of energy for life, as photosynthesis converts sunlight to energy that works its way up the food chain. However, sunlight is unlikely to provide much energy for life on Europa, for two reasons. First, because Europa is about five times as far from the Sun as Earth, sunlight is about $5^2 = 25$ times weaker at Europa than at Earth. Second, and likely more important, sunlight cannot penetrate through more than a few tens of meters of ice, so it could not directly provide energy for life in a europan ocean. The weak sunlight at Jupiter's distance from the Sun could provide limited energy for life on Europa only if pockets of liquid water exist near the top of the ice crust. In that case, organisms living in these pockets could conceivably obtain energy from photosynthesis, and this energy might then filter downward as these organisms (dead or alive) cycle between the crust and the ocean.

Before the origin of photosynthesis on Earth, biochemical reactions may have been facilitated by energy sources such as lightning, ultraviolet radiation from the Sun, and heat released by impacts of asteroids and comets. Unfortunately, none of these energy sources is likely to be useful to life in europan seas. Europa has no atmosphere for lightning, ultraviolet light does not penetrate the ice, and the time during which impacts were frequent enough to provide significant energy ended some 4 billion years ago.

There is, however, another possible scheme for producing energy on Europa's icy surface. High-energy particles accelerated and trapped in Jupiter's magnetic field, as well as ultraviolet light from the Sun, regularly slam into the surface ice. These particles and photons hit the surface with enough energy to break up molecules in the ice, leading to the production of small quantities of other molecules such as hydrogen peroxide (H_2O_2), molecular oxygen (O_2), and hydrogen (which quickly escapes); this process explains why Europa has an extremely thin atmosphere (not noticeable to the eye, but detected by instruments). These

FIGURE 9.17
On Earth, we find abundant life near deep-sea volcanic vents. However, while the vent supplies energy used by some microbes, most of the life around it—including the tube worms visible in this photo—actually gets energy from materials that filter down from above.

molecules can facilitate energy-producing reactions, and they should be mixed into the uppermost portion of the europan surface by the frequent churning caused by small meteorites. If all these molecules were ultimately to end up in the ocean below, they could provide energy to support life there. Unfortunately, we don't know how much of the outer ice actually gets cycled into the water below, or how often. In addition, the total amount of energy that might be available in this way is at least ten thousand times less than the amount of energy that photosynthesis generates on Earth, so an ocean on Europa could not support anywhere near as much life as do the oceans on our planet.

One other known process might yield some energy for life on Europa. Some of the potassium contained in the rocky material that makes up this moon would dissolve in the ocean (and be frozen in the ice above). The energy from the natural decay of one of potassium's radioactive isotopes would produce both hydrogen and oxygen molecules. Rough estimates suggest that these molecules could then facilitate chemical reactions that might support a small biomass.

THE CASE FOR POSSIBLE LIFE ON EUROPA Summarizing the case for life on Europa, we can say:

- There is strong but indirect evidence that Europa has a liquid water ocean.
- We expect the elements needed for life to be present in that ocean and on its floor.
- There are possible energy sources to support life, but the total available energy is small compared to the energy available for life on Earth.

Taken together, the evidence makes Europa seem like a good candidate for the possibility of life. That said, we should caution that it seems unlikely that adequate sources of energy could exist in the europan ocean to support macro-fauna—the complex sea creatures of our aquariums, for example. If life exists in Europa's oceans, it is probably quite simple and small, perhaps analogous to the most primitive single-celled organisms that have existed on Earth.

Think About It Given the various uncertainties about a liquid ocean and available energy on Europa, do you consider it likely or unlikely that we will find life there? Why?

The only way to find out for sure whether life exists on Europa will be to go there. NASA's planned Europa mission (mentioned earlier) would not only confirm (or deny) the existence of the hidden ocean, but also analyze Europa's very thin atmosphere, offering a small chance of finding organic molecules or other hints of life as it measures the composition of the atmosphere. Subsequent missions could land on the surface. If a lander found any water that might have been brought up from below, we could analyze it for various organic molecules such as amino acids. The lander might also melt and filter a sample of surface ice—preferably a large sample since we expect the abundance of any life on Europa to be low—in search of evidence of life. Note that we'd need to be careful to sterilize the spacecraft so that we didn't accidentally transport biology from Earth to Europa and then "discover" it there. (Indeed, in 2003, NASA deliberately ended the *Galileo* mission by causing the spacecraft to plunge into Jupiter's atmosphere so as to prevent any possibility that the spacecraft might someday crash into and contaminate Europa with hitchhiking microbes from Earth.) If the results from a lander were encouraging—for example, if they showed the presence of organic molecules in the ice—we could then dream and scheme about probes that would manage to work their way through the ice and explore the eternally dark ocean depths where alien life might swim.

Could other moons of Jupiter have life?

Europa is the most likely of the Galilean moons to be habitable. However, both Ganymede and Callisto are also composed of significant amounts of water ice. Could they, too, have underground oceans, and hence the possibility of life?

GANYMEDE Ganymede, the third from Jupiter of the four Galilean moons, is the largest moon in the solar system. Like Europa, Ganymede has a surface of hard, brittle ice. However, while Europa's impact craters have been mostly erased by other geological processes, Ganymede appears to have both young and old surface regions, sometimes separated by remarkably sharp boundaries (Figure 9.18). Some regions are dark and densely cratered, suggesting that they look much the same today as they did billions of years ago. Other regions are light-colored with few craters, suggesting that they are geologically younger and have had their ancient craters erased; the young regions also exhibit strange grooves. The most likely explanation for the younger regions of ice is that they were created by "water eruptions" that occurred when internal heat caused ice below the surface to melt, erupt as watery "lava," and then freeze. The grooves were probably made by tectonic stresses that stretched the icy crust.

The idea that water sometimes erupts onto Ganymede's surface implies that partial melting must occasionally occur in the underlying ice,

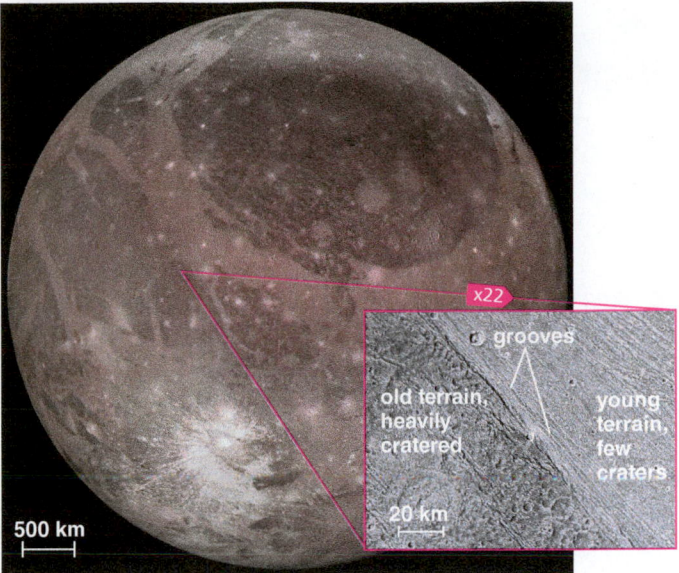

FIGURE 9.18

Ganymede, the largest moon in the solar system, has both old and young regions on its surface of water ice. The dark regions are heavily cratered and must be billions of years old, while the light regions are younger landscapes where tectonic faulting and eruptions of liquid water or slush have presumably erased ancient craters; the long grooves in the light regions were probably formed by tectonic stresses. Notice that the boundary between the two types of terrain can be quite sharp.

Movie Madness 2010: THE YEAR WE MAKE CONTACT

The monoliths are back, and they've brought plenty of buddies.

In the film *2010*, a sequel to the classic *2001*, humans are once again prompted by some enigmatic black slabs to head for the outer solar system. It seems that an unseen race of aliens is trying to make an important point about Jupiter and its moons. So we oblige them by sending yet another batch of astronauts on a billion-kilometer joy ride.

However, unlike the mission in *2001*, this crotchety crew actually returns to Earth, bringing with them a clear warning delivered by the mysterious aliens: "All these worlds are yours except Europa. Attempt no landing there."

What's the deal? After all, this is *our* solar system. So why are some unknown, unseen entities marking Europa off-limits ? That's like Mom forbidding you to go into a basement closet. Of course that's the one place you'll find most interesting, and Arthur C. Clarke, the author of the novel *2010,* knew that there was, indeed, an interesting closet in the jovian system. The *Voyager* spacecraft had shown Europa to be an ice-covered world that could have a huge liquid ocean, and maybe even life.

In *2010*, this possibility is subtly exploited. The astronauts learn that there are chlorophyll-equipped critters somewhere below the europan ice. Alas, life in such a deep, dark habitat is a bit of a drag. So just as the humans arrive, the sophisticated aliens who built the monoliths decide to reengineer Jupiter, turning it into a mini-sun. They do this for the benefit of the primitive life-forms on Europa. The new star eventually warms their moon and converts it to something that looks like Earth during the Mesozoic era. The Europans, we presume, are destined to crawl out of their formerly ice-capped seas and find a monolith that will promote them to intelligent beings, the way our simian ancestors were improved in *2001*.

We lose Jupiter, but we gain a second, dimmer sun in our skies (no doubt a headache for astronomers and a source of confusion for migratory birds). But one wonders why these unseen extraterrestrials are so keen to meddle in the biological evolution of other worlds. Perhaps they just want our distant descendants to have some intelligent company right here in the solar system. It's a nice thought, but in the meantime someone needs to tell the space agencies that it's "hands off" Europa for the next few hundred millennia!

but could Ganymede have a full-fledged ocean like that thought to exist on Europa? The *Galileo* spacecraft turned its magnetometer to Ganymede just as it did to Europa, and made two intriguing discoveries. First, Ganymede apparently has its own intrinsic magnetic field—one generated within the moon—which may indicate that it has a molten, convecting core somewhat like Earth's outer core. Second, the magnetometer data showed that a small part of Ganymede's magnetic field varies with Jupiter's 10-hour rotation, just like Europa's magnetic field. Recent observations with the Hubble Space Telescope further strengthen the case for a subsurface ocean on Ganymede. Although Hubble cannot directly measure the magnetic field, it acquired images of aurorae around Ganymede that are caused by interactions of the solar wind with the magnetic field. Careful study of how the aurorae change with time indicates variations in the magnetic field that make sense only if Ganymede has a deeply buried ocean of salty water. Overall, a liquid water ocean as deep as that on Europa now seems likely to exist beneath Ganymede's frozen surface.

Assuming the ocean is real, what heat source keeps it liquid? Because Ganymede is farther from Jupiter than Europa or Io is, its tidal heating is weaker and could not by itself supply enough heat to melt ice today. However, Ganymede's larger size means it should retain more heat from radioactive decay. Perhaps tidal heating and radioactive decay together provide enough heat to make a liquid layer beneath the icy surface. Planetary scientists also suspect that Ganymede went through a period of increased tidal heating, perhaps about 1 billion years ago, and is still cooling off from that event.

While the existence of a subsurface ocean is encouraging from the standpoint of habitability, the lesser heating on Ganymede means that the ice cover is estimated to be much thicker than on Europa—about 150 kilometers thick. This would make finding life in a subsurface ocean far more difficult and the transport of possible nutrients from the surface considerably less efficient. In addition, the pressure in Ganymede's interior is high enough to create high-density forms of ice that likely lie beneath any liquid water ocean. As a result, Ganymede probably does not have a rock/water boundary at its ocean bottom like that on either Earth or Europa, a fact that would further reduce the available energy for life. Life, if it exists on Ganymede at all, would probably be less abundant and less evolved than seems possible on Europa. It might also be so deep below the surface that we'd have little hope of gaining access to it.

CALLISTO Callisto is the farthest out of the four Galilean moons. Figure 9.19 shows that its entire surface is densely pockmarked by craters (the bright, circular patches) that must date back to the heavy bombardment. Other surface features are more difficult to interpret. For example, the close-up photo shows a dark, powdery substance that is concentrated in low-lying areas, leaving ridges and crests bright white. The dark powder may be debris left behind when ice vaporizes into gas from Callisto's surface.

Gravity measurements suggest that Callisto is mostly a ball of mixed ice and rock, overlaid by several hundred kilometers of water ice. Because its interior doesn't seem to be fully differentiated, we conclude that Callisto was never very warm inside and that neither radioactive decay nor tidal heating ever heated this moon enough to melt it through. Surprisingly, however, the *Galileo* spacecraft found an induced magnetic field

500 km

2 km

FIGURE 9.19
Callisto is heavily cratered, indicating an old surface that nonetheless may hide a deeply buried ocean. The inset shows how a dark powder appears to cover low-lying areas of the surface.

for Callisto, too, suggesting—as for Europa and Ganymede—the presence of a salty, subsurface ocean.

If a subsurface ocean exists on Callisto, what heat source could keep it liquid? Unlike the other three Galilean moons, Callisto doesn't participate in the orbital resonances that cause tidal heating, meaning that the warmth required to maintain a liquid ocean would almost surely have to come from radioactive decay. This meager heat source might be sufficient because of the insulating properties of a thick, icy skin. In addition, the water might contain salts or ammonia that acts like antifreeze to help keep it liquid at low temperatures.

If Callisto really does have a deep, unseen ocean, we arrive at the astonishing possibility that three of Jupiter's large moons could have liquid water oceans and, perhaps, life. Energy considerations make it unlikely that any of these moons has the abundance or diversity of life that we find on Earth. Callisto, which probably has the least energy available for life, is the least likely of the three moons to have life in its ocean. Nonetheless, there is the intriguing thought that any aliens that might come from afar to study biology in our solar system might find much to interest them by spending time in the vicinity of Jupiter.

Think About It Suppose we had the technology to send landers that could somehow reach subsurface oceans on any of Jupiter's moons. Which moon would you choose to explore first? Why?

9.3 Life Around Saturn, and Beyond

Prospects for habitability and life dim considerably as we go beyond Jupiter. Although a few outer solar system moons have tidal heating, we do not find more examples quite as extreme as those of, say, Io and Europa. The greater distance from the Sun also lessens the possibility of obtaining energy from sunlight, and cold temperatures would presumably slow any metabolic reactions. Nevertheless, the outer solar system offers some intriguing prospects for life—or at least for interesting organic chemistry that could be a precursor to life.

Could Titan have life?

The best-studied candidate for habitability beyond Jupiter is Saturn's moon Titan. The second-largest moon in the solar system after Ganymede, Titan is the only solar system moon to have a substantial atmosphere. In fact, Titan's atmosphere is even thicker than Earth's. The surface pressure is about 1.6 times that on Earth, which means that if you could visit Titan, the pressure would feel fairly comfortable even without a space suit. The temperature, however, would not. Here, where sunlight is nearly 100 times as weak as on Earth, the surface temperature is a frigid $-180°C$ ($-290°F$). Moreover, while the atmosphere is 90% nitrogen (N_2)—not so different from the 77% nitrogen content of Earth's atmosphere—there is no appreciable oxygen to breathe.

While Titan's low temperatures make life there seem unlikely, some interesting chemistry is taking place on this frigid, smoggy world. As a result, it was selected as a prime target for the *Cassini* mission that reached Saturn orbit in 2004. By 2008, *Cassini* had completed its original reconnaissance of the Saturn system, and soon thereafter went into "overtime," continuing to study the ringed planet and its moons. By mid-

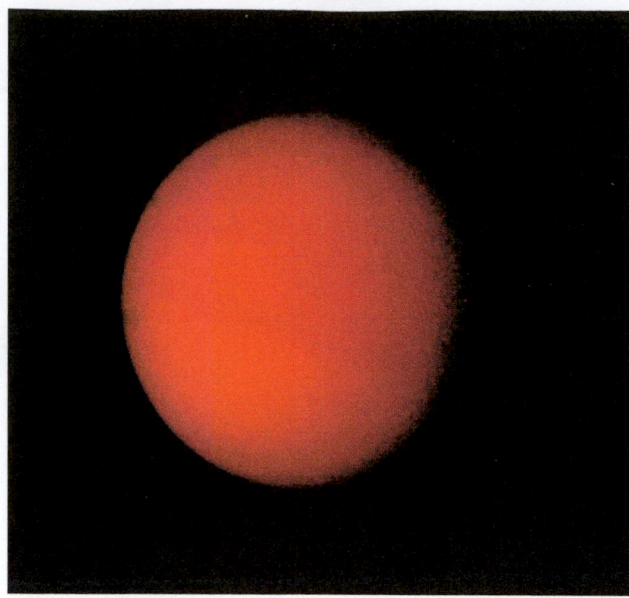

FIGURE 9.20
Titan, as photographed by *Voyager 2*. It is enshrouded by a
reddish smog, and the mystery of what lies beneath encouraged
scientists to make Titan a prime target of the *Cassini* mission.

2015, it had flown past Titan 109 times and had mapped much of the
moon's hidden landscape. In addition, *Cassini* carried a probe, called *Huygens*, that parachuted to a soft landing on Titan's surface in January 2005.
The results from *Huygens* and the ongoing observations from *Cassini* have
revolutionized our understanding of Titan.

Think About It Saturn's distance from Earth varies between about 1.2 and
1.6 billion kilometers (depending on whether Earth is on the same or the
opposite side of the Sun); in other words, the *Huygens* probe successfully
landed on a world more than *a billion kilometers away.* Does this fact tell you
anything important about human capabilities? Defend your opinion.

A MOON LONG SHROUDED IN MYSTERY Titan's atmosphere was first discovered in 1944, when spectroscopic observations from Earth showed
the presence of methane. However, the amount of atmosphere was not
immediately known, in part because methane is not the dominant gas
(nitrogen is). Because scientists originally included the hazy atmosphere
when measuring Titan's size, it was once thought to be the largest moon
in the solar system (which is in keeping with its name). The two *Voyager*
spacecraft, which passed by Saturn and its moons in 1980 and 1981,
found that Titan's girth had been overestimated because it is puffed out
by a 200-kilometer-thick atmosphere. The solid part of Titan has a radius
60 kilometers smaller than that of Ganymede. Cameras aboard the *Voyager* spacecraft were powerful enough to observe Titan in fine detail, but
they saw nothing of the surface* because their vision was blocked by the
opaque, reddish haze—in essence, smog (Figure 9.20). Most of the visible
smog is due to chemical by-products formed when ultraviolet light from
the Sun breaks apart molecules of methane.

Titan's gravitational pull on the *Voyager* spacecraft allowed researchers to accurately determine its mass, which turns out to be nearly twice
that of our Moon. Knowing its mass, combined with its size, permits
us to compute its average density, which is nearly the same as Callisto's
(about 1.9 gm/cm^3). This suggests that Titan is made up of roughly equal
volumes of rock and ice, like other large moons of the outer solar system.

Despite the fact that the *Voyager* cameras were frustrated by smog,
these spacecraft managed to tell us much about conditions within Titan's
atmosphere. *Voyager 1* was given a trajectory that allowed it to sail behind
Titan, so that its radio signal would pass through the smog on its way
back to Earth. This ingenious experiment provided data that were used
to determine both the temperature and the composition of the atmosphere. The composition measurements proved especially intriguing. Besides its 90% nitrogen content, the atmosphere is composed of (in order
of abundance) methane (CH_4), argon, and ethane (C_2H_6). There are
lesser quantities of propane (C_3H_8), acetylene (C_2H_2), hydrogen cyanide (HCN), and carbon dioxide (CO_2). What accounts for this mixture
of hydrocarbons, which reads like an oil company's product line?

To begin with, once ammonia (NH_3) from Titan's icy interior made
it into the atmosphere, energetic ultraviolet light from the Sun would
have broken it apart, allowing it to react to form nitrogen (N_2) and hydrogen (H_2). The nitrogen molecules were heavy enough to stay put and

---

*Subsequent reprocessing of *Voyager* images shows that they did just barely detect the
surface, demonstrating how improving image-processing technology can give new life
to old data.

became the principal ingredient of Titan's air. The much lighter hydrogen escaped into space. When methane (CH_4) from the interior entered the atmosphere, it too was broken apart by ultraviolet light into hydrogen (which, again, escaped) and the simpler compounds CH and CH_2. Products of the methane breakdown then reassembled themselves into more complex hydrocarbons, especially ethane (C_2H_6). Eventually, this should have led to an atmosphere so saturated with ethane that a nonstop drizzle of ethane rain began to fall.

So far, so good. But the fact that measurements from both *Voyager* and Earth-bound telescopes showed that there's still lots of methane gas in the air was surprising; we might expect that it would all have been converted into other molecules long ago. This puzzle can be solved if we assume that a replenishment source—a reservoir of methane—slowly feeds new gas into the atmosphere. The source was hypothesized to be large pools of slowly evaporating methane on the surface, where the $-180°C$ temperature should be just warm enough to allow methane and ethane to be in a liquid state (see Table 7.1). Some evaporating methane would remain in the atmosphere, some would be converted to ethane by ultraviolet light and chemical processes, and some might even rain down.

Overall, the *Voyager* studies suggested that Titan could have a drizzle of rain made up of methane or ethane droplets—in essence, liquid natural gas (which is largely made up of methane and ethane)—and perhaps even lakes or oceans of liquid methane and ethane on the surface. Clearly, this was a world that called for further study.

ORIGIN OF THE ATMOSPHERE Given that we usually think of moons as airless worlds, like our own Moon, how is it that Titan has such a thick atmosphere? In fact, the main reason moons generally have little or no atmosphere is their small size, which results in weak gravity that is insufficient to hold on to substantial amounts of gas. Titan can hold its atmosphere because of its relatively large size, along with extremely cold temperatures that make it less likely that gas molecules can attain escape velocity. A subtler question is why Titan has a thick atmosphere while Jupiter's moon Ganymede, which is even larger (and still quite cold), does not.

Two explanations for the difference have been suggested. First, recall that at Saturn's distance from the Sun, ices such as methane and ammonia should have been able to condense in the early solar system, but we expect mostly water ice at Jupiter's distance. Therefore, while the outer crusts of Europa, Ganymede, and Callisto are composed largely of water ice, Titan's outer layer should have substantial amounts of methane and ammonia ice. These compounds can vaporize into gas at lower temperatures than does water (see Table 7.1). If internal temperatures rose on Titan during differentiation, methane and ammonia ice might have turned to gas, bubbled out of the crust, and built up an atmosphere.

A second possible reason for Titan's atmosphere is that comets and asteroids hitting a moon of Saturn are traveling at lower speeds than are those that fall onto the moons of Jupiter (both because Jupiter's stronger gravity accelerates incoming objects more and because the Sun's gravity accelerates objects more at Jupiter's distance than at Saturn's). When such bodies slam into an atmosphere, they can blast away much of the atmosphere. If Ganymede once had an atmosphere, it would have

600 km

FIGURE 9.21

Titan unmasked. This picture of Titan was made with a *Cassini* camera outfitted with an infrared filter able to image wavelengths of light that can penetrate the thick smoggy atmosphere. The image uses black and white to show infrared contrast, since we cannot see infrared light. Compare the detail in this image to the featureless *Voyager* photo in Figure 9.20.

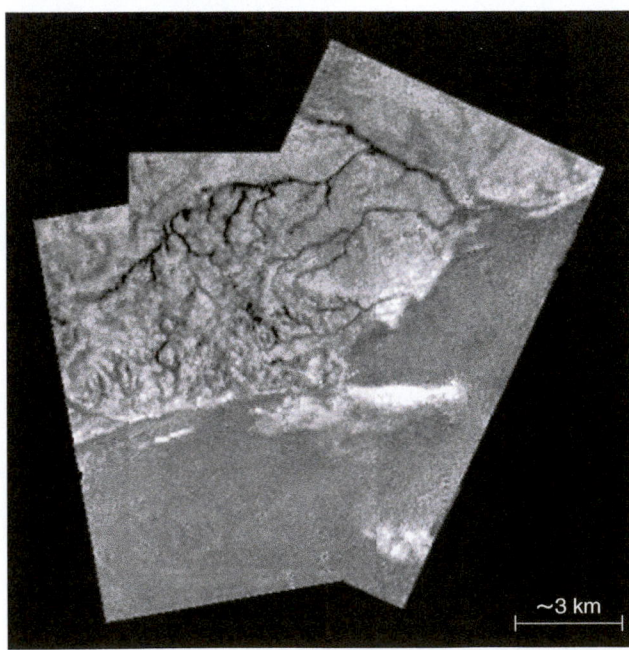

~3 km

FIGURE 9.22 INTERACTIVE FIGURE

The *Huygens* probe made the photos in this mosaic when it was still several kilometers above Titan's frigid surface. The tributary-like dark channels flowing to the flat area below were probably carved by liquid methane and ethane. The bright icy hills reach heights of approximately 100 meters.

been more likely than Titan's atmosphere to have been blown away by impacts.

THE *CASSINI–HUYGENS* MISSION The intriguing prospect of a smoggy moon with frigid seas of liquid methane was a major incentive for launching the 2-ton *Cassini–Huygens* spacecraft to Saturn in 1997 (see Figure 7.14). After a circuitous, 7-year journey that took it twice past Venus, back past Earth, and then on beyond Jupiter—with each planetary pass giving it a gravitational slingshot energy boost [Section 7.4] that was necessary because of its relatively low launch speed—*Cassini* reached Saturn orbit in July 2004. With infrared cameras and filters that allow the orbiter to "see" through the smoggy atmosphere, *Cassini* quickly gave us much clearer pictures of Titan (Figure 9.21).

Our closest views of Titan came from the *Huygens* probe. Released from the *Cassini* "mother ship" on Christmas Day, 2004, the probe spent 21 days coasting toward Titan. On January 14, 2005, the probe entered Titan's smoggy atmosphere. A series of parachutes was deployed to ease *Huygens* through its $2\frac{1}{2}$-hour descent. As the descent proceeded, the probe radioed back information about the composition and temperature of Titan's atmosphere, and also snapped hundreds of photos of the landscape below.

The aerial views from *Huygens* show fantastic topographic details. Some are clearly reminiscent of coastal areas on Earth, with hills laced by branching channels that meander down to flat areas that look like rivers or lakes (Figure 9.22). However, unlike on Earth, the liquids responsible for carving the channels are liquid methane and ethane, not water. *Huygens* found no sign of flowing or pooled liquids at the time of its descent, but this is probably because *Huygens* landed during a dry spell, or perhaps even in an area that is only very occasionally rinsed by liquids during wet seasons.

Although *Huygens* was designed to float in case it settled onto a lake of liquid hydrocarbons, it actually landed on a hard surface, reminiscent of a streambed strewn with dirty water-ice pebbles (Figure 9.23). It hit the surface at about the speed of a bicycle. For another $1\frac{1}{2}$ hours (until the mother ship was too far away to pick up signals and the lander's battery died), the parked probe continued to send back data. This was the first time that a spacecraft from Earth had landed on a moon other than our own.

The *Huygens* probe did not find any liquids in its descent region, but the *Cassini* orbiter has found lakes, particularly in the north polar region of this moon (Figure 9.24). However, *Cassini* has ruled out the idea that Titan might be largely covered with liquid seas of methane or ethane, which some scientists had deemed likely prior to the spacecraft's arrival.

Further analysis of Titan's atmosphere has provided new insight into its origin. In particular, aside from the argon previously identified by *Voyager*, *Huygens* found no other "noble gases" (gases of elements in the last column of the periodic table [Appendix D]) such as krypton, xenon, or neon. These nonreactive gases are present in Earth's atmosphere, presumably as a consequence of a rain of comets billions of years ago. Once the comets smashed into our world, these elements, which are relatively heavy, were trapped. The fact that they are absent on Titan (the exception, argon, arises from the radioactive decay of potassium [Section 4.2]) suggests that no gas was supplied by comets. Instead, we infer that Titan's smoggy atmosphere was outgassed from its interior.

Outgassing implies either some type of volcanism or vaporization of interior ices, but Titan's interior should not be warm enough to melt rock and drive volcanoes like those on Earth. Instead, Titan may have some sort of low-temperature volcanism in which the eruptions are driven by melting or vaporizing ices of water, methane, or ammonia. Some researchers suspect that Titan has "ice volcanoes," an idea that conjures up visions of "icy volcanism" (sometimes called *cryovolcanism*) looking much like volcanism on Earth aside from the far lower temperatures. Other researchers suspect that the gas is being released by vaporization rather than by anything that would look like a volcano. Either way, the release of methane from the interior may help to explain the mystery of Titan's atmospheric methane.

Perhaps the most exciting discovery to date concerns the new perspective we have gained on Titan's alien environment. Titan's landscape looks remarkably similar to Earth's, yet it is shaped by very different materials. Instead of liquid water, Titan has liquid methane and ethane. Instead of rock, Titan has ice. Instead of molten lava, Titan has a slush of water ice mixed with ammonia. Instead of dirt, Titan's surface has smog particles that rain out of the sky. Titan even has wind-sculpted dunes, found in patterns similar to those found on Earth but possibly made of organic hydrocarbons rather than sand or snow (Figure 9.25). Evidently, the similarities in the physical processes that occur on Titan and Earth are far more important in shaping the landscapes than the fact that the two worlds have different compositions and temperatures.

Think About It Visit the *Cassini* website. Is the mission still operating (it is scheduled to end in 2017)? Has it made any new discoveries that shed further light on the ideas discussed in the preceding paragraphs?

THE POSSIBILITY OF LIFE Titan is so cold that any surface water would be solid ice, but, as we've discussed, there is strong evidence for liquid hydrocarbons on Titan. In Section 7.1, we noted that water might not be the only liquid that could support life, though it has some clear advantages over its competitors. In particular, because methane and ethane liquids are colder than liquid water by some 200°C, any life using these liquids would probably have much slower chemical reaction rates, and hence a slower metabolism, than life using liquid water; many biologists doubt that such "slow" life is possible. (A possible exception would be if molecules of life could have weaker chemical bonding than we find in life on Earth, in which case the reaction rates might not be so limiting.) In addition, methane and ethane are far less able than water to dissolve other compounds or to facilitate the type of chemistry that might lead to life. Consequently, the outlook for biology on Titan is bleak.

It is not, however, completely hopeless. The ultraviolet light that hits Titan's atmosphere produces a wide range of organic molecules (the main contributors to the observed smog). Over billions of years, some of these compounds should have accumulated as a deep layer of organic sediment on the surface—these sediments may be the material in the dunes (see Figure 9.25). Occasional impacts by comets or asteroids would provide enough heat locally to melt any water ice and create pockets of warm water that might persist for a thousand years or so. While it is not clear that life could form in such a short time period, some interesting chemistry would certainly occur. In fact, *Cassini* has found possible evidence for several volcanic cones, between 1000 and 1500 meters high, on Titan's

5 cm

FIGURE 9.23 INTERACTIVE FIGURE
The view from the *Huygens* touchdown site. The small objects visible in this photo are water ice. They measure 4–15 centimeters in diameter and show some indication of having been "smoothed" by tumbling in liquid. It may be that this is a currently dry streambed for liquid methane. The inset shows an artist's conception of the *Huygens* probe when it came to rest.

100 km

FIGURE 9.24
Radar image of Ligeia Mare, near Titan's north pole, showing lakes of liquid methane and ethane at a temperature of −180°C. Most solid surfaces reflect radar well, and these regions are artificially shaded tan to suggest land. The liquid surfaces reflect radar poorly, and these regions are shaded blue and black to suggest lakes.

FIGURE 9.25

Dunes on Titan. The dark streaks in this radar image are thought to be windblown dunes, possibly made of hydrocarbon sediments.

surface. If they really are volcanic, these cones would suggest icy volcanism on Titan. The material erupted from these cold volcanoes would be liquid water from reservoirs under Titan's frozen crust—in essence, volcanic "hot springs" in which temperatures might rise slightly above 0°C. Perhaps life could have originated in such places and gained a foothold. Another possibility is suggested by the discovery of bacteria existing in an oily lake on the Caribbean island of Trinidad. The bacteria are found in extremely tiny droplets of water within the lake's sludge, and they break down the oil to supply themselves with the energy necessary to live. Perhaps water brought to the surface of Titan by volcanic action might support a similar type of microbial life in its hydrocarbon seas.

Simple life on Titan might alternatively draw energy from chemical reactions in the upper atmosphere. These reactions produce a variety of products that include *acetylene*—an energy-rich, heavy compound that accumulates on the surface (where the *Huygens* probe detected it). When acetylene reacts with oxygen, it releases a lot of energy (think of welding torches). On bitterly cold Titan, acetylene would react more slowly with hydrogen, releasing energy and producing methane as a waste product. The idea that there is life that actually feeds on acetylene in this manner is, of course, only speculation, but at least a possible energy source is available.

One further possibility for life on Titan is deep beneath the surface. *Cassini* measurements show small changes in Titan's shape as it orbits Saturn, suggesting that it may have a liquid ocean far beneath its icy crust. However, this ocean would probably consist of a very cold ammonia/water mixture, rather than just liquid water, and the cold temperature would make life seem less likely.

Even if Titan lacks biology, studying this moon will give us valuable insights into chemical processes that might have been important to the beginnings of life on the early Earth. For example, scientists would like to know whether amino acids on Titan form equally in both right- and left-handed forms or if they exist mostly in left-handed forms like those used by life on Earth (see Figure 5.10). Moreover, we often give lip service to the fact that our ideas about the conditions that would produce life might be too conservative, but we don't often discuss alternatives. If future missions find living things on this small and hostile world—a place where the liquid environment and energy sources are quite different from those on Earth today—it most assuredly will be life as we *don't* know it.

Could other moons of Saturn have life?

No other moon of Saturn is even close in size to Titan or to any of Jupiter's Galilean moons. Rhea, the next largest of Saturn's moons after Titan, has less than one-third Titan's diameter and less than one-half the diameter of Europa (smallest of the Galilean moons). Iapetus, only slightly smaller than Rhea, seems to contain even more water ice than rock. Nonetheless, we expect such relatively small moons to retain far less radioactive heat than larger moons, and any tidal heating should be at least slightly less effective. Nevertheless, several of Saturn's moons (besides Titan) show evidence of past geological activity, and one—Enceladus—has surprised scientists by showing that it is geologically active today.

Cassini photos of Enceladus show a moon with some very young surface regions. The bluish "tiger stripes" in Figure 9.26 (which are not really blue in color) show regions near the moon's south pole that are measurably warmer than the surrounding terrain, and close-up examination suggests that they are cracks or grooves through which material can well up from below. These regions appear to be covered by "fresh" ice—ice that has solidified on the surface within no more than the past few thousand years, and possibly within just the past few decades or less. Moreover, images taken by *Cassini* as it looked at Enceladus backlit by the Sun show that fountains of ice particles and water vapor are spraying out from the tiger stripe regions (see the inset in Figure 9.26). This is clear evidence of icy volcanism on Enceladus. *Cassini*'s on-board Cosmic Dust Analyzer has also found sodium compounds, such as common salt (as well as simple organic compounds), in the neighborhood of this moon. This suggested that there might be a large, underground lake or ocean surrounding a rocky core. Additional studies showed that Enceladus wobbles slightly back and forth as it rotates and orbits Saturn, and the amount of wobble indicates the presence of a global ocean. The ocean is estimated to lie 30 to 40 kilometers beneath the moon's surface and may be up to about 30 kilometers in depth.

Cassini has also detected tiny particles of silica (quartz) floating in space near Enceladus. These particles could have been made only at temperatures near 90°C, suggesting that they came from very hot water at the bottom of the ocean, where the water contacts the moon's rocky core. The heated water would dissolve a bit of the rock, thereby becoming salty. The water would slowly vaporize, becoming the source of the salty grains of ice that are seen escaping from the tiger stripes on the surface. The warm water is also a potential energy source for life, similar to the ocean vents of Earth. When you put it all together, Enceladus seems to have subsurface liquid water (probably an ammonia/water mixture) that drives its icy volcanism, at least some simple organic molecules, and enough heat to power all this activity. The astonishing conclusion: Enceladus may have subsurface habitable zones.

This unexpected finding has at least two important implications for the search for life in our solar system and the universe. First, while we know that Enceladus is tidally heated, it's not clear whether the amount of heating has remained constant or changed with time. Apparently, we don't yet fully understand the way that moons can be heated, and as a result conditions for liquid water (or lower-temperature ammonia/water mixtures) might be more widespread than we had imagined. Second, the fact that we have been so surprised by Enceladus should tell us that other surprises are likely to await us. Our basic ideas about where to look for life are probably still valid as starting points, but it might be wise to keep an open mind about other places as well.

Could moons of Uranus or Neptune have life?

As we have progressed through this chapter, we have encountered a range of potential habitability. Europa seems reasonably likely to be habitable, as does Enceladus; Ganymede and Callisto seem somewhat less likely; and Titan, with its cold temperatures, seems to stretch the prospects for habitability to the limit. Nevertheless, our surprising findings make us ask whether there could be still other habitable places in our solar system.

After the Galilean moons and Titan, the next largest moon in our solar system is Neptune's moon Triton. As we discussed earlier, Triton

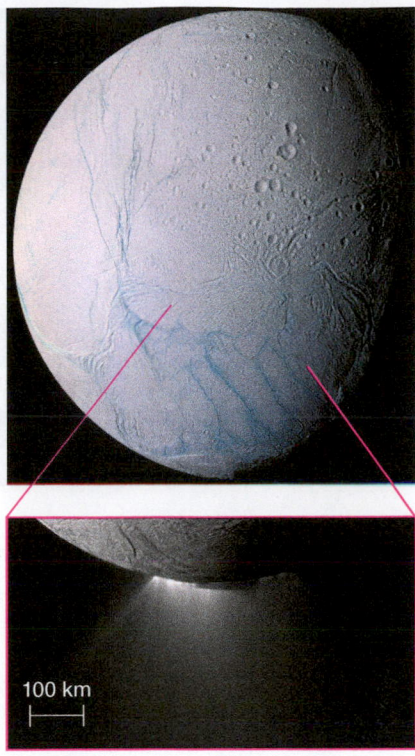

FIGURE 9.26

Active Enceladus. The blue "tiger stripes" near the bottom of the main photo are regions of fresh ice that must have recently emerged from below. The colors are exaggerated; the image is a composite made at ultraviolet, visible, and infrared wavelengths. The inset shows Enceladus backlit by the Sun, with fountains of ice particles (and water vapor) clearly visible as they spray out of the south polar region.

100 km

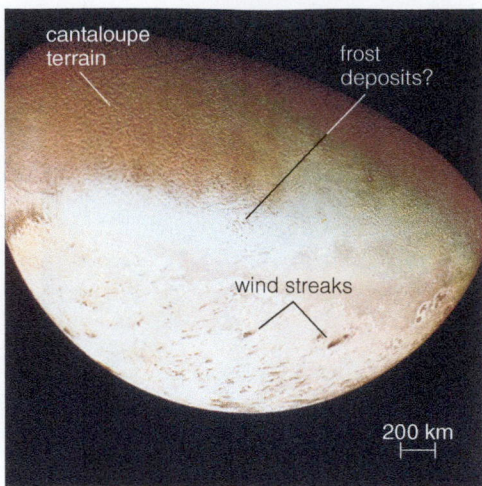

FIGURE 9.27
The southern hemisphere of Neptune's moon Triton, photographed by *Voyager 2* in 1989.

orbits Neptune "backward," suggesting that it was somehow captured from what was once an independent orbit around the Sun. The question of how such a capture might have occurred is quite interesting, but here we are more concerned with the issue of potential habitability. Triton was photographed close-up only once, by *Voyager 2* in 1989. Its icy surface is an enigmatic mix of terrain that in some places is smooth and in others is crinkled into patterns resembling the skin of a cantaloupe (Figure 9.27). There are few impact craters. The crater count on Triton today leads scientists to estimate that its last resurfacing must have been no more than 10–100 million years ago. Clearly, Triton has had internal heat, possibly left over from tidal heating that would have occurred as it was being captured into its present orbit, and probably with an additional contribution from radioactive decay. Some researchers think this heat may be sufficient to occasionally cause ice volcanoes to erupt from a liquid ocean beneath the surface. The liquid would be much colder than ordinary liquid water, and probably would consist of water mixed with nitrogen, ammonia, methane, or other melted ices.

Perhaps the most important idea emerging from our inspection of Triton is that even a moon in Neptune's distant realms, where surface temperatures are horrifically cold (colder than −230°C, or −382°F), might be geologically active and could possibly hide a liquid ocean and a refuge for life. And, given the surprise of finding a subsurface liquid mixture on a moon as small as Enceladus, we might just find similar surprises if and when we send future spacecraft to orbit and study the satellite systems of Uranus and Neptune.

Overall, while we do not yet know if any of the moons of the jovian planets in our solar system have life, we've found at least six that seem potentially habitable. The lesson is clear: If similar moons are also numerous around the planets of other stars, such moons might be the most common habitats for life in the universe.

 **THE PROCESS OF SCIENCE IN ACTION**

9.4 Chemical Energy for Life

In our discussions of Europa, we touched on the fact that simply having a heat source is not by itself enough to make energy available for life. To support an origin of life there must also be a chemical pathway by which complex molecules can be made, and to support ongoing life there must be chemical reactions that life can tap to fuel metabolism.

Not so long ago, the only known chemical pathways for metabolism were those used by plants and animals. However, as scientists have studied microbes living in "extreme" environments on Earth [Section 5.5], they have discovered many other metabolic pathways. These chemical pathways typically occur at interfaces between rock and water (such as in water-infiltrated rock underground), or through reactions with minerals in the hot water near deep-sea vents. If there is life on Europa or other jovian moons, it is likely to get its energy from the same types of chemical reactions. We therefore focus this chapter's case study in the process of science in action on the ways that life can extract chemical energy. As you'll see, this topic illustrates the interdisciplinary nature of astrobiology, in which biologists discover new forms of life on Earth, chemists and biochemists figure out how these organisms obtain energy, and planetary scientists then seek other worlds that might offer the same types of energy sources.

What is the role of disequilibrium in life?

From a chemical energy standpoint, the basic requirement for life is a situation in which chemicals naturally exist in a state that is "unbalanced," which we describe as a state of **disequilibrium.** The idea is similar to that of a scale on which you place objects on both sides to see which is heavier. If the two sides weigh the same, the scale is balanced. But if you then add a little extra to one side, the scale quickly tips because it is no longer balanced. In a similar way, if there is disequilibrium among chemicals, they will start to react just the way a scale starts to tip. With a scale, the movement quickly stops, because a scale can move only so far. But with chemicals, the reactions can continue as long as the disequilibrium remains. The idea will be clearer if we go into a bit of chemistry.

Any mixture of atoms and molecules can naturally undergo chemical reactions that may rearrange the atoms in such a way as to form or break chemical bonds. Left to themselves, however, chemical reactions will ultimately come to an **equilibrium** that represents a balance between the reacting atoms and molecules and the product atoms and molecules. For example, molecular hydrogen and oxygen can react together to make water. We can write this chemical reaction as

$$H_2 + \tfrac{1}{2} O_2 \longleftrightarrow H_2O$$

The double arrow indicates that the reaction can proceed in both directions, and the $\tfrac{1}{2}$ in front of the O_2 indicates that the reaction requires only half as many oxygen molecules as hydrogen molecules.

If we begin by mixing hydrogen and oxygen, at first the reaction will proceed only toward making water molecules, since there is no water present initially. But eventually the reaction will proceed at equal rates in both directions, at which time we will have chemical equilibrium. In some cases nearly all the hydrogen and oxygen will be converted to water, while in other cases little of it may be converted to water. The relative amounts of hydrogen, oxygen, and water at equilibrium depend on the external circumstances (such as pressure, temperature, and the presence of other chemicals). For example, the reaction between H_2 and O_2 needs a "push" to get it started. In a room filled with these two gases, this push might come from lighting a match. The energy from the match gets the reaction started; as H_2 and O_2 combine to form H_2O, additional energy is released that can then trigger more molecules to combine. This sequence can occur extremely rapidly, and the amounts of energy released can cause an explosion.

Now, imagine that the reaction between hydrogen and oxygen is occurring in a small flask that is inside a large room and that the reaction has come to equilibrium at the room's temperature. Suppose we suddenly do something that disturbs the equilibrium, such as adding excess hydrogen and oxygen molecules. Because the excess hydrogen and oxygen means that the reaction is no longer in equilibrium, we have created a state of *disequilibrium*. This disequilibrium will cause the rate of water formation to speed up until the equilibrium is restored. Note that, once we force the chemicals into disequilibrium, the rest of the process is completely natural: The reaction rate changes automatically in such a way as to bring the relative amounts of hydrogen, oxygen, and water back to their equilibrium values. Under the right circumstances, these reactions moving back toward equilibrium can release chemical energy that might be used by life to fuel metabolism.

No one is adding chemicals to the mixture of life, so the key to making chemical energy available for life lies in having some *natural* set of circumstances that can create and maintain a state of disequilibrium. In that case, the ongoing disequilibrium means that the reactions are always trying to move back toward equilibrium, thereby offering a continuous source of chemical energy. Natural processes that maintain chemical disequilibrium turn out to be quite common on Earth, and probably exist to some extent on any geologically active world.

For example, chemical disequilibrium inevitably exists near deep-sea volcanic vents, because mixing between the high-temperature vent water and the surrounding low-temperature ocean water creates conditions in which minerals and water will undergo chemical reactions. Because the vents continually release hot water, the disequilibrium can be maintained for long periods of time. Another place where we find ongoing disequilibrium is at interfaces between rock and water. The rock and water will naturally undergo chemical reactions, and as long as there is a supply of water that continuously circulates and comes in contact with the rock, the reactions will remain out of equilibrium. Thus, both deep-sea vents and rock/water interfaces offer places where chemical disequilibrium can provide ongoing energy that could be utilized to create complex molecules. This energy can also string complex molecules together into complicated structures—a process that may have been important to the origin of life—or support the metabolism of living organisms.

What types of chemical reactions supply energy for life?

Chemical disequilibrium offers the *potential* of providing chemical energy for life, but realizing this potential requires chemical reactions that life can actually use. A particular class of chemical reactions turns out to be especially important for life on Earth—reactions called **redox reactions.** Redox reactions involve an exchange or reshuffling of electric charge (which occurs through movement of electrons) between the reacting atoms or molecules.

Let's consider again what happens when hydrogen and oxygen combine to make water. Viewed on a molecular level, the reaction occurs in two steps. First, a hydrogen molecule decomposes into two protons (hydrogen nuclei) and two electrons:

$$H_2 \longrightarrow 2\,H^+ + 2\,e^-$$

(H^+ represents a single proton, which is positively charged, and e^- represents a single negatively charged electron.) Next, the two protons and two electrons combine with an oxygen atom (half of an oxygen molecule) to make a water molecule:

$$\tfrac{1}{2}\,O_2 + 2\,H^+ + 2\,e^- \longrightarrow H_2O$$

Viewed in this way, the production of water is a redox reaction, because electrons are effectively transferred from hydrogen to oxygen. Because the hydrogen gives up the electrons, we say that it is the *electron donor* for the overall reaction. Because the oxygen takes on the electrons, we say that it acts as the *electron acceptor.*

In accepting electrons, the electrical charge of the oxygen is *reduced* (because electrons are negatively charged); hence the first three letters in *red*ox refer to this process of **reduction** of electrical charge. The charge of the hydrogen is increased, but because this increase occurs as the result of action by oxygen, we say that the hydrogen has become *oxidized*. In fact, oxygen is so efficient at grabbing electrons from other chemicals (atoms or molecules) that chemists have come to use the term **oxidation** to describe the process of losing electrons in general, even when the electrons are lost to something besides oxygen. The overall process of making water from hydrogen and oxygen, $H_2 + \frac{1}{2}O_2 \longrightarrow H_2O$, is called a redox reaction because the oxygen gets *reduced* while the hydrogen gets *oxidized*.

A redox reaction always involves the transfer of one or more electrons from an electron donor (which becomes oxidized) to an electron acceptor (which becomes reduced). The transfer of electrons gives off energy that can then drive other chemical reactions, including the biochemical reactions of life.

REDOX REACTIONS ON EARTH Most of the key energy-generating chemical reactions used by life on Earth are redox reactions. For example, the basic process of aerobic respiration in animals involves combining a sugar acquired by eating, such as glucose ($C_6H_{12}O_6$), with oxygen acquired by breathing. The reaction makes carbon dioxide and water, and releases energy in the process:

$$C_6H_{12}O_6 + 6\,O_2 \longrightarrow 6\,CO_2 + 6\,H_2O + energy$$

This is a redox reaction because the glucose donates electrons (it is oxidized) while the oxygen accepts electrons (it is reduced).*

Many cellular energy-generating processes proceed through chains of redox reactions, sometimes called *electron transport chains* because a series of redox reactions means a series of electron transfers. In photosynthesis, for example, the chain begins when chlorophyll in a plant cell absorbs sunlight. The energy from the sunlight creates disequilibrium in the cell, and this disequilibrium then offers energy that the cell utilizes through a chain of redox reactions.

POSSIBLE REDOX REACTIONS FOR OTHER WORLDS Redox reactions are especially important when we consider the prospects for life in extreme environments, either on Earth or on other worlds. Many Earth organisms use fairly simple redox reactions as their primary source of energy. For example, bacteria known as *Thiobacillus ferrooxidans*, which can thrive in highly acidic conditions such as in mine tailings, obtain energy by oxidizing iron:

$$2\,Fe^{+2} + \frac{1}{2}O_2 + 2\,H^+ \longrightarrow 2\,Fe^{+3} + H_2O$$

(Fe^{+2} and Fe^{+3} represent iron atoms missing two and three electrons, respectively.) In this case, the iron is the electron donor (it is oxidized) and the oxygen is the electron acceptor (it is reduced). Note that neither preexisting organic molecules nor sunlight is needed for this reaction, which means reactions like this one could have been used by early life-forms on Earth—and might be used by life living underground on other worlds.

---

*You can often recognize what is being oxidized or reduced in redox reactions, even without knowing precisely how electrons are rearranged. In the aerobic respiration reaction, for example, the C in glucose is being oxidized because it ends up being combined with a greater number of O atoms. Glucose has equal numbers of C and O atoms, but CO_2 has twice as many O as C atoms and so is more oxidized. The oxygen is being reduced because it ends up being combined with more hydrogen; it has no H atoms on the left side of the reaction, but has two H atoms per O atom on the right. Thus, for example, the C in CH_4 is more reduced than the C in CO_2, which is more oxidized.

Many other redox reactions produce energy for various microbes on Earth, including reactions involving molecular hydrogen and sulfur. These are especially important when we consider possible energy sources for an origin of life. For example, both iron and sulfur are common in the disequilibrium environments of hot springs and deep-sea vents, and thus could be involved in redox reactions that might ultimately lead to life. At rock/water interfaces, which exist any place there is liquid water underground, chemical reactions between water and iron in rock can produce molecular hydrogen, which can then be used in redox reactions for biochemistry.

As a result of the understanding of this chemistry gained by studying life on Earth, we now have reason to think that life could exist on many other worlds. In particular, any geologically active world with liquid water may have places where chemical disequilibrium persists for long periods of time, such as near underwater volcanic vents or anywhere where rock and water come into contact. At these places, redox reactions can provide energy that could power biochemical reactions that might ultimately lead to life and that could support life once it arises. That is why places such as underground pockets of liquid water on Mars or the subsurface ocean of Europa seem so promising as potential abodes of life.

The Big Picture

PUTTING CHAPTER 9 IN PERSPECTIVE

In this chapter, we have considered the moons of the outer solar system and found that several of them might offer conditions suitable for life. As you continue in your studies, keep in mind the following "big picture" ideas:

- Our own Moon, a now-dead relic of an early collision, may have misled us into thinking that only planets can harbor life. In fact, moons exhibit enormous variety. Some of the moons of the outer solar system are as large as small planets, several might have liquid oceans, and one even has a substantial atmosphere. Life might well be possible in such places.

- The solar system moon most likely to be habitable, Europa, is kept warm inside by a mechanism quite different from the one that warms Earth's interior. Tidal heating, the result of orbital resonances that occur among three large moons of Jupiter, can provide a continuous source of heat for billions of years. Because orbital resonances can arise quite naturally, tidal heating may be common among moons of jovian planets throughout the universe.

- The icy moons of the outer solar system force us to rethink our basic concept of "habitability." If any of these moons are indeed homes to life, then the range of habitability is much broader than we might have guessed from studies of terrestrial worlds.

- From a chemical energy standpoint, life requires conditions in which there is a natural and ongoing source of chemical disequilibrium. Such conditions probably exist on almost any geologically active world that has liquid water, either at underwater volcanic vents or simply at the interfaces of rock and even tiny amounts of liquid water.

Summary of Key Concepts

9.1 The Moons of the Outer Solar System

What are the general characteristics of the jovian moons?
The moons of jovian planets range greatly in size, from a few kilometers across to somewhat larger than Mercury. They tend to have ice mixed in with their rock: water ice for all the jovian moons, plus ammonia, methane, and other ices for moons of the more distant jovian planets. Nearly all are in **synchronous rotation,** keeping one side perpetually turned toward their host planet.

Why do we think that some moons could harbor life?

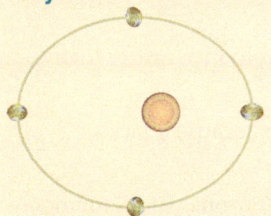

Some moons have substantial internal heat as a result of **tidal heating,** along with radioactive decay. Tidal heating explains the tremendous volcanic activity on Io and the heating thought to melt subsurface ice on Europa and perhaps Enceladus. A few other moons may also have liquid water or other liquids, and thus would seem to meet the minimum requirements for life.

9.2 Life on Jupiter's Galilean Moons

Does Europa have an ocean?

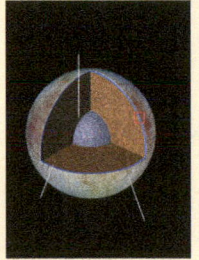

Europa's surface shows numerous features suggesting that liquid or slush from below has occasionally gushed through and re-paved the surface, and Europa's magnetic field makes sense only if we assume Europa has a salty ocean. These observational data, combined with the known tidal heating of Europa, make it likely that the moon has a subsurface ocean, which may contain twice as much water as the oceans of Earth.

Could Europa have life?
While it's probable that Europa has both a liquid water environment and the elements necessary for life, possible energy sources for life are much more limited than on Earth. Volcanic vents on the ocean floor could provide some energy, perhaps enough for life to have arisen but probably not enough to support life in great abundance. A few other energy sources may contribute additional energy, but overall we would expect any life on Europa to be simple and small.

Could other moons of Jupiter have life?

Magnetic field measurements suggest that both Ganymede and Callisto also have subsurface oceans, and Ganymede shows some evidence of water having gushed out onto parts of its surface. Thus, both Ganymede and Callisto could conceivably offer conditions for life, although energy sources are even more limited on these moons than on Europa.

9.3 Life Around Saturn, and Beyond

Could Titan have life?

Titan has a thick atmosphere, lakes of liquid methane and ethane, and numerous other surface features reminiscent of Earth. However, the bitterly cold temperatures would greatly slow chemical reactions, making metabolism difficult and decreasing the chances for life. It is also possible that Titan sometimes has surface or near-surface pockets of liquid water and a subsurface ocean of a cold ammonia/water mixture. Some energy sources for life might also be available.

Could other moons of Saturn have life?

The relatively small moon Enceladus has a warm, subsurface aquifer that drives the fountains of ice and water vapor observed to be emerging from the moon. It is therefore possible that Enceladus has zones of habitability. While we have no direct evidence for similar possibilities on other moons of Saturn, the case of Enceladus tells us not to rule them out too quickly.

Could moons of Uranus or Neptune have life?
Life seems less likely on such distant moons, but it is still possible that some could have habitable zones beneath their surfaces. Neptune's moon Triton shows evidence of tidal heating and icy volcanism, suggesting it might have liquid beneath its surface. Other moons seem like much longer shots, but we should study them further before concluding that they lack liquids and chemistry that might sustain life.

✿ THE PROCESS OF SCIENCE IN ACTION

9.4 Chemical Energy for Life

What is the role of disequilibrium in life?
Life as we know it can exist only in places where natural conditions maintain a state of chemical **disequilibrium.** This disequilibrium can cause chemical reactions that may be used to create complex molecules, to string complex molecules together into complicated structures, or to support metabolism of living organisms.

What types of chemical reactions supply energy for life?
Life on Earth gains energy from **redox reactions** in which one molecule gains electrons and another loses them. The redox reactions used by terrestrial life include those that occur near deep-sea vents and at underground rock/water interfaces, suggesting that the same types of reactions might be used by life on worlds that have similar conditions, such as Mars (underground rock/water interfaces) and Europa (deep-sea vents).

Exercises and Problems

REVIEW QUESTIONS

Short-Answer Questions Based on the Reading

1. Briefly explain how the larger *jovian moons* tend to differ in general from the smaller ones. How does the formation process of the moons explain these differences?

2. Briefly describe the cause of the tides on Earth, why they lead to *tidal friction,* and how tidal friction affects Earth's rotation and the orbit of the Moon.

3. What is *synchronous rotation,* and why is it so common among the jovian moons?

4. What is *tidal heating*? Briefly explain how it can arise and persist as a result of *orbital resonances.* How does tidal heating affect Io?

5. Describe the evidence suggesting that Europa has a liquid water ocean beneath its icy crust. How might future observations confirm this idea?

6. What energy sources might be available to life on Europa? Overall, what can we say about the likelihood and abundance of life on Europa?

7. Describe the evidence for subsurface oceans on Ganymede and Callisto. What are the prospects for life on these worlds?

8. Why was Titan chosen for such intense study by the *Cassini–Huygens* mission? Why is it surprising to find methane in Titan's atmosphere?

9. Based on recent data, describe the general nature of Titan and discuss its prospects for life.

10. What evidence suggests that Enceladus might be habitable? What lessons does Enceladus hold for our more general search for life in the universe?

11. Could Triton be habitable? Briefly discuss the possibility of finding habitable moons around Uranus or Neptune.

12. What do we mean by chemical *equilibrium* and chemical *disequilibrium*? Why is disequilibrium necessary for life?

13. What are *redox reactions*? Give a couple of examples.

14. Based on our understanding of the chemistry of life, where should we expect such chemistry to be possible? What are the implications of this idea to the search for life beyond Earth?

TEST YOUR UNDERSTANDING

Evaluate the Claims

Each of the following statements makes some claim. Evaluate the claim, writing a few sentences describing why you think it is valid or invalid (or clearly true or false). Explain clearly; because not all of these have definitive answers, your explanation is more important than your chosen answer.

15. Io is riddled with volcanoes because of its proximity to Jupiter's strong magnetic field.

16. Europa is likely to have fishlike organisms the size of whales swimming in its ocean.

17. While Europa, Ganymede, and Callisto are all candidate locations for life, we expect that the most abundant and diverse life would be found within Callisto.

18. The fact that our Moon keeps one side always facing Earth is an astonishing coincidence.

19. Titan is simply too cold to have any life.

20. Triton might have life that uses liquid ammonia, rather than liquid water, as its transport medium.

21. Io doesn't have a significant atmosphere because it lacks a source of outgassing.

22. Orbital resonances like those among Io, Europa, and Ganymede are the results of extremely rare accidents, so we would not expect tidal heating to be important in other planetary systems.

23. If there is life on Enceladus, it probably gets its energy from sunlight.

24. If our solar system is typical, other star systems might have an average of five to ten worlds on which liquid water (or a mixture of water and some other liquid) exists in at least some places.

Quick Quiz

Choose the best answer to each of the following. Explain your reasoning with one or more complete sentences.

25. The moons of Saturn may have large amounts of ammonia and methane ice, while those of Jupiter do not because (a) methane and ammonia come only from comets that exist in the Oort cloud; (b) Jupiter's strong magnetic field encourages water ice to form; (c) the greater cold at Saturn's distance from the Sun means that ices of ammonia and methane could condense there but not at Jupiter.

26. Which statement about synchronous rotation is true? (a) It can develop only on moons that are born with slow rotation. (b) It occurs commonly as a result of tidal forces exerted on moons by their parent planets. (c) It can develop only on moons with liquid oceans.

27. Io is covered in volcanoes while Europa is covered in ice because (a) Io is larger than Europa; (b) Io receives much more sunlight than Europa; (c) Io is subject to stronger tidal heating than Europa.

28. Which of the following is *not* an indication of liquid water beneath Europa's icy surface? (a) the moon's changing magnetic field (b) the moon's average density (c) surface features that look like jumbled icebergs

29. Photosynthesis is an unlikely source of energy for life in europan seas primarily because (a) the moon's ice cover is too thick; (b) sunlight at the distance of Jupiter is too weak to provide any energy at all; (c) there is no soil on which plants could grow.

30. It's assumed that, even if Europa has life, the total amount of that life will be small. That's because (a) Europa is only about as big as our own Moon, and consequently there's not much room for life; (b) the ocean will be cold, slowing down metabolism; (c) there are likely to be only limited sources of energy for life.

31. The chances for life on Titan's surface are considered slim, mainly because (a) there's little oxygen in the atmosphere; (b) the liquid methane and ethane rain would be lethal; (c) the surface temperature is far below the freezing point of water.

32. Where might we find liquid water on Titan? (a) in lakes and rivers on the surface (b) beneath the surface near sources of "icy volcanism" (c) as droplets in high-altitude clouds

33. Why were scientists so surprised to find active geology on Enceladus? (a) because it is so small (b) because it lacks any possibility of tidal heating (c) because it is so far from the Sun

34. Chemical disequilibrium is likely to be present in all the following places *except* in (a) volcanic vents on ocean floors; (b) solid ice exposed to the extreme cold of space; (c) underground aquifers where thin films of liquid water move over rock.

PROCESS OF SCIENCE

35. *How definitive is my experiment?* The *Viking* landers that searched for evidence of life on Mars in the 1970s produced ambiguous results in the minds of some scientists. While no clear proof of life was found, the craft could not persuasively rule out life either. If you were to search for life in the outer solar system, can you think of an experiment that would not be liable to return ambiguous results?

36. *Cost-Effective Exploration.* Given its budget, NASA can't afford to launch many spacecraft per year. Yet the public is interested to know if there's life in space. If you were named the NASA Administrator and had money for only one mission to the outer solar system to hunt for biology, which world would you choose to investigate? Defend your position.

GROUP WORK EXERCISE

37. *Old-School Exploration.* **Roles:** *Scribe* (takes notes on the group's activities), *Proposer* (proposes ideas to the group), *Skeptic* (points out weaknesses in proposed ideas), *Moderator* (leads group discussion and makes sure everyone contributes). **Activity:** Some researchers have suggested that we should do exploration the old-fashioned way, by using a ship—in this case, a small, robotic floating vessel—to search Titan's lakes for evidence of life. Given that the ship would have to be small, what should be the highest priority experiments? Be specific. Develop a group consensus, and write a one-page report that includes a list of instrumentation to put on board.

INVESTIGATE FURTHER

In-Depth Questions to Increase Your Understanding

Short-Answer/Essay Questions

38. *Lessons for Life.* Considering everything we've learned about the possible habitability of jovian moons, make a list of what you think are the three most important lessons that apply to the search for life in *other* solar systems. Describe the importance of each lesson clearly, and conclude by summarizing how the study of jovian moons has changed our perspective about life in the universe.

39. *Exploring Europa I.* Although Europa is a promising place to look for life, penetrating its thick, icy crust will be difficult. Suggest a possible way of making a spacecraft that could enter the europan ocean. If it is technically feasible, do you think we should do it soon, or wait until we have further evidence of life? Defend your opinion.

40. *Exploring Europa II.* One suggestion for determining whether Europa has life is to send an orbiter that passes close to the surface of the planet, drops a "bowling ball" that makes a crater on the surface, and then catches the ice thrown up by the impact. This sample ice would be brought back to Earth for analysis. Briefly describe what we would be looking for in such an experiment and what it might teach us. How do you think the cost of such a mission to Europa would compare to, say, that of a mission that orbits but does not return to Earth? Explain.

41. *Europan Fish.* On Earth, fish breathe oxygen that is dissolved in water. Do you expect that we will find dissolved oxygen in Europa's ocean? Why or why not? Based on your answer, do you think that, if we could somehow transport fish to Europa, it is possible that they could survive in the europan ocean? What other types of life from Earth might survive on Europa?

42. *Life on Titan.* Several possibilities have been suggested for the support of life on Titan, including acetylene that falls onto the surface and the possible presence of subsurface, liquid water aquifers. If you could plan one new Titan lander, how would you design it to search for life?

43. *Migrating Life.* As we discussed in Chapter 6, there is a decent likelihood that life might at some point have traveled from Earth to Mars (or vice versa) on meteorites. Discuss the likelihood that life from Earth or Mars could have made its way to, and taken root on, Europa or other jovian moons.

44. *Movie Review.* A number of science fiction movies have concerned jovian moons, including *2010, Outland,* and *Gattaca.* Watch one of these movies, and write a critical review of it. Be sure to comment on the accuracy of any scientific content in the movie.

45. *The Sirens of Titan.* The moon Titan plays the title role in Kurt Vonnegut's novel *The Sirens of Titan.* Although the book was never intended to give a realistic portrayal of Titan, it helped popularize the moon. Read the novel and write a short critical review, focusing on how Vonnegut's portrayal of Titan differs from reality.

Quantitative Problems

Be sure to show all calculations clearly and state your final answers in complete sentences.

46. *Orbital Resonances I.* Using the data in Appendix E, identify an orbital resonance relationship between Titan and Hyperion. (*Hint:* If the orbital period of one were 1.5 times that of the other, we would say that they were in a 3:2 resonance.)

47. *Orbital Resonances II.* Using the data in Appendix E, identify an orbital resonance that affects Enceladus. In light of the *Cassini*

mission findings about Enceladus, how might this resonance be important?

48. *Tidal Force on the Moon.* In Cosmic Calculations 9.1, we found the tidal force that Earth exerts on the Moon today. Following a similar procedure, calculate the tidal force Earth would have exerted on the Moon shortly after the Moon formed, when it was only about $\frac{1}{10}$ its current distance from Earth. How much greater was the tidal force than it is today? What does this tell you about how a factor-of-10 change in distance affects the tidal force?

49. *Tidal Force on Io.* Using the procedure from Cosmic Calculations 9.1, calculate the tidal force exerted on Io by Jupiter. Compare this force to the tidal force that Earth exerts on the Moon, and comment on the implications for Io's volcanism.

50. *Tidal Force on Europa and Ganymede.* Using the procedure from Cosmic Calculations 9.1, calculate the tidal force exerted on Europa and Ganymede by Jupiter. Compare this force to the tidal force exerted on Io (from Problem 49), and comment on the expected strength of tidal heating on each world.

51. *Astrology Explained (or Not).* Recall that astrology claims that the positions of the planets in the sky at a person's moment of birth forever influence the person's life. In an attempt to "explain" this claim, some astrologers have proposed that the source of the influence could be tidal effects from the planets.
    a. Using the method of Cosmic Calculations 9.1, calculate the tidal force exerted by Jupiter on a baby being born on Earth. For the purposes of the calculation, assume the baby is 50 centimeters long, so the distance of the baby's "near" side to Jupiter is Jupiter's distance *minus* 25 centimeters (0.25 meter), and the distance to the baby's "far" side from Jupiter is Jupiter's distance *plus* 25 centimeters; for Jupiter's distance, use 780 million kilometers (which is about the average).
    b. Jupiter, of course, is not the only gravitational influence on the baby. For example, there is also a gravitational force acting between the baby and the doctor supervising the delivery. Calculate the gravitational force that is pulling the baby toward the doctor. Assume that the baby's mass is 3 kilograms and the doctor's mass is 50 kilograms, and that they are 0.5 meter apart during delivery.
    c. Compare the tidal effect of Jupiter on the baby to the gravitational pull between the baby and the doctor. Based on your answer, do you find it plausible to claim that tidal effects of the planets can influence the baby's life?

Discussion Questions

52. *Importance of Life.* In coming decades, scientists hope to devote a lot of effort to searching for life on jovian moons. How important do you think it would be if we found life on any of these moons? Answer in terms of both scientific importance and philosophical importance. Overall, do you think the possible benefits justify the expense required to undertake this search?

53. *Limited Thinking.* Throughout this book, we have generally assumed that any life found elsewhere would be at least somewhat similar to Earth life; for example, we assume it would be carbon-based, most likely use liquid water, and gain energy from sunlight or chemistry. Are these assumptions still valid in light of what we've learned about jovian moons? Under what circumstances might we need to broaden our assumptions? Explain.

WEB PROJECTS

54. *Photo Journal.* Visit the website for NASA's *Galileo* or *Cassini* mission. For one of the potentially habitable moons of Jupiter or Saturn, create your own photo journal in which you include at least ten photos, along with a paragraph or two for each photo that explains how it relates to the question of habitability on that moon.

55. *Europa Orbiter.* Find out the current status of NASA's plans to send a mission to study Europa. What will the mission do? When will it be launched? What are its science goals? Write a one- to two-page summary of your findings.

10

The Nature and Evolution of Habitability

LEARNING GOALS

10.1 THE CONCEPT OF A HABITABLE ZONE
- How does a planet's location affect its prospects for life?
- Could life exist outside the habitable zone?

10.2 VENUS: AN EXAMPLE IN POTENTIAL HABITABILITY
- Why is Venus so hot?
- Is Venus in the habitable zone?

10.3 SURFACE HABITABILITY FACTORS AND THE HABITABLE ZONE
- What factors influence surface habitability?
- Where are the boundaries of the Sun's habitable zone today?

10.4 THE FUTURE OF LIFE ON EARTH
- How will the Sun's habitable zone change in the future?

- How long can life survive on Earth?

 **THE PROCESS OF SCIENCE IN ACTION**

10.5 GLOBAL WARMING
- What is the evidence for global warming?
- What are the potential consequences of global warming?

▲ **About the photo:** Earth's surface has remained habitable despite a gradual increase in the intensity of sunlight over our planet's 4½-billion-year history.

Other worlds of our solar system may have life, but none are likely to have anything like the diversity and abundance of life on Earth. Why is this?

On one level, this question seems easy to answer: Unlike any other world in our solar system, Earth has abundant liquid water on its surface. But as we probe deeper, we find that the causes underlying *why* Earth has oceans that teem with life, and other worlds do not, are more subtle. For example, Mars had surface liquid water in the past, but no longer does—implying that the potential for life can change with time.

In this chapter, we'll take a broad approach to the question of habitability in our solar system, seeking to understand the factors—such as size and distance from the Sun—that make Earth so different from its neighboring worlds. In doing so, we'll gain deep insight into the likelihood of finding Earth-like worlds around other stars. After all, we can expect Earth-like planets to be common only if they can exist over a fairly wide range of sizes and distances. Learning what makes Earth habitable will also help us understand how its habitability might change in the future, either through long-term, natural changes that will ultimately occur in our solar system or through human-induced changes that could have much more immediate and serious consequences for our own civilization.

10.1 The Concept of a Habitable Zone

We all know of urban legends—for example, that small pet alligators flushed down toilets have grown to adulthood and are roaming the sewers of New York. Urban legends occur across a wide variety of subject areas, and astrobiology is no exception. So here's another one to think about: If Earth moved a mile closer to the Sun, would we all burn up? According to an urban legend that has spread widely in recent years, the answer is yes, and the legend also holds that we'd freeze if our planet moved just a mile farther away.

But in truth, we are no more likely to burn up if Earth moved a mile closer to (or freeze if it moved a mile farther from) the Sun than New York City sanitation employees are to become an alligator's meal while at work. The legend is nonsense, as you can realize just by thinking about Earth's orbit around the Sun: The orbit is an ellipse, not a circle, and Earth's distance from the Sun varies from a minimum of about 147.1 million kilometers each January to a maximum of about 152.1 million kilometers each July. So if the legend were true, our whole planet would burn up each January and freeze each July. In reality, this 5-million-kilometer variation in distance has little effect on the weather, a fact that becomes obvious when you remember that the Northern Hemisphere has summer during the time that Earth is *farthest* from the Sun and winter when Earth is closest to the Sun. If the distance variation were important, the whole planet would be significantly warmer in January than in July, and it's not.

But like many urban legends, this one holds at least one deeply buried kernel of truth, because there must be *some* distance from the Sun at which it would be too hot on Earth for life to survive, and some distance at which it would be too cold. The distance isn't a mile, or even a few million miles, but in principle we should be able to determine what it is and use the range of acceptable distances to learn about prospects for life in the universe.

How does a planet's location affect its prospects for life?

If there are distances from the Sun at which our planet would become too hot or too cold for life, there must also be some range of distances that is "just right" for a world like Earth. This range defines what we call the Sun's **habitable zone**; that is, the habitable zone is the range of distances from the Sun within which we could in principle move our planet without fundamentally changing the characteristics that make it home to abundant life (Figure 10.1). Other stars also have their own habitable zones, meaning distances at which a world similar in size to Earth, and with a similar atmosphere, would be habitable, although the sizes of habitable zones are different for different types of stars [Section 11.1].

Note that this definition captures the essence of what we mean by a habitable zone, but there are several important caveats to keep in mind:

1. The concept of a habitable zone is based on the range of distances at which *worlds similar to Earth* could exist. In other words, a habitable zone is a zone in which it is possible for a world to have abundant liquid water on its surface.
2. Simply being in a star's habitable zone is *not* sufficient to make a world habitable. The Moon presents an obvious case in point: As a companion to our planet, it is located at essentially the same distance from the Sun as Earth, but it is not habitable.
3. Habitable zones evolve with time. In particular, because stars like the Sun tend to brighten as they age, we expect a star's habitable zone to move outward over time.

In summary, *at any particular time, a star's habitable zone is the range of distances around it at which a planet could potentially have surface temperatures that would allow for abundant liquid water.*

We have considered a number of places besides Earth in our solar system that might either have life now or have had life in the past, including Mars, Europa, and Titan. However, none of these places seem likely to have *surface* liquid water or life at present; instead, their life would be in underground locations where liquid water might be present (or perhaps in the surface lakes of methane and ethane on Titan). This distinction may not be that important for the search for life within our own solar system, because future space missions should eventually allow us to search for life beneath the surface of worlds like Mars and Europa. However, the distinction between surface life within a habitable zone and subsurface life beyond it is important when we consider the challenges of finding life in other star systems.

Think About It The surface pressure and temperature on Mars are too low for liquid water to exist there today. Does this imply that Mars is beyond the outer boundary of the Sun's habitable zone? Why or why not?

We are unlikely to be able to travel to the stars anytime soon [Section 13.1], so telescopic images and spectra offer our only realistic hope of

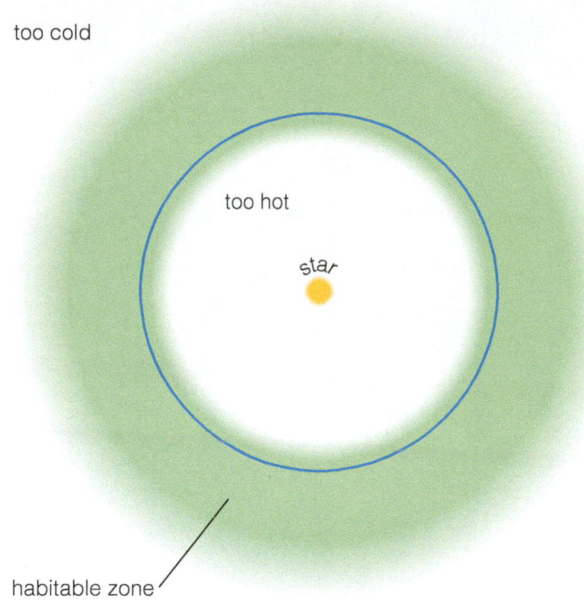

too cold

too hot

star

habitable zone

FIGURE 10.1
A planet with abundant liquid water on its surface can in principle exist only within some particular range of distances from a star. This range defines the star's habitable zone, and the size of the range depends on characteristics of the star.

finding life on extrasolar planets or moons. Surface life may well create spectral signatures in a planet's atmosphere that would allow us to detect it [Section 11.4], but subsurface life is unlikely to cause enough atmospheric change to make an unmistakable spectral signature. Therefore, while we may be able to find subsurface life (if it exists) within our own solar system, we have little hope of finding it on planets or moons around other stars. So the search for life beyond the realm of the Sun is by necessity a search for surface life on worlds within the habitable zones of their stars.

Could life exist outside the habitable zone?

There are several ways that habitability might present itself outside the habitable zone. One is in small pockets of subsurface groundwater, such as that which may still exist on Mars today [Section 8.3]. Because this water would be kept liquid primarily by geological conditions rather than solar heat, a Mars-like planet could have subsurface habitability even if the planet itself lay beyond the outer edge of the habitable zone. It's not far-fetched to imagine that Mars-like planets could be more common than Earth-like planets.

The possibility of life in subsurface oceans, such as those thought to exist on Europa, Ganymede, and Enceladus [Section 9.2], could make habitability even more common. We expect moons in the outer regions of any solar system to contain large amounts of water ice, because water is the most common of the ices that can condense in regions far from a forming star. Orbital resonances like those that contribute to the tidal heating [Section 9.1] of Io, Europa, Ganymede, and Enceladus could occur in any system where a large planet has multiple moons, and the heating could potentially melt subsurface ice. Heat from radioactive decay might also contribute to melting interior ice. Because tidal heating and radioactive decay can supply heat in the absence of sunlight, internally heated moons with subsurface oceans could exist around planets orbiting at almost any distance from a star, except close-in, where all ice would melt and evaporate away.

Another intriguing possibility concerns *surface* habitability on Earth-size planets outside habitable zones. Theoretical calculations suggest that an Earth-size planet's own internal heat could keep its surface warm enough for liquid water *if* the surface were protected against heat loss by a thick hydrogen atmosphere. Such a hydrogen atmosphere is not possible on an Earth-size planet in our solar system, because solar heat would cause the hydrogen to escape into space fairly rapidly. But, as we will discuss further in Chapter 11, it is possible that Earth-size planets are sometimes ejected from solar systems in the process of forming, and sent into interstellar space (becoming orphan planets [Section 11.4]). When an Earth-size planet is born, it might have a hydrogen atmosphere for a short time, particularly if it forms at a greater distance from its star or around a star cooler than our Sun. If such a planet is ejected into interstellar space before it loses this atmosphere, its thick hydrogen envelope might remain for many billions of years. Indeed, such "free-floating Earths" could conceivably be quite common in interstellar space, though their relatively small size and low brightness would make them extremely difficult to detect. If they exist, such planets could have surface oceans, as well as geothermal and chemical energy much like that sustaining life underground and around deep-sea vents on Earth. Although the total available energy would probably not sustain a huge amount of life, at least some might be possible.

We could also conceive of habitability outside the habitable zone if life can use a liquid medium other than water, such as liquid ethane, which has a much lower freezing point (see Table 7.1). Any liquid will evaporate rapidly when atmospheric pressure is low, so the presence of surface liquids of any type requires an atmosphere. In general, this means we can hope to find surface liquids only on worlds large enough to hold significant atmospheres, but not so large that they become giants like the jovian planets, where strong vertical winds probably preclude life. In our solar system, Titan is the only world that meets this criterion. Nevertheless, moons like Titan might be relatively common among other star systems. If life based on other liquids is possible, there could be many habitable worlds similar to Titan.

In summary, it's quite possible that the majority of habitable worlds in the universe are not located within stellar "habitable zones."* From this standpoint, the concept of a habitable zone might seem obsolete, because it doesn't account for the potential habitability of Europa-like or Titan-like moons, of Mars-like subsurface water, or of "free-floating Earths." Nevertheless, while life might be common on such worlds, it would be extremely difficult to detect. Also, it seems unlikely that such worlds could give rise to complex life, let alone advanced civilizations. The traditional concept of a habitable zone is therefore still quite useful in searching for life beyond Earth, and it is critical when we are searching for intelligent life.

10.2 Venus: An Example in Potential Habitability

One of our primary goals in this chapter is to define the boundaries of the habitable zone in our own solar system, so that we can then extend the definition to the search for planets within the habitable zones of other stars. A useful first step in defining these boundaries is to determine which planets currently lie within the Sun's habitable zone. We can rule out Mercury, which is too close to the Sun, and all the planets beyond the asteroid belt, which are too far. That narrows the possibilities down to Venus, Earth, and Mars.

We have already explored the cases of Earth and Mars in some depth. Earth is obviously within the habitable zone. Mars must be at least near the borderline of the habitable zone, since it apparently had liquid water on its surface in the past. But what about Venus?

Why is Venus so hot?

The surface of Venus is far too hot for liquid water [Section 7.2], but we can't directly attribute this heat to Venus's proximity to the Sun. Venus orbits the Sun at a distance about 72% of Earth's distance. Because the intensity of sunlight follows an inverse square law (see Figure 7.2), the intensity of sunlight at Venus is a little less than twice as great as at Earth (because $1/0.72^2 \approx 1.9$). We'd expect this extra sunlight to make Venus warmer than Earth, but not nearly by enough to account for its searing

---

*This fact has led some astrobiologists to argue that the term *habitable zone* should be redefined to account for other types of habitable worlds. However, in this book, we stick with the traditional idea of the habitable zone as the region in which it is possible to have *surface* oceans kept liquid by the heat of a star.

470°C (878°F) surface temperature. Calculations show that if Venus had the same atmosphere as Earth, its average temperature would be only about 30°C hotter than Earth's—making Venus somewhat like the tropical planet envisioned in old science fiction.

As we discussed in Chapter 7, the immediate cause of Venus's high temperature is an extreme greenhouse effect produced primarily by atmospheric carbon dioxide. Venus has about 200,000 times as much carbon dioxide in its atmosphere as Earth, and carbon dioxide is a greenhouse gas that traps heat near a planet's surface [Section 4.5]. However, given their similar sizes and compositions, we expect Venus and Earth to have had similar levels of volcanic outgassing—and the released gas ought to have had about the same composition on both worlds. We therefore are left with a deeper question about Venus's extreme heat: Why is Venus's atmosphere so different from Earth's?

Our understanding of planetary geology suggests that huge amounts of water and carbon dioxide should have been outgassed into the atmospheres of both Venus and Earth. Venus's atmosphere does indeed have huge amounts of carbon dioxide, but it has virtually no water. Earth's atmosphere has only small amounts of either gas. The big question of why the planets have such different atmospheres therefore has two smaller parts: (1) Why did Venus keep its atmospheric carbon dioxide and Earth did not? (2) Where did all the outgassed water go on each planet?

THE ABUNDANCE OF CARBON DIOXIDE ON VENUS The first question is relatively easy to answer. In fact, Venus and Earth have similar total amounts of carbon dioxide, and differ primarily in where the carbon dioxide is located. On Venus, the carbon dioxide is all in its atmosphere. On Earth, nearly all the carbon dioxide is locked up in carbonate rocks or dissolved in the oceans [Section 4.5], through the action of the carbon dioxide cycle (see Figure 4.27). This cycle is possible because carbon dioxide gas dissolves in water, where it can undergo chemical reactions to make carbonate minerals (minerals rich in carbon and oxygen) such as limestone.

Venus lacks a similar carbon dioxide cycle because it has no liquid water in which the gas can dissolve. We conclude that the difference in carbon dioxide is a direct consequence of a difference in liquid water: Earth's carbon dioxide became incorporated into rock because Earth has oceans and a carbon dioxide cycle; Venus's carbon dioxide is all in its atmosphere because it has no oceans and hence no carbon dioxide cycle.

THE LACK OF WATER ON VENUS Given that the differences between Venus and Earth in carbon dioxide are attributable to differences in water, we must next ask what happened to outgassed water on each planet. For Earth, the answer is easy: The water outgassed from volcanoes is still here, having condensed and rained down to form the oceans. Venus, however, is nearly totally lacking in water in any phase. The surface is far too hot for either ice or liquid water. It is even too hot for water to be chemically bound in surface rock, and any water deeper in its crust or mantle was probably baked out long ago. The only place where Venus could conceivably have much water is in its atmosphere, but measurements show that there's little there, either. Overall, Earth has roughly 100,000 times as much water as Venus.

Broadly speaking, there are two possible explanations for the lack of water on Venus. Either Venus never had much water in the first place or Venus once had more water but somehow lost it. We cannot absolutely

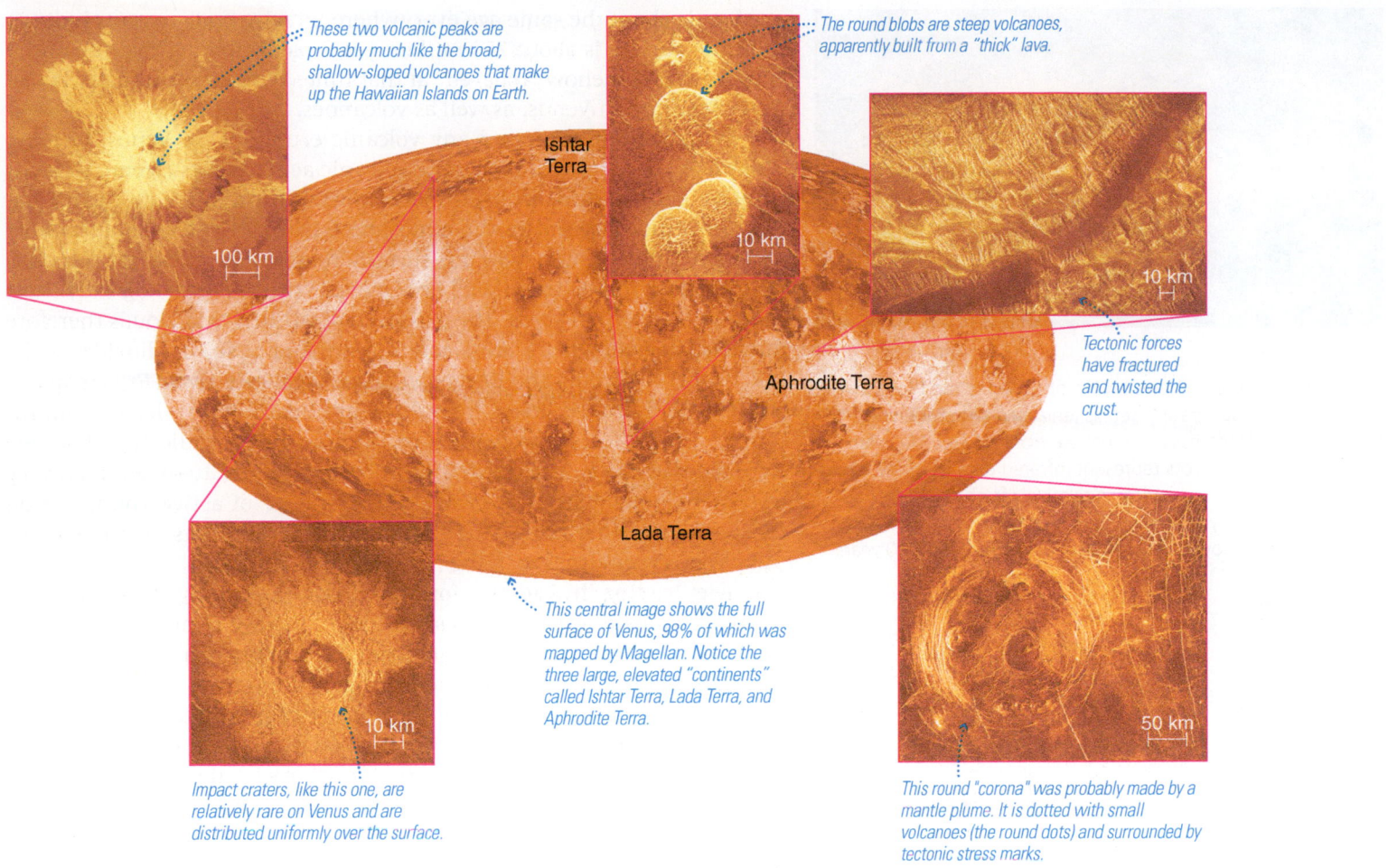

These two volcanic peaks are probably much like the broad, shallow-sloped volcanoes that make up the Hawaiian Islands on Earth.

100 km

The round blobs are steep volcanoes, apparently built from a "thick" lava.

Ishtar
Terra

10 km

Tectonic forces have fractured and twisted the crust.

10 km

Aphrodite Terra

Lada Terra

This central image shows the full surface of Venus, 98% of which was mapped by Magellan. Notice the three large, elevated "continents" called Ishtar Terra, Lada Terra, and Aphrodite Terra.

10 km

Impact craters, like this one, are relatively rare on Venus and are distributed uniformly over the surface.

50 km

This round "corona" was probably made by a mantle plume. It is dotted with small volcanoes (the round dots) and surrounded by tectonic stress marks.

FIGURE 10.2
The surface of Venus, as revealed by radar observations from the *Magellan* spacecraft. Bright regions in the radar images represent rough areas or higher altitudes.

rule out the first explanation, but it seems unlikely. At the time the planets were forming, the temperature of the protoplanetary disk would not have been very different at the orbits of Venus and Earth, so both planets should have accreted from planetesimals of similar composition. As we discussed in Chapter 3, these planetesimals probably contained little or no water, because temperatures in the inner solar system were too high for water vapor to condense into solid particles of ice. The water on Earth must have come from planetesimals that originated in more distant parts of the solar system and had their orbits perturbed in such a way that they ended up crashing into our planet. Simple physical arguments suggest that such collisions should have been equally common on Venus, so Venus should have obtained water in this same way. In other words, Venus should have started out with nearly as much water locked into its crust and mantle as Earth, which means we are left with the question of where all this water went.

As on Earth, trapped water on Venus would have been released into the atmosphere through outgassing. Radar mapping of Venus's surface shows plenty of evidence of ongoing geological activity such as tectonics and volcanism (Figure 10.2). Venus has relatively few impact craters, which tells us that the craters due to large impacts must have been erased by other geological processes. Moreover, the few impact craters are distributed fairly uniformly over the planet's surface, suggesting that the

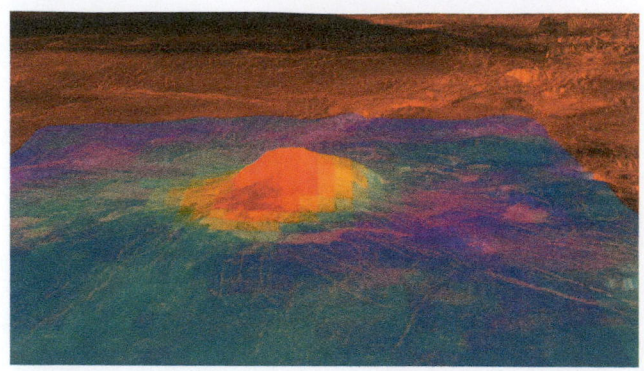

FIGURE 10.3

This composite image shows a volcano called Idunn Mons on Venus. Surface topography details (enlarged about 30 times to make the volcano easier to see) are from NASA's *Magellan* radar mapper, and colors represent infrared data from the *Venus Express* spacecraft. Red indicates relatively fresh rock that has not been chemically altered by Venus's harsh atmosphere, suggesting that lava flows occurred within the past 250,000 years.

surface is about the same age everywhere. Precise crater counts indicate that the surface is about 750 million years old, implying that the entire surface was somehow "repaved" at that time. We also see numerous tectonic features on Venus, as well as volcanoes.

We have not witnessed any volcanic eruptions on Venus, but two lines of evidence point to ongoing volcanic activity. First, Venus's clouds contain sulfuric acid, which is made from sulfur dioxide (SO_2) and water. Sulfur dioxide enters the atmosphere through volcanic outgassing, but once in the atmosphere it is steadily removed by chemical reactions with surface rocks; it would all be removed within the geologically short time of 100 million years. The existence of sulfuric acid clouds therefore implies that outgassing must continue to supply sulfur dioxide to the atmosphere. Second, the European Space Agency's *Venus Express* spacecraft (which orbited Venus from 2006 until early 2015) detected an infrared spectral feature from rocks on three volcanoes indicating that they erupted within about the past 250,000 years (Figure 10.3), which is very recent on a planetary time scale. The existence of active volcanism on Venus implies that outgassing should have released lots of water vapor into Venus's atmosphere in the past.

The leading hypothesis for the disappearance of Venus's water invokes some of the same processes thought to have removed water from Mars [Section 8.3]: Ultraviolet light from the Sun broke apart water molecules in Venus's atmosphere. The hydrogen atoms then escaped to space (through *thermal escape*), ensuring that the water molecules could never re-form. The oxygen from the water molecules was lost to some combination of chemical reactions with surface rocks and stripping by the solar wind. Venus's upper atmosphere is vulnerable to the solar wind because Venus lacks a protective magnetic field, probably because of its slow rotation. Recall that at least moderately rapid rotation is one of the requirements for a global magnetic field [Section 4.4]; Venus, which takes 243 days to complete a single rotation on its axis, does not meet this requirement. Without a magnetic field, the solar wind can strip atmospheric gas away, and probably has stripped away from Venus at least as much or more gas than it has stripped from Mars. However, because Venus has such a thick atmosphere, this gas loss would be barely noticeable.

Acting over billions of years, the breakdown of water molecules and the escape of hydrogen can easily explain the loss of an ocean's worth of water from Venus. Indeed, because Venus would have lost any surface water it once had, outgassed water could not be recycled back into the mantle the way it is recycled on Earth. Venus should therefore have lost water from its interior continuously throughout its history, explaining why its crust and mantle must by now be extremely dry—a situation quite different from that on Earth, where significant amounts of water are chemically bound to crust and mantle rock. As we discussed in Chapter 4, this dryness may have strengthened and thickened Venus's lithosphere, explaining why Venus lacks Earth-like plate tectonics.

Proving that Venus really did lose so much water is difficult, but evidence comes from the gases that didn't escape. Recall that most hydrogen nuclei contain just a single proton, but a tiny fraction of all hydrogen atoms (about 1 in 6400 measured in Earth's oceans) contain a neutron in addition to the proton, making the isotope (see Figure 3.19) of hydrogen that we call *deuterium*. Water molecules that contain an atom of deuterium instead of ordinary hydrogen (called "heavy water") behave chemically just like ordinary water and can be broken apart by ultraviolet light just as easily. However, a deuterium atom is twice as heavy as an ordinary hydrogen atom and therefore does not escape to space as easily when the water molecule is broken apart. If Venus lost a huge amount of hydrogen from water molecules to space, the rare deuterium atoms would have been more likely to remain behind than the ordinary hydrogen atoms. Recall that scientists have used this idea to estimate the amount of water lost from Mars over time [Section 8.3]. Measurements show even greater water loss for Venus: The fraction of deuterium among hydrogen atoms is a hundred times higher on Venus than on Earth, suggesting that a substantial amount of water must have been broken apart and its hydrogen lost to space.

The deuterium ratio does not allow us to determine exactly how much water Venus has lost, because Venus must be subjected to occasional comet impacts. The comets bring water that enters the atmosphere, where it ultimately gets broken down and its hydrogen escapes, just as with Venus's original water. Because comet water contains deuterium in a ratio higher than that found on Earth, the addition of comet water tends to lower the atmospheric ratio of deuterium to hydrogen on Venus. As a result, the water loss that we can infer from the current ratio of deuterium to hydrogen in the venusian atmosphere is only a *lower limit* on the actual water loss. The measured deuterium-to-hydrogen ratio implies that Venus has lost at least the equivalent of a global layer of water several meters deep. However, because this value is a lower limit, it is likely that Venus had much more water, making it plausible that Venus really did outgas as much water as Earth and then lost virtually all of it.

THE RUNAWAY GREENHOUSE EFFECT We have explained how Venus might have lost its water, but why didn't Earth lose its water in the same way? The answer is the ocean itself. On Earth, most of the outgassed water vapor condensed into rain, forming the oceans, long before ultraviolet radiation could break apart much of the water vapor. The short-wavelength ultraviolet light that tends to break apart water molecules does not penetrate far into the atmosphere, let alone penetrate the ocean surface, so water in the oceans is protected. (Today, ultraviolet light is also absorbed by the ozone layer, but the ozone layer didn't exist when our planet was young.) To understand why Venus was unable to protect its water in a similar way, let's consider what would happen if we moved Earth to Venus's distance from the Sun.

Figure 10.4 summarizes what would occur. The greater intensity of sunlight would almost immediately raise the global average temperature by about 30°C, from its current 15°C (59°F) to about 45°C (113°F). Although this is still well below the boiling point of water, the higher temperature would lead to increased evaporation of water from the oceans. The higher temperature would also allow the atmosphere to hold more water vapor before the vapor condensed to make rain (think of how much more humid hot days are than cold days). The combination of more evaporation and greater atmospheric capacity for water vapor would substantially increase the total amount of water vapor in Earth's atmosphere. Now, remember that water vapor, like carbon dioxide, is a greenhouse gas. The added water vapor would therefore strengthen the greenhouse effect, driving temperatures a little higher. The higher temperatures, in turn, would lead to even more ocean evaporation and more water vapor in the atmosphere—strengthening the greenhouse effect even further. In other words, we'd have a "positive feedback loop" in which each little

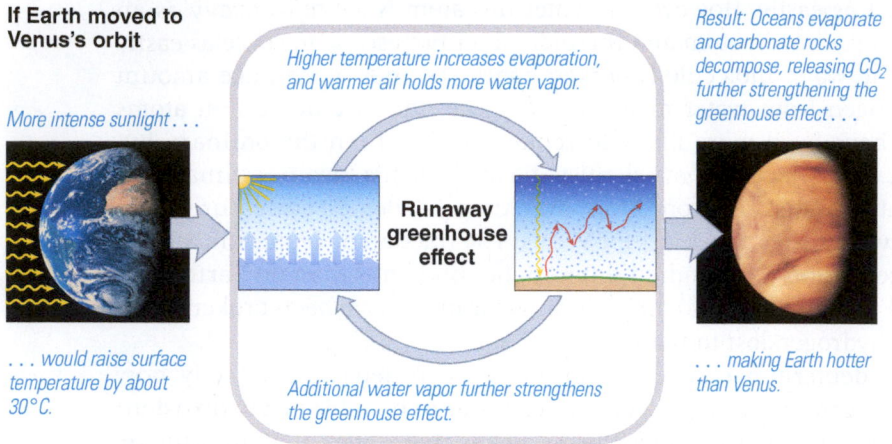

If Earth moved to Venus's orbit

More intense sunlight...

...would raise surface temperature by about 30°C.

Higher temperature increases evaporation, and warmer air holds more water vapor.

Runaway greenhouse effect

Additional water vapor further strengthens the greenhouse effect.

Result: Oceans evaporate and carbonate rocks decompose, releasing CO$_2$ further strengthening the greenhouse effect...

...making Earth hotter than Venus.

FIGURE 10.4

This diagram shows what scientists suspect would happen if Earth were placed at Venus's distance from the Sun: A runaway greenhouse effect would cause the oceans on Earth to evaporate completely.

bit of additional water vapor in the atmosphere would mean higher temperatures and even more water vapor. The process would careen rapidly out of control, resulting in what we call a **runaway greenhouse effect.**

The runaway process would cause our planet to heat up until the oceans were completely evaporated and the carbonate rocks released all their carbon dioxide back into the atmosphere. By the time the process was complete, temperatures on our "moved Earth" would be even higher than they are on Venus today, thanks to the combined greenhouse effects of the released carbon dioxide and water vapor in the atmosphere. The water vapor would then gradually disappear, as ultraviolet light broke water molecules apart and the hydrogen escaped to space. In short, moving Earth to Venus's orbit would essentially turn our planet into another Venus.

We have arrived at a simple explanation of why Venus is so much hotter than Earth. Even though Venus is only about 30% closer to the Sun than Earth, this difference was enough to be critical. On Earth, it was cool enough for water to rain down to make oceans in which carbon dioxide could dissolve and undergo chemical reactions that locked it away in carbonate rocks. As a result, our atmosphere was left with only enough greenhouse gases to make our planet pleasantly warm. On Venus, the greater intensity of sunlight made it just enough warmer that oceans either never formed or soon evaporated. Without oceans to dissolve carbon dioxide and make carbonate rock, carbon dioxide accumulated in the atmosphere, leading to a runaway greenhouse effect, which resulted in the extreme temperatures of Venus today.

Think About It We've seen that moving Earth to Venus's orbit would cause our planet to become Venus-like. If we could somehow move Venus to Earth's orbit, would it become Earth-like? Why or why not?

Is Venus in the habitable zone?

The fact that moving our own planet to Venus's orbit would lead it to the same fate as Venus implies that *any* planet at this distance would be doomed to a runaway greenhouse effect (assuming it had an atmosphere from outgassing, and therefore similar to that of the early Earth or

Venus). In other words, if our scenario is correct, it seems impossible for a planet to have a habitable surface with abundant liquid water at Venus's distance from the Sun. We conclude that Venus does *not* lie within the Sun's habitable zone today.

In contrast, Earth clearly *does* lie within the habitable zone, which is why our planet has not suffered a runaway greenhouse effect. Indeed, models show that a runaway greenhouse effect could not occur on the present-day Earth, at least through natural processes (as opposed to human-induced processes), an idea supported by our planet's climate history. Whenever the atmospheric concentration of greenhouse gases has increased on Earth in the past—as it has numerous times, perhaps most extremely during "hothouse" phases that follow snowball Earth episodes [Section 4.5]—our greater distance from the Sun has helped prevent the temperature from increasing to the point at which a runaway process would occur. Putting the lessons of Earth and Venus together, we conclude that the inner boundary of the Sun's habitable zone must currently lie beyond the orbit of Venus but within the orbit of Earth.

PAST HABITABILITY OF VENUS While Venus clearly isn't in the Sun's habitable zone today, it is possible that it was in the distant past, in which case it might once have had a much more moderate climate. Recall that the Sun has gradually brightened with age, which means that sunlight was less intense on all the planets when they were young. Models suggest that, some 4 billion years ago, the intensity of sunlight at Venus was only slightly greater than it is at Earth today. In that case, rain might have fallen on Venus, and oceans could have formed. It's even conceivable that life could have arisen in the oceans of a young Venus (or been transported to those oceans in meteorites from Earth).

We may never know whether Venus once had oceans or life, as any evidence is probably long gone. Recall that Venus's entire surface appears to have been "repaved" by tectonics and volcanism. This repaving would have covered up any shorelines or other geological evidence of past oceans. Rocks from Venus probably can't tell us anything about an oceanic past either: The high temperatures would have baked out any gases that might once have been incorporated into surface rock, eliminating evidence that might have told us whether the rocks formed in water or ever held life.

THE POSSIBILITY OF PRESENT-DAY LIFE ON VENUS If life did once take hold on Venus, could it survive anywhere on the planet today? For the reasons we've already discussed, the gradually brightening Sun would have doomed any life on Venus's surface. Even if Venus once had oceans, the onset of the runaway greenhouse effect raised the temperature so high that all the water evaporated, leaving no habitats on or below the planet's surface.

However, there is one place where it is conceivable that life might survive, if it ever took hold on Venus in the first place. As we discussed in Chapter 7, the temperature drops significantly with altitude in the atmosphere, and Venus has high-altitude clouds containing tiny amounts of liquid water (in highly acidic form). Though the prospects of finding life in these clouds are admittedly slim, it is at least within the realm of possibility to imagine that life once arose in venusian oceans and, as the runaway greenhouse effect set in, successfully adapted to life in the sky.

10.3 Surface Habitability Factors and the Habitable Zone

The case of Venus tells us that distance from the Sun is a critical factor in surface habitability, since Venus seems to have been doomed to suffer a runaway greenhouse effect no matter how habitable it may have been in its youth. But we know that distance from the Sun is not the only habitability factor, because we have seen that Mars's small size probably explains how it lost liquid water that existed on its surface in the past. In this section, we'll summarize what we have learned about surface habitability factors, and then use these lessons to consider the present-day boundaries of the Sun's habitable zone.

What factors influence surface habitability?

Our comparative studies of Venus, Earth, and Mars have given us deep insight into the factors that make these three worlds so different. Let's build on the discussions in both this chapter and prior chapters, so that we will gain additional insight into the prospects of finding worlds with habitable surfaces around other stars.

THE ROLE OF DISTANCE FROM A CENTRAL STAR As we've discussed, the first factor that affects surface habitability is a planet's distance from its home star, in our case the Sun. Subtle differences in distance can result in dramatic variations in habitability, as the case of Venus shows. A planet with an Earth-like atmosphere at Venus's distance would be only a little warmer than Earth; the reason Venus is not habitable is that "a little warmer" turns out to be warm enough to start a runaway greenhouse process that heats the planet far more. When we consider the minimum distance at which surface habitability is possible, we must therefore account not simply for solar heating but also for resultant processes that can lead to the evaporation of surface water. Similarly, when we consider the maximum distance at which surface habitability is possible, we must consider processes that might weaken greenhouse warming, causing surface water to freeze.

Another factor that influences the range of distances that define the habitable zone is the central star's luminosity (brightness). Stars come with a wide range of luminosities: A few stars are much more luminous than the Sun, about 10% of all stars have luminosities similar to that of the Sun, and the vast majority of stars are considerably dimmer than the Sun. Clearly, a brighter star must have a wider and more distant habitable zone than the Sun, while a dimmer star must have a narrower and closer-in habitable zone. The wider habitable zones of brighter stars might seem to increase the odds of finding planets within these zones, but another factor may make biology rare or nonexistent in these cases: The brightest stars turn out to have extremely short lifetimes—millions of years rather than billions of years—and hence may not offer enough time to nurture biology [Section 11.1]. The dimmer stars, with their narrow habitable zones, seem less likely to have planets in these zones. However, they may make up for this shortcoming by their sheer numbers, since they are so much more common than stars like the Sun (see Cosmic Calculations 10.1). We'll discuss the prospects for habitability around different types of stars in Chapter 11.

THE ROLE OF PLANETARY SIZE Clearly, distance from a central star is not the only factor affecting surface habitability. If it were, the Moon would be habitable because it is the same distance from the Sun as is Earth. More to the point, if distance were all that mattered, Mars should have become *more* habitable as the Sun grew brighter with time; instead, it froze over. Mars's current lack of surface habitability appears to be less the result of its distance from the Sun than of its size. As we discussed in Chapter 8, its small size allowed its interior to cool more quickly than that of a larger planet like Earth. Its core presumably cooled to the point at which it no longer had a convecting layer and therefore could no longer generate a global magnetic field. The martian atmosphere was then left vulnerable to stripping by the solar wind, and through this and other mechanisms, it lost too much gas to support a strong greenhouse effect. Without a strong greenhouse effect, Mars effectively froze. The cooler interior also means that if martian volcanoes remain active at all, they do not erupt frequently enough to resupply lost atmospheric gases.

Is there a minimum or maximum size that makes it possible for a planetary surface to remain habitable over long time periods? A minimum size clearly exists. We have traced Earth's long-term habitability directly to the climate regulation provided by the carbon dioxide cycle, which in turn depends on the cycle of plate tectonics [Section 4.5]. We therefore have at least some reason to think that plate tectonics is necessary for long-term habitability on any world, in which case the size requirement for surface habitability is a size that allows plate tectonics to exist.

We do not fully understand the minimum planetary size needed for plate tectonics, but the size must be large enough so that the planet has active tectonics *and* a relatively thin lithosphere that can fracture into plates. The Moon and Mercury are clearly too small, and we see no evidence of ongoing plate tectonics on Mars. This suggests that Mars has also had enough interior cooling to ensure that its lithosphere has thickened to the point where it cannot break into plates. But while Mars—with a radius about half that of Earth—seems too small, we don't know how much bigger a planet would need to be so that it would have both sufficient internal warmth to power tectonics and a crust thin enough to split into plates that could be pushed around by the convective currents below. The case of Venus illustrates the problem. Recall that, given the similarity in size of Venus and Earth, we might expect to find plate tectonics operating on both planets, but there is little evidence of Earth-like plate tectonics on Venus. Could it be that Venus's slightly smaller size was just enough to make the difference, much as its slightly nearer distance to the Sun led to its runaway greenhouse effect? It's possible, but as we've discussed, a more likely hypothesis holds that Venus's high temperature baked water out of the rock in its crust and upper mantle, making the crust too stiff to allow subduction and thereby suppressing plate tectonics. If this explanation is correct, Venus might well have had plate tectonics if it had not been so close to the Sun, and might have had plate tectonics in its early history when the Sun was dimmer.

At this point, all we can say with confidence is that Earth is of sufficient size for long-term surface habitability and Mars is not. Beyond that, we don't know how large a planet must be to allow for the presence of liquid water over extended periods of time. We are similarly uncertain of the maximum size that would allow for long-term habitability. As we'll discuss in the next chapter, we now know of many planets around other stars that are probably rocky in nature but larger than Earth (the so-called *super Earths*), but we do not yet know enough about their likely geology to say with confidence whether they could harbor surface oceans and life.

THE ROLE OF AN ATMOSPHERE A third crucial factor in surface habitability is the presence of an atmosphere. Without sufficient atmospheric pressure, liquid water cannot be stable and abundant regardless of other factors. Moreover, an atmosphere is necessary to protect a planetary surface against harmful solar radiation. So while a world without an atmosphere might still shelter life underground, it seems implausible that any biology would creep, crawl, or fly across its landscape.

The presence or absence of a significant atmosphere sometimes depends on size. In the case of Mars, for example, we have traced its loss of surface habitability to loss of atmospheric gas, which most likely was tied to its small size and loss of magnetic field. However, there may be several other ways in which a planet either might never get or would subsequently lose an atmosphere, even if it were of the right mass and at the right distance from its star.

Remember that the atmospheres of Venus, Earth, and Mars are all thought to have been produced largely through outgassing. If a planet lacked trapped gases in its interior, no outgassing would be possible. Given that most of the gases within the terrestrial worlds are assumed to have been brought by impacts of objects from more distant reaches of the solar system, and that these objects were presumably cast onto collision courses by gravitational interactions, terrestrial planets might form without gases in a star system that lacks gas-bearing planetesimals or in which these planetesimals for some reason never get perturbed inward. Neither situation seems likely in light of what we know about solar system formation [Section 3.4], but we can't rule them out.

For a planet that has outgassing and sufficient size to hold its gas gravitationally, we know of at least two possible ways in which it might nonetheless lose its atmosphere. First, large impacts can in principle blast significant amounts of atmospheric gas into space. This process may have played a role on Mars. If large impacts are more common in some other star systems—as may be the case in systems that lack a planet comparable in size to Jupiter at the right distance [Section 11.4]—these impacts could cause significant gas loss on Earth-size planets. Second, as we have discussed, the solar wind can strip atmospheric gas from any planet that lacks a global magnetic field, and we don't expect to find magnetic fields on slowly rotating planets. Venus has so much atmospheric gas that loss to solar wind stripping has been insignificant in comparison. Earth, however, might well have become uninhabitable (at least on its surface) if not for the protection that the magnetic field has offered against billions of years of stripping by the solar wind.

SUMMARY OF HABITABILITY FACTORS We have identified three major factors for long-term surface habitability (Figure 10.5):

1. The planet must be neither too close to nor too far from its star; that is, it must be within its star's habitable zone.
2. The planet must be large enough to retain internal heat and may also require plate tectonics for climate regulation. We don't know

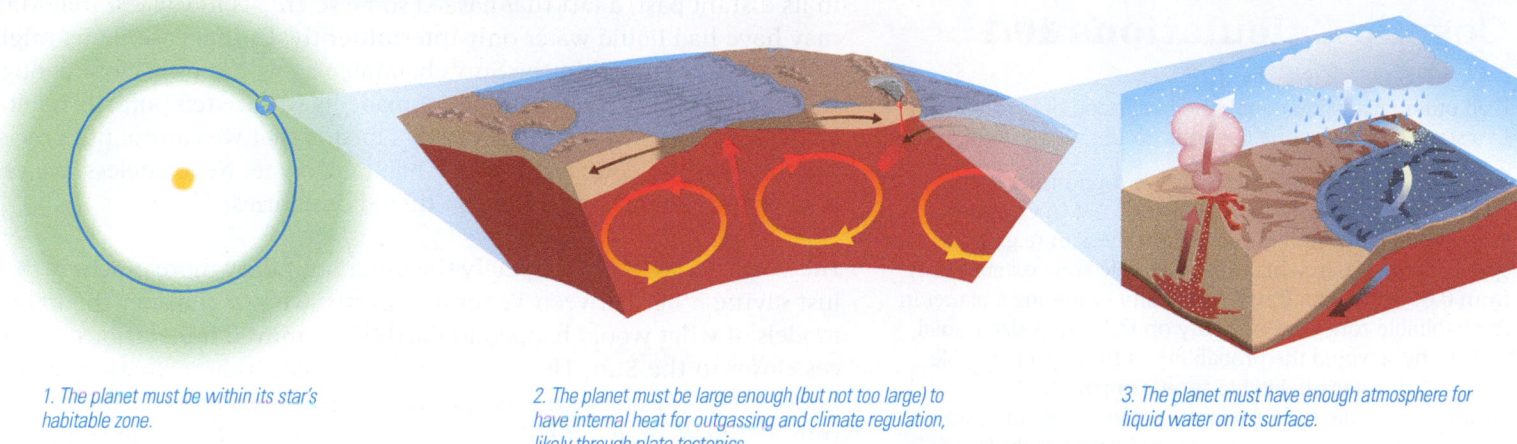

1. The planet must be within its star's habitable zone.

2. The planet must be large enough (but not too large) to have internal heat for outgassing and climate regulation, likely through plate tectonics.

3. The planet must have enough atmosphere for liquid water on its surface.

FIGURE 10.5
Summary of the factors required for long-term surface habitability.

the precise minimum size that meets these conditions, but it is certainly larger than Mars. We also do not know the maximum size, but Earth is obviously not too large.

3. The planet must have enough of an atmosphere for liquid water to be present on its surface. This probably means that it must have had gases trapped in its interior, so that an atmosphere could form through outgassing, and that it has not since lost too much of this atmospheric gas to impacts or solar wind stripping. Protection against the latter may require a global magnetic field, which in turn would require at least moderately rapid rotation.

The latter two requirements depend on factors intrinsic to the planet itself. The first depends on where it forms around its star, which brings us to the topic of habitable zone boundaries.

Think About It Using the three habitability factors, explain why the Moon is not habitable.

Where are the boundaries of the Sun's habitable zone today?

The boundaries of the Sun's present-day habitable zone depend on the range of distances at which a planet of suitable size and with sufficient atmospheric pressure could have liquid water on its surface. The inner boundary marks the place where, if a planet were any closer to the Sun, a runaway greenhouse effect would be triggered. The outer boundary marks the place where, if a planet were any farther from the Sun, surface water would freeze.

The case of Venus has shown us that the inner boundary of the habitable zone does not lie as far inward as Venus's orbit. The case of Mars tells us something about the outer boundary, though it is more ambiguous: Geological evidence makes it clear that Mars once had liquid water on its surface, in which case it must once have been within the Sun's habitable zone—and since the habitable zone moves outward with time, it would then still be in the habitable zone today. However, as we discussed in Chapter 8, models of the martian climate cannot yet fully account for the conditions that might have caused Mars to have been quite so warm

Cosmic Calculations 10.1

CHANCES OF BEING IN THE ZONE

If all other factors are equal, the likelihood of finding planets in a star's habitable zone depends on the width of the zone.

Example: The Sun's habitable zone is (optimistically) calculated to extend from about 0.84 to 1.7 AU. Consider a star much smaller and dimmer than the Sun (e.g., a star of spectral type M), in which the habitable zone extends only from 0.05 to 0.1 AU. If the probability of finding a planet in the habitable zone depends only on the zone's size (radial width), how would the probability of finding a habitable planet in this star's habitable zone compare to the probability around a Sun-like star? Given that these types of small stars outnumber Sun-like stars by approximately eight to one in our galaxy, for which class of stars would we expect more worlds in habitable zones?

Solution: The width (range of radii) of a habitable zone, R_{HZ}, is,

$$R_{HZ} = R_{outer} - R_{inner}$$

For a Sun-like star,

$$R_{HZ} = 1.7 - 0.84 = 0.86 \text{ AU}$$

For the smaller star,

$$R_{HZ} = 0.1 - 0.05 = 0.05 \text{ AU}$$

For the smaller star, the probability of being in the habitable zone is only $0.05/0.86 = 0.058$ times the probability for a Sun-like star. However, because these smaller stars are eight times as common, the *total number* of worlds in habitable zones around these small stars would be about $8 \times 0.058 = 0.46$, or 46%, times the number of worlds around Sun-like stars. (Of course, we have not considered factors besides the size of the habitable zone that may also be important.)

in its distant past, a fact that has led some scientists to suggest that Mars may have had liquid water only intermittently. In that case, Mars might never have been within the Sun's habitable zone, but only close enough that special events (such as impacts) made its surface temporarily habitable. The ambiguity in the case of Mars means that we cannot put precise numbers on the boundaries of the habitable zone. Nevertheless, we can place some general constraints on these boundaries.

THE INNER BOUNDARY To specify the inner boundary more precisely than just saying it lies between Venus and Earth, we can consider theoretical models of what would happen to Earth if we moved it to various distances closer to the Sun. These calculations suggest that a runaway greenhouse effect would occur if we placed Earth anywhere inside of 0.84 AU from the Sun, or about halfway between the orbits of Venus and Earth. (Recall that 1 AU, or astronomical unit, is Earth's average distance from the Sun, or about 150 million kilometers.)

However, another factor might cause temperatures to spin out of control even beyond 0.84 AU from the Sun. According to some models, moderate additional warming would allow water vapor to circulate to much higher altitudes in Earth's atmosphere (into the stratosphere). At these high altitudes, water molecules would be above much of the ozone layer and hence could be broken apart by ultraviolet light from the Sun. The hydrogen would then escape to space, causing Earth to lose this water and thereby allowing more water to rise into the upper atmosphere and be lost in turn. Over time, this **moist greenhouse effect**—so called because the upper atmosphere would become moist with water, at least until all the water was lost—might cause Earth to lose its oceans. Note that the moist greenhouse effect leads to water loss not because the temperature is outside the range in which liquid water could exist, but rather because water that evaporates from the surface can rise into the upper atmosphere, where it can be lost to space. This is a less dramatic version of the scenario we discussed when we imagined placing Earth at Venus's orbital distance, but it could be just as lethal in the long run. An Earth-like planet might suffer this moist greenhouse water loss anywhere within 0.95 AU of the Sun. However, models that describe the onset of the moist greenhouse effect have numerous recognized uncertainties (such as the effects of clouds) that might push this distance inward.

In summary, the inner boundary of the present-day habitable zone in our solar system may be at 0.84 AU if we allow for only a simple runaway greenhouse effect, but as far out as 0.95 AU if we allow for water loss by a moist greenhouse effect.

THE OUTER BOUNDARY The outer boundary of the present-day habitable zone is the distance from the Sun at which even a strong greenhouse effect could not warm a planet enough to keep liquid water from freezing. At first, we might guess that Mars is beyond the outer boundary, since the temperature is too cold for liquid water at its surface today. However, if Mars were larger and had retained a thick atmosphere, it might still have enough greenhouse warming to have a habitable surface. In that case, the outer boundary of the habitable zone would lie beyond the orbit of Mars.

Calculations suggest that this is indeed the case. If we allow for a thick atmosphere with a strong greenhouse effect, the outer boundary of the present-day habitable zone lies at about 1.7 AU, well outside

Mars's average orbital distance of 1.52 AU. However, there is at least one potential problem that could bring the outer boundary in closer. If the atmosphere of a planet is too cold, the atmospheric carbon dioxide that produces greenhouse warming will condense into snowflakes and fall to the surface. This carbon dioxide snow might limit how much carbon dioxide could reside in the atmosphere, preventing the atmosphere from staying thick enough for a strong greenhouse effect. This scenario might occur if the atmosphere lacked dust or any other greenhouse gas to help keep the middle atmosphere warm. In that case, the outer boundary of the present habitable zone might lie at only about 1.4 AU, or just inside the orbit of Mars. Note that this case would also imply that Mars never was in the habitable zone, so it could have had surface liquid water only intermittently in the past.

THE EXTENT OF THE HABITABLE ZONE We have found two estimates each for the distances of the inner and outer boundaries of the Sun's present habitable zone. The more optimistic estimates indicate that the present-day habitable zone extends from about 0.84 to 1.7 AU. The more conservative estimates suggest that it extends from only about 0.95 to 1.4 AU. Both sets of boundaries are shown in Figure 10.6. Note that, even in the more conservative case, the Sun's present-day habitable zone represents a fairly wide region in the inner solar system.

Keep in mind that we don't yet know enough about how planetary atmospheres work to know which estimate is correct—or if the truth lies somewhere in between. In addition, there might be processes that can affect the atmospheric temperature that we haven't yet discovered, and therefore have not accounted for in our calculations of the habitable zone boundaries. These estimates of the boundaries of the habitable zone should therefore be considered just that—estimates. They might be significantly refined as we learn more in the future.

Think About It About how much wider is the more optimistic estimate of the present-day habitable zone than the more conservative estimate? If planets form at random locations in the inner portions of other stars' solar systems, how would these two different estimates affect the likely number of planets within habitable zones around Sun-like stars in the Milky Way Galaxy?

10.4 The Future of Life on Earth

Because the boundaries of the habitable zone are calculated under the assumption that we have a planet of suitable size, they depend only on the amount of heat and light put out by the Sun. Therefore, when the Sun was dimmer in the past, the habitable zone must have been closer in. In the future, when the Sun will be brighter than it is today, the habitable zone will lie farther out. In this section, we'll discuss the way in which the Sun's habitable zone evolves with time.

How will the Sun's habitable zone change in the future?

To determine how the habitable zone moves over time, we need to know how the Sun brightens as it ages. Fortunately, the process that causes the Sun to brighten is well understood.

The Sun shines by fusing hydrogen into helium. Overall, each fusion reaction converts four independent hydrogen nuclei into a single helium

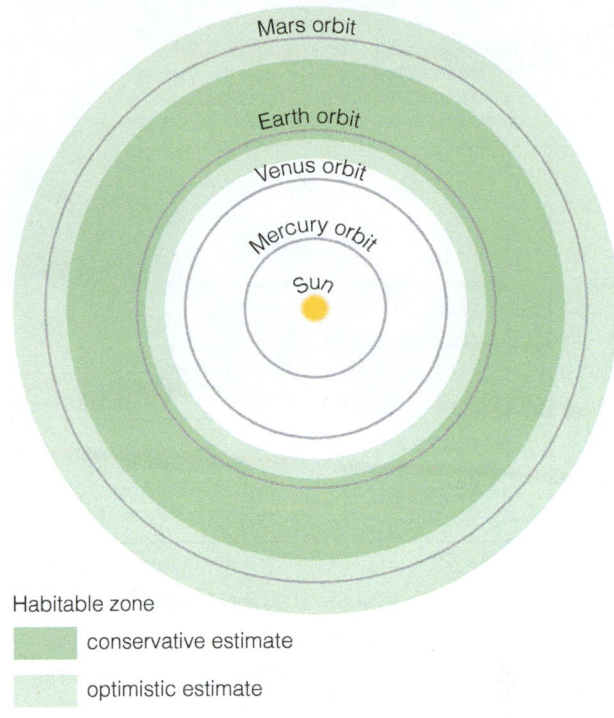

Habitable zone

■ conservative estimate

□ optimistic estimate

FIGURE 10.6
Boundaries of the Sun's habitable zone today. The narrower set of boundaries represents a model based on the more conservative assumptions, while the wider set represents the most optimistic scenarios.

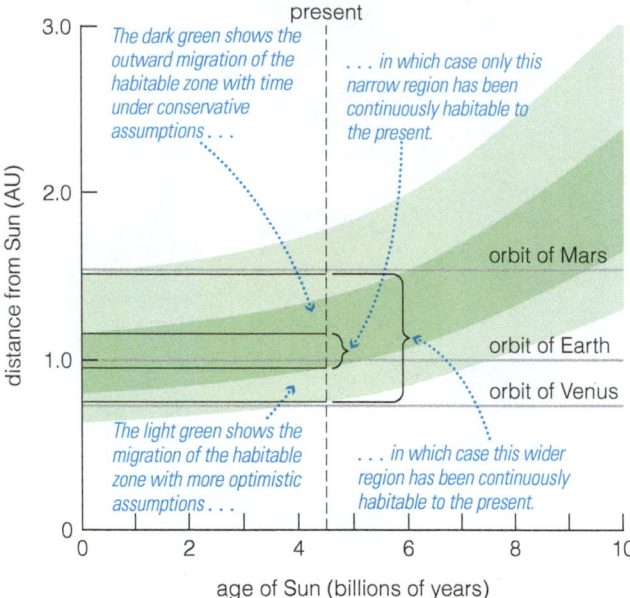

FIGURE 10.7
This graph shows the Sun's habitable zone through time. The narrower region represents the habitable zone based on the more conservative assumptions, and the wider region represents the habitable zone based on the more optimistic assumptions. The horizontal swaths represent the zone that has been continuously habitable from 4 billion years ago to the present, again under the conservative (narrower) and optimistic (wider) assumptions.

The dark green shows the outward migration of the habitable zone with time under conservative assumptions . . .

. . . in which case only this narrow region has been continuously habitable to the present.

present

orbit of Mars

orbit of Earth

orbit of Venus

The light green shows the migration of the habitable zone with more optimistic assumptions . . .

. . . in which case this wider region has been continuously habitable to the present.

nucleus (see Figure 3.10), which means that the total number of *independent* particles in the solar core gradually falls with time. This in turn causes the solar core to shrink, because for a given core size, fewer independent particles means less pressure pushing back against the weight of the Sun's overlying layers, allowing that weight to compress the core. The slow shrinkage, in turn, gradually increases the core temperature, much as a bicycle tire pump gets warm when you compress the air in it by pushing on its piston. The gradual temperature increase causes a corresponding increase in the fusion rate, which is why the Sun slowly brightens.

Theoretical models indicate that the Sun's core temperature should have increased enough to raise its fusion rate and the solar luminosity by about 30% since the Sun was born $4\frac{1}{2}$ billion years ago. The models also allow scientists to predict the Sun's future luminosity. Observational data support the models' conclusions, as stars of particular masses do indeed vary somewhat in brightness, with older stars being brighter for any particular mass.

Figure 10.7 shows the results of calculations of the boundaries of the habitable zone from the Sun's birth until it exhausts its core hydrogen fuel some 5 billion years from now. Notice how the habitable zone gradually moves outward from the Sun, just as we would expect.

The horizontal swaths in Figure 10.7 show the distances from the Sun at which conditions have remained habitable from 4 billion years ago to the present. This zone is often called the **continuously habitable zone,** because it has been habitable at all times since the end of the heavy bombardment about 4 billion years ago. The width of the continuously habitable zone is fairly narrow if we use the more conservative assumptions, and substantially wider with the more optimistic assumptions. In fact, under the optimistic assumptions, both Earth and Mars are in the continuously habitable zone. Note also that the continuously habitable zone is defined for habitability only up to the present. If we look billions of years into the future, the habitable zone will continue to move outward and the continuously habitable zone will therefore become narrower.

Think About It Was Venus ever in the habitable zone? Is it in the continuously habitable zone? Under the more conservative assumptions, about when will the continuously habitable zone move beyond Earth's orbit? Explain.

How long can life survive on Earth?

Earth has remained habitable for some 4 billion years, allowing plenty of time for life to evolve, diversify, and ultimately make our own existence possible. However, the continuing evolution of the habitable zone means that Earth's days of habitability must eventually come to an end.

It's important to note that the demise of Earth's habitability is not something worth losing sleep over. Even under the most pessimistic scenarios, our planet will remain habitable for many hundreds of millions of years to come. Under more optimistic scenarios, Earth may have billions of years of remaining habitability. Compared to the length of time our civilization has existed so far, these are incredibly long time scales. If our species survives for this long, we will have had plenty of time to find ways to move to other planets in our solar system or in other star systems, or to otherwise prevent our perishing with Earth.

Think About It Given the more immediate threats to our civilization, is it even worth thinking about what will happen to our planet millions or billions of years from now? Defend your opinion.

THE END OF EARTH'S HABITABILITY Thanks to the climate regulation provided by the carbon dioxide cycle, Earth has remained habitable even as our Sun has brightened by some 30% over the past 4 billion years. As the Sun continues to brighten, the carbon dioxide cycle should continue to keep the climate relatively pleasant for at least hundreds of millions of years to come. Eventually, however, the warming Sun will cause the cycle to break down.

If you study Figure 10.7, you'll see that under the more conservative assumptions, the inner boundary of the habitable zone will move beyond Earth's orbit in about a billion years. Recall that these conservative assumptions are based on the idea that a *moist greenhouse effect* will cause the oceans to evaporate away. In other words, this scenario envisions that about a billion years from now, water vapor will begin to circulate into Earth's upper atmosphere, where ultraviolet light will break apart water molecules, allowing the hydrogen to escape into space. As water is lost from the upper atmosphere, more water will evaporate to take its place, until the oceans have evaporated completely. At that point, it seems unlikely that any life could continue to survive on Earth. Of course, no one is yet sure whether the moist greenhouse problem will really arise in about a billion years. Many feedback mechanisms in Earth's climate are not yet well understood. For example, increased cloud cover may reduce the amount of sunlight reaching Earth's surface, preventing the onset of the moist greenhouse effect.

Under more optimistic assumptions, Earth may remain habitable until some 3 to 4 billion years from now. At that time, the Sun will have grown so bright that Earth will finally suffer the fate of Venus—a runaway greenhouse effect. The rising temperature on Earth will cause increased ocean evaporation, and the buildup of water vapor in the atmosphere will increase the greenhouse effect further (see Figure 10.4). The positive feedback won't stop until the oceans have evaporated away and all the carbon dioxide has been released from carbonate rocks. Our planet will become a Venus-like hothouse, with temperatures far too high for liquid water to exist.

We know of no natural phenomena that can prevent this runaway greenhouse effect from occurring. However, if our descendants have become sufficiently advanced in their technology, we can imagine several ways they could survive in our solar system. For example, they might protect Earth itself by building a giant sunshade in space to reduce the intensity of the light reaching Earth from the brightening Sun. Or they might simply move: When the habitable zone first moves past the orbit of Earth, Mars will be well within it, so perhaps our descendants will be able to make Mars livable with some advanced terraforming [Section 8.4]. If they are truly powerful, they might even find a way to move our planet gradually outward from the Sun to keep it within the habitable zone; by slowly moving Earth to the orbit of Mars over the next 3–4 billion years, humans could keep our planet habitable and stay home. Another possibility might be to relocate the population to large, artificial habitats constructed in space. Still, at best all these solutions can be only temporary, because the Sun itself eventually will die.

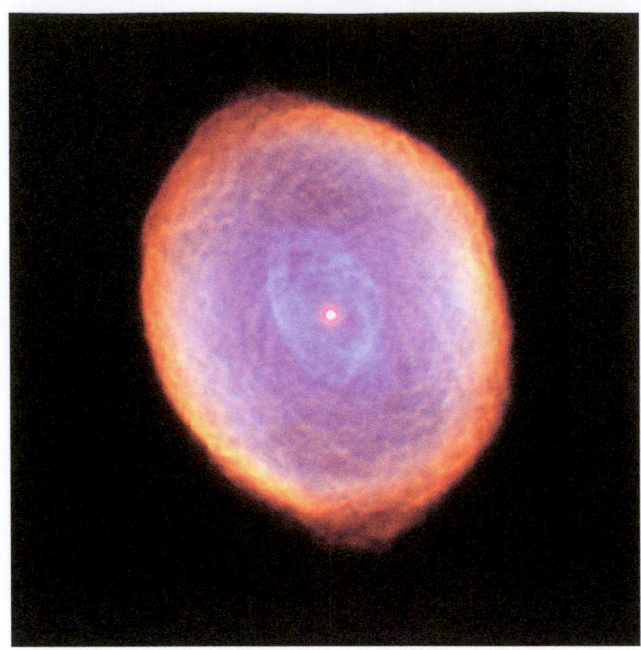

FIGURE 10.8

This Hubble Space Telescope photograph shows the Spirograph Nebula, an example of a planetary nebula. The gas of the nebula was expelled from a Sun-like star that had reached the end of its life. The central white dot is the white dwarf that remains, which is essentially just the hot core of the now-dead star. Our Sun will eventually suffer the same fate as the star that created this nebula.

DEATH OF THE SUN About 5 billion years from now, the Sun will have essentially used up the hydrogen in its central core, having fused it all into helium. At that point, the Sun will begin to undergo a series of dramatic transformations (driven by a combination of the initiation of hydrogen fusion in the layers surrounding the core and, a little later, fusion of helium into carbon in the core). Over a period of a few hundred million years (less than 10% of the Sun's total lifetime), the Sun will swell to about 100 times (or more) its present radius, its surface temperature will drop (changing its color from yellow to red), and it will become what we call a *red giant* star. Even though the Sun will have a lower surface temperature as a red giant, its much greater size will cause it to pump more energy into space than it does now. At its peak, the red giant will be about 1000 times as luminous as the Sun is today, and Earth's surface temperature will rise to 700°C (about 1300°F) or higher. Any remaining oceans will evaporate, and even underground life will be baked to death during this period as subsurface water boils away.

In its final death throes, the Sun will expel its outer layers into space, creating a *planetary nebula** (Figure 10.8). All that will remain of the Sun will be its hot central core; no more nuclear fusion will occur. This remaining core, known as a *white dwarf* star, will then gradually cool over time. The violence that accompanies the planetary nebula ejection will probably destroy Earth. Even if our planet somehow survives this event and continues to orbit the white dwarf Sun, the light from the white dwarf will eventually become so feeble that Earth's charred surface will face a future of perpetual, frigid darkness.

COULD WE STILL SURVIVE? For those undaunted by the thought of humans or other intelligent Earth beings surviving some 5 billion years into the future, the next step is to wonder whether we could also survive the death of the Sun. The obvious solution to the Sun's death is to move to another star system. Stars that are being born today might offer great homes to us in 5 billion years. When these stars died, we could move on to others born still later. As long as there are new stars with habitable planets, we could potentially survive by migrating to new homes.

However, even this type of long-term migration has its limits. The recycling of stellar material cannot continue forever, because dying stars return to space only part of the gas from which they were made. Over time, the galaxy will contain less and less interstellar gas. About 100 billion years from now, the Milky Way Galaxy will contain so little gas that new stars will no longer be born. What then?

After Edwin Hubble first discovered that our universe is expanding, scientists wondered whether the expansion would continue forever or someday stop, causing the universe to collapse back in on itself. In the past couple of decades, astronomers have learned that the expansion is actually accelerating, making it seem that the fate of the universe will be to expand forever at a greater and greater clip. However, keep in mind that *forever* is an extremely long time. It remains possible that we will someday discover new surprises that will change our view of the long-term fate of the universe.

If the universe continues to expand after all star formation has ceased, life will be able to continue only around those long-lived stars that still have habitable planets. But even the longest-lived stars will run out of

---

*Despite the name, these structures have nothing to do with planets; the term comes from the planetlike appearance of some of these nebulae when seen through a small telescope.

hydrogen to fuse within a few hundred billion years. Once all stars have died, the now-brilliant galaxies are destined to fade into darkness. As the universe continues to age, there will be enough time for relatively infrequent events to have a noticeable effect on the structure of galaxies. Close encounters will eventually send many of the burned-out stars flying into the vastness of intergalactic space, while the remaining stars will spiral in toward their galactic centers, merging into gigantic black holes. At this point, the story becomes even more speculative. If the so-called grand unified theories of physics are correct, the stellar corpses would eventually disintegrate into swarms of subatomic particles. Meanwhile, the giant black holes would slowly evaporate into energy and particles in a process first described by the noted British physicist Stephen Hawking (portrayed in the movie *The Theory of Everything*). At some point in the far distant future, the universe would consist of nothing but a dilute sea of subatomic particles and photons of light, each separated from others by immense distances that would grow larger as the universe endlessly expanded. Our current epoch of a universe filled with stars and galaxies would have been just a fleeting moment in an eternity of darkness.

This end in darkness may sound a bit depressing, but it is, after all, inconceivably far in the future. Nevertheless, it is fair to ask whether it is truly the end or instead could be followed by something else. Some serious scientists already argue that there might be ways by which an intelligent civilization could survive even as the universe died. For a more lighthearted viewpoint, we turn to science fiction. Isaac Asimov, in his story "The Last Question," begins with a couple of people asking a supercomputer whether there is a way to reverse the decline of the universe and thereby avert a cold and dark end. The computer responds that the available data are insufficient to answer the question. Over billions of years, computers advance and humankind survives, making the question ever more important as the universe approaches its cold, dark future. By the end of the story, humanity has died and the computer exists solely in hyperspace, outside the time and space of our universe. The universe has reached a state of ultimate darkness, with nothing left alive.

Special Topic 10.1 FIVE BILLION YEARS

The Sun's demise in about 5 billion years might at first seem worrisome, but 5 billion years is an extremely long time. It is longer than Earth has yet existed, and human time scales pale by comparison. A single human lifetime, if we take it to be about 100 years, is only 2×10^{-8}, or two hundred-millionths, of 5 billion years. Because 2×10^{-8} of a human lifetime is about 1 minute, we can say that the relationship of a human lifetime to the life expectancy of the Sun is roughly the same as that of 60 heartbeats to a human lifetime.

What about human creations? The Egyptian pyramids have often been described as "eternal." But they are slowly eroding because of wind, rain, air pollution, and the impact of tourists, and all traces of them will probably have vanished within a few hundred thousand years. While a hundred thousand years may seem like a long time, the Sun's remaining lifetime is some 50,000 times longer.

On a more somber note, we can gain perspective on billions of years by considering evolutionary time scales. During the past century, our species has acquired sufficient technology and power to destroy human life totally, if we so choose. However, even if we suffer that unfortunate fate, some species (including many insects) are likely to survive. Would another intelligent species ever emerge on Earth? There is no way to know, but we can look to the past for guidance. Many species of dinosaurs were biologically quite advanced, if not actually intelligent, when they were suddenly wiped out about 65 million years ago. Some small, rodentlike mammals survived, and here we are 65 million years later. We therefore might guess that another intelligent species could evolve some 65 million years after a human extinction. If these beings also destroyed themselves, another species could evolve 65 million years after that, and so on. But even at 65 million years per shot, Earth would have some 15 more chances for an intelligent species to evolve in 1 billion years—the length of time our planet will remain habitable under fairly conservative scenarios. Under more optimistic estimates for long-term habitability, there could be 60 or more periods—each as long as the period separating us from the dinosaurs—still to come before our planet dies. That is a lot of potential opportunities for the evolution of a species wise enough to avoid self-destruction and to move on to other star systems by the time the Sun finally dies. Perhaps we ourselves will prove to be so wise.

But the computer, which Asimov calls AC, whirs on in the timelessness of hyperspace until finally it learns how to reverse the decay of the universe. Asimov wrote:

> For another timeless interval, AC thought how best to do this. Carefully, AC organized the program.
>
> The consciousness of AC encompassed all of what had once been a Universe and brooded over what was now Chaos. Step by step, it must be done.
>
> And AC said, "LET THERE BE LIGHT!"
>
> And there was light—

 **THE PROCESS OF SCIENCE IN ACTION**

10.5 Global Warming

While we may not need to be concerned about a runaway greenhouse effect that is billions of years in the future, changes in the strength of the greenhouse effect can still cause dramatic shorter-term changes in Earth's climate. In particular, considerable evidence suggests that Earth's global average temperature is currently on the rise, a trend referred to as **global warming.** You're undoubtedly aware that global warming is a hot political topic that has generated debate about whether it is occurring naturally or as a result of human activity and, if the latter, what (if anything) we should do about it. For this chapter's case study in the process of science in action, we'll investigate how researchers seek to understand global warming and its potential consequences.

What is the evidence for global warming?

The basic science behind global warming is surprisingly simple: We know that carbon dioxide is a greenhouse gas that can enhance the greenhouse effect and therefore cause a rise in a planet's surface temperature, and we know that human activity (such as the burning of fossil fuels) is adding

---

Movie Madness WALL-E

It's sometime in the 28th century, and humankind has failed to live up to its nobler ambitions. Runaway consumerism, not to mention a sudden distaste for recycling, has carpeted Earth in industrial waste, making *WALL-E* the ultimate in trashy movies. Thank goodness it's a cartoon, sparing you the more odious details of a planet that everywhere mirrors the raunchier neighborhoods of *Slum Dog Millionaire.*

Humanity has decided to deal with the teeming refuse of its wretched shores by decamping to a fleet of space-based cruise ships where, as anyone who's ever spent time sailing the Caribbean knows, the greatest challenge is deciding whether to make another foray to the midnight buffet on the Lido deck. In a cynical comment on our contemporary tendency to overeat, these earthly émigrés have degenerated to the status of animated lard.

But our species hasn't given up on Earth: We've left our natal world in the hands of a fleet of robots. The thankless job of these machines is to clean the place up. Once they've finished, the humans will return and resume their profligate lifestyles.

As the film opens, this plan has reached a low ebb. Nearly all the robots have gone treads up and become scrap themselves. There's only one functioning unit left, WALL-E (Waste Allocation Load Lifter Earth-Class). He's roughly the size of a bantam fridge and has a head patterned after World War II Navy binoculars. WALL-E is a trash compactor who is self-aware and endowed with childlike innocence. He's frequently surprised, frightened, and—after a few centuries on the job—hankering for some romance.

WALL-E is an example of what roboticists call a *Von Neumann machine,* named after the mathematician who first described a self-replicating device. WALL-E has the smarts to make a copy of himself, and obviously can scavenge for the necessary parts. If he wished, he could build his own offspring, providing company and some free time. Instead, he looks for love—apparently unaware that this is the old, biological method for making children, and probably ineffective and awkward for robots.

Unlike other more realistic dystopian films, *WALL-E* shows us not how miserable our future will be, but how indolent. We're all going on an extended cruise for a time out, while a sheet-metal junkyard dog cleans up the planet, slowly returning it to its former status as a habitable zone. Who could resist this obvious appeal to our idyllic childhoods, when mom would tidy up our room while we went outside to play?

---

carbon dioxide and other greenhouse gases to Earth's atmosphere. It might therefore seem natural to conclude that our activity will cause Earth's climate to warm up. However, we also know from our study of Earth's climate regulation mechanisms that there are many feedback processes that could in principle make the reality less straightforward. As a result, the obvious starting point for the scientific study of global warming is to find out if it is indeed occurring as we might expect and, if so, whether we can attribute the warming to the carbon dioxide emissions of our civilization.

EVIDENCE OF RECENT WARMING You might think that determining whether our planet has been warming up during the past century or two would require nothing more than collecting temperature data from old newspapers. However, remember that global warming refers to an increase in the *average* temperature of our whole planet. We don't expect all localities to warm by the same average amount; indeed, it's possible that some places will get colder even as the planet as a whole warms up.

Today, orbiting satellites provide data that allow us to determine the global average temperature quite accurately, because they give us a view of our entire planet. We can validate these records with "ground truth" measurements recorded at more than 7000 weather stations around the world, along with ocean temperatures generally obtained by measuring the temperature of water collected by ships' intake valves. As a result, we have reliable temperature data for the approximately five decades for which we have satellite observations of our planet.

The data become somewhat less reliable for years prior to the satellite era. Getting a good estimate of the global average temperature requires having many local measurements from around the world, and fewer such measurements are available as we look deeper into the past. Moreover, most historic temperature records kept were for cities, which have tended to become warmer over time for reasons independent of global warming, such as through changes that occur as vegetation is paved over and as local utilities generate more heat. Scientists can often account for this "urban heat island" effect, but even then are left primarily with data on land temperatures and few records of temperatures over the oceans, which cover three-fourths of Earth's surface. As a result, scientists have devoted a lot of effort in the past few years to examining past temperature data in detail. Today, thanks to techniques of statistical analysis, it is possible to reconstruct a fairly reliable temperature history for most of the past two centuries, though the uncertainties become larger as we look further back.

Figure 10.9 shows the reconstructed history of Earth's global average temperature since 1880. Despite the uncertainties, the overall conclusion is clear: Global average temperatures have risen by about 0.85°C (1.5°F) in the past century. Moreover, most of the warming (about 0.6°C) has occurred in just the last 40 years, which is within the period for which the data are most reliable.

A temperature increase of 0.85°C may not sound like much, and it might not be so important if it were uniform everywhere. But some regions are warming much more than others because of differences in topography, wind direction, and other factors (Figure 10.10). In general, polar regions are warming much more than equatorial regions, and the Northern Hemisphere is warming more rapidly than the Southern Hemisphere. (The greater proportion of land in the north is responsible for this, because water takes a great deal of energy to either heat or cool.) As a result, glaciers are on the retreat around the world, and polar ice appears to be melting.

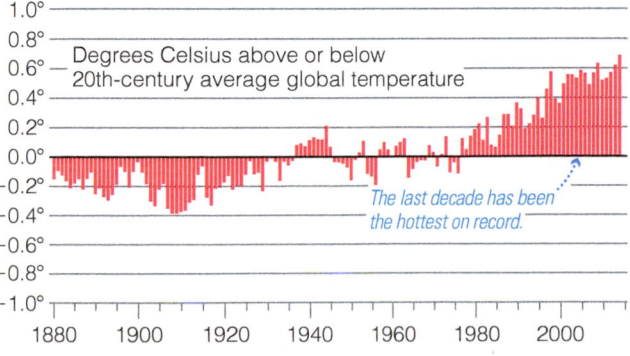

FIGURE 10.9

Average global temperatures from 1880 through 2014. Notice the clear global warming trend of the past few decades. Data from the National Climate Data Center.

FIGURE 10.10
This map shows how regional average temperatures for the period 2010–2014 (the most recent 5-year period at time of publication) compared to the averages from 1951–1980. Notice the large regional differences, and the particularly large warming in the north polar region.

Regional temperature change: 2010–2014 compared to 1951–1980

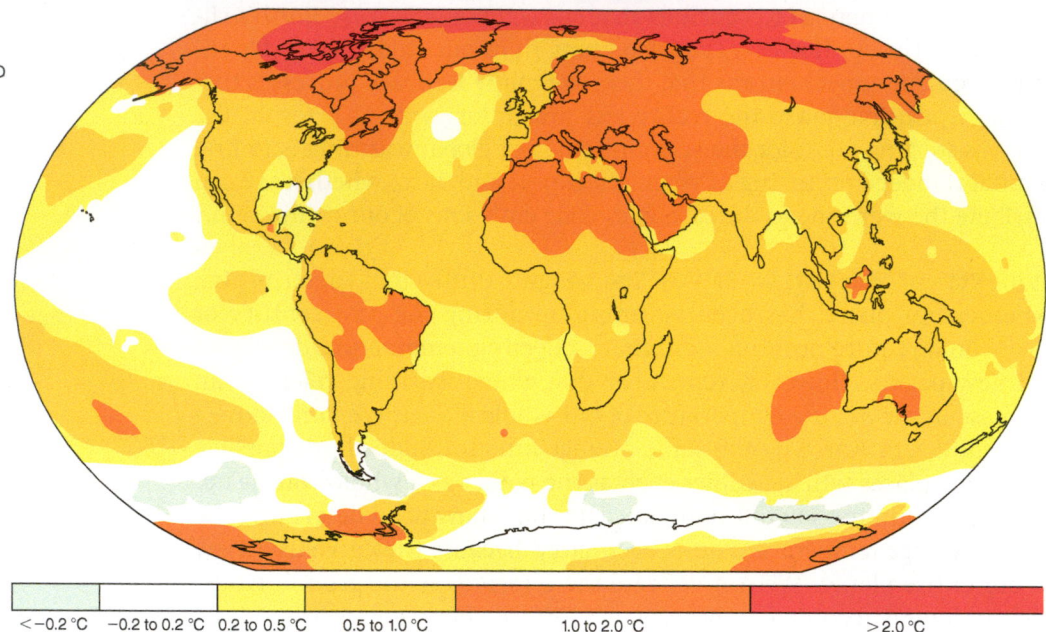

| <−0.2 °C | −0.2 to 0.2 °C | 0.2 to 0.5 °C | 0.5 to 1.0 °C | 1.0 to 2.0 °C | >2.0 °C |

CORRELATION OF WARMING WITH INCREASED CARBON DIOXIDE Given the clear evidence of warming over recent decades, the next question is whether this warming has been caused by the human release of carbon dioxide (or other greenhouse gases) into the atmosphere. To establish causality in any case like this, we generally proceed in two steps. First, we look for a *correlation*—in this instance, evidence that temperatures rise and fall with carbon dioxide concentration. If we find a correlation, we next ask whether the correlation implies causality—in this case, whether the temperature rise is *caused* by the carbon dioxide rise. Keep in mind that correlations do not automatically imply causality: Often, they are coincidental or attributable to some underlying commonality. For example, there is a correlation between the number of churchgoers in a city and the number of beer drinkers in a city, but that obviously doesn't mean that going to church causes people to drink beer (or that drinking beer causes people to go to church); instead, it just reflects the fact that both numbers tend to go up as population increases.

We can see evidence of a correlation by looking at past data for temperatures and carbon dioxide concentration. Direct measurements of carbon dioxide concentration have been made since 1958, giving us a clear record of the increase caused by human activity over a period of almost 60 years. In case you are wondering how we know that this recent rise in carbon dioxide concentration is due to human activity and not natural factors, there are at least three clear lines of evidence. First, the amount of the rise in the atmospheric concentration tracks almost perfectly with the amount of carbon dioxide being released by industry. Second, scientists can make very good estimates of changes in the amount of carbon dioxide from natural sources, such as volcanoes or release from the oceans, and these estimates show that natural factors can account for no more than about 1% of the rise in the atmospheric concentration. Third, and perhaps most convincingly, the ratio of the three carbon isotopes (carbon-12, carbon-13, and carbon-14) in fossil fuels differs from that

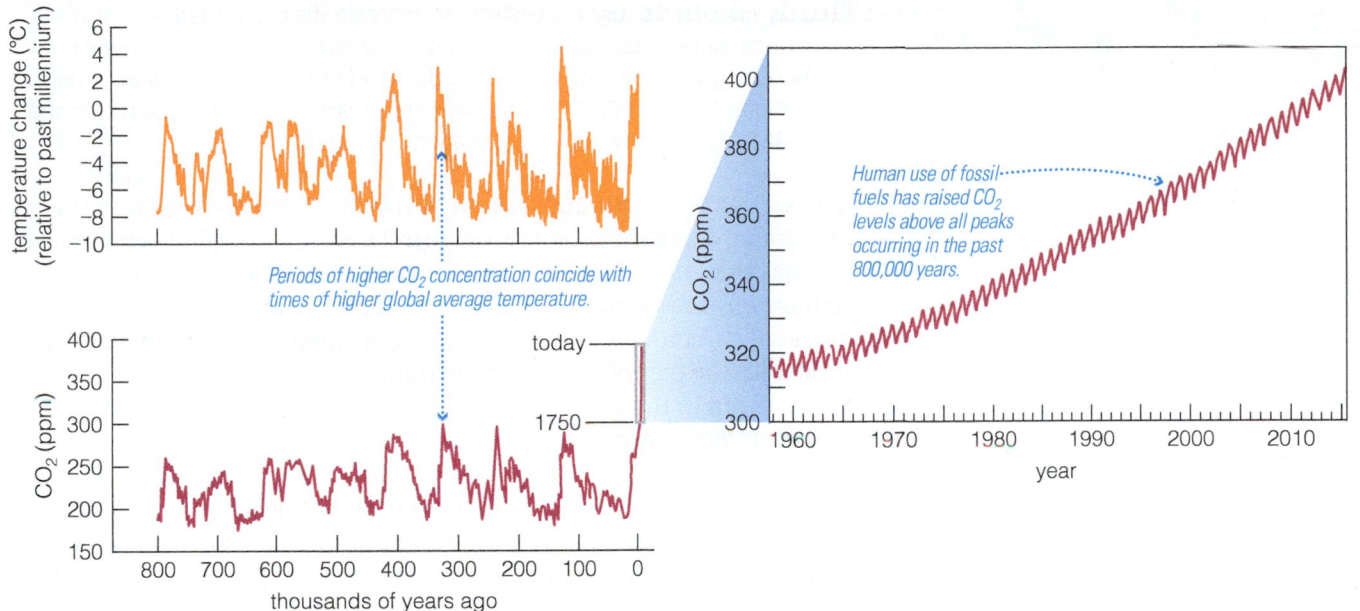

Periods of higher CO_2 concentration coincide with times of higher global average temperature.

Human use of fossil-fuels has raised CO_2 levels above all peaks occurring in the past 800,000 years.

FIGURE 10.11

This diagram shows the atmospheric concentration of carbon dioxide and global average temperature over the past 800,000 years. Notice that they vary together: More carbon dioxide goes with higher temperature, and vice versa. Data for the past few decades come from direct measurements (the up and down wiggles reflect annual season cycles); earlier data come from studies of air bubbles trapped in Antarctic ice core samples. The concentration is measured in parts per million (ppm), which is the number of CO_2 molecules among every one million air molecules. Data are from NOAA and the European Project for Ice Coring in Antarctica.

found in other carbon sources, and isotope measurements show clearly that the increase in atmospheric carbon dioxide is coming primarily from the burning of fossil fuels.

For longer-term data, scientists use ice cores drilled out of the Antarctic ice sheet. Ice cores are the frozen equivalent of sedimentary rock, made up of accumulated layers of ancient, compressed snow. Layering in ice cores shows the year-to-year history of snowfall, and counting the layers allows us to date the cores in much the same way we date trees from their rings. Trapped bubbles of air in the ice can be analyzed, and oxygen isotopes within them give clues to past temperatures. When temperatures are colder, the heavier isotopes aren't transported through the air as easily as when temperatures are warm. Therefore, by measuring the ratio of the heavier to the lighter isotopes at each layer of the ice cores, researchers can determine relative temperatures in the distant past. Similar analysis allows reconstruction of past levels of atmospheric carbon dioxide.

Today, the available ice core record allows scientists to reconstruct temperatures and carbon dioxide concentration going back about 800,000 years; Figure 10.11 shows these data along with the direct record of carbon dioxide concentration collected over the past few decades. Notice that there is indeed a correlation between temperature and carbon dioxide concentration: Periods of higher temperature tend also to be periods of higher carbon dioxide concentration. Moreover, the data show that both temperature and carbon dioxide concentration vary substantially and naturally with time. Average temperatures have risen or fallen by as much as 10°C several times during the last million years, and carbon dioxide concentration has varied naturally between about 180 and 290 carbon dioxide molecules per million molecules of air (ppm). Nevertheless, the recent rise in carbon dioxide concentration is clearly different from what has occurred naturally in the past: The atmospheric CO_2 concentration is now about 40% higher than it was before the industrial revolution began or at any other time in the past 800,000 years, and it is continuing to rise rapidly.

Think About It Use Figure 10.11 to estimate the current rate at which the carbon dioxide concentration is rising. If the rise continues at the same rate, how long will it be until we reach a doubling of the preindustrial-age concentration of 280 ppm? What if the rate of increase also rises, as it has over the past few years? Discuss the implications of your answer.

GLOBAL WARMING MODELS AND UNCERTAINTIES We have seen that a rising carbon dioxide concentration correlates with a rising temperature, but does that mean it *causes* the temperature rise? More to the point, do we have any reason to think that recent global warming is a result of the increased carbon dioxide put into the atmosphere by human activity, or could it just be part of the natural variation in temperature that we know has occurred in the past?

Given our understanding of the basic mechanism of the greenhouse effect, we cannot doubt that a continually rising concentration of greenhouse gases would eventually make our planet heat up. But for geologically short time scales—anything less than tens of thousands of years—there are feedback mechanisms that could potentially alter the cause and the effect. Scientists seek to understand the mechanisms and answer our questions about global warming by creating sophisticated computer models of the climate and comparing the model predictions to actual observations. If the models can correctly mimic the real climate, they can also be used to understand how much of the temperature increase is due to human activity rather than to natural factors, and to predict how the climate will change in the future as we continue to pump greenhouse gases into the atmosphere.

Earth's climate is incredibly complex, and many uncertainties remain in attempts to model the climate on computers. Nevertheless, today's models are the result of decades of work and refinement: Each time a model of the past failed to match real data, scientists sought to understand the missing (or incorrect) ingredients in the model and then tried again with improved models. While models may never be perfect, they now match real climate data quite well, giving scientists confidence that the models do indeed have predictive value. Figure 10.12 compares model data and real data. Notice that climate models that ignore human activity fail to match the observed rise in global temperatures. In contrast, climate models that include the enhanced greenhouse effect from human production of greenhouse gases match the observed temperature trend quite well. Figure 10.13 summarizes the evidence showing that global warming is a result of human activity.

Think About It Consider each piece of evidence summarized in Figure 10.13 individually, and then consider the evidence all together. Overall, how strong is the scientific case linking global warming to human activity? Defend your opinion.

What are the potential consequences of global warming?

As we have seen, there is virtually no scientific doubt that human activity is causing global warming. Therefore, the next question to ask is how it might affect us.

Climate models tell us that if current trends in the greenhouse gas concentration continue—that is, if we do nothing to slow our emissions of carbon dioxide and other greenhouse gases—the warming trend will

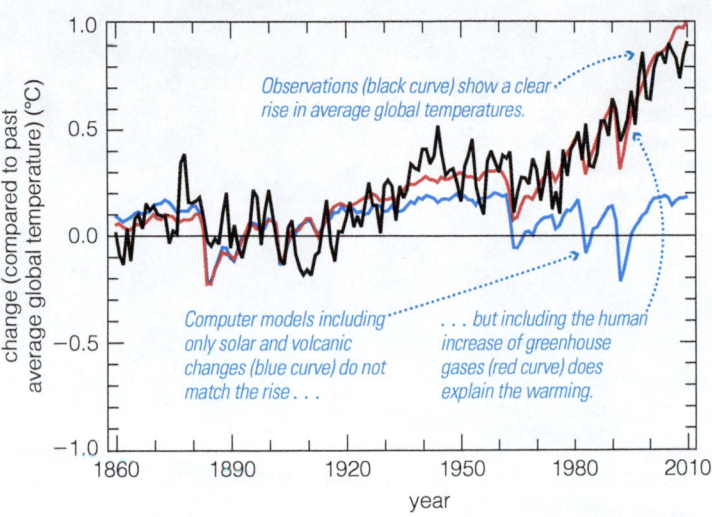

FIGURE 10.12
This graph compares observed temperature changes (black curve) with the predictions of climate models that include only natural factors such as changes in the brightness of the Sun and effects of volcanoes (blue curve) and models that also include the human contribution to increasing greenhouse gas concentration (red curve). (The red and blue model curves are averages of many scientists' independent models of global warming, which generally agree with each other within 0.1–0.2°C.)

continue to accelerate. By the end of this century, the global average temperature will be 2°C to 5°C (4°F to 10°F) higher than it is now, potentially giving our children and grandchildren the warmest climate that any generation of *Homo sapiens* has ever experienced.

The consequences of global warming are not simply hotter weather. A change in *average* global temperature is likely to mean much greater changes in local weather patterns. As we've discussed, some regions of Earth will warm more than others. We can also expect changes in rainfall patterns; for example, many scientists suspect that recent drought and wildfires in the southwestern United States are results of the changing climate, in which case we can expect these regions to experience more periods of prolonged drought and more wildfires in the future. Moreover, the general warming of the atmosphere means more total energy in the atmosphere and increased evaporation from the oceans, leading to more intense storms—perhaps including more frequent or more severe hurricanes. The same phenomena can also increase the severity of winter storms, leading to the somewhat surprising fact that global warming can cause more severe winter blizzards, a pattern that we may already be seeing in recent winters in the eastern United States.

Polar regions will warm the most, leading to increased melting of polar ice. This is clearly threatening to the species of these high-latitude regions (polar bears, who depend on an abundance of ice floes, are already under pressure), but the potentially greater threat is to sea level. Sea level has already risen about 20 centimeters in the past century because of the fact that water expands slightly as it warms. This "thermal expansion" by itself is expected to cause sea level to rise another 30 centimeters by about 2100, which could have devastating effects on coastal communities and low-lying countries such as Bangladesh. It also greatly increases the likelihood of devastating storm surges like the one that accompanied Hurricane Sandy in 2012. Melting ice may cause even greater increases in sea level. While melting of ice in the Arctic Ocean does not affect sea level—the ice is already floating—melting of landlocked ice does. Such melting appears to be occurring already. For example, the famous "snows" (glaciers) of Mount Kilimanjaro are rapidly retreating and may be gone within the next decade or so. More ominously, recent data suggest a surprisingly rapid change in Greenland's ice sheet, leading some scientists to worry that ice melt could cause sea level to rise as much as

Scientific studies of global warming apply the same basic approach used in all areas of science: We create models of nature, compare the predictions of those models with observations, and use our comparisons to improve the models. We have found that climate models agree more closely with observations if they include human production of greenhouse gases like carbon dioxide, making scientists confident that human activity is indeed causing global warming.

1 The greenhouse effect makes a planetary surface warmer than it would be otherwise because greenhouse gases such as carbon dioxide, methane, and water vapor slow the escape of infrared light radiated by the planet. Scientists have great confidence in models of the greenhouse effect because they successfully predict the surface temperatures of Venus, Earth, and Mars.

2 Human activity is adding carbon dioxide and other greenhouse gases to the atmosphere. While the carbon dioxide concentration also varies naturally, it is now much higher than it has been at any time in the previous million years, and it is continuing to rise rapidly.

The graph shows that today's CO_2 levels are higher than at any point in the past 800,000 years.

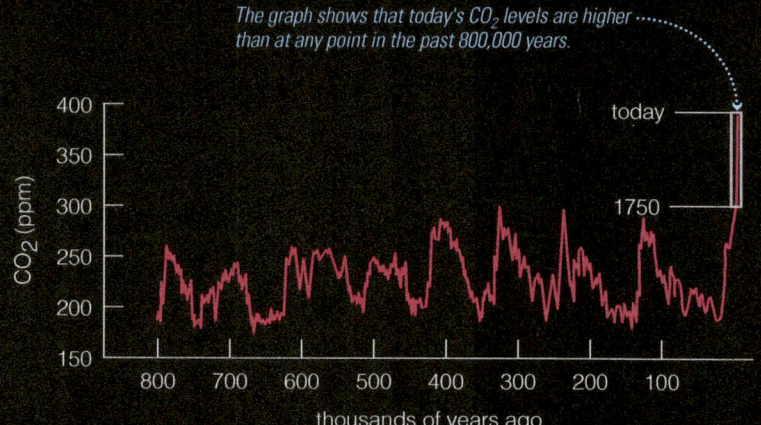

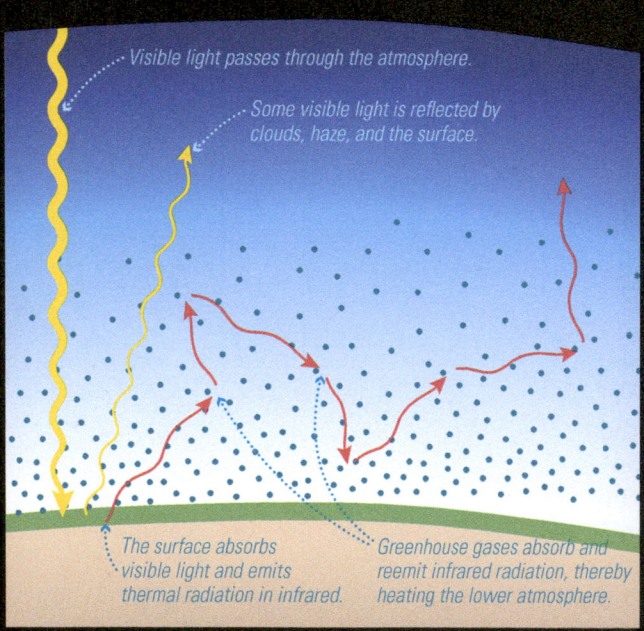

Visible light passes through the atmosphere.

Some visible light is reflected by clouds, haze, and the surface.

The surface absorbs visible light and emits thermal radiation in infrared.

Greenhouse gases absorb and reemit infrared radiation, thereby heating the lower atmosphere.

Global Average Surface Temperature

Planet	Temperature Without Greenhouse Effect	Temperature With Greenhouse Effect
Venus	−40°C	470°C
Earth	−16°C	15°C
Mars	−56°C	−50°C

This table shows planetary temperatures as they would be without the greenhouse effect and as they actually are with it (but no changes in reflectivity from ground and clouds). The greenhouse effect makes Earth warm enough for liquid water and Venus hotter than a pizza oven.

③ Observations show that Earth's average surface temperature has risen during the last several decades. Computer models of Earth's climate show that an increased greenhouse effect triggered by CO_2 from human activities can explain the observed temperature increase.

④ Models can also be used to predict the consequences of a continued rise in greenhouse gas concentrations. These models show that, without significant reductions in greenhouse gas emissions, we should expect further increases in global average temperature, rising sea levels, and more intense and destructive weather patterns.

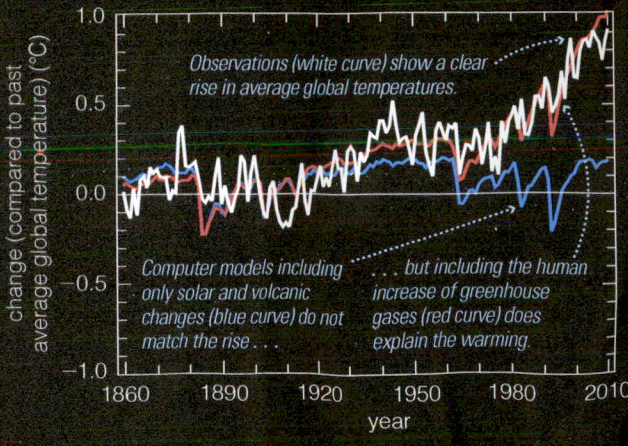

Observations (white curve) show a clear rise in average global temperatures.

Computer models including only solar and volcanic changes (blue curve) do not match the rise . . .

. . . but including the human increase of greenhouse gases (red curve) does explain the warming.

HALLMARK OF SCIENCE   **Science progresses through creation and testing of models of nature that explain the observations as simply as possible.** Observations showing a rise in Earth's temperature demand a scientific explanation. Models that include an increased greenhouse effect due to human activity explain those observations better than models without human activity.

Tallahassee

Orlando

Miami

Key West

This diagram shows the change in Florida's coastline that would occur if sea levels rose by 1 meter. Some models predict that this rise could occur within a century. The light blue regions show portions of the existing coastline that would be flooded.

several *meters*—enough to flood much of Florida—by the end of this century. Looking further ahead, scientists estimate that complete melting of the polar ice caps would increase sea level by some 70 meters (more than 200 feet). Although such melting would probably take centuries or millennia, it suggests the disconcerting possibility that future generations would have to send deep-sea divers to explore the underwater ruins of many of our major cities.

Other potential consequences of global warming are more difficult to predict. Some researchers worry that the fresh water entering the oceans from ice melting could alter major ocean currents; changes to the flow of the Gulf Stream, for example, could have drastic consequences for the climate of western Europe and parts of the United States. Ecological changes brought on by global warming could also have severe consequences for the well-being of human populations; for example, many researchers suspect that global warming would reduce agricultural production, global fish populations, fresh water availability, and the biodiversity that supports many critical forest ecosystems.

The bottom line is that we are in effect conducting a dramatic experiment on our planet in which we are increasing its greenhouse gas concentration far more than it would increase naturally. We do not know what the outcome of this experiment will be, or how easy or difficult it will be for us to deal with its consequences. The cases of Venus and Mars show that major climate change can occur even without any human intervention. While we are unlikely to do anything quite so dramatic to our own planet, we really do not know how our tampering might affect the finely balanced climate mechanisms upon which our civilization has been built.

Think About It If you were a political leader, how would *you* deal with the uncertain threat of global warming?

The Big Picture

PUTTING CHAPTER 10 IN PERSPECTIVE

In this chapter, we have tied together much of what we learned in other chapters to examine the concept of a habitable zone and its evolution over time. As you continue in your studies, keep in mind the following "big-picture" ideas:

- The *habitable zone* refers to the region around a star in which a planet of suitable size (often taken to be that of Earth) could have liquid water on its surface. Despite its name, the habitable zone is not the only region around a star where habitability is possible.

- Venus, Earth, and Mars may all have been habitable in the early history of the solar system. However, only Earth has remained habitable to this day. Understanding why the climates of Venus and Mars have changed and why

Earth's climate has remained comparatively stable can help us understand our climate and how we can protect it.

- The habitable zone is gradually migrating outward from the Sun. Eventually, it will lie beyond Earth's orbit, rendering our planet uninhabitable. Still, we can imagine our descendants, with sufficient technology, finding a way to remain within the habitable zone until the Sun ___es and then surviving by moving to other star systems.

- ___n that an advanced civilization could find a way to ___o_ as its star died and could certainly find ways to ___it s_e other natural threats such as asteroid impacts, ___on t_at nature will not impose intractable problems ___ that f__ions of years to come. If we do not survive, ___ actions __far more likely to be the result of our own ___ result of any natural catastrophe.

Summary of Key Concepts

10.1 The Concept of a Habitable Zone

How does a planet's location affect its prospects for life?
At any particular time, a star's **habitable zone** is the range of distances around it at which a planet could *potentially* have surface temperatures that would allow for abundant liquid water.

Could life exist outside the habitable zone?
There could be many worlds that have underground or underwater life fueled by energy other than sunlight, in which case these worlds need not be (and are unlikely to be) in habitable zones. However, it would be difficult to detect life on such worlds outside our solar system, and it seems unlikely that complex, intelligent life could arise in such environments.

10.2 Venus: An Example in Potential Habitability

Why is Venus so hot?

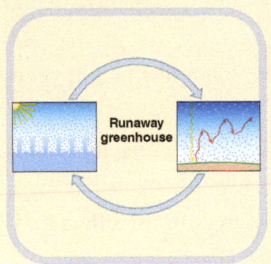

Venus's distance from the Sun ultimately led to a **runaway greenhouse effect:** Venus became too hot to develop (or keep) liquid oceans like those on Earth. Without oceans to dissolve outgassed carbon dioxide and lock it away in carbonate rocks, all of Venus's carbon dioxide remained in its atmosphere, creating its intense greenhouse effect.

Is Venus in the habitable zone?
Venus clearly is not in the habitable zone today, as any planet that once had Earth-like conditions would have suffered a runaway greenhouse effect. However, early in its history, when the Sun was some 30% dimmer than it is today, Venus may have been within the Sun's habitable zone and hence could have had rain, oceans, and perhaps even life.

10.3 Surface Habitability Factors and the Habitable Zone

What factors influence surface habitability?
According to present understanding, a planet can have a habitable surface only if it is within its star's habitable zone, is large enough to retain internal heat and have plate tectonics, and has enough of an atmosphere for liquid water to be stable on its surface.

Where are the boundaries of the Sun's habitable zone today?

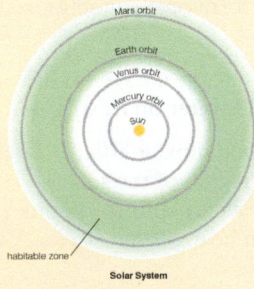

Under the most optimistic assumptions, the boundary currently extends from a distance of about 0.84 AU to 1.7 AU. Under more conservative assumptions, the boundary extends from about 0.95 AU to 1.4 AU.

10.4 The Future of Life on Earth

How will the Sun's habitable zone change in the future?
As the Sun ages, its luminosity gradually increases. As a result, the habitable zone gradually moves outward with time.

How long can life survive on Earth?

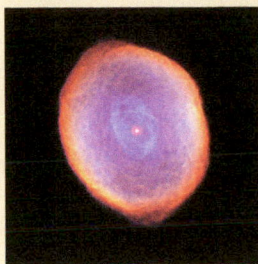

At minimum, Earth should remain habitable for another several hundred million years. By about a billion years from now, a **moist greenhouse effect** could cause Earth's oceans to evaporate away, though natural feedback processes might prevent this from occurring so soon. In 3–4 billion years, the Sun will become bright enough that our planet will certainly be subject to a runaway greenhouse effect, ending surface habitability.

✳ **THE PROCESS OF SCIENCE IN ACTION**

10.5 Global Warming

What is the evidence for global warming?

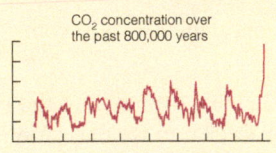

Measurements show that human activity is causing a substantial increase in the atmospheric concentration of CO_2. The well-understood mechanism of the greenhouse effect suggests that this increase should lead to an increase in the global average temperature, and such an increase has indeed been observed over the past century. Climate models indicate that this temperature increase is due primarily to the human contribution to global warming, rather than to natural factors.

What are the potential consequences of global warming?

Continued global warming could raise the average worldwide temperature by 2°C to 5°C during this century. Regional climate changes will be greater, and we can expect increased polar melting and a rise in sea level. Additional heat should increase ocean evaporation, which may lead to more numerous and more intense storms. Many other serious effects could also occur, though precise consequences are difficult to predict.

Exercises and Problems

MasteringAstronomy® *For instructor-assigned homework and other learning materials, go to MasteringAstronomy®.*

REVIEW QUESTIONS

Short-Answer Questions Based on the Reading

1. What is a *habitable zone*, and how is the idea useful? Is a planet in the habitable zone necessarily habitable? Explain.

2. Describe several ways in which it may be possible to have habitability outside the habitable zone.

3. Why do we think that Venus should have outgassed amounts of carbon dioxide and water vapor similar to those outgassed by Earth? Where is Venus's carbon dioxide today? Where is Earth's?

4. How much water is present on Venus today? How do we think that Venus lost water, and what evidence supports the idea that Venus really did lose water this way?

5. What is a *runaway greenhouse effect*, and why did it occur on Venus but not on Earth? What does this fact tell us about the inner boundary of the Sun's habitable zone?

6. Could Venus ever have had oceans and, if so, could we find geological evidence that they existed? Explain.

7. How do we expect the habitable zones of brighter stars to compare to that of the Sun?

8. Why is planetary size important to habitability? What does the case of Mars tell us about the minimum size? Can we draw any conclusions about size from the case of Venus? Explain.

9. What factors besides size and distance from the Sun might influence habitability?

10. What factors affect the location of the inner boundary of the habitable zone? Be sure to explain and consider the role of a possible *moist greenhouse effect* in such calculations.

11. What factors affect the location of the outer boundary of the habitable zone? Briefly summarize the current boundaries of our Sun's habitable zone under both the more optimistic and the more conservative scenario.

12. Why does the Sun gradually brighten, and how does this brightening affect the location of the habitable zone over time? What do we mean by a *continuously habitable zone*?

13. How and when will Earth become uninhabitable? Why? Could humans still survive? Explain.

14. Briefly describe the eventual fates of the Sun and of the universe, and what these fates might mean to our descendants (if anyone survives that long).

15. How do we determine global average temperatures from the past? What do the data show?

16. What do ice core data tell us about the past climate and the role of greenhouse gases? How does the current concentration of atmospheric carbon dioxide compare to the concentration over the past 800,000 years?

17. What is the role of climate modeling in understanding global warming?

18. Describe several potential consequences of global warming, and how they might affect our civilization.

TEST YOUR UNDERSTANDING

Does It Make Sense?

Decide whether each statement makes sense or does not make sense. Explain clearly; because not all of these have definitive answers, your explanation is more important than your chosen answer.

19. If Venus were just a little bit smaller, its climate would be Earth-like.

20. Venus is not in the habitable zone now, but it may have been in the past.

21. Venus is not in the habitable zone now, but a few billion years from now it will be.

22. If we could somehow start plate tectonics on Venus, its surface would cool and it would regain the oceans it had in the past.

23. Mars will someday undergo a runaway greenhouse effect and become extremely hot.

24. We are not yet certain, but it is quite likely that Earth has suffered through a runaway greenhouse effect at least once in the past 4 billion years.

25. While the habitable zone of the Sun migrates outward over time, the habitable zones of other Sun-like stars might instead migrate inward over time.

26. If Earth someday becomes a moist greenhouse, it will have a climate that is humid but still quite comfortable.

27. Global warming can lead to more intense winter blizzards.

28. Earth has been warmer in the past than it is today, so global warming is nothing to worry about.

Quick Quiz

Choose the best answer to each of the following. Explain your reasoning with one or more complete sentences.

29. The habitable zone refers to (a) the regions of a planet where good weather allows life to exist; (b) the range of distances from a star where a planet's surface temperature is always above the freezing point of water; (c) the range of distances from a star within which water could exist in liquid form on a suitably sized planet.

30. A planet that is *not* within a habitable zone *cannot* have (a) life; (b) subsurface oceans; (c) abundant liquid water on its surface.

31. Venus's atmosphere has much more carbon dioxide than Earth's because (a) Venus was born in a region of the solar system where more carbon dioxide gas was present; (b) Venus lacks oceans in which carbon dioxide can be dissolved;

(c) Venus has volcanoes that outgas much more carbon dioxide than volcanoes on Earth.

32. What is the likely reason for Venus's lack of water in any form? (a) The planet accreted little water during its birth. (b) The water is locked away in the crust. (c) The water was in the atmosphere, where molecules were broken apart by ultraviolet light from the Sun.

33. If Earth were to be moved to Venus's orbit, it would probably (a) stay about the same temperature, thanks to the small amount of CO_2 in Earth's atmosphere; (b) become a tropical paradise; (c) suffer a runaway greenhouse effect and become even hotter than Venus is today.

34. The inner boundary of the Sun's habitable zone today is (a) inside the orbit of Venus; (b) between Venus and Earth; (c) outside the orbit of Earth.

35. As the Sun ages, the habitable zone will (a) move outward and grow wider; (b) move outward but get narrower; (c) stay about the same as it is now.

36. Which of the following could cause Earth to become uninhabitable in about 1 billion years? (a) a moist greenhouse effect (b) a runaway greenhouse effect (c) the death of the Sun

37. Global warming means that (a) Earth's average temperature is increasing; (b) every place on Earth is getting warmer; (c) Earth will soon have a greenhouse effect.

38. The current concentration of atmospheric carbon dioxide on Earth is (a) higher than the concentration at any time during the past 800,000 years; (b) higher than the concentration in the past century, but lower than at many other times during the past 800,000 years; (c) gradually rising, but only about average for the time period during which we have ice core data.

PROCESS OF SCIENCE

39. *Science with Consequences.* A small but vocal group of people still dispute that humans are causing global warming. Do some research to find the basis of their claims. Then defend or refute their findings based on your own studies and your understanding of the hallmarks of science discussed in Chapter 3.

40. *The Habitable Zone.* Considering the possibility that life might exist on worlds outside of a star's habitable zone, do you think the concept of a "habitable zone" still makes scientific sense? Defend your opinion.

GROUP WORK EXERCISE

41. *Are We Causing Global Warming?* **Roles:** *Scribe* (takes notes on the group's activities), *Advocate* (argues in favor of the claim that human activity is causing global warming), *Skeptic* (points out weaknesses in the arguments made by the *Advocate*), *Moderator* (leads group discussion and makes sure everyone contributes). **Activity:**
    a. Work together to make a list of scientific observations that have been proposed as evidence that humans are causing global warming. Your list should include, but need not be limited to, the evidence in Figures 10.9–10.13.

b. *Advocate* presents the case that humans are causing global warming, drawing on the evidence from part a.
c. *Skeptic* attempts to refute *Advocate's* case using scientific arguments.
d. After hearing these arguments, *Moderator* and *Scribe* decide whose arguments are more persuasive and explain their reasoning.
e. Each person in the group writes up a summary of the discussion.

INVESTIGATE FURTHER

In-Depth Questions to Increase Your Understanding

Short-Answer/Essay Questions

42. *Are Habitable Zone Planets Common?* Based on what you have learned so far about solar system formation and habitable zones, do you think we should expect to find many planets in habitable zones? Explain.

43. *No Plate Tectonics.* Suppose plate tectonics magically stopped on Earth, but other geological processes (such as volcanism) continued. Would Earth's *surface* get warmer or cooler? Explain.

44. *Continuously Habitable Zone.* Is Earth in a zone that will remain continuously habitable from the Sun's birth to its death? Is any planet? Explain.

45. *Planetary Changes.* Write two or three paragraphs explaining why each of the planets described below would or would not be habitable today.
    a. A planet the size of Mars located at the distance of Venus
    b. A planet the size of Mars located at the distance of Earth
    c. A planet the size of Venus located at the distance of Earth
    d. A planet the size of Earth located at the distance of Mars

46. *A Billion Years.* At minimum, it appears that our planet will remain habitable for at least the next billion years, give or take a couple hundred million years. How long is a billion years? Think of some ways to put this time period into perspective.

47. *Venus's History.* Many people are not surprised to learn that Venus is hotter than Earth, given that it is closer to the Sun. Explain why we cannot attribute its heat to distance from the Sun alone. How do we explain Venus's high temperature?

48. *Habitable Moons.* As we'll discuss in Chapter 11, some of the newly discovered extrasolar planets are Jupiter-like in size but are located at Earth-like distances from Sun-like stars. These planets are unlikely to be habitable themselves. Could they have moons with habitable surfaces? Explain.

49. *Greenhouse Lessons.* While it seems unlikely that human activity could cause a runaway greenhouse effect on Earth, we could still cause the climate to warm substantially. Do you think we can learn anything valuable about our potential effects on Earth's climate by studying the climate histories of Venus and Mars? If so, what? Defend your opinion.

50. *Global Warming.* Briefly summarize the evidence suggesting that global warming is occurring and is a result of human activity. Then write a short essay outlining what, if anything, we should do about it.

Quantitative Problems

Be sure to show all calculations clearly and state your final answers in complete sentences.

51. *Stellar Habitable Zone.* Consider a star slightly smaller and less luminous than the Sun, with a habitable zone that extends from 0.3 to 0.5 AU. How does the size (radial width) of this star's habitable zone compare to that of the Sun, and what are the implications for the likelihood of finding planets within the zone?

52. *Massive Stellar Habitable Zone.* Consider a star that is more massive and more luminous than the Sun, with a habitable zone that extends from 2.5 to 4 AU. How does the size (radial width) of this star's habitable zone compare to that of the Sun? Should we consider life to be likely around such a star? Explain.

53. *Strength of Sunlight at Earth.* The power of sunlight reaching the top of Earth's atmosphere is 1370 watts per square meter. The amount of power flowing outward through Earth's surface caused by radioactivity within Earth is estimated to be 3 trillion watts. Which one—sunlight or radioactive heat—is providing more energy to Earth's surface, and by how much? What does your answer tell you about the relative importance of these two energy sources for life? (*Hint:* You'll need to convert the radioactive power number from a total to an amount per square meter of surface; the surface area of a sphere of radius r is $4\pi r^2$.)

54. *Strength of Sunlight at Venus and Mars.* The solar energy reaching the top of Earth's atmosphere is 1370 watts per square meter. What is the comparable energy (a) at the distance of Venus; (b) at the distance of Mars? (*Hint:* Remember that light follows an inverse square law [see Figure 7.2]; you'll need to look up distances in AU for Venus and Mars.)

55. *Energy Use by Cars.* There are approximately 1 billion automobiles in use worldwide. If the average auto runs 1 hour a day, at 50 horsepower, what is the resultant total average power use by cars, in watts? How does this amount of power compare to the amount Earth gets from sunlight? Note that 1 horsepower is 746 watts and 1 watt = 1 joule/second. (*Hint:* You'll first need to figure out the average number of cars running at any one time, and once you find the total power for the cars, you'll need to divide by Earth's surface area in square meters so that you can compare it to the solar power given in Problem 53.)

56. *Atmospheric Mass of Venus.* The atmospheric pressure on Venus is 90 bars. What is the total mass of Venus's atmosphere? You may use the fact that 1 bar is the pressure exerted by 10,000 kilograms pushing down on a square meter in Earth's gravity, and assume that Venus's gravity is essentially the same as Earth's.

57. *Melting Greenland.* Greenland is approximately 700 × 2400 km in size, with an average ice cover that's about 1.5 kilometers thick. Suppose all the Greenland ice were to melt. By approximately how much would this raise the level of Earth's oceans? Assume that oceans cover 70% of Earth, and that water and ice are approximately the same density. (*Hint:* Dividing the volume of melted water by the surface area of the oceans will tell you how much the ocean depth will increase.)

Discussion Questions

58. *The Fate of Life in the Universe.* Consider the evidence suggesting that life is just a fleeting phase in the long-term history of the universe. Assuming this to be the case, how do you think it should influence our perspective on our own place in the universe? Why?

59. *The Politics of Global Warming.* The current scientific case for global warming seems quite strong, but the topic nonetheless generates significant political controversy. Why do you think that is the case? Do you consider global warming to be primarily a scientific issue or a political issue? Explain.

60. *Dealing with Uncertainty.* One of the difficulties in deciding what to do about global warming is the fact that its precise consequences are uncertain. In general, how do you think we as a society should deal with issues whose consequences are potentially severe but highly uncertain? How would you deal with this uncertainty in the particular case of global warming? Explain.

WEB PROJECTS

61. *Global Warming Scenarios.* Research data showing how the amount of global warming might differ depending on whether we decrease, keep at current levels, or increase our future carbon dioxide emissions. Summarize your findings.

62. *Global Warming Skeptics.* Compare the arguments from websites that claim that (a) global warming is caused by fossil fuel use and (b) there is no convincing proof that this is the case. How would you evaluate these arguments?

63. *Long-Term Survival.* Read about some exotic ideas concerning how advanced civilizations might survive as the universe grows cold and dark. Do you think any of these ideas make sense, or are they just wishful thinking? Write a short essay describing one of these ideas and your opinion of it.

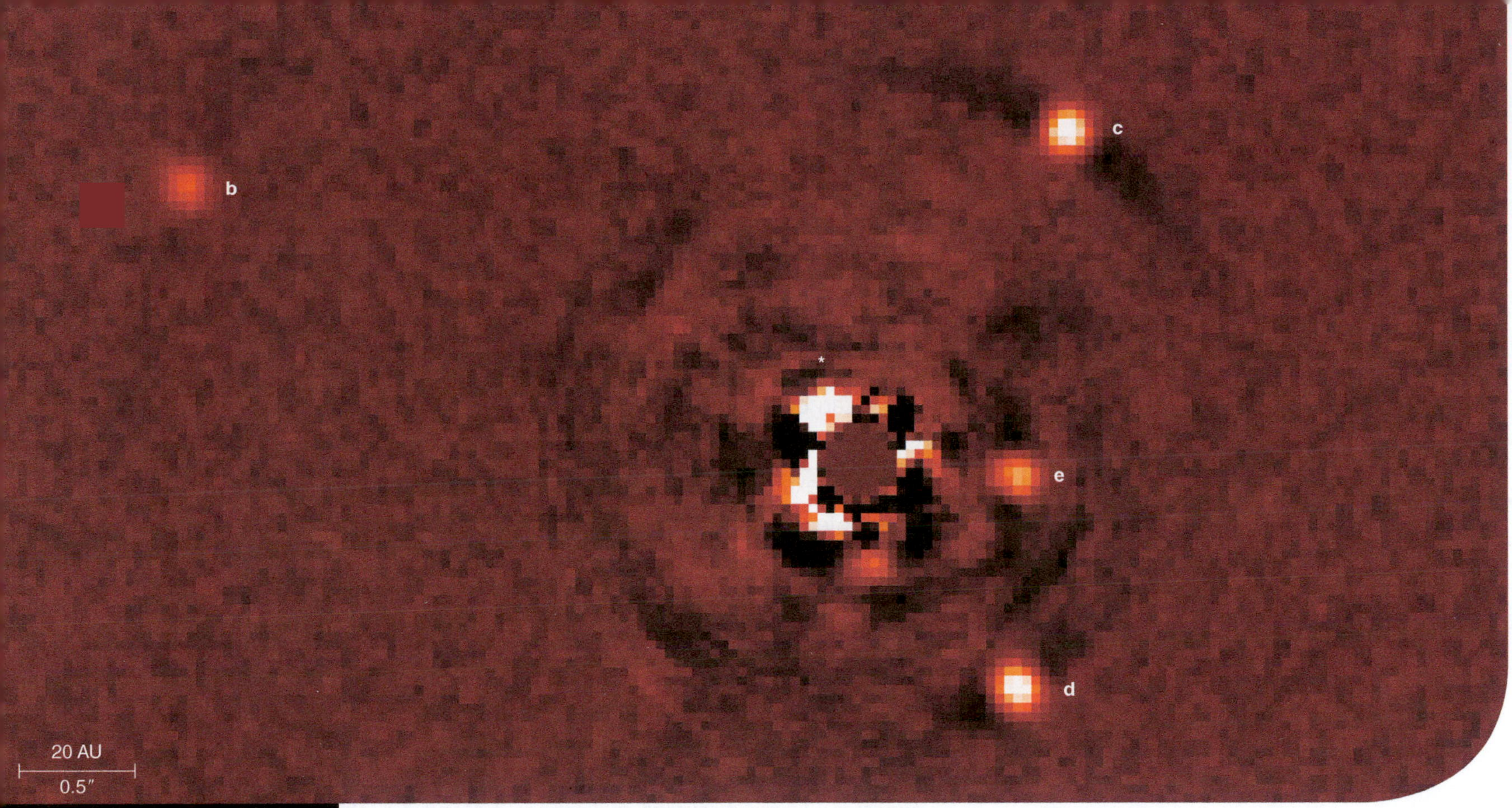

20 AU
0.5"

11 Extrasolar Planets: Their Nature and Potential Habitability

LEARNING GOALS

11.1 DISTANT SUNS
- How do other stars differ from the Sun?
- Which stars would make good suns?

11.2 DISCOVERING EXTRASOLAR PLANETS
- How do we detect planets around other stars?
- What properties of extrasolar planets can we measure?

11.3 THE NATURE OF EXTRASOLAR PLANETS
- How do extrasolar planets compare with planets in our solar system?
- Are planetary systems like ours common?

11.4 THE HABITABILILTY OF EXTRASOLAR PLANETS
- What kinds of extrasolar worlds might be habitable?

- How could we detect life on extrasolar planets?
- Are Earth-like planets rare or common?

 THE PROCESS OF SCIENCE IN ACTION

11.5 CLASSIFYING STARS
- How did we learn to classify stars?
- What is the Hertzsprung–Russell diagram?

▲ **About the photo:** This infrared image from the Large Binocular Telescope shows direct detection of four planets (marked b, c, d, e) orbiting the star HR 8799. Light from the star itself (center) was mostly blocked out during the exposure, as indicated by the solid red circle.

Earth is the only planet known to harbor life, but several other worlds in our solar system seem potentially habitable. If our solar system is typical, the total number of habitable worlds in our galaxy must be enormous. But is this really the case?

As recently as 1995, we did not know of any planets beyond those of our own solar system, and therefore had no way to address the question of whether habitable worlds are common. This situation has changed dramatically over the past two decades, which have seen the discovery of thousands of extrasolar planets, and we now have enough statistical data that we can begin to assess the prospects for finding habitable worlds around other stars.

In this chapter, we will explore the exciting and rapidly advancing science of extrasolar planets and their potential habitability. This discussion will then help us frame the questions that influence the search for extraterrestrial intelligence, a topic we'll turn to in the next chapter.

11.1 Distant Suns

The discovery in the 1990s of planets around other stars was not unexpected. As we discussed in Sections 3.4 and 3.5, the nebular theory of solar system formation predicted that extrasolar planets should be common, and their discovery provided further validation of this theory. But not all stars will necessarily have planets, and most planets are not expected to be habitable. Therefore, in a search for life beyond our solar system, the first step is to understand what types of stars might make good "suns" that could provide a long-term source of light and heat to support habitable worlds.

How do other stars differ from the Sun?

If all stars were exactly like the Sun, we might expect any star to be a possible sun to life on orbiting planets. However, while the Sun is a fairly average star in terms of size and brightness, there are enough differences among stars that we must first understand the range of stellar properties before we can know where it makes the most sense to look for habitable worlds and possible life.

STELLAR COMPOSITION Recall that all stars are born from the gravitational collapse of interstellar clouds of gas and dust (see Figures 3.8 and 3.34). The compositions of stars at birth therefore reflect the composition of interstellar clouds, and measurements show that these clouds have approximately the same composition throughout the universe: roughly three-quarters hydrogen and one-quarter helium (by mass), with other elements representing only a very small proportion of the total mass.

The reason for this general pattern of composition goes back to the history of the universe. As we discussed in Chapter 3, the early universe contained only the chemical elements hydrogen and helium, and it contained them in the proportions 75% hydrogen and 25% helium (by mass). All other elements are "star stuff" that was produced later by stars and released into space when they died, which explains why these other elements are comparatively rare. This fact also explains why stars vary in their proportions of these other elements. Very old stars were born when the universe was much younger, which means there had been few prior generations of stars to produce elements besides hydrogen and helium. Observations confirm that such stars were born with very small proportions of these other elements; for the oldest stars, this proportion was less than about 0.1% of their mass. In contrast, stars that formed more recently in the history of the universe—including our own Sun, which came into being about $4\frac{1}{2}$ billion years ago when the universe was already more than 9 billion years old—were born from gas clouds that had as much as 2% of their mass in the form of elements heavier than helium.

Despite the fact that these "heavy elements" are so rare, they are very important to gauging the prospects for planets and life. Look back at Figure 3.36, which summarizes the way planets formed in our solar system. Notice that the planet formation process begins with the condensation of solid seeds of metal, rock, or ice, all of which require material besides hydrogen and helium. In other words, there must be some minimum critical amount of other elements that is necessary for the formation of planets, particularly of rocky planets like Earth. No one knows exactly what this minimum proportion is. The roughly 2% in our own solar system is clearly sufficient, but there is great debate about how much lower the proportion can be and still allow for planets. We will return to this question shortly, but let's first look at other ways in which stars can differ from our Sun.

STELLAR TYPES The relatively small differences in the proportions of "other elements" among stars may be important to planet formation, but they do not have major effects on the general lives of stars. All stars are made mostly of hydrogen and spend most of their lives generating energy by fusing hydrogen into helium (see Figure 3.10). Because of these commonalities, the major differences among stars are traceable to their different masses.

Mass is the critical factor because it determines the strength of a star's self-gravity, and that in turn determines the central temperature and density of the star. The more massive the star, the more vigorously fusion proceeds in the core. Both observations and theory show that stars can range in mass from about 8% of the mass of the Sun to more than 150 times the mass of the Sun. The reason for the minimum mass is that below that value (0.08 solar mass), the core of the "star" never becomes hot enough to sustain hydrogen fusion; objects that are star-like but below this minimum mass are known as *brown dwarfs*. The maximum mass is less precisely known, but at above about 150 solar masses, a star generates energy at such a prodigious rate that it will essentially blow itself apart.

We cannot see into stellar cores to measure the fusion rate directly, but for stars that are in the long hydrogen-fusing stages of their lives, there are two key differences among stars of different masses that *can* be observed from outside (Figure 11.1):

1. Hydrogen-fusing stars of different masses come in different colors, and these colors reflect differences in surface temperature. Recall that higher-temperature objects emit most of their radiation at shorter wavelengths (see Figure 3.27), so hotter stars are bluer in color (and actually emit most of their light as ultraviolet) and

FIGURE 11.1 INTERACTIVE PHOTO
Stars of different masses differ in color and luminosity. Most of the stars in this Hubble Space Telescope photo are at roughly the same distance (about 25,000 light-years away), so the differences in brightness reflect real differences in luminosity. Notice that most of the yellow and red stars are far less luminous than the white and blue stars, because less massive stars are redder and less luminous during their hydrogen-fusing lifetimes. The very luminous red stars in the photo are red giants and supergiants, which are stars that have already exhausted their core hydrogen and have become much more luminous in their final stages of life.

cooler stars are redder in color. Stellar surface temperature also determines whether various atoms in the star are neutral or ionized, which means that stars of different temperatures show different sets of spectral lines. For historical reasons that we'll discuss in Section 11.5, astronomers categorize stars according to a **spectral type** that corresponds to surface temperature, using letters that go from hottest to coolest in the order OBAFGKM. Roughly speaking, stars are blue-white in color on the hot end of the spectral sequence (spectral types O, B) and red in color on the cool end (spectral type M).

2. Hydrogen-fusing stars also vary in their intrinsic brightness, or **luminosity:** More massive stars are more luminous because they generate energy through fusion at a higher rate.

Table 11.1 summarizes the properties of hydrogen-fusing stars. The first column lists the spectral types, and the next three columns list typical surface temperatures, luminosities, and masses for hydrogen-fusing stars of each spectral type. Notice that these stars come in a much larger range of luminosities than masses.

The relationship between mass and luminosity also leads us to conclusions about the lifetimes of stars of different masses. The high luminosities of the most massive stars imply that they generate energy through fusion at such a rapid rate that, even though they are born with more hydrogen than the Sun, they use it up much more quickly and hence have much shorter lifetimes than the Sun. Conversely, the least massive stars use their hydrogen to generate energy so much more slowly than the Sun that they will live far longer. The fifth column of Table 11.1 shows the typical hydrogen-fusing lifetimes of stars of different spectral types.

The fact that more massive stars live much shorter lives means that even if stars of different masses were all born in about the same numbers, lower-mass stars would be much more common (because most of the higher-mass stars would have long since died off). In fact, observations show that high-mass stars are born in much smaller numbers than low-mass stars when interstellar clouds collapse under gravity to form stars. Together, the differing lifetimes and differing initial numbers lead to the result shown in the final column of Table 11.1, which lists the approximate percentage of all stars that fall into each spectral type. Notice that although the Sun is roughly in the middle of the stellar mass range, about 90% of all stars are lower in mass—and hence are redder and less luminous—than the Sun.

TABLE 11.1 Typical Properties for Hydrogen-Fusing Stars of the Seven Major Spectral Types

Numbers given in solar units are values in comparison to the Sun; for example, a mass of 60 solar units means 60 times the mass of the Sun.
Note that the Sun is a G star. (More precisely, the Sun's spectral type is G2.)

Spectral Type	Surface Temperature (°C)	Luminosity (solar units)	Mass (solar units)	Lifetime (years)	Approximate Percentage of Stars in This Class
O	50,000	1,000,000	60	500 thousand	0.001%
B	15,000	1000	6	50 million	0.1%
A	8000	20	2	1 billion	1%
F	6500	7	1.5	2 billion	2%
G	5500	1	1	10 billion	7%
K	4000	0.3	0.7	20 billion	15%
M	3000	0.003	0.2	600 billion	75%

Think About It As we'll soon discuss, we have good reason to believe that many types of stars could host life. But suppose it turned out that only G stars (like the Sun) could be orbited by habitable worlds. Approximately how many stars in the Milky Way Galaxy would potentially be orbited by habitable worlds in that case? Explain.

STELLAR LIFE CYCLES Stars do not last forever. They are born, live, and die. A star is "born" when hydrogen fusion begins in its central core, and it lives as long as it can generate energy through fusion. A star "dies" when it finally ceases to produce energy by any kind of fusion.

All stars spend most of their lives (about 90% of the time from star birth to star death) fusing hydrogen into helium. Stars shine fairly steadily during this period, brightening gradually as they age (for the same reason the Sun is slowly brightening [Section 10.4]). As shown in Table 11.1, the most-massive stars fuse hydrogen for less than a million years, while the least-massive stars will still be fusing hydrogen when the universe is many times its current age of about 14 billion years.

What happens to a star when it finally runs out of hydrogen to fuse in its central core? The details vary with mass, but the general ideas are the same for all stars. Perhaps surprisingly, a star that exhausts its core hydrogen begins to grow larger and brighter, becoming a *giant* or *supergiant* star. For example, when our own Sun becomes a *red giant* some 4–5 billion years from now [Section 10.4], it will grow so large that it will engulf some of the inner planets. The reason an aging star grows so large and luminous is traceable to changes that occur deep in the core. During the time that the central core fuses hydrogen, the energy generation helps the core resist the crush of gravity and maintain its size. This resistance disappears when the hydrogen in the central core runs out, and as a result the core shrinks in size and increases in temperature. The core may ultimately become so hot that it begins to fuse helium or heavier elements, but in the meantime a layer surrounding the central core—a layer that still contains unused hydrogen—ignites with nuclear fusion. This layer becomes so hot that the total rate of fusion is higher than it was while the central core fused hydrogen, and the energy released by this fusion causes the star's outer layers to expand in size and emit more light.

Stars can begin to fuse heavier elements during their giant or supergiant stage. Our own Sun will someday produce carbon by fusing the helium in its core, and more-massive stars ultimately create all the other elements through a combination of core fusion reactions and reactions that occur in their final death throes. As we've discussed, this stellar manufacturing explains the existence of all the elements in the universe except the original hydrogen and helium produced by the Big Bang, which is why we say we are made of "star stuff."

Eventually, a star will reach a point at which it can fuse no other elements, and at that point the star dies. Relatively low-mass stars like our Sun end their lives comparatively gently, ejecting their outer layers into space as planetary nebulae and leaving behind the type of dead star that we call a *white dwarf* (see Figure 10.8). High-mass stars die in titanic explosions called *supernovae* (see Figure 3.11). A supernova blasts the star's outer layers into interstellar space, accompanied by enormous amounts of radiation, while the star's remaining core collapses to form a stellar remnant that may be either a *neutron star* or a *black hole*.

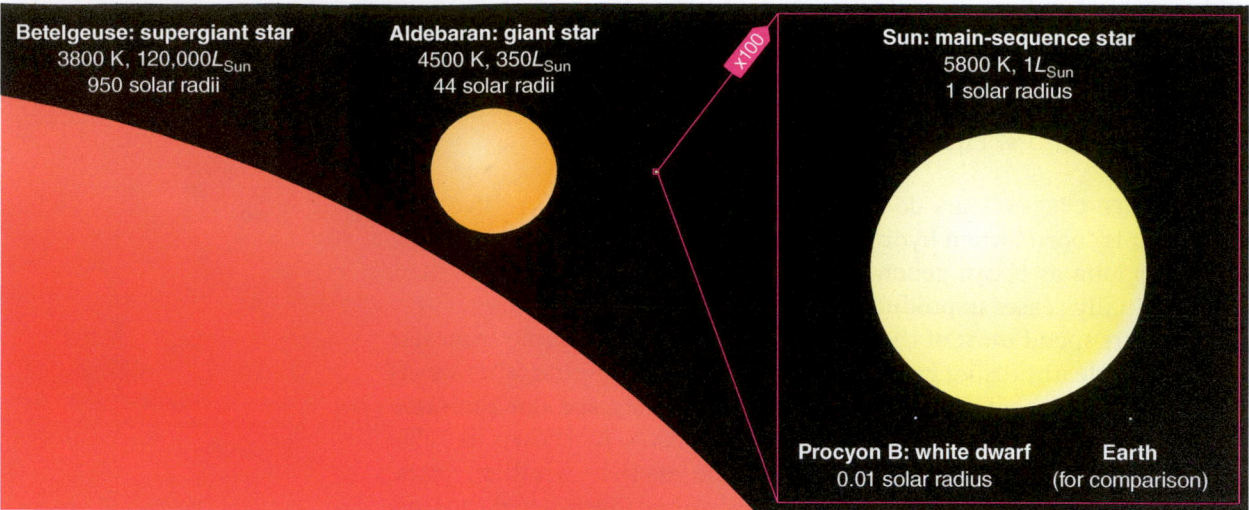

Betelgeuse: supergiant star
3800 K, 120,000L_{Sun}
950 solar radii

Aldebaran: giant star
4500 K, 350L_{Sun}
44 solar radii

x100

Sun: main-sequence star
5800 K, 1L_{Sun}
1 solar radius

Procyon B: white dwarf
0.01 solar radius

Earth
(for comparison)

FIGURE 11.2

The relative sizes of stars. A supergiant like Betelgeuse would fill the inner solar system and extend more than 80% of the way to Jupiter's orbit. A giant like Aldebaran would fill the inner half of Mercury's orbit. The Sun is a hundred times as large in radius as a white dwarf like Procyon B, which is roughly the same size as Earth.

From the standpoint of planets and life, the key idea from stellar life cycles is that stars tend to shine steadily for long periods of time only during their core hydrogen–fusing lifetimes. After that, a star goes through dramatic changes in its surface temperature, luminosity, and even size (Figure 11.2), all of which are likely to have impacts on any orbiting planets and life. Indeed, it seems almost certain that any surface life on orbiting planets would be extinguished either during a star's final stages of life or when the star ultimately died. It is possible, though, that subsurface life might survive, particularly if it obtained energy from a planet or moon's internal heat rather than sunlight (as might be possible on worlds like Europa or Enceladus, for example).

STELLAR COMPANIONS AND NEIGHBORS Our solar system has only one star, the Sun, and we live in a part of the galaxy where great distances separate us from other stars. The same is not true of all other star systems, and this fact is also relevant to planets and life.

Surveys show that about half of all stars are members of systems with two ("binary") or more stars. For example, our nearest neighbor system, Alpha Centauri, consists of three stars: Two of its stars (Alpha Centauri A and B) orbit each other once every 80 years at an average distance roughly equivalent to the distance between our Sun and Uranus; the third star (Proxima Centauri) is much farther away (about one-fourth of a light-year) and probably takes a few hundred thousand years to complete an orbit around the other two. (There is some debate as to whether the third star is gravitationally bound to the others or whether it just happens to be passing by them at this time.)

Stellar companions are relevant to planets because they can affect the possibility of having stable orbits. For example, if our solar system had a second Sun located at the distance of Mars, the competing effects of gravity from the two stars would have prevented Earth from forming in its current orbit. Nevertheless, stable orbits are sometimes possible, as we'll discuss shortly.

The general neighborhood in which a star system resides may also be relevant to planets and life. For example, stars are crowded much more closely together near the center of the galaxy and in some star clusters (particularly those known as *globular clusters*) than they are in the region of the disk where we reside. While stars in crowded neighborhoods are still far enough apart that direct collisions or planet-disrupting gravitational interactions are improbable, such stars are much more likely to be relatively near to the intense radiation released when a star dies in a supernova. Some scientists speculate that such events would therefore make life difficult or impossible on worlds in these crowded neighborhoods, though others disagree. We'll consider this debate in more detail in Section 11.4.

Which stars would make good suns?

Now that we've discussed the general differences among stars, we are ready to discuss which stars would make good "suns," meaning stars that are likely to host habitable worlds and that could potentially support life on those worlds. If you think about the various properties we've discussed, you'll recognize that any star system, in order to possibly be home to life, should meet at least the following four criteria:

- The star (and the gas cloud from which it formed, which would have had the same chemical composition as the resulting star) should contain enough elements besides hydrogen and helium to have allowed for the formation of habitable worlds.
- The star should last long enough for life to take hold and evolve.
- For surface life, at least, the star should shine with steady light and should have a habitable zone in which such life could potentially arise.
- The star or star system should allow worlds around it to have stable, nearly circular orbits.

Let's briefly explore each of these criteria.

ELEMENTS FOR BUILDING PLANETS As we've noted, the approximately 2% of "other elements" (elements other than hydrogen and helium) that comprise our own solar system are obviously sufficient for building planets. But given that many stars are much older than the Sun and therefore contain smaller proportions of elements besides hydrogen and helium, we might wonder how low the proportion can go before planets become unlikely.

Until fairly recently, many scientists suspected that heavy-element proportions much lower than that of our own solar system would make it impossible for planets to form. However, data from the *Kepler* mission and other searches for extrasolar planets now suggest otherwise. *Kepler* has discovered numerous cases in which planets orbit stars with as little as about one-quarter of the heavy-element proportion found in our solar system. For example, the star Kepler-444 is orbited by at least five planets similar in size to Earth, even though the star is estimated to be more than 11 billion years old and has less than about 0.5% of its mass in the form of elements besides hydrogen and helium. Interestingly, statistics to date suggest that such old stars tend to be less likely than younger stars to be orbited by large (jovian) planets, but there does not appear to be a significant difference in the numbers of smaller, Earth-size planets.

The bottom line is that while there is still some debate, the age and heavy-element abundance of a star do not appear likely to be major factors in determining whether the star can have habitable planets or life. In that case, the vast majority of stars could be orbited by planets.

TIME ENOUGH FOR LIFE Recall that evidence from radiometric dating indicates that it took a few tens of millions of years for the planets in our own solar system to form. This fact immediately seems to rule out stars of spectral type O as possible suns, because these stars die long before planets could form (see Table 11.1). The same is probably true for stars of spectral type B, because the star's death would probably occur before the process of accretion had settled down enough for life to take hold.

Stars of types A and F, with lifetimes of 1–2 billion years, would seem to offer enough time for both the formation of planets and the beginnings of biology. After all, evidence suggests that life took hold on our own world within several hundred million years or less of its formation [Section 6.1]. However, complex and intelligent life took much longer to arise on Earth, so if Earth is a typical case, stars of spectral types A and F may not live long enough for life to progress very far.

The above facts seem to suggest that life could not arise among stars of two of the seven spectral types (O and B) and could not evolve very far around two more (A and F). However, this is not as much of a limitation as it may seem. If you total up the numbers in the last column of Table 11.1, you'll see that these four spectral types together represent only about 3% of all stars. The other 97% live plenty long enough for life to take hold and, at least in principle, to evolve into complex and possibly intelligent forms.

HABITABLE ZONES In general, planets with oceans or life on the surface can exist only within what we've defined as a star's *habitable zone* [Section 10.3]. The basic requirement for having a habitable zone is that the star shine steadily, and this requirement seems to be easily met by almost any star in the core hydrogen–fusing stage of its existence. This stage lasts plenty long enough for life to arise on planets around stars of spectral types G, K, and M. An exception might be "flare stars," which are dim M stars that frequently emit sudden bursts of intense light and radiation that might cook any complex life on orbiting planets. However, current evidence suggests that M stars can be flare stars only when they are relatively young (perhaps in their first billion or so years of life),* and therefore that these stars eventually settle into a steadier state in which they could support a habitable zone and planets with life.

While most stars probably have habitable zones, the location and width of these zones will vary. Stars more luminous than the Sun (such as those of spectral types A and F) will have habitable zones that are wider and more distant from the star, while less-luminous stars (of spectral types K and M) will have narrower and closer-in habitable zones. To understand why, recall that the energy in starlight decreases with the square of the distance from the star (see Figure 7.2). For example, a planet orbiting a star with one-fourth the Sun's luminosity would have to orbit its star at one-half Earth's distance from the Sun $\left(\text{because } \left(\frac{1}{2}\right)^2 = \frac{1}{4}\right)$ in order to receive as much radiant energy as Earth. Figure 11.3 shows the approximate sizes of the habitable zones, in comparison to our Sun's habitable zone, for stars of one-half and one-tenth the mass of the Sun. The small size of the habitable zones around low-mass stars may decrease the probability of finding planets within this region (see Cosmic Calculations 10.1). However, because the spacing of planets seems to increase the

*The occurrence and size of flares, sunspots, and other types of surface "activity" on a star are related to the strength of the star's magnetic field. The magnetic field strength depends on the star's rotation rate and the depth of convection in its outer layers. Small stars have deeper convection, so if they rotate quickly they can have very strong magnetic fields associated with large flares and other surface activity. However, stellar rotation tends to slow with time, so the frequency and intensity of activity should diminish as a star ages.

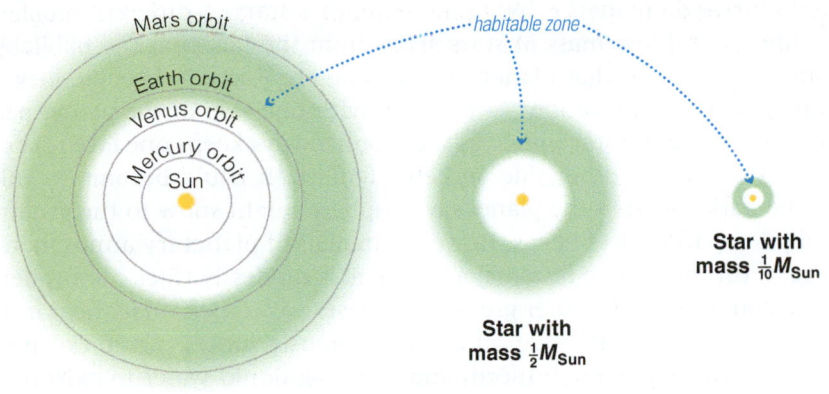

Mars orbit

habitable zone

Earth orbit

Venus orbit

Mercury orbit

Sun

Star with mass $\frac{1}{10}M_{\text{Sun}}$

Star with mass $\frac{1}{2}M_{\text{Sun}}$

Solar System

FIGURE 11.3 INTERACTIVE FIGURE
The approximate habitable zones around our Sun, a star with one-half the Sun's mass (spectral type K), and a star with one-tenth the Sun's mass (spectral type M), shown to scale. The habitable zone becomes increasingly smaller and closer-in for stars of lower mass and luminosity. (*Note:* As discussed in Section 10.4, the habitable zone around any star moves outward with time; the zones shown here are for stars similar in age to the Sun.)

farther you are from the star, the fraction of low-mass stars with a planet in the habitable zone may not be very different from that for brighter stars. In addition, the overwhelming majority of stars are low-mass stars, so their sheer numbers may mean that they are the locations of a sizable fraction of all habitable worlds.

The surface temperature and hence the wavelength of most light coming from a star may also affect prospects for life within the habitable zone. For example, stars of spectra types A and F emit many times as much ultraviolet light as does the Sun. Biology on Earth, and perhaps biology in general, is vulnerable to high-energy ultraviolet light, which easily breaks chemical bonds in complex organic molecules. Intense ultraviolet radiation might therefore keep planetary surfaces sterile. However, there are at least two ways around this problem. First, ultraviolet radiation does not penetrate far into the ground, oceans, or ice. If life on Earth emerged near volcanic vents in the deep oceans, as some evidence suggests, the same thing might happen on a watery planet around an A or F star. Similarly, worlds with a subsurface ocean, such as Europa may have, would offer life plenty of protection. Second, even though our Sun emits far less ultraviolet light than A or F stars, it still emits enough to make Earth's land surface sterile if not for the shielding provided by the ozone layer in the atmosphere. Planets around an A or F star might enjoy similar shielding if they had either a sufficiently thick atmosphere or an atmosphere containing sufficient oxygen. In the latter case, the additional ultraviolet light would split more atmospheric O_2 molecules, producing single oxygen atoms that would then combine to form ozone. In any event, even if their ultraviolet radiation makes life less likely around A and F stars, we've already discussed the fact that this rules out only about 3% of all stars as hosts for planets with life.

Think About It In view of current evidence concerning the past oxygen content of Earth's atmosphere (see Section 6.3), does it seem likely that planets around A or F stars could have ozone layers? Why or why not?

Stars lower in mass than our Sun have lower surface temperatures and therefore emit less visible light (and proportionally more infrared light) than the Sun. But this does not rule out the possibility of life within the habitable zone of such stars because, by definition, the total energy from sunlight will be the same. However, the lesser fraction of "blue" light relative to red around such stars could affect the type of plant life that

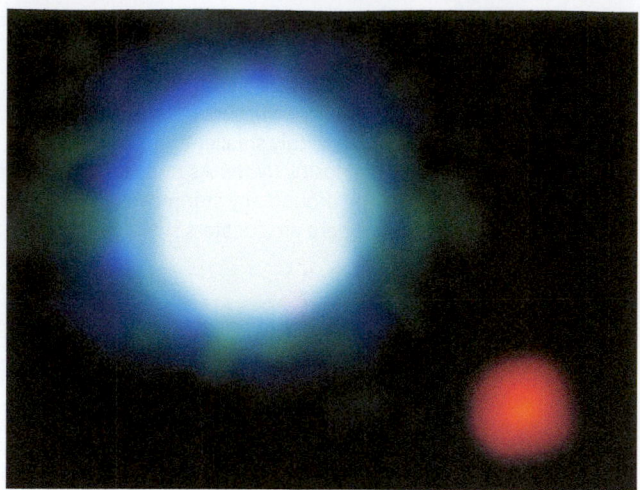

This infrared image from the European Southern Observatory's Very Large Telescope shows a brown dwarf called 2M1207 (blue) and a planet somewhat larger than Jupiter in orbit around it (red). Although a brown dwarf cannot provide the light and heat needed to support surface life on orbiting planets, a planet like the one shown here could potentially have tidally heated moons with subsurface oceans.

could thrive on planets orbiting these dimmer stars. A different problem for life around low-mass M stars arises from the fact that the habitable zone is so close-in that planets in the zone are likely to become locked into synchronous rotation [Section 9.1], with one side of the planet perpetually facing the star (much as the Moon always keeps one face turned to Earth) and the other side perpetually dark. It could become so cold on the dark side that the planet's atmosphere might snow to the ground and pile up in useless heaps. However, models of planetary atmospheres suggest this might not be a major problem: A modestly thick atmosphere of carbon dioxide (or other greenhouse gases) would circulate heat from the bright to the dark side on a synchronously locked world, keeping temperatures relatively uniform and allowing liquid water to exist over much of the planet's surface.

The bottom line is that most stars are likely to have habitable zones in which planets with surfaces able to nurture life might be found. Moreover, if we extend the prospects for habitability to subsurface zones, the possibilities may be far wider. For example, moons with subsurface liquid water (like Europa and Enceladus) could potentially exist around jovian planets wherever they might be found, even if they orbit brown dwarfs or dead stars (Figure 11.4).

STABLE ORBITS Based on our discussion so far, we have good reason to believe that the vast majority of single stars are, at least in principle, capable of having habitable planets. However, for the roughly half of all stars that are members of binary or multiple star systems, there's an additional consideration: The system must allow for stable planetary orbits. Moreover, the presence of two or more stars may change the prospects of having a habitable zone, and even when such a zone exists, the stable orbit must be nearly circular for a planet to remain within the habitable zone at all times.

Let's consider possible planetary orbits for the most common cases, which are *binary star systems*, with just two stars. (Systems with three or more stars are much less common.) Broadly speaking, there are three possible situations to consider for planetary orbits in a binary star system:

- A planet could orbit around both stars (Figure 11.5a). If the planet orbits the stellar pair at a distance considerably greater than the

Three orbital possibilities in a binary star system.

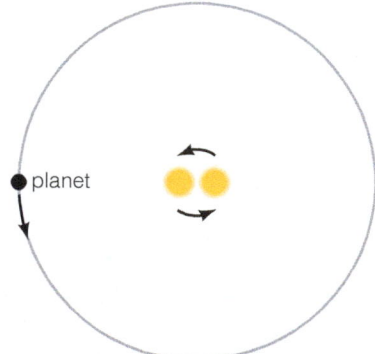

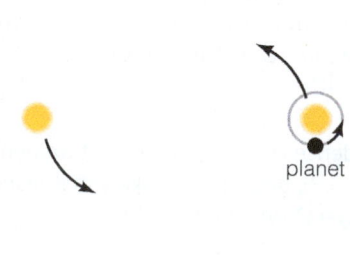

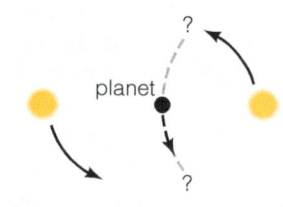

a *Potentially stable:* The planet orbits both stars; the radius of the planet's orbit is much larger than the separation between the stars. (This type of orbit is sometimes called a *P-type* or *circumbinary orbit*.)

b *Potentially stable:* The planet orbits one star; the radius of the planet's orbit is much smaller than the separation between the stars. (This type of orbit is sometimes called an *S-type orbit*.)

c *Not stable:* The orbit of the planet is neither much larger nor much smaller than the separation between the stars.

separation between the two stars, then gravitationally the two stars act much as one, and the planet can orbit without disruption around its distant twin hosts. Computer simulations indicate that stable orbits are generally possible for planets that orbit at a distance of more than about five times the separation of the stars. (The exact distance depends on the details of the system, such as the relative masses of the two stars and the eccentricity of their orbits.)

- A planet could orbit one star or the other (Figure 11.5b). If the two stars are fairly widely separated, then a planet near either one can orbit steadily because it will feel little disturbing effect from the second star. Computer simulations indicate that stability becomes possible when the two stars are separated by more than about five times the planet's orbital distance.

- We might imagine a planet trying to orbit between the two stars (Figure 11.5c). However, if the distance between the planet and at least one of the stars is not sufficiently different from the distance between the two stars, the planet's orbit will not be stable. The planet will experience competing gravitational tugs from both stars that will ultimately fling it out of the system (or send it crashing into one of the stars).

As an example of the first case, imagine that our Sun had a companion star at Mercury's position. The orbits of the inner planets would not be stable, but Jupiter's orbit could be. As an example of the second case, imagine that our Sun had a companion star at Saturn's distance (which is about 9.5 times Earth's distance). In that case, Earth and the other inner planets could have stable orbits.

Given these orbital possibilities, what can we say about habitability in binary star systems? In fact, both stable cases seem to allow for habitable planets. In the first case, the habitable zone would be a region surrounding the two stars together, with its size dictated largely by the combined luminosity of the two stars. If a planet had a nearly circular orbit within this habitable zone, it could potentially offer habitable conditions. Life on such a planet would see two suns rising and setting close together in the sky (Figure 11.6). In the second case, the more distant star would have relatively little influence on the planet's climate, so the habitable zone would be defined by the star that the planet orbited. Again, a planet with a nearly circular orbit in this zone could be po-

FIGURE 11.6
This painting shows a hypothetical planet much like Earth—with the lights of its own civilization visible on the night side—orbiting a binary star system in an orbit of the type shown in Figure 11.5a.

tentially habitable. Similar conclusions probably apply to other multiple star systems, though the situation can become more complex when more than two stars are involved.

SUMMARY: DISTANT SUNS We began this section with the goal of determining what types of stars might make suitable suns for habitable planets. We have found relatively few limits. It seems likely that most stars contain enough elements besides hydrogen and helium for planet formation to be possible. All but the most massive stars live long enough for life to take hold and evolve, and the rarity of massive stars means that stars with sufficient lifetimes for highly evolved life represent about 97% of all stars. Nearly all of these stars should have habitable zones in which planets with surface oceans and life are at least in principle possible. And even the roughly half of all stars that are members of binary systems can often make the grade by allowing stable orbits within habitable zones. The bottom line is that we have good reason to believe that habitable planets should be very common, which brings us to our next topic: how we find and study planets around other stars.

11.2 Discovering Extrasolar Planets

For more than 450 years after Copernicus recognized that Earth was merely one member of our Sun's planetary system, the study of planetary systems remained limited to our own. Then, a little more than two decades ago, a new scientific revolution began with the first discoveries of planets around other stars. This revolution is now in full bloom, with new discoveries coming at a rapid pace. In this section, we will explore the methods used to discover extrasolar planets and the planetary properties we can currently measure with these methods.

How do we detect planets around other stars?

Detecting extrasolar planets poses a huge technological challenge. Recall from Chapter 3 that detecting planets around other stars is equivalent to looking for dim ball points or marbles from a distance of thousands of kilometers away. Moreover, stars are typically *a billion times* brighter than the visible light reflected by any orbiting planets, so starlight tends to overwhelm any planetary light in photographs. This problem can be somewhat lessened—but not eliminated—by observing in infrared light, because planets emit their own infrared light and stars are usually dimmer in the infrared. Still, the great difficulty of detecting extrasolar planets should be clear.

The first unambiguous discovery of a planet around another Sun-like star—a star called 51 Pegasi—came in 1995. Thousands of other extrasolar planets have been discovered since that time, using several different planet-finding strategies. If we strip away the details, however, there are really only two general ways of learning about a distant object: *directly*, which means by obtaining images or spectra of the object, and *indirectly*, which means by inferring the object's existence or properties without actually seeing it. While scientists have achieved some success in direct detection of extrasolar planets (such as the planets in the image that opens this chapter), we will need significant advances in technology before we are able to obtain higher-resolution images or spectra. As a result, nearly all of our current understanding of extrasolar planets comes from indirect study. There are two major indirect approaches to finding and studying extrasolar planets:

1. Observing the motion of a star to detect the subtle gravitational effects of orbiting planets
2. Observing changes in a star's brightness that occur when one of its planets passes in front of the star as viewed from Earth

Both approaches are indirect because we discover the planets by observing their stars without actually seeing the planets themselves.

The earliest discoveries of extrasolar planets came primarily through the first approach, but the current champion of extrasolar planet discoveries—NASA's *Kepler* mission—used the second. We can learn even more in cases in which we can combine the two indirect approaches, because each can provide different information about a distant planet. Because these two approaches are so important to current understanding of extrasolar planets, we will devote most of this section to studying them in more detail.

Think About It Do a quick Web search on "extrasolar planets." How many are now known? How many have been discovered in just the past year?

GRAVITATIONAL TUGS—GENERAL PRINCIPLES Although we usually think of a star as remaining still while planets orbit around it, that is only approximately correct. In reality, all objects in a star system, including the star itself, orbit the system's "balance point," or **center of mass.** You can understand this concept by thinking of a waiter carrying a tray of drinks: To carry the tray, he places his hand under the spot at which it balances—its center of mass. The center of mass will not necessarily be the center of the tray; for example, if the tray has a heavy glass of water

---

Special Topic 11.1 THE NAMES OF EXTRASOLAR PLANETS

The planets in our solar system have familiar names rooted in mythology. Unfortunately, there's not yet a well-accepted scheme for naming extrasolar planets, although the International Astronomical Union has begun a process by which the public can suggest and eventually vote on planet names. In the meantime, astronomers still generally refer to extrasolar planets by the star they orbit, such as "the planet orbiting the star named ... ." Worse still, the stars themselves often have confusing or even multiple names, reflecting naming schemes used in star catalogs made by different people at different times in history.

A few hundred of the brightest stars in the sky carry proper names from ancient times. Many of these names are Arabic—such as Betelgeuse, Algol, and Aldebaran—because of the work of the Arabic scholars of the Middle Ages. In the early seventeenth century, German astronomer Johann Bayer developed a system that gave names to many more stars: Each star gets a name based on its constellation and a Greek letter indicating its ranking in brightness within that constellation. For example, the brightest star in the constellation Andromeda is called Alpha Andromedae, the second brightest is Beta Andromedae, and so on. Bayer's system worked for only the 24 brightest stars in each constellation, because there are only 24 letters in the Greek alphabet. About a century later, English astronomer John Flamsteed published a more extensive star catalog in which he used numbers once the Greek letters were exhausted. For example, 51 Pegasi gets its name from Flamsteed's catalog. (Flamsteed's numbers are based on position within a constellation rather than brightness.)

As more powerful telescopes made it possible to discover more and fainter stars, astronomers developed many new star catalogs. The names we use today usually come from one of these catalogs. For example, the star HD 209458 appears as star number 209458 in a catalog compiled by Henry Draper (HD). You may see star names consisting of numbers preceded by other catalog names, including Gliese, Ross, and Wolf; these catalogs are also named for the astronomers who compiled them. Because the same star is often listed in several catalogs, a single star can have several different names.

Objects orbiting other stars usually carry the star name plus a letter denoting their order of discovery around that star. If the second object is another star, a capital B is added to the star name; a lowercase b is added if it's a planet. For example, HD 209458b is the first planet discovered orbiting star number 209458 in the Henry Draper catalog; Upsilon Andromedae d is the third planet discovered orbiting the twentieth brightest star (because upsilon is the twentieth letter in the Greek alphabet) in the constellation Andromeda. Some recently discovered planets have been named for the observing program that discovered them. For example, Kepler 11g is the sixth planet in the eleventh planetary system announced by the *Kepler* mission.

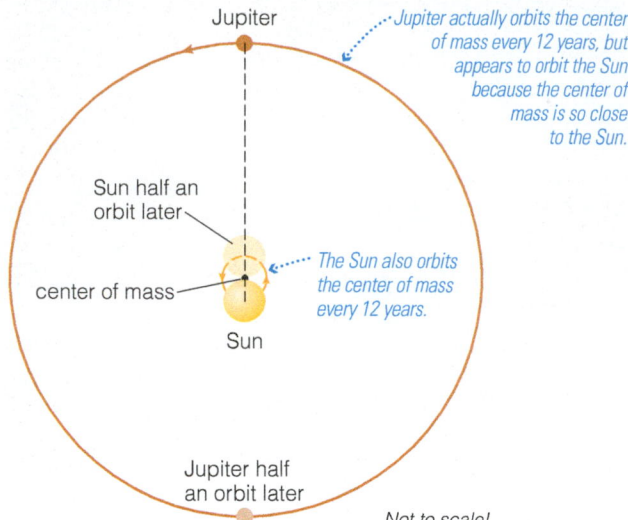

Jupiter

Jupiter actually orbits the center of mass every 12 years, but appears to orbit the Sun because the center of mass is so close to the Sun.

Sun half an orbit later

center of mass

Sun

The Sun also orbits the center of mass every 12 years.

Jupiter half an orbit later

Not to scale!

FIGURE 11.7

This diagram shows how both the Sun and Jupiter actually orbit around their mutual center of mass, which lies very close to the Sun. The diagram is not to scale; the sizes of the Sun and its orbit are exaggerated about 100 times compared to the size shown for Jupiter's orbit, and Jupiter's size is exaggerated even more.

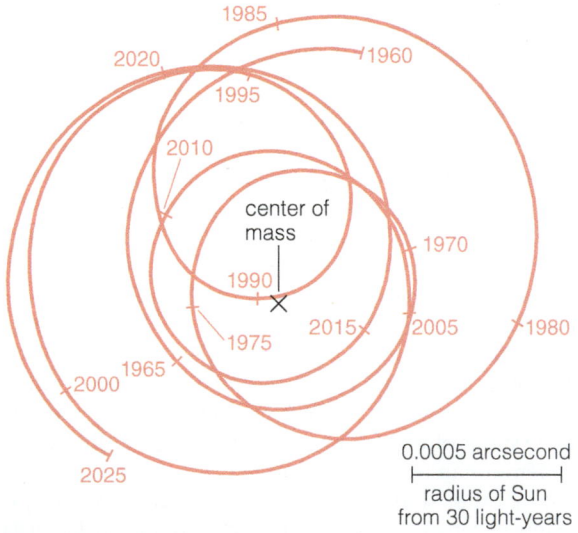

1985
2020
1960
1995
2010
center of mass
1970
1990
1975
2015
2005
1980
1965
2000
2025

0.0005 arcsecond

radius of Sun from 30 light-years

FIGURE 11.8

This diagram shows the orbital path of the Sun around the center of mass of our solar system as it would appear from a distance of 30 light-years away, for the period 1960–2025. Notice that the entire range of motion during this period is only about 0.0015 arcsecond, which is almost 100 times smaller than the angular resolution of the Hubble Space Telescope. Nevertheless, if alien astronomers could measure this motion, they could learn of the existence of planets in our solar system.

off to one side, the waiter will place his hand a little to that side of the tray's center. To understand how the center of mass idea can allow us to discover extrasolar planets, imagine the viewpoint of extraterrestrial astronomers observing our solar system from afar.

Let's start by considering only the influence of Jupiter (Figure 11.7). Jupiter is the most massive planet, but the Sun is still about a thousand times as massive. As a result, the center of mass between the Sun and Jupiter lies about one-thousandth of the way from the center of the Sun to the center of Jupiter, putting it just outside the Sun's visible surface. In other words, what we usually think of as Jupiter's 12-year orbit around the Sun is really a 12-year orbit around the center of mass. Because the Sun and Jupiter are always on opposite sides of the center of mass (otherwise it wouldn't be a "center"), the Sun must orbit this point with the same 12-year period. The Sun's orbit traces out only a small ellipse with each 12-year period, because the Sun's average orbital distance is barely larger than its own radius; that is why we generally don't notice the Sun's motion. Nevertheless, with sufficiently precise measurements, extraterrestrial astronomers could detect this orbital movement of the Sun and thereby deduce the existence of Jupiter, even without having observed Jupiter itself. They could even determine Jupiter's mass from the orbital characteristics of the Sun as it goes around the center of mass. A more massive planet at the same distance would pull the center of mass farther from the Sun's center, giving the Sun a larger orbit. Because the Sun's period around the center would still be 12 years, the larger orbit would mean the Sun would move at a faster orbital speed around the center of mass.

Now let's add in the effects of Saturn. Saturn takes 29.5 years to orbit the Sun, so by itself it would cause the Sun to orbit their mutual center of mass every 29.5 years. However, because Saturn's influence is secondary to that of Jupiter, this 29.5-year period appears as a small effect added on top of the Sun's 12-year orbit around its center of mass with Jupiter. In other words, every 12 years the Sun would return to *nearly* the same orbital position around its center of mass with Jupiter, but the precise point of return would move around with Saturn's 29.5-year period. By measuring this motion carefully over many years, an extraterrestrial astronomer could deduce the existence and masses of both Jupiter and Saturn.

The other planets also exert gravitational tugs on the Sun, each adding a small effect to the effects of Jupiter and Saturn (Figure 11.8). In principle, with sufficiently precise measurements of the Sun's orbital motion made over many decades, an extraterrestrial astronomer could deduce the existence of all the planets of our solar system. If we turn this idea around, it means we can search for planets in other star systems by carefully watching for the tiny orbital motion of a star around the center of mass of its star system.

GRAVITATIONAL TUGS—THE ASTROMETRIC METHOD Astronomers use two distinct observation methods to search for the gravitational tugs of planets on stars. The first, called the **astrometric method** (*astrometry* means "measurement of the stars"), uses very precise measurements of stellar positions in the sky to look for the slight motion caused by orbiting planets—the type of motion shown in Figure 11.8. If a star "wobbles" gradually around its average position (the center of mass), we must be observing the influence of unseen planets.

The principle behind the astrometric method is straightforward, but two practical difficulties have limited its usefulness to date. The first difficulty stems from the fact that we are searching for changes to position that are very small even for nearby stars, and these changes become smaller for more distant stars. For example, from a distance of 10 light-years, a Jupiter-size planet in a Jupiter-like orbit (5 AU from a Sun-like star) would cause its star to move slowly over a side-to-side angular distance of only about 0.003 arcsecond—approximately the width of a hair seen from a distance of 5 kilometers. At double the distance (20 light-years) the observed motion would be only half as large (0.0015 arcsecond), and at ten times the distance (100 light-years) it would be only one-tenth as large (0.0003 arcsecond). For this reason, the astrometric method is best suited to searching for relatively massive planets around relatively nearby stars.

To understand the second difficulty, consider what would happen if we could move Jupiter into an orbit farther from the Sun. This larger orbit would increase the angular extent of the Sun's side-to-side motion as seen from a distance (because moving Jupiter outward would also move the center of mass outward from the Sun), but Kepler's third law [Section 2.2] tells us that this move would also increase Jupiter's orbital period. As a result, it would take a much longer time for alien astronomers to recognize Jupiter's effect. For example, if Jupiter moved to Neptune's distance from the Sun, at which the orbital period is 165 years, it might take a century or more of patient observation to be confident that stellar motion was occurring in a 165-year cycle.

As a result of these difficulties, the astrometric method has led to few discoveries to date. However, the European Space Agency's *GAIA* mission (Figure 11.9), launched in 2013, should soon change this. *GAIA* has a goal of obtaining astrometric observations of a billion stars in our galaxy to an accuracy that in some cases will be better than 10 microarcseconds, equivalent to the angular width of a human hair viewed from a distance of 2000 kilometers. If fully successful, *GAIA* should thereby detect thousands of extrasolar planets via the astrometric method.

FIGURE 11.9
Artist's conception of the *GAIA* spacecraft. *GAIA* is capable of extremely precise measurements of stellar positions, making scientists optimistic that it will detect thousands of planets through the astrometric method.

GRAVITATIONAL TUGS—THE DOPPLER METHOD The second method of searching for gravitational tugs due to orbiting planets is the **Doppler method,** which searches a star's spectrum for telltale signs that the star is orbiting a center of mass. These signs appear in the form of small shifts in the wavelengths of spectral lines caused by what we call the *Doppler effect.*

You've probably noticed the Doppler effect on the sound of a train whistle near train tracks. (You can also notice the Doppler effect with emergency sirens or even just with the "buzz" of a fast car as it goes past you.) If the train is stationary, the pitch of its whistle sounds the same no matter where you stand (Figure 11.10a). But if the train is moving, the pitch will sound higher when the train is coming toward you and lower when it's moving away from you. Just as the train passes by, you can hear the dramatic change from high to low pitch—a sort of "weeeeeeee–ooooooooooh" sound. To understand why, think about what happens to the sound waves coming from the train (Figure 11.10b). When the train is moving toward you, each pulse of a sound wave is emitted a little closer to you. The bunching up of the waves between you and the train gives them a shorter wavelength and higher frequency (pitch). After the train passes you by, each pulse comes from farther away, stretching out the wavelengths and giving the sound a lower frequency.

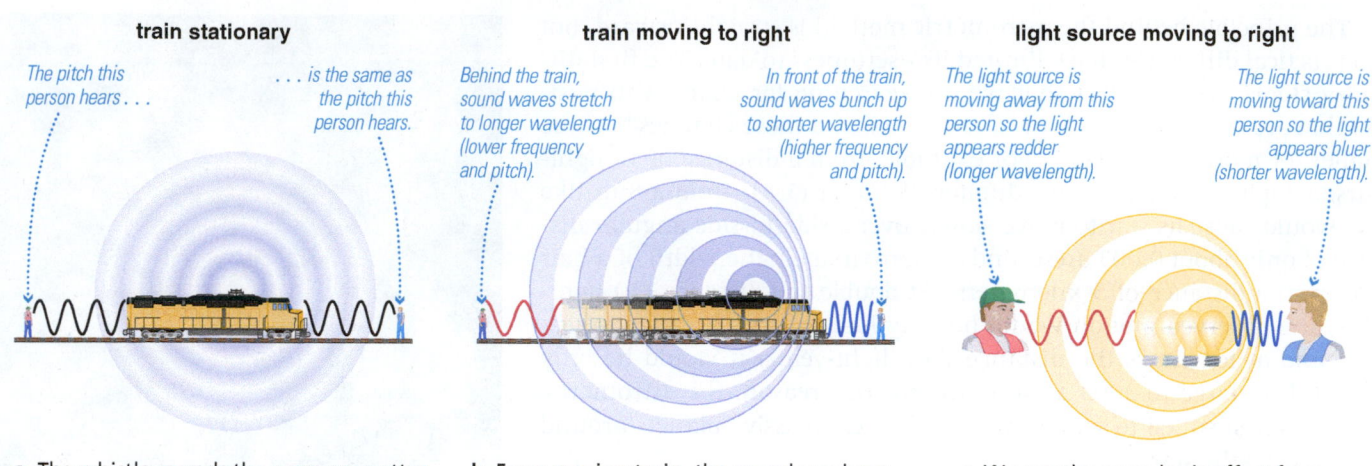

train stationary

The pitch this person hears . . .

. . . is the same as the pitch this person hears.

a The whistle sounds the same no matter where we stand near a stationary train.

train moving to right

Behind the train, sound waves stretch to longer wavelength (lower frequency and pitch).

In front of the train, sound waves bunch up to shorter wavelength (higher frequency and pitch).

b For a moving train, the sound you hear depends on whether the train is moving toward you or away from you.

light source moving to right

The light source is moving away from this person so the light appears redder (longer wavelength).

The light source is moving toward this person so the light appears bluer (shorter wavelength).

c We get the same basic effect from a moving light source (although the shifts are usually too small to notice by eye).

FIGURE 11.10

The Doppler effect. Each circle represents the crests of sound (or light) waves going in all directions from the source. For example, the circles from the train might represent waves emitted 0.001 second apart.

The Doppler effect causes similar shifts in the wavelengths of light (Figure 11.10c). If an object is moving toward us, its entire spectrum—including the spectral lines within it—is shifted to shorter wavelengths; the spectral lines therefore serve as reference lines for measuring the amount of the shift. Because shorter wavelengths of visible light are bluer, the Doppler shift of a star coming toward us is called a **blueshift.** If an object is moving away from us, its light is shifted to longer wavelengths; we call this a **redshift,** because longer wavelengths of visible light are redder. The faster the object is moving, the greater the amount of its blueshift or redshift.

The Doppler method involves looking for alternating blueshifts and redshifts (relative to a star's average Doppler shift) in a star's spectrum, because this pattern indicates that the star is orbiting a center of mass due to the gravitational tug of an unseen planet (Figure 11.11a). For example, the 1995 discovery of a planet orbiting 51 Pegasi occurred when this star was found to be moving with a rhythmic wobble that repeated about every 4 days, corresponding to an orbital speed of 57 meters per second (Figure 11.11b). The 4-day period of the star's motion must be the orbital period of the planet causing this gravitational tug on the star.

Current techniques can measure a star's velocity to a fraction of a meter per second—much slower than walking speed—which corresponds to a Doppler wavelength shift on the order of 0.0000001%. We can therefore find planets that exert a considerably smaller gravitational tug on their stars than the planet orbiting 51 Pegasi does on its star. As of 2015, the Doppler method had been used to detect more than 600 planets, including more than 100 in multiplanet systems.

Despite the success of the Doppler method, it has three major limitations. First, like the astrometric method, the Doppler method searches for gravitational tugs from orbiting planets and is therefore better for finding massive planets like Jupiter than small planets like Earth. Second, the Doppler method is best suited to identifying these planets when they

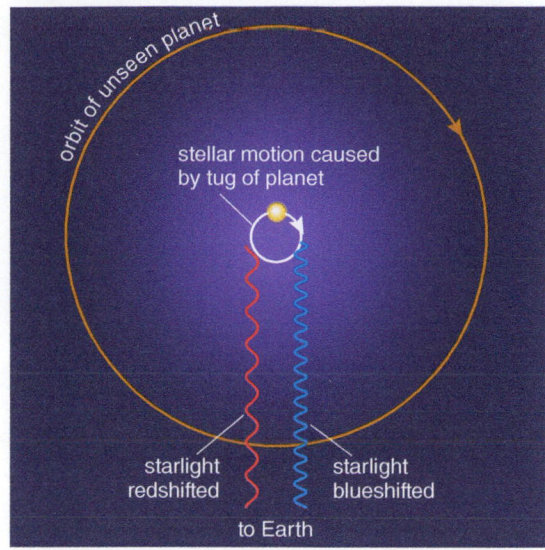

a The star's Doppler shift alternates toward the blue and toward the red, allowing us to detect its slight motion–caused by an orbiting planet–around the center of mass.

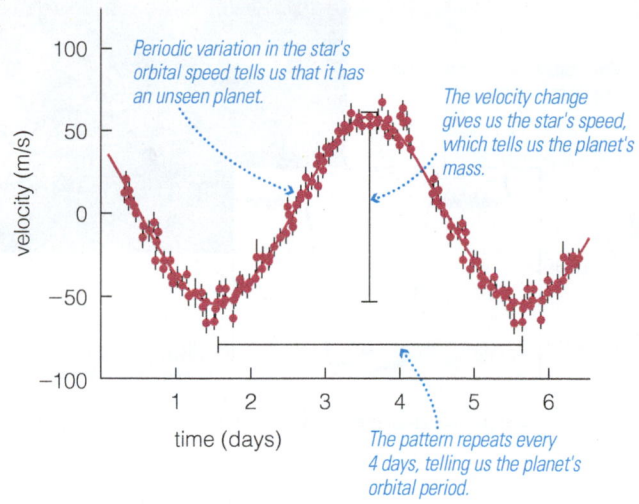

Periodic variation in the star's orbital speed tells us that it has an unseen planet.

The velocity change gives us the star's speed, which tells us the planet's mass.

The pattern repeats every 4 days, telling us the planet's orbital period.

b A periodic Doppler shift in the spectrum of the star 51 Pegasi shows the presence of a large planet with an orbital period of about 4 days. Dots are actual data points; bars through dots represent measurement uncertainty.

FIGURE 11.11 INTERACTIVE FIGURE
The Doppler method for discovering extrasolar planets.

orbit relatively *close* to their star (in contrast to the astrometric method, which works better for planets farther from their star); close-in planets are easier to detect both because being closer means a stronger gravitational tug and hence a greater velocity for the star (which is easier to measure) and because closer-in planets have shorter orbital periods that allow their orbits to be observed in shorter amounts of time. The third limitation is that the Doppler method requires a fairly large telescope (and long exposure time) in order to obtain spectra with high enough resolution to reveal very small Doppler shifts, which limits the number of stars that can be studied with this method.

CHANGES IN BRIGHTNESS—THE TRANSIT METHOD The astrometric and Doppler methods both search for planets by looking for evidence of the gravitational tugs the planets exert on their stars. The second general approach to detecting planets indirectly relies on searching for slight changes in a star's brightness caused by orbiting planets.

If we were to examine a large sample of stars with planets, a small number of these systems—about 1%—would by chance be aligned in such a way that one or more of the star's planets would pass directly between us and the star once each orbit. The result is a **transit,** in which the planet appears to move across the face of the star. We occasionally witness this effect in our own solar system when Mercury or Venus crosses in front of the Sun (see Figure 11.12).

Other star systems are so far away that we cannot actually see a planetary dot set against the face of the star as we can see Mercury or Venus against the face of the Sun. Nevertheless, a transiting planet will block a little of its star's light, allowing us to detect its passage as the star appears temporarily dimmer (Figure 11.13). The larger the planet, the more dimming it will cause. Some transiting planets also create a measurable **eclipse** as the planet goes behind the star; eclipses too cause a small dip in a system's brightness, because during the eclipse we see light only from the star rather than from both the star and the planet. Eclipse

FIGURE 11.12
This photo shows the Venus transit of June 6, 2012, as it appeared in Germany at dawn. Venus is the small black dot visible near the upper center of the Sun's face. Our orbit around the Sun makes transits of Mercury and Venus occur at uneven time intervals, and the next Venus transit will not occur until 2117. When we look to other stars, however, transits will occur at regular intervals if the planet orbits are in our line of sight.

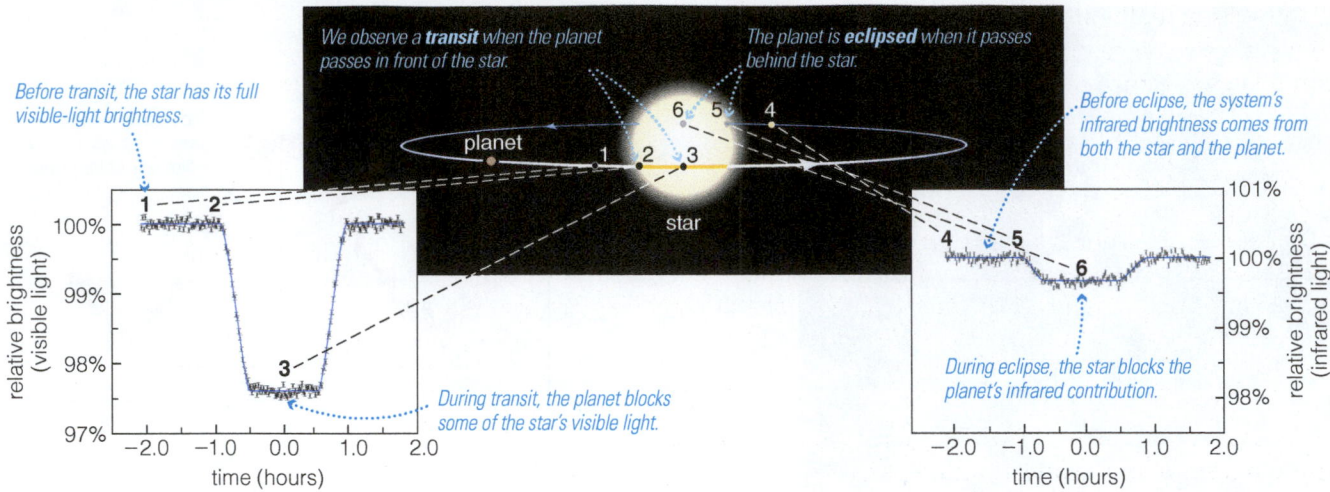

We observe a **transit** when the planet passes in front of the star.

The planet is **eclipsed** when it passes behind the star.

Before transit, the star has its full visible-light brightness.

During transit, the planet blocks some of the star's visible light.

Before eclipse, the system's infrared brightness comes from both the star and the planet.

During eclipse, the star blocks the planet's infrared contribution.

planet

star

FIGURE 11.13 INTERACTIVE FIGURE

This diagram represents transits and eclipses of the planet orbiting the star HD 189733. The left graph shows that each transit lasts about 2 hours, during which the star's visible-light brightness dips by about 2.5%. The eclipses are observable in the infrared, because the star blocks the infrared contribution of the planet. The transits and eclipses each occur once during every 2.2-day orbit of the planet.

observations are more easily accomplished in the infrared, because planets contribute a greater proportion of a system's infrared brightness than visible-light brightness.

The **transit method** searches for transits and eclipses by carefully monitoring a star system's brightness over an extended period of time. Because most stars exhibit intrinsic variations in brightness, we can be confident that we have detected a planet only if we observe the dips in brightness to repeat with a regular period, indicating that the same planet is passing in front of the star repeatedly with each orbit. (Intrinsic stellar brightness variations tend to be more random.) The period of the repeated transits is the orbital period of the planet. Scientists generally require at least three repeated transits before concluding that a planet is probably responsible for the variations in brightness. Using the transit method to discover a planet therefore requires monitoring a star's brightness for at least three full orbits of the planet. For example, we would need at least about three years of monitoring with the transit method to detect a planet in an Earth-like orbit.

Putting these ideas together, we find that the first requirement for success with the transit method is that a system have planet orbits aligned "just right" so that they pass in front of their star as seen from Earth. When that is the case, the transit method will always be able to find close-in planets with short orbital periods before it can find more distant planets, simply because less time is needed to observe repeated transits of the closer-in planets. Adding the fact that larger planets create bigger dips in brightness means that transits are easiest to measure for planets that are both large and close in.

Transits have been observed with many telescopes, but the vast majority of transit detections to date have come from NASA's *Kepler* mission. Between 2009 and 2013, *Kepler* monitored about 170,000 stars for transits, measuring their brightnesses about every 30 minutes. Based on data that had been analyzed as of mid-2015, scientists had already identified about 2000 stars being orbited by a total of more than 4000 planet "candidates," including dozens that are as small as Earth. The term "candidates" is used because even though *Kepler* data show three (or more) transits, the existence of all of these planets has not yet been confirmed by follow-up observations (such as detection by the Doppler method); as a result, there's still a small possibility that evidence for some "candidates" is actually an artifact of something besides an orbiting planet that causes a star to dim. Statistically, though, at least about 90% of the candidates will turn out to be real planets.

Think About It The *Kepler* mission was thought "dead" in 2013 when it suffered a failure of the systems that help it point steadily at stars, but engineers developed a remarkable work-around that has kept this space-based telescope in the planet-hunting game. Moreover, two new transit missions—NASA's *TESS* mission and the European Space Agency's *CHEOPS* mission—are both slated to launch in 2017. Find the current status of the *Kepler*, *TESS*, and *CHEOPs* missions, and look for any recent discoveries.

SUMMARY OF PLANET DETECTION METHODS Table 11.2 summarizes the advantages and limitations of the major methods we have discussed for studying extrasolar planets, and Cosmic Context Figure 11.14 (pages 388–389) summarizes how these methods work. Note that while these will almost certainly be the most important methods over the long term, several other strategies have met with some success. More than a dozen planets have been detected through a method called *microlensing*, in which we see the effects that occur when one star passes in front of another as seen from Earth, causing a temporary magnification of the background star's light as predicted by Einstein's general theory of relativity. (Microlensing is actually a special case of the more general phenomenon

TABLE 11.2 Major Extrasolar Planet Detection Methods

	Description	Key Advantages	Major Limitations	Confirmed or Candidate Planets (as of mid-2015)
Direct Detection	Obtain images or spectra of extrasolar planets	• Is the only method that allows direct study of the planets themselves	• Requires large telescopes and some means of blocking light from star planet is orbiting	A few
Astrometric Method	Infer planet's existence from small changes in star's position in sky	• Is now possible with *GAIA* spacecraft • Detects planets in all orbit orientations except edge-on	• Is generally possible only for relatively nearby stars • Is biased toward finding massive planets that orbit far from their stars • May require many years of observation to detect these planets (with large orbital periods)	A few
Doppler Method	Infer planet's existence from star's motion toward/away from us as revealed by Doppler shifts in its spectrum	• Is possible from ground-based telescopes • Detects planets in all orbit orientations except face-on	• Is biased toward finding massive planets with close-in orbits • Underestimates star's true motion except when system is viewed edge-on • Requires stellar spectra, which means large telescopes and long observation times	More than 600
Transit Method	Infer planet's existence from slight changes in star's brightness as planet passes in front of (or behind) it	• Allows many stars to be observed at once • Can detect very small planets • Is feasible with small telescopes • Can provide some atmospheric data in cases of measurable eclipses	• Is possible only for planets with edge-on orbits as viewed from Earth • For small planets, requires sensitivity possible only from a space observatory	More than 4000

FIGURE 11.14 DETECTING EXTRASOLAR PLANETS

The search for planets around other stars is one of the fastest growing and most exciting areas of astronomy, with the known extrasolar planets already numbering well into the thousands. This figure summarizes major techniques that astronomers use to search for and study extrasolar planets.

① Gravitational Tugs: We can detect a planet by observing the small orbital motion of its star as both the star and its planet orbit their mutual center of mass. The star's orbital period is the same as that of its planet, and the star's orbital speed depends on the planet's distance and mass. Any additional planets around the star will produce additional features in the star's orbital motion.

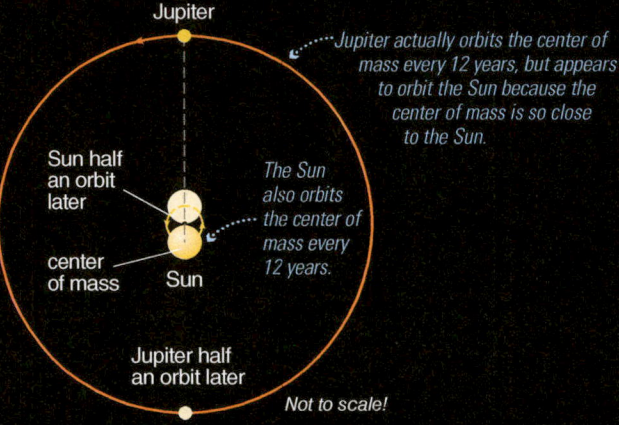

Jupiter

Jupiter actually orbits the center of mass every 12 years, but appears to orbit the Sun because the center of mass is so close to the Sun.

Sun half an orbit later

The Sun also orbits the center of mass every 12 years.

center of mass   Sun

Jupiter half an orbit later

Not to scale!

1a The Doppler Method: As a star moves alternately toward and away from us around the center of mass, we can detect its motion by observing alternating Doppler shifts in the star's spectrum: a blueshift as the star approaches and a redshift as it recedes.

1b The Astrometric Method: A star's orbit around the center of mass leads to tiny changes in the star's position in the sky. The *GAIA* mission is expected to discover many new planets with this method.

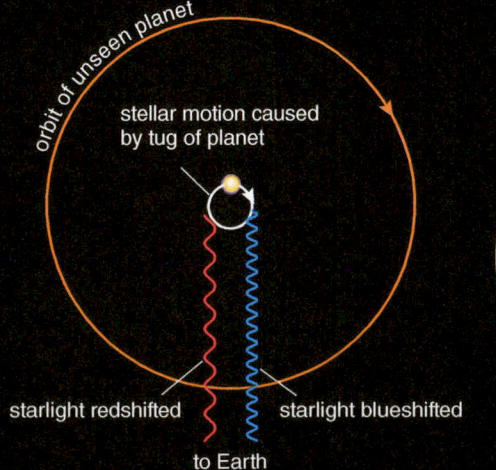

orbit of unseen planet

stellar motion caused by tug of planet

starlight redshifted   starlight blueshifted

to Earth

Current Doppler-shift measurements can detect an orbital velocity as small as 1 meter per second—walking speed.

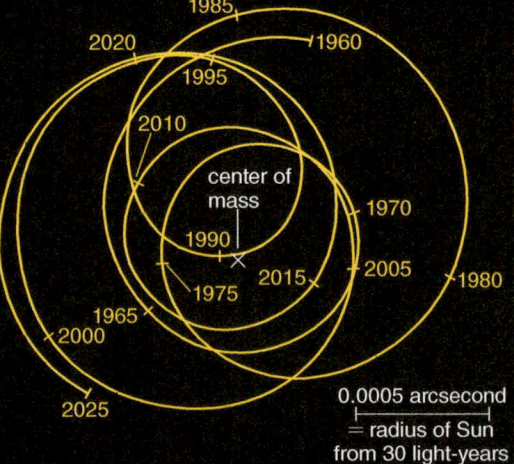

1985
2020   1960
1995
2010
center of mass
1990   1970
2015   2005
1975   1980
1965
2000
2025

0.0005 arcsecond
= radius of Sun
from 30 light-years

The change in the Sun's apparent position, if seen from a distance of 10 light-years, would be similar to the angular width of a human hair at a distance of 5 kilometers.

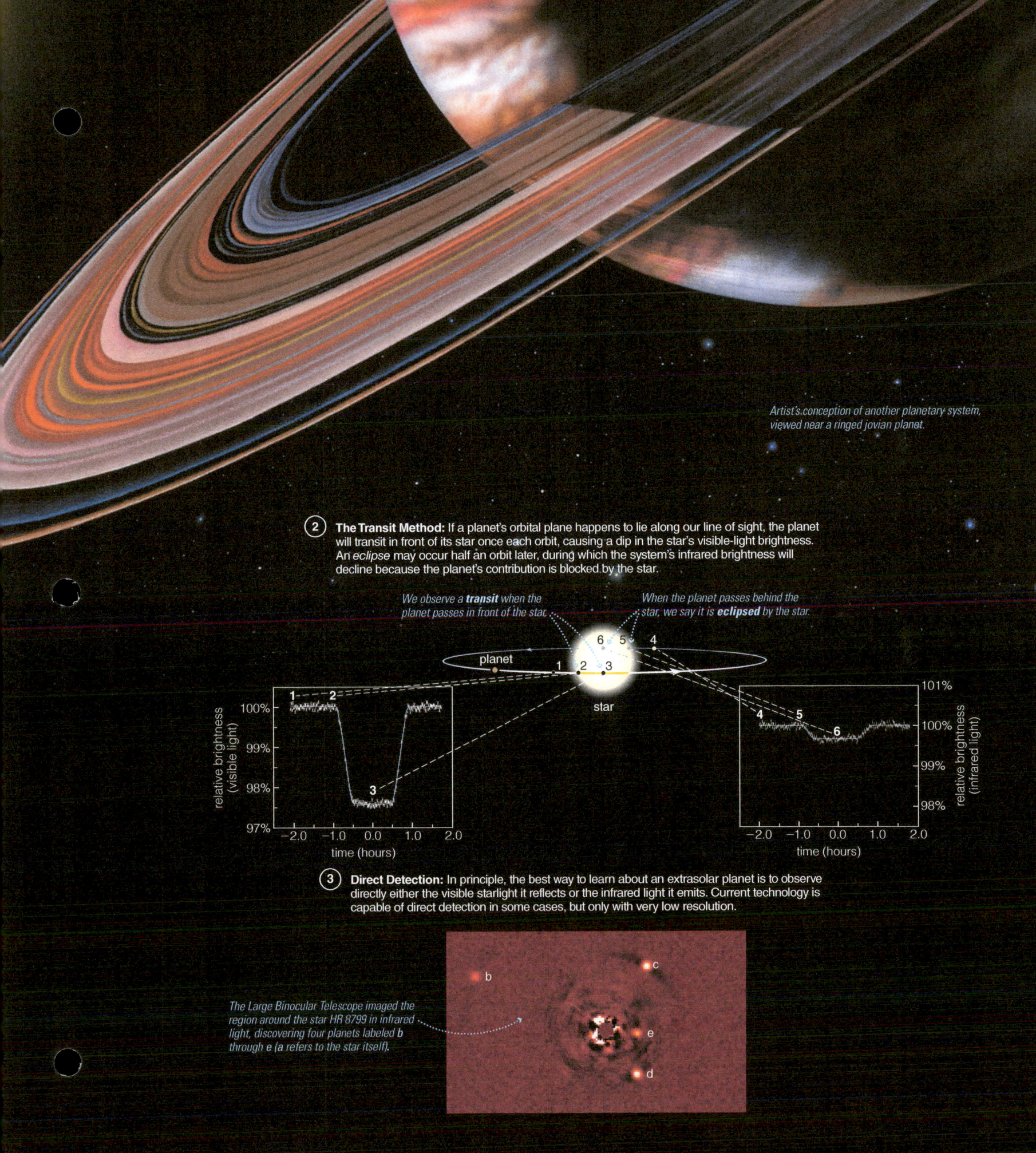

Artist's conception of another planetary system, viewed near a ringed jovian planet.

②　The Transit Method: If a planet's orbital plane happens to lie along our line of sight, the planet will transit in front of its star once each orbit, causing a dip in the star's visible-light brightness. An *eclipse* may occur half an orbit later, during which the system's infrared brightness will decline because the planet's contribution is blocked by the star.

*We observe a **transit** when the planet passes in front of the star.*

*When the planet passes behind the star, we say it is **eclipsed** by the star.*

planet

star

③　Direct Detection: In principle, the best way to learn about an extrasolar planet is to observe directly either the visible starlight it reflects or the infrared light it emits. Current technology is capable of direct detection in some cases, but only with very low resolution.

The Large Binocular Telescope imaged the region around the star HR 8799 in infrared light, discovering four planets labeled b through e (a refers to the star itself).

called *gravitational lensing*.) Careful study of a microlensing event can reveal whether the foreground star has planets. The major drawback to this method is that the special alignment necessary for microlensing is a one-time event, which generally means that there is no opportunity for confirmation or follow-up observations. A different strategy takes advantage of the fact that many stars are surrounded by disks of dust. A planet within such a disk can exert small gravitational tugs on dust particles to produce gaps, waves, or ripples that may be detectable. Some astronomers are searching for thermal emission attributable to the heat of large impacts on planets in young planetary systems, and others are searching for the special kinds of emission expected from the magnetospheres of jovian planets. As we learn more about extrasolar planets, new search methods are sure to arise.

What properties of extrasolar planets can we measure?

The mere existence of planets around other stars has changed our perception of our place in the universe, because it shows that our planetary system is not unique. Scientifically, however, we want to know much more than just that these planets exist. Despite the limitations of current methods, we can learn a surprising amount about extrasolar planets. Depending on the method or methods used, we can determine such planetary characteristics as orbital period and distance, orbital eccentricity, mass, size, density, and even a little bit about a planet's atmospheric composition and temperature.

Cosmic Calculations 11.1 FINDING ORBITAL DISTANCES FOR EXTRASOLAR PLANETS

Recall that Newton's version of Kepler's third law reads

$$p^2 = \frac{4\pi^2}{G(M_1 + M_2)} a^3$$

In the case of a planet orbiting a star, p is the planet's orbital period, a is its average orbital distance (semimajor axis), M_1 and M_2 are the masses of the star and planet, respectively, and $G = 6.67 \times 10^{-11}$ m^3/(kg $\times$ s^2) is the gravitational constant. Because a star is so much more massive than a planet, we can approximate $M_{star} + M_{planet} \approx M_{star}$ (see Cosmic Calculations 7.1). Using this approximation, we can apply simple algebra to solve for the average orbital distance a:

$$a \approx \sqrt[3]{\frac{GM_{star}}{4\pi^2} p_{planet}^2}$$

Example: The star 51 Pegasi has a mass 1.06 times that of our Sun, and Doppler measurements (shown in Figure 11.11b) indicate it has a planet with an orbital period of 4.23 days. What is the planet's average orbital distance?

Solution: We can use Newton's version of Kepler's third law, but to make the units consistent, we first convert the given stellar mass to kilograms (using the fact that the Sun's mass is 2×10^{30} kg) and the orbital period to seconds:

$$M_{star} = 1.06 \times M_{Sun} = 1.06 \times (2 \times 10^{30}\,\text{kg}) = 2.12 \times 10^{30}\,\text{kg}$$

$$p = 4.23\ \text{day} \times \frac{24\ \text{hr}}{1\ \text{day}} \times \frac{3600\ \text{s}}{1\ \text{hr}} = 3.65 \times 10^5\ \text{s}$$

We now plug these values into our equation from above to find the average distance a:

$$a \approx \sqrt[3]{\frac{GM_{star}}{4\pi^2} p_{planet}^2}$$

$$= \sqrt[3]{\frac{6.67 \times 10^{-11}\,\dfrac{\text{m}^3}{\text{kg} \times \text{s}^2} \times 2.12 \times 10^{30}\,\text{kg}}{4\pi^2}(3.65 \times 10^5\,\text{s})^2}$$

$$= 7.81 \times 10^9\ \text{m}$$

The planet orbits 51 Pegasi at an average distance of 7.8 billion meters, or 7.8 million kilometers. It's easier to interpret this number if we convert it to astronomical units (1 AU $\approx$ 1.50 $\times$ 10^{11} meters):

$$a = 7.81 \times 10^9\,\text{m} \times \frac{1\ \text{AU}}{1.50 \times 10^{11}\,\text{m}} = 0.052\ \text{AU}$$

The planet's average orbital distance is 0.052 AU—small even compared to that of Mercury, which orbits our Sun at 0.39 AU. In fact, comparing the planet's 7.8-million-kilometer distance to the size of the star itself (presumably close to the 700,000-kilometer radius of our Sun), we estimate that the planet orbits its star at a distance only a little more than 10 times the star's radius.

ORBITAL PERIOD, DISTANCE, AND ECCENTRICITY All three of the major indirect detection methods tell us a planet's orbital period. The astrometric method allows us to observe the star's orbital motion around the system's center of mass, which means we know the star's orbital period; the planet's orbital period must be the same. The orbital period for a planet detected by the Doppler method is simply the time between peaks in the velocity curve (see Figure 11.11b). With the transit method, the orbital period is the time between repeated transits.

Once we know the orbital period, we can determine average orbital distance (semimajor axis) with Newton's version of Kepler's third law. For a small object like a planet orbiting a much more massive object like a star, this law expresses a relationship between the star's mass, the planet's orbital period, and the planet's average distance (see Cosmic Calculations 7.1). We generally know the masses of the stars with extrasolar planets (thanks to our understanding of stellar types discussed in Sections 11.1 and 11.5). Therefore, using the star's mass and the planet's orbital period, we can calculate the planet's average orbital distance (see Cosmic Calculations 11.1).

In addition, the astrometric and Doppler methods can tell us the eccentricity of a planet's orbit. Recall that all planetary orbits are ellipses, but ellipses can differ in eccentricity, which is a measure of how "stretched out" they are (see Figure 2.8). A planet with a perfectly circular orbit travels at a constant speed around its star, so the star moves with constant speed around the center of mass. Any variation in the star's speed tells us that the planet is moving with varying speed along its orbit and therefore must have a more eccentric elliptical orbit. Such measurements have revealed that many extrasolar planets have highly eccentric

---

Cosmic Calculations 11.2 FINDING MASSES OF EXTRASOLAR PLANETS

An object's momentum is defined as its mass m times its velocity v; like angular momentum [Section 3.4], momentum must be conserved. Consider a star with a single planet. Because the center of mass remains stationary between them (see Figure 11.7), the system has no momentum relative to this center of mass. The star's momentum ($m_{star} \times v_{star}$) must therefore be equal in magnitude (but opposite in direction) to the planet's momentum ($m_{planet} \times v_{planet}$):

$$m_{star}v_{star} = m_{planet}v_{planet}$$

Solving, we find that the planet's mass is

$$m_{planet} = \frac{m_{star}v_{star}}{v_{planet}}$$

We generally know the star's mass from its spectral type, and the Doppler method tells us the star's velocity toward or away from us (v_{star}). We can calculate the planet's orbital velocity (v_{planet}) from its orbital period p (the time between peaks in the Doppler curve) and its average orbital distance a (calculated with the method in Cosmic Calculations 11.1). If we assume a circular orbit, the planet travels a distance $2\pi a$ during each orbit that takes time p, so its orbital velocity is $2\pi a_{planet}/p_{planet}$. Substituting this expression for the planet's velocity into the above equation for mass gives

$$m_{planet} = \frac{m_{star}v_{star}\,p_{planet}}{2\pi a_{planet}}$$

Remember that with velocity data from the Doppler method, this formula gives us the *minimum* mass of the planet.

Example: Estimate the mass of the planet orbiting 51 Pegasi.

Solution: From Cosmic Calculations 11.1, we know the planet's orbital period ($p = 3.65 \times 10^5$ s) and orbital distance ($a = 7.81 \times 10^9$ m) and the star's mass ($M_{star} = 2.12 \times 10^{30}$ kg). The graph in Figure 11.11b shows that the star's velocity is about $v_{star} = 57$ m/s. We now plug these values into the above formula to find the planet's mass:

$$m_{planet} = \frac{m_{star}v_{star}p_{planet}}{2\pi a_{planet}}$$

$$= \frac{(2.12 \times 10^{30}\ \text{kg}) \times \left(57\ \frac{\text{m}}{\text{s}}\right) \times (3.65 \times 10^5\ \text{s})}{2\pi \times (7.81 \times 10^9\ \text{m})}$$

$$\approx 9 \times 10^{26}\ \text{kg}$$

The minimum mass of the planet is about 9×10^{26} kilograms, which is about half of Jupiter's mass (given in Appendix E as 1.9×10^{27} kilograms) and about 150 times the mass of Earth. The fact that the planet has a Jupiter-like mass but orbits very close to its star (as found in Cosmic Calcualtions 11.1), where we expect it to have a very hot surface, makes this planet an example of what astronomers refer to as a *hot Jupiter*.

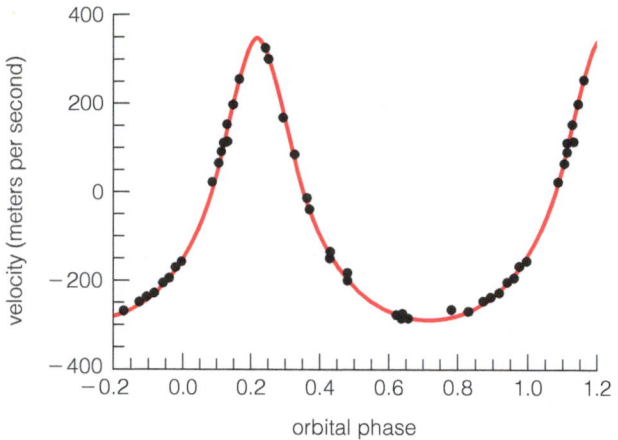

This graph shows Doppler data for the star 70 Virginis. Note the uneven nature of the change in velocity, which tells us that the planet causing the Doppler shifts is in a highly eccentric orbit.

orbits that bring them very close to their star on one side of their orbit and take them much farther from their star on the other side (Figure 11.15).

PLANETARY MASS Both the astrometric and the Doppler method measure motions caused by the gravitational tug of a planet, so both can in principle allow us to estimate planetary masses. These methods tell us about planetary masses because, for a given orbital distance, a more massive planet will cause its star to move at higher velocity around the center of mass (see Cosmic Calculations 11.2).

There is an important caveat for the Doppler method. Doppler shifts reveal only the part of a star's motion directed toward or away from us. For example, a planet whose orbit we view face-on (perpendicular to the plane of the orbit) does not cause any Doppler shift in the spectrum of its star, making it impossible to detect such a planet with the Doppler method (Figure 11.16a). We can observe Doppler shifts in a star's spectrum only if it has a planet orbiting at some angle other than face-on (Figure 11.16b), and the Doppler shift tells us a star's full orbital velocity only when we view an orbit precisely edge-on.

As a result, planetary masses that we infer from the Doppler method will be accurate only for planets in edge-on orbits, which are the same types of orbits that lead to transits. In other words, the Doppler method generally tells us a planet's precise mass only if we also observe transits of that same planet. In all other cases, the velocity inferred from Doppler shifts will be less than the full orbital velocity, which means that the mass we calculate will be the planet's *minimum* possible mass (or a "lower limit" mass). Statistically, however, we expect that the actual planetary mass should be no more than double the minimum mass in at least about 85% of all cases, so masses obtained by the Doppler method provide relatively good estimates for most planets. Moreover, in some cases we can combine Doppler data with other data (such as astrometric data) to determine a planet's precise orbital inclination, in which case we can calculate a more accurate mass.

FIGURE 11.16

The amount of Doppler shift we observe in a star's spectrum depends on the orientation of the planetary orbit as we view it from Earth.

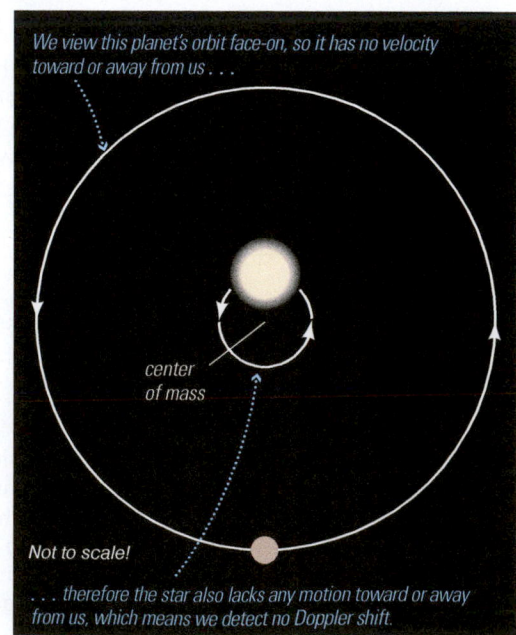

a If we view a planetary orbit face-on, we will not detect any Doppler shift at all.

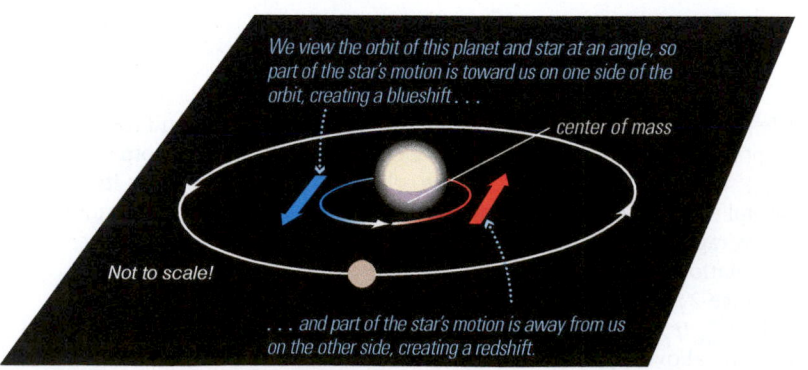

b We can detect a Doppler shift only if some part of the orbital velocity is directed toward or away from us. The more an orbit is tilted toward edge-on, the greater the shift we observe.

The transit method cannot by itself tell us masses for single planets, which is one reason scientists often try to follow up transit discoveries with the Doppler method, since the two methods together yield a precise planetary mass. However, the transit method can in some cases reveal planetary masses in multiplanet systems, because the gravitational tug of one planet on another can affect the timing of transits. For example, when another planet is tugging on a transiting planet in the direction in which it is moving, the tug will make the planet move a little faster so that the transit may begin tens of minutes to days earlier than it would otherwise; it may begin a similar amount of time later when the planet is tugging from the opposite direction. Precise measurements of timing can therefore allow us to measure slight variations in orbital speed and thereby determine the masses of the planets causing these variations. That is, we can use the timing data of one planet to determine the mass of another. This capability is particularly important for planets with masses too small to be detectable by the Doppler method with current technology.

PLANETARY SIZE Transit observations are presently the only means by which we can measure a planet's size or radius. The basic idea is easy to understand: The more of a star's light that a planet blocks during a transit, the larger the planet must be (Cosmic Calculations 11.3 provides an example of the calculations).

The Kepler 11 planetary system offers a great example. Figure 11.17 shows the brightness of the star Kepler 11 over a 110-day period. Each downward dip represents a transit in which a planet blocks a small fraction of the starlight. Careful study shows that there are six different planets (represented by the colored dots) that transit the star with periods ranging from 10 to 120 days; sometimes more than one transits at the same time, causing a larger dip. The panels show a transit of each planet in more detail, along with the planet's size calculated from the depth

FIGURE 11.17
The Kepler 11 system. The black line shows the brightness of the star Kepler 11 over a 110-day period. Colored dots indicate transits by six different planets. The panels at right show the dips due to each planet separately, and indicate the planet's size in Earth radii.

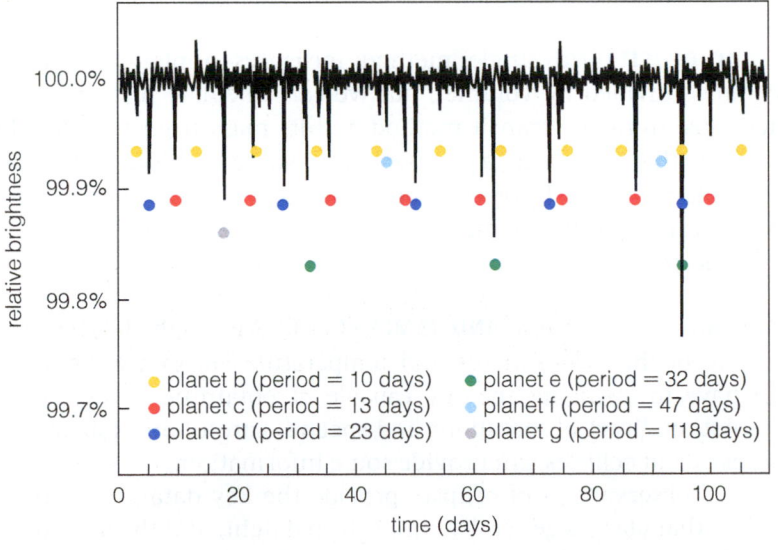

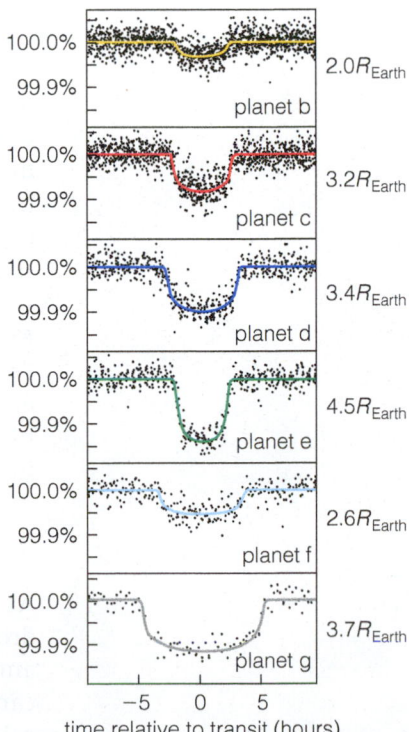

FIGURE 11.18

This diagram summarizes how we can combine data from the Doppler and transit methods to calculate a planet's average density. (The final density calculation requires converting the mass to grams and the volume to units of cubic centimeters.)

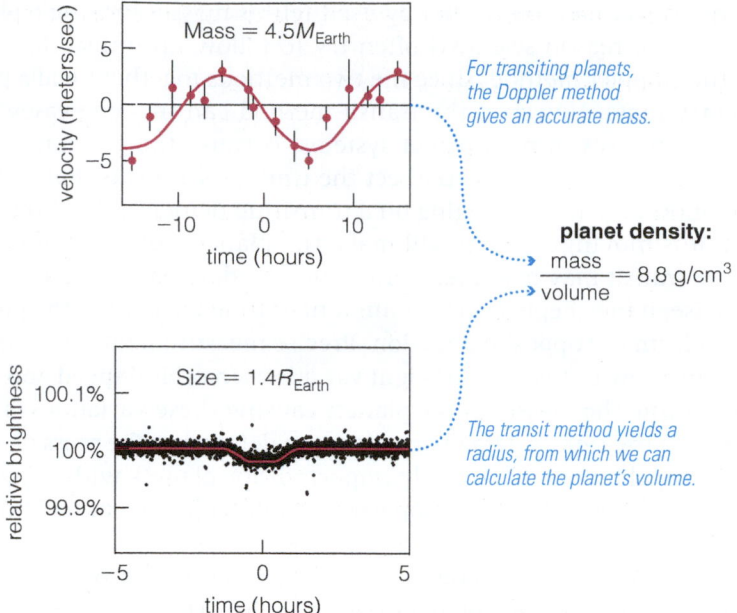

For transiting planets, the Doppler method gives an accurate mass.

planet density:

$$\frac{mass}{volume} = 8.8 \text{ g/cm}^3$$

The transit method yields a radius, from which we can calculate the planet's volume.

of the dip; note that more distant planets have wider dips because they orbit more slowly, which means their transits last longer. When we combine these sizes with the orbital periods and distances, we find that this system's planets have sizes ranging from 2 to 5 times Earth's size and all orbit between 0.1 and 0.5 AU from their star. Transported to our solar system, the six planets would lie inside Venus's orbit.

Think About It If we discover a planet and measure its size through the transit method, what follow-up observations are needed to learn its mass? If we discover a planet through the Doppler method, should we expect to be able to do follow-up observations to learn its size? Explain.

PLANETARY DENSITY No single method measures a planet's average density (mass divided by volume), but we can calculate it if we know a planet's size from the transit method and its mass from the Doppler method. Note that we can get a precise density value in this case, because transiting planets must have edge-on orbits, and the Doppler method gives an exact (rather than a minimum) mass in these cases. Figure 11.18 shows the process applied to a planet known as Kepler 10b.

ATMOSPHERIC COMPOSITION AND TEMPERATURE Although detailed understanding of the atmospheres and temperatures of extrasolar planets will have to await technology capable of obtaining moderate- to high-resolution direct observations and spectra, careful analysis of data from transits and eclipses can provide some information.

Infrared observations of eclipses provide the key data for temperature. Recall that planets generally emit infrared light, and the amount of infrared emission (per unit area) depends on the planet's temperature. As a planet goes behind its star, the system's infrared brightness will drop because we are no longer seeing the planet's infrared emission. The amount of the drop tells us how much infrared the planet emits, and we can combine this with the planet's radius (measured by the transits) to calculate an approximate temperature.

TABLE 11.3 A Summary of How We Measure Properties of Extrasolar Planets

	Planetary Property	Method(s) Used	Explanation
Orbital Properties	**period**	Doppler, astrometric, or transit	We directly measure orbital period.
	distance	Doppler, astrometric, or transit	We calculate orbital distance from orbital period using Newton's version of Kepler's third law (Cosmic Calculations 11.1).
	eccentricity	Doppler or astrometric	Velocity curves and astrometric star positions reveal eccentricity (Figure 11.15).
	inclination	transit or astrometric	Transits identify edge-on orbits; astrometric data measure any inclination angle.
Physical Properties	**mass**	astrometric or Doppler*	We calculate mass based on the amount of stellar motion caused by the planet's gravitational tug (Cosmic Calculations 11.2).
	size (radius)	transit	We calculate size based on the amount of dip in a star's brightness during a transit (Cosmic Calculations 11.3).
	density	transit and Doppler	We calculate density by dividing the mass by the volume (using size from transit method) (Figure 11.18).
	atmospheric composition and temperature	transit or direct detection	Transits and eclipses provide data on atmospheric composition and temperature.

*Transits in some multiplanet systems allow us to determine masses from the mutual gravitational tugs of the planets.

Transits and eclipses can also provide limited information about atmospheric composition. For a planet with an atmosphere, astronomers compare spectra of the system taken with the planet in front of (for transits) or behind (for eclipses) its star to spectra taken at other times. Careful analysis of differences between these spectra can reveal spectral lines caused by the planet's atmosphere. Because we generally know what gases are responsible for particular spectral lines, we can infer the existence of these gases in the planet's atmosphere.

Table 11.3 summarizes how we measure planetary properties.

11.3 The Nature of Extrasolar Planets

The number of extrasolar planets for which we have measured many key properties is now large enough that we are beginning to gain insight into how these planets compare to the planets of our own solar system. In this section, we'll explore current understanding of extrasolar planets and what that tells us about the likelihood of finding other planetary systems that are similar in nature to our own.

How do extrasolar planets compare with planets in our solar system?

Broadly speaking, we can divide comparisons of extrasolar planets and the planets of our solar system into two categories: comparisons of orbital properties (such as orbital period, distance, and eccentricity) and comparisons of physical properties (such as mass, size, and density). Together, these types of comparisons are teaching us about the nature of extrasolar planets, revealing that they exhibit a broader range of properties than we find among the planets of our own solar system.

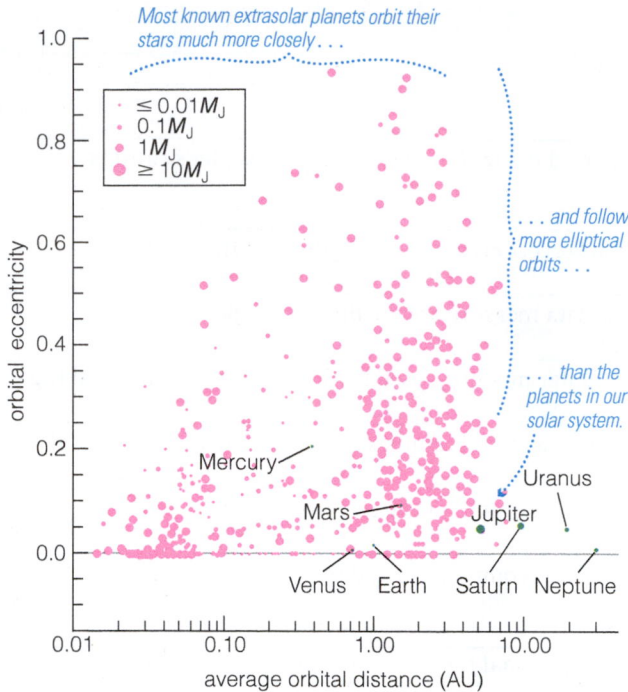

Orbital Properties of Extrasolar Planets

Most known extrasolar planets orbit their stars much more closely . . .

. . . and follow more elliptical orbits . . .

. . . than the planets in our solar system.

Key: ≤ 0.01M_J · 0.1M_J · 1M_J · ≥ 10M_J

Mercury, Mars, Venus, Earth, Jupiter, Saturn, Uranus, Neptune

(x-axis: average orbital distance (AU), 0.01 to 10.00; y-axis: orbital eccentricity, 0.0 to 1.0)

FIGURE 11.19

Orbital properties of more than 500 extrasolar planets with known masses, distances, and eccentricities. Dots closer to the left represent planets that orbit closer to their stars, and dots lower down represent planets with more circular orbits. Green dots are planets of our own solar system.

ORBITAL PROPERTIES Much as Johannes Kepler first appreciated the true layout of our own solar system [Section 2.2], we now have enough data to allow us to step back and see the layout of many other solar systems. The two key orbital properties of extrasolar planets that help us understand planetary system layout are (1) average orbital distance (semimajor axis), which allows us to compare the locations of planets in other solar systems to the locations of planets in our own, and (2) orbital eccentricity, which allows us to determine whether and by how much the orbits in other solar systems differ from the nearly circular orbits in our own solar system.

Figure 11.19 shows a graph of these two orbital properties for more than 500 extrasolar planets, with average orbital distance on the horizontal axis and orbital eccentricity on the vertical axis. By definition, eccentricity for an elliptical orbit is always between 0 and 1: An eccentricity of 0 is a perfect circle, and higher eccentricity means a more stretched-out ellipse. The sizes of the dots on the graph represent approximate masses of the planets, as shown in the key at the upper left. These planets were detected by the Doppler method, so the indicated masses are minimum masses. Be sure to note that the distance scale increases by powers of 10, which makes it easier to plot the large number of planets with relatively close-in orbits.

Look first at the green dots representing the planets in our own solar system. To be sure you are reading the graph correctly, make sure you see how it shows that the four inner planets orbit within about 1.5 AU of the Sun, while the four outer planets are spread out to about 30 AU. Also notice that all but Mercury have very small eccentricities, meaning nearly circular orbits. Now look at the dots representing extrasolar planets. Three differences between many of the extrasolar planets and our solar system's planets should jump out at you:

1. Many of the known extrasolar planets orbit their stars more closely than Mercury orbits the Sun, and almost none are located as far from their stars as are the jovian planets of our solar system.

Cosmic Calculations 11.3 FINDING SIZES OF EXTRASOLAR PLANETS

We determine planet radii from the fraction of a star's light blocked during a transit. Viewed against the sky, both the star and the planet appear as tiny circular disks. These disks are far too small for our telescopes to resolve, but the fraction of the star's light that is blocked must be equal to the area of the planet's disk (πr^2_{planet}) divided by the area of the star's disk (πr^2_{star}). We generally know the approximate radius of the star (from its spectral type), so the fractional drop in the star's light during a transit is

$$\frac{\text{fraction of light blocked}} {} = \frac{\text{area of planet's disk}}{\text{area of star's disk}} = \frac{\pi r^2_{planet}}{\pi r^2_{star}} = \frac{r^2_{planet}}{r^2_{star}}$$

Solving for the planet's radius, we find

$$r_{planet} = r_{star} \times \sqrt{\text{fraction of light blocked}}$$

Example: Figure 11.13 shows a transit of the star HD 189733. The star's radius is about 800,000 kilometers (1.15R_{Sun}), and the planet blocks 1.7% of the star's light during a transit. What is the planet's radius?

Solution: We simply plug the star's radius (800,000 km) and the fraction of its light blocked during a transit (1.7% = 0.017) into the above equation:

$$r_{planet} = r_{star} \times \sqrt{\text{fraction of light blocked}}$$

$$= 800,000 \text{ km} \times \sqrt{0.017}$$

$$\approx 100,000 \text{ km}$$

The planet's radius is about 100,000 kilometers, which is about 1.4 times Jupiter's radius of 71,500 kilometers. That is, the planet is about 40% larger than Jupiter in radius.

2. Many of the planets with close-in orbits are Jupiter-like in mass, making them examples of what astronomers call **hot Jupiters,** because they are presumed to be jovian in nature but must be hot owing to proximity to their star.
3. Many of the planets also have large orbital eccentricities, telling us that their elliptical orbits have very stretched-out shapes.

As we discussed in Section 3.5, these orbital properties can be explained by planetary migration, suggesting that this type of migration played a much larger role in many other solar systems than in our own.

Think About It Should we be surprised that we haven't found many planets orbiting as far from their stars as Saturn, Uranus, and Neptune do from the Sun? Why or why not?

We can learn even more from multiplanet systems, which allow us to make more detailed comparisons between other planetary systems and our own. Several hundred stars have so far been found to have two or more planets (or *Kepler* planet candidates), including the six-planet Kepler 11 system (see Figure 11.17). As a result, we've learned that other planetary systems can sometimes differ from ours in at least two ways. First, many have planets packed much closer to each other than our solar system's planets (as is the case for Kepler 11), which may be a consequence of planetary migration. Second, the *Kepler* mission has identified candidate planets in binary star systems, showing that planetary systems need not be like ours with a single star and confirming predictions that at least some orbits in binary systems should be stable (see Figure 11.5).

PHYSICAL PROPERTIES The key physical properties we can measure are mass and size, and for the cases in which we have measured both, we can also calculate density. These cases are particularly valuable, because they allow us to address a key question: Do extrasolar planets fall into the same terrestrial and jovian categories as the planets in our solar system, or do we find additional types of planets?

Scientists address this question by creating models that use our understanding of the behavior of different materials to determine the expected composition of a planet from its mass and radius. The results are shown in Figure 11.20 for a sample of planets for which both mass (usually from the Doppler method) and radius (from transits) are known. Be sure you understand the following key features of the figure:

- The horizontal axis shows planetary mass, in units of Earth masses. (The top of the graph shows the equivalent values in Jupiter masses.) Notice that this axis uses a scale that rises by powers of 10 because the masses vary over such a wide range.
- The vertical axis shows planetary radius in units of Earth radii. (The right side of the graph shows the equivalent values in Jupiter radii.)
- Each dot represents one planet for which both mass and radius have been measured. Planets of our solar system are marked in green.
- The paintings around the graph show artist conceptions of what representative worlds might look like.

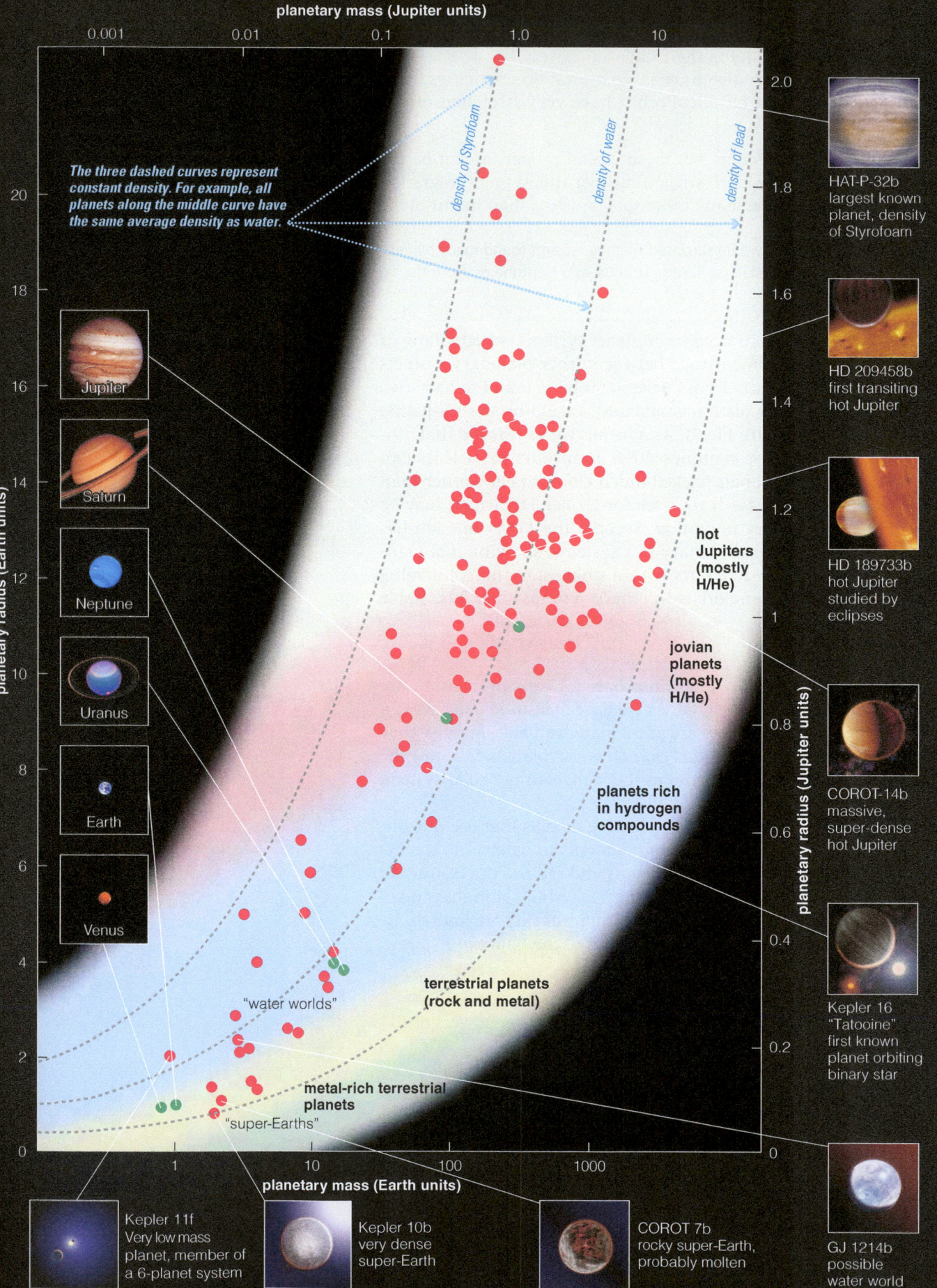

FIGURE 11.20 Masses and sizes of a sample of extrasolar planets for which both have been measured, compared to those of planets in our solar system. Each dot represents one planet. Dashed lines are lines of constant density for planets of different masses. Colored regions indicate expected planet types based on models of their compositions.

planetary mass (Jupiter units)

planetary mass (Earth units)

planetary radius (Earth units)

planetary radius (Jupiter units)

The three dashed curves represent constant density. For example, all planets along the middle curve have the same average density as water.

density of Styrofoam

density of water

density of lead

hot Jupiters (mostly H/He)

jovian planets (mostly H/He)

planets rich in hydrogen compounds

terrestrial planets (rock and metal)

"water worlds"

metal-rich terrestrial planets

"super-Earths"

Jupiter

Saturn

Neptune

Uranus

Earth

Venus

HAT-P-32b largest known planet, density of Styrofoam

HD 209458b first transiting hot Jupiter

HD 189733b hot Jupiter studied by eclipses

COROT-14b massive, super-dense hot Jupiter

Kepler 16 "Tatooine" first known planet orbiting binary star

GJ 1214b possible water world

Kepler 11f Very low mass planet, member of a 6-planet system

Kepler 10b very dense super-Earth

COROT 7b rocky super-Earth, probably molten

- Average density is easy to calculate from mass and radius, but the different scales used on the two axes make it difficult to read average density directly from the graph. To help with that, the three curves extending from the lower left to the top show three representative average densities.
- The colored regions indicate models representing the expected compositions of planets with the indicated combinations of mass and radius.

As you study Figure 11.20, notice that extrasolar planets show much more variety than the planets of our own solar system. For example, HAT-P-32b has more than twice Jupiter's radius despite having the same mass, giving it an average density of about 0.14 g/cm^3—similar to that of Styrofoam. This low average density is probably a result of the fact that this planet orbits only 0.035 AU from its star, putting it more than 10 times closer to its star than Mercury is to the Sun. This close-in orbit gives the planet a very high temperature, making it an example of a hot Jupiter; this high temperature should puff up the planet's atmosphere and may explain why the planet has such a large size relative to its mass. Near the other extreme, the planet COROT-14b is only slightly larger than Jupiter but several times as massive, giving it an average density near 8 g/cm^3, about the same as the density of iron. Although such a high average density might seem surprising, models indicate that it is actually expected for a jovian planet (made of hydrogen and helium) with such high mass. The reason is that the planet's large mass and strong gravity compress its hydrogen and helium to much higher density than is the case for Jupiter. In other words, despite the wide spread in their densities, both HAT-P-32b and COROT-14b seem clearly to fall into the jovian planet category.

We also see many planets that appear to be terrestrial in nature, with compositions of rock and metal. For example, COROT-7b has an average density near 5 g/cm^3, comparable to Earth's density. Because it has a mass about 5 times that of Earth, COROT-7b is an example of what is sometimes called a "super-Earth." It orbits very close to its star, so its surface is probably molten. This and the many other known super-Earths are likely to have a rock/metal composition similar to that of the terrestrial worlds in our solar system.

Perhaps the most surprising planets shown in Figure 11.20 are the ones that are not clearly in either the terrestrial or the jovian category. Several planets cluster in the region of Uranus and Neptune, and perhaps share their composition of hydrogen compounds shrouded in an envelope of hydrogen and helium gas. Others (perhaps GJ 1214b) appear to fit the model for "water worlds," which may be made predominantly of water, either in liquid form or as a high-pressure solid, or perhaps of other hydrogen compounds. Alternatively, some of these worlds might be composed of a dense rocky/metallic core and a thick envelope of low-density hydrogen-helium gas.

Finally, it's worth noting that while we've focused here on composition, orbital distance must also play a role in a planet's nature. In particular, because many of the planets known to date orbit fairly close to their stars, they will be much hotter than similar planets in our own solar system. We've already discussed how this could cause hot Jupiters to have puffed-up atmospheres and terrestrial worlds to have molten surfaces. Hot Jupiters might also have very different clouds than we see on the actual Jupiter, such as clouds of mineral flakes instead of clouds

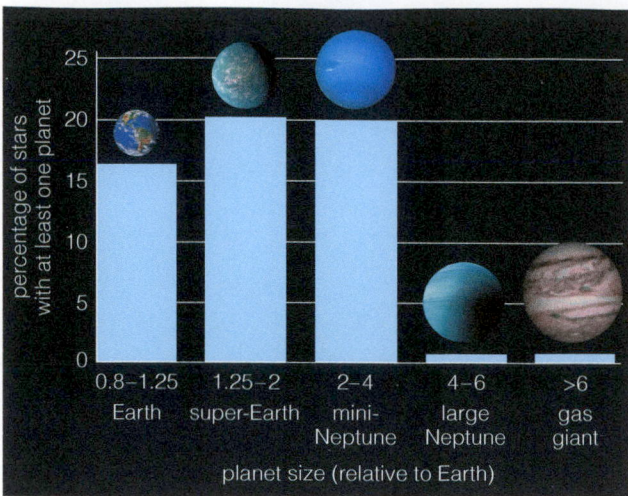

FIGURE 11.21

This bar chart shows the estimated proportions of all stars that have planets of different size categories, based on *Kepler* results. Because *Kepler* data have so far been analyzed only for planets with relatively short orbital periods, these estimates are nearly certain to increase as additional data are studied.

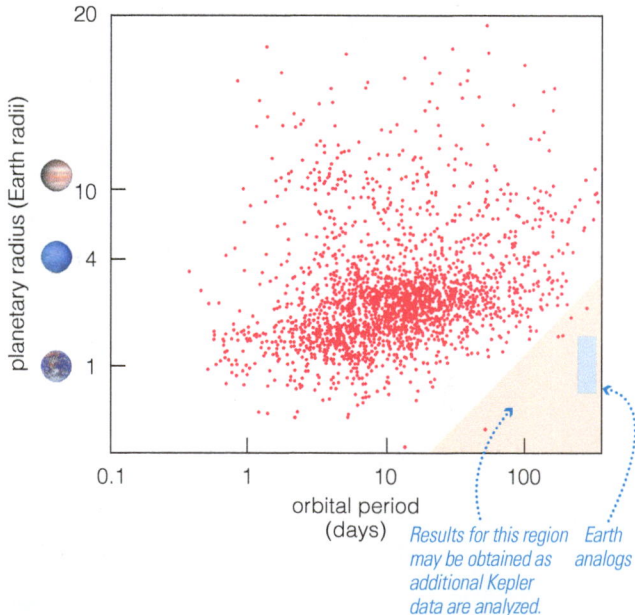

Results for this region may be obtained as additional Kepler data are analyzed. Earth analogs

FIGURE 11.22

This figure shows the orbital periods and sizes of all the candidate planets identified from Kepler data as of mid-2015. As additional data come in, we should learn more about the region in the lower right, which would include possible Earth-like planets.

of ammonia snow or water droplets. Worlds that might resemble larger versions of Ganymede or Titan in our solar system would be water worlds at closer orbital distances. With enough warmth, water worlds might become "steam planets" with vast amounts of water vapor in their atmospheres.

The bottom line is that extrasolar planets are abundant and diverse. Unlike our solar system, with only two clear types of "typical planets," these new solar systems have additional types of planets that defy easy categorization.

Are planetary systems like ours common?

We have discovered planetary systems around thousands of stars, but these represent only a tiny fraction of the more than 100 billion stars in the Milky Way Galaxy. What about the rest, which we haven't yet studied? Are planetary systems common and, if so, are planets similar to Earth common as well? Remarkably, we can now begin to answer these questions, because the *Kepler* mission observed an essentially random sample of stars in our galaxy. It is possible to use simple geometry to calculate the fraction of these stars that would be expected to have transiting planets if they all had planetary systems. We can then use the statistics of the actual *Kepler* discoveries to estimate the fraction of all stars that have planets of various sizes.

Figure 11.21 shows the results of such statistical studies through mid-2015. Two remarkable conclusions are apparent. First, planets are common. By looking across all size categories, astronomers conclude that at least 70% of all stars harbor at least one planet. Second, small planets appear to outnumber large planets by a significant margin, suggesting that Earth-size planets are also very common. Keep in mind that these statistics reflect only planets with relatively short orbital periods, both because *Kepler* observed for only four years (and required at least three transits for confidence in a planet detection) and because not all its data have yet been analyzed. The percentages shown in Figure 11.21 can only increase as we gather and process more data.

The fact that the current statistics are dominated by planets that orbit relatively close to their stars means that most of the planets represented in Figure 11.21 are probably too hot to harbor life. However, Figure 11.22 shows *Kepler* data in a different way: as a graph of planet size versus orbital period. At first glance, the empty space at lower right might seem to suggest there are few Earth-sized planets in Earth-sized orbits. But the lack of data points in this region of the graph actually stems from the extreme difficulty in detecting the infrequent, shallow transits such planets make. Accounting statistically for these effects, scientists now estimate that at least 10% to 20% of stars are likely to have a planet less than twice Earth's size in an orbit within its star's habitable zone. Although only eight such planets had been detected as of mid-2015, the statistics suggest that the number of such planets in the galaxy may be in the tens of billions.

These are remarkable results. If we put them all together, we now have statistical evidence indicating that the vast majority of stars have planetary systems, that most of these systems have multiple planets, that many of these planets are similar in size to Earth, and that many of these Earth-size planets orbit within the habitable zones around their stars. However, we do not yet know whether being Earth-size within a habitable zone necessarily leads to Earth-like conditions (such as continents and oceans) or life, a question we will discuss in the next section. So while it seems likely that many other planetary systems will turn out to be much like ours, we will need more data to know for sure. The bottom line is that, while we cannot yet answer with certainty the questions of whether planetary systems like ours and planets like Earth are common, we are rapidly learning more and will likely have definitive answers to these questions within the next decade or two.

Think About It Not long before most of today's college students were born, the only known planets were those of our own solar system. Today, the evidence suggests that many or most stars have planets. For the galaxy as a whole, that's a change in the estimated number of planets from fewer than 10 to more than 100 billion. How do you think this change should alter our perspective on our place in the universe? Defend your opinion.

11.4 The Habitability of Extrasolar Planets

The results we've discussed so far in this chapter are both exciting and daunting. Exciting in that we now know there are billions of planets in our galaxy that we might want to examine for potential habitability and life. Daunting in that having spent nearly all this book so far just exploring the potential habitability of the planets and moons of one solar system (our own), we now face the challenge of considering the habitability of billions of times as many worlds. Fortunately, we can draw on what we've learned about our own solar system and our limited understanding of extrasolar planets to make a few general statements. In this section, we'll discuss the potential habitability of extrasolar worlds, how we might find life if it exists on any of these worlds, and a controversial hypothesis suggesting that life—and especially intelligent life—might be much rarer than we would otherwise suppose.

What kinds of extrasolar worlds might be habitable?

The lessons from our own solar system tell us that we should consider habitability in more than one way. For surface habitability like we find on Earth, we need a world that is large enough to have an atmosphere and oceans and is located within its star's habitable zone. But our study of jovian moons like Europa and Enceladus tells us we need to also consider the prospects of subsurface life, which may not be limited to habitable zones and can include moons as well as planets. It's also possible that some moons could have the conditions necessary to support surface life. Let's briefly consider the types of worlds that could potentially be habitable.

SURFACE HABITABILITY When we speak of surface habitability, we are looking for worlds with oceans and atmospheres at least somewhat like those of Earth, or at least like those Mars may have had early in its history, before its surface water dried up. These worlds by definition must orbit within their star's habitable zone. This fact likely rules out planets with very eccentric orbits, since these orbits would cross in and out of the habitable zone, presumably making life unlikely. However, planets with less eccentric orbits that only briefly exit the habitable zone might be habitable if they had atmospheres or oceans, which could act as heat stores to keep the climate on these worlds tolerable for life. In any case, given that plenty of planets have relatively circular orbits and that most stars have well-defined habitable zones—including many stars in binary or multiple-star systems—we expect that we will find vast numbers of planets orbiting within habitable zones.

The key question then seems to be which types of worlds within habitable zones might actually be habitable. We can probably rule out any world much smaller than Mars, because much like our Moon, such small worlds would be unlikely to have an atmosphere and surface pressure that could allow for surface liquid water. Moreover, if we are correct in our ideas about the role of a magnetic field in protecting an atmosphere and about climate regulation through plate tectonics, the world may also have to be comparable to or larger than Earth in size, although the exact minimum size for surface habitability is still being debated [Section 10.3].

The most obvious candidates for surface habitability would be Earth-size planets in habitable zones. As we discussed previously, some preliminary analyses suggest that such planets are quite common, with planets less than about twice Earth's size in a habitable zone orbiting at least 10% to 20% of all stars. The number could go higher still through at least two other categories of worlds.

First, there are worlds somewhat larger than Earth, but still not so large that they would be subject to the kinds of vertical winds that we think make jovian planets unlikely to be habitable [Section 7.3]. For example, super-Earths likely have compositions very similar to that of Earth, even though they are up to several times as massive. The major unknown is how these larger masses would affect the likelihood of plate tectonics for climate regulation. The water worlds also offer possibilities for surface habitability if they actually have surface water. Current models are not adequate for us to say with confidence whether this is likely, but if it is, these worlds could be one of the most common types with surface habitability.

The second additional category to consider for surface habitability is moons within habitable zones that might be large enough to have surface oceans. There are no such moons in our solar system, but what we've learned from our solar system suggests three possible ways that such moons might come to exist. One possibility is that they might form as a consequence of giant impacts like that thought to have formed our own Moon. For example, while our Moon has a mass only about 1/80 that of Earth, we might imagine that a similar giant impact on a super-Earth could lead to the formation of a moon with an Earth-like mass. The second possibility is that they form as ice-rich moons of jovian planets, just like the jovian moons in our own solar system. Then, if their planet migrated inward and the moons managed to remain in orbit, they might eventually find themselves in the habitable zone. A third possibility is the formation of giant planets with large moons close to their home stars, something that models suggest may be possible in some situations; in this case, migration inward wouldn't be required. The compositions of these moons might make the criteria for surface habitability somewhat different than they are for terrestrial worlds, though we do not yet understand planetary dynamics well enough to know the specifics.

EXTENDED HABITABLE ZONES The prospects for surface habitability might be much greater if we are underestimating the size of habitable zones, and the discovery of planets like super-Earths and planets that could turn out to be water worlds suggests that this might indeed be the case. The standard way in which habitable zones are calculated—such as the zones shown in Figure 11.3—assumes rocky planets (like Venus, Earth, and Mars) with greenhouse effects from water vapor and carbon dioxide in their atmospheres. However, super-Earths and water worlds can be several times as massive as Earth, and as a result they may retain substantial atmospheres of hydrogen gas captured from a solar nebula during planet formation. This hydrogen can also act as a greenhouse gas, which means it could keep these planets warm enough to maintain surface liquid water at distances well beyond those usually assumed for habitable zones. Some estimates suggest that these planets could have habitable surfaces even at distances up to 10 or more times as far from their star as the standard calculations would suggest. For example, a super-Earth or a water world with a substantial hydrogen atmosphere might be habitable at Saturn's distance from the Sun, or even beyond. Research into this idea is continuing, and it raises the intriguing possibility that surface habitability may be possible for a greater range of distances from a home star than we have generally assumed.

SUBSURFACE HABITABILITY Just as we found in our own solar system, where we've identified Mars and several jovian moons as possibly having habitable subsurface regions, it is possible that there could be far more worlds with subsurface than surface habitability. These worlds also fall into several categories. First, there are planets, like Mars, that are too small to have the climate stability that would have allowed them to

Movie Madness STAR WARS

It's "a long time ago" in someone's far-off galaxy, and the political situation is turning ugly. Yes, we're talking about *Star Wars,* the cinematic space opera that, like relatives, just keeps on returning (with seven installments through 2015).

The premise of the original *Star Wars* ("Episode IV") is that a galaxy-wide republic has been hijacked and converted to an autocratic, evil empire by despots in gray flannel suits. This may sound vaguely reminiscent of the story of Rome, but unlike what happened to that ancient civilization, this political shift has encouraged a serious rebellion, a war among the stars. The rebels are led by Princess Leia (you can tell she's a princess because her hair is done up like twin Danish pastries), and her strategy is to take out the empire's headquarters—an enormous, spherical spacecraft known to its friends as "the Death Star." The Death Star packs weaponry that can explode a planet in seconds (calculate, if you wish, the energy required to do that). On the other hand, the rebels have "the Force" on their side—a mystical ability to change the odds of every situation based on moral merit and self-discipline.

Most of *Star Wars* is battle of the sort that's familiar to any movie fan, except that the bad guys wear brittle, white plastic suits and fly spacecraft that look like box kites. But *Star Wars* offers some interesting peeks into life as it might be elsewhere in the cosmos. The rebels have their base of operations on a large planet's moon, a not-impossible scenario since hefty moons could be habitable. Luke Skywalker, the young hero, hails from a world circling a close double star, which research suggests could be possible.

There are peculiar anachronisms in *Star Wars,* however. The Death Star is obviously extremely high-tech, and yet the principals occasionally face off using souped-up swords. Everyone jets around in spacecraft that are capable of circumventing the prohibition of faster-than-light travel by jumping into hyperspace (in essence, leaving the universe and then jumping back in, thereby avoiding the need to follow universal laws while not in the universe), and yet we often see aliens saddled up on giant, dinosaur-like creatures.

All of that can be forgiven. But *Star Wars* takes its biggest literary license in showing dozens of alien races all living contemporaneously (although it's clear that the human types are in charge). In the movie's famous cantina scene, which takes place in the wretched port city of Mos Eisley, aliens of all shapes and colors get together to do business and get drunk. In fact, the chance that any two intelligent species (let alone dozens) would arise in the galaxy within 100,000 years of each other is quite small. If there are other societies out there, they will be either far behind us or enormously beyond our level. We won't be sharing dance music and booze with them in a seedy extraterrestrial dive.

And besides that, why does a republic have a princess, anyhow?

This artist's conception shows an orphan planet—a planet that has escaped from its star system—as it might appear in the infrared. Orphan planets are too far from any star to be observable with reflected visible light and too cool to emit visible light of their own, but they would emit infrared light.

maintain habitable surfaces for billions of years, but that might have had habitable surfaces for a shorter time and might have retained subsurface liquid water for far longer. Second, there are jovian moons that might be similar to Europa, Enceladus, Titan, and other moons of our solar system that we've identified as potentially habitable. Beyond that, some of the "other" categories of extrasolar planets, such as the water worlds or even super-Earths located beyond the habitable zone, might also offer opportunities for finding subsurface liquid water. For example, because we expect super-Earths to contain much more internal heat than Earth, they might have deep subsurface zones of liquid water even if they are far enough from their star to have completely frozen surfaces.

ORPHAN PLANETS There is another category of potentially habitable worlds that we have not yet discussed: **orphan planets** (sometimes called *rogue planets*), which do not orbit a star. These "planets" pose a definitional challenge since we usually *define* a planet as something that orbits a star, but it's easy to understand where the idea comes from. Recall that one mechanism by which jovian planets may migrate inward in some planetary systems is through close gravitational encounters in which one object loses energy and moves inward while the other object gains energy and is flung outward [Section 3.5]. Given that migration appears to be quite common in planetary systems, it seems reasonable to expect that many planets have been flung outward into interstellar space in this way. Orphan planets would be extremely difficult to detect directly because they are so dim, but there are other potential ways of finding them (Figure 11.23). One observational search (using a strategy based on *microlensing* events similar to those we briefly mentioned earlier) found that such orphan planets may be about as numerous as the stars in our galaxy. In addition, some scientists have developed models suggesting that planet-size bodies can form independently (rather than being born around a star), in which case the number of such planets could be far higher.

You might at first guess that planets without a star would lack an energy source for life, but our study of jovian moons should temper that idea: If moons like Europa are indeed habitable, it is due to energy from internal heat, not to energy from the Sun. A planet like Earth retains internal heat for billions of years, and therefore might have active volcanism and tectonics—and the prospects of subsurface liquid water—even if it were not orbiting the Sun. In that case, many orphan planets might have subsurface habitable zones.

Moreover, some researchers suggest that orphan planets might even have surface oceans and habitability. Much like the idea behind extended habitable zones, this might be possible if the planets have a thick enough hydrogen atmosphere. In that case, internal heat leaking outward might be trapped by the greenhouse effect, and in some cases this might raise the surface temperature above the freezing point of water. Indeed, with a thick enough atmosphere, even a planet the size of Earth could potentially offer such conditions while floating freely in interstellar space. A key question, then, is whether a thick hydrogen atmosphere would be likely around planets similar in size to Earth. No one really knows, but remember that the main reason Earth and the other terrestrial planets did not retain hydrogen from the solar nebula is that their small sizes allowed the lightweight hydrogen to escape. Because the speed of hydrogen atoms depends on the temperature, which would be far lower for planets ejected into interstellar space, it is possible that orphan planets might retain their hydrogen atmospheres, if they acquired such atmospheres in the first place. However, although these conditions might allow surface habitability on orphan planets, these worlds would not be bathed in the high-intensity energy that you can get from a star and that is necessary for photosynthesis on Earth.

How could we detect life on extrasolar planets?

At present, we have no way to search for actual life on any of the many extrasolar worlds that may potentially be habitable, other than by using experiments, known as SETI, that could find evidence of technically accomplished life [Section 12.3]. None of our indirect methods for learning about these worlds (astrometric, Doppler, transit) can offer information that would tell us whether they have life, and our direct detection capabilities remain too rudimentary to detect the presence of life. However, we expect our direct observational capabilities to improve dramatically in coming decades, and as a result, scientists are working on strategies that might allow future telescopes to search for life on worlds around other stars.

This immediately bring us to an important caveat about the search for life beyond our solar system: While we have some hope that new telescopes may allow us to detect the presence of life on the *surfaces* of extrasolar worlds, no technology likely to arise in this century seems capable of detecting life that exists only deep in the subsurface. To understand why, consider the cases of Mars and Europa. Both might potentially have subsurface life, but the only way we are going to find it is by going to these worlds and searching for it. As we'll discuss in Chapter 13, the technology that would enable us to visit extrasolar worlds is so far beyond our current capabilities that such visits do not seem plausible in the near future. Telescopically, we might learn about the *potential* for subsurface habitability, but not about life itself.

Nevertheless, we've seen that there are likely to be vast numbers of worlds with habitable surfaces, and these will be open to study. How might we search for life on them? As with most astronomical observations, we can hope to learn through both images and spectra. Let's begin with images. Even relatively crude images of an extrasolar planet or moon—say, just a few pixels in area—might provide important clues. For example, simply watching how a world changes in brightness might tell us about the presence of oceans and the ratio of ocean to land, because seas are darker than continents. We might also find that the light from such a world changes from day to day, because of clouds, or from season to season, because of snow or ice. With higher resolution, we might imagine seeing changes that, with sufficient study, might prove to be due to life, such as changes in vegetation.

Spectra would allow us to measure many other properties. For example, even a fairly crude spectrum should allow us to gauge the surface temperature of a planet. If we could collect enough light from a distant planet, we could make a more detailed spectral analysis, one that might suggest far more convincingly that a planet was home to life. For instance, we might gain information on the types of minerals or ices on its surface. If we found that the planet had characteristics like those of Earth or Mars, we would know that it was habitable in principle, though perhaps not whether it actually harbored life.

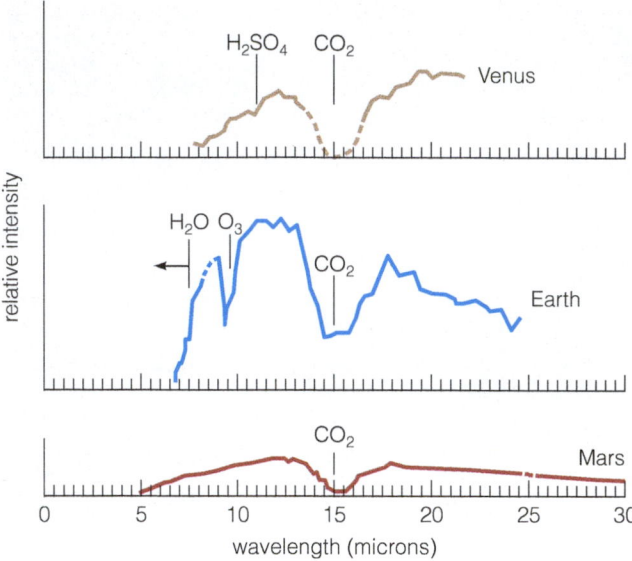

FIGURE 11.24

The infrared spectra of Venus, Earth, and Mars, as they might be seen from afar, showing absorption features that point to the presence of various gases in their atmospheres. While carbon dioxide is present in all three spectra, only our own planet has appreciable oxygen (and hence ozone)—a product of photosynthesis. If we could make similar spectral analyses of distant planets, we might possibly detect atmospheric gases that would indicate life.

With spectra from infrared telescopes, we could search for the absorption or emission features of gases in the atmosphere of a planet. Several of these gases will be easy to detect if they are present, including carbon dioxide, ozone, methane, and water vapor (Figure 11.24). While the mere presence of such gases would not necessarily point to life, their precise abundances and the combinations in which they occurred could provide stronger evidence about whether life was present. With large telescopes of the future, even the direct detection of compounds like chlorophyll would be possible in principle.

For example, Earth's atmosphere has large amounts of oxygen, the result of photosynthesis. If we found abundant oxygen in the atmosphere of another world, we would have reason to suspect the presence of life, particularly if the ratio of oxygen to the other detected gases seemed incompatible with nonbiological chemistry. To some extent, the same is true of methane, which is present in Earth's atmosphere today largely thanks to the "exhaust" gases produced by livestock and rice paddies. In early times, before photosynthesis raised the oxygen level, the atmosphere may have been altered by another metabolic process, known as *methanogenesis,* through which microbes expel methane rather than oxygen. The first billion years or so of Earth's biological history might therefore have been marked by the presence of atmospheric methane.

The bottom line is that for billions of years, life on Earth has been making its presence known to anyone with a telescope large enough to find our world and make a spectrum of its reflected light. In principle, we could identify life on other worlds in the same way. Perhaps in the next few decades we will discover abundant oxygen or methane on a distant world visible to us as no more than a dot in a telescope, providing an exciting and encouraging clue that it harbors life.

Are Earth-like planets rare or common?

Based on everything we've discussed to this point in the book, and on the fact that we now expect billions of planets in our galaxy to be at least potentially habitable, it may seem reasonable to assume that the prospects for life on extrasolar worlds—and in some cases, advanced or even intelligent life—are quite good. Most astrobiologists would probably agree with this assumption, but some have questioned it. In essence, these scientists argue that the long-term evolution of life on Earth has been possible only because our planet has been the fortunate beneficiary of several kinds of "planetary luck," without which life would have been unlikely to progress beyond simple microbes, if it got started at all. This suggestion, sometimes called the "rare Earth hypothesis," would have profound implications if true, particularly for the efforts to search for extraterrestrial intelligence, which we will discuss in the next chapter. We'll therefore conclude our discussion of extrasolar habitability by briefly examining a few of the key issues in the rare Earth hypothesis. Note that because this hypothesis is directed primarily at prospects for highly evolved life, it focuses on Earth-like planets in habitable zones, since these seem most likely to support advanced surface life.

GALACTIC CONSTRAINTS Proponents of the rare Earth hypothesis suggest that Earth-like planets can form in only a relatively small region of the Milky Way Galaxy, making the number of potential homes for life far smaller than we might otherwise expect it to be. In essence, they argue that there is a fairly narrow ring at about our solar system's distance from the center of the Milky Way Galaxy that makes up a *galactic habitable zone* analogous to the habitable zone around an individual star (Figure 11.25).

The basic argument holds that regions beyond the galactic habitable zone are unlikely to have Earth-like planets because stars farther from the galactic center tend to have lower proportions of elements besides hydrogen and helium, needed to make terrestrial worlds. With regard to regions interior to the galactic habitable zone, they argue that the increased crowding of stars near the galactic center makes supernovae more common, and supernovae will expose any habitable planets to radiation that would be detrimental to life.

However, other scientists offer counterarguments to both sets of galactic constraints. As we discussed in Section 11.1, results from the *Kepler* mission suggest that Earth-size planets can form even around stars with significantly lower heavy-element abundances than our Sun. In that case, the outer boundary of any galactic habitable zone would be much farther out than the one shown in Figure 11.25, if there is an outer boundary at all. The existence of an inner boundary is similarly uncertain, because while there would undoubtedly be more nearby supernovae for planets in the crowded inner regions of the galaxy, we do not really know whether the resulting radiation would be fatal to life. A planet's atmosphere might protect life from this radiation, and even if radiation got through the atmosphere, we can't be certain that the radiation would be detrimental rather than beneficial to the evolution of life. For example, because radiation would increase the mutation rate, it might actually accelerate rather than inhibit evolution.

IMPACT RATES AND JUPITER A second issue raised by rare Earth proponents concerns impact rates. Recall that Earth was probably subjected to numerous large impacts—some possibly large enough to vaporize the oceans—during the first few hundred million years after our planet was born [Section 4.3]. In our solar system, the impact rate lessened dramatically after that. Might the impact rate remain high much longer in other solar systems?

The most numerous small objects in our solar system are the trillion or so comets located in the distant *Oort cloud* [Section 3.4], where they pose little threat to Earth. However, recall that these objects are thought to have formed much closer to the Sun, and their current great distances are a result of close encounters with the jovian planets, especially Jupiter. If Jupiter did not exist, these objects might have remained in regions of the solar system where they were far more likely to pose an impact threat. The view of rare Earth proponents is that only the "luck" of having Jupiter as a planetary neighbor made the end of the heavy bombardment and our existence on Earth possible.

The primary question in this case is just how "lucky" this situation might be. Current data suggest that Jupiter-size planets are quite common, though so far we don't know how commonly they are found in Jupiter-like orbits, since the limitations of current detection techniques have made it much easier to detect such planets when they've migrated to close-in orbits around their stars. But it should be only a matter of time (primarily because we need more years of observation to detect a planet in a more distant orbit) until we know whether having a Jupiter as a planetary companion is actually rare. In addition, we might again argue that a higher rate of impacts could actually accelerate the evolution of life.

FIGURE 11.25
The green ring in this diagram of the Milky Way Galaxy highlights what some scientists hypothesize to be a galactic habitable zone—the only region of the galaxy in which Earth-like planets can form. However, other scientists doubt the claims that underlie this hypothesis, in which case Earth-like planets could be far more widespread.

Think About It The story of life on Earth is replete with disasters that served to stress terrestrial species, resulting in the rapid evolution of new, more complex organisms. For example, the K–T impact [Section 6.4] apparently led to the demise of the dinosaurs and opened the door for the rise of mammals. More recently, ice ages are thought to have played a major role in the evolution of modern humans. Do you think it's possible that a higher rate of impacts could be *good* rather than bad for life on another planet? Explain.

CLIMATE STABILITY Another factor affecting the rarity of Earth-like planets is climate stability. Recall that, in comparison to Venus and Mars, Earth has had a remarkably stable climate. This climate stability has almost certainly played a major role in allowing complex life to evolve on our planet. If our planet had frozen over like Mars or overheated like Venus, we would not be here today.

Advocates of the rare Earth hypothesis point to at least two pieces of "luck" with regard to Earth's stable climate. The first is plate tectonics. As we discussed in Chapters 4 and 10, plate tectonics has been very important to Earth's long-term climate stability, and we do not yet know whether this type of tectonics is common or rare. The second piece of "luck" in Earth's climate stability claimed by rare Earth proponents is our relatively large Moon. Recall from Chapter 8 that Mars undergoes dramatic climate changes because the tilt of its axis varies over a significant range (see Figure 8.29). Some models suggest that Earth would undergo similar changes in axis tilt (as a result of the small gravitational tugs from other planets) if the Moon did not exist, leading to greater seasonal changes that might in turn cause deeper ice ages and more intense periods of warmth. Because the Moon is thought to have formed as a result of a random, giant impact [Section 4.6], we might seem to be very lucky to have the Moon and the climate stability it brings.

Again, however, there are other ways to look at the issue. For one thing, as scientists have refined their models, it now appears that the changes in axis tilt in the absence of the Moon might be no more than about 10 degrees. Moreover, even these changes would occur only with Earth's current rotation rate. If Earth rotated in less than about 12 hours—as it almost certainly did before tidal friction from the Moon slowed it down (see Figure 9.8)—the axis would be fairly stable even without the Moon. Finally, even if the axis tilt changed significantly, the change might not have a major impact on life. Changes in axis tilt might cause different parts of the planet to warm or cool dramatically, but the changes might occur slowly enough for life to adapt or migrate as the climate changed.

RARE EARTH SUMMARY For each potential argument that Earth has been lucky, we have seen counterarguments suggesting otherwise. We therefore do not yet know whether the rare Earth hypothesis is valid. Nevertheless, it's worth keeping the arc of history in mind. It was little more than 400 years ago that we learned that Earth is not the center of the physical universe, and since then we have learned that it is also not at the center of our galaxy, and that our galaxy is just one of billions of galaxies in the universe. Given those discoveries, it would not be so surprising if we were to learn that Earth is not the center of the biological universe either, and that we live in a universe teeming with Earth-like planets and advanced life.

Think About It Considering all the factors we have discussed, what do *you* think of the rare Earth hypothesis? Defend your opinion.

11.5 Classifying Stars

In Section 11.1, we discussed how stellar types and stellar lifetimes affect the prospects for finding stars that might make good "suns" for habitable worlds. But how did we learn so much about stars?

As is often the case in science, efforts at understanding began with efforts at classification. Biology offers a well-known example: Long before biologists knew anything about DNA or even that cells exist, they were busy setting up categories for the life they could observe (this is called *taxonomy*, from the Greek *tassein*, meaning "to classify," and *nomos*, meaning "a science or study"). Separating animals into reptiles, amphibians, insects, mammals, marsupials, and so on, requires no more equipment than a good eye and a pencil, but doing so eventually gives insight into the evolutionary processes that produced animal life. In other words, even if you don't yet understand the underlying reasons for an observed diversity of objects, classifying them is a good first step. For this chapter's section on the process of science in action, we'll briefly explore how stellar classification efforts helped lead to our modern understanding of stars and their life cycles.

How did we learn to classify stars?

Try to put yourself in the place of an astronomer of the mid- to late nineteenth century. Telescopes were rapidly advancing, allowing much better images of star clusters and gas clouds. But what exactly *are* stars? It was clear that other stars shone brightly like the Sun, looking dim in our sky only because they are so far away. But no one yet knew how the Sun shined, nor did anyone know what stars were made of. Visual inspection through telescopes gave only limited information. No telescope of the time could see a star as anything more than a point of light. Our ability to measure distances accurately was still very limited, though the fact that stars in clusters varied significantly in brightness and color demonstrated that there were a least some inherent differences among stars. To learn more, astronomers began to look not only at images of stars, but at their spectra.

In the 1870s, astronomers at Harvard College Observatory, under the directorship of Edward Pickering (1846–1919), began a massive effort to study stellar spectra and thereby determine other characteristics of the stars. Making a detailed stellar spectrum can be a tedious process, but at the end of the nineteenth century astronomers took an important step forward when they invented a method for recording the spectra of many stars at once. To do this, astronomers mounted a glass prism in front of a telescope's objective lens and then photographed a patch of sky containing a large number of stars. On the resulting photo, the image of each star was spread out into a rainbow-like streak, and the most obvious spectral lines could be seen (Figure 11.26).

Once the technique was developed, it wasn't long before thousands of stellar spectra were in hand. The next step was to study this wealth of data and try to make sense of what it was telling us. Pickering needed help with these tasks, so he began to hire assistants whom he called

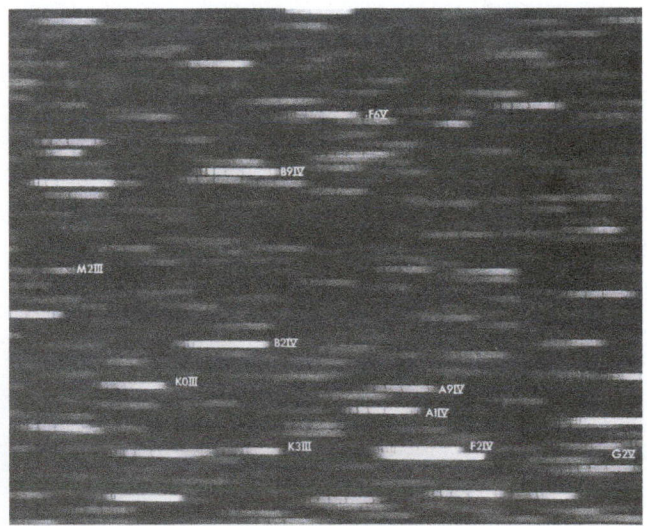

FIGURE 11.26

With photographs like this one, made by placing a prism in front of a telescope's lens, astronomers recorded the spectra of many stars at once. The individual spectra were somewhat crude, but were adequate for identifying the stronger hydrogen and helium lines that are the basis of the spectral type classification.

"computers." The job required people well trained in physics and astronomy, but, in part because the task was seen as somewhat tedious, Pickering found few takers among the men graduating from what was then the all-male Harvard College. Because at that time women with equivalent educations faced enormous obstacles to securing good positions in science, a job with Pickering represented a rare chance for career advancement for women. So Pickering recruited women who had studied physics and astronomy at colleges such as Wellesley and Radcliffe. Although the work was indeed tedious, it was also cutting-edge research, and the women astronomers of Harvard College Observatory made many great discoveries (Figure 11.27).

At first, the astronomers found it difficult to make sense of the spectra. Pickering suggested a scheme in which stars were classified by the visibility of hydrogen lines in their spectra (that is, spectral lines caused by the element hydrogen), using type A to designate stars with the strongest hydrogen lines, type B for those with slightly fainter lines, and so on down the alphabet to type O for stars showing the weakest lines. Following this suggestion, one of Pickering's first "computers," Williamina Fleming (1857–1911), had classified more than 10,000 stellar spectra by 1890. But the work was just beginning.

In 1896, Pickering hired Annie Jump Cannon (1863–1941), who in the course of her career would personally classify the spectra of more than 400,000 stars. Within a few years of being hired, she realized that Pickering's sequence of spectral types A to O included some redundancies and, more importantly, the spectra fell into a much more natural order than he had supposed. She concluded that there were only seven major spectral types, which could be logically ordered as OBAFGKM, the *spectral sequence* that legions of astronomy majors have memorized using the politically incorrect mnemonic "Oh, Be A Fine Girl, Kiss Me." Cannon also subdivided each type by number; for example, stars of spectral type G could be subclassified as G0, G1, G2, and so on to G9, with G0 being most similar to the F stars and G9 being nearly the same as a K0 star. Our Sun is now classified as spectral type G2. The astronomical community adopted Cannon's system of stellar classification in 1910.

The stellar classifications clearly were telling us something important about the nature of stars, but no one yet knew just what that was. Some suspected that the different spectral types reflected different compositions, but that did not turn out to be the case. The correct answer finally came in 1925, in the dissertation of another woman working at Harvard College Observatory, Cecilia Payne-Gaposchkin (1900–1979). Relying on insights from what was then the newly developing science of quantum mechanics, Payne-Gaposchkin showed that the differences in the spectral types reflected differences in the surface temperatures of the stars (see Table 11.1), not differences in composition. A later review of twentieth-century astronomy called her work "undoubtedly the most brilliant Ph.D. thesis ever written in astronomy."

What is the Hertzsprung–Russell diagram?

Another key step in learning the nature of stars came as astronomers tried to understand how the spectral types related to other stellar properties, such as luminosity. Early in the twentieth century, Danish astronomer Ejnar Hertzsprung, who had collected data on the distances of nearby stars, noticed something peculiar: If he arranged these nearby stars by color, from blue-white to red (spectral types B through M), the arrangement was also a sequence in intrinsic luminosity. That is, for nearby stars, the hottest (most blue) stars were more luminous than the cooler stars. However, if he included stars that were farther away, he found that the distant red stars were, on average, intrinsically much more luminous than the nearer ones. Since he realized that the more distant sample of red stars undoubtedly included intrinsically faint objects he couldn't see, this suggested to Hertzsprung that there might be two categories of red stars: small ones (the intrinsically fainter stars) and giants (the intrinsically more luminous stars).

Hertzsprung's results were surprising because we might expect that stars could have any combination of luminosity and color. Why were the red stars separating into distinct categories? In 1911, Hertzsprung began to expand this work by plotting the colors and luminosities of stars in two clusters, the Pleiades and the Hyades. Using clusters of stars allowed him to assume that differences in brightness within the cluster reflected differences in luminosity, because all the stars in a cluster are at more or less the same distance (just as the people in Chicago are all at approximately the same distance from the people in New York). He soon found that, indeed, the stars were not scattered at random across his plot, but fell into two broad groups: (1) Most were in a swath running diagonally through the diagram, which was later named the **main sequence,** and (2) nearly all of the remainder were in a part of the diagram he called the *giants.*

The American astronomer Henry Norris Russell, working independently, was making a similar analysis. However, rather than examining stars in clusters, he used stars whose distances were known from trigonometric measures. Of course, this meant that his diagram was based on relatively close stars, those for which geometric calculation could be used to determine distances. But the results were quite similar to Hertzsprung's. By the 1920s, Hertzsprung's and Russell's results were codified as what we now call the **Hertzsprung–Russell (H–R) diagram,** a graphic found in astronomy textbooks ever since. This diagram is as fundamental to astronomy as the periodic table is to chemistry.

Figure 11.28 shows a modern H–R diagram and explains how it is constructed and what it means. It is largely from this diagram, along with the work on spectral types first performed by Annie Jump Cannon and the meaning of the spectral types discovered by Cecilia Payne-Gaposchkin, that we have developed our modern understanding that stars differ primarily by mass and go through life cycles in which they are born as hydrogen-fusing stars, later grow into giants or supergiants, and then die.

The first key point to notice in Figure 11.28 is that stars do not fall randomly all over the H–R diagram; instead, they are confined to only a few regions. You might compare this to characterizing humans on the basis of two easily measured parameters: height and weight. If you actually took these measures for a large number of people and plotted them on a graph with height on one axis and weight on the other, you'd find that there's a pretty narrow range within which most of the data points fall. Very short, very heavy people are rare, as are very tall, very light folk. The clustering of the data along a diagonal swath relating height and weight is a consequence of the fact that, in a general way, all people are the same. The Hertzsprung–Russell diagram suggests that stars, too, might be generally similar.

Hertzsprung-Russell (H-R) diagrams are very important tools in astronomy because they reveal key relationships among the properties of stars. An H-R diagram is made by plotting stars according to their surface temperatures and luminosities. This figure shows a step-by-step approach to building an H-R diagram.

① An H-R Diagram Is a Graph: A star's position along the horizontal axis indicates its surface temperature, which is closely related to its color and spectral type. Its position along the vertical axis indicates its luminosity.

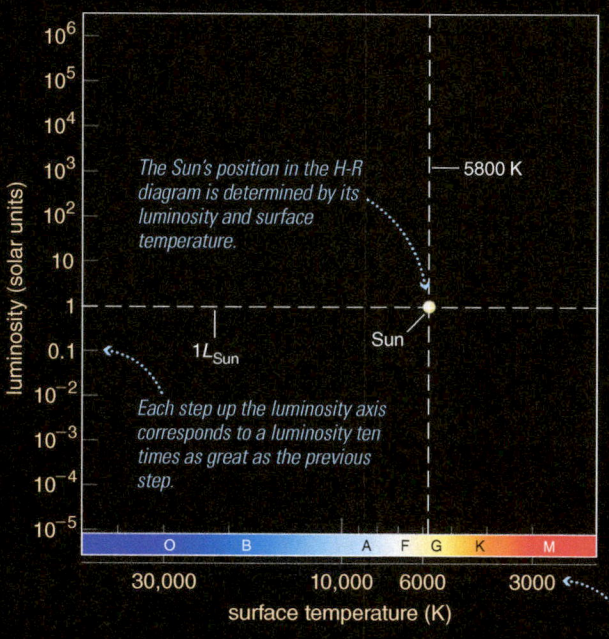

The Sun's position in the H-R diagram is determined by its luminosity and surface temperature.

5800 K

$1L_{Sun}$

Each step up the luminosity axis corresponds to a luminosity ten times as great as the previous step.

Sun

Temperature runs backward on the horizontal axis, with hot blue stars on the left and cool red stars on the right.

② Main Sequence: Our Sun falls along the main sequence, a line of stars extending from the upper left of the diagram to the lower right. Most stars are main-sequence stars, which shine by fusing hydrogen into helium in their cores.

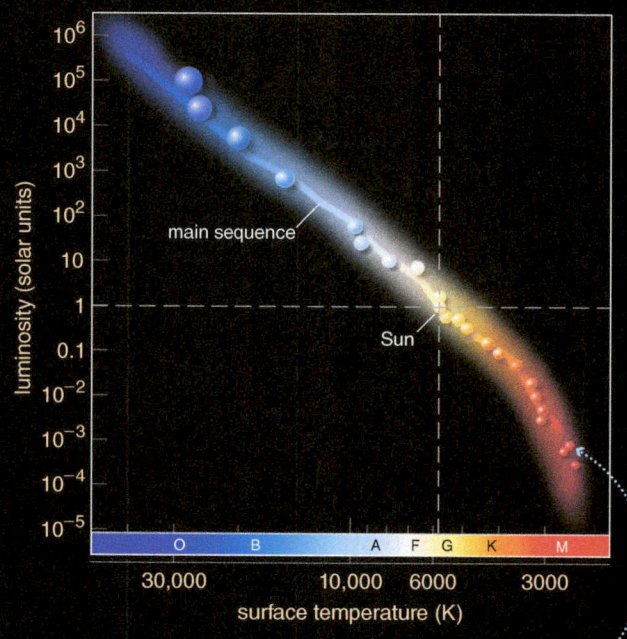

main sequence

Sun

Star sizes on these diagrams indicate the general trend, but actual size differences are far greater than shown.

③ Giants and Supergiants: Stars in the upper right of an H-R diagram are more luminous than main-sequence stars of the same surface temperature. They must therefore be very large in radius, which is why they are known as *giants* and *supergiants*.

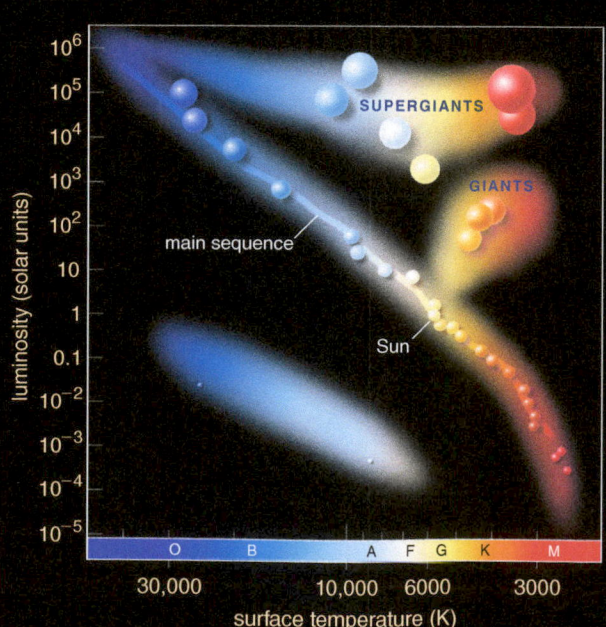

SUPERGIANTS

GIANTS

main sequence

Sun

④ White Dwarfs: Stars in the lower left have high surface temperatures, dim luminosities, and small radii. These stars are known as *white dwarfs*.

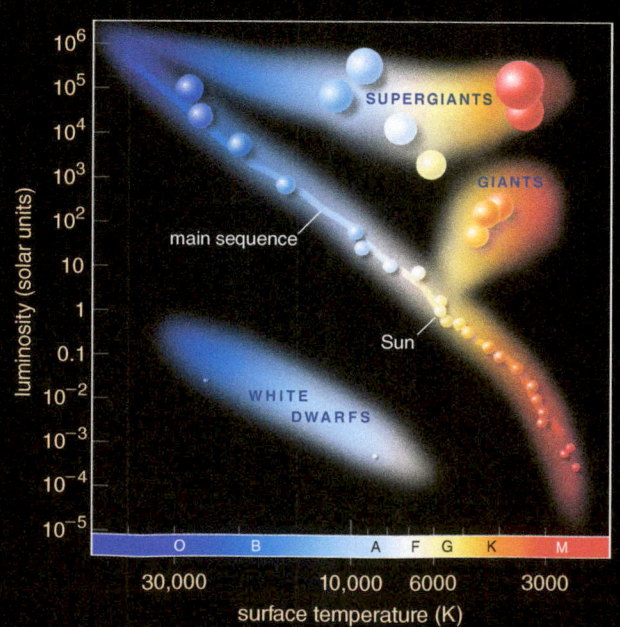

SUPERGIANTS

GIANTS

main sequence

Sun

WHITE DWARFS

5 **Masses on the Main Sequence:** Stellar masses (purple labels) decrease from the upper left to the lower right on the main sequence.

6 **Lifetimes on the Main Sequence:** Stellar lifetimes (green labels) increase from the upper left to lower right on the main sequence: High-mass stars live shorter lives because their high luminosities mean they consume their nuclear fuel more quickly.

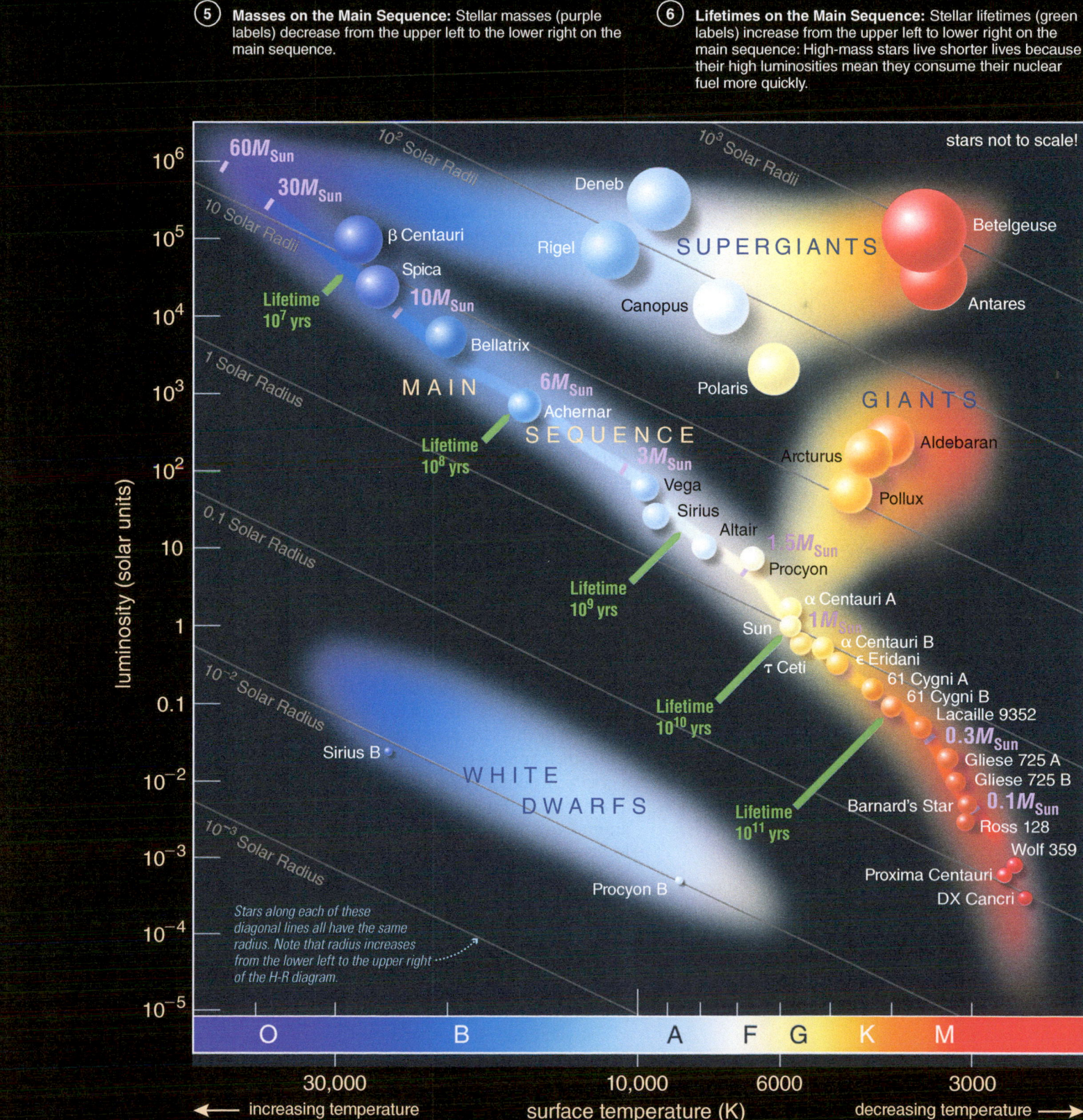

Next, look at the regions of a modern H–R diagram where large numbers of stars are found. The major occupied regions are the following:

- The main sequence is the narrow swath that curves from top left to bottom right. Stable stars fusing hydrogen in their cores—the overwhelming majority of all stars—fall on this line. Our Sun lies about midway along the main sequence, near the center of the diagram, which is why the Sun is often referred to as an "average star," even though it is actually more massive than the majority of stars (the K and M stars on the lower part of the main sequence).
- At the top right are the giants and their somewhat bigger brethren, the supergiants. These stars are mostly cool and red, but very luminous. In fact, giants and supergiants are stars that used to be on the main sequence but are now at the end stages of their lives, having already exhausted their core hydrogen (see the discussion of giants and supergiants in Section 11.1).
- At the bottom left are the hot but dim white dwarfs. White dwarfs are the collapsed remains of stars like the Sun, stars that have exhausted their fuel and are slowly turning into ashen, stellar corpses.

With a little more effort, we can see how this diagram and classification system led to the major ideas we've discussed about stars and their lives. The fact that more than nine out of ten stars are on the main sequence tells us that stars spend about 90% of their lives fusing hydrogen in their cores, and their arrangement on this line tells us something about the character of stars during this long phase of their lives: The hotter they are, the more luminous they are (that is, the more energy they pump into space). Moreover, the existence of the main sequence suggests that all the stars populating it are built basically the same way, and Russell concluded that there must be one parameter that determined where on the main sequence a star would lie. As we discussed in Section 11.1, that parameter turns out to be a star's mass, with mass increasing from bottom right to top left along the main sequence (notice the mass labels in Figure 11.28).

What about the giants and supergiants? Despite the fact that they are not on the main sequence, these luminous heavyweights do not violate the idea that mass is the most important property. They have simply changed their internal chemical composition by having reached a point at which they have fused most of their central hydrogen fuel. They have moved on to a different "engine" for producing energy: fusing helium or even heavier elements. As we discussed in Section 11.1, the temporary energy boost provided by this switch in fuel causes them to swell in size and luminosity, while their surface temperatures are lowered (they become redder). From a graphical point of view, they've moved off the main sequence to take up a rather short residence in the giant or supergiant region of the H–R diagram. Eventually they exhaust these secondary fuels and either collapse to become a white dwarf (at the bottom left of the diagram) or, if they began life as a star several times as massive as the Sun, explode in a supernova and leave behind either a neutron star or a black hole. There is no location on the H–R diagram for such pathological stellar remnants, so the heavier giants eventually, and rather suddenly, disappear from the chart.

The H–R diagram was originally just an organizational strategy, like classifying living things as plants or animals. But we have seen that organizing leads to insights. The lesson for the process of science is clear: When you're not sure what's going on, start by organizing what you do know, and it may take you down the path to discovery.

The Big Picture

PUTTING CHAPTER 11 IN PERSPECTIVE

In this chapter, we have discussed the search for habitable planets beyond our own solar system. As you continue in your studies, keep the following "big picture" ideas in mind:

- In a period of barely more than two decades, we have gone from knowing of no other planets around other stars to knowing that many or most stars have planets. As a result, there is no longer any question that planets are common in the universe.

- We are rapidly learning about the nature of extrasolar planets, with one of the key lessons being that they come in a wider range of types than the planets we find in our own solar system. This wide range seems likely only to enhance the prospects for finding habitable worlds.

- We have not yet reached the point at which we can search directly for life on extrasolar worlds, but it's likely that we'll have the necessary technology within a few decades. At that point, we may be able to definitively answer the question of whether planets like Earth—and life—are common or rare.

Summary of Key Concepts

11.1 Distant Suns

How do other stars differ from the Sun?

All stars are born from interstellar clouds made mostly of hydrogen and helium, and their compositions differ only in the small proportions of other chemical elements. Because of their similar compositions, the major differences among stars are a result of differing masses: More massive stars have higher surface temperatures, higher luminosities, and shorter lifetimes than less massive stars. Many stars are members of binary or multiple star systems.

Which stars would make good suns?

To make a good "sun" for a habitable world, a star should have a high enough proportion of elements besides hydrogen and helium to allow planet formation (most stars probably qualify); it should be low enough in mass that it will live long enough for life to take hold and evolve; it should be in the hydrogen-fusing stage of life that allows for steady sunlight and a habitable zone; and if it is in a binary or multiple-star system, it should allow for stable orbits within the habitable zone.

11.2 Discovering Extrasolar Planets

How do we detect planets around other stars?

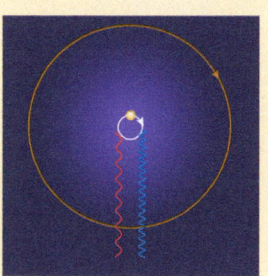

Current technology is limited in direct-detection capabilities, but we can detect planets indirectly through three major methods. We can look for a planet's gravitational effect on its star through the **astrometric method,** which looks for small shifts in stellar position, or the **Doppler method,** which looks for the back-and-forth motion of stars revealed by Doppler shifts. For the small fraction of planetary systems with orbits aligned edge-on to Earth, we can search for **transits,** in which a planet blocks a little of its star's light as it passes in front of it.

What properties of extrasolar planets can we measure?

All detection methods allow us to determine a planet's orbital period and distance from its star. The astrometric and Doppler methods can provide masses (or minimum masses), while the transit method can provide sizes and, for some multiple-planet systems, also masses. When the transit and Doppler methods can be used together, we can determine average density. In some cases, transits (and eclipses) can provide other data, including limited data about atmospheric composition and temperature.

11.3 The Nature of Extrasolar Planets

How do extrasolar planets compare with planets in our solar system?

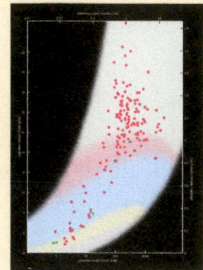

The known extrasolar planets have a much wider range of properties than the planets in our solar system. Many orbit much closer to their stars and with more eccentric orbital paths; some jovian planets, called **hot Jupiters,** are also found close to their stars. We have also observed properties indicating planetary types, such as super-Earths and water worlds, that differ from the terrestrial and jovian planets in our solar system.

Are planetary systems like ours common?

Current evidence suggests that most stars have planets, and at least some are Earth-sized and in their star's habitable zone. Nevertheless, we don't yet have enough data to know for certain whether planetary systems like ours—and planets like Earth—are common.

11.4 The Habitability of Extrasolar Planets

What kinds of extrasolar worlds might be habitable?

Surface habitability seems possible for planets or moons similar in size and composition to Earth and located within the habitable zone, and the habitable zone may extend farther for super-Earths or water worlds with thick hydrogen atmospheres. Subsurface habitability may be even more common, since it is possible on any world with enough internal heat to keep water liquid beneath the surface. **Orphan planets,** which do not orbit a star, also offer intriguing possibilities for subsurface life, and possibly even for surface life if they have thick enough atmospheres.

How could we detect life on extrasolar planets?

Future telescopes should allow us to obtain crude images or spectra of planets within stellar habitable zones. An image of an extrasolar planet—even if only a few pixels in size—might indicate the presence of snow or clouds, and would tell us the planet's rotation period. Spectroscopic analysis could tell us much more, and might reveal combinations of atmospheric gases, such as oxygen and methane, that would be evidence for life. SETI experiments might directly detect the presence of technologically sophisticated life.

Are Earth-like planets rare or common?

We don't know. Some of the key questions are whether our galaxy, like a star, has a relatively narrow habitable zone; whether the role Jupiter has played in lowering our solar system's impact rate is rare or critical to life; and whether Earth's relatively stable climate, due largely to plate tectonics and our large Moon, is likely on otherwise similar worlds. Arguments can be made on both sides of each question, and at present we lack the data to determine which side is correct.

THE PROCESS OF SCIENCE IN ACTION

11.5 Classifying Stars

How did we learn to classify stars?

The advent of spectroscopy allowed astronomers to study stars by categorizing them according to their spectra. The women astronomers of Harvard recognized the spectral sequence OBAFGKM and, later, that this sequence represents a sequence in surface temperatures.

What is the Hertzsprung–Russell diagram?

The **Hertzsprung–Russell diagram** plots stars according to their spectral type or surface temperature on the horizontal axis and luminosity on the vertical axis. Most stars fall along a continuous swath on the diagram, called the **main sequence,** that runs from hot, luminous stars at the upper left to cool, dim stars at the lower right. Other stars—giants and supergiants—clump in the part of the diagram where stars are luminous but have cool surface temperatures. The stellar corpses known as white dwarfs are dim but hot, so they are found in the lower left of the diagram. Studying H–R diagrams helped astronomers realize that mass is a star's most fundamental property, and the organizational power of the diagram makes it one of astronomy's most useful tools.

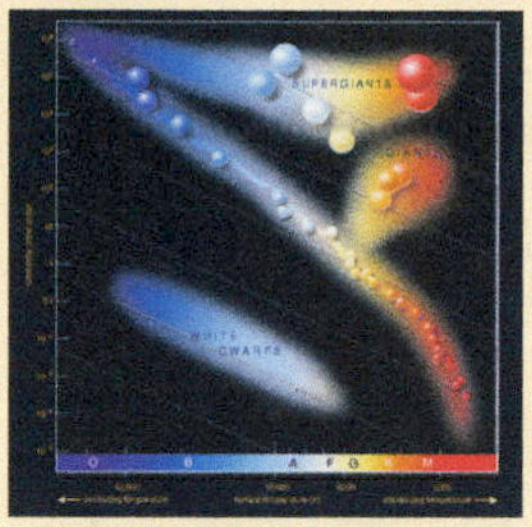

Exercises and Problems

MasteringAstronomy® *For instructor-assigned homework and other learning materials, go to MasteringAstronomy®.*

REVIEW QUESTIONS

Short-Answer Questions Based on the Reading

1. What is the general composition of all stars, and how does composition differ among stars in terms of elements important to planet formation and life?
2. Summarize how stars differ along the *spectral sequence* OBAFGKM. Why is mass the fundamental property in determining a star's type, and how does mass affect a star's life cycle?
3. Describe four criteria a star must meet to make a good "sun" that could support life, and which types of stars meet each of these criteria. Overall, how common are such stars? Be sure to include discussion of the relevant factors in binary or multiple-star systems.
4. How do habitable zones differ among stars of different mass? Explain.
5. Briefly describe the conditions under which habitable zones and stable orbits might be possible in binary or multiple-star systems.
6. Why are *extrasolar planets* hard to detect directly? What are the two general approaches to indirect detection?
7. Briefly describe the *astrometric, Doppler,* and *transit* methods for detecting extrasolar planets. Summarize the key advantages and major limitations of each method.
8. Briefly summarize the planetary properties we can in principle measure with current detection methods.
9. Why does the Doppler method generally allow us to determine only *minimum* planetary masses? In what cases can we be confident that we know precise masses? Explain.
10. How does the transit method tell us planetary size, and in what cases can we also learn mass and density?
11. How do the orbits of known extrasolar planets differ from those of planets in our solar system? What leads us to conclude that some of these planets are *hot Jupiters*?
12. Summarize the key features shown in Figure 11.20, and briefly describe the nature of planets that would fit each of the model curves shown on the graph.
13. According to current statistics, how common are planets of various masses and sizes around other stars? Explain.
14. What types of worlds seem most likely to support surface habitability? What types of worlds might have subsurface habitability?
15. How might a star's habitable zone be wider than we assume based on planets like Earth? What are *orphan planets,* and how might they potentially be habitable?
16. How might future imagery and spectroscopy allow us to determine whether distant planets are habitable or have life?
17. What is the *rare Earth hypothesis*? Briefly summarize the arguments used to advance it and the counterarguments against each of these.
18. How is classification useful in science? Briefly describe how our stellar classification scheme was discovered.
19. What is the *Hertzsprung–Russell diagram*? How does a star in the upper left section differ from one in the lower right?
20. Briefly summarize the characteristics of stars in each of the three major regions of the H–R diagram—the *main sequence,* the giants and supergiants, and the white dwarfs.

TEST YOUR UNDERSTANDING

Would You Believe It?

Suppose that, on the dates indicated, you saw the following headlines. (These are not real discoveries.) In each case, decide whether the headline is believable in light of what we currently know about extrasolar planets and our technological capabilities. Explain clearly; because not all of these have definitive answers, your explanation is more important than your chosen answer.

21. Date: February 16, 2025. Headline: Astronomers Conclude That Earth-Size Planets in Habitable Zones Don't Exist.

22. Date: January 9, 2026. Headline: Astronomers Discover Earth-Like World Orbiting Massive Star of Spectral Type O.

23. Date: June 19, 2028. Headline: Spectrum Reveals Unmistakable Evidence of Life on a "Hot Jupiter."

24. Date: November 7, 2020. Headline: New Images Show Oceans on Extrasolar Planet.

25. Date: November 7, 2050. Headline: New Images Show Oceans on Extrasolar Planet.

26. Date: July 20, 2020. Headline: Giant Planet Found in Our Solar System Just Beyond Pluto.

27. Date: September 15, 2045. Headline: Sun-Like Star Has Three Planets with Life.

28. Date: March 30, 2037. Headline: More than One-Third of Stars Have Habitable Planets.

29. Date: December 13, 2033. Headline: Orphan Planet Has Surface Oceans and Oxygen Atmosphere.

30. Date: June 1, 2040. Headline: First Spacecraft to Reach an Extrasolar Planet Discovers Evidence of Life.

Quick Quiz

Choose the best answer to each of the following. Explain your reasoning with one or more complete sentences.

31. Compared to a star of spectral type K, a star of spectral type A is generally (a) hotter, more luminous, and more massive; (b) hotter, more luminous, and less massive; (c) cooler, dimmer, and less massive.

32. Compared to the habitable zone of our Sun, the habitable zone of a lower-mass star is (a) larger and farther out; (b) hotter and much brighter; (c) smaller and closer in.

33. Which method could detect a planet in an orbit that is face-on to Earth? (a) Doppler method (b) transit method (c) astrometric method

34. To determine a planet's average density, we can use (a) the transit method alone. (b) the astrometric and Doppler methods together. (c) the transit and Doppler methods together.

35. Based on the model types shown in Figure 11.20, a planet made almost entirely of hydrogen compounds would be considered (a) a terrestrial planet; (b) a jovian planet; (c) a water world.

36. According to current statistics, about what percentage of all stars have planets? (a) 1% (b) 15% (c) 70%

37. The term *super-Earth* means a planet that is (a) the size of Earth but with more water; (b) larger than Earth but on a close-in orbit that makes it much hotter than Earth; (c) similar in composition to Earth but larger in size.

38. Our best hope for determining that life exists on an extrasolar world lies in (a) obtaining telescopic images with high enough resolution to see the life; (b) obtaining spectra that allow us to determine atmospheric composition; (c) sending spacecraft to study the worlds up close.

39. Jupiter has had an important effect on life on Earth because (a) Jupiter's heat has helped supply energy to life; (b) Jupiter's gravity helped clear the inner solar system of objects that could cause impacts; (c) without Jupiter, Earth could not have a stable orbit around the Sun.

40. The main sequence on an H–R diagram represents stars that are (a) in the final stages of their lives; (b) fusing hydrogen into helium in their cores; (c) all extremely low in mass.

PROCESS OF SCIENCE

41. *The Rare Earth Hypothesis.* We do not yet have enough data to evaluate the rare Earth hypothesis fully, but the pace of discovery suggests that we may have the needed data relatively soon. Describe how you think scientists will be able to go about deciding whether this hypothesis is valid or invalid.

42. *Unanswered Questions.* As discussed in this chapter, we are only just beginning to learn about extrasolar planets. Briefly describe one important but unanswered question related to the study of planets around other stars. Then write two or three paragraphs in which you discuss how we might answer this question in the future. Be as specific as possible, focusing on the type of evidence necessary to answer the question and how the evidence could be gathered. What are the benefits of finding answers to this question?

GROUP WORK EXERCISE

43. *Time to Move On.* **Roles:** *Scribe* (takes notes on the group's activities), *Proposer* (proposes explanations to the group), *Skeptic* (points out weaknesses in proposed explanations), *Moderator* (leads group discussion and makes sure everyone contributes). **Activity:** A common theme in science fiction is "leaving home" to find a new planet for humans to live on. Now that we know about thousands of planets, we can start imagining how to choose.
    a. Make a list of characteristics that you would look for in a planet that might make a good home.
    b. Examine the planets in Figure 11.20. Does this graph give enough information to allow you to determine which planets might make good homes or poor ones? If not, what's missing?
    c. Suppose you also knew the orbital distance for each of the planets in Figure 11.20. Would that make it easier to find potential good homes? Why or why not?

INVESTIGATE FURTHER

In-Depth Questions to Increase Your Understanding

Short-Answer/Essay Questions

44. *Explaining the Doppler Method.* Explain how the Doppler method works in terms an elementary school child would understand. It may help to use an analogy to explain the difficulty of direct detection and the general phenomenon of the Doppler shift.

45. *Explaining the Transit Method.* Explain how the transit method works in terms an elementary school child would understand. Why can't we use this method to discover *all* extrasolar planets?

46. *Comparing Methods.* What are the strengths and limitations of the Doppler and transit methods? What kinds of planets are easiest to detect with each method? Are there certain planets that each method cannot detect, even if the planets are very large? Explain. What advantages are gained if a planet can be detected by both methods?

47. *Super-Earth.* You've discovered a super-Earth orbiting a Sun-like star at the distance of Jupiter. Is it possible that it will have surface oceans and life? Is it possible that it will have subsurface life? Explain.

48. *Stars with Habitable Planets.* Based on what you've learned about stars in this chapter, make your best estimate of the fraction of all stars around which you'd expect to find planets in habitable zones. Clearly explain how you come up with your estimate, and what uncertainties still need to be addressed.

49. *Are Earth-Like Planets Common?* Based on what you have learned in this chapter, form an opinion as to whether Earth-like planets will ultimately prove to be rare, common, or something in between. Write a one- to two-page essay explaining and defending your opinion.

50. *Ages of Stars on the H–R Diagram.* The giants and supergiants on the H–R diagram are stars in the last stages of their lives. Does this mean they are *older* than most main-sequence stars? Why or why not?

51. *Nightfall.* Read the short story "Nightfall" by Isaac Asimov. If such a planet really exists, do you think the scenario described is realistic? Why or why not? Summarize and defend your opinions in a one- to two-page essay.

52. *Science Fiction Planet.* Choose one fictional planet with life depicted in a science fiction book or movie. Briefly describe the planet's fictional characteristics, and then write a few paragraphs stating whether such a planet seems plausible in light of our current understanding of extrasolar habitability.

Quantitative Problems

Be sure to show all calculations clearly and state your final answers in complete sentences.

53. *Number of Stars with Habitable Planets.* Assume that the Milky Way Galaxy has 300 billion stars (a reasonable estimate). Based on the statistics given in this chapter, determine approximately how many stars would be Sun-like. How many would be K stars? How many would be M stars? If you assume that an average G star has one planet in its habitable zone, while only one in five K stars and one in ten M stars has such a planet, how

many total planets would you expect to find in habitable zones in the Milky Way Galaxy?

54. *Lost in the Glare.* This exercise will help you explore how hard it would be for alien astronomers to detect the light from the planets of our solar system.
    a. Calculate the fraction of the total emitted sunlight that reaches Earth. (*Hint:* Imagine a sphere around the Sun the size of the planet's orbit [area = $4\pi a^2$], and then calculate the fraction of that area taken up by the disk of Earth [area = πr^2_{planet}].)
    b. Earth reflects 29% of the Sun's light. Based on this fact and your answer from part a, calculate the fraction of total sunlight reflected by Earth. (*Hint:* Your answer will simply be the overall fraction of all the Sun's light that is reflected by Earth.)
    c. Would detecting Jupiter be easier or harder than detecting Earth? Comment on whether you think Jupiter's larger size or greater distance has a stronger effect on its detectability. You may neglect any difference in reflectivity between Earth and Jupiter.

55. *Finding Orbit Sizes.* The Doppler method allows us to find a planet's semimajor axis using just the orbital period and the star's mass (Cosmic Calculations 11.1).
    a. Imagine that a new planet is discovered orbiting a $2M_{Sun}$ star with a period of 5 days. What is its semimajor axis?
    b. Another planet is discovered orbiting a $0.5M_{Sun}$ star with a period of 100 days. What is its semimajor axis?

56. *Finding a Planetary Mass.* Using the Doppler method, you discover a planet that is causing its star to move at a maximum speed of 14 meters per second. The planet has an orbital period of 56 days and an average orbital distance of 55 million kilometers from its star. What is the planet's mass? (*Hint:* See Cosmic Calculations 11.2.)

57. *Transit of TrES-1.* The planet TrES-1, orbiting a distant star, has been detected by both the transit and the Doppler method, so we can calculate its density and get an idea of what kind of planet it is.
    a. Using the method of Cosmic Calculations 11.3, calculate the radius of the transiting planet. The planetary transits block 2% of the star's light. The star TrES-1 has a radius of about 85% of our Sun's radius.
    b. The mass of the planet is approximately 0.75 times the mass of Jupiter, and Jupiter's mass is about 1.9×10^{27} kilograms. Calculate the average density of the planet. Give your answer in grams per cubic centimeter. Compare this density to the average densities of Saturn (0.7 g/cm³) and Earth (5.5 g/cm³). Is the planet terrestrial or jovian in nature? (*Hint:* To find the volume of the planet, use the formula for the volume of a sphere: $V = \frac{4}{3}\pi r^3$. Be careful with unit conversions.)

58. *The Doppler Formula.* The amount of Doppler shift for light or radio waves can be calculated from this formula:

$$\frac{\text{wavelength shift}}{\text{rest wavelength}} = \frac{v}{c}$$

The rest wavelength is the wavelength of a particular spectral line in an object that is not moving (relative to us), *v* is the velocity of the star from which we observe a wavelength shift, and c is the speed of light (c = 300,000,000 m/sec). Suppose that, in a particular star, a spectral line with a rest wavelength of 600 nm is found to be shifted by 0.1 nm (toward the blue). How fast is that star moving toward us, in meters per second? (1 nm = 10^{-9} m.)

59. *Finding a Center of Mass.* In the simple case of a two-body system—for example, a star and a single planet—the position of their center of mass can be determined from

$$m_{star} r_{star} = m_{planet} r_{planet}$$

where m_{star} and m_{planet} are the masses of the star and the planet, respectively, and r_{star} and r_{planet} are the distances of the star and the planet from their center of mass. Consider the Sun and Jupiter, which are separated by 780 million kilometers. Using the fact that the Sun's mass is about 1000 times the mass of Jupiter, determine about how far the center of mass is from the center of the Sun. How does this distance compare to the Sun's radius of 700,000 kilometers?

Discussion Questions

60. *Future Mission.* Imagine that a wealthy benefactor has just given you a large grant to search for Earth-like planets around other stars. What would you do? Explain.

61. *Is It Worth It?* Thanks to rapidly advancing technology, we could probably now build space observatories capable of obtaining images and spectra of Earth-size planets around other stars, with enough resolution to be able to determine whether they are Earth-like, and perhaps even to detect spectral signatures that would indicate the presence of life. However, such observatories would likely cost several billion dollars. Suppose you were a member of the U.S. Congress. How much would you be willing to spend on such observatories? Defend your opinion.

62. *The Copernican Principle and Rare Earth.* The Copernican revolution taught us that our planet is not the center of the universe, as had been generally believed before that time. Taking this lesson to heart, we have since assumed that our planet is not "central" or "special" in any way but rather that we are on a fairly typical planet in a fairly typical place in the universe. This principle, often called the *Copernican Principle* or the *Principle of Mediocrity,* has been borne out many times since. For example, we have learned that we are not near the center of our galaxy and that the universe has no center at all. Do you consider this principle to be in conflict with the rare Earth hypothesis? If so, does this make the rare Earth hypothesis any less scientific? Defend your opinions.

WEB PROJECTS

63. *New Planets.* Find the latest extrasolar planet discoveries. Create a personal "planet journal," complete with illustrations as needed, with a page for each of at least three recently discovered planets. On each page, be sure to note the method that was used to find the planet, give any information we have about the nature of the planet, and discuss how this information fits in with our current understanding of extrasolar planets in general.

64. *Direct Detections.* In this chapter, we saw only a few examples of direct detection of possible extrasolar planets. Search for new information on these and any other direct detections now known. Have the detections discussed in this chapter been confirmed as planets? Have we made any other direct detections and, if so, how? Summarize your findings in a short written report, including images of the directly detected planets.

65. *Extrasolar Planet Mission.* Learn about a proposed future mission to study extrasolar planets, including its proposed design, capabilities, and goals. Write a short report on your findings.

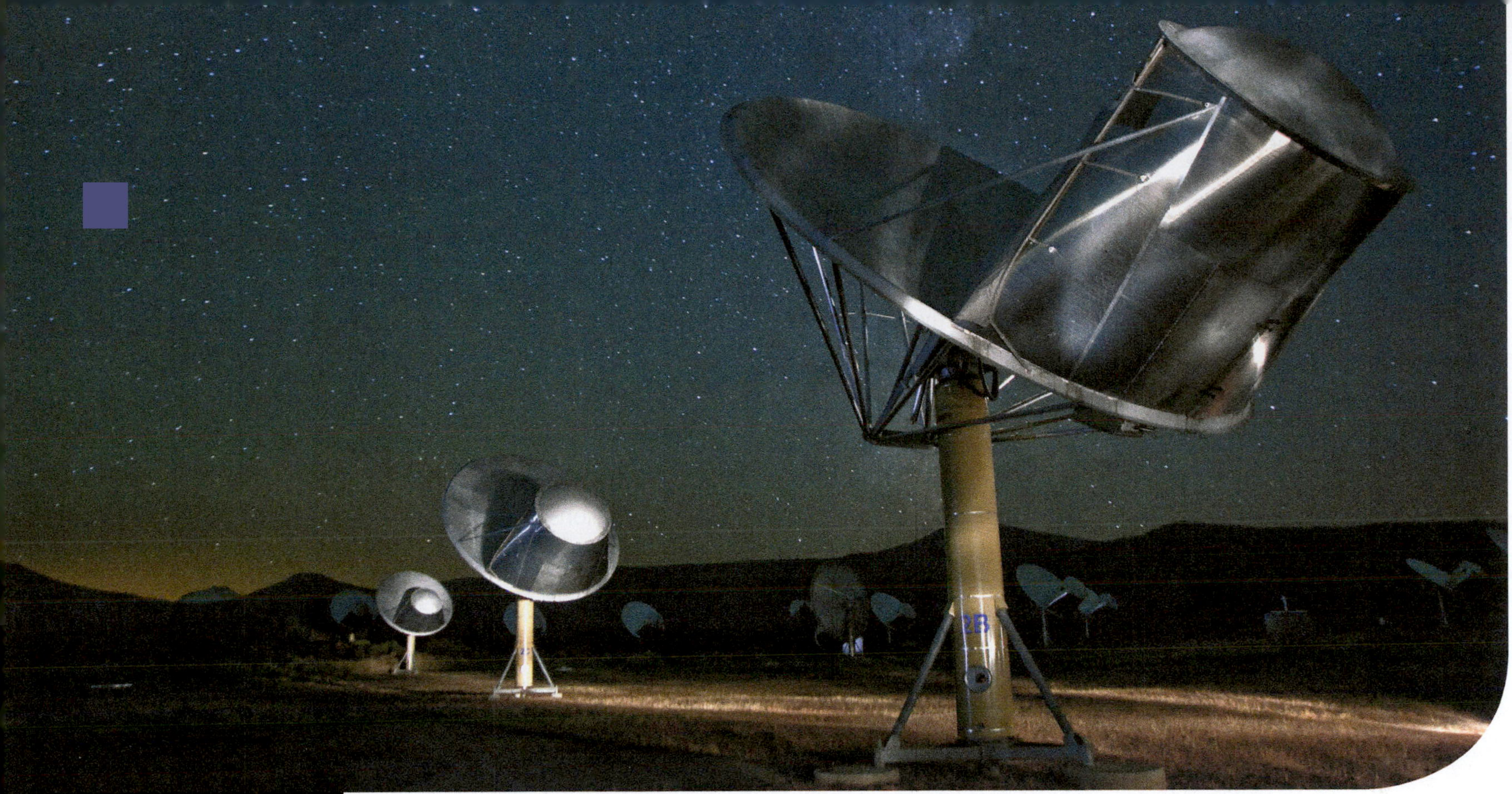

12 The Search for Extraterrestrial Intelligence

LEARNING GOALS

12.1 THE DRAKE EQUATION
- What is the Drake equation?
- How well do we know the terms of the Drake equation?

12.2 THE QUESTION OF INTELLIGENCE
- Even if life is widespread, is intelligence common?
- Will intelligence inevitably spawn technology?

12.3 SEARCHING FOR INTELLIGENCE
- How did SETI begin?
- How do we search for intelligence today?
- What happens if SETI succeeds?

THE PROCESS OF SCIENCE IN ACTION

12.4 UFOS AND ALIENS ON EARTH
- What have we learned from UFO sightings?
- Have aliens left any compelling evidence of visitation?
- Is there a case for alien visits?

▲ **About the photo:** The Allen Telescope Array (Hat Creek, California) is used in the search for extraterrestrial intelligence (SETI).

There are approximately 15,000 stars within 70 light-years of Earth. If any of these nearby stellar systems have planets with technologically sophisticated beings, the inhabitants could already know that we exist by picking up our high-frequency radio, radar, and television transmissions. These broadcasts—mostly unintentional evidence of our technological society—are moving into space at the speed of light and are currently washing over star systems at the rate of almost one new star system a day. By using sufficiently powerful radio telescopes, others could learn that we're here.

We are only beginning to signal our presence, but other civilizations may have been doing something similar for a long time. As we discussed in the prior chapter, the Milky Way Galaxy alone may contain tens of billions of Earth-like worlds, and some of these could harbor civilizations that are filling the interstellar voids with their broadcasts. Using both radio and optical telescopes, scientists are attempting to find such transmissions. These experiments, called the *search for extraterrestrial intelligence (SETI)*, are the primary topic of this chapter.

12.1 The Drake Equation

The search for extraterrestrial intelligence (SETI) differs in a fundamental way from all the other searches for life that we have discussed in this book. Those searches are concerned not just with finding life itself, but with finding evidence that might point to its existence elsewhere—such as whether habitable planets are common or rare or whether our understanding of the origin of life would allow for life to arise on Mars or the jovian moons. In contrast, SETI restricts itself to seeking clear and conclusive evidence of technologically advanced life.

Indeed, if SETI is successful in receiving and perhaps interpreting a message from a distant civilization, it might give us answers to many or most of the other questions we have discussed. To begin with, the discovery of a distant civilization would immediately prove that life is not unique to Earth. Because it is likely that any extraterrestrials we might detect with SETI experiments would be more advanced than we are, there's at least a possibility that we could learn a great deal from their transmissions, if we could understand them.

The principal goal of this chapter is to explore the methods by which scientists are now searching for evidence of other civilizations. First, however, it's worth asking whether our current understanding of life in the universe gives us any good reason to believe that the search may be successful. To some extent, the answer to this question may not matter, as stated so eloquently in the quotation that opens this chapter. That is, our innate scientific curiosity inspires us to search even if we cannot be certain that the search will ever be successful. But although we may not know the probability of finding a signal, recent advances in astrobiology allow us to say a lot about the factors that might influence this probability. As in all of science, knowledge about such factors can provide important guidance to research efforts.

What is the Drake equation?

Early in the search for evidence of other civilizations, scientists realized that it would be useful to try to estimate the chances of finding something. In 1961, the first scientific conference on the search for extraterrestrial intelligence was held in Green Bank, West Virginia, at the radio observatory where a pioneering hunt for an alien signal had recently been conducted. (We will discuss this search, called Project Ozma, in Section 12.3.) Only about a dozen people attended the conference—just about the entire world's complement of people with a professional interest in the subject at the time—but their expertise ranged across the disciplines of astronomy, biology, and engineering.

In setting the meeting's agenda, astronomer Frank Drake (Figure 12.1) tried to summarize the factors that would determine whether any attempt to detect intelligent extraterrestrials could succeed. He came up with a simple equation—now called the **Drake equation**—that in principle could be used to calculate the number of civilizations existing elsewhere in our galaxy* (or in the universe at large) *from which we could potentially get a signal*. Note that this definition limits what we mean by "civilization" in this context. For example, the ancient Greeks had a remarkable civilization, but theirs doesn't count under this definition because they never developed radio or other technologies that could be used to effectively communicate across space.

The Drake equation does not give us a definitive answer for the number of transmitting civilizations. Rather, it lays out the factors that are important in determining this number. The equation has played a guiding role in research bearing on life in the universe, because much of it deals with life in general and not just the smaller fraction of life that is intelligent enough to produce signals. Many of the factors in the Drake equation have been discussed earlier in this book as part of our search for biology on other worlds. Consequently, we expect that—with improvements in our knowledge of these factors—our estimate of the number of signaling civilizations will get better. However, we needn't await such improvements before we embark on a SETI search.

To keep our discussion simple, we will follow Drake's lead and use his formula to focus only on the number of intelligent species in our own galaxy. We can always extend our estimate to the rest of the universe by multiplying the result we find for our galaxy by 100 billion, the approximate number of galaxies in the observable universe [Section 3.2]. We limit our focus for practical reasons, too. Any signals from other galaxies will be severely weakened by distance, making them far harder for us to detect. To stay consistent with key ideas already discussed, we'll consider a slight modification of Drake's original equation. (For the original version, see Special Topic 12.1 on page 425.) In modified form, the Drake equation looks like this:

$$\text{Number of civilizations} = N_{HP} \times f_{\text{life}} \times f_{\text{civ}} \times f_{\text{now}}$$

Let's examine each term to see how this equation tells us the number of civilizations in the Milky Way Galaxy capable of interstellar communication:

FIGURE 12.1
Astronomer Frank Drake, with the equation he first wrote in 1961. He has been both on the staff and a member of the board of the SETI Institute in California.

---

*Drake's formula wasn't the first time that such a calculation had been considered; for example, astronomer Harlow Shapley made a somewhat similar attempt two years earlier. However, Drake's formula was both more specific and better constructed, and therefore caught on much more widely.

- N_{HP} is the number of habitable planets in the galaxy. It is the first term because we assume that a prerequisite to having life or a civilization is having a habitable planet on which that life can evolve.
- f_{life} is the fraction of habitable planets that actually *have* life. For example, $f_{life} = 1$ would mean that all habitable planets have life; $f_{life} = \frac{1}{1,000,000}$ would mean that only 1 in a million habitable planets has life. The product $N_{HP} \times f_{life}$ tells us the number of life-bearing planets in the galaxy.
- f_{civ} is the fraction of the life-bearing planets on which a civilization capable of interstellar communication *has at some time* arisen. For example, $f_{civ} = \frac{1}{1000}$ would mean that such a civilization has existed on 1 out of 1000 planets with life, while the other 999 out of 1000 have not had a species intelligent enough to build radio transmitters, high-powered lasers, or other devices for interstellar conversation. When we multiply this term by the first two terms to form the product $N_{HP} \times f_{life} \times f_{civ}$, we get the total number of planets on which intelligent beings have evolved and developed a civilization at some time in the galaxy's history.
- f_{now} is the fraction of the civilization-bearing planets that happen to have a civilization *now*, as opposed to, say, millions or billions of years in the past. This term is important because it tells us how many civilizations we could potentially talk to,* since there's no point in listening for civilizations that are long gone. Because the previous three terms told us the total number of civilizations that have *ever* arisen in the galaxy, multiplying by f_{now} tells us how many civilizations we could potentially make contact with today. For example, if the first three terms were to tell us that 10 million planets in the galaxy have at some time had a communicating civilization but f_{now} turns out to be 1 in 5 million, then only two civilizations could be expected to exist today. As we will see shortly, the value of f_{now} must depend largely on how long civilizations survive once they arise.

To summarize, the Drake equation gives us a way to calculate the number of civilizations capable of interstellar communication that are currently sharing the Milky Way Galaxy with us. It provides a useful way of organizing our thinking about the problem, because it tells us exactly what numbers we need to know to learn the answer. Indeed, it suffers from only one significant drawback: We don't know precise values for any of its terms!

Think About It Try the following sample numbers in the Drake equation. Suppose that there are 1 billion (1,000,000,000,000) habitable planets in our galaxy, that 1 in 10 habitable planets has life, that 1 in 100 planets with life has at some point had an intelligent civilization, and that 1 in 500 civilizations that have ever existed is in existence now. How many civilizations would exist at present? Explain.

How well do we know the terms of the Drake equation?

The only term in the Drake equation for which we can make even a reasonably educated guess is the number of habitable planets, N_{HP}. As

---

*For the purposes of the Drake equation, we'll assume that the term f_{now} takes into account the light-travel time for signals from other stars; for example, if a star with a civilization is 10,000 light-years away, it counts in determining f_{now} if the civilization existed 10,000 years ago, because signals broadcast at that time would just now be arriving at Earth.

we discussed in Chapter 11, statistics based on known extrasolar planets show that planets are quite common, and preliminary data suggest that as many as one in five star systems (and possibly more) may have one or more *habitable* planets. Given that there are several hundred billion stars in the Milky Way Galaxy, it seems entirely reasonable to suppose that there could be 100 billion or more habitable planets.

The rest of the terms in the Drake equation present more difficulty. For the moment, we have no statistics that would allow us to estimate the fraction f_{life} of habitable planets on which life actually arose. The problem is that we have only one known case in which life arose on a habitable world—our own Earth. Still, we are not completely without guidance. The fact that life apparently arose rapidly on Earth [Section 6.1] suggests that the origin of life was fairly "easy," in which case we

Special Topic 12.1 FRANK DRAKE AND HIS EQUATION

If anyone could be called the Father of SETI Research, Frank Drake is that person. As a young man, Drake learned electronics in the Navy. He then studied for an advanced degree in astronomy, and on graduation took a job at the new National Radio Astronomy Observatory (NRAO) in Green Bank, West Virginia. In the late 1950s, the Observatory was busy with the construction of a large radio telescope, a project that would take many years. Consequently, the Observatory opted to buy an "off-the-shelf" instrument that could be used right away. This telescope boasted a 26-meter (85-foot) reflector, which at the time was larger than most of the world's operating radio telescopes.

Once this telescope was up and running, the Observatory staff was encouraged to suggest interesting experiments for its use. Drake had already been thinking about the possibility of interstellar communication by radio, and he proposed a simple experiment to search for alien signals from two nearby star systems. This became Project Ozma, the first modern SETI search. A year later, in 1961, Drake organized a conference at Green Bank to discuss the possibility that such an experiment could actually find an extraterrestrial transmission. His "agenda" for that conference became known as the Drake equation.

Drake's original form for his equation is

$$N = R_* \times f_{planet} \times n_e \times f_{life} \times f_{intell} \times f_{civ} \times L$$

N is the number of transmitting civilizations in our galaxy. R_* is the galactic birthrate of stars suitable for hosting life, in stars per year. For example, if we assume there are roughly 100 billion stars in the Milky Way and the galaxy is approximately 10 billion years old, R_* is approximately 10 per year (under the rather crude assumption that all stars are suitable and the rate of star formation has been relatively constant). The term f_{planet} is the fraction of such stars having planets; n_e is the number of planets per solar system that have an environment favorable for life; f_{life} is the fraction of such planets on which life actually evolved; f_{intell} is the fraction of inhabited worlds that develop intelligent life; f_{civ} is the fraction of planets having intelligent beings that produce a civilization capable of interstellar communication; and L is the lifetime over which such civilizations are "on the air," broadcasting signals. You might want to compare these terms with the more compact factors used in the equation on page 423.

Using some admittedly optimistic estimates for the first six terms in the equation, Drake suggested that we could reasonably guess that they would multiply to approximately one per year. This is the "birthrate" of civilized societies in the Milky Way. To find out how many are broadcasting now, we need only multiply this rate by L, the number of years during which they broadcast. This is analogous to determining the number of students attending a college by multiplying the number per year who enter as freshmen (the entrance rate) times the number of years spent as a student (typically four). If the birthrate for civilizations is taken to be one per year, the Drake equation becomes simply $N = L$; that is, the number of transmitting civilizations is simply the average lifetime (in years) of a transmitting society.

Unfortunately, L is dependent on sociology rather than on astronomy or biology—making it far more difficult to determine its value. Attempts to estimate L usually involve guessing what the one technological civilization we know—our own—is likely to do. Only a half-century after inventing radio, we also developed atomic weapons. To some people, this suggests that L might be very short—only a few centuries or less. On the other hand, we can be optimistic and assume that we will survive our own technology and exist as a society for millions of years into the future. Perhaps one of the most important things we could learn from a SETI detection is that not all technologically sophisticated societies are doomed to early self-destruction.

Frank Drake's personalized license plate.

might expect that most habitable planets would also have life, making the fraction f_{life} close to 1. However, until we have solid evidence that life arose anywhere else, such as on Mars, it remains possible that Earth was somehow extremely lucky and that f_{life} is so close to 0 that life has never arisen on any other planet in our galaxy.

Similarly, we have little basis on which to guess the fraction f_{civ} of life-bearing planets that eventually develop a civilization. On the one hand, the fact that life flourished on Earth for almost 4 billion years before the rise of humans might suggest that it is very difficult to produce a civilization even when there is life. On the other hand, given that the majority of stars in the Milky Way are older than our Sun, there has been plenty of time for evolution to work on numerous planets. Any evolutionary drive toward intelligence might inevitably lead to huge numbers of civilizations, even if it takes a long time on any given world. This question of whether intelligence is a rare accident or an inevitable result of evolution is so important to the issue of the search for extraterrestrial civilizations that we will devote the next section to investigating it.

The final term in the equation, f_{now}, is particularly interesting because it is related to the survivability of civilizations. Consider our own example. Our galaxy has existed for roughly 13 billion years. Let's say that, since it takes some time for life to evolve to intelligence, technically capable societies could have arisen only in the last 10 billion years. We have been capable of interstellar communication via radio for only about 70 years. If we were to destroy ourselves tomorrow (saving students the unpleasantness of a final exam), our technological "lifetime"—the length of time we could make ourselves known to other star systems—would have been only 70 years. If this is typical of other civilizations, then our chances of finding a signal from any one of them at any random time would be only 70/10,000,000,000, or 1 part in 140 million. In that case, even if there have been hundreds of millions of civilizations in the galaxy's history, no more than a few would be detectable now.

Of course, we have not yet destroyed ourselves, so the fraction f_{now} might be significantly larger. For example, suppose civilizations stay in a technologically active state for a billion years. Then the chance of a signal reaching us from any given civilization at a random time is 1 in 10, which means $f_{now} = \frac{1}{10}$. In that case, there may be a large number of communicating civilizations out there now, even if civilizations arise rather infrequently. To take some numbers, suppose that only 1 in 10 million stars ever gets a planet with a civilization. In a galaxy of 100 billion stars, this would mean that only about 10,000 civilizations ever arise. But if $\frac{1}{10}$ of them are here *now*, then there are some 1000 civilizations we could potentially find. Four decades ago, considerations like these led Frank Drake to conclude that the typical lifetime of civilizations must be one of the primary factors—perhaps even *the* primary factor—in the potential success of SETI efforts.

The Drake equation is mathematically simple, but its chain of terms is only as strong as its weakest link. If we know one term poorly, there is no way to improve our estimate of the number of civilizations by knowing other terms well. For example, while we might get better estimates of most of the factors in the equation as our knowledge of astronomy and biology improves, f_{now} depends on sociological factors—that is, the behavior of alien civilizations. Do they quickly self-destruct, or do they survive for long periods? The only way we can make realistic estimates of f_{now} is by actually detecting extraterrestrial societies. Until then, we will face great uncertainty in the total number of signaling worlds no matter how much we learn about the other terms in the Drake equation.

As a result of this "weakest link" problem, as well as the fact that other terms in the formula are still highly uncertain, we cannot draw any definitive conclusions from the Drake equation. Indeed, some of the terms have such a large uncertainty that the numbers we enter into the equation (choosing numbers within the range consistent with our present knowledge) can give us anything from an optimistic view that a large fraction of stars in our galaxy have intelligent, communicating beings to a pessimistic view that we would have to search a large number of galaxies to find even one other example of intelligence. The main value of the Drake equation, then, is in pointing out what factors are important and underscoring the implications of our lack of knowledge about particular factors. That is, it can be used to help us recognize what the issues are and where we remain ignorant, and therefore to steer us to areas in which more research is needed.

Think About It The Drake equation assumes that each transmitting civilization has sprung up independently on its own habitable planet. Is this a reasonable assumption, or might it be too limiting? Defend your opinion.

12.2 The Question of Intelligence

There are several key factors in the Drake equation that we must understand better in order to estimate the number of civilizations. We have already discussed (in earlier chapters) the uncertainties surrounding the number of habitable planets and the origin of life. In this section, we turn our attention to the term f_{civ}, which describes the probability that life will eventually give rise to intelligence and a technologically adept civilization.

Cosmic Calculations 12.1 THE DISTANCE BETWEEN SIGNALING SOCIETIES

How far is it to the nearest other world with technologically advanced beings? We don't know, of course, but if we use the Drake equation to estimate the number of civilizations, we can then compute their average separation.

We start by estimating the volume V of space available for civilizations in the Milky Way Galaxy, assuming that civilizations are confined to the galaxy's disk:

$$V = \pi R^2 \times T$$

where R is the disk radius and T is the disk thickness.

Suppose the Drake equation tells us that there are N technological civilizations in our galaxy. If we assume that these civilizations are spread randomly, then the average volume of space that contains just one civilization must be the total volume of the galaxy divided by the number of civilizations, V/N. If we consider this volume per civilization to be a cube in which each side measures d light-years, then d is also the distance from the center of one cube to the center of the next, which means it is the average distance between civilizations. The volume of a cube with side length d is d^3, so $d^3 = V/N$. Solving

for d, we find

$$d = \left(\frac{V}{N}\right)^{1/3} = \left(\frac{\pi R^2 \times T}{N}\right)^{1/3}$$

Example:
Suppose that $N = 10{,}000$. What is the average distance between civilizations?

Solution:
The disk of the galaxy has radius $R = 50{,}000$ light-years and thickness $T = 1000$ light-years, so its volume is

$$V = \pi R^2 \times T = \pi \times (50{,}000 \text{ ly})^2 \times (1000 \text{ ly}) \approx 8 \times 10^{12} \text{ ly}^3$$

We use this volume to calculate d:

$$d = \left(\frac{V}{N}\right)^{1/3} = \left(\frac{8 \times 10^{12} \text{ ly}^3}{10{,}000}\right)^{1/3} \approx 900 \text{ ly}$$

If there are 10,000 civilizations in the disk of our galaxy, the average distance between these civilizations is nearly 1000 light-years.

FIGURE 12.2
The path of evolution on Earth was severely affected by chance events such as the K–T impact 65 million years ago, which wiped out the dinosaurs and most other species. This artist's impression depicts the 10-kilometer-wide asteroid as it approached Earth. If this rock had arrived at Earth's orbit a day earlier (or a day later), it would have missed our planet. Would intelligent beings still have eventually arisen?

Even if life is widespread, is intelligence common?

The probability of a SETI success—receiving a signal—depends on how many civilizations are out there and broadcasting, and intelligence is a prerequisite to a civilization. SETI is likely to be successful only if intelligent life is widespread. But is it? Broadly speaking, there are two opposing schools of thought on this question.

One school considers intelligence that is comparable to our own (in other words, one that is able to develop both science and technology) to be unlikely. From this point of view, biology might be widespread but the evolution of technological intelligence extremely rare. Life has existed on our planet for at least 3.5 billion years. But only in the last few million years has our genus *Homo* developed the capability to understand its environment, and only within the last half-millennium have we come to understand the nature of Earth and begun our exploration of the cosmos. At the very least, the late appearance of *Homo sapiens* on Earth suggests that a long period of evolution must precede the emergence of technologically intelligent creatures.

Moreover, as we discussed in Chapter 6, our existence seems to have resulted from a number of chance events. For example, the Cambrian explosion that gave rise to the "body plans" of all modern animals (including those of our phylum, chordata) might have been the result of environmental stress introduced by snowball Earth episodes or a massive asteroid strike, and the rise of primates might never have occurred if the K–T impact had not wiped out the dinosaurs (Figure 12.2). The chance nature of these events and the many other forks in the evolutionary road suggest to some people that the appearance of technological intelligence on Earth was an enormously improbable event.

The second school of thought holds the opposite view. It proposes that there is evolutionary pressure for intelligence; that is, various evolutionary mechanisms consistently encourage an increase in intelligence for a wide range of species. If this is true, then some technologically intelligent species would still have evolved on Earth even if a different sequence of past events had prevented the existence of humans. Because we are the only species on our planet that has ever developed an advanced civilization, no other known species directly supports the view that the evolution of technological intelligence is likely. Instead, those who adopt this viewpoint look more generally at the process of evolution for evidence that some of the workings of natural selection promote greater intellectual capability.

CONVERGENT EVOLUTION The evolutionary argument in favor of widespread intelligence is based on the phenomenon of **convergent evolution,** the tendency of organisms of different evolutionary backgrounds that occupy similar ecological niches to resemble one another. In such cases, natural selection often produces analogous adaptations. One example of convergent evolution is the shape of large marine predators: Dolphins and sharks evolved from earlier mammals and fish, respectively, but both have a similar streamlined body form. The obvious reason for this is that being shaped like a torpedo makes for greater speed underwater—and speed has clear survival value for a predator. We say that the evolution of originally quite different animals has *converged* on this optimized underwater shape.

Eyesight offers another example. Vision is a useful adaptation, and most multicellular animals have some ability to see. However, the eye was not a unique evolutionary invention. Studies of evolutionary relationships show that eyes evolved independently at least eight different times. Indeed, the design of the human eye is by no means the best. Compared to the eyes of some other animals, ours have "flaws," such as the fact that the nerves in our eyes come together in a bundle before exiting through the back of our eyeballs, resulting in a small blind spot. The several independent origins of eyes suggest that evolution tends to *converge* toward developing some kind of eye to provide vision.

Think About It All crabs may look quite similar, but DNA studies show that various crab species evolved from very different ancestors; for example, some crab ancestors were shrimplike and others were lobsterlike. Based on this information, how do you think a crablike body, with its round shape and unusual manner of walking, might be an example of convergent evolution? Why might this body type be a natural evolutionary development?

Like speed or eyesight, intelligence—the way an animal processes information—is subject to natural selection. If there are ecological niches for which keener intelligence has survival value, then we would expect a convergent evolution to greater brain power for animals in these niches. Intelligence would be just as likely to emerge as any other generally useful adaptation. In this case, evolution would tend to raise the level of intelligence to at least some degree in a great many species (much as eyes developed in many species), which in turn would increase the chance that some of these species would evolve even higher intelligence. With more players in the game, the chance of producing human-style intelligence would be greater. If we could establish the presence of an evolutionary trend favoring intelligence, we would be encouraged about the potential emergence of technological intelligence elsewhere.

MEASURING INTELLIGENCE In principle, we can test whether intelligence is evolutionarily favored by measuring the brain power of a variety of animal species over time. But how can we measure the mental ability of animals? Few creatures are eager or able to take IQ tests, but we can resort to a simpler measure of raw brain power based on brain mass.

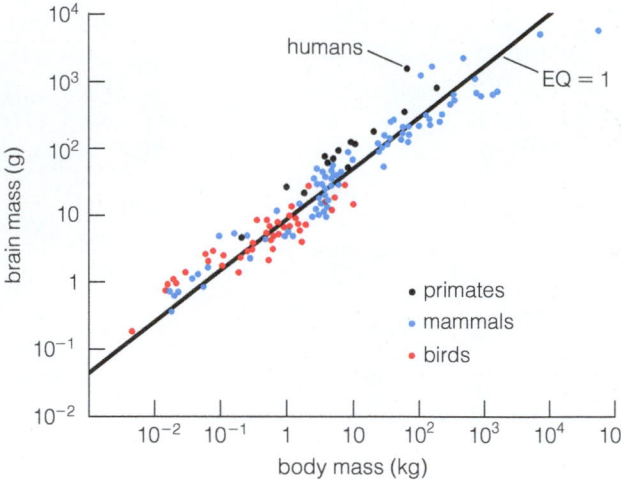

FIGURE 12.3

This graph shows how brain mass compares to body mass for some mammals (including primates) and birds. The straight line represents an average ratio of brain mass to body mass, which we define as an encephalization quotient of EQ = 1. Animals that fall above the line have an EQ greater than 1 and are presumed to be "smarter" than average. Animals that fall below the line have an EQ less than 1 and are "intellectually challenged." Note that the scale uses powers of 10 on both axes. (Data from Harry J. Jerison.)

Figure 12.3 plots brain mass against body mass for a sample of birds and mammals (including primates). There is a clear and expected trend: Heavier animals have heavier brains. By drawing a straight line that fits these data, we define an average brain mass for each body mass. Those animals whose brain mass falls on this line are said to have an **encephalization quotient (EQ)** of 1, which means a typical allotment of brain mass for creatures of their size. They have enough brain mass, we can assume, for a basic set of behaviors. If their brain mass falls above this line, then it seems reasonable to suppose that they are capable of more elaborate behavior. Creatures whose brain mass falls below the line are presumably less mentally agile than average animals in this sample. The EQ, the brain mass relative to the value on the line EQ = 1, serves as an indicator of general intelligence.

Although other indicators of intelligence have been suggested (the amount of brain folding, for example, or the number of neural connections), measuring EQ has the advantage of being fast and easy. For example, notice that the EQ = 1 line shows that the average species with body mass 1 kilogram (read off the horizontal axis) has a brain mass of about 10 grams (along the vertical axis). We'd therefore say that a species with the same body mass but a brain weight of 20 grams has EQ = 2, meaning it has twice as much brain as average for its size. Similarly, a species with body mass 1 kilogram and brain mass 5 grams has EQ = $\frac{1}{2}$, meaning it has only half as much brain as average for its size. Besides this ease of computation, EQ also has the advantage of being something we can compute for extinct species, since we can often estimate their total body masses and we can gauge their brain masses from the volume of their fossil skulls.

Admittedly, EQ is a simple measure; it might be likened to judging the computational ability of computers by weighing their CPU chips. Nonetheless, EQ seems to correlate well with complex behavior. Carnivorous animals that need to hunt down their meals generally have higher EQs than leaf eaters, and animals that lavish care on their offspring score higher than those that ignore them.

EXPLORING THE EVOLUTION OF INTELLIGENCE What happens when we use EQ to investigate the premise that, on Earth, there has been a trend toward increasing brainpower over time? If we look at contemporary species, we find that although humans don't sport the most massive brains (whales' brains, for instance, are much larger than ours), we do have the biggest brains in relation to body mass. Our species's encephalization quotient is 7, meaning that our brains are 7 times more massive than would be expected for an "average" mammal of the same body weight. This EQ not only exceeds those of chimps and dolphins, whose EQs are 2.5 and 4 or 5, respectively, but is higher than that of any other known species, alive or extinct. These numbers therefore seem to confirm that we're the cleverest critters on the planet.

However, a look to the past shows that we were not the only creatures evolving toward greater intelligence. Biologists have measured the EQs for a range of toothed whales and dolphins, and noted how these values changed during the past 50 million years (Figure 12.4). They did this by using computer tomography to determine the brain volumes of long-dead animals whose fossilized skulls were sitting in museums. The animal's weight was estimated by measuring the size of some of the bones around the eye sockets, a parameter known to be strongly correlated with body mass. With data in hand, these biologists then computed the EQs of 200 specimens, representing 62 cetacean species. What they found was that these animals experienced a major improvement in EQ starting about 35 million years ago, when they developed the navigation scheme known as echolocation (using audible "pings" as sonar), and some species underwent another big shift 15 million years ago. Not all the EQs went up, and today's cetaceans range in EQ from 0.2 to 5. But the smarter ones are not far behind our own EQ of 7. Humans are not closely related to dolphins in an evolutionary sense, so the fact that they have also developed large brains suggests that there is real survival value in cleverness, and that there are many ways that nature can produce it.

Another way to approach the question of the evolution of intelligence is to ask what factors might encourage its appearance and whether those factors are likely to be commonplace. To begin with, a good many "preadaptations" seem necessary. High-performance brains like those of birds and mammals need a vigorous metabolism, so intelligence is most likely to arise in warm-blooded animals. In addition, a relatively large body size is necessary to house a large brain (although research with insects suggests that, by miniaturizing neurons, nature might in principle be able to pack a lot of intelligence into small packages). A long period of parental care of offspring is probably also a necessary precondition for intelligence, which wouldn't be that useful without a time to learn from parents. These preconditions have existed for a wide range of species on Earth during at least the last 50 million years. Because this is the same period in which intelligence seems to have risen most dramatically, we have some hint that intelligence has indeed been evolutionarily favored.

Still, we might wonder about precisely which evolutionary mechanisms will select for intelligence. After all, having a big brain is metabolically expensive. In humans, for example, roughly one-fourth of our metabolism is for the benefit of our brain, even though that brain accounts for only about 2% of body weight. Consequently, big brains will occur only in species in which the cost is rewarded with real survival advantages. One survival benefit might come from the interactions among individuals of social species. Dolphins and primates are highly social animals. Success in a social environment is enhanced by intelligence, because there is survival value in being able to judge the mood and meaning of fellow creatures. Social position in your troop or pod depends on how savvy and canny you are. Moreover, an elevated social position often allows you to have first choice in mates, so clever, high-ranking individuals will tend to produce clever, high-ranking offspring. We might reasonably expect intelligent aliens, if they exist, to have also evolved in interactive social environments.

Competition between species can also ratchet up intellect. For example, large carnivores and their prey encourage improvements in one another's intelligence. When a lioness stalks gazelles for dinner, she's more likely to catch a gazelle that's less aware of its surroundings or less cagey in devising escape maneuvers. The lion gets a meal and in the process inadvertently raises the average intelligence of the surviving gazelles. The rise in gazelle intelligence makes getting a meal tougher for less mentally agile lions, which then preferentially drop out of the gene pool. The less alert, less cunning of both species are weeded out, and the smarter members survive.

In summary, several common circumstances in the animal world naturally select for brain power, and there is evidence that more than

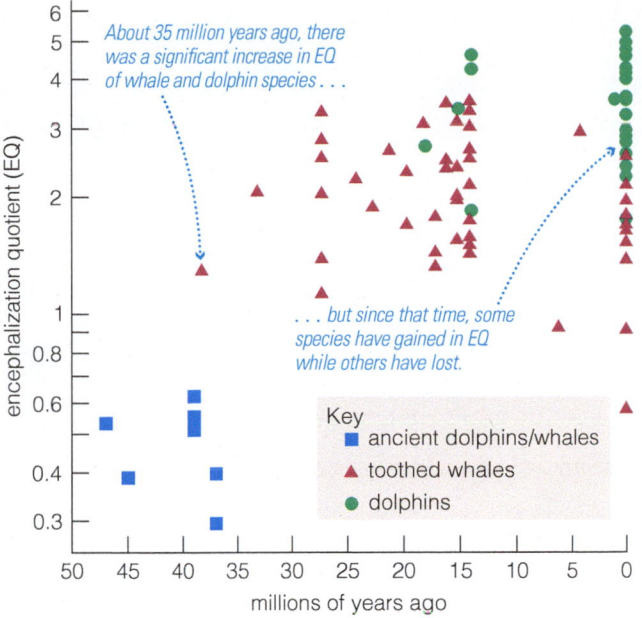

FIGURE 12.4

This graph shows the change in encephalization quotient (EQ) with time for toothed whales and dolphins. Note that over the course of the last 50 million years, some EQs increased substantially, while others remained relatively unchanged and some even went down. The fact that some dolphins have EQs comparable to our own suggests that high intelligence has survival value in many species, and may be a common evolutionary development. (Adapted from Marino, L., McShea, D., Uhen, M. D. 2004, *The Anatomical Record*. 281A: 1247–1255. Used with permission.)

one group of animals on our planet has been on the track to high general intelligence. This argues against the idea that the appearance of intelligence is some special, extraordinary accident of evolution on Earth and suggests that, given time and a competitive environment, many intelligent species will arise on any inhabited world. The existence of many candidates enhances the probability that one or more of these species will develop humanlike brainpower. On the other hand, increased intelligence comes at a cost in terms of resources to support it (such as a high metabolism or carrying around a heavy head), and it is unclear that these resources would not be better utilized by, say, evolving the ability to run faster or fight more fiercely. Moreover, many species *didn't* evolve intelligence over the last 50 or 100 million years. In the end, the example of terrestrial life alone does not tell us whether there is an evolutionary imperative toward humanlike intelligence. And, even if there is such an imperative, we still must ask whether high intelligence necessarily leads to technical competence and the ability to communicate across interstellar distances.

Will intelligence inevitably spawn technology?

It might seem natural that a species with sufficient brainpower will eventually develop science and technology. However, there are both physiological and sociological counterarguments to this idea.

On the physiological side, suppose dolphins were as intelligent as we are. Their lack of hands and their need to live in water would prevent them from building anything resembling modern technology. They might be smart and have sophisticated social structures, but without telescopes they wouldn't know much about astronomy, and without radios or lasers they wouldn't be able to talk to the dolphins of other worlds. Wolves have mouths, elephants have trunks, and ravens sport beaks and feet, but none of these animals have overall body designs that would allow them to manipulate complex tools—no matter how smart they might be.

On the sociological side, remember that *Homo sapiens* emerged about 200,000 years ago, and it's generally believed that by 20,000 or 30,000 years ago our ancestors had essentially the same level of intelligence as we do. Yet our ability to communicate through space is less than a century old, and it derives from the emergence of science—itself an endeavor that has only gradually developed over the past 2500 years. Many different cultures developed mathematics and astronomy to varying degrees, but only the cultural line that emerged from ancient Greece ultimately led to modern science [Section 2.1]. Was this a lucky accident, or was it inevitable that some culture would develop modern technology?

Again, a single example—what happened on our own planet—cannot guarantee that science and technology will be common in the cosmos, even if high intelligence is. The only way we can answer the question of how frequent or rare technological civilizations might be is to find evidence of them on other habitable worlds. We'll now turn our attention to experiments that are trying to do just that.

12.3 Searching for Intelligence

Receiving a message from another civilization could be one of the most important events of human history. We would know that we were not alone in the universe. We could conceivably learn a great deal about science and sociology. We could even learn how other civilizations have successfully survived periods in their history in which they had the power to destroy themselves. How might we receive such a message? In this section, we'll investigate the context in which we engage in SETI efforts. We'll begin with some historical background, then discuss the types of signals that SETI might be able to detect. We'll also explore a few possible ways of detecting extraterrestrial intelligence that don't involve communication signals.

How did SETI begin?

Shortly after its invention, radio was recognized as a possible means of extraterrestrial communication. After all, radio travels at the speed of light and can easily bridge the airless voids of space. If extraterrestrials are using radio, we might detect their presence without anyone having to leave the home planet.

EARLY CLAIMS OF SUCCESS As the twentieth century dawned, two pioneers of wireless technology became convinced (wrongly) that they had heard aliens on the airwaves. One of these pioneers was Guglielmo Marconi (1874–1937), generally celebrated as the man who made radio practical. The other was Nikola Tesla (1856–1943).

Tesla was both eccentric and brilliant. He was a prolific inventor (among other things, he made the first fluorescent lamp), but his most enduring legacy was the use of alternating current (AC) for distributing electrical power. Alternating current, produced by having a voltage that cycles positive and negative 60 times a second, is commonplace today, but it was a radical concept when Tesla first proposed its use for a Niagara Falls generating station.

Tesla thought that he could distribute electrical power by using alternating currents to generate low-frequency radio radiation rather than using copper wires. To demonstrate his idea, he built a 60-meter-tall transmitting tower, wrapped with wire, in Colorado Springs, Colorado. His device, intended to radiate electrical energy, was an outsized example of what is now known as a Tesla coil. One night in 1901, the inventor noted that his apparatus was picking up electrical disturbances, which he ascribed to interplanetary communication. He wrote, "The feeling is constantly growing on me that I had been the first to hear the greeting of one planet to another."* We now believe that he was listening to an atmospheric phenomenon known as "whistlers," electrical noise created by distant lightning discharges.

That same year, Marconi successfully sent a radio signal across the Atlantic, proving that this invention was useful for communication over large distances. It took little imagination to guess that radio might also serve for sending messages into space. In the early 1920s, Marconi stated that he, too, had picked up signals that came from extraterrestrial sources. These experiments culminated in 1924, during one of the periods when Earth and Mars are closest to each other in their orbits. Marconi and others encouraged anyone with a radio set to listen for martian broadcasts, and even the U.S. Army joined in the search (Figure 12.5). There was considerable optimism that something would be detected, and

Army radio operators thruout the country "listened in"; or messages from Mars last month—without much success. Corp. John H. Sadler of the Signal Corps is shown at a radio station of the War Department.

FIGURE 12.5
An Army operator listening for martian radio signals, as pictured in *Radio Age*, October 1924.

*Original source: Nikola Tesla, "Talking with the Planets," *Collier's Weekly* 26, no. 19, 4 (1901); taken from S. J. Dick, *The Biological Universe*, Cambridge University Press, 1996.

a cryptographer was standing by in case the signals required decoding. Alas, while signals were picked up, there was no evidence that they were from Mars or any other extraterrestrial source. The signals that Marconi heard may also have been "whistlers" from distant lightning, or possibly garbled U.S. Navy broadcasts of which he was unaware.

In retrospect, these early experiments were doomed. Aside from the fact that there are no Martians building radio transmitters, we now know that these pioneer listening attempts were all made at the wrong spot on the radio dial. They were conducted at relatively low frequencies, which have problems penetrating Earth's *ionosphere*—a layer of the upper atmosphere that consists of particles ionized by sunlight. The ionosphere acts as a radio "mirror," reflecting low-frequency radio emissions. Indeed, it was the ionosphere that bounced Marconi's 1901 transatlantic signal from England to Canada, allowing communication despite Earth's curved surface. The apparatus used by both Tesla and Marconi operated at frequencies too low to penetrate the ionosphere and therefore would have been insensitive to any cosmic broadcasts.

Although the radio pioneers might have mistaken thunderstorms for Martians, their enthusiasm for radio's use as a long-distance communication medium was ahead of its time. After World War II, when high-frequency radio equipment became widely available, it was possible to listen at frequencies that *could* penetrate the ionosphere. In addition, improvements in antennas and receivers made such experiments enormously more sensitive than those attempted early in the century. The stage was set for a scientific approach to SETI.

ORIGINS OF MODERN SETI The beginning of modern SETI is generally attributed to two physicists working at Cornell University. In 1959, Giuseppe Cocconi and Philip Morrison wondered how difficult it would be to send radio signals over interstellar distances. They made simple calculations showing that communication between nearby stars was possible using technology no more advanced than our own.

Cocconi and Morrison realized that the galaxy is considerably older than our solar system and consequently that there could be civilizations among the stars that have been around far longer than ours—perhaps surviving for many millions of years. For such long-lived societies, the construction of powerful radio beacons that could be used to send "hailing signals" to other star systems should be a simple matter.

Cocconi and Morrison advocated looking for such beacons using large radio telescopes aimed at nearby stars. The exact arrangement of the planets around the stars under scrutiny was irrelevant, because the area of sensitivity of any reasonable-size radio telescope would encompass all of a star's possible planetary environs. The only other technical question concerned which frequencies to monitor with the radio receivers.

The radio portion of the electromagnetic spectrum (see Figure 3.24) extends over a wide range of frequencies. When we build a radio receiver, it is sensitive only to a particular set of frequencies within this wide range, which we refer to as its **band** of operation. If you look at a typical FM radio, you'll see that it is designed to cover the band of frequencies from about 87 to 108 MHz. (Recall that the basic unit of frequency, hertz [Hz], is equivalent to waves [or cycles] per second. For example, 1 megahertz [MHz] is a frequency of one million cycles per second.)

Of course, radio stations do not broadcast their signals over this entire band—if they did, no matter where you tuned your radio you'd hear the overlapping sounds of all the radio stations that were transmitting. Instead, radio stations limit their signals to only a narrow range of frequencies. For example, if your favorite radio station broadcasts at "97.3 on your FM dial," then you must tune your receiver so that it picks up only a small range of radio frequencies centered on 97.3 MHz. That range of transmitted frequencies is called the **bandwidth** of the signal. As a practical matter, the bandwidth is governed by how much information the broadcast contains. For example, most television signals (Figure 12.6) have a bandwidth 500 times wider than that of an AM radio station, because the TV signal contains picture (video) information in addition to sound (audio).

Terrestrial broadcasts are made with a wide range of bandwidths depending on their purpose, but a hailing signal designed to get the attention of someone light-years away would likely include a very narrow bandwidth signal component, with much of the transmitter energy concentrated at one spot on the radio dial. This would make the signal easier to pick out against the background noise that is naturally produced by interstellar gas, distant galaxies, and the radio receiver itself. But in what part of the radio spectrum would we expect to find such a narrow-band hailing signal? In truth, we cannot say for certain, since a wide range of radio frequencies are serviceable. However, because searching all these frequencies would be a hopelessly difficult task, Cocconi and Morrison proposed that SETI experimenters tune their receivers near a spot on the dial that every scientifically literate society would know: 1420 MHz. This is the frequency at which neutral hydrogen gas—the major constituent of the thin material that floats between the stars—produces natural radio static. Radio astronomers often use this frequency to study the distribution of interstellar gas in galaxies. Because it is such a useful frequency, astronomers throughout the universe (of whatever species) would have it marked on the receivers of their radio telescopes, and a transmitting beacon tuned near this frequency would be likely to attract the attention of others.

Having set out all the important principles of a radio SETI search, Cocconi and Morrison tried to interest radio astronomers in their idea. They approached the director at England's Jodrell Bank Observatory but got little response. However, Frank Drake, then a young astronomer at the Green Bank radio observatory in West Virginia, had independently reached many of the same conclusions as Cocconi and Morrison. In the spring of 1960, using the 26-meter (85-foot) radio dish at Green Bank, Drake began an experiment much like the one Cocconi and Morrison had in mind (Figure 12.7). He conducted a weeks-long search for 1420 MHz radio signals from two nearby, Sun-like stars: Epsilon Eridani and Tau Ceti. (These stars are approximately 12 light-years distant.) Drake whimsically named his search Project Ozma, after the fictional princess in the books by Frank Baum. It was the first modern SETI experiment.

Drake's search failed to detect alien signals, but it fired the imaginations of many in the science community and ultimately led to a small SETI research program run by NASA. The NASA researchers spent many years studying the feasibility and technology of SETI. Their work led to the construction of specialized receiving systems that could be fitted to large, existing radio telescopes. In 1992, the NASA search began. But a year later, with the data collection barely under way, the program was canceled by the U.S. Congress. Since then, scaled-down SETI programs have continued with private funding in the United States and as small university research efforts elsewhere in the world.

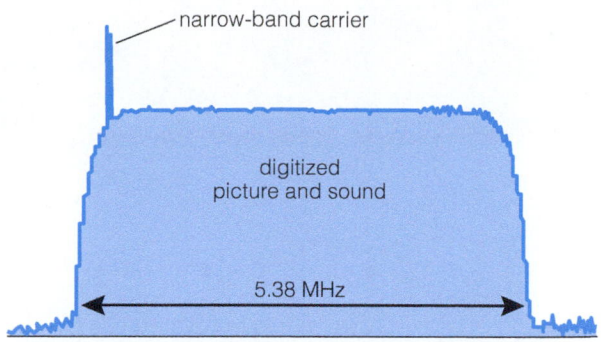

FIGURE 12.6
The spectrum of a high-definition television (HDTV) signal, showing how the energy of the transmission is spread over a nearly 6 MHz bandwidth. The large bandwidth is necessary because the signal carries a great deal of picture information. However, note that approximately 7% of the total broadcast energy is concentrated in a narrow "carrier"—the spike visible at the low frequency end of the spectrum. This carrier helps the television receiver lock onto the signal. Because the carrier concentrates a fair amount of signal power into a very narrow bandwidth, it would be the easiest part of the signal to detect from a great distance. Hence, it is a good example of the type of narrow-band signal that SETI experiments seek.

FIGURE 12.7
The 26-meter (85-foot) radio telescope used by Frank Drake in his pioneering 1960 SETI search, Project Ozma. This instrument was the first to be built at the National Radio Astronomy Observatory in Green Bank, West Virginia. Drake scrutinized two nearby star systems for signals near 1420 MHz in frequency.

How do we search for intelligence today?

Many of the SETI techniques in use today are elaborations on the earlier schemes proposed by Cocconi and Morrison, and first tried by Drake. However, SETI has also branched out to search for signs of intelligence in other ways. To understand the modern search for extraterrestrial intelligence, we first need to consider the types of signals we might be able to detect.

CATEGORIES OF SIGNALS Drake's Project Ozma searched for a deliberately broadcast signal at a specific frequency that would be known to astronomers anywhere. It was therefore a search for an interstellar hailing signal, or beacon, intentionally sent so that others might detect it. But there are other possible signals we might hope to find. Broadly speaking, alien signals could fall into any one of the following three categories:

1. *Signals used for local communication on the world where intelligent beings live.* Our own radio and television signals fall into this category, because they are designed for our own use and not for interstellar communication. Another local use for radio signals is radar. Advanced civilizations might use radar to locate comets that pose a potential threat to their planet, for example.
2. *Signals used for communication between a civilization's home world and some other site,* such as a colony on another world. We have used relatively weak signals of this type to communicate with our interplanetary spacecraft. Such signals would be far stronger if they were being used, say, to communicate between colonies on planets in different star systems light-years apart.
3. *Intentional signal beacons,* such as the type searched for by Project Ozma, purposefully designed to get the attention of other societies.

In principle, SETI can search for all three types of signals, but in practice, our ability to receive signals depends on the sensitivity of our equipment. To get a rough idea of what we might be able to detect with our current technology, let's consider our own signals as an example.

The first commonplace, high-power, high-frequency transmissions from Earth were our early television broadcasts. These transmissions began in earnest during the late 1940s, so they are now some 70 light-years away in space—followed by all television broadcasts since. In principle, any civilization within about 70 light-years could watch our old television shows (although it's unclear whether they would *like* them!). However, while television transmitters are fairly powerful, their antenna systems are designed to spread the signal over a wide angle, so that they reach homes in all directions from the transmitting station. This means that the strength of that portion of the signal that is, by chance, aimed at any given star is quite weak. And, like all light, these signals continue to weaken with distance (following an inverse square law, as shown in Figure 7.2). If another civilization has the same sort of receiving technology that we do, they could detect our television signals only if they were within about 1 light-year of us, and no stars are that close (the nearest are more than 4 light-years away). Turning the situation around, we find that we are not yet capable of detecting signals in the first category listed above, unless for some reason they are being broadcast with much more power than we use for our own television signals. (Some of our military radars, which employ large antennas to narrowly focus their transmissions, could be detected as far away as a few tens or hundreds of light-years.) The situation is no better for signals in the second category, at least if such signals are comparable in strength to those we currently use for communicating with our own interplanetary spacecraft.

Our technology is rapidly improving, and in the future we might be able to detect alien signals in any of the three categories. But for the moment, at least, our best chance of detection is to consider signals in the third category. Beacon signals should be the easiest to detect because they would deliberately be made strong enough to be heard across interstellar distances. They should also be the least difficult to interpret, because they presumably would be designed for easy decoding.

DECODING A SIGNAL How might an alien civilization design a signal so that we could decode it? In the book and movie *Contact*, author Carl Sagan supposed that the aliens might make it easy for us to recognize their message by playing back to us one of our own television transmissions. However, because our broadcasts have had only enough time to make it a relatively short distance into space (roughly 70 light-years), this strategy could work only if a civilization happened to be near enough to have already found such a signal and returned it to Earth. Fortunately, there are other ways that aliens could make it easy for us.

For example, aliens might choose to broadcast a strong, narrow-band signal. Most natural radio emissions have fairly broad bandwidths, so a signal confined to one narrow spot on the radio dial would immediately offer a hint that it might be the work of intelligent beings. If the signal was also flashing on and off or switching between two nearby frequencies, we would suspect that it was a coded message. We would undoubtedly record the pattern and try to analyze what was being "said."

The first thing to do would be to look for repetition. If aliens really wanted us to detect and decode their signal, they would repeat the entire broadcast many times, since otherwise we'd have to be listening at just the right moment to catch the signal. Knowing the total length of the message might help us greatly in figuring it out. If the total number of flashes or frequency changes was, for example, 1679, we might note that this is the product of two prime numbers, 23 and 73. (A prime number is one that can be divided only by 1 and itself without any remainder. The prime numbers are 2, 3, 5, 7, 11, ...) We could then arrange the message in a 23-by-73 grid and look for pictures or other figures.

This simple approach is one we have taken ourselves in one of the few deliberate broadcasts that we have made to the stars. In 1974, the powerful planetary radar transmitter on the Arecibo radio telescope (Figure 12.8a) was fired up and used to send a 3-minute message to the object M13, a globular cluster containing a few hundred thousand stars. The message consisted of 1679 bits, and each bit was represented by one of two radio frequencies. If aliens picked up the signal and recognized that 1679 is the prime number product 23 × 73, they could arrange the bits in a rectangular grid with 73 rows and 23 columns. The resulting graphic, shown in Figure 12.8b, represents the Arecibo radio dish, our solar system (shown with Pluto, since the signal was sent decades before Pluto's demotion to dwarf planet), a human stick figure, and a schematic of DNA and the eight simple molecules used in its construction. However, because we did not repeat the signal, any aliens living around stars in M13 will have only one chance to receive it as it passes by their planet

a Arecibo has the world's largest single-dish radio antenna, shown here, with a diameter of 305 meters (1000 feet).

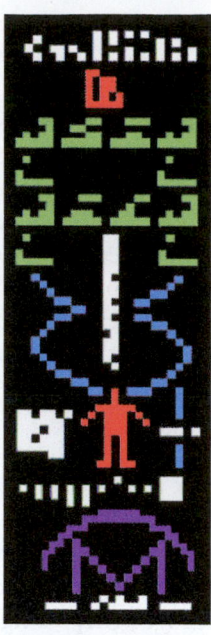

b The pictorial message broadcast in 1974, as it looks once you realize that the message is intended to be laid out as a rectangular grid with 73 rows and 23 columns (each of these numbers is a prime number, which should help enable any alien recipients to guess the layout). The colors are shown only to make the components clearer; the actual picture was sent in "black and white."

FIGURE 12.8

In 1974, the Arecibo radio telescope in Puerto Rico was used to send a 3-minute broadcast toward the globular cluster M13.

at the speed of light. That is one reason the signal was sent to a globular cluster: The cluster's several hundred thousand stars would seem to improve the odds of someone's receiving the signal. If it is received, it won't be soon. M13 is about 25,000 light-years from Earth, so it will take our signal some 25,000 years to get there and another 25,000 years for any response to make its way back to Earth.

Think About It Some people consider the 1974 broadcast to have been a dangerous exercise that might attract the unwanted attention of hostile aliens. In general, do you think it is "safe" for us to broadcast messages to the stars? Why or why not?

Of course, alien messages could be far more sophisticated and more difficult to decode. In the movie *Contact*, the pictures were three-dimensional, not flat, and in that case we should be looking for messages whose length is the product of three, rather than two, prime numbers. Alternatively, the message might be encoded in ways similar to the schemes used for sending files on the Internet. Given the enormous variety of possible ways a message could be transmitted, it might be that we would never be able to understand an alien broadcast. However, if advanced societies truly wished to get in touch, they would undoubtedly go to some trouble to make their messages simple enough to be understandable to any civilization able to build the telescopes necessary to receive them.

Finally, it's important to remember that beacon signals will exist only if other societies make a deliberate decision to broadcast them. If no one does anything more than send occasional short signals like ours in 1974, the chances of detection will be quite small, even if many societies are out there listening. As a result, some people have asked whether we should undertake our own more deliberate effort to send a beacon signal into space. While this question continues to intrigue both researchers and the public, the consensus has been that we should focus on receiving signals first. After all, as we've seen in our consideration of the Drake equation, the chance that anyone will pick up a signal depends on how long broadcasting civilizations are "on the air." There's little point in transmitting signals for only a few weeks or even a few years. A broadcasting project would require long-term investment and a great deal of patience. While some SETI researchers are considering a transmitting effort to some of the nearer star systems, we should keep in mind that we developed radio technology only in the past century. There may be galactic civilizations that have had the ability to transmit signals into space for hundreds of millennia or longer. As the new kids on the block, it may well make sense for us to listen first.

RADIO SETI TODAY Let's assume that someone really is broadcasting a signal that we could recognize and potentially decode. How should we go about searching for it? Today, just as in the early days of Drake's Project Ozma, most SETI searches are attempts to detect *radio* transmissions from other worlds.

A large radio antenna—most often a radio telescope constructed for more conventional research purposes—is used to try to pick up the hoped-for cosmic signals. A low-noise amplifier at the antenna's focus boosts the signal levels before they are further processed. The processing usually consists of digitally "slicing" the incoming, wide-band signal (typically spanning hundreds of megahertz) into many narrow-band channels. This processing is based on the assumption that some strong narrow-band component will be present in any deliberately broadcast extraterrestrial signal.

Movie Madness CONTACT

Do the aliens know we're here? They might, if they're not too far away.

Contact, the movie based on Carl Sagan's novel about getting in touch with our celestial pals, starts out with a nifty sequence in which the camera backs away from Earth, slips by the Moon and planets, and eases out into the galaxy and beyond. During this high-speed countermarch, we hear the sounds of radio programs that have reached each of these cosmic outposts. With every step outward, we move back in time. (OK, there's some cinematic license here: As the camera passes Saturn, we've regressed to 1950s rock and roll. Saturn is never more than 88 light-*minutes* from Earth!)

In fact, less than 100 light-years out, Earth really does go "silent." Easy evidence for the presence of *Homo sapiens* extends only this far—about one-tenth of 1% of the distance to the far edges of the galaxy.

Fortunately for the cash-starved SETI researchers in *Contact,* some friendly aliens have an outpost around the bright star Vega, a mere 25 light-years away. They've tuned in to one of our early TV broadcasts and, apparently intrigued, have replied. Their response is picked up by the film's heroine, Ellie Arroway, as she uses a pair of earphones to monitor the cosmic static received by a large radio telescope. (More license here: Radio receivers for SETI sport hundreds of millions of channels. Ellie should have either donned a few hundred million headsets or left the listening to computers.)

The alien reply signal, which sounds like a pile driver hitting a pod of whales, contains an original 1936 broadcast of the Berlin Olympic games (that way we know that the extraterrestrials are deliberately beaming to us) interwoven with construction details for some sort of large device.

Faced with a SETI detection, the government goes nuts, and so do a lot of the citizenry. Some are ecstatic about the possibility of alien company; others see the news as heralding the apocalypse. Meanwhile, the scientists build the device—a multibillion-dollar machine that looks like the ultimate theme park attraction, but in fact is a wormhole transporter (is there any difference?).

It's nice to think that advanced aliens would want to improve our lifestyle by sending us plans for high-tech hardware. But this seems an unlikely message from space. After all, if we could somehow contact the Neanderthals and give them plans for a personal computer, do you think they could ever, *ever* build it?

During the search, the radio telescope may be either pointed in selected directions, such as toward individual stars, or swept across the heavens to study a larger section of the sky. The former technique, known as a *targeted search*, proceeds on the assumption that not all locations in space are equally probable sites for intelligent life. For example, Drake put forth the reasonable hypothesis that civilizations were most likely to exist on planets around Sun-like stars, and consequently he targeted his search at this type of star. The sweep technique, known as a *sky survey*, makes no assumptions about where intelligent aliens might be located.

Table 12.1 summarizes key features of three major SETI surveys that are currently under way, including the number of narrow-band channels being examined and the overall band of frequencies covered in each search. The last column of the table gives an indication of each experiment's sensitivity: It lists the minimum transmitter power that alien broadcasters would need to be using for us to detect their signals, assuming the aliens are located 100 light-years away and use a transmitting antenna 100 meters in diameter. The Inner Galactic Plane Survey, for example, could detect such an extraterrestrial broadcasting setup if the transmitter had a power of 70 megawatts or more. That's more than the power of most radio and television broadcasts on Earth, but not that much more: Some of our own broadcasting stations have power greater than a megawatt, although they do not use a 100-meter antenna to narrowly focus their transmissions.

A key limitation of most SETI efforts, past and present, has been that they have "piggybacked" on telescopes used primarily for other astronomical research (Figure 12.9). As a result, the choice of where the telescopes point is dictated by the needs of these other research programs. The Inner Galactic Plane Survey is the first major project to use telescopes dedicated to SETI research. This survey is being conducted with the Allen Telescope Array, located in northern California and operated by the SETI Institute (Figure 12.10). As of 2015, the Allen Array consists of 42 antennas, but the intention (pending funding) is that it will eventually consist of 350 small (6-meter) radio dishes, giving it a combined collecting area equivalent to that of a single 100-meter telescope. One major advantage of an array, as compared to a single-dish radio telescope, is that it can be electronically focused on several positions on the

TABLE 12.1   Major Current Radio SETI Surveys

Name	Institution	Telescope	Total Number of Channels	Width of Single Channel	Band Covered	Detectable Power for 100-Meter Transmitter at 100 Light-Years
Inner Galactic Plane Survey*	SETI Institute	Allen Telescope Array	450 million	1 Hz	1390–1720 MHz	70 megawatts
SERENDIP	University of California, Berkeley**	Arecibo Radio Telescope (305 m)	168 million	0.6 Hz	1370–1470 MHz	1 megawatt
SETI Italia	Istituto di Radioastronomia, Bologna	Medicina radio telescope (32 m)	24 million	0.6 Hz	Bands centered at 1.4, 2.8, 6.4, and 22.4 thousand MHz	Typically 30 megawatts

*Values given here are for the Allen Telescope Array capabilities at the beginning of 2015, when 42 of its 6-meter antennas were in operation. If more antennas are added (the design goal is to build 350), the sensitivity will increase.

**About 2.5% of the data collected by the SERENDIP project are being distributed over the Internet for processing on a downloadable screen saver. This project (called SETI@home) has involved more than seven million home computer users.

FIGURE 12.9

The 64-meter Parkes radio telescope in New South Wales, Australia. For many years, a SETI experiment "piggybacked" on this telescope while it was engaged in other astronomical research. Piggyback schemes avoid competition for telescope time and therefore provide SETI experiments with a lot of data. On the other hand, the SETI astronomers—who are only "along for the ride"—have no say in where the telescope is aimed. Since there are stars in every direction, however, this type of sky survey could still chance on a transmitting civilization.

FIGURE 12.10

The Allen Telescope Array, constructed by the SETI Institute in Hat Creek, California. The array consists of 42 small (6-meter-diameter) dishes and is designed to be expanded to 350 antennas (inset). It is currently studying the inner parts of our galaxy as well as nearby star systems.

Cosmic Calculations 12.2

SENSITIVITY OF SETI SEARCHES

The distance from which a SETI experiment could detect an alien transmission depends on the sensitivity of the receiver and on the strength of the signal. The inverse square law for light [Section 7.1] tells us that if aliens broadcast a signal with power P, it will have weakened by a factor d^2 by the time it reaches Earth, where d is the distance of the broadcasting civilization. Therefore, the strength S of the signal we receive at Earth is this diminished power P/d^2 multiplied by the area of the receiving radio dish, A_r, and a constant of proportionality, k:

$$S = k \times A_r \times \frac{P}{d^2}$$

Example: A SETI search using the 300-meter-diameter Arecibo Radio Telescope can pick up a 10-million-watt signal from 1000 light-years away (if the transmitting antenna is also 300 m in diameter). If we were to detect a similar signal coming from 25,000 light-years away, how powerful would the alien transmitter have to be?

Solution: Since all else is held constant in this problem, the only change is that in the second case the transmitter is 25 times farther away than in the first case. Therefore, for a given transmitted power P, the signal strength S will be reduced by a factor of $\left(\frac{1}{25}\right)^2 = \frac{1}{625}$. To get the same S, the transmission power from this star would have to be 625 times stronger, or 625×10 million $= 6.25$ billion watts.

sky simultaneously, thereby speeding up the search. The Allen Array is used nearly every day for SETI observations, though it can also be used simultaneously for conventional radio astronomy studies.

A remaining limitation for all current radio SETI experiments comes from interference, especially from radar and orbiting satellites. Radio interference greatly hinders SETI searches, because telecommunications satellites, aircraft, and Earth-bound radar produce narrow-band signals of exactly the type being sought. These signals are so strong that they are picked up by radio telescopes no matter in which direction they are pointed. SETI researchers have devised various tricks to sort out terrestrial signals from possible extraterrestrial ones, but this problem will continue to worsen. Perhaps in the future we will be able to place radio telescopes on the far side of the Moon, where they would be shielded from Earth's noisy radio presence.

OPTICAL SETI The fact that most SETI searches have used radio telescopes may seem logical, since we often think of radio as something that we "listen to," but it is not the only possibility. Remember that radio is a form of light and that we "listen" to it only after the light is converted to sound by a radio receiver. In essence, the radio station uses electronics to convert the information content of the sound (for example, music) into light waves with wavelengths far longer than what the human eye can perceive. These light waves travel through the air to your receiver, where electronics converts the information back into sound. It just so happens that the wavelengths of light that we use for this process are in the radio part of the spectrum (see Figure 3.24), which is why we say we are "listening to the radio." Broadcast television, cell service, and WiFi are also encoded in radio waves, but in this case the encoding includes pictures and data in addition to sound.

The common use of radio waves for encoding information is a technological choice, not a requirement. In principle, any form of light would do—and often does. For example, if your computer or television is hooked up to a fiber-optic cable network, you are receiving information transmitted in the form of infrared or visible-light waves bouncing through the fiber optics, rather than in the form of radio waves. Alien civilizations might choose to communicate with visible light or other forms of light besides radio, and some SETI efforts are designed to search for such signals.

The idea of communicating between worlds via visible light has a long and interesting history. In fact, scientists considered using light as a communication medium even before radio's invention. During the nineteenth century, several scientists advanced proposals intended to put us in touch with sophisticated societies that were imagined to live on either the Moon or Mars. The eminent mathematician Karl Gauss is said to have suggested planting trees in the form of a right triangle in Siberia, bounded by clear-cut squares. This greenery graphic presumably would be seen by other solar system inhabitants, proving that Earthlings were at least smart enough to know the theorem of Pythagoras. While this story may be apocryphal, Gauss is definitely known to have proposed using 100 mirrors, each about 1 meter on a side, to shine the light of gas lamps into space. "This would be a discovery even greater than that of America, if we could get in touch with our neighbors on the Moon," he claimed. Another nineteenth-century suggestion was to dig trenches in the Sahara in various geometric shapes, fill them with water and oil, and ignite them at night.

Despite these old ideas, until fairly recently optical SETI did not draw much attention from researchers. Part of the reason had to do with estimates of the energy needed. In general, far less energy is required to send an interstellar signal via radio than via visible light, so researchers guessed that radio would be the aliens' technology of choice. In addition, while radio waves travel through interstellar space with ease, visible light tends to be absorbed by tiny grains of interstellar dust that float between the stars. This dust creates the dark rift through the center of the Milky Way as we see it in our sky (Figure 12.11). When we look in different directions within the galactic plane, the dust completely blocks our view of visible light beyond a few thousand light-years—a relatively short distance compared to the 100,000-light-year diameter of our galaxy. The fact that visible light can be used to send signals for only limited distances through the galaxy seemed to argue against its use as a beacon for interstellar communication.

Researchers no longer believe that either problem is a serious strike against optical signaling. The limitations imposed by interstellar dust are real, but signals that can travel up to a few thousand light-years could still be quite useful. After all, millions of stars lie within 1000 light-years of Earth. Moreover, if longer-distance communication is needed, using infrared rather than visible light would largely circumvent the problem, because infrared light penetrates the interstellar dust. The energy–cost issue argues in favor of radio only if we consider transmitters that beam their signals in all directions into space. For focused communication, such as sending a signal to a particular nearby star, visible light can easily be concentrated into a narrow beam (by using a large lens or mirror). Whereas radio is ideal for a station that wants its signal to be picked up by people living in all directions from the station, visible light from a flashlight (or a laser beam) is a better medium if you want to signal someone in a particular place. SETI researchers have concluded that visible light might well be useful for interstellar communication and that optical SETI could be a worthwhile enterprise.

How might an alien civilization communicate optically? Simply turning a continuously shining, high-powered laser on someone else's star system isn't particularly effective. A continuous beam could be lost in the glare of starlight from the transmitter's home sun unless it was extremely bright. In addition, a light that is always on doesn't convey much information. Short bursts of laser light, say a billionth of a second long, would work better, because they could momentarily outshine the starlight even with relatively modest laser power. A series of such bursts—a pulse train—could contain patterns that made up a message. Such arguments have led SETI scientists to imagine that advanced alien societies might be using automated, high-powered laser transmitters to send pulsed messages to thousands of nearby stars. The messages, perhaps only a few seconds long, could be repeated every several dozen hours. If a civilization in our galactic neighborhood is sending laser messages, we might hope to find the laser "pings" by monitoring nearby stars.

Optical SETI efforts today are attempting to do just that. Like their radio counterparts, these efforts are passive—that is, they are searching for incoming signals but are not transmitting anything outward—and are systematically checking out the vicinities of local stars. The experiments look for short pulses of light that are bright enough to outshine the background shower of photons naturally produced by the alien world's host star. Photomultiplier tubes, sensitive light detectors able to see short flash-

FIGURE 12.12
Setup for an optical SETI experiment that was carried out on the 1-meter Nickel telescope at the Lick Observatory, on Mount Hamilton near San Jose, California. The photomultiplier tube detector, which is designed to react to bursts of light, is contained in the white box attached to the back of the telescope.

es, are affixed to conventional mirror telescopes to hunt for these photon bursts. Figure 12.12 shows the setup for an optical SETI experiment at the Lick Observatory, near San Jose, California. This experiment was quite sensitive: If someone were sending out laser pulses with a transmitter no more powerful than we ourselves could build, the Lick experiment could have picked it up from as far away as about 500 light-years. Researchers at Harvard University have constructed a telescope whose sole job is to scan the sky for short, bright flashes of light sent from other worlds.

Think About It It has been suggested that one benevolent use for our nuclear weapons would be to rocket them into space, line them up in geometric patterns, and then detonate them as a "broadcast signal" to distant alien societies. Do you think this would make an effective SETI signal? Why or why not?

Beyond radio and visible light, almost any other form of light might also be used for interstellar communication. We've already noted the potential value of using infrared signals in place of visible signals in order to penetrate interstellar dust. Similarly, distant civilizations might choose to communicate with ultraviolet light or X rays. However, these other forms of light have no obvious advantages over radio, infrared, and optical options, and they are more expensive in terms of the energy cost per "bit" of information. As a result, SETI researchers feel justified in concentrating current searches on radio and optical frequencies.

What about signals using more exotic technologies? For example, some people have suggested that advanced civilizations might send messages via the ghostlike subatomic particles called *neutrinos* or via the so-called *gravity waves* that Einstein predicted but that we have not yet detected directly.* We cannot rule out these possibilities, but we might wonder what the point would be. After all, neither of these schemes would communicate any faster than light, and they would make both transmitters and receivers far harder to build. Of course, it is also possible—perhaps even likely—that advanced civilizations have developed

---

*However, we have strong indirect evidence for the existence of gravity waves, and a new instrument called the Advanced Laser Interferometer Gravitational-Wave Observatory (LIGO) may detect them within just a few years.

communication technologies that we could not detect. These technologies might involve physics that we have not yet discovered, or they might use ways of disguising signals so that our detectors could not distinguish them from natural background signals. Indeed, if highly advanced civilizations *want* to keep their communications secret from us, they probably can. SETI's hopes rest with civilizations that either are looking for contact or don't care if their communications are intercepted by others. In either of these cases, sticking with the simplest available technologies would seem to make the most sense.

ARTIFACTS AND INTERSTELLAR CRAFT There might be other ways of detecting extraterrestrial civilizations besides picking up signals traveling through interstellar space. These possibilities apply if civilizations have advanced far beyond our own capabilities.

If other civilizations have achieved interstellar travel, they might have visited our solar system. In that case, they may have left artifacts behind, either accidentally or deliberately. Some people claim that we already have such artifacts from UFOs, but as we'll discuss in Section 12.4, these claims do not meet the standards of science. There are more plausible ways to imagine aliens leaving artifacts, and many science fiction writers have considered them. For example, in the classic movie *2001—A Space Odyssey,* based on the story by Arthur C. Clarke, highly advanced aliens buried a monolith on the Moon (Figure 12.13). When humans found the monolith, it notified the extraterrestrials.

It might be a while before we can search the Moon and the planets for artifacts such as buried monoliths, but a few researchers have suggested that aliens wishing to get our attention might leave calling cards in places where we could find them somewhat more easily. In particular, some people believe we should look especially hard at the so-called Lagrange points of the Earth–Moon system, named for the eighteenth-century French mathematician Joseph-Louis Lagrange (1736–1813). At these five positions in space, the effects of gravity from Earth and the Moon "cancel" in such a way that, if you floated weightlessly at one of these five points, you wouldn't be tugged toward either body. Figure 12.14 shows the five Lagrange points for the Earth–Moon system. Note that three of them are on a line passing through the centers of Earth and the Moon and that the other two are located 60° to either side of the Moon. These latter two positions, known as L4 and L5, are of particular interest. Unlike the other three, they are "stable," which means that if you started at L4 or L5 but then drifted slightly away, the competing effects of gravity from Earth and the Moon would bring you back to your starting point. This effect is much like what happens to a marble in a shallow bowl; if the marble is moved a bit off-center, it still rolls back to the bottom of the bowl. As a result, L4 and L5 would be obvious places to leave artifacts in cold storage for possible retrieval after long periods of time. Two decades ago, a limited survey of the Lagrange points was made with small telescopes. The hope was to find bright objects that might indicate the presence of parked artifacts. In addition, a preliminary search was made using the Arecibo radar. Although neither search turned up anything clearly artificial, these locations might be worth a more thorough reconnaissance in the future.

If interstellar travelers have not left artifacts for us to find, we might still be able to detect their powerful interstellar spacecraft. As we will discuss in Chapter 13, spacecraft capable of traveling at speeds close to the speed of light would likely have enormous engines powered by energy

FIGURE 12.13

In the movie *2001—A Space Odyssey,* American astronauts dig up a clearly artificial monolith on the Moon. This is a signaling device left behind by an advanced society, placed where it could be found only when humans had reached a high enough level of technical sophistication.

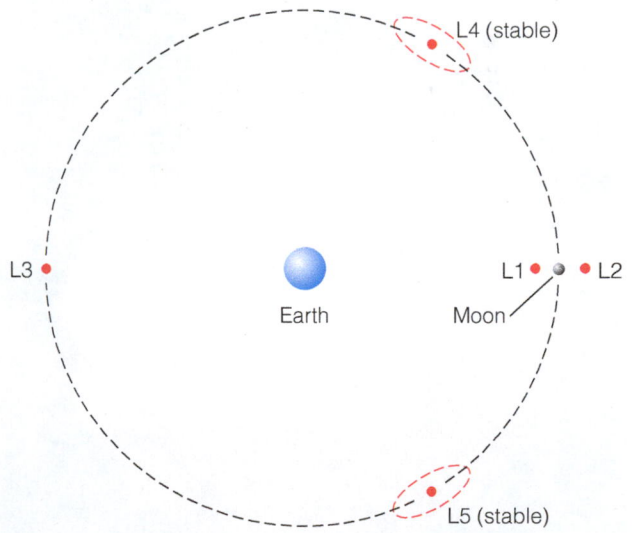

FIGURE 12.14

The five Lagrange points of the Earth–Moon system. The points L4 and L5, with the stable regions schematically indicated by the dashed ovals, are most attractive for the long-term parking of artifacts or probes. As the Moon orbits Earth, the Lagrange points also orbit, so they remain in the same relative positions with respect to Earth and the Moon.

sources such as nuclear fission, nuclear fusion, or matter–antimatter drives. These engines would leave telltale signs of their operation—signs that might be detectable at distances of hundreds or even thousands of light-years away. Although only a limited search for distant rockets has been made, this type of phenomenon might be inadvertently discovered in the course of more conventional astronomical research.

ASTRO-ENGINEERING Sufficiently advanced civilizations might also betray their presence through tremendous feats of "astro-engineering." In terms of their ability to exploit natural energy resources, the twentieth-century Russian physicist Nikolai Kardashev suggested that civilizations might be divided into three distinct categories:

1. **Planetary (or Type I) civilizations,** which use the resources of their home planet
2. **Stellar (or Type II) civilizations,** which corral the resources of their home star
3. **Galactic (or Type III) civilizations,** which employ the resources of their entire galaxy

We are in the first category, since we exploit (often with abandon) the resources of our home planet. The lights of our cities and the heat from our homes and factories, while considerable, are feeble on a cosmic scale and would be extraordinarily difficult to detect at the distances of the stars. However, we might be able to detect civilizations in the second or third categories.

A Type II civilization, for example, might decide to capitalize on solar energy in a big way. A star like the Sun puts out a great deal of power. If we could capture just 1 second's worth of the Sun's total energy output, it would be enough to supply today's world demand for energy for approximately the next million years. But nearly all the Sun's energy escapes into space, and Earth intercepts only the tiny bit that heads our way. In principle, a technologically adept civilization could capture *all* of its star's energy by fashioning a large, thin-walled sphere (possibly built from a dismantled outer planet) around its solar system and covering the inner surface with solar cells or their equivalent. Such spheres are called **Dyson spheres,** after physicist Freeman Dyson, who proposed their possible existence (Figure 12.15).* You might think a Dyson sphere would be invisible, because it would absorb all the light of its interior star. However, the laws of physics dictate that waste heat must escape from the sphere. This heat would be radiated as infrared light (coming off the outside of the sphere) that we could detect with specialized telescopes. We could in principle discover the presence of a stellar civilization by finding the infrared signature of a Dyson sphere. Limited searches have already been undertaken, so far to no avail.

Civilizations in the third category would be so advanced that they are hard for us to imagine. As Arthur C. Clarke has stated, "Any sufficiently advanced technology is indistinguishable from magic." Nevertheless, one recent experiment used an optical telescope to search for the waste heat that might be produced by Type III civilizations by looking for an unexpectedly large excess of infrared light coming from entire

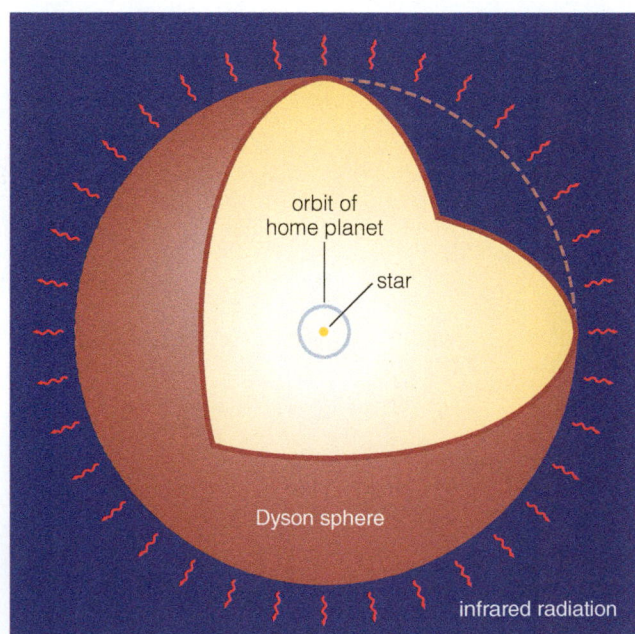

FIGURE 12.15
Schematic representation of a Dyson sphere, constructed exterior to an inhabited planet, with waste heat escaping from its outer surface in the form of infrared light.

*In fact, a sphere would be gravitationally unstable, and wouldn't stay in place. However, this idea could be implemented with a large swarm of orbiting satellites, each covered in solar cells, thus producing a "Dyson swarm." The idea and the method of detection are the same as for the sphere.

galaxies. Preliminary observations of 100,000 galaxies have so far failed to find any that show clear signs of such excesses, but note that only societies using a *trillion trillion* times as much energy as we do would produce enough infrared to be detected in this survey.

What happens if SETI succeeds?

Until now, no SETI experiment has turned up a confirmed extraterrestrial transmission. Several intriguing signals have been reported, however. Perhaps the best known was found in 1977 by the (now-defunct) automated SETI search at the Ohio State Radio Observatory. When an Ohio State astronomer examined the data from the night's observing, in the form of a computer printout, he found a signal so impressively strong that he wrote "Wow" on the printout margin. The "Wow" signal, made popular by its appealing nomenclature, was never seen again despite repeated observations at the same frequency and the same sky position. Consequently, SETI researchers do not consider it a detection of extraterrestrial intelligence but instead presume it was some sort of terrestrial interference.

Despite their continued failure to find a persistent, verifiable signal, those engaged in SETI research remain optimistic (Figure 12.16). The motivation for this optimism is the rapid improvement in the abilities of the telescopes and detectors used in SETI. Because much of the hardware used in this research is built with digital electronics, SETI benefits from the rapid growth in computer capability. The density of transistors placed on silicon chips has been doubling approximately every 18 months for decades (a phenomenon sometimes called *Moore's law*, after Gordon Moore, a cofounder of Intel Corporation). Suppose Frank Drake is correct in suggesting that the number of transmitting civilizations, N, is around 10,000; if 10% of all stars have a planet suitable for habitation, then one in a million such stars will have a transmitting society. The Allen Telescope Array, which is designed to take advantage of technological improvements, will be able to examine approximately a million targets by the year 2035. This suggests that, if Drake's assumptions are correct, a signal might be detected within a few decades. Continued progress in astronomy—and, in particular, the possibility of learning which extrasolar planets might have atmospheres that indicate the presence of life—could shorten this time by improving our choice of star systems to examine.

What will happen if we really do receive an artificial signal from the stars? A signal detection would need to be thoroughly verified by continued observations both with the discovering telescope and at other observatories. The nature of the signal itself would be evidence of artificiality, but scientists would need to be sure it wasn't due to terrestrial interference, equipment failure, or a college prank. It might take many weeks of careful work by excited astronomers before they could be fully confident of a genuine detection.

Once a detection had been verified, the news would be released to the outside world. Given the potential implications of the discovery of an extraterrestrial civilization, this step would have to be taken with some care. For this reason, SETI research groups have agreed to follow a protocol known as the *Declaration of Principles Concerning Activities Following the Detection of Extraterrestrial Intelligence.* The protocol has no force of law, but it lays out a reasonable course of action. In particular, if a signal had been found and verified as extraterrestrial, astronomers around the

FIGURE 12.16

A SETI experiment at the Arecibo radio telescope in Puerto Rico. The observer, Jill Tarter (often said to be the model for the fictional Ellie Arroway in the novel and movie *Contact*), monitors the progress of the observations on computer workstations. It is the computers that do the "listening," since many millions of channels are monitored simultaneously.

world would be notified so that they could swing their telescopes in the direction of the signal and learn as much as possible. Governments and the public would also be informed directly. The open nature of research and the necessity to confirm any detection by using other telescopes dictate that the discovery not be kept secret or covered up. The evidence, after all, can be verified by anyone with access to a large telescope. In addition, the protocol addresses the matter of sending a reply to any detected signal. The SETI community has suggested that any deliberate response represent a consensus of the world's population, not just the wishes of whatever group made the detection.

Suppose we found not just a signal, but also a message—bits of information. Might we be able to decipher it? In science fiction, messages from aliens often have a mathematical bent, because we assume that mathematics would be developed by any technological society. The fictional aliens send us the value of pi or numbers from a well-known mathematical series so that we will recognize that the signal is from intelligent beings. But this sort of labeling wouldn't be needed, because we are seeking narrow-band radio signals or pulsed laser light that clearly would be artificial. In addition, using mathematics as a "language" might be a cumbersome way to convey information about such things as political systems, religious beliefs, art, or even physical appearance. Our own messages to space have been pictorial (see Figure 12.8b and Figure 13.2), and pictures might be a better way to communicate to unfamiliar societies.

Any SETI success in itself would be an astonishing event, because it would prove that we are not alone in the universe. Depending on the nature of the signal, however, we might learn far more. If a civilization is broadcasting to us, it is likely to be hundreds or even thousands of light-years distant, and two-way communication would be tedious at best. A broadcasting society therefore might not expect a reply and might simply send a one-way message—perhaps some version of their encyclopedia or information from their internet—knowing that conversation would be impractical. Such a message might take us a while to figure out, but doing so would be worth the trouble. From our discussion of the Drake equation, we learned that the chances of detecting another society increase with the average age of technological civilizations. Consequently, if we do find a signal, the chances are great that it will be from a civilization far older than our own. The knowledge we might gain from such an advanced society would be enormous.

Finally, we should also be aware of the possibility that the signal we find will not be a beacon intended for societies like ours, and might contain a message that we can never decode. Even in that case, we would at least have learned that we are neither alone nor particularly special.

✳ **THE PROCESS OF SCIENCE IN ACTION**

12.4 UFOs and Aliens on Earth

A fundamental assumption of all SETI efforts is that our best hope of detecting an alien civilization lies in looking beyond Earth for signals or other evidence of its existence. But what if the aliens are already here on our planet? Remarkably, public opinion polls since the 1960s have consistently shown that about one-third of the American public believes that we are being visited by alien spaceships, and many are convinced that hard evidence of such visits exists. If this were true, then in principle we could learn about extraterrestrial intelligence without ever resorting to the use of expensive telescopes and complex receivers. Unfortunately, when we examine these claims of alien visits using the accepted principles of science, they prove to be far from convincing.

What have we learned from UFO sightings?

The bulk of the claimed evidence for alien visitation consists of UFO sightings—many thousands of UFOs are reported each year. No one doubts that unidentified objects are being seen. Even many astronomers have seen objects in the sky that they cannot identify, and some of the most interesting UFO reports have come from seasoned pilots or astronauts whose credibility is not in question. However, the mere fact that something is unidentified does not automatically make it an alien spacecraft. The real question is not whether UFOs are seen, but whether these sightings are actually of visitors from the stars.

We can gain some insight into this question by examining both the history of UFO sightings and the history of scientific examination of UFO claims. The first modern report of a UFO as an alien spacecraft was made by businessman Kenneth Arnold in June 1947. Arnold was flying a private plane near Mount Rainier, in the state of Washington, when he saw nine shiny objects having wings, but no tails, that appeared to be streaking across the sky at nearly 2000 km/hr. A reporter for the United Press wrote up Arnold's experience as a sighting of "flying saucers," and the story became front-page news throughout America.

Within a decade, "flying saucers" had invaded popular culture, if not our planet. They were often seen in books and movies, and the term is still used today. This is ironic, given that Arnold didn't actually describe the objects he saw as saucer shaped, or even round. Three decades after the sighting, he explained that this impression was the result of a newspaperman's error. When, in 1947, the United Press reporter asked him how the objects moved, Arnold answered that they "flew erratic, like a saucer if you skip it across the water." He was describing their motion, not their shape. In fact, several later investigators have suggested that the objects seen by Arnold were meteors, streaks of light caused by small particles entering Earth's atmosphere.

Despite the reporter's misunderstanding, the idea of alien disks buzzing the countryside caught the imagination of the public. It also intrigued the U.S. Air Force, which spent two decades conducting investigations into the nature of UFOs. The military interest was prompted largely by Cold War concerns that UFOs might be new types of aircraft being developed by the Soviet Union.

In the 1950s and 1960s, teams of academics met to study the most interesting of the UFO reports. In the overwhelming majority of cases, these experts were able to plausibly identify the UFOs. They included bright stars and planets, aircraft and gliders, rocket launches, balloons, birds, ball lightning, meteors, atmospheric phenomena, and the occasional hoax. For a minority of the sightings, the investigators could not deduce what had been seen, but their overall conclusion was that there was no reason to believe that UFOs were either highly advanced Soviet craft or visitors from other worlds. The Air Force ultimately dropped its investigation of the UFO phenomenon.

However, some believers felt that the inquiries were either incomplete or part of a ruse organized by the government to put peo-

FIGURE 12.17
A UFO. The object in this photo, which shows far more detail than most pictures made of supposed visitors from other worlds, has the familiar saucer shape made popular by a reporter's error. In fact, this photo is faked, and the saucer is only a lampshade.

ple off the scent of alien visitation. The number of UFO sightings increased (of course, the number of human-made objects in the sky was also growing during these years), and countless books, photos, and film clips were offered as evidence (Figure 12.17). Yet little of the photographic material was compelling. Some of it was ambiguous (is that a distant spacecraft or a nearby bird or bug?), and some was clearly faked. Moreover, the fact that about 90% of the well-documented UFO cases were explainable as earthly phenomena only encouraged some people to point to the 10% that weren't clearly explained. However, this is not a valid argument for the idea that the unexplained cases represent alien spaceships. After all, if a metropolitan police department solved 90% of the murder cases in a large city—all committed by humans against other humans—this wouldn't suggest that the other 10% were committed by, say, aliens.

We can see the problems with UFO "evidence" even more clearly when we evaluate the sightings using the methods of science [Section 2.3]. Recall that modern science seeks natural causes to explain observed phenomena. Although aliens presumably would be "natural" if they really exist, invoking them to explain unidentified objects is no different from invoking ghosts, spirits, or Greek gods and goddesses. Any of these things might be the real explanation for UFOs, but simply saying so does not give us any insight into their true nature. We can learn something only if we can make a model to describe the phenomenon of UFOs, and then test the model against additional observations. Reports of UFO sightings do not allow us to do either: Various UFO sighting reports are so different from one another that it's difficult to create any testable hypothesis about the designs of the supposed alien spacecraft, the types of engines that they use to traverse interstellar distances, or the behavior of their occupants. We are left with just individual, eyewitness reports—and as we discussed in Chapter 2, these reports offer no way for other people to evaluate their validity.

The bottom line is that UFO reports tell us nothing more than the fact that people sometimes see things in the sky that they are unable to identify or explain. But the fact that an object has not been identified does not even make it unidentifiable—it might just be that the person making the report did not recognize something that another person might have considered obvious—and it certainly does not make it an alien spacecraft. If we want to establish evidence of alien visitation on Earth, we'll need evidence that's much more solid.

Think About It Suppose you had a year and an unlimited budget with which to investigate the claim that UFOs represent alien spacecraft. What sorts of experiments would you conduct? What types of evidence might convince you—and the scientific community—that fast-moving lights in the sky are nonterrestrial craft?

Have aliens left any compelling evidence of visitation?

If aliens landed in Paris, held press conferences, shook hands with crowds, and provided artifacts from their home world, scientists would surely accept their visit as a fact. But short of that, what could convince scientists that aliens have been here? According to the hallmarks of science, we'd need something that anyone could examine, at least in principle, and that stood up to intense and continued scrutiny. While it's conceivable that photographic evidence might do—for example, if the same UFO were photographed from many different observatories—some sort of artifact or other hard evidence would be even better. Some people claim that such evidence already exists, but to date none of it has withstood serious scientific scrutiny. Let's investigate a few of the more famous claims.

CRASHED ALIENS IN ROSWELL Perhaps the most celebrated case of supposedly alien artifacts concerns the so-called Roswell incident of 1947. The story began in early July 1947, only a few weeks after the nationwide coverage of Kenneth Arnold's "flying saucers." A rancher found some crash remnants in a pasture close to the city of Roswell, New Mexico. He reported the debris to the local sheriff, who then passed on the information to the Roswell Army Air Field, which was nearby. Several military personnel drove out to the ranch, picked up the debris, and explained to the local papers that they had recovered the remains of a "flying disk."

This dismaying story was quickly quashed. Only a day later, an Air Force officer from Fort Worth, Texas, where the debris had been flown, held a press conference in which he dismissed the flying disk notion (Figure 12.18). He stated that the debris was merely a crashed weather balloon, substantially ending interest in the incident at the time. But in 1978, UFO investigator Stanton Friedman began looking into the events at Roswell. Friedman claimed that the debris was from a spacecraft and that alien occupants had been picked up as well. He believed that the government was covering up both the fact of the mishap and its extraterrestrial victims.

The possibility of alien bodies lent this story an appeal beyond that of routine sightings of lights in the sky. However, despite the claim of hard evidence, most of the story is based on testimony recorded by Friedman and others—and remember that they conducted their interviews more than *three decades* after the event. Given that long time lag, it's perhaps unsurprising that different witnesses contradicted one another, that some supposed witnesses to the crashed "saucer" had originally claimed not to have seen it, and that a few were caught in flat-out lies. Clearly, the eyewitness testimony by itself could not be considered rigorous enough to prove something as profoundly important as alien visitation.

But what of the crash remains and alien bodies? Something *did* crash at Roswell, and despite the Air Force claim, it was not a weather balloon. Instead, declassified government records tell us that the crash was of a then-secret military balloon experiment designed to detect Soviet nuclear tests. The highly classified operation, known as Project Mogul, used balloon trains to carry the detection devices aloft; the idea was that the balloons would hover at constant altitude near the borders of the Soviet Union and listen for sound waves produced by distant nuclear blasts. Project Mogul was being tested at Roswell in the late 1940s, and the principal scientist behind it has confirmed that the debris recovered by the rancher was from his experiment. Moreover, photos taken of the recovered materials by the Air Force match contemporary photos of Project Mogul balloon trains.

As for the alien bodies or more convincing spacecraft debris, believers can claim only that the government has maintained a tight cover-up, storing the evidence securely in some secret location (such as "Area 51" in Nevada) since the 1947 crash. But consider what that would entail: Over almost 70 years, through administrations with widely varying political agendas, hundreds of scientists and other military personnel would have had access to such a secret. Yet not a shred of actual evidence has

FIGURE 12.18
Debris recovered in Roswell, New Mexico, in 1947. In this photo, made by newspaper photographer James Johnson, General Roger Ramey is showing the debris to reporters in Fort Worth, Texas.

emerged to public view, despite the fact that anyone who produced such evidence would become instantly famous on the talk-show circuit. At minimum, this conspiracy claim seems highly improbable.

In short, claims that the Roswell debris was from a downed alien craft are based on eyewitness accounts made long after the fact, and the supposed evidence of both craft and alien bodies could be real only if you believe the government capable of a highly efficient, six-decade cover-up. The nature of these claims renders them untestable and therefore unscientific, and they seem even more difficult to believe once you realize that we have an alternative explanation—the military experiment—backed by verifiable data in the form of the declassified records and photographs. It might be more interesting to believe that small people from another world just happened to make a navigational error a few weeks after flying saucer mania first swept the country, but the evidence suggests only a crashed Air Force project.

CROP CIRCLES, ABDUCTIONS, AND MORE UFOs and crashed aliens can nearly always be readily explained in terms that don't involve visitors from other worlds. The same is true for other highly publicized events interpreted as proof of an extraterrestrial presence in the neighborhood, including abductions, crop circles, and miscellaneous phenomena such as mutilated cattle and goat-eating chupacabras.

Crop circles, geometric patterns made in wheat fields, have generated a great deal of interest among the public (Figure 12.19). For several decades, such patterns have appeared each summer in England. According to some, they are the work of visiting extraterrestrials who are using this method to communicate to us. The proof offered for their nonhuman origin is the speed with which the crop circles appear (overnight) and the condition of the matted-down wheat, which is claimed to be inconsistent with simple trampling by humans. But there are many indications that these patterns are merely pranks, not alien messages. The continuing increase in complexity of the patterns over the years suggests that their creators are getting better at their craft—something we would not expect if aliens were using technology brought from distant stars. The fact that the crop patterns are made only at night also suggests human subterfuge. Moreover, many of the crop circles have been acknowledged as pranks, and the human pranksters have demonstrated for news reporters how the crop circles can be made quickly and easily. (Competitions in crop circle making have even been held!) In one interesting case reported in a documentary about crop circles, a self-proclaimed "expert" on the alien origins of crop circles acknowledged that some crop circles were pranks but claimed that not all of them could be. He explained to interviewers why a particular set of circles could not have been made by humans—and then watched in dismay as the interviewers showed him a video of experienced pranksters making the very ones he had just labeled "alien."

Cattle mutilations and the reputed killings of various small animals by short, bipedal creatures called chupacabras have also been attributed to alien activity, despite a centuries-long history of attacks on farm animals by both earthly predators and humans. The only evidence offered to connect these attacks to extraterrestrial beings is uncertainty over who or what is doing the killing. As always, the lack of a confirmed culprit does not mean we are seeing the work of aliens.

A more interesting phenomenon involves claims of alien abduction of humans. Although they make up a small percentage of the population,

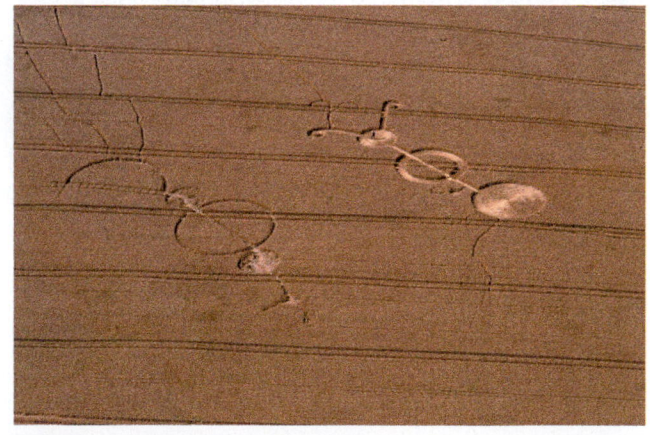

FIGURE 12.19

Crop circles. These relatively simple patterns have been "cropping up" in southern England for several decades and are claimed by some to be manifestations of alien activity. However, the designs are easily produced in a few hours by a small group of motivated students using nothing more than boards and ropes, which suggests a terrestrial origin. In addition, we might wonder why sophisticated extraterrestrials would travel many light-years simply to carve graffiti in our wheat.

a substantial number of Americans claim to have been alien abductees, often supposedly stolen away by extraterrestrials while asleep (far fewer claim to have been disturbed during waking hours). The victims state that they were either taken aboard spacecraft for observation and unwholesome experiments or simply watched while they lay immobilized in their beds. These claims may seem difficult to explain away, but a clue to their likely origin comes from research showing that similar phenomena have been described since ancient times in a multitude of cultures. In the past, the accused culprits have been witches, ghost babies, and goblins. Today, the molester of choice is frequently a visitor from the stars.

According to many psychologists, such abduction experiences are probably attributable to **sleep paralysis,** which occurs during REM (rapid eye movement) sleep. During REM sleep, the body is naturally paralyzed, so that we don't thrash physically along with the movements we make in our dreams. Sometimes, this paralysis persists for a few minutes after the brain has started waking up, and it can give a person the alarming sensation of being awake in a paralyzed body. Visions and other sensations often occur in this state—including strange imaginings that might plausibly include witches, ghosts, goblins, or wide-eyed extraterrestrials.

Surveys suggest that sleep paralysis is experienced by approximately half of all people at some point in their lives. It therefore seems quite reasonable to imagine that a small percentage might believe they had experienced some type of alien abduction. Proponents of alien abduction dispute this idea, noting that a few victims were fully awake and alert during their experience. However, in some cases, daydreams are known to produce sensations similar to those produced by sleep paralysis. Once again, we are left to accept either a simple and earthly explanation (sleep paralysis) or an extraordinary one (abduction by aliens). If you recall our discussion of Occam's razor in Chapter 2, you'll understand why scientists prefer the simple explanation, and would accept the extraordinary one only if presented with very strong evidence. As Carl Sagan said, "Extraordinary claims require extraordinary evidence."

ANCIENT VISITATIONS The vast majority of those who believe in extraterrestrial visitation claim that the aliens and their craft are among us now. However, some people have suggested that there is evidence for alien visits in the dim past of human history, or even earlier.

The idea that Earth might have been visited seems plausible, especially if civilizations turn out to be common in our galaxy. Our planet might well be of interest to alien scientists, and we'd have no way of knowing whether alien ships studied Earth in the billions of years that preceded the evolution of *Homo sapiens*. Such visits might even have continued after the rise of civilization. But while a few people have claimed that this has happened, the evidence they offer is far from compelling.

This evidence consists of things such as ancient drawings that supposedly show alien visitors or their spacecraft, and archaeological wonders supposedly too sophisticated for our ancestors to have made on their own. These archaeological claims generally make little sense. For example, desert markings on the Nazca plains of Peru (Figure 12.20), claimed by some to be landing strips and markers for alien craft, could just as well be the work of the Nazca Indian culture that occupied the area about 2000 years ago. No special technology would have been needed to make

FIGURE 12.20
Hundreds of lines and patterns, generally obvious only from the air, are etched in the sand of the Nazca desert in Peru. This aerial photo shows the large figure of a hummingbird. Some people have claimed that these patterns must have been created or inspired by aliens, though it would not have been difficult for the local people to have made them on their own.

FIGURE 12.21

This 15-meter-long etching in the Nazca plains shows an "owl man" that some people claim to be a drawing of an astronaut—presumably one of the astronauts that some people believe used the "runways" also found etched in the desert.

the markings, and patterns visible only from the air seem more likely to have been intended for gods than for aliens. That some of these drawings look somewhat like astronauts hardly improves the case (Figure 12.21), as they could as easily represent a god or a person wearing a ceremonial headdress.

Claims of alien origin for structures such as the Egyptian pyramids are nothing less than an insult to ancient people, because they seem to suggest that people of these cultures were incapable of constructing such impressive structures on their own. In fact, while the pyramids were a remarkable achievement, their construction was well within the technical expertise of the kingdom of the pharaohs. All other known ancient structures—including Stonehenge, the Mayan pyramids of Central America, and the massive stone heads of Easter Island—were also within the technical capabilities of the cultures that created them.

Perhaps the one thing going for the claims of ancient visitation is that they at least attempt to rely on archaeological artifacts that can in principle be studied by anyone. In that sense, they can be subjected to scientific study. The problem with these claims is that when scientists *have* studied them, the claims have failed to withstand scrutiny.

Is there a case for alien visits?

Despite the lack of evidence of alien visits past or present, many people continue to believe they have occurred. Champions of alien visitation generally explain away the lack of compelling evidence in one of two ways: government cover-ups or a failure of the mainstream scientific community to take the relevant phenomena seriously. Let's consider both of these arguments in a little more detail.

Government conspiracy and cover-up are popular notions, particularly in the United States. It's certainly conceivable that a secretive government might try to put a lid on the best spacecraft videos and reconnaissance images from orbiting satellites, although the motivation for doing so is unclear. The usual explanation is that the public couldn't handle the news or that the government is taking secret advantage of the alien materials to design (via "reverse engineering") new military hardware. But both explanations are silly. A third of the population already believes in alien visitors and would hardly be shocked if newspapers announced tomorrow that aliens were stacked up in government warehouses. As for the reverse engineering of extraterrestrial spacecraft, we should keep in mind the difficulty of traveling from star to star. Any society that could do so would be technologically far superior to our own. Our reverse engineering their spaceships is as unlikely as Neanderthals' constructing personal computers just because a laptop somehow landed in their cave. In addition, while a government might successfully hide evidence for a short time, the evidence is unlikely to remain secret for decades (more than six decades, in the case of the Roswell claims). And, unless the aliens landed only in the United States, can we seriously believe that *every* government would cooperate in hiding the evidence?

The alleged lack of interest among the scientific community is also an unimpressive claim. Scientists are continually competing with one another to be first with a great discovery and are aggressive in pursuing any important new phenomenon. Clear evidence that aliens exist and are (or have been) on our planet would be hailed as one of the most important discoveries of all time. Countless researchers would work evenings and weekends if they thought such a discovery were possible. The fact that few scientists are engaged in such study reflects not a lack of interest but a lack of evidence worthy of study.

Indeed, lack of physical evidence is the foremost reason given by scientists to justify their skepticism regarding UFOs and alien visits. Airport radar and orbiting satellites have never yielded proof of the passage of an alien spacecraft. No one has ever walked into a research lab with an alien artifact, such as a piece of material that we could not manufacture with our current technology—not even a fragment. In Earth orbit, the U.S. Air Force tracks thousands of pieces of "space junk" from our own satellites and shuttles, including discarded rockets, exploding bolts, a Hasselblad camera, paint chips, and the occasional astronaut glove. But no one has ever found a single piece of space junk that can be attributed to aliens. Most scientists are open to the possibility that we might someday find evidence of alien visits, and many would welcome aliens with open arms. But so far, the evidence is simply lacking. Extraterrestrial visitors to Earth remain a routine fixture of cinema and television, but in real life they are like ghosts—a pervasive and attractive idea that the vast majority of scientists treat with skepticism.

Of course, "absence of evidence is not evidence of absence," meaning that a lack of evidence for some claim doesn't make the claim untrue. In the case of alien visitation, it's conceivable that aliens really are here, and that we lack the evidence to prove it because they don't want us to know. If you wish to believe that, no one will stop you—just don't claim that your belief is based on science.

Finally, it is important to distinguish between claims that aliens are visiting us now (or visited Earth in the past) and the possibility that alien civilizations might exist. When, after World War II, space travel moved from the theoretical to the practical, it was only natural to assume that what we were trying to do—travel to other worlds—was routinely done by other civilizations. However, as we'll discuss in the next chapter, there is a great difference between journeys within the solar system and jaunts to the stars. While the former are straightforward, the latter are both enormously difficult and extremely costly in terms of energy. Nonetheless, interstellar travel doesn't violate physics, and if civilizations are common, it is certainly possible and perhaps even likely that some might have been inspired to voyage from star to star.

The Big Picture

PUTTING CHAPTER 12 IN PERSPECTIVE

In this chapter, we have explored the rationale and the methods of the search for extraterrestrial intelligence. As you continue your study, keep in mind the following "big picture" ideas:

- SETI is both a part of and distinct from other efforts in astrobiology research. Its justifications and methods depend on what we learn more generally about life in the universe. However, whereas other astrobiology research makes slow and steady progress, SETI offers the potential to give us absolute proof in one fell swoop that we are not alone—but only if we receive a clear signal from another civilization.

- We do not yet know enough about life in the universe to make a reasonable estimate either of the number of civilizations that might exist or of our odds of achieving success in SETI efforts. Nevertheless, we have no hope of success unless we try, and contact with another civilization would surely be one of the greatest discoveries in human history.

- The telescopic search for distant civilizations would be rather superfluous if aliens were already here among us. However, despite the many sightings of UFOs and other phenomena supposedly caused by aliens visiting Earth, no compelling evidence for such visits has ever been found. Scientifically, SETI represents our only current hope of detecting other civilizations.

Summary of Key Concepts

12.1 The Drake Equation

What is the Drake equation?

The **Drake equation** gives us a way to organize our thinking about the question of the number of civilizations in the Milky Way Galaxy. In its modified form, it says that the number of civilizations with which we could potentially communicate is $N = N_{HP} \times f_{life} \times f_{civ} \times f_{now}$, where N_{HP} is the number of habitable planets in the galaxy, f_{life} is the fraction of habitable planets that actually have life on them, f_{civ} is the fraction of life-bearing planets on which a civilization capable of interstellar communication has at some time arisen, and f_{now} is the fraction of all these civilizations that exist now.

How well do we know the terms of the Drake equation?

We don't know the values of any of the terms well. We have some data from extrasolar planets that allow us to make at least an educated guess about the first term, N_{HP}; it seems likely to be quite large, perhaps 100 billions or more habitable planets in our galaxy. For the rest of the terms, we have only the example of Earth to look to, making any guesses far more uncertain.

12.2 The Question of Intelligence

Even if life is widespread, is intelligence common?

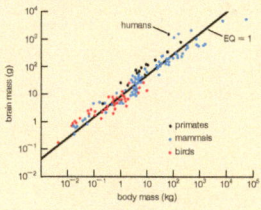

We really don't know, because we have only the example of Earth. Nevertheless, evolutionary studies indicate some drive toward intelligence, so it is plausible to imagine intelligence appearing on any planet with life, at least if given enough time.

Will intelligence inevitably spawn technology?

Certainly, there are some species with physical limitations, such as a lack of hands, that would seem to prevent the development of technology. But our own case is ambiguous: It took us a long time to develop technology, but we do not know if this means the development was a fortunate accident or something destined to happen eventually.

12.3 Searching for Intelligence

How did SETI begin?

Although there were some attempts at radio contact with aliens in the early twentieth century, in retrospect we know that these efforts were doomed because they used frequencies that are blocked by Earth's ionosphere and because they focused on nearby worlds like Mars, where complex life is unlikely to ex-

ist. The origin of modern SETI is generally credited to ideas proposed by physicists Giuseppe Cocconi and Philip Morrison. Frank Drake's Project Ozma was the first organized search.

How do we search for intelligence today?

SETI today is conducted primarily by searching for either radio or optical signals transmitted by distant civilizations. There may be other means of interstellar communication, but it seems reasonable to suppose that radio or optical signals would be used by at least some, if not all, other technological societies. Current signal detection efforts are probably sensitive enough to find only deliberately broadcast beacon signals.

What happens if SETI succeeds?

The scientific and technological aspects of SETI are important but may well pale in comparison to the societal issues that might arise if a signal were found. As a result, an important part of SETI work involves thinking about what will happen if the search ultimately proves successful.

❋ **THE PROCESS OF SCIENCE IN ACTION**

12.4 UFOs and Aliens on Earth

What have we learned from UFO sightings?

We've learned that people sometimes see things in the sky that they cannot identify or explain, but we have not found any convincing evidence pointing toward an alien origin for such sightings.

Have aliens left any compelling evidence of visitation?

Although many people have made claims of hard evidence of alien visits, none of these claims has ever withstood scientific scrutiny.

Is there a case for alien visits?

Based on the tenets of science, there is no current case for alien visits to Earth, either past or present. Keep in mind, however, that absence of evidence is not evidence of absence. If civilizations really are common, then it is conceivable that some aliens have come our way.

Exercises and Problems

MasteringAstronomy® *For instructor-assigned homework and other learning materials, go to MasteringAstronomy®.*

REVIEW QUESTIONS

Short-Answer Questions Based on the Reading

1. What is the purpose of the *Drake equation*? Define each of its terms, and describe the current state of understanding regarding the possible values of each term.

2. What is *convergent evolution*? How does this idea suggest that intelligence would tend to be an evolutionary imperative?

3. Briefly explain the idea of the *encephalization quotient* (*EQ*). How does it suggest that humans are indeed intelligent? What does it tell us about intelligence among other animal species?

4. Describe a few physiological and sociological factors that might influence whether an intelligent species could develop technology for interstellar communication.

5. Briefly describe early attempts at interplanetary communication by Marconi and Tesla. Why were these attempts doomed from the start?

6. Briefly discuss early SETI efforts. What do we mean by the *bandwidth* of a signal, and why does SETI concentrate on a search for narrow-bandwidth signals?

7. What are the three general categories of signals that might be detected at great distance? What are the current prospects for detecting each type of signal through SETI efforts?

8. Why do SETI researchers assume that beacon signals would be designed for easy decoding, and how might we recognize them?

9. Summarize the current techniques of radio SETI and some of the major current projects.

10. Explain why it is reasonable to imagine optical or other signals, and the method behind current optical SETI efforts.

11. Briefly discuss the possibilities of finding other civilizations via artifacts or signs of interstellar craft.

12. What are the three distinct categories of civilizations (as outlined by Kardashev)? Which one(s) can we imagine detecting through their use of resources and why?

13. Briefly discuss some of the issues that would surround an actual SETI detection.

14. Discuss several types of claims about alien visitation on Earth. Why, so far at least, do they seem not to reach the level of scientific evidence?

TEST YOUR UNDERSTANDING

Evaluate the Opinions

Each of Problems 15–24 makes a clear statement of opinion. Evaluate each statement and write a few sentences describing why you agree or disagree with it. Explain clearly; not all of these have definitive answers, so your explanation is more important than your chosen answer.

15. Humans are the "crown of creation" and an inevitable result of billions of years of evolution.

16. If, for some reason, we humans were to suddenly wipe out our species, another species—possibly the raccoons—would soon evolve greater intelligence than we possessed.

17. Sea creatures, no matter how clever they are, could never master the technology required to communicate with other worlds.

18. Most of the intelligence in the universe is not biological, but artificial ("machine intelligence").

19. Because SETI researchers are "listening" to star systems that are hundreds of light-years distant, there's a good chance that by the time we hear a signal, the civilization that sent it will have disappeared.

20. No advanced society would ever construct a beacon transmitter, because it would inevitably attract attention and might be dangerous. Similarly, we should not make deliberate transmissions to the stars.

21. We should consider including in the space program an "artifact hunt" that would search on the Moon for objects left behind by advanced extraterrestrial societies.

22. Looking for signals from star systems is a poor approach, because any truly advanced civilization will have moved beyond its home planet and populated interstellar space.

23. If 10,000 people saw the same UFO, scientists would be forced to conclude that an alien visit really occurred.

24. The absence of any scientific evidence for alien visitation on Earth implies that civilizations are rare and that SETI efforts are doomed to failure.

Quick Quiz

Choose the best answer to each of the following. Explain your reasoning with one or more complete sentences.

25. The end result of a calculation with the Drake equation is intended to be an estimate of (a) the number of worlds in the galaxy on which life has arisen; (b) the number of worlds in the galaxy on which intelligence has arisen; (c) the number of worlds in the galaxy on which civilizations are transmitting signals now.

26. Which of the following statements is true about the terms in the Drake equation? (a) Astronomical research will soon give us firm values for all of the terms. (b) Some of the terms depend on sociology and cannot be determined by astronomers alone. (c) We already know the terms of the equation to an accuracy of within a factor of two.

27. The fact that marine predators like dolphins and sharks have similar shapes despite different ancestry is an example of (a) convergent evolution; (b) narrow bandwidth; (c) spontaneous creation.

28. Which of the following would lead an animal to a higher encephalization quotient (EQ) as it evolved? (a) growth in both body size and brain size (b) growth in body size but not in brain size (c) growth in brain size but not in body size

29. The *bandwidth* of a radio signal is a measure of (a) its frequency; (b) the range of frequencies that carry information; (c) the amount of power carried by the signal.

30. Why are we more likely to be able to detect a deliberately broadcast "beacon" signal than, say, the television broadcasts of a distant civilization? (a) because we expect beacon signals to be far more common (b) because our current technology is probably sensitive enough to detect beacons but not much weaker television transmissions (c) because television is a sociological phenomenon and beacons are not, so we'd expect all civilizations to have beacons but not all to have television

31. What is the distinguishing characteristic that those doing radio SETI experiments look for? (a) a signal containing the value of pi and other mathematical constants (b) a signal that is an echo of an earthly broadcast (c) a signal that extends over only a narrow band of frequencies

32. Two-way conversation with other societies is probably unlikely, even if we make contact. This is mainly because (a) aliens won't speak our language; (b) it might be dangerous to get in touch; (c) the time it takes for signals to cross the distance to other societies could be centuries or more.

33. According to the best available evidence, the famous Roswell crash of 1947 involved (a) an alien spacecraft; (b) an Air Force balloon experiment; (c) There is no evidence that gives us any information about the crash.

34. One reason scientists doubt that crop circles have alien origin is that (a) they are always beautiful; (b) they can be easily made by humans; (c) their appearance is not correlated with sightings of bright lights.

PROCESS OF SCIENCE

35. *Measuring Intelligence.* In judging the intelligence of animals, we use the *encephalization quotient* (EQ), which depends on the ratio of brain weight to body weight. Can you think of situations for which this might be a poor way to gauge intelligence? For example, could animals have some special processing needs (such as the navigation mechanism of bats) that would make their brains larger without contributing to their intelligence? Alternatively, could animals exist (on Earth or elsewhere) whose brains were relatively lightweight but who were still highly intelligent? Defend your opinion.

36. *Alien Visits.* Learn more about a particular claim of alien visitation to Earth, past or present. Evaluate the claim according to the hallmarks of science. Does it meet the standards of science? Explain.

GROUP WORK EXERCISE

37. *The Drake Equation.* **Roles:** *Scribe* (takes notes on the group's activities), *Proposer* (proposes ideas to the group), *Skeptic* (points out weaknesses in proposed ideas), *Moderator* (leads group discussion and makes sure everyone contributes). **Activity:** Using all the knowledge you have acquired during your astrobiology course, come to a group consensus on a "best estimate" (you may also choose to give a range of estimates) for each of the terms in the Drake equation. Clearly defend each estimate, and then use your estimates to calculate the expected number of civilizations if your estimates are correct.

INVESTIGATE FURTHER

In-Depth Questions to Increase Your Understanding

Short-Answer/Essay Questions

38. *Drake Values.* Make your own estimate of a value for each of the four terms in the modified Drake equation used in this chapter. Explain how you arrived at each estimate, and then use your estimates to calculate N.

39. F_{now}. Suppose that the number of civilizations in the galaxy has been quite large—say, one million. Does that necessarily mean that other civilizations should exist right now? Explain why or why not, and describe the factors that would influence the answer.

40. *Evolution of Intelligence.* Based on your understanding of natural selection and of the evolution of humans as discussed earlier in the book, describe at least three distinct environmental factors that have contributed to the evolution of human intelligence. Explain clearly.

41. *Intelligence on Other Worlds.* Consider again the three factors you identified in Problem 40. For each one, decide whether you think the same factor would be likely to arise and select for intelligence on another world with life, and clearly explain how you reached your conclusions.

42. *Communication.* Imagine that you had to fashion a short message that would tell extraterrestrials something about human society. What would you "say" using only three simple pictures? What would you write if you could use only a half-page of English text?

43. *Talking Back.* Suppose SETI were to find a signal coming from a star system 200 light-years away. Write a one- to two-page essay describing what, if anything, we should do to establish contact. You should think about how quickly we should respond, what the response should be, and what possible dangers might be involved.

44. *Contact.* Watch the movie *Contact*, paying careful attention to the SETI experiment described in the first third of the film. How accurately does this experiment reflect any of the current SETI search programs? Did you spot any obvious scientific or technical errors? Write a one- to two-page essay comparing *Contact* to the reality of SETI efforts.

45. *Invasions of Movie Aliens.* Choose a science fiction movie in which aliens are presumed to be visiting Earth. Identify at least three ideas in the movie that either do or do not meet the standards of being testable by science. Describe each in detail.

Quantitative Problems

Be sure to show all calculations clearly and state your final answers in complete sentences.

46. *How Many Stars to Search?* The number of star systems that a SETI search would have to investigate before achieving success depends on how common signaling societies are in the galaxy. This is the number estimated by the Drake equation. Suppose this number is $N = 1$ million. How many star systems must be checked out by SETI in order to find one signal? What if $N = 1000$? Assume that there are roughly 100 billion stars in our galaxy.

47. *Distance to E.T.* Suppose there are 100,000 signaling societies in our galaxy. What is the average distance between civilizations, assuming that civilizations are spread evenly throughout the galactic disk?

48. *Actual SETI Searches.* Project Phoenix, the largest search of individual star systems for radio signals before 2015, trained its antennas on Sun-like stars up to about 150 light-years away. How many civilizations would there have to be in order for the average distance between civilizations to be 150 light-years? If the actual value of N is 10,000, would we expect Project Phoenix to have made a detection? What if the actual value of N is 10 million? Explain.

49. *Power Used by E.T.* A modern SETI search using the 300-meter-diameter Arecibo Radio Telescope in Puerto Rico could pick up a 10-million-watt signal from 1000 light-years away (assuming that the broadcasting aliens had a transmitting antenna that was also 300 meters in diameter). Suppose we used Arecibo to search the far side of the Milky Way Galaxy (roughly 80,000 light-years away). Making the same assumptions about our set-up and the transmitting antenna, determine how powerful the alien transmitter would have to be for us to detect the signal.

50. *Transmitter Used by E.T.* A modern SETI search using the 300-meter-diameter Arecibo radio telescope in Puerto Rico could pick up a 10-million-watt signal from 1000 light-years away (assuming that the aliens had a transmitting antenna that was also 300 meters in diameter). Suppose an alien civilization is using this same transmitter setup but is on the other side of the Milky Way Galaxy (roughly 80,000 light-years away). How large an antenna would we need to hear the signal?

Discussion Questions

51. *Detecting Signals.* SETI scientists are sometimes criticized for using "old technology" in their search for signals. Perhaps extraterrestrials have moved beyond radio and light signaling and are using something much more sophisticated. Discuss (a) the advantages of radio and light for interstellar communication and (b) any reasonable alternatives you can think of. There

is always the possibility that "new physics" will provide faster or more efficient methods for signaling. Do you think this is a reason to limit current SETI efforts?

52. *Societal Reaction.* It is frequently said that the detection of a signal by SETI would revolutionize human society. Does this statement seem reasonable? Some researchers have tried to find historical events, such as the Copernican revolution or the publishing of Darwin's theories of evolution, whose impacts might compare to that of a SETI detection. Are such examples likely to be accurate in predicting how we would react? How likely do you think it is that a SETI discovery would cause either mass panic or an outbreak of universal brotherhood?

53. *Dealing with UFO Claims.* Given the large number of people who claim to have seen a UFO, you are likely to know at least one such person, now or in the future. Perhaps *you* have seen a UFO. Suppose someone who has seen a UFO believes deeply that it was an alien spacecraft. What, if anything, would you say to that person? Why?

WEB PROJECTS

54. *Current SETI Research.* Go to the SETI Institute's website and use links listed there to make an inventory of current SETI projects worldwide. Organize these projects according to whether they are radio or optical, and then separate targeted searches from sky surveys. Prepare a one-page summary of this information in which you discuss how thorough the current searches for extraterrestrial signals are. You should consider how many star systems have been looked at carefully, how wide a band (for radio searches) has been covered, and how sensitive the searches are.

55. *SETI@home.* This is a project organized by researchers at the University of California, Berkeley, to process radio SETI data on home computers. Download the free SETI@home screen saver onto your computer, and use it to analyze data collected by Project SERENDIP. Write a one-page description of the general processing scheme used by SETI@home, as well as the types of signals it is searching for.

56. *What to Do in Case of a Signal Detection?* Download the text of the protocol *Principles Concerning Activities Following the Detection of Extraterrestrial Intelligence* from the SETI Institute's website. Write a short discussion of what SETI scientists expect will happen if they detect an extraterrestrial signal and whether this expectation is realistic. In particular, do you think a discovery could be "covered up," or would it leak out to the public before the scientists themselves were sure of the discovery?

13 Interstellar Travel and the Fermi Paradox

LEARNING GOALS

13.1 THE CHALLENGE OF INTERSTELLAR TRAVEL
- Why is interstellar travel so difficult?
- Could we travel to the stars with existing rockets?

13.2 SPACECRAFT FOR INTERSTELLAR TRAVEL
- How might we build interstellar spacecraft with "conventional" technology?
- How might we build spacecraft that could approach the speed of light?
- Are there ways around the light-speed limitation?

13.3 THE FERMI PARADOX
- Where is everybody?
- Would other civilizations really colonize the galaxy?
- What are possible solutions to the Fermi paradox?
- What are the implications of the Fermi paradox for human civilization?

✸ THE PROCESS OF SCIENCE IN ACTION

13.4 EINSTEIN'S SPECIAL THEORY OF RELATIVITY
- What is "relative" about relativity?
- What evidence supports Einstein's theory?

▲ **About the photo:** Artist's conception of a hypothetical interstellar spacecraft.

In an age of rapid technological progress, it may seem inevitable that our rockets will soon reach the depths of interstellar space. The reality, however, is that interstellar travel is much more challenging than bridging the distances to nearby moons and planets. There are engineering and physical constraints, not least of which is the cosmic speed limit—the speed of light. Nevertheless, we can envision at least some ways by which our descendants might someday rocket to the stars.

The idea that humans might someday travel throughout the galaxy should make us wonder whether other civilizations have already achieved this ability. Indeed, if civilizations are common, it seems reasonable to expect that some societies—perhaps many—began colonizing the galaxy long before the earliest humans walked the Earth, and maybe even before Earth was born. This idea leads directly to the so-called *Fermi paradox:* If someone could have colonized the galaxy by now, why don't we see any evidence of a galactic civilization?

We will begin this chapter by discussing both the challenges and the possibilities of interstellar travel as we understand them today. Then, with that understanding in mind, we will confront the Fermi paradox and see why, despite the seeming innocence of the question, its solution will undoubtedly have profound implications for the future of our own civilization.

Provide ships or sails adapted to the heavenly breezes, and there will be some who will not fear even that void ...

Johannes Kepler in a letter to Galileo, 1593

13.1 The Challenge of Interstellar Travel

Science fiction routinely portrays our descendants hurtling through the galaxy, wending their way from one star system to another as easily as we now travel from one country to the next. We have already used rockets to explore other worlds in our solar system. Could future generations travel among the stars just by building larger versions of the rockets we use today? Perhaps surprisingly, the answer is no. The chemical rockets that have sent people to the Moon are wholly inadequate for taking people to the stars.

Why is interstellar travel so difficult?

The fact that interstellar travel is a daunting enterprise is due to a simple circumstance: the tyranny of distance. The stars are so remote that only in the nineteenth century did astronomers develop instruments sufficiently precise to measure the distances of the closest stars besides the Sun. When it was realized just how far away these pinpoints of light are, the French philosopher Blaise Pascal was moved to write that "the eternal silence of infinite spaces" left him terrified.

SPACECRAFT BOUND FOR THE STARS We've considered the vast distances to the stars in earlier chapters, using the scale model of the solar system introduced in Chapter 3. Here we consider the distances from the point of view of travel. Five interplanetary probes—*Pioneers 10* and *11*, *Voyagers*

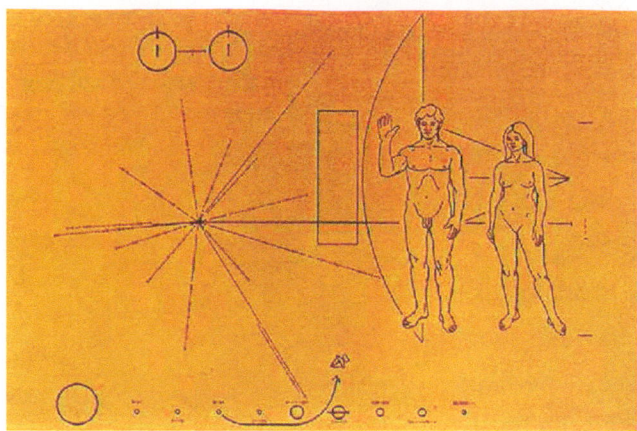

FIGURE 13.1

The *Pioneer* plaque, carried on both the *Pioneer 10* and the *Pioneer 11* spacecraft, is about the size of an automobile license plate. The human figures are shown in front of a drawing of the spacecraft to give them a sense of scale. The "starburst" to their left shows the Sun's position relative to nearby stellar remnants known as *pulsars*, which are rapidly rotating neutron stars, and Earth's location around the Sun is shown below. Binary code indicates the pulsar periods. Because pulsars slow with time, the periods will allow someone reading the plaque to determine when the spacecraft was launched.

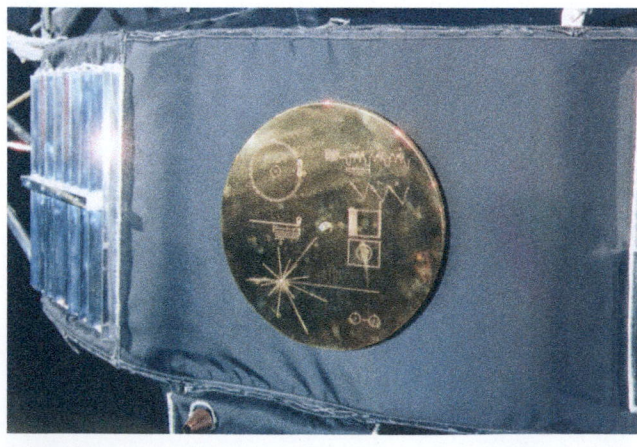

FIGURE 13.2

Voyagers 1 and *2* carry a phonograph record—a 12-inch gold-plated copper disk containing music, greetings, and images from Earth. One of the etchings on the disk surface gives instructions on how to play it.

1 and *2,* and the *New Horizons* spacecraft that flew past Pluto in 2015—are currently on their way out of the solar system. How long will it take these spacecraft to reach the stars?

Let's take the first, *Pioneer 10,* as an example. This spacecraft, launched in the early 1970s, took 21 months to reach its first target, the planet Jupiter. This might seem speedy enough in view of the fact that the giant planet is never closer than 628 million kilometers from Earth. But our nearest stellar neighbor, the Alpha Centauri star system, is 70,000 times as far away. If *Pioneer 10* were to cover the 4.4 light-years to Alpha Centauri at the same average speed at which it traveled to Jupiter, the journey would take 115,000 years. But *Pioneer 10* was not aimed at Alpha Centauri or at any other deliberate target—its trajectory was designed to reach Jupiter and Saturn, not any particular stars beyond. If we plot its trajectory along with the motions of nearby stars, we find that *Pioneer 10* will come no nearer than 3.3 light-years from any star during the next million years. About 2 million years from now, the probe will reach the general neighborhood of the bright star Aldebaran, in the constellation Taurus.

You can now see why we did not equip *Pioneer 10* (or any of our other probes) with instruments for studying planets in other star systems. Nevertheless, because the spacecraft themselves should survive unscathed for millions of years in the near-vacuum of interstellar space, we included messages in case any extraterrestrial beings someday find them. The *Pioneer* probes each carry a small engraved plaque bearing a drawing of a man and a woman as well as diagrams giving the layout of the solar system and our general location in the galaxy (Figure 13.1). The *Voyager* craft, launched about 5 years after the *Pioneer* craft, carry a somewhat more sophisticated message consisting of pictures, multilingual greetings, and two dozen musical selections (ranging from Chuck Berry to Bach) on a gold-plated copper record (Figure 13.2). Although you might wonder how intelligible these earthly calling cards would be to any aliens, the chances that they will ever be found are slim. They are like messages in a bottle thrown into the ocean surf, and they were intended more as a statement to Earthlings than to extraterrestrials. (*New Horizons* does not carry a physical message, but the "One Earth Message" project hopes to upload a globally developed message to its computers, a project still being planned as this book went to press in 2015.)

Think About It The *Pioneer* and *Voyager* "messages" are in the form of sounds, music, and pictures. But this assumes that any aliens finding these craft would have sensory organs similar to ours. Is it possible that our messages are too anthropocentric to even be recognized, or are there good reasons to think that E.T. will have eyes and ears, with characteristics similar to ours? Defend your opinion.

THE COSMIC SPEED LIMIT The *Pioneer 10* example makes the problem of interstellar travel quite clear. But it also seems to offer an obvious solution: Build spacecraft that can travel a lot faster. If it would take a little more than a hundred thousand years for *Pioneer 10* to reach the nearest stars, then a spacecraft that travels 100,000 times faster should be able to make the trip in only a little over a year.

However, this seemingly obvious solution is not allowed by the laws of physics. In particular, we know from Einstein's *special theory of relativity* that it is impossible to travel through space faster than the speed of light.

(We'll discuss this theory and why it imposes a cosmic speed limit in Section 13.4.) This might not seem too limiting, given that light travels incredibly fast—about 300,000 kilometers per second (186,000 miles per second), fast enough to reach the Moon in barely more than 1 second. But even at this remarkable speed, light takes time to travel the vast distances between the stars, which is why we measure stellar distances in light-years [Section 3.2]. As noted earlier, the nearest star system, Alpha Centauri, is about 4.4 light-years away, which means it takes light 4.4 years to reach us from this system. Because that is the fastest possible speed of travel, the best spacecraft we could hope to build would take longer than 4.4 years for the one-way trip and hence at least 8.8 years for a round trip. To make a trip across our entire galaxy—a distance of 100,000 light-years—would take any spacecraft a minimum of 100,000 years.

Could it be that Einstein's theory is wrong and that we will someday find a way to break this cosmic speed limit? Probably not. Special relativity merits the status of being a scientific theory because it is supported by an enormous body of evidence. Its predictions have been carefully tested and verified in countless experiments, so it cannot simply be "wrong." While it might someday be augmented by a more comprehensive theory, the verified results cannot simply disappear, and the cosmic speed limit will almost certainly remain in place.

ENERGY ISSUES Another challenge of interstellar travel is the tremendous amount of energy it would require, particularly if we wanted to send people and not just lightweight robotic probes.

Imagine that we wanted to colonize an extrasolar planet. To get a decent-size colony started, we'd need to send a fair number of people with many different sets of skills. For the sake of argument, suppose we wanted to send 5000 people, meaning we would need a starship with a capacity similar to that of the large starships shown in the *Star Trek* television shows and movies. How much energy would such a ship require?

Interestingly, the minimum energy requirement doesn't depend on the fuel source or the ship design. Sending a bowling ball flying through space takes more energy than sending a baseball flying at the same speed, regardless of whether the energy comes from your arm, from a catapult, or from some kind of gas-powered launcher. Similarly, sending either ball flying at a faster speed takes more energy. That is, the energy required to put an object in motion depends on only two things: the object's mass and the speed with which you want it to move.

We can estimate the mass of the starship by comparing it to other ships that transport large numbers of passengers. For example, the *Titanic* weighed about 18,000 kilograms per passenger (although accommodations for most passengers were hardly roomy and it carried provisions for only a couple of weeks, not many years). If we conservatively adopt the same per-person weight for our starship, we expect its total mass to be about 100 million kilograms. Let's assume further that our starship travels at a modest 10% of light speed, which means it will take more than 40 years to reach the nearest stars. Now that we have estimated both the mass and the speed of our starship, a simple physics formula allows us to calculate the energy needed. (The required formula is the one used to compute kinetic energy, which is equal to $\frac{1}{2}mv^2$, where m is the mass of the moving object and v is its velocity.) The energy needed to get this ship to cruising speed is calculated to be 4.5×10^{22} joules, roughly equivalent to 100 times the world's current annual energy use.

In fact, we should double this value, because the amount of energy required to slow the ship down for a soft landing once we arrive at the colony is the same as that required to accelerate it to cruising speed. In other words, the total energy bill for the trip would be equivalent to at least two centuries' worth of current world energy usage. Let's put this in monetary terms: At a typical price for home electricity in the United States (10¢ per kilowatt-hour), the energy cost of sending our craft to another star would be about $2,500,000,000,000,000,000. (To this you can add the cost of food and fresh towels for 40 years.) Unless and until we find a way to produce enormously more energy at vastly lower prices, large-scale interstellar travel will remain out of reach.

Could we travel to the stars with existing rockets?

Practical ideas for traveling through space were considered only following the Renaissance, when modern scientific thought first took hold. By 1687, Isaac Newton had produced a treatise on universal mechanics that not only described the workings of the heavens but also explained the physics required to reach them.

Newton's third law of motion states that "for every action there is an opposite and equal reaction." Envision the recoil of a gun when it is fired. The bullet moves in one direction, and the gun moves in the opposite direction. Squids, octopuses, and some other mollusks employ a similar technique in their movements. A squid takes in water that it then squirts out at higher speed behind it, thus propelling itself forward. A rocket operates slightly differently, vaporizing on-board fuel that is shot out the back. However, in both cases it is Newton's third law (or, equivalently, the law of conservation of momentum) that accounts for the forward motion.

DEVELOPMENT OF THE ROCKET A rocket has been described as the simplest type of engine, and even scientists in Newton's time realized it could work in empty space. Serious thought about travel to other worlds began in the nineteenth century. In the 1860s and 1870s, the French author Jules Verne wrote influential stories describing travel to the Moon. His propulsion scheme used an oversize artillery shell specially constructed for the task. While this scheme was hardly practical (the enormous acceleration of the shell when fired would turn the passengers to pancakes), Verne's writings stimulated investigation of space travel by three giants of early rocketry: Konstantin Tsiolkovsky (1857–1935) in Russia, Hermann Oberth (1894–1989) in Germany, and the American Robert Goddard (1882–1945). These fledgling rocket scientists explored many of the theoretical possibilities of this type of propulsion. In particular, they worked out the so-called **rocket equation,** which describes how a vehicle's final speed depends on the propellant's velocity (see Cosmic Calculations 13.1). Both Tsiolkovsky and Goddard realized that it would be difficult for a single rocket to reach **escape velocity,** the speed necessary to overcome gravity and leave Earth behind (about 11 km/s, or 25,000 mi/hr), so they proposed the use of multistage vehicles for space flight. The three pioneers also envisioned space stations, intercontinental ballistic missiles, and ion engines.

At first, few people saw much benefit in turning these ideas into working hardware. In a 1920 technical publication, Goddard mused about the possibility that a sufficiently large rocket could reach the Moon.

Cosmic Calculations 13.1

THE ROCKET EQUATION

The rocket equation tells us how a spacecraft's final velocity, v, depends on the velocity of the exhaust gas expelled out the back, v_e, and the rocket's *mass ratio*. The mass ratio is M_i/M_f, where M_i is the mass of the rocket (including any payload—such as a rover or craft it is taking into space) with all its fuel and M_f is the mass of the rocket after the fuel has been burned (that is, the payload and any still-attached but empty fuel tanks). We can write the rocket equation in the following two equivalent forms:

$$v = v_e \ln\left(\frac{M_i}{M_f}\right) \quad \Leftrightarrow \quad \frac{M_i}{M_f} = e^{v/v_e}$$

In the equation at left, "ln" is the natural logarithm; your calculator should have a key for computing this. In the equation at right, e represents a special number with value $e \approx 2.718$; your calculator should also have a key for computing e to any power. If you are familiar with the algebra of logarithms, you can confirm that the two equations are equivalent.

Example: Suppose you want a rocket to achieve escape velocity from Earth (11 km/s) and its engines produce an exhaust velocity of 3 km/s. What mass ratio is required?

Solution: We set the rocket's final velocity to $v = 11$ km/s and its exhaust velocity to $v_e = 3$ km/s, and use the second form of the equation to find the mass ratio:

$$\frac{M_i}{M_f} = e^{v/v_e} = e^{(11 \text{ km/s})(3 \text{ km/s})} = e^{11/3} \approx 39$$

The required mass ratio is about 39. As discussed in the text, this mass ratio cannot be achieved with a single-stage rocket but can be reached with a multistage rocket.

FIGURE 13.3
Early rocketry.

a Robert Goddard stands by his pioneering liquid-fuel rocket in 1926. This craft reached a modest altitude: the height of a four-story building.

b The German V-2 rocket, first launched in 1942, was used as a weapon during World War II. Note that only 16 years separate the V-2 from Goddard's first rocket.

His speculation was ridiculed by the *New York Times,* which claimed that no lunar-bound rocket could ever work since there is no air for a rocket to "push against" between Earth and the Moon. But this claim from the *Times* was wrong: Rockets do not operate by pushing against air or anything else. They simply employ Newton's third law, firing hot gas in one direction so that the rocket moves in the opposite direction. In fact, atmospheres actually hinder the performance of rockets, because they create drag that slows them down.

A few years later, in 1926, Goddard launched his first liquid-fueled rocket from a field in Auburn, Massachusetts (Figure 13.3a). It reached a height of 13 meters. This heroic, build-it-in-a-garage phase of rocketry was soon surpassed. The 1930s brought larger-scale efforts, particularly in Germany, where the military saw value in guided missiles. The German work culminated in the development of the V-2 rocket (Figure 13.3b), used primarily against England during World War II (more than 1300 V-2 rockets hit London alone). The rapid development of rocketry that followed the war was driven largely by German scientists who had been recruited by both the Russians and the Americans. The space age truly began in October 1957 with the launch of the Soviet Union's *Sputnik I,* an 84-kilogram beeping metal ball—the world's first artificial satellite.

Although spurred primarily by national rivalries and military considerations, rocket development over the past five decades has allowed humans and our robot proxies to enter those tantalizing realms that had so long been beyond our grasp. What countless generations could only dream of, we can now do. Today's rockets—the direct descendants of those first envisioned nearly a century ago—are fast enough and powerful enough to allow us to explore the nearby worlds of our solar system. But could they ever be improved enough to take us to the stars?

LIMITATIONS OF CHEMICAL ROCKETS Even today, every rocket we use to launch spacecraft from Earth's surface works in basically the same way as Goddard's first rocket. The engines ignite and burn a chemical fuel,

FIGURE 13.4

The Saturn V rocket, which was used to carry the *Apollo* astronauts to the Moon. The most powerful rocket yet built, this now-four-decade-old, three-stage design weighed about 30% more than the Space Shuttle at liftoff and was capable of sending a 45,000-kilogram (50-ton) payload to the Moon. With a launch pad mass of 2.8 million kilograms for the Moon trips, it had an overall mass ratio of 62.

such as a mixture of oxygen and kerosene. The chemical burning creates very hot gas, which is expelled through a narrow nozzle, propelling the spacecraft into orbit or to other worlds. These chemical rockets serve our current purposes fairly well (though many people dream of new technologies that would allow us to leave Earth at far lower cost). Unfortunately, they are completely inadequate for interstellar travel.

The largest chemical rocket built to date was the Saturn V, the rocket that carried the *Apollo* astronauts to the Moon (Figure 13.4). This vehicle burned liquid oxygen and kerosene, with water the major combustion product. The hot water vapor was expelled out the back at a speed of about 3 kilometers per second—roughly three times the speed of a rifle bullet. The Saturn V consisted of three separate "stages"—that is, three distinct rockets perched atop one another so that each lower stage could drop away after exhausting its fuel. We can see why these multiple stages were useful—and why chemical rockets are limited—by investigating rocket mechanics in a bit more depth.

Let's start by imagining that a rocket like the Saturn V had only a single stage. In order to leave Earth behind, we need to reach Earth's escape velocity of 11 kilometers per second. Using the rocket equation developed by the pioneers of rocketry, we can calculate the **mass ratio** required to attain this speed. The mass ratio is defined as the mass of the fully fueled rocket (including any spacecraft it is carrying) divided by the rocket (and spacecraft) mass after all the fuel is burned. As shown in Cosmic Calculations 13.1, reaching escape velocity requires a mass ratio of 39, meaning that the fueled rocket on the launch pad must weigh 39 times as much as the empty rocket and spacecraft alone. That is, the fuel weight must be about 38 times the weight of the spacecraft and the engines. This is clearly a discouraging requirement and one that's just about impossible to meet given the weight of tanks, fuel pumps, fins, and astronauts. Indeed, the best single-stage rockets have mass ratios of only 15 or less.

If it takes a mass ratio of 39 to leave Earth and our best rockets have mass ratios of only 15, how can we ever succeed? As the early rocket pioneers realized, the trick is to use multiple stages. If each stage is discarded as its fuel is used up, the upper stages don't need to accelerate the dead weight of those below. The rocket as a whole—with all of its stages—still must weigh 39 times as much as the parts that will actually reach space. But each stage requires a much lower mass ratio, because the weight of the rocket will decrease as stages are discarded. For example, escaping Earth by means of a three-stage rocket (assuming the stages have identical mass ratios and exhaust velocities) would require that each stage have a mass ratio of only 3.4—well within our capabilities. (See Problem 54 at the end of the chapter.)

Think About It The Space Shuttle did not use stacked rockets like the Saturn V, but it still used staging. Explain how. (*Hint:* It should be obvious if you look at a picture of the Shuttle on its launch pad.)

In principle, adding more stages can propel chemical rockets to higher speeds, but not high enough for convenient interstellar travel. For example, an oxygen-kerosene rocket like the Saturn V consisting of a stack of *100* stages (each with a mass ratio of 3.4) would reach a speed of 370 km/sec—33 times faster than the Saturn V but barely more than 0.1% of the speed of light. A trip to Alpha Centauri at this speed would still take some 4000 years. Using more efficient chemical fuels (such as oxygen and hydrogen, rather than kerosene) could help, but by no more than about a factor of 2. Indeed, no matter what engineering refinements we consider, chemical rockets simply are not powerful enough to deliver large payloads to the stars in a reasonable length of time. Interstellar travel requires a different approach.

13.2 Spacecraft for Interstellar Travel

Chemical rockets may be insufficient for travel to the stars, but other technologies hold greater promise. Generally speaking, we can break these technologies into two groups: "conventional" technologies—that is, technology that seems within our grasp (at least if we disregard cost), even if we don't yet have it—and technologies that are theoretically possible but far beyond our present capabilities. In this section, we'll begin by investigating a few conventional technologies that would allow at least a modest degree of interstellar travel, then move on to explore more far-out ideas.

How might we build interstellar spacecraft with "conventional" technology?

Conventional technologies for interstellar travel are based on the idea that we could adapt existing technologies to the task. None of these technologies would make interstellar travel "easy"—at best, they might reduce the travel time to nearby stars to centuries or decades. Still, if we had unlimited funds, we could in principle begin work on these technologies today.

NUCLEAR ROCKETS Chemical reactions involve shuffling the outer electrons of atoms. While these reactions can seem quite powerful (consider the drama of a Space Shuttle launch), the energy they release is insignificant compared to the amount of energy at least potentially available in the reacting materials. According to Einstein's famous formula, $E = mc^2$, any piece of matter contains an amount of energy E equivalent to its mass m multiplied by the speed of light c squared [Section 3.3], which represents an enormous amount of energy. For example, if you could turn a 1-kilogram (2.2-pound) rock completely into energy, the energy released would be equivalent to that contained in nearly 8 billion liters of gasoline—or as much gasoline as is used by all the cars in the United States in a week. However, while this energy is "there" in any piece of matter, it is very difficult to extract. Chemical reactions extract so little of it that we do not notice any change in the mass of the reacting materials.

Nuclear reactions, in contrast, can noticeably affect the mass of reacting materials. They involve changes in the dense atomic nucleus. Two basic types of nuclear reactions can be used to generate power: fission and fusion. Nuclear **fission** involves the splitting of large nuclei such as uranium or plutonium. When a uranium nucleus is split, approximately 0.07% of its mass is turned into energy. For example, if 1000 grams of uranium underwent fission, the fission products (the material left over after the fission has occurred) would weigh a total of only 999.3 grams, 0.07% less than the starting weight. Although this mass loss may sound fairly small, the energy it releases dwarfs that released by chemical reactions. Nuclear fission bombs are what destroyed the Japanese cities of Hiroshima and Nagasaki at the end of World War II, and all current nuclear power plants get their energy from fission.

Nuclear *fusion*, the power source of the Sun and other stars [Section 3.2], is about ten times as efficient as fission. Fusion of hydrogen into helium converts about 0.7% of the hydrogen fuel mass into energy. The Sun, for example, fuses 600 million tons of hydrogen into helium *each second*. The resulting helium weighs 0.7% less than the original hydrogen, or about 596 million tons. The other 4 million tons of mass simply "disappears" as it becomes the energy that makes our Sun shine. We humans have managed to achieve nuclear fusion here on Earth, but only in thermonuclear bombs (or "H-bombs") and as yet not in a well-controlled, commercially useful way. This is unfortunate; not only is the fuel for fusion (hydrogen) readily available in water, but the efficiency of fusion is so great—at least compared to that of current energy sources such as oil, coal, and hydroelectric power—that fusion power would seem almost unlimited if we were able to tap it. For example, if we could somehow hook up a nuclear fusion plant to your kitchen sink, then by continuously fusing the hydrogen in the water flowing from the faucet we could generate more than enough power to meet all the current energy needs of the United States.* That is, with your kitchen faucet fusion plant, we could stop the drilling and importing of oil, dismantle all hydroelectric dams, shut down all coal-burning power stations, get rid of all fission power plants, and still have power to spare. And there'd be no more worries about ongoing contributions to global warming, because fusion does not release any greenhouse gases into the atmosphere.

Think About It Scientists have been working for decades in hopes of developing the technology for viable nuclear fusion power plants, but so far without success. How much effort do you think we should put into attempts to develop fusion power? If we achieved it, how do you think it would change our world?

The tremendous efficiency of nuclear energy over chemical energy was bound to appeal to rocket scientists. In 1955, the U.S. Atomic Energy Commission and the U.S Air Force (and later NASA) embarked on an experiment called *Project Rover* to develop nuclear fission reactors that could be flown in a rocket. The idea was to use the fission reactor to generate enormous heat, which would be used to bring hydrogen gas to a temperature of millions of degrees before expelling it out the engine nozzles. (Note that the hydrogen was being used as a propellant, not for fusion.) At its peak, Project Rover employed 1800 people and ultimately tested six fission engines. The program made substantial progress and showed that fission-powered rockets could achieve speeds at least two to three times those of similar-size chemical rockets. By the late 1960s, NASA officials were confident that the Project Rover rockets could be used to send humans to Mars in what they hoped would be an immediate follow-up to the *Apollo* Moon landings (Figure 13.5). However, the political climate changed, and the United States abandoned its early plans for a human mission to Mars. Project Rover was terminated in 1973.

FIGURE 13.5
President John F. Kennedy departing the Nevada Test Site after viewing a full-scale mock-up of a nuclear-powered engine for Project Rover, December 8, 1962.

*Actual attempts to generate fusion power use deuterium (the isotope of hydrogen with one neutron), which is present naturally in the ratio of about 1 part deuterium to 50,000 parts ordinary hydrogen. Thus, with deuterium, the needed water flow would be about 50,000 times greater than that of your kitchen faucet—but this flow (about 130,000 liters per minute) is still only about that of a small stream.

FIGURE 13.6
Artist's conception of the Project Orion starship, showing one of the small H-bomb detonations that would propel it. Debris from the detonation impacts the flat disk, called the pusher plate, at the back of the spaceship. The central sections (enclosed in a lattice) hold the bombs, and the front sections house the crew.

Another experimental approach, dubbed *Project Orion,* was more radical. Physicists at Nevada's Los Alamos Scientific Laboratory realized that one way to get a rocket up to much higher speed would be to toss small nuclear (fusion) bombs out the rear and let the resulting explosions push the craft forward. The bombs, released at a rate of one every few seconds or more, would drop back about 50 meters and then detonate behind a large metal "pusher plate" affixed to the tail of the rocket (Figure 13.6). This would provide an impulse to move the rocket forward. Despite suffering obvious abuse, the pusher plate wouldn't vaporize because it would be exposed to these searing explosions for only a few milliseconds at a time. The Los Alamos scientists calculated that a spaceship 1 mile long accelerated by the rapid-fire detonation of a million H-bombs could reach Alpha Centauri in just over a century. In this sense, Project Orion represented the first true "starship" design to be fashioned by humans. No actual construction ever began, although in principle we could build a Project Orion–type starship with existing technology. However, this kind of starship would be very expensive and would require an exception to the international treaty banning nuclear detonations in space. Project Orion ended in 1965, because of both budget cuts and the nuclear test ban treaty.

Another nuclear rocket design was developed in the 1970s by the British Interplanetary Society under the name *Project Daedalus* (Figure 13.7). The idea was to shoot frozen fuel pellets of deuterium and helium-3 into a reaction chamber where they would undergo nuclear fusion. The fuel pellets, about the size of gravel, would be shot into the chamber at a rate of 250 pellets per second. There they would be encouraged to fuse by electron beams, producing a rapid-fire series of explosions that would propel the ship. Because we cannot yet build nuclear fusion reactors, this design remains beyond our current technological capabilities. Nevertheless, the proponents of Project Daedalus developed a plan for sending a robotic spacecraft to Barnard's star, a dim, type M star 6 light-years distant and the next closest star to Earth beyond the Alpha Centauri system. After 4 years of firing the engine, the craft would reach about one-tenth the speed of light and then spend the next four decades coasting to its destination. Once there, it would deploy probes and sensors to relay back

FIGURE 13.7
Artist's conception of a robotic Project Daedalus starship. The front section (upper right) holds the scientific instruments. The large spheres hold the fuel pellets for the central fusion reactor.

photos and other data, giving us our first close-up view of another stellar system only about 50 years after its launch.

In 2010, NASA, together with the Defense Advanced Research Projects Agency (DARPA), announced an initiative called the 100 Year Starship. The intention of this initiative is to devise a plan for sending a crewed spacecraft to a nearby star, with a travel time of one century or less. The 100 Year Starship project awarded $500 thousand to a private foundation to start thinking about the technical, organizational, sociological, and even fund-raising problems that would confront any such future undertaking. Note that this initiative is not currently trying to build an interstellar rocket, but to design a project that could potentially achieve success over the long time period required for both the development and the voyage.

Nuclear-powered rockets are undoubtedly feasible in some form. Still, at best they would achieve speeds of about one-tenth the speed of light. Interstellar journeys would be possible, but it would take decades to reach even the nearest stars.

IONS, SUNLIGHT, AND LASERS The propulsion schemes described so far involve a relatively quick acceleration of the rocket to high speed, after which the engines shut down and the craft cruises for whatever length of time it takes to reach its target. Another approach is to use a low-powered rocket whose engines keep firing continuously. The **ion engine** is an example of this approach. It works something like an old-fashioned television picture tube in that it accelerates charged particles (ions). In a television tube, electrons are fired from the back of the tube to the phosphor screen that faces the viewer. An ion rocket engine does the same, with charged particles fired rearward as the rocket exhaust. Both NASA and the European Space Agency (ESA) have already used low-power ion engines successfully. These engines can be started only in space (they don't have enough thrust to lift off Earth, and they work best in a vacuum), but they can keep firing for long periods because the mass expelled per unit of time is small. Moreover, the exhaust ions are shot from the craft at tremendous speeds, and a powerful ion rocket could in principle reach speeds approaching 1% or so of the speed of light.

FIGURE 13.8
Artist's conception of a spaceship propelled by a solar sail, shown as it approaches a forming planet in a young solar system. The sail is many kilometers across. The scientific payload is at the central meeting point of the four ladder-like structures.

Other schemes envision spacecraft that overcome the limitations of the rocket equation by not taking along bulky fuel. One possibility that's been considered for nearly a century is to use sunlight as power. Large, highly reflective, very thin (to minimize mass) **solar sails** could be pushed by the pressure exerted by sunlight. This pressure is so slight that we normally don't notice it, but in the vacuum of space, where friction is absent, the steady pressure of sunlight impinging on a mirrored sail could push a spacecraft to impressive speed, particularly with sails hundreds of kilometers in size. Solar sailing might well prove to be a fairly inexpensive way of navigating within the solar system, and it could even be useful for interstellar travel (Figure 13.8). Although the push from the Sun would slowly fade once such craft reached the outer solar system (at Saturn, the light intensity is less than 1% of its value near Earth), a solar sailing vehicle that was started very near the Sun might achieve speeds of a few percent of light speed. It could then coast to neighboring stars in less than a century.

The fact that sunlight weakens so much with distance limits the ultimate speed of a solar sailing spacecraft. However, we could get around this problem by using a powerful laser on Earth as an energy source, instead of sunlight. In principle, the laser could provide a steady and continuous "push" for the solar sail, all the way to its destination if necessary. If building large sails proved too difficult, the laser could be used to vaporize propellant on the rocket that would then propel the craft forward in the usual rocket-like manner. As these craft moved light-years away, a large focusing mirror hundreds of kilometers in size would be needed at the laser base to concentrate the beam on the pinpoint target that the spacecraft had become. The primary drawback to these schemes is the power requirement. For example, accelerating a ship to half the speed of light within a few years would require a laser that uses 1000 times all current human power consumption. Nevertheless, this approach would allow us to travel to nearby stars in a decade or two rather than many decades.

A laser-powered rocket would leave the passengers dependent on the efforts of those at the base to keep the laser shining so that they could accelerate to their desired final velocity. This might be somewhat risky given the fact that even the fastest of these transports would be en route and accelerating for decades. What if the laser crew went on strike? In addition, with no laser shining in the *opposite* direction, slowing the spacecraft to a halt at its destination (let alone returning home) would be a problem. One possibility is to use on-board propellant heated by the laser. The propellant could then be fired out the front of the craft to slow it down. Alternative braking schemes that use natural magnetic fields in space have also been suggested.

INTERSTELLAR ARKS Another, less demanding approach to interstellar travel is often featured in science fiction. Forget the high-tech rocketry and accept relatively low speeds. Then deal with the resulting long travel times by putting the crew into suspended animation—hibernation, if you will—and letting them doze their way to the stars. A challenging variation on this idea is to somehow increase human lifetimes long enough for travelers to cruise from star to star. A third suggestion is to build enormous craft that can accommodate a very large crew: in essence, an "ark." Many generations would live out their lives aboard this slow-moving vehicle before it finally reached its destination.

The difficulty with the first suggestion is that no one yet knows how to put humans to sleep for hundreds or thousands of years (and then have them wake up). However, genetic researchers have identified genes that control hibernation in animals, so it is possible that genetic engineering techniques could someday allow humans to hibernate as well.

The second suggestion, to allow the crew to live the many thousands of years necessary for interstellar travel at conventional speeds, depends on advances in medical technology. Could we somehow stop the aging process, enabling people to live such long lives that centuries or millennia of travel might seem like a walk to the corner store? We simply don't know.

The idea of building interstellar arks usually gets a skeptical reaction from sociologists. They point out that long voyages on Earth (even those that last only a few months) often end badly. Crews splinter into antagonistic factions and frequently fight for control of the ship. In addition, we might justifiably fear a deterioration in the level of expertise of the crew, with the result that the generation of folk who finally reached the target star system would neither remember why they journeyed there nor have the technical skills required to land on or colonize a world.

How might we build spacecraft that could approach the speed of light?

We have seen that conventional approaches to interstellar travel—such as chemical, nuclear, or laser-powered rockets or a solar sail—will not bring us to the stars in anything less than decades. What we really need for interstellar travel are ships that can travel at speeds close to the speed of light. We could then reach the nearest stars in years and explore space within a few tens of light-years of the Sun in just a few decades. Moreover, such ships would be traveling fast enough for on-board passengers to benefit from some astonishing effects of high-speed travel.

THE ROLE OF RELATIVITY The fact that we cannot exceed the speed of light might at first make distant stars seem forever out of reach. However, the same theory that imposes the cosmic speed limit—Einstein's special theory of relativity—also tells us that time is different for high-speed travelers than for people who stay at home. We'll discuss the reason this occurs in Section 13.4; here, we'll consider its implications for interstellar travel.

Imagine a trip to the star Vega, about 25 light-years away, in a spaceship traveling at a constant speed of 90% of the speed of light (0.9c). Because light takes 25 years to travel the distance to Vega, and a ship traveling at 0.9c is going 90% as fast as light, the ship's travel time to Vega should be $\frac{25}{0.9} \approx 28$ years. This is indeed the time that would be measured by people staying home on Earth; that is, if the ship made the round trip at this speed, it would return home $28 \times 2 = 56$ years after it left. However, it is *not* the time that would be measured by the ship's crew.

Einstein's theory tells us that when a spaceship (or any other object) travels at close to the speed of light, its length becomes noticeably shorter in the direction of movement, its mass becomes noticeably greater, and time measured aboard proceeds noticeably more slowly than time measured by clocks at rest; the latter phenomenon is called **time dilation.** These changes to time and space are not just speculation—they have all been carefully measured in experiments with subatomic particles that move at speeds close to the speed of light. Note that, while the ship's time was running slow according to people back on Earth, time would feel perfectly "normal" to the crew. But less time really would pass for them. As shown in Table 13.1 (and calculated in Cosmic Calculations 13.2), only about 24 years would pass on the spaceship during the round-trip voyage to Vega at 0.9c. In other words, this is what would happen if a ship left on this journey in the year 2100: The ship would return in the year 2156, but the crew would have aged only 24 years. If a crew member was 20 years old when she left, she'd be 44 on her return, but her twin brother, who stayed home, would be 76.

Table 13.1 shows the benefits of relativistic travel for a hypothetical trip from Earth to Vega at various speeds. At low speeds, there's no noticeable difference between ship time and Earth time, and the journey takes a very long time. The closer the ship gets to the speed of light, the less time that passes for the crew. Indeed, from the crew's standpoint, the trip can be made arbitrarily short simply by getting ever closer to the speed of light. But for friends left behind on Earth, the rocket can never return less than 50 years after it left.

If you study the table carefully, you might wonder if the crew of a very-high-velocity rocket would conclude that they were traveling faster than the speed of light—which would violate the cosmic speed limit of relativity. For example, at a speed of 99.99% that of light (0.9999c), their trip would take only 8 months for a distance we said was 50 light-years. This would seem to imply a speed some 75 times the speed of light. However, special relativity also tells us that distances shrink at high speed. Once traveling at high speed, the crew would find that the distance to Vega was not 25 light-years, as we measure it on Earth, but instead a little under 0.4 light-year. Therefore, they could cover this short distance in a short time, and they'd never think they were traveling faster than the speed of light.

Think About It Suppose you were offered the opportunity to take a trip to Vega and back at a speed of 0.9999c. How long would the round trip take, according to you? How much time would pass on Earth while you were gone? All things considered, are there any circumstances under which you would agree to take such a trip? Explain.

TABLE 13.1 Round-Trip Travel Time to Vega

This table shows the time that passes on Earth and the time that passes for the crew of a spaceship on round-trip journeys at various speeds to the star Vega, a trip of 25 light-years in each direction. Speeds are given as fractions of the speed of light, c = 300,000 km/s. Note that the first-row speed of 0.00005c is equivalent to 54,000 kilometers per hour, which is roughly the speed of our fastest chemical rockets today.

Speed	Time Measured on Earth	Time Measured on Ship
0.00005c	1,000,000 yr	1,000,000 yr
0.1c	500 yr	498 yr
0.5c	100 yr	86 yr
0.7c	72 yr	52 yr
0.9c	56 yr	24 yr
0.99c	50 yr	7 yr
0.999c	50 yr	2.2 yr
0.9999c	50 yr	8 mo

Cosmic Calculations 13.2

TIME DILATION

The effects of time dilation on a fast spaceship can be calculated with a simple formula if we assume that the spaceship travels at constant speed:

$$t_{ship} = t_{Earth}\sqrt{1 - \left(\frac{v}{c}\right)^2}$$

where t_{ship} is the amount of time that passes on the rocket, t_{Earth} is the amount of time that passes on Earth, v is the rocket's velocity (speed), and $c = 3 \times 10^8$ m/s is the speed of light.

Example: Consider a spaceship that travels round-trip to Vega at 90% of the speed of light. As noted in the text, the round-trip travel time measured by people on Earth is 56 years. How much time passes for passengers on the spaceship?

Solution: Because we are given that the ship travels at 90% of the speed of light, or $0.9c$, we know that $v/c = 0.9$. We plug in this value along with the time that passes on Earth, $t_{Earth} = 56$ yr:

$$t_{ship} = t_{Earth}\sqrt{1 - \left(\frac{v}{c}\right)^2}$$

$$= 56 \text{ yr} \times \sqrt{1 - 0.9^2}$$

$$= 56 \text{ yr} \times \sqrt{1 - 0.81}$$

$$= 56 \text{ yr} \times \sqrt{0.19}$$

$$= 24.4 \text{ yr}$$

This is the approximately 24-year round-trip time shown for the ship in Table 13.1.

INCREDIBLE JOURNEYS In fact, if we could somehow boost our spacecraft to speeds arbitrarily close to the speed of light, we could go anywhere in the universe within a human lifetime. Astronomer Carl Sagan considered a hypothetical rocket that accelerates at a steady $1g$ (or "1 gee")—an acceleration that would feel comfortably like gravity on Earth—to the halfway point of its voyage. This constant acceleration would bring the ship closer and closer to the speed of light, though it would never exceed it. (Note that the acceleration of $1g$ is constant, but the speed is not!) The ship then reverses and decelerates at $1g$ to its destination. During most of the trip, the ship would be traveling at speeds quite close to the speed of light, so time would pass quite slowly on the ship compared to time on Earth. Longer trips would mean longer periods of acceleration, bringing the ship even closer to the speed of light for most of the journey.

Calculations show that such a continuously accelerating ship could make a trip to a star 500 light-years away in only about 12 years according to those on board the ship. However, 500 years would pass on Earth. If a crew of 20-year-olds left Earth in the year 2100, they would be merely 32 years old when they reached their destination; but it would be the year 2600 on Earth (actually a bit later, since the ship would be traveling at not quite the speed of light). If the travelers sent a radio message back to Earth, the message would take 500 years to travel the 500-light-year distance. More than 1000 years after the crew had left, we'd get a message from people who had aged only 12 years since they'd last been seen on Earth.

Even longer trips would be possible in principle. For example, in a craft with a constant $1g$ acceleration, only about 21 years of ship-board time would be required to bridge the 28,000-light-year distance to the center of the Milky Way Galaxy, where the crew could observe firsthand the mysterious black hole that resides there. The 2.5-million-light-year distance to the Andromeda Galaxy could be traveled in only about 29 years of the ship's time. The passengers could travel to the Andromeda Galaxy, spend 2 years studying one of its star systems and taking our first pictures of the Milky Way as it appears from afar, and return only 60 years older than when they had left. However, they would not exactly be returning "home," since 5 million years would have passed on Earth. In this sense, special relativity offers sufficiently fast travelers only a one-way "ticket to the stars." No place is out of reach—but you cannot return home to the same people and places you left behind.

Although such incredible trips are allowed by the laws of physics, the energy costs would be extraordinary. Because special relativity also tells us that an object's mass increases as the object approaches the speed of light, the energy cost rises just as much as time slows down. Indeed, that is one explanation for why the speed of light cannot be reached: As the ship gets closer and closer to the speed of light, its mass becomes greater and greater, so the same rocket thrust generates ever less additional speed. The mass approaches infinity as the ship's speed nears the speed of light—and no force in the universe can give a push to an infinite mass. That is why the ship can never reach the speed of light, no matter how powerful its engines might be, and even getting close to that speed would require amounts of energy far beyond anything we will be able to muster in the near future. Nevertheless, such practical difficulties can't stop us from speculating, and at least two potential ways of approaching the speed of light are known to exist in principle.

MATTER–ANTIMATTER ROCKETRY The most efficient energy source we have discussed so far is nuclear fusion. But fusion converts only 0.7% of the mass of the fusing hydrogen into energy; 99.3% of the mass still remains, as helium. Is there a way to turn more of the mass, or even all of it, into energy? The answer is yes, and it is called **matter–antimatter annihilation.**

Antimatter might sound like the stuff of science fiction, but it really exists. All material things are composed of "ordinary" matter, but physicists have discovered particles that are in some ways the mirror images of normal particles, differing principally in their electrical charge. The first known such particle, christened the *positron*, was discovered in 1932. It is the antimatter twin of the electron; that is, it is identical to an electron except that it has a positive rather than a negative charge. In 1955, the proton's antimatter partner was found: the antiproton. If you were to introduce an antiproton to a positron, they would form an atom of antihydrogen. This antimatter atom would behave chemically just like ordinary hydrogen, except for one thing: You wouldn't want to get near it. When matter and antimatter meet, the result is total annihilation, with 100% of the mass turning into energy. (Note that antimatter still has mass just like ordinary matter; there is no such thing as "antimass.")

The annihilation of matter and antimatter can create energy in a variety of forms. For example, annihilation of positrons with electrons produces energy as a burst of gamma rays. The annihilation of heavier particles, such as antiprotons with protons, produces a gush of particles that soon decay into neutrinos and gamma rays. These might be able to power a rocket, because the flood of reaction particles could be directed out a rearward-facing nozzle. A matter–antimatter rocket could, in principle, achieve speeds of 90% of the speed of light with modest mass ratios.

While such numbers are seductive, the problem lies in rounding up and storing the required antimatter. No practical reservoirs of this material are known. Instead, we would have to manufacture the antimatter, as physicists now do with high-energy particle accelerators. However, current worldwide production of antimatter amounts to only a few billionths of a gram per year, and the energy that would be needed to manufacture larger amounts is staggering. For example, with present technology, manufacturing 1 ton of antimatter—far less than would be needed for an interstellar trip—would take more energy than humankind has used in all of history. Moreover, even if we could make the antimatter, we don't yet know of a good way to store it aboard our rockets, which would require some type of container in which it never touched any ordinary matter at all, since otherwise the result would be premature annihilation. Containment using very powerful magnets has been considered, but it's not clear that such massive devices could ever be practical.

INTERSTELLAR RAMJETS Another approach to achieving relativistic velocities circumvents the problems involved in carrying highly energetic fuel on board. The idea is that a starship could collect its fuel as it goes, using a giant scoop to sweep up interstellar gas. Because this gas would be mostly hydrogen, it could be funneled to a nuclear reactor, fused into helium, and then expelled out the back to propel the starship. Such propulsion systems are known as **interstellar ramjets** (Figure 13.9). In principle, interstellar ramjets can accelerate continuously by collecting and using fuel nonstop, getting ever closer to the speed of light.

FIGURE 13.9
Artist's conception of a spaceship powered by an interstellar ramjet. The giant scoop in the front (left) collects interstellar hydrogen for use as fusion fuel.

Of course, there are practical difficulties. The typical density of the gas between the stars is only a few atoms per cubic centimeter, so the scoop would need to be hundreds of kilometers across to collect adequate supplies of fuel. As Carl Sagan said, we are talking about "spaceships the size of worlds." Another problem facing an interstellar ramjet—or any ship traveling at relativistic speeds—comes from the interstellar gas and dust itself. At 99% of the speed of light (0.99c), a particle the size of a sand grain packs energy equivalent to an explosion of about 100 kilograms of TNT. Even individual atoms encountered at this speed would be deadly, so the ship would need substantial shielding to protect both its structure and its crew.

Any type of relativistic travel, whether with matter–antimatter engines, interstellar ramjets, or some as-yet-unthought-of method, remains far beyond our current technological capabilities. But we have at least imagined ways by which an extremely advanced civilization *might* be able to travel among the stars. Whether anyone has actually done so remains unknown.

All of the schemes we've considered for rapid interstellar travel assume that we wish to send humans into space. But our sensor technologies—the high-resolution cameras and other devices we use to measure an environment—are improving much more rapidly than our rocket technology. Perhaps the most practical way to go to the stars is not to go ourselves, but to send lightweight probes that could map in detail another planetary system, returning the data to Earth via radio. Because these payloads could be enormously smaller than a spacecraft designed to support humans, even fairly conventional rockets could launch them at high speed. In this way we could explore distant worlds while comfortably sitting in front of our computer screens (or using virtual reality devices), traveling no farther than our own living rooms.

Are there ways around the light-speed limitation?

If you are a science fiction fan, our discussion of interstellar travel may depress you. Interstellar tourism and commerce seem out of the question, even with ships that travel at speeds close to the speed of light, because of the long times involved (at least as seen from home planets and colonies). If we are ever to travel about the galaxy the way we now travel about Earth, we will need spacecraft that can somehow get us from here to there much faster than the cosmic speed limit would seem to allow. Could such spacecraft be possible?

No one really knows. However, there just might be a "loophole" in the law limiting cosmic speed. In particular, while Einstein's special theory of relativity showed that we can't travel *through* space faster than the speed of light, his general theory of relativity suggests that there might be "shortcuts" that, in effect, let us travel *outside* ordinary space in a way that greatly reduces the distances to be traveled. If so, then we might reach far-off places by taking a shortcut that lessens the distance to them.

HYPERSPACE In 1915, Einstein enlarged his earlier work (special relativity) with his announcement of the general theory of relativity [Section 2.4]. Einstein had already shown that we live in a four-dimensional universe, with three dimensions of space and one of time; the four dimensions together are usually called **spacetime.** With general relativity, he showed that gravity arises from curvature of spacetime, and that this curvature explains everything from why planets orbit the Sun to the pathways taken by light as it travels through the universe.

How can we tell that space (or spacetime) is "curved"? One simple test is to consider the paths of light beams, which travel through space in what we call straight lines. If space had no curvature, then two parallel light beams—for example, from two laser pointers taped side by side—would never cross. However, general relativity insists that the presence of matter can cause a warping of space that will lead these parallel beams to cross. The matter produces a distortion of the ordinary three dimensions of space into other, hypothetical dimensions that we can't see or "get into." These additional dimensions are called **hyperspace.**

To visualize this idea, physicists often resort to "embedding diagrams," such as that depicted in Figure 13.10. Space is reduced from three dimensions to two, and the resulting diagram resembles a rubber sheet; if you inhabited the world shown in the figure, you would be flat, infinitely thin, and incapable of appreciating that any dimension exists on either side of the sheet. Embedded in the sheet is a large mass, such as a star, that causes the sheet to distort into hyperspace. You cannot see the hyperspace dimensions, but you can make measurements on the rubber sheet that will tell you whether or not it is curved. In fact, during solar eclipses, we have measured changes in the apparent positions of stars whose light passes near the Sun. As Figure 13.10 shows, these changes are what we would expect if our space was curved through hyperspace.

The warping of space is usually quite small. Even the bending of starlight passing close to the Sun amounts to only a fraction of a thousandth of a degree. But if space could be warped more dramatically, the distortion might offer us shortcuts to distant destinations.

BLACK HOLES, WORMHOLES, AND WARP DRIVE The bizarre objects called black holes are, in fact, holes in spacetime—places where space becomes so distorted that it in effect becomes a bottomless pit. As a result, science fiction writers have sometimes imagined using black holes as shortcuts to other places. Unfortunately, this idea suffers from at least two major drawbacks. First, the only known black holes are themselves extremely far away, so getting to them in the first place would be a problem. Second, we do not know of any way we could survive a close encounter with a black hole.

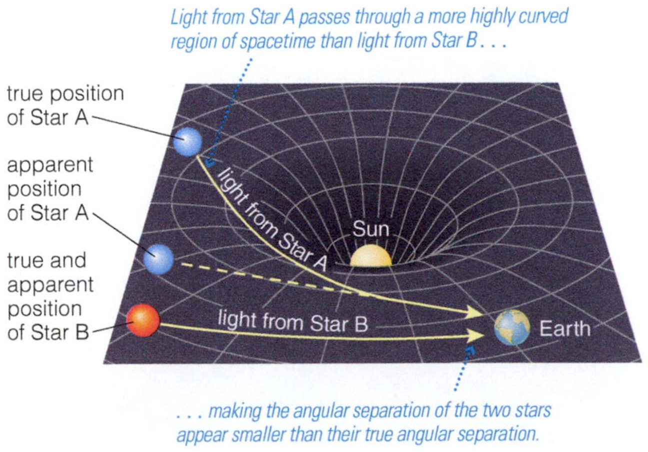

Light from Star A passes through a more highly curved region of spacetime than light from Star B . . .

true position of Star A

apparent position of Star A

light from Star A

Sun

true and apparent position of Star B

light from Star B

Earth

. . . making the angular separation of the two stars appear smaller than their true angular separation.

FIGURE 13.10
This embedding diagram shows how starlight is bent as it passes near the Sun, causing stars to appear slightly offset from their true positions in space. This effect (exaggerated in this diagram) has been measured during solar eclipses, proving that our space really is curved through hyperspace.

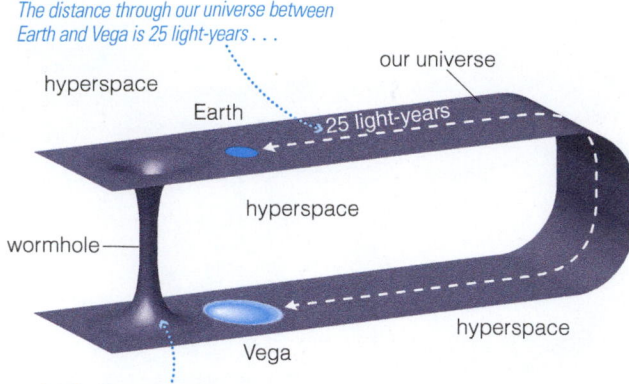

The distance through our universe between Earth and Vega is 25 light-years . . .

hyperspace

our universe

Earth

25 light-years

hyperspace

wormhole

hyperspace

Vega

hyperspace

. . . but the distance would be much shorter if we could travel through a wormhole.

FIGURE 13.11

Illustration of the idea of a wormhole. Once again, we have reduced our three spatial dimensions to the flat, two-dimensional realm of a rubber sheet. Two distortions of space, one near Earth and one near Vega, have met up in hyperspace, forming a connecting tunnel. Going from Earth to Vega in ordinary space would be a 25-light-year trip. But the wormhole offers a radically shorter route—one that might be traversed in minutes without ever exceeding the speed of light. (Adapted from a drawing by Caltech physicist Kip Thorne.)

However, a related phenomenon, called a **wormhole,** might be more useful. Just as a worm might shorten its trip from one side of an apple to the other by tunneling through it, so might a wormhole provide a hyperspace shortcut to a distant part of the universe. Imagine a dense mass floating somewhere near Earth, distorting the surrounding space into hyperspace. Now imagine a similar distortion occurring somewhere else in the cosmos, many light-years away. If these two distortions somehow met up in hyperspace, they could connect two distant places in ordinary space via a short, hyperspace tunnel (Figure 13.11). Traversing this tunnel might take little time (perhaps minutes) and could short-circuit the necessity of traveling those many light-years.

While this is clearly an appealing idea, could it actually be made to work? In particular, how do we arrange for the wormhole's opportune existence? Quantum physics suggests that on the tiniest scales of the universe, in regions of space far smaller than an atomic particle, spacetime is a seething foam, constantly punctured by distortions into hyperspace. In this highly microscopic world, wormholes might be forming (and self-destructing) all the time. It's conceivable that a highly advanced society might have learned how to capture one of these natural wormholes—one that connected two places of interest—and how to quickly enlarge it to a size that would permit its use for travel. However, such wormhole construction would seem to carry an impossibly large energy cost—estimated to be a thousand times the energy released by an exploding massive star (a supernova). Moreover, even if the energy could be found, we do not yet know of a way to stabilize a wormhole against immediate collapse. All in all, we do not know enough about physics to say for sure whether wormhole travel is even possible. Given this uncertainty, we can but imagine and hope. Carl Sagan's book and movie *Contact* postulated a network of wormhole tunnels permitting fast travel throughout the universe, and the more recent movie *Interstellar* made use of one as well—but even the fictional movie characters did not know how it had been built.

Another possible way to travel great distances in a short time might be to exploit the warping of space by placing a dense mass in front of a spacecraft. Hanging like bait on a fishing line, this black hole on a stick would allow the shortening of spatial distances in front of the rocket. Much as you might move across a floor by scrunching a rug in front of you and straightening it out behind you, this highly unusual craft would bend space in front and leave it unaltered behind. You would continually fall into the warped space in front of the craft. This concept comes closest to what we know as the "warp drive" of science fiction. However, it would require either capturing a black hole (difficult, to be charitable) or creating one using enormous amounts of energy. In either case, the black hole would have to be carried along for the ride.

Could there be simpler ways to take advantage of hyperspace? We do not know. This leaves the door wide open for science fiction writers. In the *Star Wars* movies, a simple flip of a lever takes a ship outside our universe and into hyperspace, allowing nearly instantaneous travel to anywhere. In *Star Trek,* a command from the captain sends the ship into warp drive, apparently without the need for a black hole in front of the ship. If such schemes are at all possible, we have no inkling of how they might work. But who knows what an advanced civilization might have discovered? After all, we have been studying physics in earnest for only a few centuries. Others out there may have been studying it for millions or billions of years.

13.3 The Fermi Paradox

We have found that practical interstellar travel is well beyond our capabilities today, but we can envision ways that more advanced civilizations might achieve it. In earlier chapters, we found good reason to think that habitable planets could be common, some with civilizations, and that many of these civilizations could be much older than ours. This type of reasoning leads to an idea first stated in 1950 by the Nobel Prize–winning Italian-American physicist Enrico Fermi (Figure 13.12). During a lunch at the Los Alamos National Laboratory in northern New Mexico, the conversation drifted to the possibility of extraterrestrial intelligence. The physicists present at the lunch were considering the likelihood that sophisticated cosmic societies might exist in great abundance. Fermi replied to these speculations with a disarmingly simple question: "So where is everybody?" Although serious scientific discussion of his query did not get under way for many years, its central idea is now known as the **Fermi paradox.**

Where is everybody?

The essence of the Fermi paradox is almost as simple as Fermi's original question. It begins with the idea that neither we nor our planet should be in any way special, in which case other Earth-like planets and other advanced civilizations (meaning civilizations capable of space travel) ought to be fairly common in the galaxy. This is more or less what we conclude from the Drake equation [Section 12.1], unless the rare Earth hypothesis turns out to be correct [Section 11.4]. However, a large number of civilizations would necessarily mean many civilizations with the opportunity to develop advanced technology and interstellar travel long before we came on the scene with our rockets and radio telescopes. In that case (for

---

Movie Madness STAR TREK

If you want to boldly go where no one has gone before, without spending a few hundred centuries doing it, you need warp drive.

As almost everyone knows, the various incarnations of the U.S.S. Enterprise, Star Trek's famous interstellar transport, high-tail it from one part of the galaxy to another in short order thanks to a futuristic propulsion system. But what is warp drive, anyway?

According to the show's technical manuals, the term is merely slang for "continuum distortion propulsion"—a Latinate mouthful that describes a scheme by which powerful fields are used to distort space and allow speeds faster than that of light. As we discuss in the chapter, rapid travel by warping space is not an entirely nutty idea. It might be possible. And if physics were to allow it in *principle*, could our clever descendants do it in practice?

Maybe yes, maybe no. A major problem is that even if you could warp space, to do so would take enormous amounts of energy. *Star Trek* deals with this small technical detail by fueling the field-generating warp engines with antimatter (antihydrogen, to be precise). Combining antimatter with ordinary matter is the most efficient combustion imaginable, as the entire mass of both is converted to energy.

Of course, there's still the problem of making the antimatter, not to mention shipping it to service stations around the galaxy (being careful to keep it out of the hands of pirates—antimatter is costly). But the truly interesting thing about warp drive is the range of speeds attained. At "Warp 1" you're loping along at the speed of light. By "Warp 9"—near the top of the *Enterprise*'s speedometer—you're streaking through space at 1000 times light speed. This means you can traverse the galaxy in a century, which is short enough to be possible, but long enough to allow you to get effectively stranded and interfere with "Starfleet Command's Prime Directive" to avoid disturbing alien cultures—which, as discussed in Section 13.3, sounds remarkably similar to one of the possible solutions of Fermi's paradox.

Mind you, you could forget warp drive entirely, and stick with the physics we know by building starships that go at 99+% of light speed. Special relativity would guarantee that travel times as perceived by the ship's crew would be short. They could cross the galaxy overnight, according to their own watches. But *Star Trek* has opted out of the relativistic approach for good reason, for otherwise the *Enterprise* crew would return home to find all their family and friends long dead and forgotten. Starfleet headquarters would probably be just an archaeological dig.

Better to call up Scotty in the engine room and tell him to put the pedal to the space-bending metal.

---

reasons we'll discuss shortly), it seems that someone else should have colonized the galaxy already. But we see no evidence of such a galactic colonization effort.

STATEMENT OF THE PARADOX Summarizing the above, we are led to the following two ideas:

1. The idea that neither we nor our planet is in any way special suggests that someone should have colonized the galaxy by now.
2. The idea of a galactic civilization implies that we should be surrounded by evidence of this civilization—but aside from unconvincing claims of extraterrestrial UFOs [Section 12.4], no such evidence exists.

By definition, the existence of two such seemingly contradictory ideas constitutes a *paradox.* But unlike some logical paradoxes (e.g., statements such as "This statement is false"), the Fermi paradox must have some solution. After all, either there is a galactic civilization out there or there isn't.

THE AGE OF CIVILIZATIONS If you look closely at the first premise of the Fermi paradox, you'll notice that it depends on the idea that, if civilizations are at all common, many should have arisen long before our own arrival.

Recall that the universe is about 14 billion years old [Section 3.2], while Earth is only $4\frac{1}{2}$ billion years old [Section 4.2]. In other words, the universe predates our planet and solar system by more than 9 billion years. Stars began to form shortly after the Big Bang—a fact we know because the ages of the oldest stars are only a few hundred million years short of the age of the universe itself. There's some debate as to whether these early generations of stars had enough heavy elements to make Earth-like planets [Section 11.3], but little doubt that the heavy-element abundance was high enough to make Earth-like planets within a few billion years of the universe's birth. We can play with more precise estimates of these times in a variety of ways, but the bottom line is this: Unless we are misunderstanding some fundamental piece of star and planet formation, it should have been possible for Earth-like planets to have been born starting *at least 5 billion years before* our own planet was born. In other words, some other Earth-like worlds should have had a 5-billion-year head start on ours.

This 5-billion-year head start means that, if intelligent life arose on these planets in the same amount of time that it took for intelligent life to arise here on Earth, the first civilizations in our galaxy should have appeared on the scene at least 5 billion years ago. In other words, if these civilizations have survived to the present day, they should be technologically ahead of us by 5 billion years.

We can take the idea a little further by making some guesses about the number of civilizations that have arisen over time. As we discussed in Chapter 12, our current understanding of the Drake equation does not allow any definitive conclusions about the number of civilizations, but it at least seems plausible to imagine that 1 in a million stars would eventually give rise to a civilization on an orbiting planet. Using a conservative estimate of 100 billion stars in the Milky Way—most far older than the Sun—this means there would have been some 100,000 civilizations in our galaxy by now. This is an astonishing idea, and one that becomes

FIGURE 13.12

Enrico Fermi (1901–1954), one of the leading physicists of the twentieth century, received the Nobel Prize in 1938 for work in understanding radioactive decay and predicting the existence of the particles known as neutrinos. By the time he was awarded his Nobel Prize, the Fascists had risen in Italy and the Nazis were in power in Germany. Abhorring these ideologies, Fermi chose not to return to Italy after attending the Nobel Prize ceremony in Sweden. Instead, he moved to the United States, where he became a prominent figure in the Manhattan Project, which was developing the atomic bomb. Element 100 in the periodic table, fermium, was named in his honor.

even more incredible when we put the number together with the time. If we assume that these 100,000 civilizations have arisen at random times over the past 5 billion years, then on average a civilization arises every 5 billion ÷ 100,000 = 50,000 years.

Think about what all this means. First, even with the odds of finding a civilization at only 1 in a million stars, there should still have been 100,000 civilizations that arose before we came on the galactic scene. Second, it means that we are almost certainly the youngest civilization in the galaxy at present, and that on average we'd expect the next-youngest civilization to have arisen 50,000 years ago, the third-youngest to have arisen 100,000 years ago, and so on.

Although the numbers we use for civilizations are essentially wild guesses, things don't change all that much even if we're much more conservative. If we assume that civilizations arise around only 1 in 100 million stars, rather than 1 in 1 million, we still end up with 1000 civilizations having arisen over the past 5 billion years. And in that case, civilizations would arise about every 5 million years on average, giving the next-youngest civilization an even greater head start on us.

Think About It Consider a couple of other variations on the above theme. How do the numbers change if the fraction of stars with civilizations is higher—say, 1 in 10,000? How do they change if the fraction of stars with civilizations is only 1 in 1 billion? Explain.

These ideas are of course speculative, but you can probably now see why Fermi asked, "So where is everybody?" And you should also be able to see why, if we ever actually meet up with another civilization, it is extremely unlikely that it will be at a technological level of development anywhere near as low as our own.

MACHINES DEEPEN THE PARADOX So far we have considered the idea that living beings should be out and about in our galaxy, but another possibility is that living beings could send machines into space, much as we have already begun to send robot spacecraft to explore other worlds in our solar system. Consideration of machine technology only deepens the Fermi paradox.

Beyond simply sending out individual robots such as those we can build now, we can imagine that in the future we might build much more sophisticated robots. For example, we might send robots to other worlds with programming instructions to dig up resources on arrival, use the resources to build factories, and use the factories to build spacecraft and more robots. These new robots would then go on to the next world, where they would do the same thing. Thus, these robots would be self-replicating, though in a way quite different from the self-replication of biological beings. The general idea of such self-replicating machines was first proposed by the American mathematician and computer pioneer John Von Neumann (1903–1957), so they are often called **Von Neumann machines.**

The use of Von Neumann machines would allow us to explore much farther and wider than we could by going to other worlds ourselves. Moreover, while interstellar travel poses huge barriers to us due to our limited life spans, these machines could presumably still function after journeys through space that take centuries or millennia. Once we sent the first wave of these machines to a few nearby star systems, they would gradually spread from star system to star system.

In 1981, physicist Frank Tipler used this idea of "colonization" by self-replicating Von Neumann machines to extend the Fermi paradox. In essence, Tipler argued that civilizations could effectively make their presence felt throughout the galaxy even without achieving the ability to send themselves on interstellar journeys. As soon as a civilization reached a level that allowed it to build Von Neumann machines, these machines would begin to spread through the galaxy. Because such colonization would require technology only slightly beyond our own, Tipler argued that if civilizations were common, the galaxy would already be overrun by self-replicating machines. Because it isn't, Tipler concluded that we are alone and thus that the search for extraterrestrial intelligence (SETI) is a waste of time. Needless to say, other scientists have found plenty of fault with this conclusion, and SETI researchers are still listening hopefully for a signal from the stars. Nevertheless, it's clear that the Fermi paradox is leading us to some deep, philosophical questions about our own civilization and the possible nature of others.

Would other civilizations really colonize the galaxy?

One obvious question built into the Fermi paradox is whether other civilizations really could or would colonize the galaxy. After all, if other civilizations are content to remain quietly on their home planets, or at least in their home star systems, then there might be lots of civilizations out there without any having come our way. We must therefore consider both the capabilities and the motives that might lead other civilizations to colonize the galaxy.

COLONIZATION MODELS Let's start by assuming that another civilization decided to start sending out spacecraft to colonize other habitable planets. How long would it take this civilization to colonize the entire galaxy?

The answer clearly depends on the civilization's technological capabilities. For example, if it had the technology to build spacecraft that could travel at speeds close to the speed of light, then it could add colonies throughout the galaxy fairly quickly, since trips between nearby stars would take only a few years. Perhaps surprisingly, the conclusion is not that much different if we assume much slower speeds.

Consider a civilization that has nuclear rockets such as the Project Orion or Project Daedalus rockets. As we've discussed, such rockets do not seem that far beyond our own current technological grasp. Recall that such rockets might attain speeds of about 10% of the speed of light (0.1c). Given that a typical distance between star systems in our region of the galaxy is about 5 light-years (the average distance is actually slightly lower than this), a nuclear spacecraft traveling at 10% of the speed of light could journey from one star system to the next in about 50 years. This trip would be possible within a human lifetime and might be practical if the colonizers had found ways to hibernate during the voyage or if they had somewhat longer life spans than we do (either naturally or through medical intervention).

After arriving at a new star system, the colonists establish themselves and begin to increase their population. Once the population has grown sufficiently, these colonists send their own pilgrims into space, adding yet more star systems to the growing civilization. Figure 13.13 shows how such colonization would gradually spread through the galaxy. The process starts at the home star system. The first few colonies are located within just a few light-years. These colonies then lead to other colonies at greater distances, as well as at unexplored locations in between. The growth tends to expand the empire around the edges of the existing empire, much like the growth of coral in the sea. For this reason, this type of colonization model is often called a *coral model*.

The overall result is a gradually expanding region in which all habitable planets are colonized. The colonization rate depends on the speed of spacecraft and the time it takes each colony to start sending spacecraft to other stars. For travel at 10% of the speed of light, assuming that it takes 150 years before each colony's population grows enough to send out more colonists, calculations show that the inhabited region of the galaxy expands outward from the home world at about 1% of the speed of light (see Problem 58 at the end of the chapter). If the home star is near one edge of the galactic disk so that colonizing the entire galaxy means inhabiting star systems 100,000 light-years away, then with these assumptions the civilization could colonize the entire galaxy in about 10 million years. The required time would be a few million years less if the home star was in a more central part of the galaxy.

For an even more conservative estimate, suppose that the colonists have rockets that travel at only 1% of the speed of light and that it takes each new colony 5000 years until it is ready to send out additional colonists. Even in this case, the region occupied by this civilization would grow at a rate of roughly $\frac{1}{1000}$ (0.1%) of the speed of light, and the entire galaxy would be colonized in 100 million years. This is still a very short time compared to the time that has been available for civilizations to arise, further deepening the mystery of why we see no evidence that anyone else has done any colonizing by now.

MOTIVES FOR COLONIZATION In developing a colonization model, we assume that other civilizations would *want* to send out colonists and colonize the galaxy. Is this a reasonable assumption?

We can address this question by considering ourselves as an example, since one of the premises of the Fermi paradox is that we are not special in any way. That is, if *we* would be likely to colonize the galaxy, then we should assume that others would probably act in the same way. Because we have not yet reached the technological level needed to start interstellar colonization, it's impossible to know with certainty whether we would try it if and when we achieved the capability. However, the history of the human species strongly suggests a predisposition to colonize any new territory available to us.

In many ways our entire history has been one of colonization. Modern humans arose in Africa and eventually began expanding around the world. Indeed, the expansion of the human species on Earth probably looked much like the coral model we have described for galactic colonization. Early humans moved outward, gradually encompassing a larger and larger region of our planet. By about 10,000 years ago, our ancestors lived in almost every place on Earth with an environment suitable for human life. Even after humans had effectively colonized the entire planet, attempts at colonization did not stop. For example, Europeans colonized the Americas, with devastating consequences for the people already living there.

Might our inclination to colonize subside in the future? Possibly, but recent history suggests otherwise. Already there are organizations dedicated to colonizing Mars. It doesn't matter if most people would

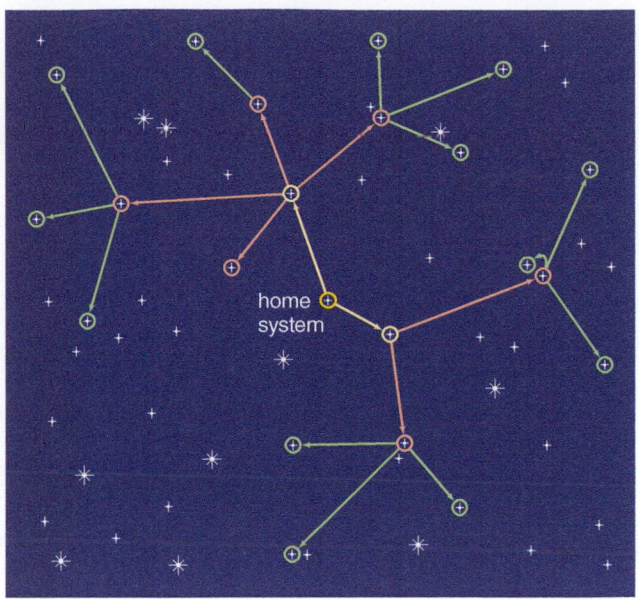

FIGURE 13.13
The coral model of galactic colonization. Colonization begins when the inhabitants of one star system send a few craft to nearby stars. After a time during which the new colonies grow and mature, each new colony sends a few ships with colonists to yet more distant stars, and so on. The colonization "frontier" expands at the edges, much like the way coral grows in the sea.

have no interest in going, because a tiny fraction of the human population would be more than sufficient to start a new colony on another planet. Overall, it seems that if other civilizations are at all like ours, they would take advantage of technological opportunities to colonize the galaxy.

Moreover, even if other civilizations don't have an inherited predisposition toward colonization, many other motives might serve to encourage it. For example, some members of an alien civilization might choose to leave their home to escape war or persecution (as did many Europeans and other immigrants coming to America, for example). Or a society might deliberately send out colonists in an attempt to make its civilization "extinction proof." For example, while our civilization could easily be wiped out in a variety of ways—from nuclear warfare to environmental catastrophe—a civilization spread among star systems would have a much more difficult time self-destructing, because a particular environmental problem affects only one planet, and the long travel times between stars would make it almost impossible to wage war on multiple planets at once. Finally, if a civilization survived long enough for its star to reach an age at which its home planet was no longer in the habitable zone and life would soon be extinguished [Section 10.4], its members might have no choice but to move on in search of a new home.

Think About It Consider these and other possible reasons for a civilization to choose to start colonizing other star systems. In general, do you think it is reasonable to assume that other civilizations would attempt galactic colonization? Defend your opinion.

NON-MOTIVES FOR COLONIZATION Before we leave this topic, it's worth looking at a few ideas that are sometimes suggested as motives for colonization but that break down on closer examination. The most notable example is the alleviation of population pressure.

After growing quite slowly for thousands of years, human population began a dramatic upward swing a few hundred years ago. Figure 13.14 shows human population over the past 12,000 years. If the current trend were to continue, today's human population of more than 7 billion (a threshold reached in 2011) would double to 14 billion by about 2070, double again to 28 billion by 2130, and double again to 56 billion by 2190. Indeed, within just a few more centuries of such growth, we would not fit on Earth even if we all stood elbow to elbow (and of course we would have long since exhausted our ability to provide enough food for ourselves by that point). Clearly, our rapid population growth on Earth must stop soon, or we will face an unparalleled catastrophe (see Cosmic Calculations 6.2). Could colonization be the answer to our population problem?

Not a chance. Let's suppose we wanted to stabilize the population at its current size, but through colonization rather than through changes in the population growth rate. Currently, we add about 80 million people to Earth each year. Therefore, to keep our population stable, we'd need to move 80 million people per year off the planet. According to NASA, the average cost of a Space Shuttle launch during the 30-year program was about $450 million, and it took only six or seven people into low-Earth orbit—the Shuttle couldn't reach other planets, let alone other stars. If we assume the same costs in the future, with the added capability of taking passengers to other worlds, the cost of sending 80 million people into

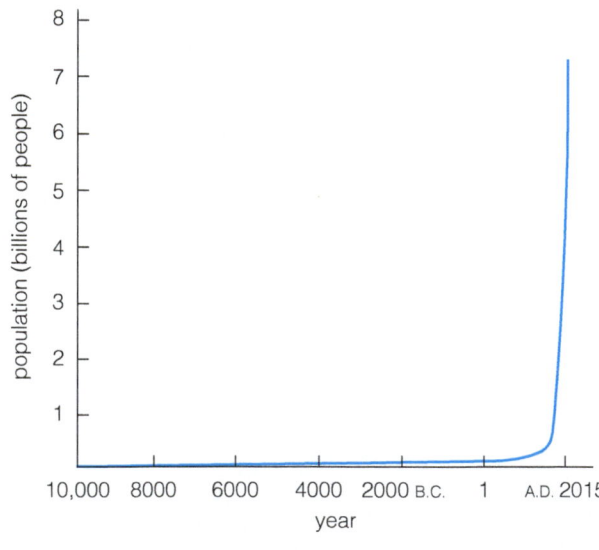

FIGURE 13.14
This graph shows human population over the past 12,000 years. Note the tremendous population growth that has occurred in just the past few centuries.

space each year would be more than $3 quadrillion, or almost 100 times the gross national product of the United States. There are not too many things that we can say are outright impossible, but solving population problems through colonization is one of them.

Another less-than-viable colonization motive is conquest. As we've discussed, if we find other inhabited planets with intelligent beings, technologically speaking we are likely to be at least either many thousands of years ahead of them or many thousands of years behind them. If we are thousands of years ahead, conquering them would be like the United States conquering cave dwellers—there hardly seems to be much to gain. If we are thousands of years behind them, we would not be likely to prevail.

Some people suggest that we would colonize not so much through a motivation for direct conquest but rather as the consequence of our species's general tendency toward aggression. This is harder to rule out as a viable motive, but many science fiction writers (including Gene Roddenberry, creator of the *Star Trek* series) have pointed out a potential flaw in this idea: If we continue to be as aggressive and warlike as we have been in the past, we are unlikely to survive long into the future because our capacity for destruction has risen along with our level of technology. As a result, these writers have argued that our surviving long enough to achieve the technology for interstellar travel would necessarily mean that we found ways to overcome our aggressive and warlike tendencies. In that case, colonization would occur only because of curiosity and a desire to explore, not because of any desire for empire building.

What are possible solutions to the Fermi paradox?

We have now seen why the Fermi paradox is real; that is, it really does seem that if civilizations are at all common, then the galaxy should have been colonized long ago. So why don't we see any evidence of a galactic civilization? There are many possible explanations, but broadly speaking, we can group them into three major categories:

1. *We are alone.* There is no galactic civilization because civilizations are extremely rare—so rare that ours is the first to have arisen on the galactic scene.
2. *Civilizations are common, but no one has completely colonized the galaxy.* There are at least three possible reasons why this might be the case:
    i. *Technological difficulties.* Interstellar travel is much harder or vastly more expensive than we have guessed, so civilizations are unable to venture far from their home worlds.
    ii. *Sociological considerations.* Our desire to explore is unusual, and other societies choose not to leave their home star systems. Or, for one reason or another, colonizers might run out of steam before they've conquered large tracts of galactic real estate.
    iii. *Self-destruction.* Many civilizations have arisen, but they have all destroyed themselves before achieving the ability to colonize the stars.
3. *There is a galactic civilization,* but it has deliberately avoided revealing its existence to us.

Let's examine each of these categories in more depth.

WE ARE ALONE The idea that we are alone is certainly the simplest solution to the Fermi paradox. However, many people object to this solution on philosophical grounds, because it would suggest that our circumstances are very special compared to those that have arisen around any other of the more than 100 billion star systems in the galaxy. If this were true, it would go against almost everything else we have learned since the time of Copernicus. That is, while our ancestors might have imagined our planet to be the center of the universe, more recent astronomical discoveries all seem to suggest that we are not particularly special. Earth is merely a planet orbiting the Sun, our Sun is a rather ordinary star in the Milky Way Galaxy, and our galaxy is much like many other galaxies in the universe.

Of course, while the idea that we are alone might be philosophically unappealing, we cannot rule it out on scientific grounds. Indeed, proponents of the rare Earth hypothesis [Section 11.4] would not be surprised to learn that we are alone. Recall that, according to this hypothesis, the combination of circumstances that allowed intelligent life to arise on Earth is so rare that we are likely to be the only civilization in the galaxy. The ideas that underlie the rare Earth hypothesis are controversial, but even if they prove incorrect, there may be other reasons we are alone. For example, perhaps some undiscovered law of nature has rendered civilizations impossible until quite recently. In that case, it would not be so strange to imagine that we are the first civilization, even if many others may follow.

CIVILIZATIONS BUT NO COLONIZATION—TECHNOLOGICAL DIFFICULTIES The second category of explanations offers the possibility that civilizations are common but colonization is not. Three possible reasons why this might be the case were listed: technological difficulties, sociological considerations, and self-destruction. Let's consider each of these, starting with technological difficulties.

We've seen that interstellar travel seems difficult but not impossible. But could we be underestimating the challenge, and could it actually be so difficult as to be essentially impossible? The energy cost of interstellar travel is sometimes suggested as an impasse. Recall that a large interstellar starship traveling at only about 10% of the speed of light would require energy comparable to what the world currently uses in a hundred years. Clearly, this requirement is prohibitive for us today, and the costs of interstellar travel might be so high compared to the costs of building habitats in our own solar system that migration to other stars might always seem untenable. However, it's possible that advanced civilizations could overcome this problem. The ability to produce power through nuclear fusion, for example, might allow them to generate the needed energy with relative ease. In an extreme case, a civilization capable of building a Dyson sphere [Section 12.3] would have access to all the energy produced by its star. It seems unlikely that the energy requirement alone could preclude all interstellar colonization.

A related possibility is that some other, unknown biological or physical barrier to interstellar travel exists. For example, we have assumed that in the future we'll be able to find a way to keep crews alive for the decades required to go from one star system to the next, but perhaps this is actually much more difficult than we have imagined. Possibly, some type of unknown danger lurking in space prevents intelligent beings from traveling among the stars. Science fiction writers have certainly considered such possibilities—for example, a mysterious effect that causes interstellar travelers to go insane—but they seem far-fetched in light of what we presently know about interstellar space.

It's worth noting that neither energy considerations nor lurking dangers would be enough to stop a civilization from sending out self-replicating Von Neumann machines. But there might be other reasons no such machines are out there. For example, such machines would tend to grow in number at a rapid (exponential) rate and could in principle use up all the resources in the galaxy in just a few million years. Carl Sagan (in a paper co-written with William Newman) addressed this idea by suggesting that any civilization smart enough to build such machines would also be smart enough to recognize their dangers and therefore would not construct them in the first place.

CIVILIZATIONS BUT NO COLONIZATION—SOCIOLOGICAL CONSIDERATIONS

Let's next turn to sociological considerations. As we've discussed, it seems quite likely that, given the technological and economic opportunity, we would choose to engage in interstellar travel and galactic colonization. But could it be that we are somehow exceptional in having this desire and that other civilizations are perfectly content to stay at home? Like the "we are alone" idea, this idea suggests that we are somehow special rather than typical of intelligent beings, and therefore it goes against our usual assumption that there's nothing special about our circumstances. After all, we are products of the competitive forces that drive evolution by natural selection, and these forces ought to be similar on any world with life. Moreover, our colonization models show that it would take only *one* other civilization to colonize the entire galaxy in a few million years, so the lack of interest in space travel would have to apply to *every other* civilization that has ever arisen. If civilizations are common—say, if there are the 100,000 civilizations expected if 1 in a million stars has one—it's difficult to believe that not one other civilization has had interests similar to ours.

On the other hand, even with fast rockets, colonization of the entire galaxy would still take millions of years. Perhaps no civilization can maintain enthusiasm for an effort that lasts this long. The individual colonies, separated by many light-years, might evolve along different lines (either biological or cultural), shattering the unity of the empire and bringing further expansion to a halt. While these possibilities are not unreasonable, it remains true that only one civilization needs to persevere with its colonization efforts in order to bring the entire galaxy under its wing.

One other sociological consideration suggests that advanced societies might start out like us but then "engineer" themselves in such a way as to shut down their drive to colonize. On our planet, and presumably on others, the development of rocketry occurred at roughly the same time as the invention of nuclear weapons, chemical weapons, and other methods of mass destruction. Societies that remain aggressive are in constant danger of self-destruction. Therefore, civilizations might be motivated to find ways to reduce or channel their aggressive tendencies, perhaps through some type of genetic engineering. The oldest alien cultures, according to this line of reasoning, would have managed to rid themselves of dangerous aggression. It's conceivable that in the process they would have also chosen to focus on improving life on their home planet rather than on moving out into space.

CIVILIZATIONS BUT NO COLONIZATION—SELF-DESTRUCTION A much more sobering possibility for why the galaxy might remain uncolonized even if many civilizations have arisen—one that assumes we are completely typical of intelligent beings—is that societies inevitably self-destruct before attaining the capability for interstellar travel. While this idea is horribly tragic, it is not far-fetched. Nuclear weapons provide clear proof that the technology needed for interstellar travel can also be used to destructive ends. Similarly, any society that learns to tap energy resources would almost certainly use the most accessible energy first, which on any Earth-like planet is likely to be fossil fuels. Like us, other civilizations must face the dangers posed by global warming and other environmental problems. Population growth probably also poses similar problems for all civilizations. Rapid population growth is a natural consequence of biological reproduction for a species that is no longer subject to the whims of predators or childhood diseases, which tend to hold population growth in check. From this perspective, a society can survive long enough to achieve interstellar travel only if it successfully navigates what amounts to a very difficult obstacle course—one in which each obstacle could mean the end of its civilization. Could it be that it simply can't be done?

Think About It What odds would you give for humanity's surviving long enough to achieve interstellar travel? Defend your choice.

THERE *IS* A GALACTIC CIVILIZATION The third category of explanations for the Fermi paradox in essence suggests that there is no paradox at all: The galactic civilization is out there, but we do not yet recognize it. Indeed, UFO buffs might claim that scientists are blind to the obvious proof of this suggestion. While we can't rule out the possibility, no evidence of alien visitation has yet withstood scientific scrutiny. Nevertheless, there are many ways by which a galactic civilization could avoid our detection.

One idea simply assumes that a galactic civilization would have no particular interest in us. After all, to a society that is millions or billions of years ahead of us, we might seem too simple to warrant attention. Of course, if civilizations are communicating or traveling among the stars, we might be able to discover them. Signaling is precisely what SETI experiments look for. The fact that we have not yet received a clearly extraterrestrial broadcast may simply be a consequence of not having yet looked at a sufficient number of stars.

Another possibility is that civilizations are aware of our presence but have deliberately chosen to keep us in the dark about their existence. This idea is sometimes called the **zoo hypothesis,** although it might better be called the "wildlife refuge" hypothesis. Just as we set aside nature reserves that are supposed to be left alone as places where wildlife can thrive without our intervention, a galactic civilization might declare planets like ours off-limits to exploration. (*Star Trek* fans might think of this as the "Prime Directive" [to avoid interfering with alien cultures] solution to the Fermi paradox.) One objection to the zoo hypothesis is that even if civilizations wanted to hide from us, we would still be able to intercept their communications among themselves. On the other hand, as we've already noted, SETI searches for signals have so far carefully investigated only a small amount of cosmic real estate and could easily have missed such communications. Alternatively, the communications might involve a technology that we have not yet developed and that therefore is undetectable by us at present.

A closely related idea suggests that a sophisticated galactic civilization might reveal itself to new societies only after they reach a certain level of technology. Perhaps the extraterrestrials place monitoring devices near star systems that show promise of emerging intelligence and patiently wait until these devices record the presence of civilization. This idea is sometimes called the **sentinel hypothesis,** after a science fiction story by Arthur C. Clarke titled "The Sentinel." The story became the basis of the book and movie *2001—A Space Odyssey,* in which a monolith buried on the Moon signals our presence when we finally dig it up (see Movie Madness in Chapter 7). Carl Sagan used a similar idea for the book and movie *Contact* (see Movie Madness in Chapter 12), in which a signaling station around the star Vega amplifies and beams back our own television broadcasts to us, leading us to our first glimpse of a galactic civilization.

What are the implications of the Fermi paradox for human civilization?

The Fermi paradox may have its origins in a simple question—Where is everybody?—but we have seen that finding an answer is much more complex than asking the question. In fact, if we consider our possible answers in more depth, we find that each leads to astonishing implications for our own species.

Consider the first solution—that we are alone. If this is true, then our civilization is a remarkable achievement. It implies that through all of cosmic evolution, among countless star systems, we are the first piece of the universe ever to know that the rest of the universe exists. Through us, the universe has attained self-awareness. Some philosophers and many religions argue that the ultimate purpose of life is to become truly self-aware. If so, and if we are alone, then the destruction of our civilization and the loss of our scientific knowledge would represent an inglorious end to something that took the universe some 14 billion years to achieve. From this point of view, humanity becomes all the more precious, and the collapse of our civilization would be all the more tragic. Knowing this to be the case might help us learn to put petty bickering and wars behind us so that we might preserve all that is great about our species.

The second category of solutions has much more terrifying implications. If thousands of civilizations before us have all failed to achieve interstellar travel on a large scale, what hope do we have? Unless we somehow think differently than all previous civilizations, we will never go far in space. Given that we have always explored when the opportunity arose, this solution almost inevitably leads to the conclusion that failure will come about because we destroy ourselves. We can only hope that this answer is wrong.

The third solution is perhaps the most intriguing. It says that we are newcomers on the scene of a galactic civilization that has existed for millions or billions of years before us. Perhaps this civilization is deliberately leaving us alone for the time being and will someday decide the time is right to invite us to join it. If so, our entire species might be on the verge of beginning a journey every bit as incredible as that of a baby emerging from the womb and coming into the world.

You can probably now see why the Fermi paradox involves far more than a simple question. No matter what the answer turns out to be, learning it is sure to mark a turning point in the brief history of our species. Moreover, this turning point is likely to be reached within the next few decades or centuries. We already have the ability to destroy our own civilization. If we do so, then our fate is sealed. But if we survive long enough to develop technology that can take us to the stars, the possibilities seem almost limitless.

Imagine for a moment the grand view, a gaze across the centuries and millennia from this moment forward. Picture our descendants living among the stars, having created or joined a great galactic civilization. They will have the privilege of experiencing ideas, worlds, and discoveries far beyond our wildest imagination. Perhaps, in their history lessons, they will learn of our generation—the generation that history placed at the turning point and that managed to steer its way past the dangers of self-destruction and onto the path to the stars.

❁ THE PROCESS OF SCIENCE IN ACTION

13.4 Einstein's Special Theory of Relativity

In this chapter, we have seen that Einstein's special theory of relativity has important implications for possibilities of interstellar travel. But that is not why Einstein came up with the theory. Rather, as with the creation of other scientific theories, Einstein developed the theory of relativity to explain one of the outstanding scientific mysteries of the time, which is why we will take it as this chapter's case study in the process of science in action.

The mystery concerned the speed and nature of light. In 1873, Scottish mathematician and physicist James Clerk Maxwell (1831–1879) published a paper in which he showed that light is an electromagnetic wave [Section 3.3]. His paper included a set of equations—now known as *Maxwell's equations*—that describe the nature of electromagnetic waves. These equations form the basis of our modern understanding of electricity and magnetism, and in essence you are confirming their validity every time you flip a light switch, use your cell phone, or watch a television show. However, the idea that light is a wave left a fundamental question: All other types of waves—for example, sound waves, water waves, or waves on a violin string—are carried by some type of medium; what medium carries light waves through "empty" space? Maxwell and other physicists presumed that space must be filled with some medium of unknown composition that could carry the light waves—an idea that had actually been around for some time already—and this medium was known as the *ether*.

If the ether really existed, then Earth's motion around the Sun would make us move in different directions *relative* to the ether at different times of year. In 1887, A. A. Michelson (1852–1931) and E. W. Morley (1838–1923) conducted an elegant experiment—later known as the *Michelson–Morley experiment*—designed to measure Earth's motion through the ether. The experiment relied on the idea that the speed of light would be slightly faster when light was moving in the same direction as the ether (relative to Earth), slightly slower when it was moving in the opposite direction, and at speeds in between for other directions. However, their experiment failed to measure any directional difference in the speed of light. In hindsight, the obvious implication was that the ether does not exist. At the time, however, this idea seemed so preposterous that Michelson and Morley went to great lengths to explain how nature might "hide" the ether's existence from human experimenters.

Einstein came up with his theory because he took the results of the Michelson–Morley experiment at face value. Instead of assuming that nature was hiding something from us, he assumed that the speed of light showed no directional variation because none exists; that is, Einstein assumed that the speed of light (through space) is a physical constant, one that will always be measured the same no matter what the motion of the light source or the observer. In doing so, he not only explained the perplexing results of the Michelson–Morley experiment but also cleared up some mysteries that had been associated with mathematical implications of Maxwell's equations. In other words, Einstein's theory did not come about by magic or just because he was a smart guy—it came about because there were known problems that needed to be solved, and Einstein was the first to come up with their solution.

Einstein's special theory of relativity is often portrayed as being difficult, but its basic ideas are actually quite easy to understand—though admittedly mind-boggling in their consequences. Because they are so important to our modern understanding of the nature of the universe, let's take a look at the basic ideas and their astonishing implications.

What is "relative" about relativity?

Imagine a supersonic airplane that flies at a speed of 1670 km/hr from Nairobi, Kenya, to Quito, Ecuador. How fast is the plane going? At first, this question sounds trivial—we have just said that the plane is going 1670 km/hr.

But wait. Nairobi and Quito are both nearly on Earth's equator, and the equatorial speed of Earth's rotation is the same 1670 km/hr at which the plane is flying. Moreover, the east-to-west flight from Nairobi to Quito is opposite the direction of Earth's rotation (Figure 13.15). Therefore, if you lived on the Moon, the plane would appear to stay put *while Earth rotated beneath it.* When the flight began, you would see the plane lift off the ground in Nairobi. The plane would then remain stationary while Earth's rotation carried Nairobi away from it and Quito toward it. When Quito finally reached the plane's position, the plane would drop back down to the ground.

We have two alternative viewpoints about the plane's flight. People on Earth say that the plane is traveling westward across the surface of Earth. Observers in space say that the plane is stationary while Earth rotates eastward beneath it. Both viewpoints are equally valid. In fact, there are many other equally valid viewpoints about the plane's flight. Observers looking at the solar system as a whole would see the plane moving at a speed of more than 100,000 km/hr—Earth's speed in its orbit around the Sun. Observers living in another galaxy would see the plane moving at about 800,000 km/hr with the rotation of the Milky Way Galaxy. The only thing all these observers would agree on is that the plane is traveling at 1670 km/hr *relative to* the surface of Earth.

This example shows that questions like "Who is really moving?" and "How fast are you going?" have no absolute answers. Einstein's special theory of relativity gets its name from telling us that measurements of time and space, as well as measurements of motion, make sense only when we describe whom or what they are being measured relative to.

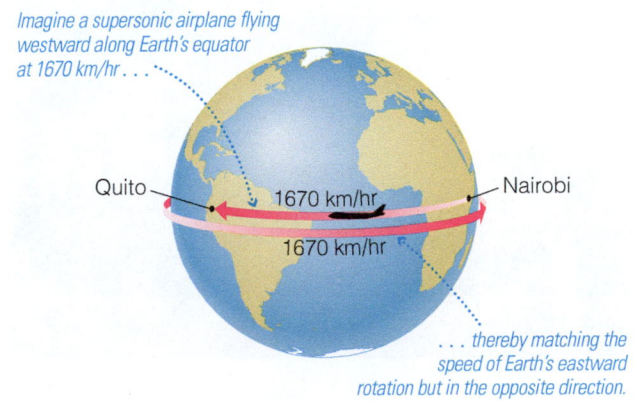

Imagine a supersonic airplane flying westward along Earth's equator at 1670 km/hr . . .

Quito · 1670 km/hr · Nairobi

1670 km/hr

. . . thereby matching the speed of Earth's eastward rotation but in the opposite direction.

FIGURE 13.15

A plane flying at 1670 km/hr from Nairobi to Quito (westward) travels precisely opposite Earth's eastward rotation. Viewed from afar, the plane remains stationary while Earth rotates underneath it.

Think About It Suppose you are running on a treadmill and the readout says you are going 10 kilometers per hour. What is the 10 kilometers per hour measured relative to? How fast are you going relative to the ground? How fast would an observer on the Moon see you going? Describe a few other possible viewpoints on your speed.

THE ABSOLUTES OF RELATIVITY The theory of relativity tells us that motion is always relative, but it does *not* say that *everything* is relative. In fact, the theory claims that two things in the universe are absolute:

1. The laws of nature are the same for everyone.
2. The speed of light is the same for everyone.

The first absolute, that the laws of nature are the same for everyone, is a more general version of the idea that all viewpoints on motion are equally valid. If they weren't, different observers would disagree about the laws of physics. The second absolute, that the speed of light is the same for everyone, is much more surprising. Ordinarily, we expect speeds to add and subtract. If you watch someone throw a ball forward from a moving car, you see the ball traveling at the speed at which it is thrown plus the speed of the car. But if a person shines a light beam from a moving car, you see it moving at precisely the speed of light (about 300,000 kilometers per second), no matter how fast the car is going. This strange fact explains why the Michelson–Morley experiment found no differences in the speed of light, and this has been experimentally verified countless times.

THE SPEED OF LIGHT The cosmic speed limit follows directly from this fact about the speed of light. To see why, imagine that you have just built the most incredible rocket possible, and you are taking it on a test ride. You push the acceleration button and just keep going faster and faster and faster. With enough fuel, you might expect that you'd eventually be moving faster than the speed of light. But can you?

Before we answer this question, the fact that all motion is relative forces us to answer another question: What is your speed being measured relative to? Let's begin with *your* point of view. Imagine that you turn on your rocket's headlights. Because the speed of light is the same for everyone, you must see the headlight beams traveling at the speed of light—which means they are racing away from your rocket at a speed you'll measure to be 300,000 km/s. The fact that you'll see your headlight beams racing away is true no matter how long you have been firing your rocket engines. This shouldn't be too surprising, as it's just another way of saying that you can't catch up with your own light.

Now, remember that everyone always measures the same speed of light. This means that people back on Earth—or anyplace else—will also say that your headlight beams are moving through space at 300,000 kilometers per second. In other words, according to anyone watching you from any place in the universe, your headlight beams are traveling at $c = 300,000$ km/s, and they're moving out ahead of you (Figure 13.16). Clearly, this implies that you must be traveling *slower* than the headlight beams, which means slower than the speed of light.

In case you are still not convinced, let's turn the situation around. Imagine that, as you race by some planet, a person on the planet turns on a light beam. Because the speed of light is absolute, you will see the light beam race past you at $c = 300,000$ km/s. The person on the planet

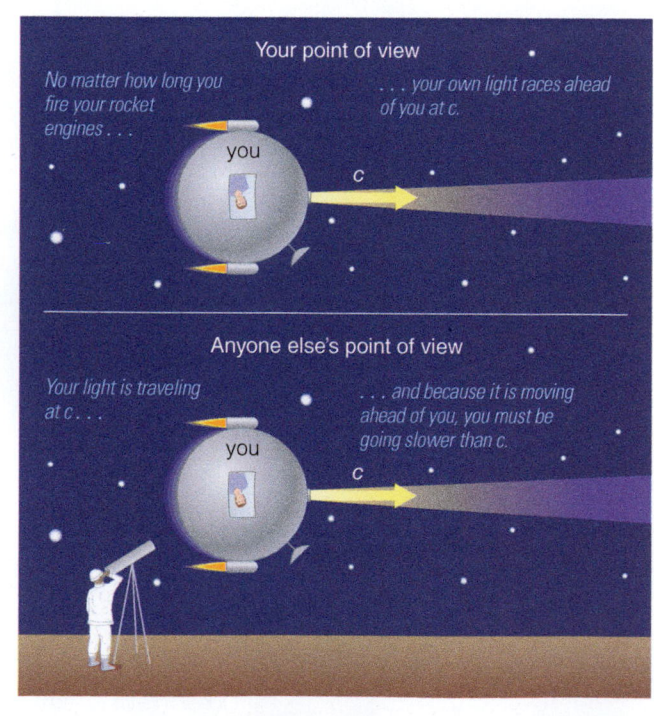

FIGURE 13.16
The fact that everyone always measures the same speed of light means that you cannot keep up with your own light (top), and therefore no matter what you do, other people must always conclude that you are traveling slower than light. In other words, there is no way for you ever to reach or exceed the speed of light.

will also see the light traveling at $c = 300,000$ km/s and will see the light outrace you. Again, everyone will agree that you are traveling slower than the speed of light.

The same argument applies to any moving object, and it is true with or without headlights. All light travels at the speed of light, including the light that reflects off an object (allowing us to see it) and the infrared light that even cool objects like people emit. As long as the speed of light is absolute, no material object can ever keep up with the light it emits or reflects, which means no material object can reach or exceed the speed of light. Building a spaceship to travel faster than the speed of light is not a mere technological challenge—it simply cannot be done.

THE RELATIVITY OF TIME AND SPACE Like the cosmic speed limit, the other strange consequences of relativity also follow from the absoluteness of the speed of light. We will not go into details here, but you can understand the general ideas by thinking about the fact that a speed is always equal to a distance divided by a time. For example, a speed of 60 miles per hour means that you travel a distance of 60 miles in a time of 1 hour, and light's speed of 300,000 km/s means that light travels a distance of 300,000 kilometers in a time of 1 second. In our ordinary slow-moving lives, we think of times and distances as absolutes, while speed is relative—as is the case with the speed of the airplane in our Nairobi–Quito example. But Einstein's theory tells us that when it comes to light, speed is absolute—everyone always measures the same speed of light. The consequence is that time and distance must become relative.

We have already seen how time and distance are affected. Time runs slower for high-speed travelers (the phenomenon of *time dilation* that we discussed earlier in the chapter), which is why relativistic starships could allow passengers to make long trips if they traveled fast enough. Distances are also shrunk as measured by the high-speed travelers, which is why high-speed travelers to Vega would see a shorter distance than the distance we measure from Earth. Mass is also affected, so we would measure high-speed objects to have higher mass than they do when they are stationary relative to us.* Another direct consequence of these ideas—and one that can be derived with nothing more than high school algebra (though we won't do it here)—is Einstein's famous formula, $E = mc^2$.

What evidence supports Einstein's theory?

Although we haven't gone into details here, we have stated that all the amazing consequences of relativity, including time dilation and $E = mc^2$, follow logically from the absoluteness of the speed of light. However, remember that logic alone is not good enough in science; conclusions must always remain tentative until they pass observational or experimental tests. Does relativity meet the test?

THE ABSOLUTENESS OF THE SPEED OF LIGHT The first thing we might wish to test is the surprising premise of relativity: the absoluteness of the

---

*You can see why mass must be affected if you think about Newton's second law of motion, which states that force = mass × acceleration (see Figure 2.13). The units of acceleration are a distance divided by a time squared (such as m/s²), so the relativity of time and distance means that even if the force (e.g., from the rocket engines) stays constant, a rocket's acceleration must decrease with increasing speed. The only way to account for this decrease when the force stays the same is to realize that the mass of the rocket must be increasing.

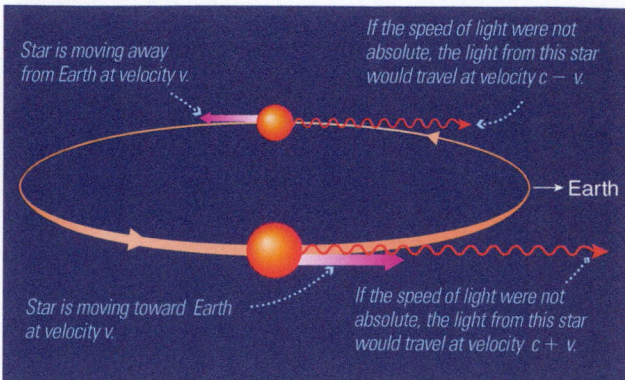

FIGURE 13.17

If the speed of light were *not* absolute, the speed at which light from a star in a binary system came toward Earth would depend on the star's position and velocity toward us in the binary orbit. Therefore, we would not see the star as a distinct point of light. Because we *do* see the star as a distinct point, the speed of light must be absolute.

speed of light. In principle, we can test this premise by measuring the speed of light coming from many different objects and going in many different directions and verifying that the speed is always the same. As we've already discussed, the Michelson–Morley experiment was in essence one such test, and it showed that the speed of light is indeed absolute.

We have verified this fact many other times. For example, some distant galaxies are moving away from us at speeds close to the speed of light, yet their light still arrives here traveling at the speed of light. Perhaps even more convincingly, if the speed of light were not always the same, we could not see distinct stars in binary star systems. Consider the star moving toward us in Figure 13.17. If the speed of light depended on the star's motion, its light would be coming toward us somewhat faster than the "normal" speed of light. Half an orbit later, when it was moving away from us, its light would travel to us at less than the "normal" speed of light. Therefore, the light from the "fast" side of the orbit would tend to catch up with the light emitted earlier from the "slow" side, reaching us at the same time and smearing the star's image so that we'd see it in different positions all at once, instead of seeing it as a distinct star. In other words, the simple fact that we see distinct stars in binary star systems proves that the speed of light does not depend on the stars' motions; light always travels at the same speed. Moreover, the speed of light from the two orbiting stars in binary systems, as well as from opposite sides of the rotating Sun, has been measured, again confirming the same result: The speed of light is always the same.

EXPERIMENTAL TESTS OF RELATIVITY Although *we* cannot yet travel at speeds at which the effects of relativity should be obvious, tiny subatomic particles can reach such speeds, thereby allowing us to test the precise predictions of the formulas of special relativity. Let's first consider one way of testing time dilation. In machines called *particle accelerators* (such as the Large Hadron Collider), physicists accelerate subatomic particles to speeds near the speed of light and study what happens when the particles collide. The colliding particles have a great deal of kinetic energy, and the collisions convert some of this kinetic energy into mass-energy that emerges as a shower of newly produced particles. Many of these particles have very short lifetimes, at the end of which they decay (change) into other particles. For example, a particle called the π^+ ("pi plus") meson has a lifetime of about 18 nanoseconds (billionths of a second) when produced at rest. But π^+ mesons produced at speeds close to the speed of light in particle accelerators last much longer than 18 nanoseconds—and by precisely the amount predicted by the time dilation formula.

The same experiments also confirm the mass increase predicted by relativity. The amount of energy released when high-speed particles collide depends on the particle masses and speeds. Just as relativity predicts, these masses are greater at high speed than they are at low speed—again by just the amount predicted by Einstein's formulas.

Particle accelerators even offer experimental evidence that nothing can reach the speed of light. It is relatively easy to get particles traveling at 99% of the speed of light in particle accelerators. However, no matter how much more energy is put into the accelerators, the particle speeds get only fractionally closer to the speed of light. Some particles have been accelerated to speeds within 0.00001% of the speed of light, but none have ever reached the speed of light.

Although the effects of relativity are obvious only at very high speeds, modern techniques of measuring time are so precise that effects can be measured even at ordinary speeds. For example, a 1975 experiment compared the amount of time that passed on an airplane flying in circles to the time that passed on the ground. Over 15 hours, the airborne clocks lost a bit under 6 nanoseconds relative to the ground clocks, matching the result expected from relativity. More recent experiments have confirmed the very tiny time dilation that relativity predicts even at walking speeds.

Nuclear energy also provides a test of relativity. Remember that $E = mc^2$ is a direct consequence of Einstein's theory. Every time you see film of an atomic bomb, or use electrical power from a nuclear power plant, or feel the energy of sunlight that the Sun generated through nuclear fusion, you are witnessing direct experimental evidence of relativity.

BEYOND SPECIAL RELATIVITY All in all, special relativity is one of the best-tested theories in physics, and it has passed every experimental test to date with flying colors. That is why it merits status as a true, scientific *theory*. Of course, like any scientific theory, it is always open to further testing and refinement. Indeed, Einstein himself realized that special relativity was missing an important ingredient: It dealt successfully with motion through empty space, but not with motion affected by gravity. This missing ingredient provided much of the motivation that led Einstein to press on and publish his general theory of relativity about a decade later.

The general theory expanded on the special theory by including gravity, and in the process Einstein found that it also turned out to be an improvement on Newton's theory of gravity [Section 2.4]. It is likely that the general theory of relativity will someday need its own refinements to make it compatible with the theory of quantum mechanics. If so, it is possible that some of the ideas of special relativity may also change. But the evidence that supports the theory as we know it today is real and cannot be made to disappear. If anyone ever comes up with a better theory, the new theory will still have to explain the many experimental results that support relativity so well.

The Big Picture

PUTTING CHAPTER 13 IN PERSPECTIVE

In this chapter, we have explored the possibilities for and challenges of interstellar travel, and the surprising Fermi paradox that arises when we think about what other civilizations might already have achieved. As you continue in your studies, keep in mind the following "big picture" ideas:

- Interstellar travel may be a staple of science fiction, but it remains well beyond our current capabilities. Nevertheless, it is possible that the challenges can be surmounted and that other civilizations might already have overcome them.

- The idea that other civilizations might already have achieved interstellar travel leads to the Fermi paradox—the question of why we see no evidence of galactic colonization. This simple question is not easily dismissed, and no matter what its solution turns out to be, it has profound implications for human civilization.

- One key lesson of the Fermi paradox is that we live at a unique moment in the history of the human species. We have the ability to destroy our civilization and perhaps even to drive our species to extinction. But if we survive, our descendants might have a boundless future. From this perspective, no generation has ever borne such great responsibility.

Summary of Key Concepts

13.1 The Challenge of Interstellar Travel

Why is interstellar travel so difficult?

Convenient interstellar travel remains well beyond our technological capabilities. Current spacecraft would take more than 100,000 years just to traverse the distance to the nearest stars. The energy requirements for sending people on interstellar trips are enormous, far greater than all current world energy usage.

Could we travel to the stars with existing rockets?

Nearly all rockets built to date are powered by chemical rocket engines. The **rocket equation** shows that these types of engines could not possibly get us to speeds of even 1% of the speed of light, making them impractical for interstellar travel.

13.2 Spacecraft for Interstellar Travel

How might we build interstellar spacecraft with "conventional" technology?

Technologies such as nuclear rocket engines, **solar sails,** and **ion engines** could in principle allow us to build starships that could travel at speeds up to about 10% of the speed of light—fast enough to reach nearby stars in less than a century.

How might we build spacecraft that could approach the speed of light?

Several technologies that are well beyond us at present but are allowed by the laws of physics could allow starships to reach speeds close to the speed of light. These include **matter–antimatter** engines and **interstellar ramjets** that scoop up fuel as they go.

Are there ways around the light-speed limitation?

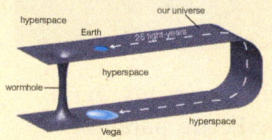

No known physical laws prevent hyperspace, **wormholes,** or warp drive from offering "loopholes" that could allow us to get from one place to another in less time than we could by traveling through ordinary space. However, we do not yet know if any of these are really possible.

13.3 The Fermi Paradox

Where is everybody?

This seemingly simple question, known as the **Fermi paradox,** comes about because our general assumption that Earth is not unique leads us to expect that many other civilizations should

by now have arisen and had the opportunity to colonize the Milky Way Galaxy. Yet we see no evidence of a galactic civilization.

Would other civilizations really colonize the galaxy?

Based on the idea that other beings would have evolved in response to evolutionary pressures similar to those that led to human evolution on Earth, we expect that other civilizations would have the same inherent drive to colonize that we seem to possess. If so, a civilization should be able to colonize the galaxy in a time that is short compared to the age of the universe, even with technology not much beyond our own.

What are possible solutions to the Fermi paradox?

There are three general categories of solution to the Fermi paradox: (1) We are alone. (2) Civilizations are common, but no one has colonized the galaxy. (3) There is a galactic civilization, but it has deliberately avoided revealing its existence to us.

What are the implications of the Fermi paradox for human civilization?

The first solution implies that we are the only piece of the universe that has ever attained self-awareness. The second suggests that civilizations may either change or destroy themselves before attaining the ability to travel to the stars. The third implies that we might someday meet up with a galactic civilization that predates us by millions or billions of years.

✻ THE PROCESS OF SCIENCE IN ACTION

13.4 Einstein's Special Theory of Relativity

What is "relative" about relativity?

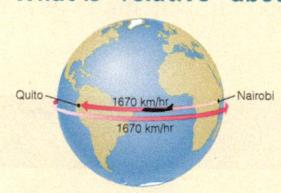

Special relativity tells us that motion is relative, but everyone always agrees on the speed of light. From this it follows that different observers can measure time, distance, and mass differently, and that no material object can reach or exceed the speed of light.

What evidence supports Einstein's theory?

Experiments with light confirm that its speed is always the same. Experiments with subatomic particles in particle accelerators confirm the predictions of **time dilation** and mass increase at speeds close to the speed of light, and time dilation has been verified at relatively low speeds in aircraft and spacecraft. Nuclear power plants and nuclear bombs release energy in accordance with the formula $E = mc^2$, which is also a prediction of special relativity.

Exercises and Problems

REVIEW QUESTIONS

Short-Answer Questions Based on the Reading

1. Briefly describe the journey of *Pioneer 10*. How does it illustrate the challenge of interstellar travel?

2. How does the speed of light affect the possibility of interstellar travel?

3. About how much energy would it take to send enough people on a trip to start a colony around another star? Explain how you arrive at the answer.

4. How do rockets work? Briefly describe the history of rocketry.

5. What is the *rocket equation* used for? Based on the rocket equation and the *mass ratio*, briefly explain why chemical rockets are inadequate for sending people or large robotic probes to the stars.

6. Describe the proposed fusion-powered starships of Project Orion and Project Daedalus. How quickly could such ships reach the stars?

7. Discuss a few ways of reaching the stars (other than nuclear rockets) that are, at least in principle, within our current technological reach.

8. How would *time dilation* affect space travel at speeds close to the speed of light? Discuss possible ways of achieving such speeds, including *matter–antimatter engines* and *interstellar ramjets*.

9. Briefly discuss how Einstein's general theory of relativity might allow "shortcuts" by which we could reach distant stars in shorter times than we would expect from their measured distances. Do we know whether these shortcuts are really possible?

10. What is the *Fermi paradox*? What two seemingly contradictory ideas underlie the paradox?

11. Why does it seem that other civilizations, if they exist, should be significantly older than ours? Explain clearly.

12. What are *Von Neumann machines*? How do they affect the Fermi paradox?

13. Describe the *coral model* of galactic colonization. Why do we conclude that civilizations could have colonized the galaxy by now even with technology not much more advanced than ours?

14. Briefly discuss possible motives for galactic colonization, as well as "motives" that don't hold up.

15. Summarize the three general categories of possible solutions to the Fermi paradox, and discuss each category in some detail.

16. Briefly discuss the profound implications of the Fermi paradox and how the answer to the paradox affects our civilization.

17. What known problems were solved when Einstein discovered the special theory of relativity?

18. Explain how the idea of an absolute speed of light leads automatically to the conclusion that no one can travel faster than light.

19. Besides the idea that you cannot reach the speed of light, what other consequences follow from the absoluteness of the speed of light?

20. Describe at least three tests that have confirmed the validity of the special theory of relativity.

TEST YOUR UNDERSTANDING

Science or Nonscience?

Each of the following describes some futuristic scenario that, while perhaps entertaining, may or may not be plausible. In each case, decide whether the scenario is plausible according to our present understanding of science or whether it is unlikely to be possible. Explain clearly; because not all of these have definitive answers, your explanation is more important than your chosen answer.

21. A brilliant teenager working in her garage discovers a way to build a rocket that burns coal as its fuel and can travel at half the speed of light.

22. Using beamed energy propulsion from a laser powered by energy produced at a windmill farm in the California desert, NASA engineers are able to send a solar sailing ship on a journey to Alpha Centauri that will take only 50 years.

23. Human colonization of the moons of Saturn occurs using spaceships powered by dropping nuclear bombs out the back of the ships.

24. In the year 2750, we receive a signal from a civilization around a nearby star telling us that the *Voyager 2* spacecraft recently crash-landed on its planet.

25. The General Rocket Corporation (a future incarnation of General Motors) unveils a new personal interstellar spacecraft that works as an interstellar ramjet with a scoop about 10 meters across.

26. Members of the first crew of the matter–antimatter spacecraft *Star Apollo*, which left Earth in the year 2165, return to Earth in the year 2450 looking only a few years older than when they left.

27. In the year 2030, we finally uncover definitive evidence of alien visits to Earth when a flying saucer crashes in the Rocky Mountains and its oxygen-kerosene fuel ignites a forest fire.

28. Aliens from a distant star system invade Earth with the intent to destroy us and occupy our planet, but we successfully fight them off when their technology proves no match for ours.

29. Aliens arrive on Earth but virtually ignore our presence, finding the diversity of earthly bacteria to be much more scientifically interesting.

30. A single great galactic civilization exists. It originated on a single planet long ago but is now made up of beings from civilizations on many different planets, each of which was assimilated into the galactic culture in turn.

Quick Quiz

Choose the best answer to each of the following. Explain your reasoning with one or more complete sentences.

31. The *New Horizons* spacecraft is currently on its way out of our solar system. About how long will it take to travel the distance to the nearest stars? (a) 100 years (b) 1000 years (c) 100,000 years

32. The amount of energy that would be needed to accelerate a large spaceship to a speed close to the speed of light is (a) about 100 times as much energy as was needed to launch the Space Shuttle; (b) more than the total amount of energy used by the entire world in a year; (c) more than the amount of energy that our Sun emits into space in a year.

33. The rocket engines of our current spacecraft are powered by (a) chemical energy; (b) nuclear energy; (c) matter–antimatter annihilation.

34. Suppose that a spaceship was launched in the year 2120 on a round-trip journey of 100 light-years, traveling at 99.99% of the speed of light. If one of the crew members was 30 years old when she left, about how old would you expect her to be on her return? (a) 31 (b) 130 (c) 29

35. Suppose that a spaceship was launched in the year 2120 on a round-trip journey of 100 light-years, traveling at 99.99% of the speed of light. In approximately what year would the ship return to Earth? (a) 2121 (b) 2170 (c) 2220

36. Which of the following best describes our current understanding of the possibility of fast interstellar travel through hyperspace? (a) Hyperspace travel is the method of choice for all advanced civilizations. (b) We do not know enough to say whether such travel is really possible. (c) The idea of hyperspace is pure fantasy and has no basis in reality.

37. Which of the following questions best represents the Fermi paradox? (a) Why can't we travel faster than the speed of light? (b) Why haven't we found any evidence of a galactic civilization? (c) Why haven't aliens invaded Earth and stolen our resources?

38. According to current scientific understanding, the idea that the Milky Way Galaxy might be home to a civilization millions of years more advanced than ours is (a) a virtual certainty; (b) extremely unlikely; (c) one reasonable solution to Fermi's paradox.

39. Which of the following is *not* relative in the special theory of relativity? (a) motion (b) time (c) the speed of light

40. What does the famous formula $E = mc^2$ have to do with special relativity? (a) Nothing; it comes from a different theory. (b) It is one of the two starting assumptions of special relativity. (c) It is a direct consequence of the theory, and hence a way of testing the theory's validity.

PROCESS OF SCIENCE

41. *The Copernican Principle.* In discussing Fermi's paradox, we generally assume that neither our planet nor our species is "special"

in any way. Explain why the history of science makes this seem a viable assumption. Do you think the assumption can be used to strongly support the argument that there is other intelligent life in the universe? Defend your opinion.

42. *Relativity as a Theory.* Look back at the hallmarks of science in Chapter 2. Evaluate the special theory of relativity in terms of each of the three hallmarks, showing why it meets the test of being science. Then explain why special relativity qualifies as a scientific *theory*, rather than just a hypothesis.

GROUP WORK EXERCISE

43. *Humanity at the Turning Point.* **Roles:** *Scribe* (takes notes on the group's activities), *Proposer* (proposes ideas to the group), *Skeptic* (points out weaknesses in proposed ideas), *Moderator* (leads group discussion and makes sure everyone contributes). **Activity:** Consider the idea, discussed in this chapter, that we are living at a critical turning point in human history, in that the decisions we make in coming decades will determine whether we destroy ourselves or whether we endure long enough that our descendants can someday journey to the stars. Make two lists: one of threats to our future, and one of opportunities for our future. Then come to a group consensus on at least three important actions that we should take today to help ensure that our civilization survives.

INVESTIGATE FURTHER

In-Depth Questions to Increase Your Understanding

Short-Answer/Essay Questions

44. *Distant Dream or Near-Reality?* Considering all the issues surrounding interstellar flight, when, if ever, do you think we are likely to begin traveling among the stars? Write a few paragraphs defending your opinion.

45. *What's Wrong with This Picture?* Many science fiction stories have imagined the galaxy divided into a series of empires, each having arisen from a different civilization on a different world, that hold each other at bay because they are all at about the same level of military technology. Is this a realistic scenario? Explain.

46. *Large Rockets.* Suppose we built a rocket that worked much like the Space Shuttle but was 1000 times as large. Could this rocket get us to speeds close to the speed of light? Explain.

47. *Ticket to the Stars.* In this chapter we stated that relativity offers only a one-way "ticket to the stars." Explain why.

48. *Solution to the Fermi Paradox.* Among the various possible solutions we have discussed for the Fermi paradox, which do you think is most likely? (Or, if you have no opinion as to their likelihood, which do you like best?) Write a one- to two-page essay in which you explain why you favor this solution.

49. *Interstellar Travel in the Movies.* Choose a science fiction movie in which aliens (or future humans) are engaged in some type of interstellar travel. In a one- to two-page essay, briefly describe how they accomplish the travel and evaluate in depth whether the scheme seems plausible.

50. *The "Relative" in Relativity.* Many people have claimed that Einstein showed that "everything is relative." Is this true? Explain.

Quantitative Problems

Be sure to show all calculations clearly and state your final answers in complete sentences.

51. *Cruise Ship Energy.* Suppose we have a spaceship about the size of a typical ocean cruise ship today, which means it has a mass of about 100 million kilograms, and we want to accelerate the ship to a speed of 10% of the speed of light.
    a. How much energy would be required? (*Hint:* You can find the answer simply by calculating the kinetic energy of the ship when it reaches its cruising speed; because 10% of the speed of light is still small compared to the speed of light, you can use the formula that tells us that kinetic energy $= \frac{1}{2} \times m \times v^2$.)
    b. How does your answer compare to total world energy use at present, which is about 5×10^{20} joules per year?
    c. Suppose the cost of energy is 3¢ per 1 million joules. At this price, how much would it cost to generate the energy needed by this spaceship?

52. *The Rocket Equation I.* Suppose a rocket with mass ratio $M_i/M_f = 15$ has engines that produce an exhaust velocity of 3 km/s. What is its final velocity? Is it sufficient to escape Earth? (*Hint:* See Cosmic Calculations 13.1.)

53. *The Rocket Equation II.* Suppose you want a rocket to achieve escape velocity from Earth (11 km/s) and its engines produce an exhaust velocity of 3 km/s. What mass ratio is required? Briefly explain the meaning of this mass ratio.

54. *The Multistage Rocket Equation.* The rocket equation takes a slightly different form for a multistage rocket:

$$v = n v_e \ln\left(\frac{M_i}{M_f}\right)$$

    where n is the number of stages.
    a. Suppose a rocket has three stages with mass ratio $M_i/M_f = 3.4$ and engines that produce an exhaust velocity of 3 km/s. What is its final velocity? Is it sufficient to escape Earth?
    b. Suppose a rocket has 100 stages with mass ratio $M_i/M_f = 3.4$ and engines that produce an exhaust velocity of 3 km/s. What is its final velocity? Compare it to the speed of light.

55. *Relativistic Time Dilation.* Use the time dilation equation from Cosmic Calculations 13.2 to answer each of the following questions.
    a. Suppose a rocket travels at the escape velocity from Earth (11 km/s). Will time on the rocket differ noticeably from time on Earth? Explain.
    b. Suppose a rocket travels at a speed of $0.9c$. If the rocket is gone from Earth for 10 years as measured on Earth, how much time passes on the rocket?
    c. Suppose a rocket travels at a speed of $0.9999c$. If the rocket is gone from Earth for 10 years as measured on Earth, how much time passes on the rocket?

56. *Testing Relativity.* A π^+ meson produced at rest has a lifetime of 18 nanoseconds (1.8×10^{-8} s). Thus, in its own reference frame, a π^+ meson will always "think" it is at rest and therefore will decay after 18 nanoseconds. Suppose a π^+ meson is produced in a particle accelerator at a speed of $0.998c$. How long will scientists see the particle last before it decays? Briefly explain how an experiment like this helps verify the special theory

of relativity. (*Hint:* Use the time dilation equation from Cosmic Calculations 13.2.)

57. *Long Trips at Constant Acceleration.* Consider a spaceship on a long trip with a constant acceleration of $1g$. Although the derivation is beyond the scope of this book, it is possible to show that, as long as the ship is gone from Earth for many years, the amount of time that passes on the spaceship during the trip is approximately

$$t_{ship} = \frac{2c}{g} \ln\left(\frac{g \times D}{c^2}\right)$$

    where D is the distance to the destination and ln is the natural logarithm. If D is in meters, $g = 9.8$ m/s^2, and $c = 3 \times 10^8$ m/s, the answer will be in units of seconds. Use this formula as needed to answer the following questions. Be sure to convert the distances from light-years to meters and final answers from seconds to years. Useful conversions: 1 light-year $\approx 9.5 \times 10^{15}$ m; 1 yr $\approx 3.15 \times 10^7$ s.
    a. Suppose the ship travels to a star that is 500 light-years away. How much time will pass on the ship? Approximately how much time will pass on Earth? Explain.
    b. Suppose the ship travels to the center of the Milky Way Galaxy, about 28,000 light-years away. How much time will pass on the ship? Compare this to the amount of time that passes on Earth.
    c. The Andromeda Galaxy is about 2.2 million light-years away. Suppose you had a spaceship that could constantly accelerate at $1g$. Could you go to the Andromeda Galaxy and back within your lifetime? Explain. If you could make the journey, what would you find when you returned to Earth?

58. *The Coral Model of Colonization.* We can estimate the time it would take for a civilization to colonize the galaxy. Imagine that a civilization sends colonists to stars that are an average distance D away and sends them in spacecraft that travel at speed v. The time required for travel, t_{travel}, is then $t_{travel} = D/v$. Suppose that the colonists build up their colony for a time t_{col}, at which point they send out their own set of colonists to other star systems (with the same average distance and same spacecraft speed). Then the speed at which the civilization expands outward from the home star, v_{col} (for the speed of colonization), is $v_{col} = D/(t_{travel} + t_{col})$. However, this is true only if the colonization is always directed straight outward from the home star. In reality, the colonists will sometimes go to uncolonized star systems in other directions, so we will introduce a constant k that accounts for this zigzag motion. Our equation for the speed at which the civilization expands outward from the home star is

$$v = k\frac{D}{(t_{travel} + t_{col})}$$

$$= k\frac{D}{\left(\dfrac{D}{v} + t_{col}\right)}$$

    For the purposes of this problem, assume that $k = \frac{1}{2}$ and that the average distance between star systems is $D = 5$ light-years.
    a. How fast (as a fraction of the speed of light) does the civilization expand if its spacecraft travel at $0.1c$ and each colony builds itself up for 150 years before sending out the next wave of colonists? How long would it take the colonists to expand a distance of 100,000 light-years from their home star at this rate?

b. Repeat part a, but assume that the spacecraft travel at 0.01c and that each colony builds itself up for 1000 years before sending out more colonists.

c. Repeat part a, but assume that the spacecraft travel at 0.25c and that each colony builds itself up for 50 years before sending out more colonists.

Discussion Questions

59. *Seeding the Galaxy.* If interstellar travel is forever impractical, are there other ways an advanced civilization might spread its culture? Clearly, communication is possible, although the speed of light makes conversations between star systems maddeningly tedious. Could a society send the information required to assemble members of its species (its "DNA," for instance) and therefore spread through the galaxy at the speed of light? Can you imagine other ways of spreading a culture without starships? Explain.

60. *Sociology of Interstellar Travel.* Suppose we somehow built a spaceship capable of relativistic travel and volunteers were being recruited for a journey to a star 15 light-years away. Would you volunteer to go? Do you think others would volunteer? In light of the effects of time dilation, discuss the benefits and drawbacks of such a trip.

61. *The Turning Point.* Discuss the idea that the responsibility of our generation to future humans is greater than that of any previous generation. Do you agree with this assessment? If so, how should we deal with this responsibility? Defend your opinions.

62. *We'll All Live in Space.* Consider the practical requirements of creating unlimited living space for future generations by building artificial habitats in orbit around the Earth. One proposal put forth in the 1970s called for constructing cylindrical habitats 20 km long and 2 km in diameter, which would slowly rotate so that the millions of residents living inside them would experience artificial gravity. Discuss potential problems and advantages of such artificial habitats. Could they reasonably be used to "save" Earth's population in the event of disaster?

WEB PROJECTS

63. *Starship Design.* Find more details about a proposal for starship propulsion or design. How would the starship work? What new technologies would be needed, and what existing technologies could be applied? Summarize your findings in a one- to two-page report.

64. *Advanced Spacecraft Technologies.* NASA supports many efforts to incorporate new technologies into spaceships. Although few of them reach the level of being suitable for interstellar colonization, most are innovative and fascinating. Learn about one such NASA project, such as the 100 Year Starship initiative, and write a short summary of your findings.

65. *Solutions to the Fermi Paradox.* Learn more about someone's pet solution to the Fermi paradox. Write a short summary of that solution, and discuss how it fits with the ideas we have discussed in this chapter.

Epilogue

Contact—Implications for the Search and Discovery

The known is finite, the unknown is infinite; intellectually we stand on an islet in the midst of an illimitable ocean of inexplicability. Our business in every generation is to reclaim a little more land.

Thomas H. Huxley (1825–1895)

We've covered a lot of ground in this book. We've studied life on Earth and learned about the prospects for life elsewhere in our solar system and on planets around other stars. Our scientific discussion mirrors a broader cultural phenomenon—an intense interest in ideas relating to extraterrestrial life. This interest shows up in many aspects of contemporary culture, including TV shows and movies, countless blogs and websites, and a notable public interest in the exploration of space and the search for life beyond Earth.

LEARNING GOALS

- Is there life elsewhere?
- How does the search for extraterrestrial life affect the human condition?
- What is the significance of the search itself?

▲ **About the photo:** Earth, photographed from the outskirts of our solar system by the *Voyager 1* spacecraft. The "sunbeam" surrounding Earth is an artifact of light scattering in the camera.

In this Epilogue, we'll briefly consider a few possible reasons why humans are so deeply interested in the question of whether life is present elsewhere. We'll begin with a short review of why the idea of life beyond Earth seems scientifically reasonable and then explore how the search for it helps us revisit age-old questions about the nature of humanity. We'll also discuss the philosophical and cultural consequences of finding life elsewhere—life of any kind, whether microbial or intelligent or somewhere in between.

• Is there life elsewhere?

Throughout this book we have discussed many reasons why it seems reasonable to think that life might be common in the universe, but we still don't know for sure that it exists anywhere besides Earth. Because of this uncertainty, all we can do today is discuss the issues that could determine whether extraterrestrial life is likely (or unlikely) to exist and how we might search for evidence of it.

Despite these limitations, we are at a unique point in the long debate over the possibility of extraterrestrial life. We have the technological capability to explore Mars and much of the rest of our solar system, and we are rapidly developing technology that should in principle allow us to seek evidence for life on planets around other stars. After millennia of speculation about life beyond Earth, we are finally reaching the point at which we could really discover it, if it exists. This remarkable prospect prompts us to discuss the philosophical and cultural consequences of finding life elsewhere. But first let's summarize the key issues in our discussion about the search for life in the universe.

WHY LIFE SEEMS LIKELY Although we don't yet know whether life exists beyond Earth, we've seen that current science offers reasons to think it should. We can examine the nature of life on Earth—its building blocks and how it originated—and understand the environmental conditions in which life can exist. We can look at the other planets and moons in our solar system and determine whether they offer conditions conducive to biology. And we can look for planets around other stars, learn how abundant they are and how they form, and consider how many of them might be capable of supporting living things.

When we do these things, we find three key pieces of evidence that point to the idea that life should be common in the universe:

1. The chemical elements that make up life on Earth are common throughout the universe, and complex, carbon-bearing molecules important to terrestrial life appear to form easily and naturally under conditions that should be common on many worlds.
2. Life on Earth thrives under a wide range of environmental conditions that we once considered too extreme to be capable of supporting life, and many of these types of environments are likely to be found on other planets in our own solar system and beyond.
3. Life appeared on Earth quite early in its history, making it seem plausible that an origin of life is "easy."

Let's start with the first item in the above list. The elements from which life is constructed (carbon, hydrogen, oxygen, nitrogen, and almost two dozen other elements) are found nearly everywhere in the universe. Hydrogen is the most abundant element by far, but the other elements used for life exist in at least modest quantities, because they have been created by nuclear fusion in earlier generations of stars and during supernova explosions of those stars. The supernovae ejected these manufactured elements into interstellar space, where they were incorporated into the gas and dust clouds out of which later stars and planets formed. Moreover, experiments in laboratories on Earth and spectroscopic observations of distant objects show that the elements used by life combine readily into the molecular building blocks of life. We have found molecules as complex as amino acids in meteorites, and we have detected numerous organic molecules even in interstellar space. We therefore expect that such molecules should form abundantly and naturally under a range of conditions that includes those that were present on the early Earth and are likely to be present on other geologically active worlds. A wide availability of organic molecules would make it likely that the starting points for an origin of life exist on many worlds.

If the starting points for life are commonplace, the next question concerns whether available environments can allow life to arise and thrive. This is the issue addressed by the second point on our list. We have learned that life on Earth is incredibly diverse and that some organisms can thrive in a wide variety of environments that seem "extreme" to humans. For example, we have found life on Earth in the hot water near deep-sea vents, in the dry and frigid deserts of Antarctica, and inside rocks deep underground. The diversity of environments in which we find terrestrial life suggests that life could survive on any world that meets relatively simple environmental requirements—the presence of liquid water (or possibly another liquid), access to the requisite elements and molecules, and an energy source to drive metabolism. These conditions are likely to be met on any geologically active world with a rocky interior and liquid water, whether a planet or a moon. Such worlds would have the necessary elements and the potential for energy to be available via water/rock chemical reactions.

The first two points on our list tell us that the starting points for life are widely available and that life, once started, can survive in a wide range of environments. The third point tells us that getting life to arise may not be difficult. Even with the present uncertainties in interpreting the geological record, life must have originated on a time scale that is very short compared to the age of Earth and to the expected ages of other habitable planets. Unless Earth was somehow atypical, it seems that life might be the natural, straightforward consequence of the types of chemical reactions that can occur in planetary environments. When we put all three key pieces of evidence together, it seems reasonable to imagine life existing in at least a few other places in our own solar system and on many similar worlds throughout the universe.

PROSPECTS FOR FINDING LIFE IN OUR SOLAR SYSTEM If life is indeed as common and wide-ranging as the evidence suggests it could be, then the first place to look for life is within our own solar system. Here in the Sun's neighborhood, Mars and Europa seem the most likely places besides Earth to harbor life, but we've also found a few other candidates.

Mars shows evidence of liquid water having been present at its surface early in its history and within its crust throughout its history. The martian atmosphere contains several of the key elements of life—notably carbon (in the form of gaseous carbon dioxide), hydrogen and oxygen (in the form of water), and nitrogen. The other necessary elements are found in surface and near-surface rocks. Energy to drive metabolism could come from chemical reactions between the water and the rocks, for example, allowing the possibility of organisms much like those that live within rocks on Earth.

Jupiter's moon Europa may have large amounts of liquid water. Although we are not yet absolutely certain, it seems likely that Europa has a global, 100-kilometer-thick ocean lying beneath an ice surface. We also expect that Europa has heavier elements in its rocky interior, so all the elements needed for life should be abundant. Energy for life could be supplied by chemical reactions between the water and the underlying rock, or by chemical compounds created by the impact of high-energy particles onto surface ice (driven by Jupiter's magnetic field). It seems plausible that life could exist on Europa if its postulated ocean proves to be real.

Other places in our solar system could also potentially support life. Two other moons of Jupiter, Ganymede and Callisto, show evidence of subsurface liquid oceans similar to Europa's but at greater depths beneath their surfaces. Saturn's moon Titan has lakes of liquid methane and ethane, and may have liquid water underground in the form of a cold, ammonia/water mixture. Saturn's moon Enceladus also seems likely to have similar liquids in its interior, and Neptune's moon Triton might have the same. Whether these worlds have a mechanism that could offer enough energy to sustain life is an open question, but it is at least possible that some or all of them could host life.

PROSPECTS FOR FINDING LIFE AMONG THE STARS Efforts to search for life beyond our own solar system are already under way. We have discovered thousands of planets orbiting other stars, and the statistics indicate that

Movie Madness E.T.

When cinema aliens come to Earth, it's usually wise policy to head for the storm cellar. But when wrinkly little E.T. is accidentally left behind by his planetary pals, it turns out to be good news—at least for a few suburban kids.

E.T. is the quintessential alien film—a movie that long held the record for being the most successful picture of all time. There's good reason. Appealing little E.T., who has the stature and gait of a penguin, is every kid's dream. After all, he's a friend (and really useful because with his super powers he can help you outsmart adults), and he's exclusively *your* friend. Other kids will stop kicking verbal sand in your face when you show up at the playground with a guy from another galaxy, even if he has a face like a polished redwood burl.

Aside from this childlike wish fulfillment, E.T. encapsulates everything we hope or think is true about intelligent extraterrestrials. To begin with, E.T. is benign and nonthreatening. Unlike evil aliens, who look like reptiles or insects, E.T. resembles a baby, with his short nose, big eyes, and wrinkled skin. He's only 2 feet high, weighs 35 pounds, and has a 25-watt fingertip. He clearly comes from a kinder, gentler planet, since his only interest in Earth's biota is its plants. (No insects had to

die for this film.) Scientists who have given any thought to the true nature of advanced aliens (and those that can come to Earth are clearly advanced) have often equated technological prowess with peaceful behavior. The aliens will be friendly. This doesn't entirely square with our experience on Earth, but one can hope.

In addition, E.T. not only looks a lot like us; he acts like us, functions like us (he can get blotto on beer), and is interested in our personal lives. None of this is likely to be true, of course. He's also well adapted to terrestrial conditions, waddling around without a space suit and, indeed, without any clothes at all. And nefarious federal agents wearing jackets, ties, and drab personalities are busily trying to keep the cuddly creature's visit under wraps, something that many people believe is happening in real life.

All of this may be in keeping with the public's perception of what aliens would be like. But if the search for extraterrestrial intelligence succeeds and we eventually learn the true nature of extraterrestrials, it is far more likely that they will be of a construction and temperament that are far beyond our most fevered imaginings.

our galaxy may have billions of planets with Earth-like sizes and orbits. Moreover, we've discovered planets with a wider range of characteristics than the planets of our own solar system, and some of these planetary types might also conceivably harbor life.

If conditions for an origin of life are indeed common, and if life really does get started easily, then at minimum, microbial life would be widespread throughout the universe. The possible prevalence of microbial life begs the question of the potential for more complex or intelligent life. Our one example of intelligence here on Earth does not allow us to extrapolate or even hazard a guess as to whether intelligence should be widespread or rare; the only way to find out whether other civilizations exist is to search for them. SETI searches for radio or light signals from possible extraterrestrial civilizations have not yet met with success, but only a small sample of the possible homes for civilization has been searched so far. Like the question of extraterrestrial life in general, the question of intelligent extraterrestrial life remains open.

• How does the search for extraterrestrial life affect the human condition?

We have reviewed the key scientific issues pertaining to life elsewhere. Now we turn our attention to issues that we have generally neglected to this point, including philosophical and societal issues that touch on reasons the search for life beyond Earth is of such broad intellectual interest to scientists and the public alike.

Why are so many people so interested in the question of whether life exists on other worlds, and what would it mean to find evidence—or a lack of evidence—for such life? We'll discuss a few possible answers to these questions, but you should recognize that the ideas we discuss may not resonate the same way with everyone. As you read, think about what issues might be driving your own interests and how you might continue to pursue these interests in the future.

OUR CHANGING PERSPECTIVE ON THE WORLD Many people are fascinated by the question of extraterrestrial life because knowing if such life exists would likely have a profound influence on our perspective about our place in the universe. Much of what we have learned about the world over the past several thousand years has changed the way we interact with it. For example, a mere 10,000 years ago—a blink of an eye in the history of our planet—humans were primarily a species of hunter gatherers, living out their lives with little knowledge about what was beyond the next mountain or valley. Today, people in nearly every corner of the world recognize our species as part of an enormous cosmos. We have learned that what we do in one place on our planet affects the environment, people, and societies in all other places around the world. In addition, we now see Earth as just one of many worlds in our solar system, our solar system and the Sun as one of more than 100 billion star systems in our galaxy, and our galaxy as just one of billions of similar galaxies in the observable universe. This expansion of our world (or contraction, depending on how you perceive it!) has brought with it a need to rethink our views both of ourselves and of our relationship to the rest of the planet and universe.

Perhaps the first and most significant major transition toward a modern scientific view of ourselves began nearly 500 years ago with the Copernican revolution. The work of Copernicus, Galileo, Kepler, and Newton allowed us to recognize that Earth is not at the center of the universe but instead orbits the Sun. This displacement of Earth from the center of the universe had profound philosophical and psychological effects. Earth, and by extension humanity, could no longer be viewed as the center of everything, and the world could no longer be viewed as obeying only the laws of Providence rather than the laws of physics. This shift in thinking had little impact on people's day-to-day lives—we would be hard-pressed even today to think of a way in which whether Earth goes around the Sun or vice versa matters to our daily activities. However, it made a fundamental difference to our worldview. As a result, the idea initially met widespread resistance in Western society and took centuries to gain full acceptance.

A second major shift occurred in the mid-nineteenth century with the recognition by Charles Darwin and Alfred Russel Wallace of the processes that drive the evolution of species. The basic idea of evolution was not new. By the time of Darwin, scientists already recognized that Earth was old, that fossils represented organisms that had lived in prior times, that older fossils were different from younger ones, and that both were different from current living organisms. These discoveries had already convinced many scientists that the nature of living organisms had changed over time. What Darwin and Wallace discovered was a way to explain *why and how* species change: through competition for survival and the process we refer to as *natural selection*. We now recognize that all the species present on Earth represent the product of some 4 billion years of evolution, traceable back to the earliest history of life on Earth.

Like the Copernican revolution, recognizing the nature of evolution sparked a change in our view of ourselves and of our world. Copernicus and others showed that we are not located at the physical center of the universe, and Darwin and Wallace showed that we are not at the biological center either. Rather, we represent just one of millions of distinct species on Earth, and while we may be "dominant" over many species in the plant and animal kingdoms, these kingdoms represent just a small part of the great diversity of life on our planet. This shift in perspective is profound, and may well explain why many people have difficulty accepting the idea of evolution by natural selection. The intensity of the public controversy only underscores the tremendous impact of the theory of evolution on Western society.

A third shift in perspective is taking place today with the discovery of other solar systems and the modern understanding of the potential for life to be widespread in the universe. Although astronomers had long suspected that planets ought to be common around other stars, the first definitive evidence of extrasolar planets is less than three decades old. If we ultimately confirm our guess that other Earth-like worlds with life exist or are common, we will no longer be able to regard Earth as special in any essential way. Throughout the history of scientific thought, new discoveries have increasingly displaced us from the center of the universe, both physically and metaphorically. Finding that life is common throughout the universe would lead to yet another major shift in perspective.

Even aside from these larger considerations of understanding our place in the cosmos, there are undoubtedly more immediate reasons for our interest in alien life. Extraterrestrial beings figure in many film and television dramas. They are easy villains—exotic, and with capabilities limited only by the imagination of the director. Their prevalence in film may reflect a natural interest in beings that could be potential competitors or even mates. This interest might be similar to our fascination with predators. Early humans paid attention to anything with big teeth—there was survival value in doing so—and these animals still command our attention today. It may be that our curiosity about large creatures (dinosaurs, for example) and potential peers (aliens) has origins that extend beyond our modern, more science-based motivations.

THE IMPACT OF EXTRATERRESTRIAL LIFE ON HUMAN PERSPECTIVE People have long speculated about life beyond Earth, and many people—including some scientists—have at times been convinced that life exists on the Moon, Mars, or other worlds. Why, then, would the actual discovery of extraterrestrial life have a major impact on human perspective? The answer lies in the difference between guessing and knowing. As long as there is uncertainty about the existence of life on other worlds, people are free to hold a wide range of opinions. People living before the time of Copernicus could continue to believe in an Earth-centered universe, but it was quite hard to continue to do so after Galileo, Kepler, and Newton offered convincing proof to the contrary. An actual discovery of life beyond Earth would force us, both as individuals and as a society, to reconsider the place of our planet and our species in the cosmos.

In contemplating the significance of finding life elsewhere, let's begin by considering what would happen if we found microbial life on Mars. The first question we would probably ask is whether the life was genetically related to terrestrial life (suggesting that it had migrated between planets on meteorites) or instead represented an independent origin of life on Mars. We could answer this question by determining the structure of the molecules that make up the martian life. Does it use DNA and RNA molecules similar to those used by terrestrial life? Does it use the same amino acids or the same proteins to carry out enzymatic reactions? Do the molecules that participate in life have the same "handedness" as those we know on Earth? It seems unlikely that there would be only one solution to the problems of containing and passing on the genetic information required for life to reproduce, of catalyzing the chemical reactions that make up life, and of storing and using energy in metabolism. We would therefore expect life that had an origin independent from life on Earth to have a different chemical structure.

While the discovery of microbial life on Mars that was genetically related to terrestrial life would be exciting, it probably would not have as great an impact as a discovery of life that showed evidence of an independent origin. A discovery of life with independent beginnings would provide clear proof that the origin of life was not a unique event. With proof that life had sprung up twice in just our own solar system, we would have every reason to think that it has originated on many worlds around other stars and that life is widespread in the universe. A discovery of alien microbial life would also help us better understand life in general. We would learn more about the conditions under which life can arise and persist, as well as the conditions under which it can evolve into more complex forms. And we could begin to engage in *comparative biology*, in which the study of alien life would help us better understand the life on our own world.

While many people who have studied the issue of life in the universe expect that we will find microbial life to be widespread, we cannot truly envision the consequences of an actual discovery unless and until it occurs. Moreover, the history of science tells us to be prepared for surprises. For example, the actual discovery of extrasolar planets proved surprising, despite their predicted existence: We learned that other solar systems can be very different from our own. If and when we do find life elsewhere, we should expect to be equally surprised about its nature.

EXTRATERRESTRIAL INTELLIGENCE AND THE NATURE OF HUMANITY The discovery of extraterrestrial life of any kind would have a profound effect on our perception of our place in the universe, but it is the potential discovery of extraterrestrial intelligence that generates the greatest public interest. What would finding intelligent life elsewhere mean to us? We discussed some of the philosophical implications in Section 13.3; here we focus on how it might affect our lives more directly.

It is difficult to predict how we would respond to such a discovery, and it is likely that people would exhibit a wide range of responses. At one extreme, some people might look to extraterrestrial intelligence as a source of help in solving our problems. This view has been portrayed in many books and movies, including *Contact* and *2001—A Space Odyssey*. Interestingly, many of these stories suggest, first, that we are not able to solve our own problems without outside help and, second, that extraterrestrials will want to help us. At the other extreme, some people imagine that extraterrestrials would come to Earth and destroy our civilization, either deliberately or by accident. The Martians of H. G. Wells's *War of the Worlds* came here bent on our destruction (Figure E.1). In Douglas Adams's *Hitchhiker's Guide to the Galaxy*, aliens destroy Earth as an accidental consequence of their need to build a new "interstellar bypass." Note also that all these stories assume that we would be able to understand communications from societies far more advanced than our own.

Neither extreme seems especially likely. As we've discussed, intelligent aliens capable of visiting our planet would almost certainly be far more technologically advanced than we are. It's not clear that they would have any interest in us at all (making the first extreme unlikely), and they'd seemingly have little to gain by destroying us (making the second extreme unlikely). If and when intelligent life is discovered, the reality may lie somewhere in between these two extremes—or it might be completely different.

• What is the significance of the search itself?

A discovery of even microbial extraterrestrial life would undoubtedly bring important practical benefits. For example, studying it would help us understand what characteristics of terrestrial life are unique to Earth and what characteristics apply generally to life anywhere. If we found intelligent life elsewhere, we would learn much more about the nature of intelligence and might be exposed to cultures and societies extremely different from those of humans. If we could communicate with more advanced beings, we could possibly learn the secrets of the universe, the nature of consciousness and the mind, and technological marvels that could dramatically change our life here on Earth.

However, for many people excited by the scientific search for life in the universe, the search itself is much more than a means to an end. From this viewpoint, the search is just one more critical component in our exploration of the world around us. Other components of astronomy and space science include exploring the planets and moons in our solar

FIGURE E.1
A 1906 drawing used to illustrate H. G. Wells's book *The War of the Worlds*, in which Earth is invaded by hostile Martians.

system as a way to understand how planets work and exploring stars and galaxies as a way to determine the nature of our universe. Other components of biology include exploring the origin and evolution of life on Earth so that we can understand how we ourselves came to exist.

In all these cases, our exploration does not seem to be driven solely by the desire to find specific answers to the scientific questions we are asking, because in each case we end up asking *more* questions. Instead, we seem to be driven by our inherent curiosity, our desire to understand the world around us (Figure E.2). Sometimes our curiosity leads to discoveries with practical applications, while at other times it simply helps us understand how or why the world is as it is. Understanding the world around us means learning about the broader environment in which humans exist. Understanding the occurrence of planets orbiting other stars helps us understand the significance of the occurrence of planets orbiting the Sun, including Earth. Understanding the occurrence of life elsewhere allows us to understand the significance of the occurrence of life on Earth. And understanding the potential for intelligent life beyond Earth brings with it an understanding of the meaning of the existence of intelligent life here on Earth. In essence, by learning about the universe around us, we are learning about ourselves and about what it means to be human.

With that, we'll turn to a few final thoughts from one of the pioneers of astrobiology, written as he looked at the "pale blue dot" of Earth as photographed by the *Voyager 1* spacecraft, which you can see in the photo that opens this Epilogue.

Look again at that dot. That's here. That's home. That's us. On it everyone you love, everyone you know, everyone you ever heard of, every human being who ever was, lived out their lives. The aggregate of our joy and suffering, thousands of confident religions, ideologies, and economic doctrines, every hunter and forager, every hero and coward, every creator and destroyer of civilization, every king and peasant, every young couple in love, every mother and father, hopeful child, inventor and explorer, every teacher of morals, every corrupt politician, every "superstar," every "supreme leader," every saint and sinner in the history of our species lived there—on a mote of dust suspended in a sunbeam.

The Earth is a very small stage in a vast cosmic arena. Think of the rivers of blood spilled by all those generals and emperors so that, in glory and triumph, they could become the momentary masters of a fraction of a dot. Think of the endless cruelties visited by the inhabitants of one corner of this pixel on the scarcely distinguishable inhabitants of some other corner, how frequent their misunderstandings, how eager they are to kill one another, how fervent their hatreds.

Our posturings, our imagined self-importance, the delusion that we have some privileged position in the Universe, are challenged by this point of pale light. Our planet is a lonely speck in the great enveloping cosmic dark. In our obscurity, in all this vastness, there is no hint that help will come from elsewhere to save us from ourselves.

The Earth is the only world known so far to harbor life. There is nowhere else, at least in the near future, to which our species could migrate. Visit, yes. Settle, not yet. Like it or not, for the moment the Earth is where we make our stand.

It has been said that astronomy is a humbling and character-building experience. There is perhaps no better demonstration of the folly of human conceits than this distant image of our tiny world. To me, it underscores our responsibility to deal more kindly with one another, and to preserve and cherish the pale blue dot, the only home we've ever known.

— Carl Sagan (1934–1996)

FIGURE E.2

Earth as viewed from the *Apollo 17* spacecraft. Looking from this perspective, we recognize the strong connections between the terrestrial ecosystem and the planet itself, and we recognize that life, indeed, is a planetary phenomenon. As we explore the universe, we will learn whether the formation of planets around stars and the occurrence of life on planets are rare or commonplace. We may then finally answer the question of whether we are alone.

Exercises and Problems

DISCUSSION QUESTIONS

1. *Is There Life Elsewhere?* After considering all the evidence to date about the possibility of extraterrestrial life, do you believe it is likely that we'll find microbial life elsewhere? Do you believe it is likely that we'll find intelligent life elsewhere? Defend your opinions, using arguments based on the full range of scientific issues discussed in this book.

2. *Microbial or Intelligent?* Do you think the implications of discovering microbial life elsewhere would be any more or less profound than those of discovering extraterrestrial intelligence? Explain your reasoning.

3. *Extraterrestrial Life and Your Religion.* Would the discovery of extraterrestrial life have any important implications for your own personal religious beliefs? Would it affect the current "official" beliefs (if any) of your religion? Explain.

4. *Extraterrestrial Life and the Debate on Evolution.* Do you think the discovery of extraterrestrial life would affect the current status of the debate over science and religion in the United States? For example, would it alter the controversy surrounding the teaching of evolution in public schools? Why or why not?

5. *Aliens and Everyday Life.* While the discovery of extraterrestrial life would surely be profound, do you think it would alter any aspect of our everyday lives? If so, how? If not, why not?

6. *The Search Itself.* Suppose we spend a fair amount of money and effort searching for life over the next few decades and ultimately find no evidence for life beyond Earth. Will the search have been a waste, a success, or something in between? Defend your opinion.

7. *Pale Blue Dot.* Read and discuss the quotation from Carl Sagan at the end of the chapter. Do you agree with its sentiments? Defend your opinion.

Appendix A   Useful Numbers

Astronomical Distances

1 AU $\approx 1.496 \times 10^8$ km $= 1.496 \times 10^{11}$ m

1 light-year $\approx 9.46 \times 10^{12}$ km $= 9.46 \times 10^{15}$ m

1 parsec (pc) $\approx 3.09 \times 10^{13}$ km ≈ 3.26 light-years

1 kiloparsec (kpc) $= 1000$ pc $\approx 3.26 \times 10^3$ light-years

1 megaparsec (Mpc) $= 10^6$ pc $\approx 3.26 \times 10^6$ light-years

Universal Constants

Speed of light: $c = 3.00 \times 10^5$ km/s $= 3 \times 10^8$ m/s

Gravitational constant: $G = 6.67 \times 10^{-11} \dfrac{m^3}{kg \times s^2}$

Planck's constant: $h = 6.63 \times 10^{-34}$ joule $\times$ s

Stefan-Boltzmann constant: $\sigma = 5.67 \times 10^{-8} \dfrac{watt}{m^2 \times K^4}$

Mass of a proton: $m_p = 1.67 \times 10^{-27}$ kg

Mass of an electron: $m_e = 9.11 \times 10^{-31}$ kg

Useful Sun and Earth Reference Values

Mass of the Sun: $1 M_{Sun} \approx 2 \times 10^{30}$ kg

Radius of the Sun: $1 R_{Sun} \approx 696{,}000$ km

Luminosity of the Sun: $1 L_{Sun} \approx 3.8 \times 10^{26}$ watts

Mass of Earth: $1 M_{Earth} \approx 5.97 \times 10^{24}$ kg

Radius (equatorial) of Earth: $1 R_{Earth} \approx 6378$ km

Acceleration of gravity on Earth: $g = 9.8$ m/s^2

Escape velocity from surface of Earth: $v_{escape} = 11.2$ km/s $= 11{,}200$ m/s

Astronomical Times

1 solar day (average) $= 24^h$

1 sidereal day $\approx 23^h 56^m 4.09^s$

1 synodic month (average) ≈ 29.53 solar days

1 sidereal month (average) ≈ 27.32 solar days

1 tropical year ≈ 365.242 solar days

1 sidereal year ≈ 365.256 solar days

Energy and Power Units

Basic unit of energy: 1 joule $= 1 \dfrac{kg \times m^2}{s^2}$

Basic unit of power: 1 watt $= 1$ joule/s

Electron-volt: 1 eV $= 1.60 \times 10^{-19}$ joule

Appendix B   Useful Formulas

- Universal law of gravitation for the force between objects of mass M_1 and M_2, with distance d between their centers:

$$F = G\frac{M_1 M_2}{d^2}$$

- Newton's version of Kepler's third law, which applies to any pair of orbiting objects, such as a star and planet, a planet and moon, or two stars in a binary system; p is the orbital period, a is the distance between the centers of the orbiting objects, and M_1 and M_2 are the object masses:

$$p^2 = \frac{4\pi^2}{G(M_1 + M_2)}a^3$$

- Escape velocity at distance R from center of object of mass M:

$$v_{escape} = \sqrt{\frac{2GM}{R}}$$

- Relationship between a photon's wavelength (λ), frequency (f), and the speed of light (c):

$$\lambda \times f = c$$

- Energy of a photon of wavelength λ or frequency f:

$$E = hf = \frac{hc}{\lambda}$$

- Stefan-Boltzmann law for thermal radiation at temperature T (on the Kelvin scale):

$$\text{emitted power per unit area} = \sigma T^4$$

- Wien's law for the peak wavelength (λ_{max}) thermal radiation at temperature T (on the Kelvin scale):

$$\lambda_{max} = \frac{2{,}900{,}000}{T}\ \text{nm}$$

- Doppler shift (radial velocity is positive if the object is moving away from us and negative if it is moving toward us):

$$\frac{\text{radial velocity}}{\text{speed of light}} = \frac{\text{shifted wavelength} - \text{rest wavelength}}{\text{rest wavelength}}$$

- Angular separation (α) of two points with an actual separation s, viewed from a distance d (assuming d is much larger than s):

$$\alpha = \frac{s}{2\pi d} \times 360°$$

- Inverse square law for light (d is the distance to the object):

$$\text{apparent brightness} = \frac{\text{luminosity}}{4\pi d^2}$$

- Parallax formula (distance d to a star with parallax angle p in arcseconds):

$$d\ (\text{in parsecs}) = \frac{1}{p\ (\text{in arcseconds})}$$

$$\text{or } d\ (\text{in light-years}) = 3.26 \times \frac{1}{p\ (\text{in arcseconds})}$$

- The orbital velocity law, to find the mass M_r contained within the circular orbit of radius r for an object moving at speed v:

$$M_r = \frac{r \times v^2}{G}$$

Appendix C   A Few Mathematical Skills

This appendix reviews the following mathematical skills: powers of 10, scientific notation, working with units, the metric system, and finding a ratio. You should refer to this appendix as needed while studying the textbook.

C.1 Powers of 10

Powers of 10 indicate how many times to multiply 10 by itself. For example:

$$10^2 = 10 \times 10 = 100$$
$$10^6 = 10 \times 10 \times 10 \times 10 \times 10 \times 10 = 1,000,000$$

Negative powers are the reciprocals of the corresponding positive powers. For example:

$$10^{-2} = \frac{1}{10^2} = \frac{1}{100} = 0.01$$
$$10^{-6} = \frac{1}{10^6} = \frac{1}{1,000,000} = 0.000001$$

Table C.1 lists powers of 10 from 10^{-12} to 10^{12}. Note that powers of 10 follow two basic rules:

1. A positive exponent tells how many zeros follow the 1. For example, 10^0 is a 1 followed by no zeros, and 10^8 is a 1 followed by eight zeros.

2. A negative exponent tells how many places are to the right of the decimal point, including the 1. For example, $10^{-1} = 0.1$ has one place to the right of the decimal point; $10^{-6} = 0.000001$ has six places to the right of the decimal point.

Multiplying and Dividing Powers of 10

Multiplying powers of 10 simply requires adding exponents, as the following examples show:

$$10^4 \times 10^7 = \underbrace{10,000}_{10^4} \times \underbrace{10,000,000}_{10^7}$$
$$= \underbrace{100,000,000,000}_{10^{4+7} = 10^{11}} = 10^{11}$$

$$10^5 \times 10^{-3} = \underbrace{100,000}_{10^5} \times \underbrace{0.001}_{10^{-3}}$$
$$= \underbrace{100}_{10^{5+(-3)} = 10^2} = 10^2$$

$$10^{-8} \times 10^{-5} = \underbrace{0.00000001}_{10^{-8}} \times \underbrace{0.00001}_{10^{-5}}$$
$$= \underbrace{0.0000000000001}_{10^{-8+(-5)} = 10^{-13}} = 10^{-13}$$

TABLE C.1   Powers of 10

	Zero and Positive Powers			Negative Powers	
Power	Value	Name	Power	Value	Name
10^0	1	One			
10^1	10	Ten	10^{-1}	0.1	Tenth
10^2	100	Hundred	10^{-2}	0.01	Hundredth
10^3	1000	Thousand	10^{-3}	0.001	Thousandth
10^4	10,000	Ten thousand	10^{-4}	0.0001	Ten-thousandth
10^5	100,000	Hundred thousand	10^{-5}	0.00001	Hundred-thousandth
10^6	1,000,000	Million	10^{-6}	0.000001	Millionth
10^7	10,000,000	Ten million	10^{-7}	0.0000001	Ten-millionth
10^8	100,000,000	Hundred million	10^{-8}	0.00000001	Hundred-millionth
10^9	1,000,000,000	Billion	10^{-9}	0.000000001	Billionth
10^{10}	10,000,000,000	Ten billion	10^{-10}	0.0000000001	Ten-billionth
10^{11}	100,000,000,000	Hundred billion	10^{-11}	0.00000000001	Hundred-billionth
10^{12}	1,000,000,000,000	Trillion	10^{-12}	0.000000000001	Trillionth

Dividing powers of 10 requires subtracting exponents, as in the following examples:

$$\frac{10^5}{10^3} = \underbrace{\frac{100,000}{10^5} \div \frac{1000}{10^3}}$$

$$= \underbrace{\frac{100}{10^{5-3} = 10^2}} = 10^2$$

$$\frac{10^3}{10^7} = \underbrace{\frac{1000}{10^3} \div \frac{10,000,000}{10^7}}$$

$$= \underbrace{\frac{0.0001}{10^{3-7} = 10^{-4}}} = 10^{-4}$$

$$\frac{10^{-4}}{10^{-6}} = \underbrace{\frac{0.0001}{10^{-4}} \div \frac{0.000001}{10^{-6}}}$$

$$= \underbrace{\frac{100}{10^{-4-(-6)} = 10^2}} = 10^2$$

Powers of Powers of 10

We can use the multiplication and division rules to raise powers of 10 to other powers or to take roots. For example:

$$(10^4)^3 = 10^4 \times 10^4 \times 10^4 = 10^{4+4+4} = 10^{12}$$

Note that we can get the same end result by simply multiplying the two powers:

$$(10^4)^3 = 10^{4\times3} = 10^{12}$$

Because taking a root is the same as raising to a fractional power (e.g., the square root is the same as the $\frac{1}{2}$ power, the cube root is the same as the $\frac{1}{3}$ power, etc.), we can use the same procedure for roots, as in the following example:

$$\sqrt{10^4} = (10^4)^{1/2} = 10^{4\times(1/2)} = 10^2$$

Adding and Subtracting Powers of 10

Unlike multiplying and dividing powers of 10, there is no shortcut for adding or subtracting powers of 10. The values must be written in longhand notation. For example:

$$10^6 + 10^2 = 1,000,000 + 100 = 1,000,100$$

$$10^8 + 10^{-3} = 100,000,000 + 0.001 = 100,000,000.001$$

$$10^7 - 10^3 = 10,000,000 - 1000 = 9,999,000$$

Summary

We can summarize our findings using n and m to represent any numbers:

- To *multiply* powers of 10, *add* exponents: $10^n \times 10^m = 10^{n+m}$
- To *divide* powers of 10, *subtract* exponents: $\frac{10^n}{10^m} = 10^{n-m}$
- To *raise* powers of 10 to other powers, multiply exponents: $(10^n)^m = 10^{n \times m}$
- To add or subtract powers of 10, first write them out longhand.

C.2 Scientific Notation

When we are dealing with large or small numbers, it's generally easier to write them with powers of 10. For example, it's much easier to write the number 6,000,000,000,000 as 6×10^{12}. This format, in which a number *between* 1 and 10 is multiplied by a power of 10, is called **scientific notation.**

Converting a Number to Scientific Notation

We can convert numbers written in ordinary notation to scientific notation with a simple two-step process:

1. Move the decimal point to come after the *first* nonzero digit.
2. The number of places the decimal point moves tells you the power of 10; the power is *positive* if the decimal point moves to the left and *negative* if it moves to the right.

Examples:

$$3042 \xrightarrow[\text{3 places to left}]{\text{decimal needs to move}} 3.042 \times 10^3$$

$$0.00012 \xrightarrow[\text{4 places to right}]{\text{decimal needs to move}} 1.2 \times 10^{-4}$$

$$226 \times 10^2 \xrightarrow[\text{2 places to left}]{\text{decimal needs to move}} (2.26 \times 10^2) \times 10^2 = 2.26 \times 10^4$$

Converting a Number from Scientific Notation

We can convert numbers written in scientific notation to ordinary notation by the reverse process:

1. The power of 10 indicates how many places to move the decimal point; move it to the *right* if the power of 10 is positive and to the *left* if it is negative.
2. If moving the decimal point creates any open places, fill them with zeros.

Examples:

$$4.01 \times 10^2 \xrightarrow[\text{2 places to right}]{\text{move decimal}} 401$$

$$3.6 \times 10^6 \xrightarrow[\text{6 places to right}]{\text{move decimal}} 3,600,000$$

$$5.7 \times 10^{-3} \xrightarrow[\text{3 places to left}]{\text{move decimal}} 0.0057$$

Multiplying or Dividing Numbers in Scientific Notation

Multiplying or dividing numbers in scientific notation simply requires operating on the powers of 10 and the other parts of the number separately.

Examples:

$$(6 \times 10^2) \times (4 \times 10^5) = (6 \times 4) \times (10^2 \times 10^5)$$
$$= 24 \times 10^7 = (2.4 \times 10^1) \times 10^7$$
$$= 2.4 \times 10^8$$

$$\frac{4.2 \times 10^{-2}}{8.4 \times 10^{-5}} = \frac{4.2}{8.4} \times \frac{10^{-2}}{10^{-5}} = 0.5 \times 10^{-2-(-5)} = 0.5 \times 10^{3}$$
$$= (5 \times 10^{-1}) \times 10^{3} = 5 \times 10^{2}$$

Note that, in both these examples, we first found an answer in which the number multiplied by a power of 10 was *not* between 1 and 10. We therefore followed the procedure for converting the final answer to scientific notation.

Addition and Subtraction with Scientific Notation

In general, we must write numbers in ordinary notation before adding or subtracting.

Examples:

$$(3 \times 10^{6}) + (5 \times 10^{2}) = 3{,}000{,}000 + 500$$
$$= 3{,}000{,}500 = 3.0005 \times 10^{6}$$

$$(4.6 \times 10^{9}) - (5 \times 10^{8}) = 4{,}600{,}000{,}000 - 500{,}000{,}000$$
$$= 4{,}100{,}000{,}000 = 4.1 \times 10^{9}$$

When both numbers have the *same* power of 10, we can factor out the power of 10 first.

Examples:

$$(7 \times 10^{10}) + (4 \times 10^{10}) = (7 + 4) \times 10^{10}$$
$$= 11 \times 10^{10} = 1.1 \times 10^{11}$$

$$(2.3 \times 10^{-22}) - (1.6 \times 10^{-22}) = (2.3 - 1.6) \times 10^{-22}$$
$$= 0.7 \times 10^{-22} = 7.0 \times 10^{-23}$$

C.3 Working with Units

Showing the units of a problem as you solve it usually makes the work much easier and also provides a useful way of checking your work. If an answer does not come out with the units you expect, you probably did something wrong. In general, working with units is very similar to working with numbers, as the following guidelines and examples show.

Five Guidelines for Working with Units

Before you begin any problem, think ahead and identify the units you expect for the final answer. Then operate on the units along with the numbers as you solve the problem. The following five guidelines may be helpful when you are working with units:

1. Mathematically, it doesn't matter whether a unit is singular (e.g., meter) or plural (e.g., meters); we can use the same abbreviation (e.g., m) for both.
2. You cannot add or subtract numbers unless they have the *same* units. For example, 5 apples + 3 apples = 8 apples, but the expression 5 apples + 3 oranges cannot be simplified further.

3. You *can* multiply units, divide units, or raise units to powers. Look for key words that tell you what to do.
   - *Per* suggests division. For example, we write a speed of 100 kilometers per hour as

$$100 \, \frac{\text{km}}{\text{hr}} \quad \text{or} \quad 100 \, \frac{\text{km}}{1 \text{ hr}}$$

   - *Of* suggests multiplication. For example, if you launch a 50-kg space probe at a launch cost *of* $10,000 per kilogram, the total cost is

$$50 \text{ kg} \times \frac{\$10{,}000}{\text{kg}} = \$500{,}000$$

   - *Square* suggests raising to the second power. For example, we write an area of 75 square meters as 75 m^2.
   - *Cube* suggests raising to the third power. For example, we write a volume of 12 cubic centimeters as 12 cm^3.

4. Often the number you are given is not in the units you wish to work with. For example, you may be given that the speed of light is 300,000 km/s but need it in units of m/s for a particular problem. To convert the units, simply multiply the given number by a *conversion factor*: a fraction in which the numerator (top of the fraction) and denominator (bottom of the fraction) are equal, so that the value of the fraction is 1; the number in the denominator must have the units that you wish to change. In the case of changing the speed of light from units of km/s to m/s, you need a conversion factor for kilometers to meters. Thus, the conversion factor is

$$\frac{1000 \text{ m}}{1 \text{ km}}$$

Note that this conversion factor is equal to 1, since 1000 meters and 1 kilometer are equal, and that the units to be changed (km) appear in the denominator. We can now convert the speed of light from units of km/s to m/s simply by multiplying by this conversion factor:

$$\underbrace{300{,}000 \, \frac{\text{km}}{\text{s}}}_{\text{speed of light in km/s}} \times \underbrace{\frac{1000 \text{ m}}{1 \text{ km}}}_{\substack{\text{conversion from} \\ \text{km to m}}} = \underbrace{3 \times 10^{8} \, \frac{\text{m}}{\text{s}}}_{\text{speed of light in m/s}}$$

Note that the units of km cancel, leaving the answer in units of m/s.

5. It's easier to work with units if you replace division with multiplication by the reciprocal. For example, suppose you want to know how many minutes are represented by 300 seconds. We can find the answer by dividing 300 seconds by 60 seconds per minute:

$$300 \text{ s} \div 60 \, \frac{\text{s}}{\text{min}}$$

However, it is easier to see the unit cancellations if we rewrite this expression by replacing the division with multiplication by the reciprocal (this process is easy to remember as "invert and multiply"):

$$300 \text{ s} \div 60 \frac{\text{s}}{\text{min}} = 300 \cancel{\text{s}} \times \underbrace{\frac{1 \text{ min}}{60 \cancel{\text{s}}}}_{\substack{\text{invert} \\ \text{and multiply}}} = 5 \text{ min}$$

We now see that the units of seconds (s) cancel in the numerator of the first term and the denominator of the second term, leaving the answer in units of minutes.

More Examples of Working with Units

Example 1. How many seconds are there in 1 day?

Solution: We can answer the question by setting up a *chain* of unit conversions in which we start with 1 *day* and end up with *seconds*. We use the facts that there are 24 hours per day (24 hr/day), 60 minutes per hour (60 min/hr), and 60 seconds per minute (60 s/min):

$$\underbrace{1 \text{ day}}_{\substack{\text{starting} \\ \text{value}}} \times \underbrace{\frac{24 \text{ hr}}{\text{day}}}_{\substack{\text{conversion} \\ \text{from} \\ \text{day to hr}}} \times \underbrace{\frac{60 \text{ min}}{\text{hr}}}_{\substack{\text{conversion} \\ \text{from} \\ \text{hr to min}}} \times \underbrace{\frac{60 \text{ s}}{\text{min}}}_{\substack{\text{conversion} \\ \text{from} \\ \text{min to s}}}$$

$$= 86{,}400 \text{ s}$$

Note that all the units cancel except *seconds*, which is what we want for the answer. There are 86,400 seconds in 1 day.

Example 2. Convert a distance of 10^8 cm to km.

Solution: The easiest way to make this conversion is in two steps, since we know that there are 100 centimeters per meter (100 cm/m) and 1000 meters per kilometer (1000 m/km):

$$\underbrace{10^8 \text{ cm}}_{\substack{\text{starting} \\ \text{value}}} \times \underbrace{\frac{1 \text{ m}}{100 \text{ cm}}}_{\substack{\text{conversion} \\ \text{from} \\ \text{cm to m}}} \times \underbrace{\frac{1 \text{ km}}{1000 \text{ m}}}_{\substack{\text{conversion} \\ \text{from} \\ \text{m to km}}}$$

$$= 10^8 \cancel{\text{cm}} \times \frac{1 \cancel{\text{m}}}{10^2 \cancel{\text{cm}}} \times \frac{1 \text{ km}}{10^3 \cancel{\text{m}}} = 10^3 \text{ km}$$

Alternatively, if we recognize that the number of kilometers should be smaller than the number of centimeters (because kilometers are larger), we might decide to do this conversion by dividing as follows:

$$10^8 \text{ cm} \div \frac{100 \text{ cm}}{\text{m}} \div \frac{1000 \text{ m}}{\text{km}}$$

In this case, before carrying out the calculation, we replace each division with multiplication by the reciprocal:

$$10^8 \text{ cm} \div \frac{100 \text{ cm}}{\text{m}} \div \frac{1000 \text{ m}}{\text{km}}$$

$$= 10^8 \text{ cm} \times \frac{1 \text{ m}}{100 \text{ cm}} \times \frac{1 \text{ km}}{1000 \text{ m}}$$

$$= 10^8 \cancel{\text{cm}} \times \frac{1 \cancel{\text{m}}}{10^2 \cancel{\text{cm}}} \times \frac{1 \text{ km}}{10^3 \cancel{\text{m}}}$$

$$= 10^3 \text{ km}$$

Note that we again get the answer that 10^8 cm is the same as 10^3 km, or 1000 km.

Example 3. Suppose you accelerate at 9.8 m/s² for 4 seconds, starting from rest. How fast will you be going?

Solution: The question asked "how fast?" so we expect to end up with a speed. Therefore, we multiply the acceleration by the amount of time you accelerated:

$$9.8 \frac{\text{m}}{\text{s}^2} \times 4 \text{ s} = (9.8 \times 4) \frac{\text{m} \times \cancel{\text{s}}}{\text{s}^{\cancel{2}}} = 39.2 \frac{\text{m}}{\text{s}}$$

Note that the units end up as a speed, showing that you will be traveling 39.2 m/s after 4 seconds of acceleration at 9.8 m/s².

Example 4. A reservoir is 2 km long and 3 km wide. Calculate its area, in both square kilometers and square meters.

Solution: We find its area by multiplying its length and width:

$$2 \text{ km} \times 3 \text{ km} = 6 \text{ km}^2$$

Next we need to convert this area of 6 km² to square meters, using the fact that there are 1000 meters per kilometer (1000 m/km). Note that we must square the term 1000 m/km when converting from km² to m²:

$$6 \text{ km}^2 \times \left(1000 \frac{\text{m}}{\text{km}} \right)^2 = 6 \text{ km}^2 \times 1000^2 \frac{\text{m}^2}{\text{km}^2}$$

$$= 6 \cancel{\text{km}^2} \times 1{,}000{,}000 \frac{\text{m}^2}{\cancel{\text{km}^2}}$$

$$= 6{,}000{,}000 \text{ m}^2$$

The reservoir area is 6 km², which is the same as 6 million m².

C.4 The Metric System (SI)

The modern version of the metric system, known as *Système Internationale d'Unites* (French for "International System of Units") or **SI,** was formally established in 1960. Today, it is the primary measurement system in nearly every country in the world with the exception of the United States. Even in the United States, it is the system of choice for science and international commerce. The basic units of length, mass, and time in the SI are

- The **meter** for length, abbreviated m
- The **kilogram** for mass, abbreviated kg
- The **second** for time, abbreviated s

Multiples of metric units are formed by powers of 10, using a prefix to indicate the power. For example, *kilo* means 10^3 (1000), so a kilometer is 1000 meters; a microgram is 0.000001 gram, because *micro* means 10^{-6}, or one millionth. Some of the more common prefixes are listed in Table C.2.

TABLE C.2 SI (Metric) Prefixes

Small Values			Large Values		
Prefix	Abbreviation	Value	Prefix	Abbreviation	Value
Deci	d	10^{-1}	Deca	da	10^1
Centi	c	10^{-2}	Hecto	h	10^2
Milli	m	10^{-3}	Kilo	k	10^3
Micro	μ	10^{-6}	Mega	M	10^6
Nano	n	10^{-9}	Giga	G	10^9
Pico	p	10^{-12}	Tera	T	10^{12}

Metric Conversions

Table C.3 lists conversions between metric units and units used commonly in the United States. Note that the conversions between kilograms and pounds are valid only on Earth, because they depend on the strength of gravity.

TABLE C.3 Metric Conversions

To Metric	From Metric
1 inch = 2.540 cm	1 cm = 0.3937 inch
1 foot = 0.3048 m	1 m = 3.28 feet
1 yard = 0.9144 m	1 m = 1.094 yards
1 mile = 1.6093 km	1 km = 0.6214 mile
1 pound = 0.4536 kg	1 kg = 2.205 pounds

Example 1. International athletic competitions generally use metric distances. Compare the length of a 100-meter race to that of a 100-yard race.

Solution: Table C.3 shows that 1 m = 1.094 yd, so 100 m is 109.4 yd. Note that 100 meters is almost 110 yards; a good "rule of thumb" to remember is that distances in meters are about 10% longer than the corresponding number of yards.

Example 2. How many square kilometers are in 1 square mile?

Solution: We use the square of the miles-to-kilometers conversion factor:

$$(1 \text{ mi}^2) \times \left(\frac{1.6093 \text{ km}}{1 \text{ mi}}\right)^2 = (1 \text{ mi}^2) \times \left(1.6093^2 \frac{\text{km}^2}{\text{mi}^2}\right)$$
$$= 2.5898 \text{ km}^2$$

Therefore, 1 square mile is 2.5898 square kilometers.

C.5 Finding a Ratio

Suppose you want to compare two quantities, such as the average density of Earth and the average density of Jupiter. The way we do such a comparison is by dividing, which tells us the *ratio* of the two quantities. In this case, Earth's average density is 5.52 g/cm^3 and Jupiter's average density is 1.33 g/cm^3 (see Figure 7.9), so the ratio is

$$\frac{\text{average density of Earth}}{\text{average density of Jupiter}} = \frac{5.52 \text{ g/cm}^3}{1.33 \text{ g/cm}^3} = 4.15$$

Notice how the units cancel on both the top and the bottom of the fraction. We can state our result in two equivalent ways:

- The ratio of Earth's average density to Jupiter's average density is 4.15.
- Earth's average density is 4.15 times Jupiter's average density.

Sometimes, the quantities that you want to compare may each involve an equation. In such cases, you could, of course, find the ratio by first calculating each of the two quantities individually and then dividing. However, it is much easier if you first express the ratio as a fraction, putting the equation for one quantity on top and the other on the bottom. Some of the terms in the equation may then cancel out, making any calculations much easier.

Example 1. Compare the kinetic energy of a car traveling at 100 km/hr to that of the same car traveling at 50 km/hr.

Solution: We do the comparison by finding the ratio of the two kinetic energies, recalling that the formula for kinetic energy is $\frac{1}{2}mv^2$. Since we are not told the mass of the car, you might at first think that we don't have enough information to find the ratio. However, notice what happens when we put the equations for each kinetic energy into the ratio, calling the two speeds v_1 and v_2:

$$\frac{\text{K.E. car at } v_1}{\text{K.E. car at } v_2} = \frac{\frac{1}{2} m_{\text{car}} v_1^2}{\frac{1}{2} m_{\text{car}} v_2^2} = \frac{v_1^2}{v_2^2} = \left(\frac{v_1}{v_2}\right)^2$$

All the terms cancel except those with the two speeds, leaving us with a very simple formula for the ratio. Now we put in 100 km/hr for v_1 and 50 km/hr for v_2:

$$\frac{\text{K.E. car at 100 km/hr}}{\text{K.E. car at 50 km/hr}} = \left(\frac{100 \text{ km/hr}}{50 \text{ km/hr}}\right)^2 = 2^2 = 4$$

The ratio of the car's kinetic energies at 100 km/hr and 50 km/hr is 4. That is, the car has four times as much kinetic energy at 100 km/hr as it has at 50 km/hr.

Example 2. Compare the strength of gravity between Earth and the Sun to the strength of gravity between Earth and the Moon.

Solution: We do the comparison by taking the ratio of the Earth–Sun gravity to the Earth–Moon gravity. In this case, each quantity is found from the equation of Newton's law of gravity. (See Section 2.4.) Thus, the ratio is

$$\frac{\text{Earth–Sun gravity}}{\text{Earth–Moon gravity}} = \frac{G\frac{M_{\text{Earth}}M_{\text{Sun}}}{(d_{\text{Earth–Sun}})^2}}{G\frac{M_{\text{Earth}}M_{\text{Moon}}}{(d_{\text{Earth–Moon}})^2}}$$

$$= \frac{M_{\text{Sun}}}{(d_{\text{Earth–Sun}})^2} \times \frac{(d_{\text{Earth–Moon}})^2}{M_{\text{Moon}}}$$

Note how all but four of the terms cancel; the last step comes from replacing the division with multiplication by the reciprocal (the "invert and multiply" rule for division). We can simplify the work further by rearranging the terms so that we have the masses and distances together:

$$\frac{\text{Earth–Sun gravity}}{\text{Earth–Moon gravity}} = \frac{M_{\text{Sun}}}{M_{\text{Moon}}} \times \frac{(d_{\text{Earth–Moon}})^2}{(d_{\text{Earth–Sun}})^2}$$

Now it is just a matter of looking up the numbers (see Appendix E) and calculating:

$$\frac{\text{Earth–Sun gravity}}{\text{Earth–Moon gravity}} = \frac{1.99 \times 10^{30} \text{ kg}}{7.35 \times 10^{22} \text{ kg}} \times \frac{(384.4 \times 10^3 \text{ km})^2}{(149.6 \times 10^6 \text{ km})^2}$$

$$= 179$$

In other words, the Earth–Sun gravity is 179 times stronger than the Earth–Moon gravity.

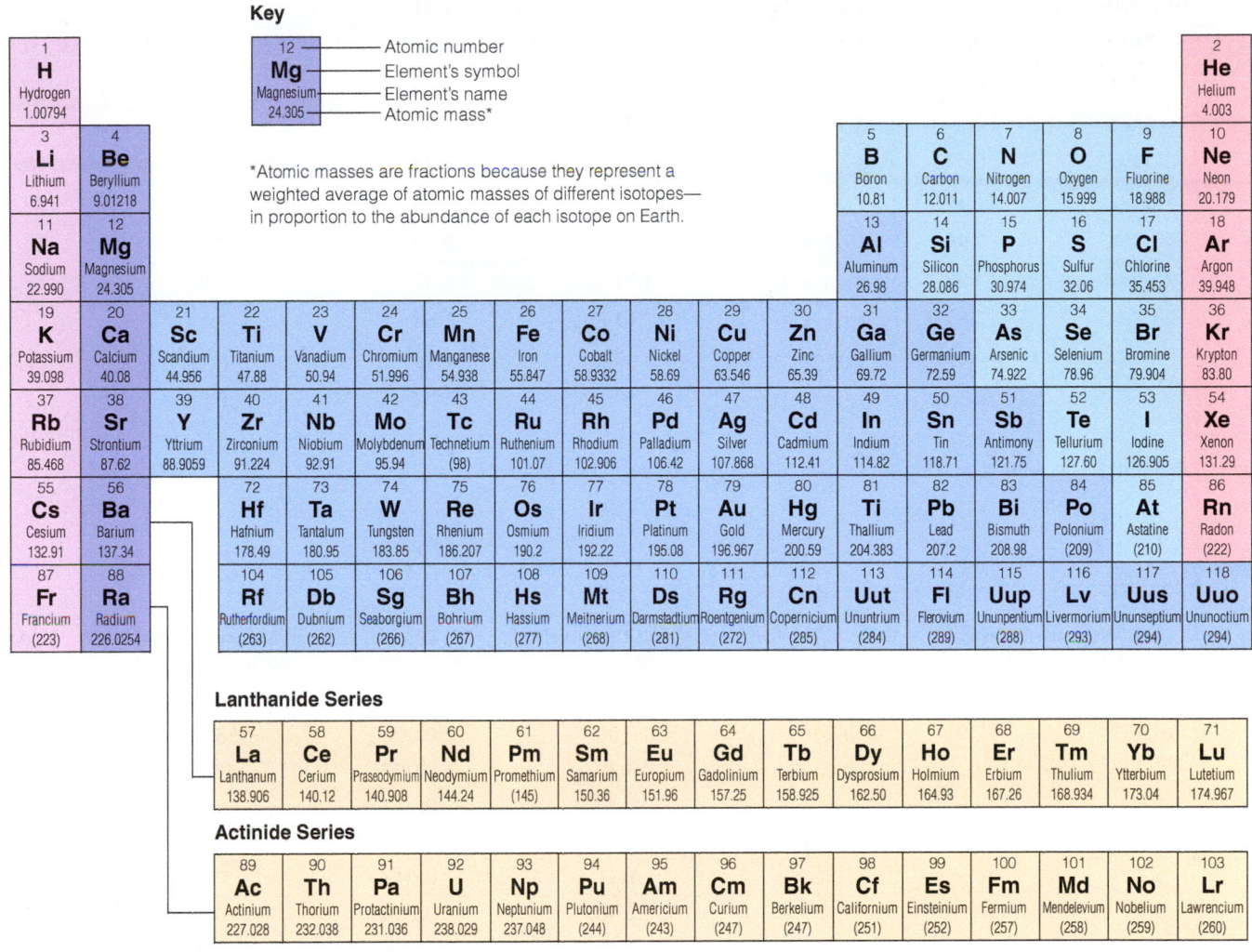

Key

12	← Atomic number
Mg	← Element's symbol
Magnesium	← Element's name
24.305	← Atomic mass*

*Atomic masses are fractions because they represent a weighted average of atomic masses of different isotopes—in proportion to the abundance of each isotope on Earth.

1																	2
H Hydrogen 1.00794																	**He** Helium 4.003
3 **Li** Lithium 6.941	4 **Be** Beryllium 9.01218											5 **B** Boron 10.81	6 **C** Carbon 12.011	7 **N** Nitrogen 14.007	8 **O** Oxygen 15.999	9 **F** Fluorine 18.988	10 **Ne** Neon 20.179
11 **Na** Sodium 22.990	12 **Mg** Magnesium 24.305											13 **Al** Aluminum 26.98	14 **Si** Silicon 28.086	15 **P** Phosphorus 30.974	16 **S** Sulfur 32.06	17 **Cl** Chlorine 35.453	18 **Ar** Argon 39.948
19 **K** Potassium 39.098	20 **Ca** Calcium 40.08	21 **Sc** Scandium 44.956	22 **Ti** Titanium 47.88	23 **V** Vanadium 50.94	24 **Cr** Chromium 51.996	25 **Mn** Manganese 54.938	26 **Fe** Iron 55.847	27 **Co** Cobalt 58.9332	28 **Ni** Nickel 58.69	29 **Cu** Copper 63.546	30 **Zn** Zinc 65.39	31 **Ga** Gallium 69.72	32 **Ge** Germanium 72.59	33 **As** Arsenic 74.922	34 **Se** Selenium 78.96	35 **Br** Bromine 79.904	36 **Kr** Krypton 83.80
37 **Rb** Rubidium 85.468	38 **Sr** Strontium 87.62	39 **Y** Yttrium 88.9059	40 **Zr** Zirconium 91.224	41 **Nb** Niobium 92.91	42 **Mo** Molybdenum 95.94	43 **Tc** Technetium (98)	44 **Ru** Ruthenium 101.07	45 **Rh** Rhodium 102.906	46 **Pd** Palladium 106.42	47 **Ag** Silver 107.868	48 **Cd** Cadmium 112.41	49 **In** Indium 114.82	50 **Sn** Tin 118.71	51 **Sb** Antimony 121.75	52 **Te** Tellurium 127.60	53 **I** Iodine 126.905	54 **Xe** Xenon 131.29
55 **Cs** Cesium 132.91	56 **Ba** Barium 137.34		72 **Hf** Hafnium 178.49	73 **Ta** Tantalum 180.95	74 **W** Tungsten 183.85	75 **Re** Rhenium 186.207	76 **Os** Osmium 190.2	77 **Ir** Iridium 192.22	78 **Pt** Platinum 195.08	79 **Au** Gold 196.967	80 **Hg** Mercury 200.59	81 **Ti** Thallium 204.383	82 **Pb** Lead 207.2	83 **Bi** Bismuth 208.98	84 **Po** Polonium (209)	85 **At** Astatine (210)	86 **Rn** Radon (222)
87 **Fr** Francium (223)	88 **Ra** Radium 226.0254		104 **Rf** Rutherfordium (263)	105 **Db** Dubnium (262)	106 **Sg** Seaborgium (266)	107 **Bh** Bohrium (267)	108 **Hs** Hassium (277)	109 **Mt** Meitnerium (268)	110 **Ds** Darmstadtium (281)	111 **Rg** Roentgenium (272)	112 **Cn** Copernicium (285)	113 **Uut** Ununtrium (284)	114 **Fl** Flerovium (289)	115 **Uup** Ununpentium (288)	116 **Lv** Livermorium (293)	117 **Uus** Ununseptium (294)	118 **Uuo** Ununoctium (294)

Lanthanide Series

57	58	59	60	61	62	63	64	65	66	67	68	69	70	71
La Lanthanum 138.906	**Ce** Cerium 140.12	**Pr** Praseodymium 140.908	**Nd** Neodymium 144.24	**Pm** Promethium (145)	**Sm** Samarium 150.36	**Eu** Europium 151.96	**Gd** Gadolinium 157.25	**Tb** Terbium 158.925	**Dy** Dysprosium 162.50	**Ho** Holmium 164.93	**Er** Erbium 167.26	**Tm** Thulium 168.934	**Yb** Ytterbium 173.04	**Lu** Lutetium 174.967

Actinide Series

89	90	91	92	93	94	95	96	97	98	99	100	101	102	103
Ac Actinium 227.028	**Th** Thorium 232.038	**Pa** Protactinium 231.036	**U** Uranium 238.029	**Np** Neptunium 237.048	**Pu** Plutonium (244)	**Am** Americium (243)	**Cm** Curium (247)	**Bk** Berkelium (247)	**Cf** Californium (251)	**Es** Einsteinium (252)	**Fm** Fermium (257)	**Md** Mendelevium (258)	**No** Nobelium (259)	**Lr** Lawrencium (260)

Appendix E Planetary Data

TABLE E.1 Physical Properties of the Sun and Planets

Name	Radius (Eq[a]) (km)	Radius (Eq) (Earth units)	Mass (kg)	Mass (Earth units)	Average Density (g/cm^3)	Surface Gravity (Earth = 1)	Escape Velocity (km/s)
Sun	695,000	109	1.99×10^{30}	333,000	1.41	27.5	—
Mercury	2440	0.382	3.30×10^{23}	0.055	5.43	0.38	4.43
Venus	6051	0.949	4.87×10^{24}	0.815	5.25	0.91	10.4
Earth	6378	1.00	5.97×10^{24}	1.00	5.52	1.00	11.2
Mars	3397	0.533	6.42×10^{23}	0.107	3.93	0.38	5.03
Jupiter	71,492	11.19	1.90×10^{27}	317.9	1.33	2.36	59.5
Saturn	60,268	9.46	5.69×10^{26}	95.18	0.70	0.92	35.5
Uranus	25,559	3.98	8.66×10^{25}	14.54	1.32	0.91	21.3
Neptune	24,764	3.81	1.03×10^{26}	17.13	1.64	1.14	23.6
Pluto[b]	1187	0.186	1.31×10^{22}	0.0022	1.86	0.06	1.21
Eris[b]	1163	0.183	1.67×10^{22}	0.0028	2.52	0.08	1.38

[a]Eq = equatorial.

[b]Under the IAU definitions of August 2006, Pluto and Eris are officially designated "dwarf planets."

TABLE E.2 Orbital Properties of the Sun and Planets

Name	Distance from Sun[a] (AU)	Distance from Sun[a] (10^6 km)	Orbital Period (years)	Orbital Inclination[b] (degrees)	Orbital Eccentricity	Sidereal Rotation Period (Earth days)[c]	Axis Tilt (degrees)
Sun	—	—	—	—	—	25.4	7.25
Mercury	0.387	57.9	0.2409	7.00	0.206	58.6	0.0
Venus	0.723	108.2	0.6152	3.39	0.007	−243.0	177.3
Earth	1.00	149.6	1.0	0.00	0.017	0.9973	23.45
Mars	1.524	227.9	1.881	1.85	0.093	1.026	25.2
Jupiter	5.203	778.3	11.86	1.31	0.048	0.41	3.08
Saturn	9.54	1427	29.5	2.48	0.056	0.44	26.73
Uranus	19.19	2870	84.01	0.77	0.046	−0.72	97.92
Neptune	30.06	4497	164.8	1.77	0.010	0.67	29.6
Pluto	39.48	5906	248.0	17.14	0.248	−6.39	112.5
Eris	67.67	10,120	557	44.19	0.442	15.8	78

[a]Semimajor axis of the orbit.

[b]With respect to the ecliptic.

[c]A negative sign indicates rotation is backward relative to other planets.

TABLE E.3 Satellites of the Solar System (as of 2015)[a]

Planet Satellite	Radius or Dimensions[b] (km)	Distance from Planet (10³ km)	Orbital Period[c] (Earth days)	Mass[d] (kg)	Density[d] (g/cm³)	Notes About the Satellite
Earth						
Moon	1738	384.4	27.322	7.349×10^{22}	3.34	*Moon:* Probably formed in giant impact.
Mars						
Phobos	13 × 11 × 9	9.38	0.319	1.3×10^{16}	1.9	*Phobos, Deimos:* Probable captured asteroids.
Deimos	8 × 6 × 5	23.5	1.263	1.8×10^{15}	2.2	
Jupiter						
Small inner moons (4 moons)	8–83	128–222	0.295–0.674	—	—	*Metis, Adrastea, Amalthea, Thebe:* Small moonlets within and near Jupiter's ring system.
Io	1821	421.6	1.769	8.933×10^{22}	3.57	*Io:* Most volcanically active object in the solar system.
Europa	1565	670.9	3.551	4.797×10^{22}	2.97	*Europa:* Possible oceans under icy crust.
Ganymede	2634	1070.0	7.155	1.482×10^{23}	1.94	*Ganymede:* Largest satellite in solar system; unusual ice geology.
Callisto	2403	1883.0	16.689	1.076×10^{23}	1.86	*Callisto:* Cratered iceball.
Irregular group 1 (7 moons)	4–85	7500–17,000	130–457	—	—	*Themisto, Leda, Himalia, Lysithea, Elara, and others:* Probable captured moons with inclined orbits.
Irregular group 2 (52 moons)	1–30	17,000–29,000	490–980	—	—	*Ananke, Carme, Pasiphae, Sinope, and others:* Probable captured moons in inclined backward orbits.
Saturn						
Small inner moons (12)	3–89	117–212	0.5–1.2	—	—	*Pan, Atlas, Prometheus, Pandora, Epimetheus, Janus, and others:* Small moonlets within and near Saturn's ring system.
Mimas	199	185.52	0.942	3.70×10^{19}	1.17	*Mimas, Enceladus, Tethys:* Small and medium-size iceballs, many with interesting geology.
Enceladus	249	238.02	1.370	1.2×10^{20}	1.24	
Tethys	530	294.66	1.888	6.17×10^{20}	1.26	
Calypso and Telesto	8–12	294.66	1.888	—	—	*Calypso and Telesto:* Small moonlets sharing Tethys's orbit.
Dione	559	377.4	2.737	1.08×10^{21}	1.44	*Dione:* Medium-size iceball, with interesting geology.
Helene and Polydeuces	2–16	377.4	2.737	1.6×10^{16}	—	*Helene and Polydeuces:* Small moonlets sharing Dione's orbit.
Rhea	764	527.04	4.518	2.31×10^{21}	1.33	*Rhea:* Medium-size iceball, with interesting geology.
Titan	2575	1221.85	15.945	1.35×10^{23}	1.88	*Titan:* Dense atmosphere shrouds surface; ongoing geological activity.
Hyperion	180 × 140 × 112	1481.1	21.277	2.8×10^{19}	—	*Hyperion:* Only satellite known not to rotate synchronously.
Iapetus	718	3561.3	79.331	1.59×10^{21}	1.21	*Iapetus:* Bright and dark hemispheres show greatest contrast in the solar system.
Phoebe	110	12,952	−550.4	1×10^{19}	—	*Phoebe:* Very dark; material ejected from Phoebe may coat one side of Iapetus.
Irregular groups (37 moons)	2–16	11,300–25,200	450–930 −550 to −150	—	—	Probable captured moons with highly inclined and/or backward orbits.

(continued)

Planet Satellite	Radius or Dimensions[b] (km)	Distance from Planet (10³ km)	Orbital Period[c] (Earth days)	Mass[d] (kg)	Density[d] (g/cm³)	Notes About the Satellite
Uranus						**Uranus**
Small inner moons (13 moons)	5–81	49–98	0.3–0.9	—	—	Cordelia, Ophelia, Bianca, Cressida, Desdemona, Juliet, Portia, Rosalind, Cupid, Belinda, Perdita, Puck, Mab: Small moonlets within and near Uranus's ring system.
Miranda	236	129.8	1.413	6.6×10^{19}	1.26	Miranda, Ariel, Umbriel, Titania, Oberon: Small and medium-size iceballs, with some interesting geology.
Ariel	579	191.2	2.520	1.35×10^{21}	1.65	
Umbriel	584.7	266.0	4.144	1.17×10^{21}	1.44	
Titania	788.9	435.8	8.706	3.52×10^{21}	1.59	
Oberon	761.4	582.6	13.463	3.01×10^{21}	1.50	
Irregular group (9 moons)	5–95	4280–21,000	260–2800	—	—	Francisco, Caliban, Stephano, Trinculo, Sycorax, Margaret, Prospero, Setebos, Ferdinand: Probable captured moons; several in backward orbits.
Neptune						**Neptune**
Small inner moons (5 moons)	29–96	48–74	0.30–0.55	—	—	Naiad, Thalassa, Despina, Galatea, Larissa: Small moonlets within and near Neptune's ring system.
Proteus	$218 \times 208 \times 201$	117.6	1.121	6×10^{19}	—	
Triton	1352.6	354.59	−5.875	2.14×10^{22}	2.0	Triton: Probable captured Kuiper belt object—largest captured object in solar system.
Nereid	170	5588.6	360.125	3.1×10^{19}	—	Nereid: Small, icy moon; very little known.
Irregulars (6 moons)	12–27	16,600–49,300	1880–9750	—	—	2002 N1, N2, N3, N4, 2003 N, 2004 N1: Possible captured moons in inclined or backward orbit.
Pluto						**Pluto**
Charon	606	19.6[e]	6.38	1.59×10^{21}	1.702	Charon: Unusually large compared to Pluto; may have formed in giant impact.
Styx	1.8–9.8	42.4	20.2	—	—	Styx, Nix, Kerberos, Hydra: Newly discovered moons outside Charon's orbit.
Nix	$54 \times 41 \times 36$	48.7	24.9	—	—	
Kerberos	2.6–14	57.8	32.2	—	—	
Hydra	43×33[f]	64.7	38.2	—	—	
Eris						**Eris**
Dysnomia	50	37.4	15.8	—	—	Dysnomia: Approximate properties determined in June 2007.

[a]*Note:* Authorities differ substantially on many of the values in this table.

[b]$a \times b \times c$ values for the dimensions are the approximate lengths of the axes (center to edge) for irregular moons.

[c]Negative sign indicates backward orbit.

[d]Masses and densities are most accurate for those satellites visited by a spacecraft on a flyby. Masses for the smallest moons have not been measured but can be estimated from the radius and an assumed density.

[e]Distance to system center of mass is 17.5×10^3 km.

[f]Third dimension not measured.

Glossary

absolute zero The coldest possible temperature, which is 0 K = −273.15°C.

absorption (of light) The process by which matter absorbs radiative energy.

absorption line spectrum A spectrum that contains absorption lines.

accelerating universe A model of the universe in which a repulsive force (see *cosmological constant*) causes the expansion of the universe to accelerate with time. Its galaxies will recede from one another increasingly faster, and it will become cold and dark more quickly than a coasting universe.

acceleration The rate at which an object's velocity changes. Its standard units are m/s^2.

acceleration of gravity The acceleration of a falling object. On Earth, the acceleration of gravity, designated by g, is 9.8 m/s^2.

accretion The process by which small objects gather together to make larger objects.

adaptive optics A technique in which telescope mirrors flex rapidly to compensate for the bending of starlight caused by atmospheric turbulence.

aerobic organisms Organisms that require molecular oxygen to survive.

albedo The fraction of sunlight reflected by a surface; albedo = 0 means no reflection at all (a perfectly black surface), and albedo = 1 means all sunlight is reflected (a perfectly white surface).

Amazonian era The present era on Mars, which began about 1.0 billion years ago.

amino acids The building blocks of proteins. (More technically, an amino acid is a molecule containing both an *amino group* [NH or NH$_2$] and a *carboxyl group* [COOH].)

anaerobic organisms Organisms that do not require (and may even be poisoned by) molecular oxygen.

Andromeda Galaxy (M13; the Great Galaxy in Andromeda) The nearest large spiral galaxy to the Milky Way.

angular momentum Momentum attributable to rotation or revolution. The angular momentum of an object moving in a circle of radius r is the product $m \times v \times r$.

angular resolution (of a telescope) The smallest angular separation that two pointlike objects can have and still be seen as distinct points of light (rather than as a single point of light).

angular size (or **angular distance**) A measure of the angle formed by extending imaginary lines outward from our eyes to span an object (or between two objects).

annihilation See *matter–antimatter annihilation.*

Antarctic Circle The circle on Earth with latitude 66.5°S.

anthropic principle The idea that our existence is possible only because a great number of aspects of the universe are "fine-tuned" for life.

antielectron See *positron.*

antimatter Any particle with the same mass as a particle of ordinary matter but whose other basic properties, such as electrical charge, are precisely opposite.

aphelion The point at which an object orbiting the Sun is farthest from the Sun.

apogee The point at which an object orbiting Earth is farthest from Earth.

apparent brightness The amount of light reaching us *per unit area* from a luminous object; often measured in units of watts/m^2.

apparent magnitude A measure of the apparent brightness of an object in the sky, based on the ancient system developed by Hipparchus.

apparent retrograde motion The apparent motion of a planet, as viewed from Earth, during the period of a few weeks or months when it moves westward relative to the stars in our sky.

archaea One of the three domains of life; the others are eukarya and bacteria.

arcminute (or **minute of arc**) $\frac{1}{60}$ of 1°.

arcsecond (or **second of arc**) $\frac{1}{60}$ of an arcminute, or $\frac{1}{3600}$ of 1°.

Arctic Circle The circle on Earth with latitude 66.5°N.

Aristotelians Ancient Greek followers of Aristotle, who held that there could be only one Earth and that the heavens were a realm distinct from Earth.

asteroid A relatively small and rocky object that orbits a star; asteroids are officially considered part of a category known as "small solar system bodies."

asteroid belt The region of our solar system between the orbits of Mars and Jupiter in which asteroids are heavily concentrated.

astrobiology The study of life on Earth and beyond; it emphasizes research into questions of the origin of life, the conditions under which life can survive, and the search for life beyond Earth.

astrometric method The detection of extrasolar planets through the side-to-side motion of a star caused by gravitational tugs from the planet.

astronomical unit (AU) The average distance (semimajor axis) of Earth from the Sun, which is about 150 million kilometers.

atmosphere A layer of gas that surrounds a planet or moon, usually very thin compared to the size of the object.

atmospheric pressure The surface pressure resulting from the overlying weight of an atmosphere.

atomic mass number The combined number of protons and neutrons in an atom.

atomic number The number of protons in an atom.

atomists Ancient Greek scholars who held that the universe was made from an infinite number of indivisible atoms.

atoms Consist of a nucleus made from protons and neutrons, surrounded by a cloud of electrons.

ATP (adenosine triphosphate) The molecule that stores and releases energy for nearly all cellular processes among life forms on Earth.

aurora Dancing lights in the sky caused by charged particles entering our atmosphere; called the *aurora borealis* in the Northern Hemisphere and the *aurora australis* in the Southern Hemisphere.

autotroph An organism that gets its carbon directly from the atmosphere in the form of carbon dioxide.

axis tilt (of a planet in our solar system) The amount by which a planet's axis is tilted with respect to a line perpendicular to the ecliptic plane.

bacteria One of the three domains of life; the others are eukarya and archaea.

band (of sensitivity) The set of frequencies that a particular radio receiver can pick up.

bandwidth (of a transmitted signal) The range of frequencies over which a communication signal is transmitted.

bar The standard unit of pressure, approximately equal to Earth's atmospheric pressure at sea level.

basalt A type of dark, high-density volcanic rock that is rich in iron and magnesium-based silicate minerals; it forms a runny (easily flowing) lava when molten.

Big Bang The name given to the event thought to mark the birth of the universe.

Big Bang theory The scientific theory of the universe's earliest moments, stating that all the matter in our observable universe came into being at a single moment in time as an extremely hot, dense mixture of subatomic particles and radiation.

Big Crunch The name given to the event that would presumably end the universe if gravity ever reverses the universal expansion and the universe someday begins to collapse.

binary star system A star system that contains two stars.

biochemistry The chemistry of life.

biosphere The "layer" of life on Earth.

blackbody radiation See *thermal radiation*.

black smokers Structures around seafloor volcanic vents that support a wide variety of life.

blueshift A Doppler shift in which spectral features are shifted to shorter wavelengths, observed when an object is moving toward the observer.

brown dwarf An object that forms much like a star but is too low in mass to sustain nuclear fusion in its core; brown dwarfs have masses much greater than that of Jupiter but always less than $0.08M_{Sun}$.

Cambrian explosion The dramatic diversification of life on Earth that occurred between about 540 and 500 million years ago.

carbohydrates Molecules such as sugars and starches that provide energy to cells and make important cellular structures.

carbonate rock A carbon-rich rock, such as limestone, that forms underwater from chemical reactions between sediments and carbon dioxide. On Earth, most of the outgassed carbon dioxide currently resides in carbonate rocks.

carbon-based life Life that uses molecules containing carbon for its most critical functions. All life on Earth is carbon-based.

carbon dioxide (CO_2) cycle The process that cycles carbon dioxide between Earth's atmosphere and surface rocks.

catalysis The process of causing or accelerating a chemical reaction by involving a substance or molecule that is not permanently changed by the reaction.

catalyst The unchanged substance or molecule involved in catalysis.

celestial sphere The imaginary sphere on which objects in the sky appear to reside when observed from Earth.

cell The basic structure of all life on Earth, in which the living matter inside is separated from the outside world.

Celsius (temperature scale) The temperature scale commonly used in daily activity internationally, defined so that, on Earth's surface, water freezes at 0°C and boils at 100°C.

center of mass (of orbiting objects) The point at which two or more orbiting objects would balance if they were somehow connected; it is the point around which the orbiting objects actually orbit.

charged particle belts Zones in which ions and electrons accumulate and encircle a planet.

chemical bond The linkage between atoms in a molecule.

chemical element See *element*.

chemical enrichment The process by which the abundance of heavy elements (heavier than helium) in the interstellar medium gradually increases over time as these elements are produced by stars and released into space.

chemical potential energy Potential energy that can be released through chemical reactions; for example, food contains chemical potential energy that your body can convert to other forms of energy.

chemoautotroph An organism that gets its carbon directly from the atmosphere and its energy from chemical reactions involving inorganic molecules.

chemoheterotroph An organism that gets both its energy and its carbon by consuming preexisting organic molecules; all animals are chemoheterotrophs.

chloroplasts Structures in plant cells that produce energy by photosynthesis.

chromosome A large molecule that contains DNA and carries genetic information in the form of genes.

civilization types A way of categorizing civilizations by whether they use resources of their planet, their star, or their galaxy. See also *galactic civilization*, *planetary civilization*, and *stellar civilization*.

clay Any of a variety of common silicate minerals with particular physical structures.

climate The long-term average of weather.

cluster of galaxies A collection of a few dozen or more galaxies bound together by gravity; smaller collections of galaxies are called *groups*.

color-coded image An image that represents information or forms of light in any way that makes an object appear different than it would appear if we looked at its true, visible-light colors. Sometimes called a *false-color image*.

comet A relatively small, icy object that orbits a star. Like asteroids, comets are officially considered part of a category known as "small solar system bodies."

comparative planetology The study of the solar system by examining and understanding the similarities and differences among worlds.

compound (chemical) A substance made from molecules consisting of two or more atoms with different atomic numbers.

condensates Solid or liquid particles that condense from a cloud of gas.

condensation The formation of solid or liquid particles from a cloud of gas.

conduction (of energy) The process by which thermal energy is transferred by direct contact from warm material to cooler material.

conservation of angular momentum (law of) The principle that, in the absence of net torque (twisting force), the total angular momentum of a system remains constant.

conservation of energy (law of) The principle that energy (including mass-energy) can be neither created nor destroyed, but can change only from one form to another.

conservation of momentum (law of) The principle that, in the absence of net force, the total momentum of a system remains constant.

continental crust The thicker, lower-density crust that makes up Earth's continents. It is made when remelting of seafloor crust allows lower-density rock to separate and erupt to the surface. Continental crust ranges in age from extremely young to as old as about 4 billion years (or more).

continental drift The way the continents slowly move around on Earth, now known to be a result of plate tectonics.

continuously habitable zone The region around a star in which conditions could allow for surface habitability throughout the history of the star system.

continuous spectrum A spectrum (of light) that spans a broad range of wavelengths without interruption by emission or absorption lines.

convection The energy transport process in which warm material expands and rises while cooler material contracts and falls.

convection cell A small individual region of convecting material.

convergent evolution The tendency of organisms of different evolutionary backgrounds to come to resemble one another because they occupy similar ecological niches.

Copernican revolution The dramatic change, initiated by Copernicus, that occurred when we learned that Earth is a planet orbiting the Sun rather than the center of the universe.

coral model (of colonization) A model of how a civilization might colonize the galaxy, based on growth much like that of coral in the sea.

core (of a planet) The dense central region of a planet that has undergone differentiation.

core (of a star) The central region of a star, in which nuclear fusion can occur.

cosmic microwave background The remnant radiation from the Big Bang, which we detect using radio telescopes sensitive to microwaves (which are short-wavelength radio waves).

cosmic rays Particles such as electrons, protons, and atomic nuclei that zip through interstellar space at close to the speed of light.

cosmos An alternative name for the universe.

crust (of a planet) The low-density surface layer of a planet that has undergone differentiation.

crystal A substance made from atoms arranged in precise geometrical patterns, such as in a mineral.

cultural evolution Changes that arise from the transmission of knowledge accumulated over generations.

cyanobacteria Photosynthetic bacteria thought to have been responsible for making most of the oxygen that gradually built up in Earth's atmosphere.

cycles per second Units of frequency for a wave; describes the number of peaks (or troughs) of a wave that pass by a given point each second. Equivalent to *hertz*.

dark energy Name sometimes given to energy that could be causing the expansion of the universe to accelerate.

dark matter Matter that we infer to exist from its gravitational effects but from which we have not detected any light; dark matter apparently dominates the total mass of the universe.

decay (radioactive) See *radioactive decay*.

December solstice Both the point on the celestial sphere where the ecliptic is farthest south of the celestial equator and the moment in time when the Sun appears at that point each year (around December 21).

density (mass) The amount of mass per unit volume of an object. The average density of any object can be found by dividing its mass by its volume. Standard metric units are kilograms per cubic meter, but density is more commonly stated in units of grams per cubic centimeter.

deuterium A form of hydrogen in which the nucleus contains a proton and a neutron, rather than only a proton (as is the case for most hydrogen nuclei).

differentiation The process by which gravity separates materials according to density, with high-density materials sinking and low-density materials rising.

disequilibrium (chemical) A state in which a mixture undergoing chemical reactions is not in equilibrium.

DNA (deoxyribonucleic acid) The molecule that constitutes the genetic material of life on Earth.

DNA bases Adenine (A), cytosine (C), guanine (G), and thymine (T); the four DNA bases can be paired across the two DNA strands only so that A goes with T and C goes with G.

DNA replication The process of copying DNA molecules.

domain (of life) The highest level at which we currently classify life; the three domains are *eukarya*, *bacteria*, and *archaea*.

Doppler effect (or shift) The effect that shifts the wavelengths of spectral features in objects that are moving toward or away from the observer.

Doppler method The detection of extrasolar planets through the motion of a star toward and away from the observer caused by gravitational tugs from the planet.

Drake equation An equation that lays out the factors that play a role in determining the number of communicating civilizations in our galaxy.

dust (or **dust grains**) Tiny, solid flecks of material; in astronomy, we often discuss *interplanetary dust* (found within a star system) or *interstellar dust* (found between the stars in a galaxy).

dwarf planet An object that orbits the Sun and is massive enough for its gravity to make it nearly round in shape, but that does not qualify as an official planet because it has not cleared its orbital neighborhood. The dwarf planets of our solar system include the asteroid Ceres and the Kuiper belt objects Pluto, Eris, Haumea, and Makemake.

Dyson sphere A hypothesized type of large, thin-walled sphere built to surround a star so that an advanced civilization could capture all the energy flowing out from the star; named after physicist Freeman Dyson, who proposed their possible existence.

Earth-orbiters (spacecraft) Spacecraft designed to study Earth or the universe from Earth orbit.

eccentricity A measure of how much an ellipse deviates from a perfect circle; defined as the center-to-focus distance divided by the length of the semimajor axis.

ecliptic The Sun's apparent annual path among the constellations.

ecliptic plane The plane of Earth's orbit around the Sun.

ejecta (from an impact) Debris ejected by the blast of an impact.

electrical charge A fundamental property of matter that is described by its amount and as either positive or negative; more technically, a measure of how a particle responds to the electromagnetic force.

electromagnetic radiation Another name for light of all types, from radio waves through gamma rays.

electromagnetic spectrum The complete spectrum of light, including radio waves, infrared light, visible light, ultraviolet light, X rays, and gamma rays.

electromagnetic wave A synonym for *light*, which consists of waves of electric and magnetic fields.

electron acceptor (in a redox reaction) The atom or molecule that gains electrons in an overall chemical reaction.

electron donor (in a redox reaction) The atom or molecule that gives up electrons in an overall chemical reaction.

electrons Fundamental particles with negative electric charge; the distribution of electrons in an atom gives the atom its size.

element (chemical) A substance made from individual atoms of a particular atomic number.

ellipse A type of oval that happens to be the shape of bound orbits. An ellipse can be drawn by moving a pencil along a string whose ends are tied to two tacks; the locations of the tacks are the foci (singular, *focus*) of the ellipse.

elliptical galaxies Galaxies that appear round in shape, often longer in one direction, like a football. They have no disks and contain little cool gas and dust compared to spiral galaxies, though they often contain extremely hot, ionized gas.

emission (of light) The process by which matter emits energy in the form of light.

emission line spectrum A spectrum that contains emission lines.

encephalization quotient (EQ) A rough measure of animal intelligence based on the ratio of an animal's brain size to its body mass.

endolith An organism that lives inside of rock; also known as *lithophile*.

endospore A special "resting" cell that allows some organisms to remain dormant for long periods of time.

energy Broadly speaking, what makes matter move. The three basic types of energy are *kinetic*, *potential*, and *radiative*.

enzyme A protein that serves as a catalyst.

eons (geological) The largest divisions of time in Earth's geological history. The four eons are the *Hadean*, *Archean*, *Proterozoic*, and *Phanerozoic*.

equilibrium (chemical) A state of balance between the reacting atoms and molecules and the product atoms and molecules in a mixture undergoing chemical reactions.

equinox See *March equinox* and *September equinox*.

eras (geological) The second-largest divisions of time in Earth's geological history, after eons. The Phanerozoic eon is subdivided into three eras: the *Paleozoic, Mesozoic,* and *Cenozoic*.

erosion The wearing down or building up of geological features by wind, water, ice, and other phenomena of planetary weather.

eruption The process of releasing hot lava onto a planet's surface.

escape velocity The speed necessary for an object to completely escape the gravity of a large body such as a moon, planet, or star.

eukarya One of the three domains of life, and the one in which all plants and animals are found; the other domains are bacteria and archaea.

eukaryote A living organism that is a member of the domain eukarya, and therefore is made from one or more eukaryotic cells.

eukaryotic cell A cell that contains a distinct nucleus that is separated from the rest of the cell by its own membrane.

evaporation The process by which atoms or molecules escape into the gas phase from the liquid phase.

evolution (biological) The gradual change in populations of living organisms that is responsible for transforming life on Earth from its primitive origins to its great diversity today.

evolutionary adaptation An inherited trait that enhances an organism's ability to survive and reproduce in a particular environment.

exoplanet See *extrasolar planet*.

expansion (of universe) The idea that the space between galaxies or clusters of galaxies is growing with time.

extrasolar planet A planet orbiting a star other than our Sun.

extraterrestrial life Life that does *not* live on Earth.

extremophiles Living organisms that are adapted to conditions that are "extreme" by human standards, such as very high or low temperature or a high level of salinity or radiation.

Fahrenheit (temperature scale) The temperature scale commonly used in daily activity in the United States; defined so that, on Earth's surface, water freezes at 32°F and boils at 212°F.

false-color image See *color-coded image*.

fault (geological) A place where lithospheric plates slip sideways relative to one another.

feedback processes Processes in which a small change in some property (such as temperature) leads to changes in other properties, which then either amplify or diminish the original small change.

fermions Particles, such as electrons, neutrons, and protons, that obey the exclusion principle.

Fermi's paradox The question posed by Enrico Fermi about extraterrestrial intelligence—"So where is everybody?"—which asks why we have not observed other civilizations even though simple arguments would suggest that some ought to have spread throughout the galaxy by now.

field An abstract concept used to describe how a particle would interact with a force. For example, the idea of a *gravitational field* describes how a particle would react to the local strength of gravity, and the idea of an *electromagnetic field* describes how a charged particle would respond to forces from other charged particles.

fission See *nuclear fission*.

flybys (spacecraft) Spacecraft that fly past a target object (such as a planet), usually just once, as opposed to entering a bound orbit of the object.

force Anything that can cause a change in momentum.

fossil Any relic of an organism that lived and died long ago.

fossil record (or **geological record**) The information about Earth's past that is recorded in fossils (*fossil record*) and rocks (*geological record*). Note that the terms are often used synonymously.

frequency The rate at which peaks of a wave pass by a point, measured in units of 1/s, often called *cycles per second* or *hertz*.

frost line The boundary in the solar nebula beyond which ices could condense; only metals and rocks could condense within the frost line.

fundamental forces There are four known fundamental forces in nature: *gravity,* the *electromagnetic force,* the *strong force,* and the *weak force*.

fusion See *nuclear fusion*.

galactic civilization A civilization that employs the resources of its entire galaxy.

galaxy A huge collection of anywhere from a few hundred million to more than a trillion stars, all bound together by gravity.

galaxy cluster See *cluster of galaxies*.

Galilean moons The four moons of Jupiter that were discovered by Galileo: Io, Europa, Ganymede, and Callisto.

gamma-ray burst A sudden burst of gamma rays from deep space; such bursts apparently come from distant galaxies, but their precise mechanism is unknown.

gamma rays Light with very short wavelengths (and hence high frequencies)—shorter than those of X rays.

gas phase The phase of matter in which atoms or molecules can move essentially independently of one another.

gas pressure The force (per unit area) pushing on any object due to surrounding gas. See also *pressure*.

gene The basic functional unit of an organism's heredity. A single gene consists of a sequence of DNA bases (or RNA bases, in some viruses) that provides the instructions for a single cell function (such as building a protein).

general theory of relativity Einstein's generalization of his special theory of relativity so that the theory also applies when we consider effects of gravity or acceleration.

genetic analysis The analysis of an organism's genes or genome.

genetic code The "language" that living cells use to read the instructions chemically encoded in DNA.

genetic engineering Making deliberate changes to an organism's genome.

genome The complete sequence of DNA bases in an organism, encompassing all of the organism's genes.

genus The next most precise level of classification after *species*; it is a "generic" category to which multiple species may belong.

geocentric model Any of the ancient Greek models of the universe that had Earth at the center of a celestial sphere.

geocentric universe The ancient belief that Earth is the center of the entire universe.

geological activity Processes that change a planet's surface long after its formation, such as volcanism, tectonics, and erosion.

geological processes The four basic geological processes are *impact cratering, volcanism, tectonics,* and *erosion*.

geological record The information about Earth's past that is recorded in both fossils and rocks.

geological time scale The time scale used by scientists to describe major eras in Earth's past. It is divided into four *eons* (the Hadean, Archaean, Proterozoic, and Phanerozoic). The Phanerozoic eon is subdivided into three *eras* (the Paleozoic, Mesozoic, and Cenozoic), which in turn are subdivided into several *periods*. (The periods are further subdivided into *epochs* and *ages*.)

geology The study of surface features (on a moon, planet, or asteroid) and the processes that create them.

giant impact A collision between a forming planet and a very large planetesimal, such as is thought to have formed our Moon.

giants (among stars) Stars that are near the ends of their lives and that have expanded in radius to extremely large sizes.

global average temperature The average surface temperature of a planet.

global warming An expected increase in Earth's global average temperature caused by human input of carbon dioxide and other greenhouse gases into the atmosphere.

globular cluster A spherically shaped cluster of up to a million or more stars; globular clusters are found primarily in the halos of galaxies and contain only very old stars.

granite A light-colored and low-density igneous rock common in mountain ranges on Earth; it gets its name from its grainy appearance and it is composed largely of quartz and feldspar minerals.

gravitation (law of) See *universal law of gravitation.*

gravitational constant The experimentally measured constant G that appears in the law of universal gravitation:

$$G = 6.67 \times 10^{-11} \frac{\text{m}^3}{\text{kg} \times \text{s}^2}$$

gravitational encounter An encounter in which two (or more) objects pass near enough so that each can feel the effects of the other's gravity and they can therefore exchange energy.

gravitational lensing The magnification or distortion (into arcs, rings, or multiple images) of an image caused by light bending through a gravitational field, as predicted by Einstein's general theory of relativity.

gravity One of the four fundamental forces; it is the force that dominates on large scales.

Great Red Spot A large, high-pressure storm on Jupiter.

greenhouse effect The process by which greenhouse gases in an atmosphere make a planet's surface temperature warmer than it would be in the absence of an atmosphere.

greenhouse gases Gases, such as carbon dioxide, water vapor, and methane, that are particularly good absorbers of infrared light but are transparent to visible light.

habitable world A world with environmental conditions under which life could *potentially* arise or survive.

habitable zone The region around a star in which planets could potentially have surface temperatures at which liquid water could exist.

Hadean eon The earliest eon in Earth's history, corresponding to times before about 4.0 billion years ago.

half-life The time it takes for half of the nuclei in a given quantity of a radioactive substance to decay.

hallmarks of science The following three general characteristics of science: (1) Modern science seeks explanations for observed phenomena that rely solely on natural causes. (2) Science progresses through the creation and testing of models of nature that explain the observations as simply as possible. (3) A scientific model must make testable predictions about natural phenomena that would force us to revise or abandon the model if the predictions did not agree with observations.

halo (of a galaxy) The spherical region surrounding the disk of a spiral galaxy.

handedness The property of some molecules, such as amino acids, that gives them two distinct forms that are mirror images of each other.

heavy bombardment The period in the first few hundred million years after the solar system formed, during which the tail end of planetary accretion created most of the craters found on ancient planetary surfaces.

heavy elements In astronomy, generally all elements *except* hydrogen and helium.

heredity The characteristics of an organism passed to it by its parent(s), which it can pass on to its offspring. The term can also apply to the transmission of these characteristics from one generation to the next.

hertz (Hz) The standard unit of frequency for light waves; equivalent to units of 1/s.

Hertzsprung–Russell (H–R) diagram A graph plotting individual stars as points, with stellar luminosity on the vertical axis and spectral type (or surface temperature) on the horizontal axis.

Hesperian era The middle history of Mars, dating from about 3.8 to 1.0 billion years ago.

heterotroph An organism that gets its carbon by consuming preexisting organic molecules.

hot Jupiter A class of planet that is Jupiter-like in size but orbits very close to its star, causing it to have a very high surface temperature.

hot spot (geological) A place within a plate of the lithosphere where a localized plume of hot mantle material rises.

Hubble's law A mathematical expression of the idea that more distant galaxies move away from us faster.

hydrogen compounds Compounds that contain hydrogen and were common in the solar nebula, such as water (H_2O), ammonia (NH_3) and methane (CH_4).

hydrosphere The "layer" of water on Earth consisting of oceans, lakes, rivers, ice caps, and other liquid water and ice.

hyperspace Any space with more than three dimensions.

hyperthermophile An organism that thrives under conditions of extremely high temperature compared to what most organisms can tolerate.

hypothesis A tentative model that is proposed to explain some set of observed facts but that has not yet been rigorously tested and confirmed.

ice ages Periods of global cooling during which polar caps, glaciers, and snow cover extend closer to the equator.

ices (in solar system theory) Materials that are solid only at low temperatures, such as the hydrogen compounds water, ammonia, and methane.

igneous rock Rock made when molten rock cools and solidifies.

image A picture of an object made by focusing light.

impact The collision of a small body (such as an asteroid or comet) with a larger object (such as a planet or moon).

impact crater A bowl-shaped depression left by the impact of an object that strikes a planetary surface (as opposed to burning up in the atmosphere).

impactor The object responsible for an impact.

impact sterilization The process by which a planet is sterilized as a result of a large impact.

infrared light Light with wavelengths that fall in the portion of the electromagnetic spectrum between radio waves and visible light.

inner solar system Generally considered to encompass the region of our solar system out to about the orbit of Mars.

inorganic Not pertaining to life or the chemistry of carbon molecules.

intensity (of light) A measure of the amount of energy coming from light of a specific wavelength in the spectrum of an object.

interferometry A telescopic technique in which two or more telescopes are used in tandem to produce much better angular resolution than the telescopes could achieve individually.

interstellar cloud A cloud of gas and dust between the stars.

interstellar ramjet A hypothesized type of spaceship that uses a giant scoop to sweep up interstellar gas for use in a nuclear fusion engine.

inverse square law A law followed by any quantity that decreases with the square of the distance between two objects.

ion engine (rocket) A rocket engine that works by accelerating charged particles and expelling them out its back.

ionization The process of stripping one or more electrons from an atom.

ionization nebula A colorful, wispy cloud of gas that glows because neighboring hot stars irradiate it with ultraviolet photons that can ionize hydrogen atoms.

ionosphere A portion of the thermosphere in which ions are particularly common (because of ionization by X rays from the Sun).

ions Atoms with a positive or negative electrical charge.

isotopes Forms of an element that have the *same* number of protons but *different* numbers of neutrons.

joule The international unit of energy, equivalent to about $\frac{1}{4000}$ of a Calorie.

jovian moons The moons of jovian planets.

jovian planets Giant gaseous planets similar in overall composition to Jupiter.

Julian calendar The calendar introduced in 46 B.C. by Julius Caesar and used until the Gregorian calendar replaced it.

June solstice Both the point on the celestial sphere where the ecliptic is farthest north of the celestial equator and the moment in time when the Sun appears at that point each year (around June 21).

Kelvin (temperature scale) The most commonly used temperature scale in science, defined such that absolute zero is 0 K and water freezes at 273.15 K.

Kepler's first law Law stating that the orbit of each planet about the Sun is an ellipse with the Sun at one focus.

Kepler's laws of planetary motion Three laws discovered by Kepler that describe the motion of the planets around the Sun.

Kepler's second law The principle that, as a planet moves around its orbit, it sweeps out equal areas in equal times. This tells us that a planet moves faster when it is closer to the Sun (near perihelion) than when it is farther from the Sun (near aphelion) in its orbit.

Kepler's third law The principle that the square of a planet's orbital period is proportional to the cube of its average distance from the Sun (semimajor axis), which tells us that more distant planets move more slowly in their orbits. In its original form, it is written $p^2 = a^3$. See also *Newton's version of Kepler's third law*.

kinetic energy Energy of motion, given by the formula $\frac{1}{2}mv^2$.

kingdoms (biological) Except for the three domains, the highest classification grouping of living organisms.

K–T boundary The thin layer of dark sediments that marks the division between the Cretaceous and Tertiary periods in the fossil record (the *K* comes from the German word for "Cretaceous," *Kreide*).

K–T event (impact) The collision of an asteroid or comet 65 million years ago that caused the mass extinction best known for wiping out the dinosaurs. *K* and *T* stand for the geological layers above and below the thin one produced by the event.

Kuiper belt The comet-rich region of our solar system that resides between about 30 and 100 AU from the Sun. Kuiper belt comets have orbits that lie fairly close to the plane of planetary orbits and travel around the Sun in the same direction as the planets.

Kuiper belt objects Any object orbiting the Sun within the region of the Kuiper belt, although the term is most often used for relatively large objects. For example, Pluto and Eris are considered large Kuiper belt objects.

Lagrange points (of the Earth–Moon system) The five positions in space where the effects of gravity from Earth and the Moon "cancel" in such a way that, if you floated weightlessly at one of these five points, you wouldn't be tugged toward either body.

late heavy bombardment An apparent increase in the impact rate near the end of the heavy bombardment, about 3.9 billion years ago.

latitude The angular north-south distance between Earth's equator and a location on Earth's surface.

light-year The distance that light can travel in 1 year, which is 9.46 trillion kilometers.

lipids Complex molecules in cells, also known as *fats*, that play a variety of roles including being key components of membranes.

liquid phase The phase of matter in which atoms or molecules are held together but move relatively freely.

lithophile An organism that lives inside rock; also known as *endolith*.

lithosphere The relatively rigid outer layer of a planet; generally encompasses the crust and the uppermost portion of the mantle.

Local Group The group of about 40 galaxies to which the Milky Way Galaxy belongs.

Local Supercluster The supercluster of galaxies to which the Local Group belongs.

longitude The angular east-west distance between the prime meridian (which passes through Greenwich, England) and a location on Earth's surface.

luminosity The total power output of an object, usually measured in watts or in units of solar luminosities ($L_{Sun} = 3.8 \times 10^{26}$ watts).

lunar maria The regions of the Moon that look smooth from Earth and are actually impact basins.

magma Underground molten rock.

magnetic field The region surrounding a magnet in which it can affect other magnets or charged particles.

magnetic field lines Lines that represent how the needles on a series of compasses would point if they were laid out in a magnetic field.

magnetosphere The region surrounding a planet in which charged particles are trapped by the planet's magnetic field.

main sequence The prominent line of points (representing *main-sequence stars*) running from the upper left to the lower right on an H–R diagram.

main-sequence stars Stars whose temperature and luminosity place them on the main sequence of the H–R diagram. Main-sequence stars release energy by fusing hydrogen into helium in their cores.

mantle (of a planet) The rocky layer that lies between a planet's core and crust.

mantle convection The flow pattern in which hot mantle material expands and rises while cooler material contracts and falls.

March equinox Both the point in Pisces on the celestial sphere where the ecliptic crosses the celestial equator and the moment in time when the Sun appears at that point each year (around March 21).

martian meteorites Meteorites found on Earth that are thought to have originated on Mars.

mass A measure of the amount of matter in an object.

mass-energy The potential energy of mass, which has an amount $E = mc^2$.

mass extinction An event in which a large fraction of the species living on Earth go extinct, such as the event in which the dinosaurs died out about 65 million years ago.

mass increase (in relativity) The effect in which an object moving past you seems to have a mass greater than its rest mass.

mass ratio (of a rocket) The ratio of the initial (launch) mass of the rocket M_i, including its fuel and any spacecraft it is carrying, to its mass M after all the fuel is burned.

matter–antimatter annihilation An event that occurs when a particle of matter and a particle of antimatter meet and convert all of their mass-energy to photons.

membrane (cell) A barrier that separates the inside of a cell (or cell nucleus) from the outside.

metabolism The many chemical reactions that occur in living organisms.

metals (in solar system theory) Elements, such as nickel, iron, and aluminum, that condense at fairly high temperatures.

metamorphic rock Rock made from igneous or sedimentary rock that gets transformed (but not melted) by high heat or pressure.

meteor A flash of light caused when a particle from space burns up in our atmosphere.

meteorite A rock from space that lands on Earth.

microwaves Light with wavelengths in the range from micrometers to millimeters. Microwaves are generally considered to be a subset of the radio wave portion of the electromagnetic spectrum.

mid-ocean ridges Long ridges of undersea volcanoes on Earth, along which mantle material erupts onto the ocean floor and pushes apart the existing seafloor on either side. These ridges are essentially the source of new seafloor crust, which then makes its way along the ocean bottom for millions of years before returning to the mantle at a subduction zone.

Milankovitch cycles The cyclical changes in Earth's axis tilt and orbit that can change the climate and cause ice ages.

Milky Way Used both as the name of our galaxy and to refer to the band of light we see in the sky when we look into the plane of our galaxy.

Miller–Urey experiment An experiment first performed in the 1950s that was designed to learn how organic molecules might have formed naturally on the early Earth.

mineral A rocky substance with a particular chemical composition and crystal structure.

mitochondria The cellular organs in eukaryotic cells in which oxygen helps produce energy (by making molecules of ATP).

model (scientific) A representation of some aspect of nature that can be used to explain and predict real phenomena without invoking myth, magic, or the supernatural.

moist greenhouse effect A process by which a planet could lose water when the atmospheric circulation allows water vapor to rise high enough to be broken apart by ultraviolet light from the Sun.

molecule Technically, the smallest unit of a chemical element or compound; in this text, the term refers only to combinations of two or more atoms held together by chemical bonds.

momentum The product of an object's mass and velocity.

moon An object that orbits a planet.

multiple star system A star system that contains two or more stars.

mutations Errors in the copying process when a living cell replicates itself.

natural selection The process by which mutations that make an organism better able to survive get passed on to future generations.

nebula A cloud of gas in space, usually one that is glowing.

nebular theory The detailed theory that describes how our solar system formed from a cloud of interstellar gas and dust.

neutrons Particles with no electrical charge that are found in atomic nuclei.

newton The standard unit of force in the metric system: $1 \text{ newton} = 1 \frac{\text{kg} \times \text{m}}{\text{s}^2}$.

Newton's first law of motion Principle that, in the absence of a net force, an object moves with constant velocity.

Newton's laws of motion Three basic laws that describe how objects respond to forces.

Newton's second law of motion Law stating how a net force affects an object's motion. Specifically, force = rate of change in momentum, or force = mass × acceleration.

Newton's third law of motion Principle that, for any force, there is always an equal and opposite reaction force.

Newton's universal law of gravitation See *universal law of gravitation*.

Newton's version of Kepler's third law A generalization of Kepler's third law used to calculate the masses of orbiting objects from measurements of orbital period and distance; usually written as $p^2 = \frac{4\pi^2}{G(M_1 + M_2)} a^3$.

Noachian era The era on Mars before 3.8 billion years ago.

nonscience As defined in this book, any way of searching for knowledge that makes no claim to follow the scientific method, such as seeking knowledge through intuition, tradition, or faith.

nuclear fission The process in which a larger nucleus splits into two (or more) smaller particles.

nuclear fusion The process in which two (or more) smaller nuclei slam together and make one larger nucleus.

nucleus (of an atom) The compact center of an atom made from protons and neutrons.

nucleus (of a cell) The membrane-enclosed region of a eukaryotic cell that contains the cell's DNA.

nucleus (of a comet) The solid portion of a comet—the only portion that exists when the comet is far from the Sun.

observable universe The portion of the entire universe that, at least in principle, can be seen from Earth.

Occam's razor A principle often used in science, holding that scientists should prefer the simpler of two models that agree equally well with observations; named after the medieval scholar William of Occam (1285–1349).

Oort cloud A huge, spherical region centered on the Sun, extending perhaps halfway to the nearest stars, in which trillions of comets orbit the Sun with random inclinations, orbital directions, and eccentricities.

orbit The path followed by a celestial body because of gravity; an orbit may be *bound* (elliptical) or *unbound* (parabolic or hyperbolic).

orbital energy The sum of an orbiting object's kinetic and gravitational potential energies.

orbital resonance A situation in which one object's orbital period is a simple ratio of another object's period, such as $\frac{1}{2}$, $\frac{1}{4}$, or $\frac{5}{3}$. In such cases, the two objects periodically line up with each other, and the extra gravitational attractions at these times can affect the objects' orbits.

orbiters (of other worlds) Spacecraft that go into orbit of another world for long-term study.

organic chemistry The chemistry of organic molecules (whether or not the molecules are involved in life).

organic molecule Generally, any molecule containing carbon and associated with life. Note that we do not generally consider molecules such as carbon dioxide (CO_2) and carbonate minerals to be organic, since they are commonly found independent of life.

outer solar system Generally considered to encompass the region of our solar system beginning at about the orbit of Jupiter.

outgassing The process of releasing gases from a planetary interior, usually through volcanic eruptions.

oxidation Chemical reactions, often with rocks on the surface of a planet, that remove oxygen from the atmosphere.

ozone The molecule O_3, which is a particularly good absorber of ultraviolet light.

ozone depletion The decline in levels of atmospheric ozone found worldwide on Earth, especially in Antarctica, in recent years.

ozone hole A place where the concentration of ozone in the stratosphere is dramatically lower than is the norm.

Pangaea A "supercontinent" that existed prior to 225 million years ago, in which all Earth's current continents were linked together.

panspermia The idea that life migrated to Earth from some extraterrestrial location.

paradigm (in science) A general pattern of thought that tends to shape scientific study during a particular time period.

paradox A situation that, at least at first, seems to violate common sense or contradict itself. Resolving paradoxes often leads to deeper understanding.

parallax The apparent shifting of an object against the background, due to viewing it from different positions. See also *stellar parallax*.

perigee The point at which an object orbiting Earth is nearest to Earth.

perihelion The point at which an object orbiting the Sun is closest to the Sun.

periodic table of the elements A table that lists properties of all the known elements in an organized way.

period–luminosity relation The relation that describes how the luminosity of a Cepheid variable star is related to the period between peaks in its brightness; the longer the period, the more luminous the star.

periods (geological) The third-largest divisions of time in Earth's geological history, after eons and eras.

phase (of matter) The state determined by the way in which atoms or molecules are held together; the common phases are *solid, liquid,* and *gas*.

photoautotroph An organism that gets its carbon directly from the atmosphere and gets its energy from sunlight through photosynthesis; plants are photoautotrophs.

photoheterotroph An organism that gets its carbon by consuming preexisting organic molecules and gets its energy from sunlight through photosynthesis.

photon An individual particle of light, characterized by a wavelength and a frequency.

phyla (singular, *phylum*) The next level of biological classification below kingdoms.

pixel An individual "picture element" in a digital image.

planet A moderately large object that orbits a star and shines primarily by reflecting light from its star. More precisely, according to a definition approved in 2006, a planet is an object that (1) orbits a star (but is itself neither a star nor a moon); (2) is massive enough for its own gravity to give it a nearly round shape; and (3) has cleared the neighborhood around its orbit. Objects that meet the first two criteria but not the third, including Ceres, Pluto, and Eris, are designated *dwarf planets*.

planetary civilization A civilization that uses the resources of its home planet; we are a planetary civilization by this definition.

planetary geology The extension of the study of Earth's surface and interior to apply to other solid bodies in the solar system, such as terrestrial planets and jovian planet moons.

planetary migration A process through which a planet can move from the orbit on which it is born to a different orbit that is closer to or farther from its star.

planetary nebula The glowing cloud of gas ejected from a low-mass star at the end of its life.

planetesimals The building blocks of planets, formed by accretion in the solar nebula.

plasma A gas consisting of ions and electrons.

plates (on a planet) Pieces of a lithosphere that apparently float on the denser mantle below.

plate tectonics The geological process in which plates are moved around by stresses in a planet's mantle.

positron The antimatter equivalent of an electron. It is identical to an electron in all respects except that it has a positive rather than a negative electrical charge.

potential energy Energy stored for later conversion into kinetic energy; includes *gravitational potential energy, electrical potential energy,* and *chemical potential energy*.

power The rate of energy usage, usually measured in watts (1 watt = 1 joule/s).

precession The gradual wobble of the axis of a rotating object around a vertical line.

precipitation Condensed atmospheric gases that fall to the surface in the form of rain, snow, or hail.

pressure The force (per unit area) pushing on an object. In astronomy, we are generally interested in pressure applied by surrounding gas (or plasma).

prokaryote A living organism made from cells in which DNA is *not* confined to a distinct, membrane-enclosed nucleus. Most prokaryotes are single-celled. Prokaryotes include all the organisms in two of the three domains of life: bacteria and archaea.

prokaryotic cell A cell that lacks a distinct nucleus.

protein A large molecule assembled from amino acids according to instructions encoded in DNA. Proteins play many roles in cells; a special category of proteins, called *enzymes*, catalyzes nearly all of the important biochemical reactions that occur within cells.

protons Particles with positive electrical charge found in atomic nuclei; they are built from three quarks.

protoplanetary disk A disk of material surrounding a young star (or protostar) that may eventually form planets.

protoplanets Planetesimals that have grown quite large, to planet size.

pseudoscience Something that purports to be science or may appear to be scientific but that does not adhere to the testing and verification requirements of the scientific method.

Ptolemaic model The geocentric model of the universe developed by Ptolemy in about A.D. 150.

radar mapping Imaging of a planet by bouncing radar waves off its surface, especially important for Venus and Titan, where thick clouds mask the surface.

radial motion The component of an object's motion directed toward or away from us.

radial velocity The portion of any object's total velocity that is directed toward or away from us. This part of the velocity is the only part that we can measure with the Doppler effect.

radiative energy Energy carried by light; the energy of a photon is Planck's constant times its frequency, or $h \times f$.

radioactive decay The spontaneous change of an atom into a different element, in which its nucleus breaks apart or a proton turns into an electron. This decay releases heat to a planet's interior.

radioactive element (or **radioactive isotope**) A substance whose nucleus tends to fall apart spontaneously.

radiometric dating The process of determining the age of a rock (i.e., the time since it solidified) by comparing the present amount of a radioactive substance to the amount of its decay product.

radio waves Light with very long wavelengths (and hence low frequencies)—longer than those of infrared light.

rare Earth hypothesis A hypothesis holding that the specific circumstances that have made it possible for complex creatures (such as birds or humans) to evolve on Earth might be so rare that ours may be the only inhabited planet in the galaxy that has anything but the simplest life.

red giant A giant star that is red in color; red giants are a late stage in the life of a star, occurring after the star has exhausted its supply of core hydrogen.

redox reactions Chemical reactions that involve an exchange or reshuffling of electric charge between the reacting atoms or molecules. A redox reaction always involves the transfer of one or more electrons from an *electron donor* (which becomes oxidized) to an *electron acceptor* (which becomes reduced).

redshift (Doppler) A Doppler shift in which spectral features are shifted to longer wavelengths, observed when an object is moving away from the observer.

reduction (chemical) The process of gaining electrons—which reduces the electrical charge (because electrons carry negative charge)—in a chemical reaction.

reference frame (frame of reference) In the theory of relativity, what two people (or objects) share if they are *not* moving relative to each other.

resonance See *orbital resonance.*

RNA (ribonucleic acid) A molecule closely related to DNA—but with only a single strand and a slightly different backbone and set of bases—that plays critical roles in carrying out the instructions encoded in DNA.

RNA world The hypothesized period during which life on Earth first evolved and used RNA, rather than DNA, as its genetic material.

rock (in solar system theory) Material common on the surface of Earth, such as silicon-based minerals, that is solid at temperatures and pressures found on Earth but typically melts or vaporizes at temperatures of 500–1300 K.

rock cycle The idea that rocks can be transformed between the three basic types: igneous, metamorphic, and sedimentary.

rotation The spinning of an object around its axis.

runaway greenhouse effect A positive feedback cycle in which heating caused by the greenhouse effect causes more greenhouse gases to enter the atmosphere, which further enhances the greenhouse effect.

rybozymes RNA molecules that function as catalysts.

sample return mission A space mission designed to return to Earth a sample of another world.

satellite Any object orbiting another object.

science The search for knowledge that can be used to explain or predict natural phenomena in a way that can be confirmed by rigorous observations or experiments.

scientific method An organized approach to explaining observed facts through science.

scientific theory A model of some aspect of nature that has been rigorously tested and has passed all tests to date.

seafloor crust On Earth, the thin, dense crust of basalt created by seafloor spreading.

seafloor spreading On Earth, the creation of new seafloor crust at mid-ocean ridges.

search for extraterrestrial intelligence (SETI) The name given to observing projects designed to search for signs of intelligent life beyond Earth.

second law of thermodynamics The law stating that, when left alone, the energy in a system undergoes conversions that lead to increasing disorder.

sedimentary rock A rock that formed from sediments created and deposited by erosional processes. The sediments tend to build up in distinct layers, or *strata.*

seismic waves Earthquake-induced vibrations that propagate through a planet.

selection effect (also called **selection bias**) A type of bias that arises from the way in which objects of study are selected and that can lead to incorrect conclusions. For example, when you are counting animals in a jungle it is easiest to see brightly colored animals, which could mislead you into thinking that these animals are the most common.

semimajor axis Half the distance across the long axis of an ellipse; in this text, it is usually referred to as the *average* distance of an orbiting object, abbreviated *a* in the formula for Kepler's third law.

sentinel hypothesis A possible solution to the Fermi paradox that suggests that extraterrestrials place monitoring devices near star systems that show promise of emerging intelligence, and patiently wait until these devices record the presence of civilization.

September equinox Both the point in Virgo on the celestial sphere where the ecliptic crosses the celestial equator and the moment in time when the Sun appears at that point each year (around September 21).

silicate rock A silicon-rich rock.

sleep paralysis The natural paralysis of the body that occurs during REM sleep; it may occasionally persist for a few minutes after the brain has started waking up, giving a person the alarming sensation of being awake in a paralyzed body. Visions and other sensations often occur in this state.

small solar system body An asteroid, comet, or other object that orbits a star but is too small to qualify as a planet or dwarf planet.

snowball Earth Name given to a hypothesis suggesting that, some 600–700 million years ago, Earth experienced a period in which it became cold enough for glaciers to exist worldwide, even in equatorial regions.

solar luminosity The luminosity of the Sun, which is approximately 4×10^{26} watts.

solar nebula The piece of interstellar cloud from which our own solar system formed.

solar sail A large, highly reflective (and thin, to minimize mass) piece of material that can "sail" through space using pressure exerted by sunlight.

solar system (or **star system**) A star (sometimes more than one star) and all the objects that orbit it.

solar wind A stream of charged particles ejected from the Sun.

solar wind stripping The stripping away of a planet's atmospheric gas by the solar wind; generally affects only planetary atmospheres that are unprotected by a global magnetic field.

solid phase The phase of matter in which atoms or molecules are held rigidly in place.

solstice See *December solstice* and *June solstice.*

special theory of relativity Einstein's theory that describes the relativity of time and space based on the fact that the laws of nature are the same for everyone and that everyone always measures the same speed of light.

species A group of organisms that is genetically distinct from other groups; species is the most precise level of biological classification among organisms.

spectral lines Bright or dark lines that appear in an object's spectrum, which we can see when we pass the object's light through a prismlike device that spreads out the light like a rainbow.

spectral resolution The degree of detail that can be seen in a spectrum; the higher the spectral resolution, the more detail we can see.

spectral type A way of classifying a star by the lines that appear in its spectrum; it is related to surface temperature. The basic spectral types are designated by letters (OBAFGKM, with O for the hottest stars and M for the coolest) and are subdivided with numbers from 0 through 9.

spectroscopy (in astronomical research) The process of obtaining spectra from astronomical objects.

spectrum (of light) See *electromagnetic spectrum.*

speed The rate at which an object moves. Its units are distance divided by time, such as m/s or km/hr.

speed of light The speed at which light travels, which is about 300,000 km/s.

spiral galaxies Galaxies that look like flat, white disks with yellowish bulges at their centers. The disks are filled with cool gas and dust, interspersed with hotter ionized gas, and usually display beautiful spiral arms.

star A large, glowing ball of gas that generates energy through nuclear fusion in its core. The term *star* is sometimes applied to objects that are in the process of becoming true stars (e.g., protostars) and to the remains of stars that have died (e.g., neutron stars).

star system See *solar system.*

stellar civilization A civilization that employs the resources of its home star (that is, not only the resources available on its home planet).

stellar evolution The formation and development of stars.

stellar parallax The apparent shift in the position of a nearby star (relative to distant objects) that occurs as we view the star from different positions in Earth's orbit of the Sun each year.

sterilizing impact An impact large enough that it would have fully vaporized Earth's oceans and killed off any life existing on Earth.

strata (rock) Layers in sedimentary rock.

stromatolites Rocks thought to be fossils made by ancient microbes.

subduction (of tectonic plates) The process in which one plate slides under another.

subduction zones Places where one plate slides under another.

sublimation The process by which atoms or molecules escape into the gas phase from the solid phase.

superclusters The largest known structures in the universe, consisting of many clusters of galaxies, groups of galaxies, and individual galaxies.

super Earth A class of planet that is rocky like Earth, but with greater mass than Earth.

supergiants The very large and bright stars that appear at the top of an H–R diagram.

supernova The explosion of a star.

surface area–to–volume ratio The ratio defined by an object's surface area divided by its volume; this ratio is larger for smaller objects (and vice versa).

symbiotic relationship A relationship in which both an invading organism and a host organism benefit from living together.

synchronous rotation The rotation of an object that always shows the same face to the object that it is orbiting because its rotation period and orbital period are equal.

technological evolution Change driven by the rapid development of technology.

tectonics The disruption of a planet's surface by internal stresses.

temperature A measure of the average kinetic energy of particles in a substance.

terraforming Changing a planet in such a way as to make it more Earth-like.

terrestrial planets Rocky planets similar in overall composition to Earth.

theories of relativity (special and general) Einstein's theories that describe the nature of space, time, and gravity.

theory (in science) See *scientific theory*.

theory of evolution The theory, first advanced by Charles Darwin, that explains *how* evolution occurs through the process of *natural selection*.

thermal energy The collective kinetic energy, as measured by temperature, of the many individual particles moving within a substance.

thermal escape The process in which atoms or molecules in a planet's exosphere move fast enough to escape into space.

thermal radiation The spectrum of radiation produced by an opaque object that depends only on the object's temperature; sometimes called *blackbody radiation*.

thermophile An organism that thrives under conditions of high temperature compared to what most organisms can tolerate.

tidal force A force that occurs when the gravity pulling on one side of an object is larger than that on the other side, causing the object to stretch.

tidal friction Friction within an object that is caused by a tidal force.

tidal heating A source of internal heating created by tidal friction. It is particularly important for satellites with eccentric orbits such as Io and Europa.

time dilation (in relativity) The effect in which you observe time running more slowly in reference frames moving relative to you.

transit An event in which a planet passes in front of a star (or the Sun) as seen from Earth. Only Mercury and Venus can be seen in transit of our Sun. The search for transits of extrasolar planets is an important planet detection strategy.

tree of life A diagram that shows relationships among different species as inferred from genetic comparisons; its three main branches are the three *domains*: bacteria, archaea, and eukarya.

troposphere The lowest atmospheric layer, in which convection and weather occur.

turbulence Rapid and random motion.

ultraviolet light Light with wavelengths that fall in the portion of the electromagnetic spectrum between visible light and X rays.

universal law of gravitation The law expressing the force of gravity (F_g) between two objects, given by the formula

$$F_g = G\frac{M_1 M_2}{d^2}$$

$$\left(G = 6.67 \times 10^{-11} \frac{m^3}{kg \times s^2} \right)$$

universe The sum total of all matter and energy.

vaporization The process by which atoms or molecules escape into the gas phase from the liquid or solid phase; more technically, vaporization from a liquid is called *evaporation* and vaporization from a solid is called *sublimation*.

velocity The combination of speed and direction of motion; it can be stated as a speed in a particular direction, such as 100 km/hr due north.

viscosity The "thickness" of a liquid described in terms of how rapidly it flows; low-viscosity liquids flow quickly (e.g., water), while high-viscosity liquids flow slowly (e.g., molasses).

visible light The light our eyes can see, ranging in wavelength from about 400 to 700 nm.

volatiles Substances, such as water, carbon dioxide, and methane, that are usually found as gases, liquids, or surface ices on the terrestrial worlds.

volcanism The eruption of molten rock, or lava, from a planet's interior onto its surface.

Von Neumann machines Self-replicating machines first proposed by the American mathematician and computer pioneer John Von Neumann (1903–1957).

watt The standard unit of power in science; defined as 1 watt = 1 joule/s.

wavelength The distance between adjacent peaks (or troughs) of a wave.

weather The ever-varying combination of winds, clouds, temperature, and pressure in a planet's troposphere.

white dwarf The hot, compact corpse of a low-mass star, typically with a mass similar to that of the Sun compressed to a volume the size of the Earth.

worldline A line that represents an object on a spacetime diagram.

wormholes The name given to hypothetical tunnels through hyperspace that might connect two distant places in the universe.

X rays Light with wavelengths that fall in the portion of the electromagnetic spectrum between ultraviolet light and gamma rays.

zircons Tiny mineral grains of zirconium silicate, usually found embedded in sedimentary rock.

zoo hypothesis A possible explanation for the Fermi paradox holding that alien civilizations are aware of our presence but have chosen to deliberately avoid contact with us.

Index

Page references preceded by "t" refer to tables. Page references followed by "n" refer to footnotes.

Lava, 126
LCROSS spacecraft, 250
Lead, early differentiation of Earth and, 127
Leeuwenhoek, Anton Van, 170
Leverrier, Urbain, 42, 42n
Library of Alexandria, 24
Lick Observatory (California), 444
Life. *See also* Life in the solar system; Life in the universe;
    Life on Earth; Origin of life
    advantages of water for, 246–248
    artificial, 164, 233–238
    building blocks of, 206, 210, 244–246
    chemical energy for, 326–330
    conditions for, 7, 156
    defining, 3, 154, 164–165
    domains of, 170–171
Life in the solar system, 4, 503–504
    on asteroids, 257
    on Callisto, 318–319, 325
    on comets, 257–258
    death of the Sun and, 354
    on Enceladus, 325
    energy for, 246
    on Europa, 314–317
    on Ganymede, 317–318
    on jovian moons, 255–256, 257, 308–310, 317–318,
        324–326
    on jovian planets, 253–255
    liquids for, 247–248, t247
    on Mars, 251, 252, 287–292, 294–297
    on Mercury, 250–251
    migration of, 169, 210–212
    on the Moon, 250
    requirements for, 244–249, 346–349
    on Titan, 319–320, 323–324
    on Venus, 212, 251–252, 345
Life in the universe. *See also* Habitability; Search for extra-
    terrestrial intelligence (SETI)
    ancient Greeks on, 3, 16, 22–24
    Bruno on, 31
    context for study of, 51
    detecting, on extrasolar planets, 405–406
    Drake equation and, 422–427, 424n
    energy for, 246
    evolution of intelligence and, 233
    Herschel on, 31
    importance of oxygen to, 220–221
    Kepler on, 30
    likelihood of, 502–505
    liquids for, 175–176, 246–248, t247
    nebular theory and, 94–95
    Nicholas of Cusa on, 31
    philosophical and social issues relating to, 505–508
    redox reactions and, 329–330
    requirements for, 172–176, 244–249
    silicon-based, 167
    stellar recycling and, 60, 62–63
    theory of, 39–40
Life on Earth, 164, 165, 165n. *See also* Evolution; Origin
    of life
    amino acid handedness and, 169, 169n
    basic metabolic needs of, 172–173
    building blocks of, 206, 210, 244–246
    Cambrian explosion and, 143, 215–217, 216n, 218, 219,
        220, 428
    carbon-based, 166–167, 245
    classification of, 169–172, 173–175
    common ancestor for, 168–169
    domains of, 159, 170–171
    dominant form of, 171
    earliest, 198–201
    end of, 353–355
    energy for, 155, 157, 246, 328–329
    in extreme environments, 183–188
    functions of water in, 175–176
    general properties of, 154–157
    greenhouse effect and, 137–138
    heavy bombardment and, 122–124, 203, 205–206
    mass extinctions and, 221–227

reproduction by, 155, 156
    role of disequilibrium in, 327–328
    sterilizing impacts and, 125
Ligeia Mare (Titan), 323
Light, 74–76
    cellular energy from, 174, t174, 175
    Doppler effect and, 384
    electromagnetic spectrum of, 75–76
    frequency of, 74–75, A-2
    inverse square law for, 246, A-2
    radiative energy of, 73, 74
    spectroscopy and, 76–77
    speed of, 56, 74, 75, 472–473, 474, 490–491, 492–495
    visible, 75, 76
    wavelength of, 74–76, A-2
Light-year, 55, 56, 58, A-1
LIGO (Advanced Laser Interferometer Gravitational-Wave
    Observatory), 444n
Limestone, 108, 120, 139, 340
Lipids, 168, 208
Liquid(s). *See also* Water
    life and, 175–176, 246–248, t247
    as phase of matter, 71–72
Lithosphere
    of Earth, 120, 126, 128–129
    of Venus, 133, 342
Living stromatolites, 199
Local Group, 52, 53, 61
Local Supercluster, 52, 53
Loihi (Hawaii), 132
Lookback time, 66n
Lowell, Percival, 31, 37, 269–270, 283, 288
Lowell Observatory, 269
Luminosity, 372
    color and, 411, 412
    in Hertzsprung–Russell diagrams, 411, 412
    relationship of, to stellar mass, 372, t372
    spectral type and, t372
    star's habitable zone and, 346
    of Sun over time, 352
    transits and, 385–387
Lunar highlands, 123
Lunar maria, 123
Lunar rocks, 119, 120, 123, 145, 147
Lyell, Charles, 163

Mad cow disease, 159
Magellan orbiter, t262, 341, 342
Magellanic clouds, 86
Magnetic field(s)
    of Callisto, 318–319
    of Earth, 106, 134–136, 137, 226, 348
    of Europa, 312–313
    of Ganymede, 318
    of Mars, 135, 284–285
    solar wind and, 134, 136
    of stars, 376n
    variation in, and mutations, 226
    of Venus, 135, 342
Magnetite, in martian meteorite, 295
Magnetosphere, 136
Main-sequence stars, 411, 412–413, 414
Makemake (dwarf planet), 256
Malaria resistance, sickle-cell disease and, 181n
Malthus, Thomas, 163
Mammals, rise of, on Earth, 223
Mammoths, 116, 124
Mangalyaan (Mars Orbiter Mission), 271, 290
Mantle, of Earth, 120, 126
March equinox, 272
Marconi, Guglielmo, 433–435
Maria, lunar, 123
Mariner spacecraft, 270, 277, 278
Mars, 15
    apparent retrograde motion of, 20, 21
    atmosphere of, 105, 134, 136, 271, 273, 278, 284–285,
        289–290, 289n
    axis tilt of, 148, 272, 285–286, 287
    "canals" on, 3, 31, 182, 269

climate history of, 283–286, 364
    color of, 76, 273
    evidence of life on, 287–290
    floods and oceans on, 281–282
    geology of, 273–275, t275
    greenhouse effect on, 283, 284, 286, 293
    habitability of, 252, 282, 286–287, 297, 339, 347
    internal heat of, 128, 133, 284–285
    lack of plate tectonics on, 276–277
    loss of water from, 284–285
    magnetic field of, 135, 284–285
    missions to, 260, 261, t262, 270–271, 287–289,
        292–293, 468
    moons of, 144, 257, A-11
    orbit of, 272, 290
    plausibility of life on, 173, 212, 252, 277, 294–297,
        503–504, 507
    polar ice caps of, 273, 274, 286, 293
    in popular culture, 268–270
    possibility of endoliths on, 186
    preventing contamination of, 291–292
    properties of, t81, t271, A-10
    searching for life on, 290–292
    seasons on, 272–273
    spectrum of, 78–79, 406
    surface of, 9, 15, 105, 124, 271, 274–275
    terraforming on, 289, 293–294, 353
    volcanism on, 275–277
    water on, 9, 252, 271–272, 277–283, 286–287, 330
    winds on, 273
Mars Express mission, t262, 281, 289
Mars Global Surveyor, 273, 274
Mars Insight lander, 274, 290
Mars Observer mission, 271
Mars Orbiter Mission (Mangalyaan), 271, 290
Mars Reconnaissance Orbiter (MRO), 257, 260, 261, t262,
    278, 279, 283, 286
Martian Chronicles, The (Bradbury), 294
Martian meteorites, 210, 277, 291, 294–297, t295, 296
Mass(es)
    center of, 381–382, 388, 391, 419
    of distant objects, 257
    of extrasolar planets, 391, 392–393, t395
    of high-speed objects, 493, 493n
    relativity of, 493n
    role of, in stellar life cycles, 371, 372, t372, 373–374
    of solar system planets, A-10
    as stored energy, 73
    of Sun, A-10
    universal law of gravitation and, 41–42
Mass-energy, 73
Mass extinctions
    causes of, 224–227
    human activity and, 226–227
    impacts and, 221–223
Mass ratio, in rocket equation, 464, 466
Mathematics, A-3–A-8
Matter, 69–71
    atomic structure of, 70–71
    dark, 57–58
    phases of, 71–73
    recycling of, by stars, 62–63
Matter–antimatter annihilation, 475
Matter–antimatter rockets, 475
MAVEN mission, t262, 271, 285
Maxwell, James Clerk, 490
Maxwell's equations, 490
Mayan astronomy, 17
Mayan pyramids, 454
McKay, David, 294
Medicina radio telescope (Italy), t440
Mediocrity, Principle of, 420
Megahertz, 434
Megaparsec, A-1
Melting point, 72
Mercury, 105
    cooling of, 128, 133
    core of, 135, 147
    lack of atmosphere on, 134, 250